WITHDRAWN
WRIGHT STATE UNIVERSITY LIBRARIES

SpringerWienNewYork

Acta Neurochirurgica
Supplements

Editor: H.-J. Steiger

Brain Edema XIII

Edited by
J.T. Hoff, R.F. Keep, G. Xi, and Y. Hua (eds.)

Acta Neurochirurgica
Supplement 96

SpringerWienNewYork

Julian T. Hoff
Richard F. Keep
Guohua Xi
Ya Hua

University of Michigan, Department of Neurosurgery, Ann Arbor, Michigan, USA

This work is subject to copyright.
All rights are reserved, whether the whole or part of the material is concerned, specifically those of translation, reprinting, re-use of illustrations, broadcasting, reproduction by photocopying machines or similar means, and storage in data banks.

Product Liability: The publisher can give no guarantee for all the information contained in this book. This also refers to that on drug dosage and application thereof. In each individual case the respective user must check the accuracy of the information given by consulting other pharmaceutical literature. The use of registered names, trademarks, etc. in this publication does not imply, even in the absence of specific statement, that such names are exempt from the relevant protective laws and regulations and therefore free for general use.

© 2006 Springer-Verlag/Wien
Printed in Austria
SpringerWienNewYork is a part of Springer Science+Business Media
springeronline.com

Typesetting: Asco Typesetters, Hong Kong
Printing and Binding: Druckerei Theiss GmbH, St. Stefan, Austria, www.theiss.at

Printed on acid-free and chlorine-free bleached paper

SPIN: 11594406

Library of Congress Control Number: 2005937333

With partly coloured Figures

ISSN 0065-1419
ISBN-10 3-211-30712-5 SpringerWienNewYork
ISBN-13 978-3-211-30712-0 SpringerWienNewYork

Preface

The XIII International Symposium on Brain Edema and Tissue Injury was held June 1–3, 2005, in Ann Arbor, Michigan, USA. This volume includes papers presented at the symposium as well as papers that were presented at a satellite Intracerebral Hemorrhage Conference on June 4, 2005. In keeping with the outstanding XII Symposium held in Hakone, Japan in 2002, we chose to include brain tissue injury as well as brain edema as the subject matter for this meeting. Brain edema, in many respects, is a marker of underlying pathological processes which include tissue injury from many diseases.

The scientific sessions included invited speakers, oral presentations, poster sessions, and panel discussions. The meeting emphasized scientific excellence in a congenial atmosphere focusing on both basic and clinical science.

The symposium featured basic science research presentations as well as clinical observations in a variety of categories, including traumatic brain injury, cerebral hemorrhage, cerebral ischemia, hydrocephalus, intracranial pressure, water channels, and blood-brain barrier disruption. The recent increase of interest in intracerebral hemorrhage, including the primary event and the secondary injury that follows, prompted a one-day satellite conference on the subject. The conference was held immediately after the Brain Edema Symposium. Most participants in the Brain Edema Symposium stayed an extra day to learn about the latest developments in intracerebral hemorrhage research, including ongoing clinical trials and basic research investigation focusing primarily on the secondary events which develop after the hemorrhage.

There was considerable enthusiasm to continue the Brain Edema Symposium series at the conclusion of the thirteenth meeting. The Advisory Board chose Warsaw, Poland as the next site for the meeting under the direction of Professor Zbigniew Czernicki and his colleagues. Symposium attendees look forward to a successful meeting in that city in 2008.

The editors wish to thank Ms. Kathleen Donahoe, Ms. Holly Wagner, and the staff of Springer-Verlag for the commitment and editorial skills necessary to prepare this volume for publication.

Julian T. Hoff, Richard Keep, Guohua Xi, and *Ya Hua*

Acknowledgments

The Editors would like to express their sincere thanks to those who made the Brain Edema XIII Symposium and the satellite Intracerebral Hemorrhage Conference possible. Thanks are due especially to the International and Local Advisory Boards for both meetings:

Brain Edema XIII

International Advisory Board

A. Baethmann
Z. Czernicki
U. Ito
Y. Katayama
T. Kuroiwa
A. Marmarou
A. D. Mendelow

Local Organizing Committee

J. T. Hoff (Chair)
R. F. Keep
Y. Hua
G. Xi

Intracerebral Hemorrhage Conference

International Advisory Board

J. Aronowski
K. J. Becker
J. R. Carhuapoma
D. F. Hanley
S. A. Mayer
A. D. Mendelow
S. Nagao
R. J. Traystman
K. R. Wagner
M. Zuccarello

Local Organizing Committee

J. T. Hoff (Chair)
W. G. Barsan
R. F. Keep
L. B. Morgenstern
Y. Hua
G. Xi

The meeting would not have been possible without the hard work of Kathleen Donahoe and Heidi Zayan as Secretariat, Pamela Staton and members of University of Michigan Conference Management Services, Drs. Yangdong He, Shuijiang Song, and Wenquan Liu (Department of Neurosurgery, University of Michigan) for their expertise with the audio/visual presentations, and Drs. John Cowan and Jean-Christophe Leveque (Department of Neurosurgery, University of Michigan) for design and maintenance of the Brain Edema 2005 website.

We would also like to thank the National Institutes of Heath and NovoNordisk for providing educational grants in support of the meeting.

Contents

Human Brain Injury

Kawamata, T., Katayama, Y.:
Surgical management of early massive edema caused by cerebral contusion in head trauma patients 3

Chambers, I. R., Barnes, J., Piper, I., Citerio, G., Enblad, P., Howells, T., Kiening, K., Mattern, J., Nilsson, P., Ragauskas, A., Sahuquillo, J. and Yau, Y. H. for the BrainIT Group:
BrainIT: a trans-national head injury monitoring research network 7

Timofeev, I., Kirkpatrick, P. J., Corteen, E., Hiler, M., Czosnyka, M., Menon, D. K., Pickard, J. D., Hutchinson, P. J.:
Decompressive craniectomy in traumatic brain injury: outcome following protocol-driven therapy 11

Hutchinson, P. J., Corteen, E., Czosnyka, M., Mendelow, A. D., Menon, D. K., Mitchell, P., Murray, G., Pickard, J. D., Rickels, E., Sahuquillo, J., Servadei, F., Teasdale, G. M., Timofeev, I., Unterberg, A., Kirkpatrick, P. J.:
Decompressive craniectomy in traumatic brain injury: the randomized multi center RESCUEicp study (www.RESCUEicp.com) 17

Ng, S. C. P., Poon, W. S., Chan, M. T. V.:
Cerebral hemisphere asymmetry in cerebrovascular regulation in ventilated traumatic brain injury 21

Marmarou, A., Signoretti, S., Aygok, G., Fatouros, P., Portella, G.:
Traumatic brain edema in diffuse and focal injury: cellular or vasogenic? 24

Beaumont, A., Gennarelli, T.:
CT prediction of contusion evolution after closed head injury: the role of pericontusional edema 30

Utagawa, A., Sakurai, A., Kinoshita, K., Moriya, T., Okuno, K., Tanjoh, K.:
Organ dysfunction assessment score for severe head injury patients during brain hypothermia 33

Kinoshita, K., Sakurai, A., Utagawa, A., Ebihara, T., Furukawa, M., Moriya, T., Okuno, K., Yoshitake, A., Noda, E., Tanjoh, K.:
Importance of cerebral perfusion pressure management using cerebrospinal drainage in severe traumatic brain injury 37

Mori, T., Katayama, Y., Kawamata, T.:
Acute hemispheric swelling associated with thin subdural hematomas: pathophysiology of repetitive head injury in sports 40

Kinoshita, K., Utagawa, A., Ebihara, T., Furukawa, M., Sakurai, A., Noda, A., Moriya, T., Tanjoh, K.:
Rewarming following accidental hypothermia in patients with acute subdural hematoma: case report 44

Furukawa, M., Kinoshita, K., Ebihara, T., Sakurai, A., Noda, A., Kitahata, Y., Utagawa, A., Moriya, T., Okuno, K., Tanjoh, K.:
Clinical characteristics of postoperative contralateral intracranial hematoma after traumatic brain injury ... 48

Human Intracranial Hemorrhage

Compagnone, C., Tagliaferri, F., Fainardi, E., Tanfani, A., Pascarella, R., Ravaldini, M., Targa, L., Chieregato, A.:
Diagnostic impact of the spectrum of ischemic cerebral blood flow thresholds in sedated subarachnoid hemorrhage patients. ... 53

Dohi, K., Jimbo, H., Ikeda, Y., Fujita, S., Ohtaki, H., Shioda, S., Abe, T., Aruga, T.:
Pharmacological brain cooling with indomethacin in acute hemorrhagic stroke: antiinflammatory cytokines and antioxidative effects ... 57

Prasad, K. S. M., Gregson, B. A., Bhattathiri, P. S., Mitchell, P., Mendelow, A. D.:
The significance of crossovers after randomization in the STICH trial ... 61

Bhattathiri, P. S., Gregson, B., Prasad, K. S. M., Mendelow, A. D.:
Intraventricular hemorrhage and hydrocephalus after spontaneous intracerebral hemorrhage: results from the STICH trial. ... 65

Ebihara, T., Kinoshita, K., Utagawa, A., Sakurai, A., Furukawa, M., Kitahata, Y., Tominaga, Y., Chiba, N., Moriya, T., Nagao, K., Tanjoh, K.:
Changes in coagulative and fibrinolytic activities in patients with intracranial hemorrhage ... 69

Okuda, M., Suzuki, R., Moriya, M., Fujimoto, M., Chang, C. W., Fujimoto, T.:
The effect of hematoma removal for reducing the development of brain edema in cases of putaminal hemorrhage ... 74

Wu, G., Xi, G., Huang, F.:
Spontaneous intracerebral hemorrhage in humans: hematoma enlargement, clot lysis, and brain edema ... 78

Fainardi, E., Borrelli, M., Saletti, A., Schivalocchi, R., Russo, M., Azzini, C., Cavallo, M., Ceruti, S., Tamarozzi, R., Chieregato, A.:
Evaluation of acute perihematomal regional apparent diffusion coefficient abnormalities by diffusion-weighted imaging. ... 81

Tagliaferri, F., Compagnone, C., Fainardi, E., Tanfani, A., Pascarella, R., Sarpieri, F., Targa, L., Chieregato, A.:
Reperfusion of low attenuation areas complicating subarachnoid hemorrhage ... 85

Human Cerebral Ischemia

Nanda, A., Vannemreddy, P., Willis, B., Kelley, R.:
Stroke in the young: relationship of active cocaine use with stroke mechanism and outcome ... 91

Sakurai, A., Kinoshita, K., Inada, K., Furukawa, M., Ebihara, T., Moriya, T., Utagawa, A., Kitahata, Y., Okuno, K., Tanjoh, K.:
Brain oxygen metabolism may relate to the temperature gradient between the jugular vein and pulmonary artery after cardiopulmonary resuscitation ... 97

Contents

Imaging/Monitoring

Daley, M. L., Leffler, C. W., Czosnyka, M., Pickard, J. D.:
Intracranial pressure monitoring: modeling cerebrovascular pressure transmission 103

Guendling, K., Smielewski, P., Czosnyka, M., Lewis, P., Nortje, J., Timofeev, I., Hutchinson, P. J., Pickard, J. D.:
Use of ICM+ software for on-line analysis of intracranial and arterial pressures in head-injured patients. 108

Czosnyka, M., Hutchinson, P. J., Balestreri, M., Hiler, M., Smielewski, P., Pickard, J. D.:
Monitoring and interpretation of intracranial pressure after head injury 114

Experimental Brain Injury

O'Connor, C. A., Cernak, I., Vink, R.:
The temporal profile of edema formation differs between male and female rats following diffuse traumatic brain injury... 121

James, H. E.:
The effect of intravenous fluid replacement on the response to mannitol in experimental cerebral edema: an analysis of intracranial pressure, serum osmolality, serum electrolytes, and brain water content 125

Shigemori, Y., Katayama, Y., Mori, T., Maeda, T., Kawamata, T.:
Matrix metalloproteinase-9 is associated with blood-brain barrier opening and brain edema formation after cortical contusion in rats 130

Nakamura, T., Miyamoto, O., Yamashita, S., Keep, R. F., Itano, T., Nagao, S.:
Delayed precursor cell marker response in hippocampus following cold injury-induced brain edema...... 134

Sakowitz, O. W., Schardt, C., Neher, M., Stover, J. F., Unterberg, A. W., Kiening, K. L.:
Granulocyte colony-stimulating factor does not affect contusion size, brain edema or cerebrospinal fluid glutamate concentrations in rats following controlled cortical impact..................................... 139

Li, S., Kuroiwa, T., Katsumata, N., Ishibashi, S., Sun, L., Endo, S., Ohno, K.:
Unilateral spatial neglect and memory deficit associated with abnormal β-amyloid precursor protein accumulation after lateral fluid percussion injury in Mongolian gerbils 144

Ohsumi, A., Nawashiro, H., Otani, N., Ooigawa, H., Toyooka, T., Yano, A., Nomura, N., Shima, K.:
Alteration of gap junction proteins (connexins) following lateral fluid percussion injury in rats 148

Vannemreddy, P., Ray, A. K., Patnaik, R., Patnaik, S., Mohanty, S., Sharma, H. S.:
Zinc protoporphyrin IX attenuates closed head injury-induced edema formation, blood-brain barrier disruption, and serotonin levels in the rat 151

Uchino, H., Morota, S., Takahashi, T., Ikeda, Y., Kudo, Y., Ishii, N., Siesjö, B. K., Shibasaki, F.:
A novel neuroprotective compound FR901459 with dual inhibition of calcineurin and cyclophilins....... 157

Ishikawa, Y., Uchino, H., Morota, S., Li, C., Takahashi, T., Ikeda, Y., Ishii, N., Shibasaki, F.:
Search for novel gene markers of traumatic brain injury by time differential microarray analysis 163

Zhao, F. Y., Kuroiwa, T., Miyasakai, N., Tanabe, F., Nagaoka, T., Akimoto, H., Ohno, K., Tamura, A.:
Diffusion tensor feature in vasogenic brain edema in cats 168

Beaumont, A., Fatouros, P., Gennarelli, T., Corwin, F., Marmarou, A.:
Bolus tracer delivery measured by MRI confirms edema without blood-brain barrier permeability in diffuse traumatic brain injury 171

Experimental Intracranial Hemorrhage

Wagner, K. R., Beiler, S., Beiler, C., Kirkman, J., Casey, K., Robinson, T., Larnard, D., de Courten-Myers, G. M., Linke, M. J., Zuccarello, M.:
Delayed profound local brain hypothermia markedly reduces interleukin-1β gene expression and vasogenic edema development in a porcine model of intracerebral hemorrhage 177

Shao, J., Xi, G., Hua, Y., Schallert, T., Felt, B. T.:
Alterations in intracerebral hemorrhage-induced brain injury in the iron deficient rat 183

Ostrowski, R. P., Colohan, A. R. T., Zhang, J. H.:
Neuroprotective effect of hyperbaric oxygen in a rat model of subarachnoid hemorrhage 188

Nakamura, T., Keep, R. F., Hua, Y., Nagao, S., Hoff, J. T., Xi, G.:
Iron-induced oxidative brain injury after experimental intracerebral hemorrhage 194

Wan, S., Hua, Y., Keep, R. F., Hoff, J. T., Xi, G.:
Deferoxamine reduces CSF free iron levels following intracerebral hemorrhage 199

Yang, S., Hua, Y., Nakamura, T., Keep, R. F., Xi, G.:
Up-regulation of brain ceruloplasmin in thrombin preconditioning 203

Lodhia, K. R., Shakui, P., Keep, R. F.:
Hydrocephalus in a rat model of intraventricular hemorrhage 207

Kawai, N., Nakamura, T., Nagao, S.:
Early hemostatic therapy using recombinant factor VIIa in collagenase-induced intracerebral hemorrhage model in rats 212

Nakamura, T., Xi, G., Keep, R. F., Wang, M., Nagao, S., Hoff, J. T., Hua, Y.:
Effects of endogenous and exogenous estrogen on intracerebral hemorrhage-induced brain damage in rats 218

Cannon, J. R., Nakamura, T., Keep, R. F., Richardson, R. J., Hua, Y., Xi, G.:
Dopamine changes in a rat model of intracerebral hemorrhage 222

Yang, S., Nakamura, T., Hua, Y., Keep, R. F., Younger, J. G., Hoff, J. T., Xi, G.:
Intracerebral hemorrhage in complement C3-deficient mice 227

Gong, Y., Tian, H., Xi, G., Keep, R. F., Hoff, J. T., Hua, Y.:
Systemic zinc protoporphyrin administration reduces intracerebral hemorrhage-induced brain injury 232

Experimental Cerebral Ischemia

Ito, U., Kawakami, E., Nagasao, J., Kuroiwa, T., Nakano, I., Oyanagi, K.:
Restitution of ischemic injuries in penumbra of cerebral cortex after temporary ischemia 239

Luo, J., Chen, H., Kintner, D. B., Shull, G. E., Sun, D.:
Inhibition of Na^+/H^+ exchanger isoform 1 attenuates mitochondrial cytochrome C release in cortical neurons following in vitro ischemia 244

Ohtaki, H., Nakamachi, T., Dohi, K., Yofu, S., Hodoyama, K., Matsunaga, M., Aruga, T., Shioda, S.:
Controlled normothermia during ischemia is important for the induction of neuronal cell death after global ischemia in mouse... 249

Kuroiwa, T., Yamada, I., Katsumata, N., Endo, S., Ohno, K.:
Ex vivo measurement of brain tissue viscoelasticity in post ischemic brain edema 254

Kleindienst, A., Dunbar, J. G., Glisson, R., Okuno, K., Marmarou, A.:
Effect of dimethyl sulfoxide on blood-brain barrier integrity following middle cerebral artery occlusion in the rat 258

Turner, R. J., Blumbergs, P. C., Sims, N. R., Helps, S. C., Rodgers, K. M., Vink, R.:
Increased substance P immunoreactivity and edema formation following reversible ischemic stroke 263

Pluta, R., Ulamek, M., Januszewski, S.:
Micro-blood-brain barrier openings and cytotoxic fragments of amyloid precursor protein accumulation in white matter after ischemic brain injury in long-lived rats 267

Sun, L., Kuroiwa, T., Ishibashi, S., Katsumata, N., Endo, S., Mizusawa, H.:
Time profile of eosinophilic neurons in the cortical layers and cortical atrophy 272

Ennis, S. R., Keep, R. F.:
Forebrain ischemia and the blood-cerebrospinal fluid barrier 276

Katsumata, N., Kuroiwa, T., Yamada, I., Tanaka, Y., Ishibashi, S., Endo, S., Ohno, K.:
Neurological dysfunctions versus apparent diffusion coefficient and T2 abnormality after transient focal cerebral ischemia in Mongolian gerbils 279

Ohtaki, H., Fujimoto, T., Sato, T., Kishimoto, K., Fujimoto, M., Moriya, M., Shioda, S.:
Progressive expression of vascular endothelial growth factor (VEGF) and angiogenesis after chronic ischemic hypoperfusion in rat 283

Sharma, H. S., Wiklund, L., Badgaiyan, R. D., Mohanty, S., Alm, P.:
Intracerebral administration of neuronal nitric oxide synthase antiserum attenuates traumatic brain injury-induced blood-brain barrier permeability, brain edema formation, and sensory motor disturbances in the rat 288

Ennis, S. R., Keep, R. F.:
Effects of 2,4-dinitrophenol on ischemia-induced blood-brain barrier disruption 295

Ishibashi, S., Kuroiwa, T., LiYuan, S., Katsumata, N., Li, S., Endo, S., Mizusawa, H.:
Long-term cognitive and neuropsychological symptoms after global cerebral ischemia in Mongolian gerbils 299

Kleindienst, A., Fazzina, G., Dunbar, J. G., Glisson, R., Marmarou, A.:
Protective effect of the V1a receptor antagonist SR49059 on brain edema formation following middle cerebral artery occlusion in the rat 303

Experimental Spinal Cord Injury

Sharma, H. S., Nyberg, F., Gordh, T., Alm, P.:
Topical application of dynorphin A (1–17) antibodies attenuates neuronal nitric oxide synthase up-regulation, edema formation, and cell injury following a focal trauma to the rat spinal cord............ 309

Sharma, H. S., Vannemreddy, P., Patnaik, R., Patnaik, S., Mohanty, S.:
Histamine receptors influence blood-spinal cord barrier permeability, edema formation, and spinal cord blood flow following trauma to the rat spinal cord 316

Sharma, H. S., Sjöquist, P. O., Mohanty, S., Wiklund, L.:
Post-injury treatment with a new antioxidant compound H-290/51 attenuates spinal cord trauma-induced c-*fos* expression, motor dysfunction, edema formation, and cell injury in the rat 322

Sharma, H. S.:
Post-traumatic application of brain-derived neurotrophic factor and glia-derived neurotrophic factor on the rat spinal cord enhances neuroprotection and improves motor function 329

Gordh, T., Sharma, H. S.:
Chronic spinal nerve ligation induces microvascular permeability disturbances, astrocytic reaction, and structural changes in the rat spinal cord 335

Hydrocephalus

Kiefer, M., Meier, U., Eymann, R.:
Gravitational valves: relevant differences with different technical solutions to counteract hydrostatic pressure 343

Aygok, G., Marmarou, A., Fatouros, P., Young, H.:
Brain tissue water content in patients with idiopathic normal pressure hydrocephalus 348

Meier, U., Lemcke, J., Neumann, U.:
Predictors of outcome in patients with normal-pressure hydrocephalus 352

Meier, U., Kiefer, M., Neumann, U., Lemcke, J.:
On the optimal opening pressure of hydrostatic valves in cases of idiopathic normal-pressure hydrocephalus: a prospective randomized study with 123 patients 358

Kiefer, M., Eymann, R., Steudel, W. I.:
Outcome predictors for normal-pressure hydrocephalus 364

Meier, U., Lemcke, J.:
First clinical experiences in patients with idiopathic normal-pressure hydrocephalus with the adjustable gravity valve manufactured by Aesculap (proGAV$^{Aesculap®}$) 368

Meier, U., Lemcke, J., Reyer, T., Gräwe, A.:
Decompressive craniectomy for severe head injury in patients with major extracranial injuries 373

Meier, U., Lemcke, J.:
Clinical outcome of patients with idiopathic normal pressure hydrocephalus three years after shunt implantation 377

Meier, U., Lemcke, J.:
Is it possible to optimize treatment of patients with idiopathic normal pressure hydrocephalus by implanting an adjustable Medos Hakim valve in combination with a Miethke shunt assistant? 381

Aquaporins

Binder, D. K., Yao, X., Verkman, A. S., Manley, G. T.:
Increased seizure duration in mice lacking aquaporin-4 water channels 389

Kleindienst, A., Fazzina, G., Amorini, A. M., Dunbar, J. G., Glisson, R., Marmarou, A.:
Modulation of AQP4 expression by the protein kinase C activator, phorbol myristate acetate, decreases ischemia-induced brain edema 393

Suzuki, R., Okuda, M., Asai, J., Nagashima, G., Itokawa, H., Matsunaga, A., Fujimoto, T., Suzuki, T.:
Astrocytes co-express aquaporin-1, -4, and vascular endothelial growth factor in brain edema tissue associated with brain contusion 398

Ghabriel, M. N., Thomas, A., Vink, R.:
Magnesium restores altered aquaporin-4 immunoreactivity following traumatic brain injury to a pre-injury state 402

Neuroprotection and Neurotoxicity

Dohi, K., Jimbo, H., Abe, T., Aruga, T.:
Positive selective brain cooling method: a novel, simple, and selective nasopharyngeal brain cooling method 409

Yue, S., Li, Q., Liu, S., Luo, Z., Tang, F., Feng, D., Yu, P.:
Mechanism of neuroprotective effect induced by QingKaiLing as an adjuvant drug in rabbits with *E. coli* bacterial meningitis 413

Kinoshita, K., Furukawa, M., Ebihara, T., Sakurai, A., Noda, A., Kitahata, Y., Utagawa, A., Tanjoh, K.:
Acceleration of chemokine production from endothelial cells in response to lipopolysaccharide in hyperglycemic condition 419

Li, F., Zhu, G., Lin, J., Meng, H., Wu, N., Du, Y., Feng, H.:
Photodynamic therapy increases brain edema and intracranial pressure in a rabbit brain tumor model ... 422

Sharma, H. S., Duncan, J. A., Johanson, C. E.:
Whole-body hyperthermia in the rat disrupts the blood-cerebrospinal fluid barrier and induces brain edema 426

ICP, CSF, and the Cerebrovasculature

Nemoto, E. M.:
Dynamics of cerebral venous and intracranial pressures 435

Zhu, Y., Shwe, Y., Du, R., Chen, Y., Shen, F. X., Young, W. L., Yang, G. Y.:
Effects of angiopoietin-1 on vascular endothelial growth factor-induced angiogenesis in the mouse brain . 438

Stamatovic, S. M., Dimitrijevic, O. B., Keep, R. F., Andjelkovic, A. V.:
Inflammation and brain edema: new insights into the role of chemokines and their receptors 444

Johanson, C. E., Donahue, J. E., Spangenberger, A., Stopa, E. G., Duncan, J. A., Sharma, H. S.:
Atrial natriuretic peptide: its putative role in modulating the choroid plexus-CSF system for intracranial pressure regulation 451

Author index 457

Index of keywords 459

Listed in Current Contents

Human Brain Injury

Surgical management of early massive edema caused by cerebral contusion in head trauma patients

T. Kawamata and Y. Katayama

Japan Neurotrauma Databank, Japan Society of Neurotraumatology, and Department of Neurological Surgery, Nihon University School of Medicine, Tokyo, Japan

Summary

Early massive edema caused by severe cerebral contusion results in elevation of intracranial pressure (ICP) and clinical deterioration within 24–72 hours post-trauma. Previous studies indicate that cells in the central area of the contusion undergo shrinkage, disintegration, and homogenization, whereas cellular swelling is predominant in the peripheral area, suggesting that early massive edema is attributable to high osmolality within necrotic brain tissue and may generate an osmotic potential across central and peripheral areas.

We analyzed the effects of surgical excision of necrotic brain tissue in 182 patients with cerebral contusion registered with Japan Neurotrauma Data Bank; 121 patients (66%; Group I) were treated conservatively, and 61 (34%; Group II) were treated surgically. Most Group II cases (90%) underwent complete excision of necrotic brain tissue and evacuation of clots. Group I demonstrated higher mortality at 6 months post-trauma compared to Group II (48% vs. 23%; $p = 0.0001$; $n = 182$). Striking differences were observed in patients scoring 9 or more on Glasgow Coma Scale at admission (56% vs. 17%; $p = 0.017$; $n = 45$) and demonstrated "talk-and-deteriorate" (64% vs. 22%; $p = 0.026$; $n = 29$), supporting our hypothesis that early massive edema is caused by cerebral contusion accompanied by necrotic brain tissue, indicating that surgical excision of necrotic brain tissue provides satisfactory control of progressive elevation in ICP and clinical deterioration in many cases.

Keywords: Cerebral contusion; brain edema; necrosis; osmolality.

Introduction

In patients with severe cerebral contusions, early massive edema occurs within the period of 24–72 hours post-trauma [11]. This type of edema results in progressive elevation of intracranial pressure (ICP) and clinical deterioration [7], giving rise to a clinical course termed "talk-and-deteriorate" [12]. Despite intensive medical therapy, the elevated ICP in patients with early massive edema is often uncontrollable and fatal.

The classic histopathological study by Freytag and Lindenberg [4] demonstrated the presence of 2 components of cerebral contusions; one is the central (core) area [3] in which cells undergo necrosis as the primary consequence of mechanical injury (contusion necrosis proper), and the other is the peripheral (rim) area in which cellular swelling occurs as a consequence of ischemia. A clear demarcation line separates these 2 components. The cellular elements in the central area, both neuronal as well as glial cells, uniformly undergo shrinkage, and then disintegration, homogenization, and cyst formation. In contrast, cellular swelling, which is largely attributable to ischemia [e.g., 1], is predominant in the peripheral area [4].

Our previous clinical studies [5–10], including diffusion magnetic resonance imaging, have provided several lines of evidence to suggest that a large amount of edema fluid is accumulated in necrotic brain tissue within the central area of contusion, and this contributes to early massive edema. We hypothesized that this early massive edema is attributable to a high osmolality within the necrotic brain tissue, which may generate an osmotic potential across the central and peripheral areas. In the present study, the effects of surgical excision of the necrotic brain tissue were analyzed in patients with severe cerebral contusion registered with the Japan Neurotrauma Data Bank (JNTDB).

Materials and methods

We analyzed data from the JNTDB in which a total of 1002 patients suffering severe traumatic brain injury (TBI) were registered during the period 1998 through 2001. A total of 10 high-volume emergency centers in Japan specializing in the management of severe TBI were involved in this registry. Patients with severe TBI were defined as those who scored 8 or less on the Glasgow Coma Scale

(GCS) during their clinical course or who underwent craniotomy. Five-year-old and younger children were excluded. The data sheets contained 392 items covering information on the injury characteristics, pre-hospital care, diagnosis, treatment, and outcome. Among these patients, 182 (18%) demonstrated severe cerebral contusions as the major cause of their clinical status.

Results

Among the 182 patients with severe cerebral contusion, 121 (66%; Group I) were treated conservatively and the remaining 61 (34%; Group II) underwent surgery. The ratio of selecting surgical management was far lower in patients with cerebral contusion (34 ± 19%), compared to those with acute epidural hematoma (88 ± 11%) or acute subdural hematoma (68 ± 18%) who were registered on the JNTDB during the same period. There was a huge variation in the ratio of selecting surgical management (9–77%) among the contributing centers (n = 10). Older patients, especially those between 40 and 60 years old, tended to undergo surgery more frequently, while younger patients tended to be treated conservatively. There was, however, no significant difference in age between Groups I and II (47.8 ± 23.8 vs. 54.4 ± 19.5 years). Surgical management involved internal decompression (complete excision of the necrotic brain tissue and evacuation of clots) with or without external decompression in most patients (90%) of Group II. The remaining patients underwent external decompression alone. Surgery was performed at 1.8–86.1 hours (19.5 ± 24.2), most (73%) within 24 hours post-trauma.

Many of the Group I patients (87%) and only half of the Group II patients (52%) were scored at 8 or less on the GCS at time of admission. The remaining patients of Group I comprised those who were scored at 9 or more at the time of admission and deteriorated later to a score of 8 or less. The remaining patients in Group II included those who scored 9 or more at time of admission and deteriorated later to a score of 8 or less (48%), and those who continued to score 9 or better but required craniotomy to prevent clinical deterioration (52%).

Group I demonstrated a poorer outcome on the Glasgow Outcome Scale at 6 months post-trauma (Table 1). Mortality was higher in Group I compared to Group II (48% vs. 23%; p = 0.0001; n = 182). A difference in mortality between the 2 groups was noted in patients who scored 8 or less on the GCS at time of admission, but did not reach a statistically significant level (Table 1). The most striking difference was observed in patients who scored 9 or better on the GCS at time of admission (Table 1). Mortality was clearly higher in Group I compared to Group II (56% vs. 17%; p = 0.017; n = 45). A clear difference in mortality between the 2 groups was observed even when the analysis was restricted to patients who definitely demonstrated "talk-and-deteriorate" (64% vs. 22%; p = 0.026; n = 29; Table 2).

Table 1. *Outcome (6 months post-trauma)*

	GOS (%)					
	n	GR	MD	SD	VS	D
GCS on admission: 3–5						
– Conservative	47	7	2	11	11	70
– Surgical	11	9	9	27	0	55
GCS on admission: 6–8						
– Conservative	58	29	21	10	10	29
– Surgical	21	24	29	24	10	14
GCS on admission: 9–15						
– Conservative	16	19	13	13	0	56*
– Surgical	29	28	28	17	10	17
Total						
– Conservative	121	19	12	12	9	48**
– Surgical	61	23	25	21	8	23

GCS Glasgow Coma Scale; *GOS* Glasgow Outcome Scale; *GR* good recovery; *MD* moderate disability; *SD* severe disability; *VS* vegetative state; *D* death. * p = 0.017, ** p = 0.0001.

Table 2. *Outcome (6 months post-trauma) in patients demonstrating "talk-and-deteriorate"*

	GOS (%)					
	n	GR	MD	SD	VS	D
Conservative	11	18	9	9	0	64*
Surgical	18	11	22	39	6	22

GOS Glasgow Outcome Scale; *GR* good recovery; *MD* moderate disability; *SD* severe disability; *VS* vegetative state; *D* death. * p = 0.026.

Discussion

Indications for surgical intervention in the case of severe cerebral contusion remain controversial [2]. The huge variation in the ratio of centers selecting surgical management reflects diversity in management policy and an absence of consensus regarding the indications for surgery. It may also be assumed that such diversity of management policy could have resulted in a randomization-like allocation of patients to the 2 groups.

The higher mortality in Group I as compared to Group II suggests that the surgery performed in the Group II patients helped to prevent their clinical deterioration and death. Effects of surgery were most striking in those patients who scored 9 or better at the time of admission. It is by no means certain whether patients in Group II who scored 9 or better before surgery would have deteriorated or not had surgery not been carried out. The effect of surgery on mortality is evident, however, since a difference in mortality was observed even when the analysis was restricted to patients who demonstrated "talk-and-deteriorate." This finding suggests that the surgery itself was the major reason for improved mortality in Group II. In other words, death was probably prevented by the surgery in many patients of Group II.

We previously examined the evolution of cerebral contusion by diffusion magnetic resonance imaging and apparent diffusion coefficient (ADC) mapping in head trauma patients [9]. This study demonstrated a low-intensity core in the central area and a high-intensity rim in the peripheral area of the contusion beginning at approximately 24 hours post-trauma, which corresponds with the timing of ICP elevation and clinical deterioration. During the period of 24–72 hours post-trauma, the ADC value increases in the central area and decreases in the peripheral area [11], so that there is maximal dissociation of ADC values between the central and peripheral areas during the period of 24–72 hours post-trauma. These changes appear to represent necrosis in the central area and cellular swelling in the peripheral area.

We also demonstrated that a marked increase in tissue osmolality occurs within the central area [8]. It remains uncertain whether or not such a marked increase in osmolality is osmotically active and causes edema fluid accumulation. Expansion of the extracellular space in the central area would appear to increase the capacitance for edema fluid accumulation. In contrast, shrinkage of the extracellular space in the peripheral area increases the resistance for edema fluid propagation or resolution. Edema fluid accumulation within the necrotic brain tissue is suggested by the formation of a fluid-blood interface within cerebral contusions, which is not an uncommon finding [7]. Such fluid-blood interfaces are not formed within a cavity but represent layering of red blood cells in the softened necrotic brain tissue, which has accumulated voluminous edema fluid.

We hypothesize that the barrier formed by the swollen cells in the peripheral area may prevent edema fluid propagation and also help to generate osmotic potentials across the central and peripheral areas. Since blood flow is greatly reduced but is not completely interrupted in the contused brain tissue, water is supplied from the blood vessels into the central area. We suggest that a combination of these events may facilitate edema fluid accumulation in the central area and contribute to the early massive edema of cerebral contusion.

At present, there is no established medical treatment which effectively inhibits edema fluid accumulation within cerebral contusions. Our results indicate that the most effective therapy for ameliorating the potentially fatal edema may be surgical excision of the necrotic brain tissue. The effects of surgical excision of necrotic brain tissue have commonly been accounted for on the basis of an increased space compensation for mass lesions. It is possible, however, that excision of the necrotic brain tissue eliminates the cause of edema fluid accumulation. If ICP is elevated by early massive edema due to cerebral contusion and is medically uncontrollable, surgical excision of the necrotic brain tissue would appear to represent the therapy of choice, regardless of the size of the associated hemorrhages. Our results suggest that surgery should be considered in patients who score 9 or better at the time of admission, as soon as they deteriorate to a score of 8 or less on the GCS.

Conclusion

The present findings support our hypothesis that early massive edema is caused by cerebral contusion through the presence of necrotic brain tissue, and indicate that surgical excision of the necrotic brain tissue is the only therapy which can provide satisfactory control of the progressive elevation of the ICP and clinical deterioration in many cases. Surgical intervention should be considered in patients with severe cerebral contusion who demonstrate "talk-and-deteriorate."

Acknowledgments

This work was supported by a Grant-in-Aid for Scientific Research from the Japanese Council of Traffic Science, and a program grant from the Ministry of Health, Labor and Welfare, Japan.

References

1. Alexander MJ, Martin NA, Khanna R, Caron M, Becker DP (1994) Regional cerebral blood flow trends in head injured patients with focal contusions and cerebral edema. Acta Neurochir [Suppl] 60: 479–481
2. Bullock R, Golek J, Blake G (1989) Traumatic intracerebral hematoma – which patients should undergo surgical evacuation? CT scan features and ICP monitoring as a basis for decision making. Surg Neurol 32: 181–187
3. Eriskat J, Schurer L, Kempski O, Baethmann A (1994) Growth kinetics of a primary brain tissue necrosis from a focal lesion. Acta Neurochir [Suppl] 60: 425–427
4. Freytag E, Lindenberg R (1957) Morphology of cortical contusions. AMA Arch Pathol 63: 23–42
5. Katayama Y, Kawamata T (2003) Edema fluid accumulation within necrotic brain tissue as a cause of the mass effect of cerebral contusion in head trauma patients. Acta Neurochir [Suppl] 86: 323–327
6. Katayama Y, Tsubokawa T, Miyazaki S, Kawamata T, Yoshino A (1990) Oedema fluid formation within contused brain tissue as a cause of medically uncontrollable elevation of intracranial pressure: the role of surgical therapy. Acta Neurochir [Suppl] 51: 308–310
7. Katayama Y, Tsubokawa T, Kinoshita K, Himi K (1992) Intraparenchymal blood-fluid levels in traumatic intracerebral hematomas. Neuroradiology 34: 381–383
8. Katayama Y, Mori T, Maeda T, Kawamata T (1998) Pathogenesis of the mass effect of cerebral contusions: rapid increase in osmolality within the contusion necrosis. Acta Neurochir [Suppl] 71: 289–292
9. Kawamata T, Katayama Y, Aoyama N, Mori T (2000) Heterogeneous mechanisms of early edema formation in cerebral contusion: diffusion MRI and ADC mapping study. Acta Neurochir [Suppl] 76: 9–12
10. Kushi H, Katayama Y, Shibuya T, Tsubokawa T, Kuroha T (1994) Gadolinium DTPA-enhanced magnetic resonance imaging of cerebral contusions. Acta Neurochir [Suppl] 60: 472–474
11. Marmarou A (2003) Pathophysiology of traumatic brain edema: current concepts. Acta Neurochir [Suppl] 86: 7–10
12. Marshall LF, Toole BM, Bowers SA (1983) The National Traumatic Coma Data Bank. Part 2: Patients who talk and deteriorate: implications for treatment. J Neurosurg 59: 285–288

Correspondence: Tatsuro Kawamata, Department of Neurological Surgery, Nihon University School of Medicine, 30-1 Oyaguchi Kamimachi, Itabashi-ku, Tokyo 173-8610, Japan. e-mail: kawamata@med.nihon-u.ac.jp

BrainIT: a trans-national head injury monitoring research network

I. R. Chambers[1], J. Barnes[1], I. Piper[2], G. Citerio[3], P. Enblad[4], T. Howells[4], K. Kiening[5], J. Mattern[5], P. Nilsson[4], A. Ragauskas[6], J. Sahuquillo[7], and Y. H. Yau[8] for the BrainIT Group

[1] Critical Care Physics, Regional Medical Physics Department, Newcastle General Hospital, Newcastle Upon Tyne, UK
[2] Department of Clinical Physics, Institute of Neurological Sciences, Southern General Hospital, Glasgow, Scotland
[3] Rianimazione H San Gerardo, Monza, Italy
[4] Department of Clinical Neurosciences, Section of Neurosurgery, Uppsala University Hospital, Uppsala, Sweden
[5] Neurochirurgie Kopfklinik, Heidelberg, Germany
[6] Telematics Science Laboratory, Kaunas University of Technology, Kaunas, Lithuania
[7] Department of Neurosurgery, Neurotraumatology Research Unit, Institut Catala de la Salut, Barcelona, Spain
[8] Department of Neurosurgery, Western General Hospital, Edinburgh, Scotland, UK

Summary

Background. Studies of therapeutic interventions and management strategies on head injured patients are difficult to undertake. BrainIT provides validated data for analysis available to centers that contribute data to allow post-hoc analysis and hypothesis testing.

Methods. Both physiological and intensive care management data are collected. Patient identification is eliminated prior to transfer of data to a central database in Glasgow. Requests for missing/ambiguous data are sent back to the local center. Country coordinating centers provide advice, training, and assistance to centers and manage the data validation process.

Results. Currently 30 centers participate in the group. Data collection started in January 2004 and 242 patients have been recruited. Data validation tools were developed to ensure data accuracy and all analysis must be undertaken on validated data.

Conclusion. BrainIT is an open, collaborative network that has been established with primary objectives of i) creating a core data set of information, ii) standardizing the collection methodology, iii) providing data collection tools, iv) creating and populating a data base for future analysis, and v) establishing data validation methodologies. Improved standards for multi-center data collection should permit the more accurate analysis of monitoring and management studies in head injured patients.

Keywords: Head injury; data collection; validation; analysis; network.

Introduction

There have been a number of projects that have collected high quality monitoring data from severely head injured patients. Patient recruitment to these projects can be restricted as the number of patients admitted to any one unit at any given time is limited. Collection of this data varies from hospital to hospital and there are no set standards; therefore, it is difficult to make comparisons of results from different centers. Physiological monitoring of these patients is essential so that the incidence and effects of secondary insults can be determined, as both play a significant role in patient recovery and outcome [1]. Accurate data collection is also required to detect subtle differences that new therapeutic and management strategies may produce in the care of these patients.

The aim of the BrainIT group is to coordinate with a number of European neurotrauma centers to collect high quality minute-by-minute physiological monitoring data and also clinical management data using a previously defined core data set [3] and standardized data collection equipment. The study started in September 2002 and the initial target for the group is to recruit 300 patients from 30 prospective centers. Data that is free of patient identification is then transferred over the internet to the BrainIT website where it is converted in a common format and entered into a database. Data is validated by country-specific data validators and only this data may be used in formative analysis. Data which has not been validated may be used for hypothesis testing. Anyone may register with the BrainIT website. Access to the database is permitted from those centers who recruit at least 5 patients per year.

Materials and methods

The BrainIT European Coordinating Center is located in Glasgow, Scotland, and a steering group of healthcare professionals leads and advises the group. There are currently 30 European centers participating, with 20 centers actively recruiting patients. Each country has a Country Coordinating Center, where a data validator assists and advises the centers taking part. The data validator's role is to assist in the training of nursing and/or medical staff who collect and transfer the data, and to act as the first point of contact for centers with any queries. A previously-defined core data set is used, of which there are 4 constituent parts: physiological monitoring data, demographic and clinical information, intensive care management data, and secondary insult management data. Minute-by-minute monitoring data are collected from the bedside monitors by either a bedside laptop computer using commercially developed software, or via a network and locally developed systems. Intensive care treatment and management data are collected using a handheld computer (personal digital assistant; PDA). Commercial software developed by Kelvin Connect Ltd. enables staff to collect demographic, clinical, and treatment data. Computed tomography (CT) data is also collected using the Traumatic Core Data Bank criteria (Marshall score) [2] and CT images devoid of patient identification are transferred to Glasgow via the BrainIT website for independent assessment.

Patients included in this project may be of any age as long as there is evidence of traumatic brain injury and the patient has both arterial and intracranial monitoring. Written consent is obtained from the relatives, although with agreement from the Multi-Research Ethics Committee in Scotland and local research ethics committees in the U.K., data collection may begin before consent is formally obtained. If consent is not given, any data collected will not be retained, as is stated in information sheets provided to the relatives.

Patients are followed-up at 6 months post-injury using the Extended Glasgow Outcome Scale, either by face-to-face or telephone interview. Data is collected for as long as the intracranial pressure and arterial monitoring are in place. Daily intensive care management data are collected by nursing staff and entered into a PDA; the data is then transferred to a local computer where it is stored in a database. From this database, the patient files are exported to Glasgow via the BrainIT website. Patient confidentiality is ensured by removing patient identification data before the transfer occurs, which is in keeping with local and national data protection policies. Patient identification for local and coordinating center staff is by means of a unique 8-digit number, which is obtained from the BrainIT website and attached to the patient's file prior to sending the data. The same identification number is used for the physiological monitoring data, which is again sent via the internet. Once the data has been received in Glasgow, it is converted to a common file format and areas of missing or ambiguous data are highlighted. A missing data list is then created and sent back to the contributing center to look for the missing data, who attempts to complete the file as much as possible and return it to Glasgow. This process is repeated until as much of the data can be found as is possible.

A random sample of 20% of the data is then selected for validation against the available source documents (Fig. 1). The validation list may include any physiological, clinical, and treatment data. For example, a request may be sent for a record from the nursing chart or all of the blood pressure and cerebral perfusion pressure readings

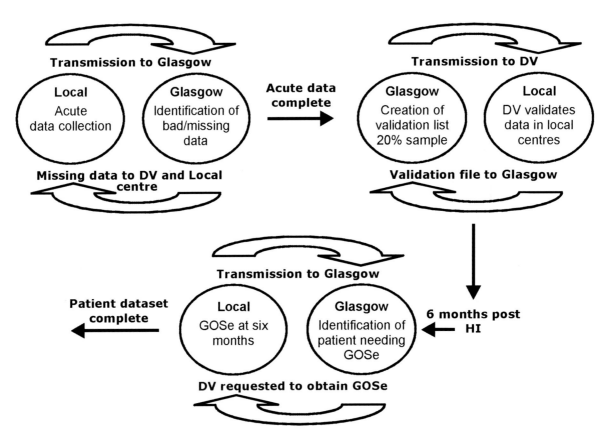

Fig. 1. Data validation cycle. *DV* Data Validator, *GOSe* Extended Glasgow Outcome Scale, *HI* Head Injury

from a 24-hour period. These readings are taken from the chart using the time nearest to that on the validation request. Simultaneously, a request is also sent for the actual number of specific episodic events during a known period; for example, the number of arterial blood gas samples taken in a given time span. A validation file is created for each patient using the BrainIT Core Data Collection tool and the file is then returned to Glasgow. The data validator is responsible for the validation of all the samples of data within a given country (Fig. 1).

Data validation is carried out following a set standard of procedures and can be done at 4 different levels. Only validated data is saved in the common database. This ensures that the 20% sample of data which is stored is the most accurate for each patient.

– Level 1 ensures the data conversion stage functions correctly; in particular, the time-stamp format (YYYY-MM-DD).
– Level 2 checks all non-numeric categorical core data set data for transcribing errors. This level of validation differs according to whether monitoring or non-monitoring data is being validated.
– Level 3 is the conversion of locally used units to BrainIT units.
– Level 4 requires intervention of the data validator.

There are 3 types of Level 4 data validation: Type 1 or self-validation where the principal investigator validates his own data; type 2 is cross-validation where local colleagues may validate each other's data; type 3 validation is where the data validator has no connection with the center from which the data has been collected.

Results

There are currently 30 centers participating in the group and 20 of these centers are actually recruiting patients and supplying data. To date, 257 patients have been recruited to the study with monitoring data sent to Glasgow from 251 patients (Fig. 2). Of these, 53 patients have currently been validated.

There are several projects planned to make use of the data collected and stored in the database. Work is currently underway to assess intracranial pressure and cerebral perfusion pressure variability analysis, and to assess the frequency of missing data and ascertain which types of data are missing most frequently. Both quantitative and qualitative analysis methods will be used in this project. Another project for the future is the BrainIT network clinical evaluation of the Raumedic Neurovent intraparenchymal probe to test its long-term clinical performance.

Conclusion

BrainIT is an open, collaborative network and, thus far, the group has demonstrated that it is possible to standardize the collection methodology of high resolution neurointensive care data. By providing country-specific data validators who are responsible for staff coordination and training, the participating centers have been able to record intensive care treatment and

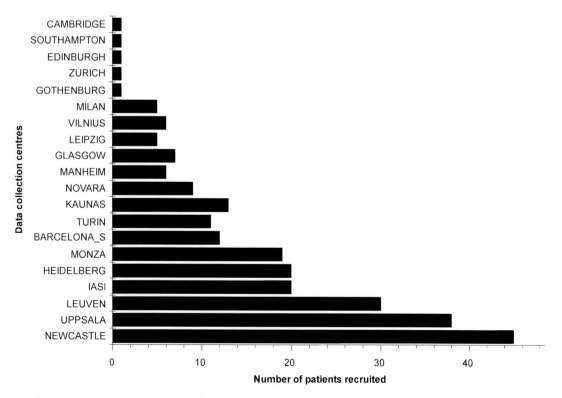

Fig. 2. Recruitment graph as at May 2005

management data using a defined core data set. The provision of standard equipment and assistance obtained from industry has enabled centers to collect data using standard data collection methods. Transfer of data has proven successful, and a populated database has provided data for future analysis by those who contribute data. A minimum of 5 patients per year is the requirement for those centers participating to have access to the data. Access to the data by personnel within the contributing centers is controlled by the principal investigator within each center. Development of software tools has enabled missing and ambiguous data to be selected from the data set, and data validators have collaborated with participating centers to find missing data. Data validation methodologies have been established and, with the help of the data validators, integrity of the data has been ensured.

References

1. Chesnut RM (2004) Management of brain and spine injuries. Crit Care Clin 20: 25–55
2. Marshall LF, Toole BM, Bowers SA (1983) The National Traumatic Coma Data Bank. Part 2: Patients who talk and deteriorate: implications for treatment. J Neurosurg 59: 285–288
3. Piper I, Citerio G, Chambers I, Contant C, Enblad P, Fiddes H, Howells T, Kleining K, Nilsson P, Yau YH; The BrainIT Group (2003) The BrainIT group: concept and core dataset definition. Acta Neurochir (Wien) 145: 615–629

Correspondence: I. R. Chambers, Critical Care Physics, Regional Medical Physics Department, Newcastle General Hospital, Westgate Road, Newcastle Upon Tyne, NE4 6BE, UK. e-mail: i.r.chambers@ncl.ac.uk

Decompressive craniectomy in traumatic brain injury: outcome following protocol-driven therapy

I. Timofeev[1], P. J. Kirkpatrick[1], E. Corteen[1], M. Hiler[1], M. Czosnyka[1], D. K. Menon[2,3], J. D. Pickard[1,3], and P. J. Hutchinson[1]

[1] Department of Neurosurgery, Addenbrooke's Hospital, Cambridge, UK
[2] Department of Anesthesia, Addenbrooke's Hospital, Cambridge, UK
[3] Wolfson Brain Imaging Centre, University of Cambridge, UK

Summary

Although decompressive craniectomy following traumatic brain injury is an option in patients with raised intracranial pressure (ICP) refractory to medical measures, its effect on clinical outcome remains unclear. The aim of this study was to evaluate the outcome of patients undergoing this procedure as part of protocol-driven therapy between 2000–2003. This was an observational study combining case note analysis and follow-up. Outcome was assessed at an interval of at least 6 months following injury using the Glasgow Outcome Scale (GOS) score and the SF-36 quality of life questionnaire. Forty-nine patients underwent decompressive craniectomy for raised and refractory ICP (41 [83.7%] bilateral craniectomy and 8 [16.3%] unilateral). Using the Glasgow Coma Scale (GCS), the presenting head injury grade was severe (GCS 3–8) in 40 (81.6%) patients, moderate (GCS 9–12) in 8 (16.3%) patients, and initially mild (GCS 13–15) in 1 (2.0%) patient. At follow-up, 30 (61.2%) patients had a favorable outcome (good recovery or moderate disability), 10 (20.4%) remained severely disabled, and 9 (18.4%) died. No patients were left in a vegetative state. Overall the results demonstrated that decompressive craniectomy, when applied as part of protocol-driven therapy, yields a satisfactory rate of favorable outcome. Formal prospective randomized studies of decompressive craniectomy are now indicated.

Keywords: Head injury; traumatic brain injury; decompressive craniectomy; ICP; brain edema; intracranial hypertension; Glasgow Outcome Scale.

Introduction

Severe traumatic brain injury is associated with high mortality and morbidity. Treatment of patients who present in coma following severe head injury aims to protect the brain from further insults, optimize cerebral metabolism, and prevent secondary injury. Intracranial pressure (ICP) monitoring is now well-established in neuro-intensive care. Recent guidelines [10] recommend target levels of ICP < 25 mmHg and cerebral perfusion pressure (CPP) ≥ 60–70 mmHg, and a number of therapeutic approaches are employed to reduce ICP and augment CPP in order to achieve these targets.

In patients with post-traumatic cerebral swelling resistant to optimal medical therapy, decompressive craniectomy may be considered. Despite observations that craniectomy leads to reduction in ICP [3, 30, 31, 35], it is still unclear how this translates into clinical outcome. Although prospective randomized evaluation of the effects of decompressive craniectomy is required, this operation continues to be used empirically in the management of patients with traumatic brain injury. In Cambridge, decompressive craniectomy is used as a part of protocol-driven intensive care management of patients with severe head injury [18] (Fig. 1), when other means of controlling elevated ICP are exhausted. It is also used in selected cases where malignant post-traumatic cerebral swelling is evident from the outset. Encouraged by our previous observations [31], we have evaluated the outcome following decompressive craniectomy in another consecutive cohort of 49 patients.

Materials and methods

Study design

This study is a retrospective observational cohort study with cross-sectional analysis of outcome. Hospital records, intensive care charts, and computed tomography (CT) scans of patients who underwent decompressive craniectomy following traumatic brain injury consecutively during the period 2000–2003 were retrospectively analyzed. Physiological parameters recorded for 24 hours before and

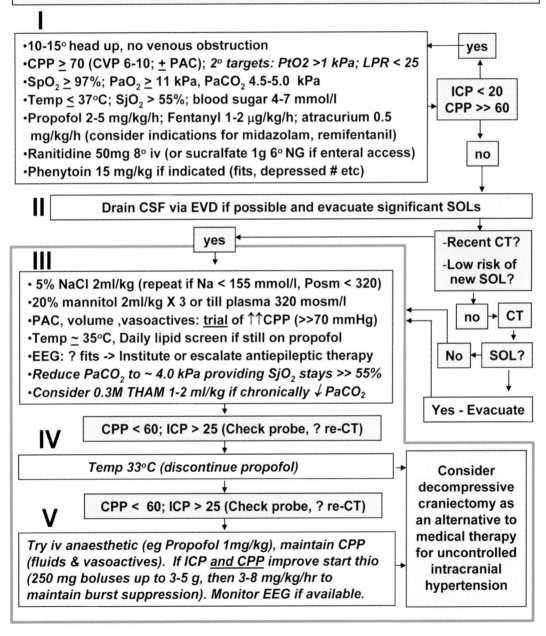

Fig. 1. Protocol of intensive care management of head injured patients

after operation were compared. Outcome was assessed using the Glasgow Outcome Score (GOS) [14] recorded at follow-up assessment at least 6 months following the injury. Data were obtained from patients' medical records and a prospectively collected head injury outcome database. SF-36 quality of life survey [29] questionnaires were mailed to surviving patients with favorable outcome (good recovery or moderate disability) for more detailed evaluation of their social and physical rehabilitation. The response rate for the postal questionnaire was 60%.

Statistical analysis

Data were analyzed using SPSS 13.0 for Windows (SPSS Inc., Chicago, IL, USA). Following distribution analysis (Kolmogorov-Smirnov and Shapiro-Wilk tests), mean values ± SD were used for data following normal distribution, and median values ± interquartile range (IQR) for non-parametric data. Means and medians of physiological variables before and after craniectomy were compared using paired *t*-test, Wilcoxon, and sign tests, respectively. P value of <0.05 was considered significant. Pearson and Spearman rank coefficients were used to test the strength of correlations.

Results

Patient characteristics

Demographic data from the study population as well as mechanisms and severity of injury are summarized in Table 1. The majority of patients had a post-resuscitation Glasgow Coma Scale (GCS) score of less than 8. The only patient who presented with mild head injury (GCS 13) deteriorated later and required intubation and intensive care management. Two patients had bilateral dilated unreactive pupils on admission; both subsequently died. A significant proportion of patients had a major intracranial injury (37%), and 25% of patients underwent non-cranial surgical procedures. Based on preoperative CT scan appearance, most patients had a diffuse brain injury (Marshall grade II–III) [16].

Initial ICP control measures

All patients were managed according to an ICP- and CPP-driven protocol [18] (Fig. 1), which aims to maintain ICP < 25 mmHg and CPP ≥ 60 mmHg. Therapeutic hypothermia was used in 44 (93.6%) patients, mannitol in 42 (89.4%) patients, hypertonic saline in 9 (19.1%) patients, external ventricular drainage in 13 (27.7%) patients, and barbiturate-induced coma in 15 (31.9%) patients. Decompressive craniectomy was considered when all medical measures to control ICP failed to achieve sustained optimal levels of ICP and CPP, and occasionally it was considered as an alternative to barbiturate therapy.

Surgical details

Decompressive craniectomy was performed on median day 2 (IQR 1;3) after injury. In 24 (49%) patients the operation was performed within 24 hours, and in 32 (65.3%) patients within 48 hours from the time of admission. Forty-one (83.7%) patients underwent bifrontal decompressive craniectomy, and 8 (16.3%) had a unilateral procedure. In all cases the dura was opened, and in cases of bifrontal decompressive craniectomy the falx cerebri was divided anteriorly if deemed necessary by the operating surgeon.

Complications associated with the operation were observed in 4 (8%) patients. Two patients suffered intracranial hemorrhage and 1 patient had an acute subdural hematoma in the early postoperative period, which required evacuation. One patient developed a cerebral abscess, which was treated with needle aspiration and antibiotics.

ICP data

Decompressive craniectomy led to a reduction in ICP pressure in all patients for whom pre- and post-

Table 1. *Summary of patient characteristics (n = 49)*

Age	28 (range 9–67)
Gender	
– Male	37 (75%)
– Female	12 (25%)
Mechanism of injury	
– Road Traffic Accident	29 (60%)
– Fall	11 (22%)
– Assault	5 (10%)
– Other	4 (8%)
Severity of injury	
GCS-based head injury grade:	
– Severe (3–8)	40 (82%)
– Moderate (9–12)	8 (16%)
– Mild (≥ 13)	1 (2%)
Preoperative CT Marshall grade (n = 27):	
– I	—
– II	4 (15%)
– III	13 (48%)
– IV	5 (18%)
– V	1 (4%)
– VI	4 (15%)
Injury severity score: median (IQR), n = 30	25 (16–27)
APACHE score: mean (SD), n = 30	17 (6.7)
Major extracranial injury	18 (37%)
Non-cranial surgery	12 (25%)
Median (IQR) ICU stay (days)	19 (range 14–26)

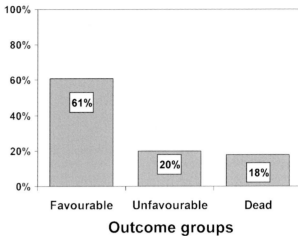

Fig. 2. Clinical outcome of patients at 6+ months assessed by Glasgow Outcome Scale score

operative ICP monitoring data was available (n = 27). Mean ± SD ICP during the 24-hour period prior to surgical decompression was 25 ± 6 mmHg as compared to 16 ± 6 mmHg for the 24 hours following an operation, and the difference is statistically significant ($p < 0.01$).

Outcome

The GOS showed favorable recovery in 30 (61.2%) patients (24 [49%] patients with good recovery and 6 [12.2%] patients with moderate disability), 10 (20.4%) patients remained severely disabled, and 9 (18.4%) patients died (Fig. 2). No patient was left in persistent vegetative state. Four out of 9 patients presenting with an initial GCS of 3 were severely disabled and 2 died; however, 3 (33.3%) patients in this group had a favorable outcome. In general, outcome was related to initial GCS score ($p = 0.004$), but there was no association with age, gender, or timing of operation with the GOS. Interestingly, in the 5 patients over 50 years of age, the rate of favorable outcome was not different than in younger patients (60%), although the numbers were too small to perform further statistical analysis. Mortality in this group accounted for the remaining 40% of patients, with no patients left severely disabled.

Eighteen out of 30 patients with favorable outcome responded to the SF-36 quality of life survey (60% response rate). The summary for all 8 domains is presented in Table 2 and compares general population levels [13]. At least 10 out of 40 surviving patients returned to work (information on re-employment was not available for 14 patients). Cranioplasty was performed in 30 patients (75% of survivors) at the time of analysis. Post-traumatic hydrocephalus, which required a shunting procedure, was observed in 6 (15%) of the 40 survivors.

Table 2. *SF-36 quality of life survey following decompressive craniectomy (n = 18). Numbers represent mean ± SD scores for each SF-6 domain in decompressive craniectomy patients and general population controls. The latter values were obtained from the results of a recent large-scale population survey [13]*

SF-36 Domain	Decompressive Craniectomy mean ± SD	Controls mean ± SD
Physical Functioning (PF)	67.5 ± 32	87.9 ± 20
Role limitation due to Physical problems (RP)	45.8 ± 46	87.2 ± 22
General Health perception (GH)	60.1 ± 22	71.0 ± 20
Bodily Pain (BP)	64.8 ± 35	78.8 ± 23
Role limitation due to Emotional problems (RE)	51.9 ± 47	85.8 ± 21
Energy Vitality score (EV)	50.3 ± 22	58.0 ± 20
Social Functioning (SF)	48.8 ± 28	82.8 ± 23
Mental Health (MH)	62.9 ± 27	71.9 ± 18

Discussion

Decompressive craniectomy continues to be used as a treatment measure for advanced cerebral edema and intractable intracranial hypertension. It has been applied to patients with malignant cerebral swelling following middle cerebral artery thrombosis [5, 19, 22, 25] and in other conditions leading to brain edema [2,

7, 8, 20, 26, 28]. The indications for decompressive craniectomy following traumatic brain injury are not universally defined. Currently, the procedure is considered when medical treatment measures fail to control ICP or when malignant intracranial hypertension is present on admission.

Although decompressive craniectomy may lead to an improvement in physiological parameters [4, 12, 32], the impact of this procedure on clinical outcome and social rehabilitation has not been clearly established. Present evidence is based on observational studies or case reports, with the single exception of a small pilot prospective randomized trial [27]. Outcome data from the published case series varies significantly. Although mortality rates quoted by different centers fall in the range of 20% ± 5%, the rates of reported favorable outcome and severe disability are considerably different. The proportion of patients achieving a favorable outcome (GOS: good recovery/moderate disability) after decompressive craniectomy has been reported to be as low as 30–40% [1, 6, 21, 23] and as high as 60–70% [9, 15, 27, 31]. One of the other serious concerns raised with regard to decompressive craniectomy is that the operation may reduce mortality, but increase the subset of patients with severe disability and persistent vegetative state. Indeed, some authors report high levels of severe disability [21, 23] and persistent vegetative state [6]. The discrepancy in published outcome data may, to some extent, be explained by difference in patient selection, indications, timing and technique of surgery, as well as an assessment of outcome.

In our unit, decompressive craniectomy is used as part of a protocol-based management of severe head injury (Fig. 1). The multidisciplinary decision to perform surgical decompression involves liaison between a senior neurosurgeon and neuro-intensivist. We have previously reported high rates of favorable outcome in a smaller series of patients [31], and the results of this study are consistent with our earlier findings. The current study shows that favorable outcome can be achieved in a large proportion of patients presenting with severe and potentially fatal head injury, with at least one-quarter returning to work. A high rate of severe disability or vegetative state after surgical decompression is not supported by our findings, as no patient was left in a vegetative state and the severe disability rate was not higher than generally accepted. In terms of quality of life, the response to the SF-36 questionnaire indicates that patients with favorable clinical outcome continue to experience physical and emotional consequences of their injury. The mean scores for patients' quality of life differ from a general population in all SF-36 domains; however, the large SDs reflect differences in response between individuals and suggest that some patients rate their quality of life as high.

Although the complication rate associated with surgery [17, 24, 33, 34] is low, the impact on outcome and the exact relationship with craniectomy requires further evaluation.

In conclusion, despite the results from this and other observational studies, consensus on the indications and timing of decompressive craniectomy following traumatic brain injury has not been achieved. There is now a need to obtain Class I evidence by proceeding with randomized, multi-center, prospective studies [11].

Acknowledgments

P. J. Hutchinson is supported by an Academy of Medical Science/Health Foundation Senior Surgical Scientist Fellowship. I. Timofeev is supported by a grant from Codman Division, Johnson and Johnson Corporation.

References

1. Albanese J, Leone M, Alliez JR, Kaya JM, Antonini F, Alliez B, Martin C (2003) Decompressive craniectomy for severe traumatic brain injury: Evaluation of the effects at one year. Crit Care Med 31: 2535–2538
2. Ausman JI, Rogers C, Sharp HL (1976) Decompressive craniectomy for the encephalopathy of Reye's syndrome. Surg Neurol 6: 97–99
3. Berger S, Schwarz M, Huth R (2002) Hypertonic saline solution and decompressive craniectomy for treatment of intracranial hypertension in pediatric severe traumatic brain injury. J Trauma 53: 558–563
4. Bor-Seng-Shu E, Teixeira MJ, Hirsch R, Andrade AF, Marino R Jr (2004) Transcranial doppler sonography in two patients who underwent decompressive craniectomy for traumatic brain swelling: report of two cases. Arq Neuropsiquiatr 62: 715–721
5. Cho DY, Chen TC, Lee HC (2003) Ultra-early decompressive craniectomy for malignant middle cerebral artery infarction. Surg Neurol 60: 227–232; discussion 232–233
6. De Luca GP, Volpin L, Fornezza U, Cervellini P, Zanusso M, Casentini L, Curri D, Piacentino M, Bozzato G, Colombo F (2000) The role of decompressive craniectomy in the treatment of uncontrollable post-traumatic intracranial hypertension. Acta Neurochir [Suppl] 76: 401–404
7. Dierssen G, Carda R, Coca JM (1983) The influence of large decompressive craniectomy on the outcome of surgical treatment in spontaneous intracerebral haematomas. Acta Neurochir (Wien) 69: 53–60
8. Fisher CM, Ojemann RG (1994) Bilateral decompressive craniectomy for worsening coma in acute subarachnoid hemorrhage. Observations in support of the procedure. Surg Neurol 41: 65–74

9. Guerra WK, Gaab MR, Dietz H, Mueller JU, Piek J, Fritsch MJ (1999) Surgical decompression for traumatic brain swelling: indications and results. J Neurosurg 90: 187–196
10. Brain Trauma Foundation, Inc, American Association of Neurological Surgeons, Congress of Neurological Surgeons, Joint Section on Neurotrauma and Critical Care (2003) Guidelines for the management of severe traumatic brain injury: cerebral perfusion pressure. Brain Trauma Foundation, Inc, New York (14 pages)
11. Hutchinson PJ, Menon DK, Kirkpatrick PJ (2005) Decompressive craniectomy in traumatic brain injury – time for randomised trials? Acta Neurochir (Wien) 147: 1–3
12. Jaeger M, Soehle M, Meixensberger J (2003) Effects of decompressive craniectomy on brain tissue oxygen in patients with intracranial hypertension. J Neurol Neurosurg Psychiatry 74: 513–515
13. Jenkinson C, Stewart-Brown S, Petersen S, Paice C (1999) Assessment of the SF-36 version 2 in the United Kingdom. J Epidemiol Community Health 53: 46–50
14. Jennett B, Bond M (1975) Assessment of outcome after severe brain damage. Lancet 1: 480–484
15. Kontopoulos V, Foroglou N, Patsalas J, Magras J, Foroglou G, Yiannakou-Pephtoulidou M, Sofianos E, Anastassiou H, Tsaoussi G (2002) Decompressive craniectomy for the management of patients with refractory hypertension: should it be reconsidered? Acta Neurochir (Wien) 144: 791–796
16. Marshall LF, Marshall SB, Klauber MR, Van Berkum Clark M, Eisenberg HM, Jane JA, Luerssen TG, Marmarou A, Foulkes MA (1991) A new classification of head injury based on computerized tomography. J Neurosurg [Suppl] 75: S14–S20
17. Mazzini L, Campini R, Angelino E, Rognone F, Pastore I, Oliveri G (2003) Posttraumatic hydrocephalus: a clinical, neuroradiologic, and neuropsychologic assessment of long-term outcome. Arch Phys Med Rehab 84: 1637–1641
18. Menon DK (1999) Cerebral protection in severe brain injury: physiological determinants of outcome and their optimisation. Br Med Bull 55: 226–258
19. Mori K, Nakao Y, Yamamoto T, Maeda M (2004) Early external decompressive craniectomy with duroplasty improves functional recovery in patients with massive hemispheric embolic infarction: timing and indication of decompressive surgery for malignant cerebral infarction. Surg Neurol 62: 420–429; discussion 429–430
20. Ong YK, Goh KY, Chan C (2002) Bifrontal decompressive craniectomy for acute subdural empyema. Childs Nerv Syst 18: 340–343; discussion 344
21. Polin RS, Shaffrey ME, Bogaev CA, Tisdale N, Germanson T, Bocchicchio B, Jane JA (1997) Decompressive bifrontal craniectomy in the treatment of severe refractory posttraumatic cerebral edema. Neurosurgery 41: 84–92; discussion 92–94
22. Robertson SC, Lennarson P, Hasan DM, Traynelis VC (2004) Clinical course and surgical management of massive cerebral infarction. Neurosurgery 55: 55–61; discussion 61–62
23. Schneider GH, Bardt T, Lanksch WR, Unterberg A (2002) Decompressive craniectomy following traumatic brain injury: ICP, CPP and neurological outcome. Acta Neurochir [Suppl] 81: 77–79
24. Schwab S, Erbguth F, Aschoff A, Orberk E, Spranger M, Hacke W (1998) "Paradoxical" herniation after decompressive trephining [in German]. Nervenarzt 69: 896–900
25. Schwab S, Steiner T, Aschoff A, Schwarz S, Steiner HH, Jansen O, Hacke W (1998) Early hemicraniectomy in patients with complete middle cerebral artery infarction. Stroke 29: 1888–1893
26. Stefini R, Latronico N, Cornali C, Rasulo F, Bollati A (1999) Emergent decompressive craniectomy in patients with fixed dilated pupils due to cerebral venous and dural sinus thrombosis: report of three cases. Neurosurgery 45: 626–629; discussion 629–630
27. Taylor A, Butt W, Rosenfeld J, Shann F, Ditchfield M, Lewis E, Klug G, Wallace D, Henning R, Tibballs J (2001) A randomized trial of very early decompressive craniectomy in children with traumatic brain injury and sustained intracranial hypertension. Childs Nerv Syst 17: 154–162
28. Wada Y, Kubo T, Asano T, Senda N, Isono M, Kobayashi H (2002) Fulminant subdural empyema treated with a wide decompressive craniectomy and continuous irrigation – case report. Neurol Med Chir (Tokyo) 42: 414–416
29. Ware JE (1993) SF-36 health survey: manual and interpretation guide. The Health Institute, New England Medical Center, Boston, pp 1–22
30. Whitfield P, Guazzo E (1995) ICP reduction following decompressive craniectomy. Stroke 26: 1125–1126
31. Whitfield PC, Patel H, Hutchinson PJ, Czosnyka M, Parry D, Menon D, Pickard JD, Kirkpatrick PJ (2001) Bifrontal decompressive craniectomy in the management of posttraumatic intracranial hypertension. Br J Neurosurg 15: 500–507
32. Yamakami I, Yamaura A (1993) Effects of decompressive craniectomy on regional cerebral blood flow in severe head trauma patients. Neurol Med Chir (Tokyo) 33: 616–620
33. Yamaura A, Makino H (1977) Neurological deficits in the presence of the sinking skin flap following decompressive craniectomy. Neurol Med Chir (Tokyo) 17: 43–53
34. Yang XJ, Hong GL, Su SB, Yang SY (2003) Complications induced by decompressive craniectomies after traumatic brain injury. Chin J Traumatol 6: 99–103
35. Yoo DS, Kim DS, Cho KS, Huh PW, Park CK, Kang JK (1999) Ventricular pressure monitoring during bilateral decompression with dural expansion. J Neurosurg 91: 953–959

Correspondence: Peter Hutchinson, Department of Neurosurgery, Addenbrooke's Hospital, Box 167, Hills Road, Cambridge, UK, CB2 2QQ. e-mail: pjah2@cam.ac.uk

Decompressive craniectomy in traumatic brain injury: the randomized multicenter RESCUEicp study (www.RESCUEicp.com)

P. J. Hutchinson[1,2], E. Corteen[1,2], M. Czosnyka[1,2], A. D. Mendelow[2], D. K. Menon[1,2], P. Mitchell[2], G. Murray[2], J. D. Pickard[1,2], E. Rickels[2], J. Sahuquillo[2], F. Servadei[2], G. M. Teasdale[2], I. Timofeev[2], A. Unterberg[2], and P. J. Kirkpatrick[1,2]

[1] University of Cambridge, Addenbrooke's Hospital, Cambridge, UK
[2] European Brain Injury Consortium, Cambridge, UK

Summary

The RESCUEicp (Randomized Evaluation of Surgery with Craniectomy for Uncontrollable Elevation of intracranial pressure) study has been established to determine whether decompressive craniectomy has a role in the management of patients with traumatic brain injury and raised intracranial pressure that does not respond to initial treatment measures. We describe the concept of decompressive craniectomy in traumatic brain injury and the rationale and protocol of the RESCUEicp study.

Keywords: Head injury; traumatic brain injury; intracranial pressure; decompressive craniectomy.

Introduction

Severe head injury is a major cause of morbidity and mortality worldwide [1, 10]. Of the pathophysiological processes implicated in traumatic brain injury, one of the most fundamental is an escalating cycle of brain swelling, increase in intracranial pressure (ICP), reduction in blood and oxygen supply, energy failure, and cell death. The goal for management of these patients is, therefore, to prevent secondary insults, reduce swelling and ICP, and maintain blood supply, oxygen delivery, and the energy status of the brain. This management goal applies at the trauma scene, during transportation, in the emergency department, in the operating room, and in the intensive care unit.

In order to assist in the management of patients with traumatic brain injury on the intensive care unit, protocols have been developed providing a step-wise approach to the control of brain swelling and raised ICP [2, 7, 8]. Targets have been established for ICP (<20–25 mmHg) and cerebral perfusion pressure (CPP) (CPP = mean arterial pressure – ICP [>60–70 mmHg]). These protocols commence with relatively simple therapeutic maneuvers including sedation, ventilation, and head-up position for nursing. If these basic measures fail to control brain swelling, more advanced medical treatment can be applied including the application of inotropes, hypertonic saline, mannitol, and hypothermia. External ventricular drainage of cerebrospinal fluid may also be feasible, depending on the size of the lateral ventricles. While brain swelling and ICP can be controlled in most patients with these basic measures, in others ICP is refractory and more advanced therapy needs to be considered. Options for these patients include the implementation of barbiturate coma and/or a decompressive surgical procedure such as decompressive craniectomy. Evidence of benefit for both of these therapies is currently sparse [4, 9].

Decompressive craniectomy is not a new operation. The phenomenon of brain decompression has been in existence for hundreds of years, commencing with trephination by the ancient Greeks, followed by Kocher approximately 100 years ago. Most recently, the operation has been performed in the context of modern, sophisticated neuro-intensive care.

In terms of controlling raised ICP, decompressive craniectomy is performed in several neurosurgical units world-wide, but there is no consensus on if and when to proceed with the surgery. There are several concerns regarding the application of decompressive craniectomy. These include failure to control ICP despite decompression, and herniation of brain through the craniectomy defect. In addition, the operation may be being performed unnecessarily in patients destined for a good prognosis, and pertinent to patients with poor prognosis, may save life at the expense of creating vegetative state and severe disability.

In terms of the current literature on the application of decompressive craniectomy to patients with traumatic brain injury, there have been several studies published (for review, see [3]). These are predominantly observational studies and demonstrate a wide range of outcomes (good recovery 29%–69%; mortality 11%–40%). There has been one attempt at randomization in the pediatric age group, which showed a significant difference in outcome, favoring patients randomized to surgery [11]. This study had a number of limitations, which were recognized by the authors, including sample size, outcome evaluation, and prolonged study duration.

In view of the widely differing results in these published studies, consensus on the role of decompressive craniectomy has not been achieved. This has resulted in proposals for randomized controlled trials. Three such trials have been proposed: an American trial, an Australasian trial (International Multicenter Randomized Controlled Trial on Early Decompressive Craniectomy – DECRA), and a European study (Randomized Evaluation of Surgery with Craniectomy for Uncontrollable Elevation of intracranial pressure – RESCUEicp).

This manuscript outlines the protocol for the RESCUEicp study – a randomized control trial comparing the efficacy of decompressive craniectomy versus optimal medical management for the treatment of refractory intracranial hypertension following brain trauma. This is a multi-center trial organized as a collaboration between the University of Cambridge Departments of Neurosurgery/Neuro-intensive Care and the European Brain Injury Consortium.

Methodology

The RESCUEicp study is a randomized trial comparing optimal medical management with surgery (decompressive craniectomy) for the management of intracranial hypertension following head injury, refractory to first-line treatment. The trial aims to recruit patients from centers experienced in the intensive care management of head injury. The target study group includes ventilated, ICP-monitored patients with refractory intracranial hypertension. Post-randomization, the trial comprises 2 arms: continuation of optimal medical management versus surgery (decompressive craniectomy).

Hypotheses

The principal hypothesis for the RESCUEicp study is that the application of decompressive craniectomy to head-injured patients with raised ICP refractory to medical treatment results in improved outcome. Specifically, 1) that decompressive craniectomy results in an improvement in the Extended Glasgow Outcome Score compared to optimal medical treatment and, 2) that decompressive craniectomy results in an improvement in surrogate endpoint measures (including specific outcome measures [SF-36 questionnaire], control of ICP, time in intensive care, and time to discharge from the neurosurgical unit) compared to optimal medical treatment.

Protocol

Patient inclusion and exclusion criteria for the trial are shown in Table 1. Approval for the study has been obtained from the UK National Ethics Committee, with ethical approval in other countries on-going. To avoid delays in treatment, it is recommended that consent for participation in the study is obtained from next of kin at time of patient admission to the neurosurgical unit, with randomization performed following failure to control ICP with initial treatment. Patients are managed on intensive care units using the trial protocol (Fig. 1). The major objective of this protocol is to maintain ICP < 25 mmHg by applying treatment measures in 3 stages.

Table 1. *Inclusion and exclusion criteria for the RESCUEicp study*

Inclusion criteria	Exclusion criteria
Head injury requiring ICP monitoring	bilateral fixed and dilated pupils
Age 10–65 years	bleeding diathesis
Abnormal CT scan	follow up not possible
Patients may have had an immediate operation for a mass lesion but not a decompressive craniectomy	devastating injury not expected to survive 24 hours

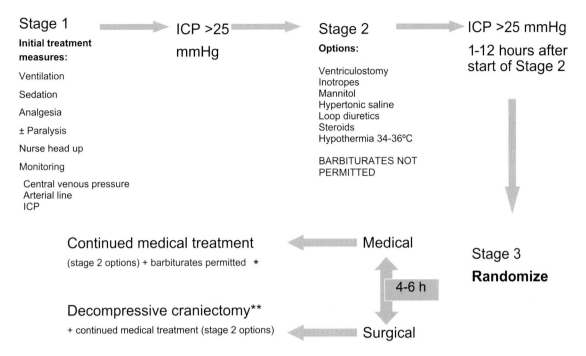

Fig. 1. Summary of RESCUEicp decompressive craniectomy protocol. The protocol is divided into 3 stages. Stage 1 comprises initial treatment measures. If the ICP > 25 mmHg after the implementation of Stage 1, the patient enters Stage 2 comprising a number of further options to control ICP. If the ICP is >25 mmHg after the implementation of Stage 2, the patient enters Stage 3 and is randomized to on-going medical treatment or decompressive craniectomy

Stage 1: initial treatment measures

Patients are sedated, relieved of pain, and ventilated (with the option for paralysis), in the head-up position for nursing and invasive monitoring (ICP, central venous pressure, and arterial lines as a minimum). ICP is assessed at this stage. If the ICP < 25 mmHg, the above medical treatment continues. If the ICP > 25 mmHg, Stage 2 is implemented (with the option of a repeat scan to investigate the possibility of an evolving mass lesion).

Stage 2: advanced treatment measures

In Stage 2, the following measures can be considered, all of which are optional: an external ventricular drain (depending on the size of the lateral ventricles), mannitol, inotropes (to increase the mean arterial pressure to maintain a CPP of >60 mmHg), arterial carbon dioxide 3.5 kPa to 4.5 kPa (can be monitored with jugular venous oxygen saturation sensors maintaining $SjvO_2 > 55\%$), hypertonic saline, moderate cooling (34–36 °C) but not severe hypothermia, loop diuretics, and steroids. Barbiturates are not implemented as part of Stage 2, but are reserved as part of continued medical treatment following randomization. This clause enables a direct comparison between the efficacy of decompressive craniectomy and extended medical treatment including the introduction of barbiturate coma.

Stage 3: randomization

If ICP remains > 25 mmHg for 1–12 hours despite Stage 1 and 2 measures, then the patient is randomized to either continued Stage 2 medical treatment including barbiturates (e.g. thiopentone boluses + infusion 4–8 mg/kg/hr), or to surgical treatment (decompressive craniectomy). Treatment following randomization should be implemented within 4–6 hours.

The surgical treatment comprises 2 options: (a)

for unilateral hemisphere swelling, a large unilateral fronto-temporo-parietal craniectomy or, (b) for bilateral diffuse hemisphere swelling, a large bilateral fronto-temporo-parietal craniectomy from the frontal sinus anteriorly to the coronal suture posteriorly and pterion laterally with a wide dural opening (pedicles based on the superior sagittal sinus medially) with the option of division of the falx anteriorly.

If the patient is randomized to continued medical treatment, no decompressive surgery is performed at the time of randomization, but may be performed later at the clinician's discretion if the patient subsequently deteriorates (for example, prolonged and unacceptably high ICP > 40 mmHg with compromised CPP). This clause is required if a situation arises whereby the treating physician feels that withholding surgery is acting against the best interest of the individual patient; "the interests of the patient always prevails over those of science and society" [6]. The same principle applies to barbiturates in the decompressive craniectomy group.

Outcome is assessed at discharge using the Glasgow Coma Score, and at 6 months using the Extended Glasgow Outcome Score [12]. Secondary endpoints are assessment of outcome using the SF-36 quality of life questionnaire [5], assessment of ICP control, time in intensive care, and time to discharge from the neurosurgical unit. For patients undergoing decompressive craniectomy, it is recommended that the bone flap is replaced within 6 months of the initial injury.

Recruitment of patients commences with a pilot phase comprising 50 patients. Overall the study aims to recruit 400 patients (200 in each arm of the study) for a 15% difference in outcome (difference in favorable outcome from 45%–60%) (power = 80%, p = 0.05). The pilot phase of the RESCUEicp study has now commenced, with 19 of out the pilot study target of 50 patients recruited.

Conclusion

Despite considerable interest in the concept of decompressive craniectomy for patients with traumatic brain injury, including several peer-reviewed publications, current opinion on the role of this operation is divided. Randomized studies are now indicated to obtain good quality Class I evidence. The RESCUEicp study has now commenced and aims to establish whether decompressive craniectomy results in improvement in outcome compared to contemporary optimal medical management. Contact with the RESCUEicp group in Cambridge, UK, can be made via the website www.RESCUEicp.com or by e-mail pjah2@cam.ac.uk.

Acknowledgments

This RESCUEicp protocol has evolved as a result of discussions between representatives of the University of Cambridge and the European Brain Injury Consortium. P. J. Hutchinson is supported by an Academy of Medical Sciences/Health Foundation Senior Surgical Scientist Fellowship. We wish to acknowledge the members of the data monitoring committee, Mr. M. D. M. Shaw, Dr. H. K. Richards, Dr. M. Smith, Prof. Lennart Persson.

References

1. American College of Surgeons Committee on Trauma (1997) Advanced trauma life support program for doctors, 6th edn. American College of Surgeons, Chicago
2. Elf K, Nilsson P, Enblad P (2002) Outcome after traumatic brain injury improved by an organized secondary insult program and standardized neurointensive care. Crit Care Med 30: 2129–2134
3. Hutchinson PJ, Kirkpatrick PJ (2004) Decompressive craniectomy in head injury. Curr Opin Crit Care 10: 101–104
4. Hutchinson PJ, Menon DK, Kirkpatrick PJ (2005) Decompressive craniectomy in traumatic brain injury – time for randomised trials? Acta Neurochir (Wien) 147: 1–3
5. Jenkinson C, Wright L, Coulter A (1993) Quality of life measurement in health care: a review of measures, and population norms for the UK SF-36. Health Services Research Unit, Oxford
6. Medical Research Council London (1998) MRC guidelines for good clinical practice in clinical trials. www.mrc.ac.uk/pdf-ctg.pdf
7. Menon DK (1999) Cerebral protection in severe brain injury: physiological determinants of outcome and their optimisation. Br Med Bull 55: 226–258
8. Polderman KH, Tjong Tjin Joe R, Peerdeman SM, Vantertop WP, Girbes AR (2002) Effects of therapeutic hypothermia on intracranial pressure and outcome in patients with severe head injury. Intensive Care Med 28: 1563–1573
9. Roberts I (2000) Barbiturates for acute traumatic brain injury. Cochrane Database Syst Rev 2:CD000033
10. Royal College of Surgeons of England (1999) Report of the working party on the management of patients with head injuries. Royal College of Surgeons, London
11. Taylor A, Butt W, Rosenfeld J, Shann F, Ditchfield M, Lewis E, Klug G, Wallace D, Henning R, Tibballs J (2001) A randomized trial of very early decompressive craniectomy in children with traumatic brain injury and sustained intracranial hypertension. Childs Nerv Syst 17: 154–162
12. Wilson JT, Pettigrew LE, Teasdale GM (1998) Structured interviews for the Glasgow Outcome Scale and the extended Glasgow Outcome Scale: guidelines for their use. J Neurotrauma 5: 573–585

Correspondence: P. J. Hutchinson, Academic Department of Neurosurgery, University of Cambridge, Box 167, Addenbrooke's Hospital, Cambridge, UK, CB2 2QQ. e-mail: pjah2@cam.ac.uk

Cerebral hemisphere asymmetry in cerebrovascular regulation in ventilated traumatic brain injury

S. C. P. Ng[1], W. S. Poon[1], and M. T. V. Chan[2]

[1] Division of Neurosurgery, Department of Surgery, Prince of Wales Hospital, The Chinese University of Hong Kong, Shatin, Hong Kong
[2] Department of Anesthesia and Intensive Care, Prince of Wales Hospital, The Chinese University of Hong Kong, Shatin, Hong Kong

Summary

Disturbances in cerebrovascular regulation in the form of diminished cerebral vasoreactivity (CVR) to carbon dioxide and an altered pressure autoregulatory response (PAR) are common after traumatic brain injury (TBI) and correlate with clinical outcome. Daily assessment of the state of cerebrovascular regulation may assist in the clinical management of TBI patients. This study examined 20 ventilated TBI patients. We employed blood flow velocity (BFV) measurement using transcranial Doppler ultrasonography to assess the impact of injury type (focal and diffuse) on cerebral hemisphere asymmetry in cerebrovascular regulation and to examine whether impairment in CVR and PAR correlate with clinical outcomes. Significant hemisphere asymmetries were found in BFV and PAR. Impairment in CVR was associated with unfavorable outcomes and bilateral CVR impairment predicted mortality.

Keywords: Blood flow velocity; cerebral vasoreactivity; pressure autoregulatory response; traumatic brain injury.

Introduction

Cerebrovascular regulation, in the form of cerebral vasoreactivity (CVR) to carbon dioxide and the pressure autoregulatory response (PAR), is affected after traumatic brain injury (TBI). Impairment in regulation is associated with severity of brain injury and correlates with neurological condition and clinical outcome [1, 2]. Knowledge of the state of cerebrovascular regulation is fundamental to understanding the pathophysiology of TBI and its sequelae in individual patients. Daily assessment of CVR and PAR may assist in the clinical management of TBI patients. However, the asymmetry of CVR and PAR between cerebral hemispheres has not been investigated systematically [4].

The aim of this study was to evaluate the cerebral hemisphere asymmetry of CVR and PAR in relation to focal brain injury and to correlate impairment in such regulation with clinical outcome in ventilated TBI patients.

Materials and methods

Patients aged less than 70 years with moderate to severe TBI, as defined by post-resuscitation Glasgow Coma Scale (GCS), with a score less than or equal to 12, were included in this study. After surgical intervention, all patients were managed in an intensive care unit under a standard protocol including artificial ventilation and sedation.

Physiological parameters including heart rate, arterial blood pressure (ABP), intracranial pressure (ICP), arterial oxygen saturation, end-tidal carbon dioxide concentration ($EtCO_2$), and jugular oxygen saturation were continuously monitored. Blood flow velocities (BFV) in both middle cerebral arteries were determined by transcranial Doppler as soon as possible after surgery.

To determine CVR, the initial $EtCO_2$ was increased by 1 kiloPascal (7.5 mmHg) using moderate hypoventilation. CVR was quantified as the percent change in BFV per unit change in $EtCO_2$, the CVR ratio. Impaired CVR was defined as a less than 1% change in BFV per unit change in $EtCO_2$. To evaluate PAR, blood pressure at normocapnia level was increased to 20 to 25% above the baseline. PAR was measured as the percent change in BFV per unit change in mean ABP, the PAR ratio. There was deemed to be a loss of PAR if there was more than a 1.5% change in velocity per unit change in mean blood pressure [3].

Intracranial lesions were classified according to CT findings as focal (with intracranial hematoma or unilateral contusion with or without brain swelling) or diffuse axonal injury. Clinical outcome was assessed at 6 months after injury using a Glasgow Outcome Score (GOS). Patients were defined as having a favorable outcome if they had GOS grades indicating good recovery or moderate disability.

Results

Sixty-six CVR and 68 PAR tests were performed on 20 patients (16 males and 4 females; mean age: 39.3 years, range: 2 to 69 years; median GCS score: 6.5).

Table 1. Side-to-side differences (in relation to pathology) in BFV, CVR ratio, and PAR ratio in focal and diffuse brain injured patients.

	Pathology right side (mean ± SD)	Non-pathology left side (mean ± SD)	Difference (mean ± SD)	p-value
Diffuse injury group				
BFV (cm/s)	91.59 ± 39.22	97.53 ± 47.55	5.94 ± 32.87	NS
CVR ratio (%/mmHg)	2.28 ± 2.46	2.58 ± 2.29	0.30 ± 1.57	NS
PAR ratio (%/mmHg)	0.79 ± 0.81	0.95 ± 0.86	0.16 ± 0.58	NS
Focal injury group				
BFV (cm/s)	66.08 ± 20.31	81.78 ± 24.17	15.71 ± 15.17	<0.0005*
CVR ratio (%/mmHg)	2.75 ± 2.83	3.40 ± 2.61	0.66 ± 2.27	NS
PAR ratio (%/mmHg)	1.55 ± 2.00	1.02 ± 0.94	−0.53 ± 1.69	0.033*

BFV Blood flow velocity; *CVR* cerebral vasoreactivity; *PAR* pressure autoregulatory response.

Fourteen patients had focal brain injury. Mortality in this group of moderate to severe brain-injured patients was 40%. Seven patients achieved a favorable outcome at 6 months after TBI.

During the CVR tests, moderate hypoventilation ($EtCO_2$: before vs. after: 30.3 ± 3.5 vs. 38.2 ± 4.0 mmHg, $p < 0.0005$) resulted in raised ICP (16.7 ± 9.3 vs. 22.4 ± 11.2 mmHg, $p < 0.0005$) while ABP remained constant (mean ABP: 84.6 ± 16.7 vs. 86.8 ± 17.4 mmHg, $p > 0.05$). During the PAR tests at normocapnia ($EtCO_2$: 34.2 ± 3.4 vs. 34.4 ± 3.8 mmHg, $p > 0.05$), the induced increase in ABP (mean ABP: 80.9 ± 14.7 vs. 101.6 ± 16.5 mmHg, $p < 0.0005$) did not alter ICP (21.3 ± 11.8 vs. 19.7 ± 12.2 mmHg, $p > 0.05$).

In relation to pathology, significant side-to-side differences were found in BFV (lesion side vs. non-lesion side: 73.8 ± 29.5 vs. 86.5 ± 33.4 cm/s, $p < 0.0005$) and CVR ratio (2.6 ± 2.7 vs. 3.2 ± 2.5%/mmHg, $p = 0.035$) but not in PAR ratio (1.3 ± 1.8 vs. 1.0 ± 0.9%/mmHg, $p > 0.05$). Data were stratified into diffuse and focal brain injury groups (Table 1). In the focal injury group, side-to-side differences were found in BFV and PAR. Although CVR ratio was lower on the lesion side, this did not reach statistical significance. No side-to-side differences were found in the diffuse injury group.

Regarding clinical outcome at 6 months after injury, the mean CVR ratios at different GOS were different ($p = 0.001$), and the CVR ratio of non-survivors was significantly lower ($p < 0.005$). No differences were found in the mean PAR ratio in relation to GOS. Further analysis of the dichotomized CVR and PAR ratios (Table 2) indicated a significant correlation between CVR and clinical outcome and mortality whereas there was no such correlation with PAR.

Table 2. Correlation between CVR and PAR ratios with clinical outcome and mortality (all measurements).

CVR ratio	Preserved	Impaired	p-value
Outcome: favorable	15/42	0/24	0.002*
Mortality: death	1/42	13/24	<0.0005*
PAR ratio	Intact	Impaired	p-value
Outcome: favorable	11/51	4/17	NS
Mortality: death	12/51	4/17	NS

CVR Cerebral vasoreactivity; *PAR* pressure autoregulatory response.

Table 3. Association of impaired CVR and PAR with clinical outcome and mortality. CVR and PAR were determined by TCD within 24 hours of admission.

CVR ratio	Preserved	Impaired		p-value
		One side	Both sides	
Outcome: favorable	7/9	0/3	0/6	0.003*
Mortality: death	2/9	2/3	6/6	0.003*
PAR ratio	Intact	Impaired		p-value
		One side	Both sides	
Outcome: favorable	1/10	1/4	2/3	NS
Mortality: death	3/10	2/4	1/3	NS

CVR Cerebral vasoreactivity; *PAR* pressure autoregulatory response; *TCD* transcranial Doppler.

CVR and PAR tests were performed within 24 hours of admission in 18 and 17 patients, respectively. Impairment in CVR was significantly associated with worse clinical outcome and mortality (Table 3). All 6 patients with bilateral CVR impairment were non-survivors. Such an association was absent for the PAR tests.

Discussion

CVR to carbon dioxide and the pressure autoregulation response are mediated through different mechanisms. The PAR is very sensitive to brain damage whereas CVR is more resistant. An asymmetry in the PAR was found in the present study and this is consistent with a previous study [4]. However, no association between PAR impairment and unfavorable outcome was found. The fact that some patients with impaired PAR can be treated may explain the dissociation between PAR impairment and unfavorable outcome. Patients with diminished CVR during the first days after TBI were more likely to have a worse outcome than patients with preserved reactivity. In our group of patients, the CVR ratio of non-survivors was lower and all of them had bilateral CVR impairment.

In conclusion, this study supports the utility of BFV measurement using transcranial Doppler coupled with CO_2 and blood pressure challenge in the assessment of cerebrovascular regulation. Cerebral hemisphere asymmetry in cerebrovascular regulation was demonstrated and CVR impairment correlated with unfavorable outcome. In contrast, PAR impairment did not necessarily mean a poor prognosis. Daily assessment of the state of CVR and PAR may help in the clinical management and outcome prediction in moderate-to-severe TBI patients.

References

1. Enevoldsen EM, Jensen FT (1978) Autoregulaton and CO_2 responses of cerebral blood flow in patients with acute severe head injury. J Neurosurg 48: 689–703
2. Kelly DF, Marin NA, Kordestani R, Counelis G, Hovda DA, Bergsneider M, McBride DQ, Shalmon E, Herman D, Becker DP (1997) Cerebral blood flow as a predictor of outcome following traumatic brain injury. J Neurosurg 86: 633–641
3. Ng SC, Poon WS, Chan MT, Lam JM, Lam WW (2002) Is transcranial Doppler ultrasonography (TCD) good enough in determining CO_2 reactivity and pressure autoregulation in head-injured patients? Acta Neurochir [Suppl] 81: 125–127
4. Schmidt EA, Czosnyka M, Steiner LA, Balestreri M, Smielewski P, Piechnik SK, Matta BF, Pickard JD (2003) Asymmetry of pressure autoregulation after traumatic brain injury. J Neurosurg 99: 991–998

Correspondence: Wai S. Poon, Division of Neurosurgery, Department of Surgery, The Chinese University of Hong Kong, Shatin, Hong Kong. e-mail: wpoon@surgery.cuhk.edu.hk

Traumatic brain edema in diffuse and focal injury: cellular or vasogenic?

A. Marmarou[1], S. Signoretti[1], G. Aygok[1], P. Fatouros[2], and G. Portella[1]

[1] Department of Neurosurgery, Medical College of Virginia Commonwealth University, Richmond, VA
[2] Department of Neuroradiology, Medical College of Virginia Commonwealth University, Richmond, VA

Summary

The objective of this study was to confirm the nature of the edema, cellular or vasogenic, in traumatic brain injury in head-injured patients using magnetic resonance imaging techniques. Diffusion-weighted imaging methods were quantified by calculating the apparent diffusion coefficients (ADC). Brain water and cerebral blood flow (CBF) were also measured using magnetic resonance and stable Xenon CT techniques. After obtaining informed consent, 45 severely injured patients rated 8 or less on Glasgow Coma Scale (32 diffuse injury, 13 focal injury) and 8 normal volunteers were entered into the study. We observed that in regions of edema, the ADC was reduced, signifying a predominantly cellular edema. The ADC values in diffuse injured patients without swelling were close to normal and averaged 0.89 ± 0.08. This was not surprising, as ICP values for these patients were low. In contrast, in patients with significant brain swelling ADC values were reduced and averaged 0.74 ± 0.05 ($p < 0.0001$), consistent with a predominantly cellular edema. We also found that the CBF in these regions was well above ischemic threshold at time of study. Taking these findings in concert, it is concluded that the predominant form of edema responsible for brain swelling and raised ICP is cellular in nature.

Keywords: Head injury; traumatic brain edema; cellular edema; vasogenic edema.

Introduction

It has been determined that the predominant cause of brain swelling and subsequent rise in intracranial pressure (ICP) in head-injured patients is brain edema [8]. However, the type of edema, vasogenic or cellular, has not been completely resolved and this knowledge is critical to the development of new methods for treatment of brain swelling. Fortunately, magnetic resonance imaging (MRI) techniques allow us to identify predominant edema using the apparent diffusion coefficient (ADC) [6]. Using these methods, experimental studies have shown that cellular edema plays an important role in traumatic brain injury [2, 4] and that secondary insult exacerbates the cellular edema development [2]. However, there have been few studies on human head injury that address this important issue. The objective of our study was to identify the type of edema during the acute stage of severe traumatic brain injury, and also to determine if ischemia played a role in the type of edema that developed.

Methods

Informed consent from family or caregivers was obtained to allow the patient to participate in this study. Following stabilization in the intensive care unit (ICU) and with approval of the attending neurosurgeon, severely head-injured patients (Glasgow Coma Scale [GCS] 8 or less) were transported to MRI suites for measurement of brain tissue water and diffusion-weighted imaging (DWI). The transport team consisted of an ICU nurse, respiratory technician, and neurosurgical resident. The research team consisted of the investigator, head-injury fellow, and radiology technician. After stabilization in the magnet, the head was positioned carefully to insure computed tomography (CT) slice compatibility. First, T1- and T2-weighed pulse sequences were used to produce images in the axial, coronal, and sagittal planes in order to identify the traumatic lesions present.

Measurement of brain water

From these images, an anatomical slice equivalent to that used in CT/cerebral blood flow (CBF) measurements was selected for quantitative brain water measurements. Five inversion recovery scans of the selected slice were obtained with a repetition time (TR) of 2.5 seconds, echo time of 28 msec and inversion times (TI) of 150, 400, 800, 1300, and 1700 ms, respectively. From these images, brain water content f_w (expressed as a fraction of water per gram of tissue) was derived as described previously [3]. The selection of an inversion recovery pulse sequence as well as particular values of the parameters for measurement of brain water have been derived after extensive studies with calibration standards of known relaxation times and in vivo infusion edema animal models. Most importantly, the technique has been validated in man. When stated in terms of water content, the percentage swelling was calculated from the equation $100(f_w - f_{wn})/(1 - f_w)$ where f_w represents the water content of the edematous brain and f_{wn} represents the water content of the normal tissue. Based on these computations, a fractional increase in water of 2% is equivalent to a volume increase of 8.8%, which emphasizes the importance of edema for the swelling process and subsequent rise in ICP.

Diffusion-weighted imaging

DWI was performed on a 1.5 T whole-body clinical imager (Siemens Vision MR system, Siemens Medical Solutions USA, Inc., Malvern, PA) equipped with 15 mT/m gradients using spin-echo sequences with and without diffusion sensitizing gradients. These pulse sequences generated an ADC trace image using a single-shot technique with 3 b values: 0, 500, and 1000 s/mm². Typically, 20 slices were generated with 5 mm slice thickness, 1.5 mm gap, 230 mm field of view, 96 × 128 matrix, and TE 100 ms. The ADC trace was generated to obviate issues arising from tissue anisotropy and head orientation. From the ADC maps, regional measurements were extracted.

Measurement of CBF

After completion of MRI studies, the patient was transported to the CT suite for measurement of CBF. After a diagnostic CT scan, a stable xenon-CT CBF study was performed on a Siemens CT/Plus scanner (Siemens, Erlangen, Germany) equipped with a xenon gas delivery system and a CBF software analysis package. Scans were performed at 4 axial planes with a thickness of 10 mm each, 15 mm apart. Two baseline scans were performed at each level followed by multiple enhanced scans during inhalation of 30% xenon and 70% oxygen. From measurements of the CT enhancement and the end-tidal curve, CBF maps were calculated by means of the Kety-Schmidt equation using a commercially available package (Diversified Diagnostic Products, Inc., Houston, Texas).

Results

Description of study cohort

A total of 45 severely injured patients (GCS < 8) and 8 normal volunteers were entered into the study, 9 females and 35 males. The mean age was 33 years, ranging from 16 to 70 years. Of these, 32 were classified as having diffuse injury and 12 with focal lesions based on initial scan assessment. In diffuse injury, 20 were classified as CT type II and 12 were type III–IV according to the Marshall Classification. The 12 patients with focal lesions were classified as type V–VI.

Patient response to imaging protocols

All patients included in this report were transported safely from ICU to imaging suite and returned safely after protocols were completed without adverse event. Comparison of blood pressures and ICP measured prior to and after return indicated no major change throughout the course of study.

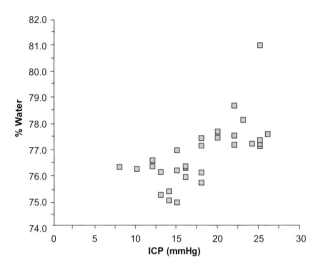

Fig. 1. ICP and water content in diffuse TBI. ICP increased to levels of 20 mmHg or greater above 77% water, which was only 1% above normal. In terms of swelling, ICP increased when brain swelling exceeded 4.3% increase in brain volume. (p < 0.001)

Water content, ICP, and CT classification in diffuse TBI

The normal hemispheric water content assessed by MRI in volunteers averaged 76.0% ± 0.66. The brain water content of diffusely injured patients in CT classification II was close to normal (76.2 ± 0.76) and was consistent with the low ICP of this group which averaged 14.8 ± 3.31 mmHg. In contrast, the water content of diffuse injury patients with CT classification III/IV was elevated and averaged 77.75 ± 1.15 g H₂O/g tissue (p < 0.001). This increase in edema is equivalent to an 8.16% increase in brain swelling according to the Elliot Jasper equation. Correspondingly, ICP at time of study for this group was also elevated and averaged 23.2 ± 2.48 mmHg (p < 0.001). Interestingly, the ICP increased to levels of 20 mmHg or higher as total water increased above 77% water, which is only 1% above normal. In terms of swelling, the ICP increased when brain swelling reached values above 4.3% (p < 0.001) (Fig. 1).

ADC in normal and head-injured patients with diffuse injury

The ADC computed from hemispheric regions of interest in 8 normal volunteers averaged (0.82 ± 0.05) × 10⁻³ mm²/sec. Henceforth, we will drop the common multiplicative factor 10⁻³. ADC values in diffusely injured patients was close to normal and averaged 0.89 ± 0.08 in the non-swelling group. This was

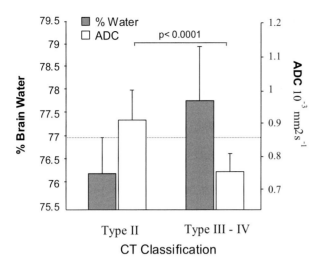

Fig. 2. Brain edema and ADC after diffuse TBI. In patients with significant brain swelling, ADC values were reduced and averaged 0.74 ± 0.05 ($p < 0.0001$), consistent with a predominantly cellular edema. Water in CT type III/IV was markedly elevated compared to less diffuse injury in CT type II

not surprising, as the ICP values of these patients were low. In contrast, the ADC values were reduced in patients with significant brain swelling and averaged 0.74 ± 0.05 ($p < 0.0001$), consistent with a predominantly cellular edema. The distribution of ADC among normal controls, swelling, and non-swelling patients correlated with CT classification (Fig. 2).

CBF in diffuse injury with reduced ADC

It is possible that the reductions in ADC seen in diffuse injury were caused by frank ischemic damage. This was the primary motivating factor to measure CBF at the same time DWI was measured. In patients with diffuse injury where ADC was significantly reduced, the corresponding hemispheric CBF of the non-swelling group averaged 50.04 ± 11.30 mL/min/100 g tissue, and in the patients with swelling averaged 45.33 ± 7.99. This reduction in CBF seen in patients with swelling did not reach significance. In absolute terms, the values of CBF in all patients studied with reduced ADC were well above ischemic threshold considered as 18 mL/min/100 g tissue. When CBF values were corrected and normalized for a PaCO2 of 34 mmHg, using 3% change per mmHg PaCO2, the CBF equaled 46.37 ± 9.63 mL/min/100 g tissue in the swelling group and 47.74 ± 11.35 mL/min/100 g tissue in the non-swelling group. These values were not statistically significant from the non-normalized CBF.

Water content, ICP, and CT classification in focal TBI

The methods for analysis in patients with focal injury were different than those of diffuse injury. First, we selected only patients for analysis which had a well-defined unilateral non-evacuated contusion, which are considered to be type VI in the Marshall classification. Second, we evaluated water content, ADC, and CBF in specific regions of interest, which included: contusion core, perilesional area, a region distant from the contusion on the ipsilateral side, and finally, a symmetrical site contralateral to the lesion. Of the initial cohort, 20 patients were selected that fit the inclusion criteria.

The water content of the contused site averaged $81.81\% \pm 2.36$, well above the brain water content of the symmetrical site of the contralateral hemisphere ($75.83 \pm 1.44\%$) ($p < 0.0001$). As water content increased, ICP increased in an exponential fashion. Similar to diffuse injury, the ICP increase was substantially above a level of 77.0% water. ICP at this level was 22 mmHg and corresponded to a 4.3% swelling.

ADC in head-injured patients with focal injury

The ADC in the core of the contusion and ring of hyperdensity surrounding the contusion was high and averaged 1.25 ± 0.37. This value is well above normal and was consistent among all of the contusions studied indicating a predominantly vasogenic edema. Beyond the core in the perilesional area, the ADC was below normal and averaged 0.72 ± 0.14 SD. The value was highly significant compared to the contused site and also to normal brain tissue ($p < 0.001$), signifying a predominantly cellular edema. The ADC of the perilesional area was also low compared to the symmetrical site in the contralateral hemisphere ($p = 0.01$). Distant from the contusion in the ipsilateral hemisphere, the ADC was approximately equal to the ADC observed in the non-contused hemisphere and averaged 0.84 ± 0.08. With the exception of the site of contusion, the ADC of the perilesional brain tissue was reduced. The ADC from tissue distant to the lesion and in the contralateral hemisphere was also reduced or near normal.

CBF in focal injury

The CBF of the non-injured hemisphere of the focal injury cohort averaged 45.86 ± 9.43, which is in the normal range for comatose patients. In contrast, the CBF in the perilesional area was low, averaging 34.01 ± 8.52. This reduction in CBF adjacent to the lesion compared to the non-injured hemisphere was statistically significant ($p < 0.001$). A similar reduction was seen when these CBF values were corrected and normalized to a PaC02 of 34 mmHg, assuming a similar 3% change in CBF per mmHg PaC02 as was used in the diffuse injury group. In both cases, CBF was above ischemic threshold levels supporting the notion that the edema distant from the contusion core is cellular in nature.

Discussion

This report provides supportive evidence that traumatic brain edema is predominantly a cellular phenomenon in both focal and diffuse injury. Further, it indicates that at time of study, when CBF measurements were made in conjunction with MRI studies, the blood flow was well above ischemic thresholds thereby obviating an ischemic-induced cellular component to the edematous process. The areas of edema were signified by zones of increased water content, measured by our MRI technique and, as might be expected, the highest levels of edema were measured in those patients with raised ICP. The ICP relationship to percentage water in patients with focal injury was exponential, illustrating a significantly reduced level of reserve in the contused hemisphere. Taken in concert, these results suggest that alternative therapies targeted to reduce the cellular edema component of swelling in both diffuse and focal injury must be considered.

For the past several decades, it has been generally accepted that the swelling process accompanying traumatic brain injury is mainly due to vascular engorgement, and it is the increase in blood volume that is mainly responsible for increase in brain bulk and subsequent rise in ICP. Edema was thought to play a minor role, particularly in a contusion where the blood-brain barrier (BBB) was thought to be transiently compromised. This controversy was resolved by studies in head-injured patients in whom both tissue water and cerebral blood volume were measured [8]. In both diffuse and focal injury, it was found that edema and not vascular engorgement was responsible for brain swelling and that blood volume was actually reduced following severe TBI. The problem remained as to what type of edema was involved.

Experimental studies to characterize the type of edema in the laboratory have used DWI to identify the extent of the cellular edema present in an impact acceleration model of diffuse injury. These studies showed that the rise in ICP following experimental TBI and concomitant reduction in ADC following injury was caused by a predominantly cellular edema. When coupled with a secondary insult, ischemia was also considered an important factor contributing to cellular swelling. Similarly, studies of focal injury using the cortical contusion model coupled with DWI indicated that the type of edema in the injured area, with or without superimposed hypotension secondary insult, was predominantly cellular [10]. However, it would seem reasonable that the formation of vasogenic and cellular edema would also be time dependent. Barzo et al. [1] studied the type of edema that forms during both the acute and chronic stage of diffuse impact acceleration injury using DWI and water content measures. The authors found a significant increase in the ADC during the first 60 minutes post-injury, which was consistent with development of a vasogenic edema secondary to BBB compromise. This transient increase in ADC was followed by a continual decrease in ADC that began 40 to 60 minutes post-injury and continued for up to 7 days post-injury. The study by Barzo et al. [2] illustrated that a vasogenic component of edema can develop soon after injury, but as the BBB closes, the cellular edema predominates. These studies implied an early closure to the BBB following traumatic brain injury. Changes in barrier opening were measured in a subsequent experimental study by the same group, using MRI monitoring with contrast agent. The authors found that closed head injury was associated with a rapid and transient BBB opening that lasted only 30 minutes. Secondary insult tended to prolong the duration of the barrier opening. Thus, the increase in barrier permeability in diffuse injury was short-lived despite the continued development of brain swelling, further supporting a predominantly cellular edema. Taken in concert, the experimental findings provide compelling evidence that the BBB opening is brief and the subsequent swelling of the brain is due to an increase in cellular water.

Human studies of DWI

The information regarding ADC changes in severely brain-injured patients is scant. Liu studied 9 patients ranging from 26 to 78 years of age with head trauma [7]. No data was available about the severity of injury.

Despite the small number of patients in their study, the authors found decreased ADC in patients with diffuse axonal injury in the acute setting. In some cases, the reduced ADC persisted into the sub-acute period, beyond that described for cytotoxic edema. This could not be confirmed because water measurements were not made. The ADC values ranged from a low of 0.334 at 9 days post-injury to 0.551. A more extensive study of DWI in head-injured patients was conducted by Kawamata *et al.* [5]. These authors studied 20 patients within 24 to 48 hours post-trauma. They expressed an ADC ratio of contusion site to normal brain. Using this ratio, the ADC value increased in the central area from 1.13 ± 0.13 and decreased in the peripheral area to 0.83 ± 0.81. The authors concluded that a combination of events facilitates edema accumulation in the tissue and contributes, together with the cellular edema in the peripheral area, to the mass effect of contusion edema. Nakahara *et al.* [9] studied 4 severely brain-injured patients and compared the ADC values with 4 reference subjects. They found that ADC values in the more severely injured patients were lower compared to the remaining patients.

In summary, investigations of severely brain-injured patients by others are consistent with the findings of our study, namely that ADC is reduced following severe TBI.

ADC and brain edema

As mentioned earlier, most studies attempt to relate the increased T2 relaxation to increased edema. In our study, where water was quantified in absolute terms, the regions of ADC change from normal values of 0.82 ± 0.05 correlated with water increase. For example, in patients with diffuse injury in the non-swelling group, the ADC values were close to normal (0.89 ± 0.08). However, in patients with significant brain swelling, ADC values were reduced to 0.74 ± 0.05, a reduction that was highly significant ($p < 0.0001$), and consistent with cellular edema. It must be emphasized that the measured ADC combines contributions from both vasogenic and cellular influences with the diffusion of water through the tissue. For example, if the vasogenic and cytotoxic influence were equal, the ADC would show no change from normal. That is the reason why we state that cellular edema is the *predominant* form of edema that is present.

In summary, the ADC, brain water, and CT classifications were consistent and changed in the expected direction. In CT classification type II where swelling was minimal, the ADC change was also minimal. However, in type III and IV with increased swelling and increased water content, the ADC was markedly reduced.

Changes in CBF and ADC

As mentioned earlier, CBF was measured at the same time that DWI was performed. The primary motivating factor was to eliminate frank ischemic damage as one possibility affecting the change in ADC. We found that CBF was reduced in patients with brain swelling and reduced ADC, but this reduction did not reach significance. All values were above ischemic thresholds.

Water content, ICP, and CT classification in focal TBI

The type of edema that surrounds a contusion is of great interest because it is the expanding lesion that leads to brain shift, elevated ICP, and eventually ischemia. According to the results of our study, we can subdivide the contusional area into 3 zones: the lesion core, the perilesional area, and the surrounding tissue. Our results show that water and the ADC in the lesion core were elevated, consistent with a zone of necrotic tissue coupled with a mixture of blood and water. Interestingly, in the perilesional area the ADC was reduced, as well as the ADC in tissue distant from the lesion. As water content increased in the contused hemisphere, ICP increased exponentially, similar to the shape of an ICP/volume curve. This suggests that small increases in water content, when reserve is exhausted, will result in dramatic increases in ICP.

The CBF changes followed a distinct pattern. In the center of the lesion, CBF was reduced below ischemic levels. In the perilesional area, CBF was higher in the vicinity of 35 mL/min/100 g tissue. The CBF in the perilesional area, although above ischemic threshold, was associated with reduced ADC indicating that the edema that formed around the contusion was predominantly cellular. The highest CBF was found, as expected, in the contralateral hemisphere and was near expected levels for the head-injured patient.

It has always been thought that changes in CT density surrounding a contusion were representative of a vasogenic edema exuding from the lesion site and migrating through the tissue. Although a vasogenic component cannot be excluded, our results provide compelling evidence that cellular edema predominates in both diffuse and focal injury.

Conclusion

The brain swelling seen in severely brain-injured patients is predominantly cellular as signaled by a reduction in ADC in regions of high tissue water. The cellular form of edema dominates in both focal and diffuse injury. The CBF seen in these regions is above ischemic levels, indicating that other mechanisms resulting in a tissue energy crisis must be involved.

Acknowledgments

This work was supported in part by grants from the National Institutes of Health, NS12587 and NS19235.

References

1. Barzo P, Marmarou A, Fatouros P, Hayasaki K, Corwin F (1997) Biphasic pathophysiological response of vasogenic and cellular edema in traumatic brain swelling. Acta Neurochir [Suppl] 70: 119–122
2. Barzo P, Marmarou A, Fatouros P, Hayasaki K, Corwin F (1997) Contribution of vasogenic and cellular edema to traumatic brain swelling measured by diffusion-weighted imaging. J Neurosurg 87: 900–907
3. Fatouros PP, Marmarou A (1999) Use of magnetic resonance imaging for in vivo measurements of water content in human brain: method and normal values. J Neurosurg 90: 109–115
4. Ito J, Marmarou A, Barzo P, Fatouros P, Corwin F (1996) Characterization of edema by diffusion-weighted imaging in experimental traumatic brain injury. J Neurosurg 84: 97–103
5. Kawamata T, Katayama Y, Aoyama N, Mori T (2000) Heterogeneous mechanisms of early edema formation in cerebral contusion: diffusion MRI and ADC mapping study. Acta Neurochir [Suppl] 76: 9–12
6. Kuroiwa T, Nagaoka T, Ueki M, Yamada I, Miyasaka N, Akimoto H, Ichinose S, Okeda R, Hirakawa K (1999) Correlations between the apparent diffusion coefficient, water content, and ultrastructure after induction of vasogenic brain edema in cats. J Neurosurg 90: 499–503
7. Liu AY, Maldjian JA, Bagley LJ, Sinson GP, Grossman RI (1999) Traumatic brain injury: diffusion-weighted MR imaging findings. AJNR Am J Neuroradiol 20: 1636–1641
8. Marmarou A, Fatouros PP, Barzo P, Portella G, Yoshihara M, Tsuji O, Yamamoto T, Laine F, Signoretti S, Ward JD, Bullock MR, Young HF (2000) Contribution of edema and cerebral blood volume to traumatic brain swelling in head-injured patients. J Neurosurg 93: 183–193
9. Nakahara M, Ericson K, Bellander BM (2001) Diffusion-weighted MR and apparent diffusion coefficient in the evaluation of severe brain injury. Acta Radiol 42: 365–369
10. Portella G, Beaumont A, Corwin F, Fatouros P, Marmarou A (2000) Characterizing edema associated with cortical contusion and secondary insult using magnetic resonance spectroscopy. Acta Neurochir [Suppl] 76: 273–275

Correspondence: Anthony Marmarou, Department of Neurosurgery, Virginia Commonwealth University Medical Center, 1001 East Broad Street, Suite 235, Richmond, VA 23298-0508, USA. e-mail: amarmaro@vcu.edu

CT prediction of contusion evolution after closed head injury: the role of pericontusional edema

A. Beaumont and T. Gennarelli

Department of Neurosurgery, Medical College of Wisconsin, Milwaukee, WI, USA

Summary

Background. Cerebral contusions have a 51% incidence of evolution in the first hours after injury. Evolution is associated with clinical deterioration and is the reason for ICP monitoring or surgical intervention. We sought to define CT features that predict cerebral contusion evolution.

Methods. Patients treated for cerebral contusion who had 2 CT scans within 24 hours after injury were evaluated (n = 21). CT scans were analyzed for area of contusion, hemorrhagic components, and edema. Increase (%) in contusion size was recorded. Contusion evolution was defined as >5% size increase. Ratios of hemorrhagic components to surrounding edema were calculated.

Results. Ten patients (47.6%) showed contusion evolution and 11 (52.4%) did not. Age, sex ratio, or injury severity between the 2 groups did not differ. Eight of 10 patients with evolving contusions had minimal or no perilesional edema on first CT; only 2 of 11 non-evolution patients had perilesional edema ($p < 0.005$). Mean ratio of area of surrounding edema to area of hemorrhagic products on first CT was 0.770 in evolution group versus 2.22 in non-evolution group ($p = 0.055$).

Conclusions. A higher proportion of patients without contusion evolution had perilesional edema present on first CT scan. The absence of pericontusional edema on early CT may be a useful marker to predict contusion evolution.

Keywords: Traumatic brain injury; contusion; computed tomography; brain edema.

Introduction

It is well recognized that cerebral contusions associated with traumatic brain injury (TBI) can increase in size after an initial computed tomography (CT) scan has been obtained [4]. This contusion evolution can be associated with clinical deterioration of the patient, including obtundation, loss of airway protection, elevated intracranial pressure (ICP), and herniation. Frequently contusions will be serially imaged until they are seen to be stable in size, or an ICP monitor will be placed to warn of worsening ICP problems. The pathological reasons why some contusions expand and some remain stable is not clear, and currently no formal criteria have been defined that can help predict the behavior of any particular contusion. Any clinical modality used to help make such determinations must be readily obtainable in the setting of acute injury. This study aims to define CT scan parameters that might help predict contusion expansion after TBI.

Materials and methods

Patients with TBI admitted to Froedtert Memorial Lutheran Hospital between March and December 2004 were retrospectively analyzed. Patients were selected according to the following criteria: presence of 1 or more contusions on CT scan, 2 CT scans performed within the first 24 hours after admission, and the presence of at least 1 reactive pupil. Clinical data regarding injury mechanism, patient demographics, and neurological function were collected. CT scan data was obtained and analyzed using ImageJ (National Institutes of Health, Bethesda, MD) and Analyze (Mayo Clinic, Rochester, MN) software. Contusion size changes, pixel intensity profiles, pixel intensity histograms, pixel mean and standard deviation, size of pericontusional edema, ratio of edema to contusion, and fractal signature [2] were calculated and recorded from regions of interest. All segmentation for contusion definition was performed manually. Statistical analysis included Chi square distribution and *t*-tests according to data type. Statistical significance was defined as $p < 0.05$.

Results

Twenty-one patients fulfilled the entry criteria: 10 with contusion expansion and 11 with stable contusion size. Contusion expansion was defined as a greater than 5% increase in size of hemorrhage between the study CT scans. The age range of the patients was 19 to 75 years. Injury mechanisms included fall, motor vehicle accident, and assault. The median age of the expanding contusion group (EC) was 41.5 years (range:

Table 1. *Comparison between imaging parameters in expanding and stable contusions.*

	Expanding contusion (n = 10)	Stable contusion (n = 11)
Ratio of Edema:No Edema	8:2	2:9***
Increase in Contusion Size (%)	439	0.39**
Mean Signal Intensity (±SD) (units)	142.1 ± 16.9	125.1 ± 7.8**
Mean Signal Intensity SD (±SD) (units)	20.9 ± 10.4	22.2 ± 6.0
Edema:Contusion Size Ratio	0.77 ± 1.79	2.22 ± 2.15
Fractal Signature	2.44 ± 0.18	2.29 ± 0.10*
Bimodal:Unimodal Ratio	1:9	11:0

Ratio of Edema:No Edema is the number of cases in each group with defined perilesional edema on CT scan compared with the number of cases without edema.
Mean Signal Intensity is the mean pixel value in the region of interest (ROI).
Mean Signal Intensity SD, is the mean standard deviation of the pixel values for the ROI, a marker of heterogeneity in the ROI.
Fractal signature is a dimensionless number reflecting the complexity of data in the ROI.
Bimodal:Unimodal Ratio is the number cases in each group where the signal intensity/frequency histogram is bimodal/skewed or unimodal, as described in Fig. 1. *** $p < 0.005$, ** $p < 0.01$, * $p < 0.05$.

21 to 75). The median age of the stable contusion group (SC) was 35 years (range 19 to 72). The male to female ratio was 7:3 and 9:2 in EC and SC, respectively. The median score on the Glasgow Coma Scale at admission was 5 (range 4 to 15) with no difference between the groups. The mean time interval between the first 2 CT scans was 10.4 hours in EC and 15.0 hours in SC. The EC group consisted of 6 frontal, 2 temporal, and 2 parietal contusions. The SC group consisted of 9 frontal, 1 temporal, and 1 parietal contusion. Overall there was no difference in the age distribution, contusion location, clinical state of the patient, or time interval between the CT scans between the 2 groups.

Table 1 shows the proportions of EC and SC that had an identifiable rim of pericontusional edema. EC clearly had a much higher proportion of contusions with no perilesional edema compared with SC where most contusions had a rim of surrounding edema ($p < 0.005$). Table 1 also shows other imaging characteristics between the 2 groups. The mean signal intensity was higher in the EC group ($p < 0.01$), consistent with the presence of hypointense edema signal in the SC group. The percentage contribution of edema signal to lesion size was higher in the SC group ($p < 0.01$), and the ratio of the size of contusion to edema was higher in this same group. Fractal signature, a measure of image texture or variability, was lower in SC ($p < 0.05$), consistent with higher variability in pixel intensity.

Analysis of histogram plots for each of the contusions demonstrated apparent differences (Fig. 1). In the SC group, where perilesional edema was seen more commonly, the histogram profiles were bimodal or unimodal with skew (100%). This observation is consistent with 2 overlapping populations of pixels. In the EC group, histogram profiles were more clearly unimodal (90%), consistent with a single population of pixel intensities, i.e., no surrounding edema.

Discussion

Prior studies have demonstrated a 23% to 51% rate of progression of hemorrhagic contusion after closed head injury [3–5]. Clinical and demographic factors that have been shown to independently correlate with risk of lesion expansion include injury severity, sex, age, and coagulopathy. No clear risk factors have been previously identified beyond injury severity that can predict lesion expansion.

In this study, CT scan features were identified that correlated with the risk of contusion expansion after TBI. Specifically, the presence of a rim of perilesional edema is associated with a lower risk of contusion expansion. The lack of surrounding edema is also reflected in a lower mean signal intensity and morphological differences in the signal intensity/frequency histograms.

Fractal signature is a dimensionless parameter applied in image processing and calculated by a process of dilation and erosion. Variation in the image fractal dimension with resolution is then tested. Fractal signature measures the deviation of the sample image from a fractal surface. Increases in the fractal signature are associated with reductions in image variability. Prior studies have demonstrated alterations in fractal signature with osteoarthritis of the knee [2]. In this study, the stable contusion group had a significantly lower fractal signature, implying less homogeneity. Care should be taken not to apply too much weight to measures of complexity in this setting with a relatively small study group. However, differences in objective image parameters between EC and SC are apparent, and these may prove useful in further study.

The pathological processes and time course of con-

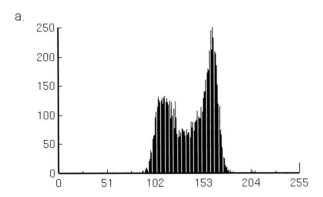
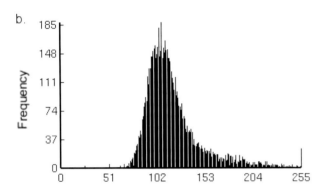
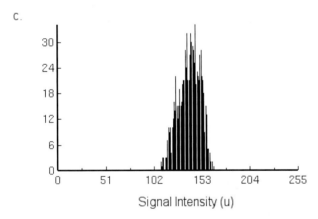

Fig. 1. Signal intensity/frequency histograms from selected contusions. (a) Histogram from a non-EC, showing bimodal distribution with peaks corresponding to contused tissue and surrounding edema. (b) Histogram from a non-EC showing skew, consistent with 2 overlapping populations of pixels corresponding to contused tissue and surrounding edema. (c) Histogram from initial CT scan of EC, showing a more clearly unimodal distribution, consistent with a single population of pixel density from contused tissue

tusions is not completely understood at present. It is certainly known that as a contusion matures it is associated with the development of a rim of pericontusional edema that gradually increases as hemorrhagic tissue undergoes necrosis. This process often takes days to weeks. However, this study suggests that the presence of pericontusional edema early after injury is associated with a lower risk of contusion expansion. As such, the edema is most likely a marker of lesion maturity that represents a state where the pathological process does not favor lesion expansion.

The lack of pericontusional edema on an early CT scan may be a useful indicator of a relatively high risk of contusion expansion, and may therefore be useful in guiding the early management of contusional injury, including decisions regarding repeat CT scanning, intensity and frequency of neurological observation, and the need for ICP monitoring. This may be particularly important given recent recommendations to rationalize the use of routine repeat CT scanning after closed head injury [1]. More extensive study is required to better determine the predictive values of some of the radiological features outlined in our study.

References

1. Brown CV, Weng J, Oh D, Salim A, Kasotakis G, Demetriades D, Velmahos GC, Rhee P (2004) Does routine serial computed tomography of the head influence management of traumatic brain injury? A prospective evaluation. J Trauma 57: 939–943
2. Lynch JA, Hawkes DJ, Buckland-Wright JC (1991) Analysis of texture in macroradiographs of osteoarthritic knees using the fractal signature. Phys Med Biol 36: 709–722
3. McBride DQ, Patel AB, Caron M (1993) Early repeat CT scan: importance in detecting surgical lesions after closed head injury. J Neurotrauma 10 [Suppl] 1: S227
4. Oertel M, Kelly DF, McArthur D, Boscardin WJ, Glenn TC, Lee JH, Gravori T, Obukhov D, McBride DQ, Martin NA (2002) Progressive hemorrhage after head trauma: predictors and consequences of the evolving injury. J Neurosurg 96: 109–116
5. Stein SC, Spettell C, Young G, Ross SE (1993) Delayed and progressive brain injury in closed-head trauma: radiological demonstration. Neurosurgery 32: 25–31

Correspondence: Andrew Beaumont, Dept. Neurosurgery, Medical College of Wisconsin, 9200 W Wisconsin Ave., Milwaukee, WI 53226, USA. e-mail: abeaumont@neuroscience.mcw.edu

Organ dysfunction assessment score for severe head injury patients during brain hypothermia

A. Utagawa[1], A. Sakurai[1], K. Kinoshita[1], T. Moriya[1], K. Okuno[2], and K. Tanjoh[1]

[1] Department of Critical Care and Emergency Medicine, Nihon University School of Medicine, Tokyo, Japan
[2] Department of Neurological Surgery, Tokyo Jikeikai University School of Medicine, Tokyo, Japan

Summary

The purpose of this study was to evaluate the utility of a novel organ dysfunction assessment score developed for patients with severe traumatic brain injury during therapeutic brain hypothermia.

The Brain Hypothermia Organ Dysfunction Assessment (BHODA) score is calculated through the combined assessment of 6 indices: central nervous system (CNS) function, respiratory function, cardiovascular function, hepatosplanchnic circulation, coagulation, and metabolism. The CNS, hepatosplanchnic circulation, and metabolic indices were based on measurements of cerebral perfusion pressure, gastric tonometry, and blood glucose, respectively. Thirty-nine patients with severe closed head injuries (scores of 3 to 8 on the Glasgow Coma Scale) were enrolled. Seven patients (18%) died during hospitalization. Outcome was favorable in 20 patients and unfavorable in 19. The BHODA score proved useful in describing sequences of complications during therapeutic brain hypothermia. A total maximum BHODA score of more than 13 points corresponded to a mortality of 70%. In a multivariate model, the total maximum BHODA score was independently associated with neurological outcome (odds ratio for unfavorable neurological outcome, 2.590; 95% confidence interval, 1.260, 5.327). In conclusion, the BHODA score can help assess multiple organ dysfunction/failure during therapeutic hypothermia and may be useful for predicting outcome.

Keywords: Traumatic brain injury; hypothermia; organ dysfunction syndrome; subdural hematoma.

Introduction

Growing evidence suggests that therapeutic brain hypothermia (THT) confers neuroprotective effects in subgroups of patients with neurological damage. Yet THT remains controversial as a strategy for severe traumatic brain injury (TBI), in part because of its many side effects during the cooling phase. The systemic organ function of TBI patients and the pathophysiological response to hypothermia are important to recognize and evaluate. The Sequential Organ Failure Assessment (SOFA) score is useful for assessing organ dysfunction and failure over time, as well as predicting outcome in critically ill patients. When TBI patients undergo THT, however, the SOFA score has only limited usefulness in evaluating cardiovascular response and central nervous system (CNS) due to sedation. As an alternative, we developed a novel organ dysfunction assessment score suitable for TBI patients undergoing THT. The purpose of this study was to evaluate the utility of this score.

Material and methods

Severe TBI patients with a Glasgow Coma Score (GCS) ≤ 8 on admission and suffering from acute subdural hematoma with cerebral contusion were enrolled in our study between January 1996 and December 2000. All patients underwent THT immediately after surgery, as previous described [6, 7]. The target brain temperature was 33 to 34 °C maintained for 48 to 72 hours. Basic treatment included sedation with midazolam (1–3 μg/kg/min), analgesic with buprenorphine (1–2 μg/kg/min), and muscle relaxant. PaCO2 concentrations were maintained at 35 to 40 mmHg and PaO2 levels were maintained above 100 mmHg. A ventricular pressure monitoring probe (Camino, Integra Neurosciences, Plainsboro, NJ) was inserted to measure ICP and ventricular temperature in each patient. Mean arterial pressure was kept above 90 mmHg and the ICP was kept below 20 mmHg. No patient underwent combined barbiturate coma therapy or active hyperventilation. Insulin generally was not administered in patients with glucose levels of less than 200 mg/dl. After THT, the patients were gradually rewarmed at a rate of 1 °C per day.

The BHODA score (Table 1) was calculated on admission and once for every 24-hour period until discharge. The worst value for each parameter was used to calculate the score for each 24-hour period. The total maximum BHODA score was calculated by summing the worst scores for each of the components. Neurological outcome was assessed at hospital discharge by the Glasgow Outcome Scale, and was dichotomized into favorable outcome (good recovery or mild disability) or unfavorable outcome (severe disability, vegetative state, or death).

Table 1. Brain Hypothermia Organ Dysfunction Assessment (BHODA) score

Variables	BHODA score				
	0	1	2	3	4
CNS					
– CPP (mmHg)	>70	≤70	≤65	≤60	≤55
Respiratory					
– PaO2/FiO2 (mmHg)	>400	≤400	≤300	≤200	≤100
Cardiovascular					
– PAR	7–10	10.1–15.0 or <7	≤20	≤30	30<
Hepatosplanchnic					
– CO2 gap (mmHg)	<10	≤15	≤20	≤25	25<
Coagulation					
– Platelet ($\times 10^3/mm^3$)	>150	≤150	≤100	≤50	≤20
Metabolism					
– Blood glucose (g/dl)	<170	170≤	200≤	230≤	260≤

CNS Central nervous system; *CPP* cerebral perfusion pressure; *PAR* pressure-adjusted heart rate defined as the product of the heart rate multiplied by the ratio of the right atrial (central venous) pressure to the mean arterial pressure. CO2 gap was defined as the difference between gastric mucosal PCO2 and arterial PCO2.

Analyses of continuous, normally-distributed variables within and between groups were undertaken using the appropriate Student *t*-test. Non-normally distributed continuous variables were analyzed using the Mann-Whitney *U*-test. Categorical variables were analyzed using Fisher exact test. A p value of <0.05 was considered significant. The results are presented as mean ± SD, except where otherwise indicated.

Results

Overall outcome

This study investigated a population of 39 patients with a mean age of 50.4 ± 15.4 years (range: 18 to 72 years). The post-resuscitation GCS of non-survivors (GCS 4) was significantly lower than that of survivors (GCS 6) (p = 0.0009). Seven patients (18%) died during hospitalization.

Organ dysfunction/failure during THT

The BHODA scores were analyzed at 4 time points for each index to identify significant differences: at the beginning of cooling, during the cooling phase, during the rewarming phase, and at the end point of THT. Scores for the respiratory system, hepatosplanchnic circulation, coagulation system, and metabolic system were significantly higher in non-survivors at all points. The score for the CNS was also significantly higher in non-survivors, but only at 3 time points (cooling phase, rewarming phase, and end point of THT).

Table 2. Hospital mortality and neurological outcome by quartile of the total maximum BHODA score

Total maximum BHODA score	0–6	7–8	9–12	13–24
Survivors, n (%)	6 (100)	14 (100)	9 (100)	3 (30)
Nonsurvivors, n (%)	0 (0)	0 (0)	0 (0)	7 (70)
Favorable, n (%)	3 (50)	14 (100)	3 (33)	0 (0)
Unfavorable, n (%)	3 (50)	0 (0)	6 (67)	10 (100)

p < 0.0001 for the difference between survivors and non-survivors; p = 0.0002 for the difference between patients with favorable and unfavorable outcome.

Hospital mortality and neurological outcome

Hospital mortality and neurological outcome by quartile of the total maximum BHODA score are shown in Table 2. The quartile of patients with the highest score had a hospital mortality of 70%, while the quartile of patients with the lowest scores had a hospital mortality of 0%. All of the patients in the quartile with the highest total maximum BHODA score had an unfavorable outcome. A multivariate model was created to assess the independent association of multiple organ dysfunction/failure and neurological outcome. After controlling for age, post-resuscitation GCS, and the BHODA score at the cooling phase, the neurological outcome was independently associated with the degree of multiple organ dysfunction/failure (odds ratio for unfavorable neurological outcome, 2.590; 95% C.I., 1.260, 5.327 for maximum BHODA score).

Discussion

Non-neurological organ dysfunction is common in patients with severe TBI and is independently associated with worse outcome. Respiratory and/or cardiovascular failure are more frequent in severe TBI than failures in other organs. Zygun [17] suggested that the potential mechanisms of organ dysfunction in severe TBI may be divided into neurogenic causes and complications from therapies. The primary etiological theory singles out catecholamine release as the neurogenic cause of myocardial dysfunction and pulmonary edema. According to a report by Dujardin et al. [2], 41% of TBI patients showed echocardiographic evidence of myocardial dysfunction after brain death.

Pulmonary edema fluid analysis indicated that both hydrostatic edema and permeability edema might be present in patients with neurogenic pulmonary edema [11]. Pulmonary constriction due to catecholamine surge increases capillary pressure and hydrostatic edema, thereby disrupting the basement membrane and ultimately producing permeability edema.

Barbiturate coma therapy is one of the main ICP-oriented therapies. The use of barbiturates has been associated with an increased incidence of pneumonia. Thiopental inhibits tumor necrosis factor and alpha-induced activation of nuclear factor kappaB by suppressing kappaB kinase activity [8]. It should be noted that cerebral perfusion pressure (CPP)-oriented therapies administered to optimize cerebral blood flow may be linked with an increased occurrence of respiratory failure. Robertson et al. [10] reported a five-fold increase in the occurrence of acute respiratory distress syndrome in a group managed with a higher CPP protocol (70 mmHg vs. 50 mmHg). The updated CPP guidelines from the Brain Trauma Foundation and the American Association of Neurological Surgeons recommend the following: "CPP should be maintained at a minimum of 60 mmHg. In the absence of cerebral ischemia, aggressive attempts to maintain CPP above 70 mmHg with fluids and pressors should be avoided because of the risk of acute respiratory distress syndrome."

The following concepts prompted the development of the SOFA score and formed its basis: 1) Organ dysfunction/failure is a process rather than an event. As such, it should be seen as a continuum rather than a condition which is simply "present" or "absent"; 2) The evaluation of organ dysfunction/failure should be based on a limited number of simple yet objective variables which can be easily and routinely measured in every institution [16]. The SOFA score can help quantify the degree of organ dysfunction/failure and is useful to predict outcome in critically ill patients [3]. The score has only limited usefulness in evaluating systemic organ dysfunction during THT, however. When severe TBI patients undergo THT, the concomitant treatment with sedatives, analgesics, muscle relaxants, and other medications generally makes it impossible to evaluate CNS function by the SOFA score. As an alternative, we created a novel organ dysfunction assessment score specialized in THT.

The BHODA score is a set of indices covering 6 physiological functions: CNS function, respiratory function, cardiovascular function, hepatosplanchnic circulation, coagulation, and metabolism. The CNS function, hepatosplanchnic circulation, and metabolic indices are based on measurements of CPP, gastric tonometry, and blood glucose, respectively. The inability of patients to respond to stimulation under THT makes the GCS inappropriate as an index of CNS function. Accordingly, the BHODA score uses CPP as the CNS index. Hypothermia is initially associated with sinus tachycardia and the subsequent development of bradycardia. The index of cardiovascular function in the BHODA score is based on the pressure-adjusted heart rate (PAR), defined as the product of the heart rate multiplied by the ratio of the right atrial (central venous) pressure to the mean arterial pressure. Hypothermia may decrease peripheral blood flow because of vasoconstriction, thereby decreasing the transfer of heat from the core to the periphery. In this case, it is difficult to maintain the target temperature by a water blanket. Dobutamine, an agent proven effective in controlling body temperature without inducing hypotension, can be used to increase the peripheral blood flow and improve heat conduction when administered concomitantly with fluid replacement therapy. For these reasons, PAR was selected as the index of cardiovascular function. PAR is also used as an index of cardiovascular function in the Multiple Organ Dysfunction (MOD) score [9]. Application of PAR in the BHODA score differs slightly, however, as 1 point is added when the PAR is calculated at less than 7 points.

Kinoshita et al. [6] showed that patients run the risk of impairing hemodynamics during THT. Systemic vasoconstriction under hypothermia may obscure the hypovolemic condition and hypoperfusion. At the point of oxygen metabolism, "masked hypoperfusion" can increase oxygen debt. Lactate concentration usually increases under hypothermia due to the fat and glucose metabolic alteration. This makes it difficult to estimate the oxygen debt by monitoring the lactate concentration alone. The pulmonary artery catheter is invasive and carries a risk of complications and higher cost. Gastric tonometry is a novel method used for indirect evaluation of regional blood flow within the splanchnic vascular bed. The measurement of the PCO2 gap (the difference between gastric mucosal PCO2 and arterial PCO2) may make it possible to accurately evaluate oxygen demand/supply status at the level of microcirculation [4]. Tonometric measurement by the air-gas method is likely to be reliable at around 34 °C [1]. The BHODA score also incorporates the hepatosplanchnic circulation calculated by PCO2 gap, as the tissue oxygen debt has been established as a common pathophysiological process leading to multiple organ failure. Venkatesh et al. [15] suggested that TBI patients developed splanchnic ischemia. A prospective randomized study by Van den Berghe et al. [13] recently identified a reduced mortality in critically ill patients under aggressive control of blood glucose levels (target < 110 mg/dl). Van den Berghe's group also reported that intensive insulin therapy reduces mean and maximal ICP in patients with isolated brain injury [14]. Kinoshita et al. [5] found that posttraumatic hyperglycemia in acute phase aggravates histopathological outcome and increases the accumulation of polymorphonuclear leukocytes. Patients with unfavorable outcome in our study had higher blood glucose concentrations than the patients with favorable outcome. Hypothermia decreases insulin sensitivity and insulin secretion, thereby facilitating the development of hyperglycemia. Hypothermia may modulate the physiological role of insulin in the regulation of target cell metabolism [12]. It remains unknown whether THT combined with intensive insulin therapy actually improves neurological outcome.

We developed the BHODA score by closely monitoring clinical experience in the management of complications during THT, and evaluated its utility in a small population of patients. The total maximum BHODA score in this multivariable logistic regression model may contribute to the prediction of hospital mortality, assuming that all other factors are held constant. Further clinical prospective studies will be required to clarify the utility of the BHODA score.

Conclusion

Our study shows that the BHODA score may be useful for assessing the evolution of organ failure over time in patients with TBI. The progress of multiple organ dysfunction/failure in this patient population was associated with worse outcome independent of age and post-resuscitation GCS.

References

1. Dohgomori H, Arikawa K, Kanmura Y (2003) Effect of temperature on gastric intramucosal PCO2 measurement by saline and air tonometry. J Anesth 17: 284–286
2. Dujardin KS, McCully RB, Wijdicks EF, Tazelaar HD, Seward JB, McGregor CG, Olson LJ (2001) Myocardial dysfunction associated with brain death: clinical, echocardiographic, and pathologic features. J Heart Lung Transplant 20: 350–357
3. Ferreira FL, Bota DP, Bross A, Melot C, Vincent JL (2001) Serial evaluation of the SOFA score to predict outcome in critically ill patients. JAMA 286: 1754–1758
4. Hurley R, Chapman MV, Mythen MG (2000) Current status of gastrointestinal tonometry. Curr Opin Crit Care 6: 130–135
5. Kinoshita K, Kraydieh S, Alonso O, Hayashi N, Dietrich WD (2002) Effect of posttraumatic hyperglycemia on contusion volume and neutrophil accumulation after moderate fluid-percussion brain injury in rats. J Neurotrauma 19: 681–692
6. Kinoshita K, Hayashi N, Sakurai A, Utagawa A, Moriya T (2003) Importance of hemodynamics management in patients with severe head injury and during hypothermia. Acta Neurochir [Suppl] 86: 373–376
7. Kinoshita K, Hayashi N, Sakurai A, Utagawa A, Moriya T (2003) Changes in cerebrovascular response during brain hypothermia after traumatic brain injury. Acta Neurochir [Suppl] 86: 377–380
8. Loop T, Humar M, Pischke S, Hoetzel A, Schmidt R, Pahl HL, Geiger KK, Pannen BH (2003) Thiopental inhibits tumor necrosis factor alpha-induced activation of nuclear factor kappaB through suppression of kappaB kinase activity. Anesthesiology 99: 360–367
9. Marshall JC, Cook DJ, Christou NV, Bernard GR, Sprung CL, Sibbald WJ (1995) Multiple organ dysfunction score: a reliable descriptor of a complex clinical outcome. Crit Care Med 23: 1638–1652
10. Robertson CS, Valadka AB, Hannay HJ, Contant CF, Gopinath SP, Cormio M, Uzura M, Grossman RG (1999) Prevention of secondary ischemic insults after severe head injury. Crit Care Med 27: 2086–2095
11. Smith WS, Matthay MA (1997) Evidence for a hydrostatic mechanism in human neurogenic pulmonary edema. Chest 111: 1326–1333
12. Torlinska T, Perz M, Madry E, Hryniewiecki T, Nowak KW, Mackowiak P (2002) Effect of hypothermia on insulin-receptor interaction in different rat tissues. Physiol Res 51: 261–266
13. van den Berghe G, Wouters P, Weekers F, Verwaest C, Bruyninckx F, Schetz M, Vlasselaers D, Ferdinande P, Lauwers P, Bouillon R (2001) Intensive insulin therapy in critically ill patients. N Engl J Med 345: 1359–1367
14. Van den Berghe G, Schoonheydt K, Becx P, Bruyninckx F, Wouters PJ (2005) Insulin therapy protects the central and peripheral nervous system of intensive care patients. Neurology 64: 1348–1353
15. Venkatesh B, Townsend S, Boots RJ (1999) Does splanchnic ischemia occur in isolated neurotrauma? A prospective observational study. Crit Care Med 27: 1175–1180
16. Vincent JL, Moreno R, Takala J, Willatts S, De Mendonca, Bruining H, Reinhart CK, Suter PM, Thijs LG (1996) The SOFA (Sepsis-related Organ Failure Assessment) score to describe organ dysfunction/failure on behalf of the Working Group on Sepsis-Related Problems of the European Society of Intensive Care Medicine. Intensive Care Med 22: 707–710
17. Zygun DA, Kortbeek JB, Fick GH, Laupland KB, Doig CJ (2005) Non-neurologic organ dysfunction in severe traumatic brain injury. Crit Care Med 33: 654–660

Correspondence: Akira Utagawa, Department of Emergency and Critical Care Medicine, Nihon University School of Medicine, 30-1 Oyaguchi Kamimachi Itabashi-ku, Tokyo 173-8610, Japan. e-mail: utakira@med.nihon-u.ac.jp

Importance of cerebral perfusion pressure management using cerebrospinal drainage in severe traumatic brain injury

K. Kinoshita, A. Sakurai, A. Utagawa, T. Ebihara, M. Furukawa, T. Moriya, K. Okuno, A. Yoshitake, E. Noda, and K. Tanjoh

Department of Emergency and Critical Care Medicine, Nihon University School of Medicine, Tokyo, Japan

Summary

Objective. To evaluate hemodynamics in patients with severe traumatic brain injury (TBI) after cerebral perfusion pressure (CPP) management using cerebrospinal fluid (CSF) drainage.

Methods. Twenty-six patients with TBI (Glasgow Coma Score = 8 or less) were investigated. Mean arterial blood pressure, CPP, cardiac index (CI), systemic vascular resistance index (SVRI), and central venous pressure were measured. The patients were divided into 2 groups after craniotomy: the intraparenchymal ICP (IP-ICP) monitoring group (n = 14) and ventricular ICP (V-ICP) monitoring group (n = 12). Patient hemodynamics were investigated on the second hospital day to identify differences. Measurements indicated a target CPP above 70 mmHg and a central venous pressure of 8–10 mmHg in both groups. Mannitol administration (IP-ICP group) or CSF drainage (V-ICP group) was performed whenever the CPP remained below 70 mmHg.

Results. High SVRI and low CI ($p < 0.05$) were observed in the IP-ICP group. The V-ICP group exhibited a reduction in the total fluid infusion volume of crystalloid ($p < 0.01$) and a reduction in the frequency of hypotensive episodes after the mannitol infusion.

Conclusions. CPP management using CSF drainage decreases the total infusion volume of crystalloid and may reduce the risk of aggravated brain edema after excess fluid resuscitation.

Keywords: Traumatic brain injury; CPP; CSF drainage; fluid management.

Introduction

The main management goals in patients with traumatic brain injury (TBI) are to reduce the intracranial pressure (ICP), provide an adequate cerebral perfusion pressure (CPP), and stabilize hemodynamics systemically. Clinical studies have documented the importance of intensive care unit (ICU) management in maintaining the CPP of traumatically brain-injured patients [9]. Fluid management is also a critical factor during intensive care in patients with TBI [5]. A recent study reported that the early use of diuresis is associated with significant side effects, including significant fluid loss and unexpected hypotension [4] during ICU management. Now that systemic arterial blood pressure can be maintained by neurogenic hypertension, it has become difficult to identify hypovolemia in the presence of increased sympathetic tone after brain injury. Hypovolemia could not be determined if the frequent use of diuresis led to a depletion in the intravascular volume after TBI. As a consequence, it has become important to maintain euvolemia in CPP management as a precondition during ICU care [6]. The drainage of cerebrospinal fluid (CSF) for ICP management has been recommended in patients with TBI in light of its efficacy in reducing ICP [2]. The changes in hemodynamics during CPP management by CSF drainage after TBI are not well understood, however. This study evaluates the hemodynamics in patients with TBI during CPP management by CSF drainage.

Materials and methods

Twenty-six patients with TBI (Glasgow Coma Score = 8 or less) were investigated in this study. Mean arterial blood pressure (MAP), CPP (CPP = MAP − ICP), cardiac index (CI), systemic vascular resistance index (SVRI), and central venous pressure were measured.

The patients were divided into 2 groups after craniotomy: the intraparenchymal ICP monitoring group (CAMINO; REF 110-4BT; IP-ICP group; n = 14) and the ventricular ICP monitoring group (CAMINO; REF 110-4BT; V-ICP group; n = 12). The hemodynamics of the patients were investigated on the second hospital day to identify differences between the groups.

Measurements indicated a target CPP above 70 mmHg and a central venous pressure of 8–10 mmHg in both groups. Mannitol (0.25–1.0 g/kg) was administered whenever the CPP remained below 70 mmHg for over 20 minutes in the IP-ICP group. The patients in the V-ICP group initially underwent CPP management by CSF drainage, followed by mannitol treatment if CSF drainage failed to bring about the desired effects. Crystalloid was administered at a basic infusion volume of 2000 mL/24 h, with adjustment of the volume every eighth hour if the water balance (total infusion volume minus urine volume) for the previous 8-hour period was calculated to be negative.

Nine cases in the IP-ICP group and 8 cases in the V-ICP group underwent therapeutic hypothermia immediately after the craniotomy. No patient underwent combined barbiturate coma therapy. Arterial PaCO2 was maintained at about 35 mmHg and never allowed to fall below 30 mmHg.

Statistical analysis

Data were expressed as mean values ± SD. The non-paired t test and Fisher exact test were employed for comparison of 2 groups. Statistical significance was defined as $p < .05$.

Results

Age and Glasgow Coma Score on admission

No differences were seen among the IP-ICP or V-ICP patients in age (IP-ICP; 51.9 ± 17.2 vs. V-ICP; 55.3 ± 15.4) or Glasgow Coma Score (IP-ICP; 6.2 ± 1.6 vs. V-ICP; 6.1 ± 1.6) on admission.

Changes in ICP and CPP

Measurements on the second hospital day indicated no change between the groups in MAP (IP-ICP 99.0 ± 11.0 vs. V-ICP 100.9 ± 9.1 mmHg) or CPP (IP-ICP 84.9 ± 16.1 vs. V-ICP 86.3 ± 8.8 mmHg).

Changes in SVRI and CI (Fig. 1a)

SVRI increased ($p < 0.05$) in the IP-ICP group compared to the V-ICP group (IP-ICP 2505.8 ± 202.5 vs. V-ICP 1934.4 ± 145.8 dyne.s/cm^5/m^2). A decrease ($p < 0.05$) in CI was seen on the second hospital day in the IP-ICP group (3.3 ± 0.7 ml/min/m^2) compared to the V-ICP group (4.2 ± 0.4 ml/min/m^2).

Total infusion volume of crystalloid and mannitol (Fig. 1b)

The total fluid infusion volume of crystalloid was lower ($p < 0.01$) in the V-ICP group (4273.0 ± 362.4 mL) than in the IP-ICP group (5036.4 ± 403.8 mL). The volume of administered mannitol was higher ($p < 0.01$) in the IP-ICP group (625.0 ± 143.8 mL) than in the V-ICP (83.0 ± 103.0 mL) group. The hypotensive episodes after mannitol infusion tended to decrease in frequency during intensive

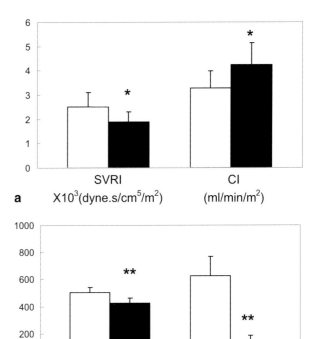

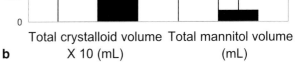

Fig. 1. (a) Changes in SVRI and CI. (b) Total infusion volumes of crystalloid and mannitol. *Open bars*: IP-ICP group, *Closed bars*: V-ICP group; *$p < 0.05$. **$p < 0.01$

care management in the V-ICP group (IP-ICP, 6 episodes in 5 cases vs. V-ICP, 1 episode in 1 case).

Discussion

Patients who suffer severe brain injury face the risk of developing secondary brain damage after the injury due to the low CPP [8, 9] stemming from either intracranial hypertension or arterial hypotension [1, 3, 7]. This study demonstrated that CPP management using CSF drainage reduced the total intravenous infusion volume after TBI. Head-injured patients administered mannitol might have unexpectedly developed a combination of hypotension and hypovolemia [4]. This study failed to uncover direct evidence of any relationship between hypotension and the early use of mannitol as a diuretic. It did demonstrate, however, that patients undergoing CPP management without CSF drainage exhibit high SVRI and low CI that is clinically significant. The effective maintenance of systemic arterial blood pressure now makes it difficult to identify hypovolemia in patients with increased systemic vascular resistance. If the depletion in the intravascular volume after craniotomy cannot be determined, the failure to maintain cardiac stroke volume may accelerate hemodynamic impairment and lead to unexpected hypotension during intensive care.

Fluid replacement should be conducted based on the blood pressure, heart rate, urine output, and central venous pressure, if available, during intensive care after craniotomy. If hypovolemia is not properly corrected, the condition will trigger hypotension and aggravate cerebral ischemia. Controversy has emerged regarding the potential danger of seriously compromised ICP as an adverse response to the large volumes of crystalloid fluids required by patients with vasogenic edema after TBI.

Small-volume fluid resuscitation is particularly attractive for use in patients with brain injuries. Hemodynamic stability can be achieved with smaller volumes of fluid, and this can theoretically reduce the risk of brain edema after TBI.

Conclusion

Our study suggests that patients with severe TBI run the risk of a masked imbalance of hemodynamics during intensive care. Hypotension may occur due to inadequate sedation, inadequate analgesia, and excessive use of diuretic agents. CPP management by CSF drainage may decrease the total infusion volume of crystalloid and reduce the risk of aggravated brain edema from excess fluid resuscitation.

References

1. Bouma GJ, Muizelaar JP, Stringer WA, Choi SC, Fatouros P, Young HF (1992) Ultra-early evaluation of regional cerebral blood flow in severely head-injured patients using xenon-enhanced computerized tomography. J Neurosurg 77: 360–368
2. Bullock MR, Chesnut RM, Clifton GL, Ghajar J, Marion DW, Narayan RK, Newell DW, Pitts LH, Rosner MJ, Wilberger JE (1995) Management and prognosis of severe traumatic brain injury. Part I: guideline for the management of severe traumatic head injury. Brain Trauma Foundation, http://www2.braintrauma.org/guidelines/index.php
3. Chesnut RM, Marshall SB, Piek J, Blunt BA, Klauber MR, Marshall LF (1993) Early and late systemic hypotension as a frequent and fundamental source of cerebral ischemia following severe brain injury in the Traumatic Coma Data Bank. Acta Neurochir [Suppl] 59: 121–125
4. Chesnut RM, Gautille T, Blunt BA, Klauber MR, Marshall LF (1998) Neurogenic hypotension in patients with severe head injuries. J Trauma 44: 958–963; discussion 963–964
5. Clifton GL, Miller ER, Choi SC, Levin HS (2002) Fluid thresholds and outcome from severe brain injury. Crit Care Med 30: 739–745
6. Graf CJ, Rossi NP (1978) Catecholamine response to intracranial hypertension. J Neurosurg 49: 862–868
7. Marmarou A, Anderson RL, Ward JD (1991) Impact of ICP instability and hypotension on outcome in patients with severe head trauma. J Neurosurg 75: S59–S66
8. Rosner MJ, Daughton S (1990) Cerebral perfusion pressure management in head injury. J Trauma 30: 933–940; discussion 940–941
9. Rosner MJ, Rosner SD, Johnson AH (1995) Cerebral perfusion pressure: management protocol and clinical results. J Neurosurg 83: 949–962

Correspondence: Kosaku Kinoshita, Department of Emergency and Critical Care Medicine, Nihon University School of Medicine, 30-1 Oyaguchi Kamimachi, Itabashi ku, Tokyo 173-8610, Japan. e-mail: kosaku@med.nihon-u.ac.jp

Acute hemispheric swelling associated with thin subdural hematomas: pathophysiology of repetitive head injury in sports

T. Mori, Y. Katayama, and **T. Kawamata**

Department of Neurological Surgery, Nihon University School of Medicine, Tokyo, Japan

Summary

Introduction. The most common head injury in sports is concussion, and repeated concussions occurring within a short period occasionally can be fatal. Acute subdural hematoma is the most common severe head injury and can be associated with severe neurologic disability and death in sports. We investigated severe brain damage resulting from repetitive head injury in sports, and evaluated the pathophysiology of sports-related repetitive injury.

Methods. We reviewed the literature containing detailed descriptions of repetitive severe sports-related head injury. In total, 18 cases were analyzed with regard to age, gender, type of sports, symptoms before second injury, and pathology of brain CT scans.

Results. The majority of cases involved young males aged 16 to 23 years old, who sustained a second head injury before symptoms from the first head injury had resolved. Ten of 15 cases did not suffer loss of consciousness at insult. Eight cases were confirmed on brain CT scans after the second injury, and all 8 cases revealed brain swelling associated with a thin subdural hematoma.

Conclusions. Second impact syndrome is thought to occur because of loss of autoregulation of cerebral blood flow, leading to vascular engorgement, increased intracranial pressure, and eventual herniation. Our investigation suggests that the existence of subdural hematoma is a major cause of brain swelling following sports-related, repetitive head injury.

Keywords: Hemispheric swelling; thin subdural hematoma; repetitive head injury; sports.

Introduction

The most common head injury in sports is concussion. Furthermore, repeated concussions occurring within a short period occasionally lead to a fatal outcome. Once athletes have suffered a concussion, they may be 4 to 6 times more likely to suffer a second head injury than someone who has never had one [7].

Subdural hematoma is the most common cause of death or severe disability in the sports-related head injury patient [11]. It has been reported that subdural hematoma can result from repeated minor head trauma [17]. In this study, we focused on the acute phase of repetitive injuries in an attempt to clarify the pathophysiology of repetitive severe sports-related head injury.

Materials and methods

We reviewed the published papers on repetitive severe sports-related head injury [1, 2, 5, 6, 10, 11, 14, 16]. In total, 18 cases, including our own 4 cases, were analyzed with regard to age, gender, type of sport, symptoms before receiving second injury, and computed tomography (CT) findings, if available.

Results

All 18 cases were male adolescents or young adults and they returned to play before the symptoms from their first injury had resolved. At the initial insult, 10 of 15 cases did not lose consciousness. After the second injury, we could confirm CT findings in 8 cases, and all 8 cases revealed subdural hematoma (Table 1).

Case reports

The first case was a 22-year-old man, who was an American college football player with an unremarkable medical history. He received a strong tackle in the first game. He did not lose consciousness but suffered partial amnesia during the game. The player stated that he was suffering from a headache and nausea, but continued to participate in the game. At 14 days after the first game, his headache still persisted.

In the second game, the player received a hard helmet-to-helmet hit that knocked him to the ground. He came off the field of play complaining of headache

Table 1. *Summary of repetitive severe sports-related head injury cases (cases with CT scan imaging available after second injury)*

Ref.	Age/sex	Sport	1st injury	Ongoing symptoms	Delay to 2nd injury	2nd injury	Pathology	GOS
16	19/M	AF	LOC–	headache	4 days	deep coma	ASDH	D
9	17/M	AF	LOC–	headache	7 days	deep coma	ASDH	D
11	18/M	AF	LOC–	headache	2 hours	headache nausea	ASDH	GR
5	22/M	Boxing	LOC–	headache vertigo	2 months	headache vomiting	ASDH	GR
*	22/M	AF	LOC–	headache	2 weeks	collapsed	ASDH	GR
*	23/M	Boxing	LOC–	headache	2 weeks	collapsed	ASDH	GR
*	20/M	Karate	LOC+	headache	4 days	headache	ASDH	GR
*	22/M	Skiing	LOC+	headache	2 days	headache	ASDH	GR

* = Our cases.
AF American Football, *ASDH* acute subdural hematoma, *D* dead, *GOS* Glasgow Outcome Scale, *GR* Good Recovery, *LOC+* with loss of consciousness, *LOC–* without loss of consciousness, *Ref* references.

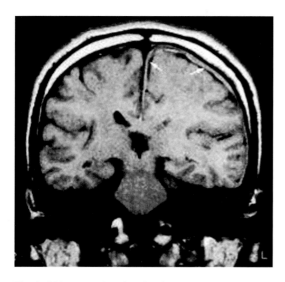

Fig. 1. MRI coronal section showing a subdural hematoma over the cerebral convexity and interhemispheric (arrows)

and vomiting, and then collapsed. He recovered consciousness immediately, and he was oriented to time, person, and place. The patient was then examined at an emergency department. Coronal magnetic resonance imaging revealed a subdural hematoma over the cerebral convexity and interhemispheric (Fig. 1).

The second case was a 20-year-old man, a beginner at karate. After his first competition, he suffered headache and dizziness, but did not undergo a medical examination. After an interval of 4 days, he attempted to participate in practice. He received a blow to the head and developed severe headache and vomiting. His initial CT scan revealed extensive cerebral swelling of the right hemisphere compared to the contralateral side, and a thin rim of subdural hematoma in the right temporal convexity (Fig. 2).

Discussion

Repetitive head injury in sports

Concussions are among the mildest forms of sports-related head injury. Such sports-related head injuries are very different from the typical severe traumatic brain injuries such as those sustained in traffic accidents and falls. Because sports-related brain injury tends to be repetitive, prevention is feasible. A player who has received a minor head injury is 4 to 6 times as likely to sustain a subsequent head injury [7]. Repetitive concussive injury induces acute and chronic damage to the brain. Repeated concussions that occur within a short period can occasionally lead to a fatal outcome.

Second impact syndrome

Second impact syndrome is a widely feared complication of sports-related traumatic brain injury, and is characterized by rapid cerebral edema following a second impact prior to recovery from the first. Loss of consciousness is not always present, but mortality and morbidity are extremely high. Seventeen cases were reported in the 3 years from 1992 to 1995 [1].

The pathophysiology of second impact syndrome is thought to involve a loss of autoregulation in the brain's blood supply. This leads to vascular engorgement within the cranium which can, in turn, markedly increase intracranial pressure and eventually result in herniation [2]. McCrory reviewed and analyzed the published cases of second impact syndrome, but described the term "second impact syndrome" as misleading because the etiology and pathology are not entirely clear [12, 13].

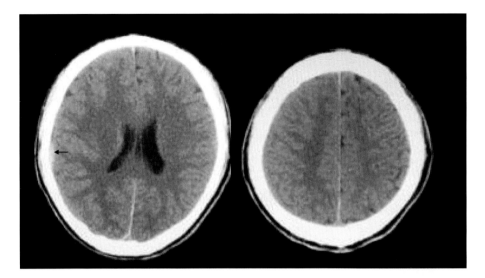

Fig. 2. CT scan showing a thin rim of acute subdural hematoma (arrow) associated with hemispheric swelling

Based on previous case reports [5, 11] and our cases, it is conceivable that the pathology of repetitive severe brain injury could be as follows. At the initial impact, the insult spreads to the bridging veins, and these veins become stretched and occasionally torn. The surface of the cortex then adheres to the dura matter. At the next impact, a thin rim of subdural hematoma compresses the bridging veins. The brain becomes swollen due to venous congestion, and fatal hemorrhage occasionally occurs. Therefore, concussion injury patients should receive close medical attention.

Management of concussion in sports

In general, a concussion can be defined as a head injury involving a traumatic alteration in mental status, commonly followed by confusion and amnesia [9]. It has been suggested that most sports-related concussions do not result in a loss of consciousness [3, 4]. Players often decide to return to play after a head injury without seeking medical attention. Kawamata *et al.* [8] have indicated that a decrease in incidence of concussion leads to a decrease in fatal injury. Both players and coaches need to understand the risks of multiple head injuries and how to apply return-to-play guidelines in their decision-making. For prevention of further injury, systems have been developed to determine when an athlete who has suffered a concussion can safely return to competition. Various grading schemas and guidelines exist [1, 10, 15, 18], and appropriate education and training for players and coaches concerning the care of head injuries are needed in order to prevent potentially catastrophic events.

Conclusion

The existence of subdural hematoma is one of the major causes of brain swelling following repetitive sports-related head injury. Further studies are needed to clarify the detailed pathophysiology of repetitive severe head injury.

References

1. Cantu RC (1998) Second-impact syndrome. Clin Sports Med 17: 37–44
2. Cantu RC, Voy R (1995) Second impact syndrome: a risk in any contact sport. Physician Sportsmed 23: 27–34
3. Collins MW, Grindel SH, Lovell MR, Dede DE, Moser DJ, Phalin BR, Nogle S, Wasik M, Cordry D, Daugherty KM, Sears SF, Nicolette G, Indelicato P, McKeag DB (1999) Relationship between concussion and neurophysical performance in college football players. JAMA 282: 964–970
4. Collins MW, Lovell MR, Mckeag DB (1999) Current issues in managing sports-related concussion. JAMA 282: 2283–2285
5. Cruikshank JK, Higgens CS, Gray JR (1980) Two cases of acute intracranial haemorrhage in young amateur boxers. Lancet 1: 626–627
6. Fekete JF (1968) Severe brain injury and death following minor hockey accidents: the effectiveness of the "safety helmets" of amateur hockey players. Can Med Assoc J 99: 1234–1239
7. Gerberich SG, Priest JD, Boen JR, Straub CP, Maxwell RE (1983) Concussion incidences and severity in secondary school varsity football players. Am J Public Health 73: 1370–1375
8. Kawamata T, Katayama Y (2002) Concussion in sports [in Japanese]. Rinsho Sports Igaku 19: 637–643

9. Kelly JP, Rosenberg JH (1997) Diagnosis and management of concussion in sports. Neurology 48: 575–580
10. Kelly JP, Nichols JS, Filley CM, Lillehei KO, Rubinstein D, Kleinschmidt-DeMasters BK (1991) Concussion in sports. Guidelines for the prevention of catastrophic outcome. JAMA 266: 2867–2869
11. Logan SM, Bell GW, Leonard JC (2001) Acute subdural hematoma in a high school football player after 2 unreported episodes of head trauma: a case report. J Athl Train 36: 433–436
12. McCrory P (2001) Does second impact syndrome exist? Clin J Sport Med 11: 144–149
13. McCrory PR, Berkovic SF (1998) Second impact syndrome. Neurology 50: 677–683
14. Miele VJ, Carson L, Carr A, Bailes JE (2004) Acute on chronic subdural hematoma in a female boxer: a case report. Med Sci Sports Exerc 36: 1852–1855
15. Report of the Sports Medicine Committee (1990) Guidelines for the management of concussion in sports. Colorado Medical Society, Denver
16. Saunders RL, Harbaugh RE (1984) The second impact in catastrophic contact-sports head trauma. JAMA 252: 538–539
17. Shell D, Carico GA, Patton RM (1993) Can subdural result from repeated minor head injury? Physician Sportsmed 21: 74–84
18. Sturmi JE, Smith C, Lombardo JA (1998) Mild brain trauma in sports. Diagnosis and treatment guidelines. Sports Med 25: 351–358

Correspondence: Tatsuro Mori, Department of Neurological Surgery, 30-1 Oyaguchi-kamimachi, Itabashi-ku, Tokyo 173-8610, Japan. e-mail: tmori-nsu@umin.ac.jp

Rewarming following accidental hypothermia in patients with acute subdural hematoma: case report

K. Kinoshita, A. Utagawa, T. Ebihara, M. Furukawa, A. Sakurai, A. Noda, T. Moriya, and K. Tanjoh

Department of Emergency and Critical Care Medicine, Nihon University School of Medicine, Tokyo, Japan

Summary

A 57-year-old man was admitted to the Emergency and Critical Care Department with accidental hypothermia (31.5 °C) after resuscitation from cardiopulmonary arrest (CPA). Brain CT revealed an acute subdural hematoma. Active core rewarming to 33 °C was performed using an intravenous infusion of warm crystalloid. The patient underwent craniotomy soon after admission, with bladder temperature maintained at 33 to 34 °C throughout the surgery. Therapeutic hypothermia (34 °C) was continued for 2 days, followed by gradual rewarming. After rehabilitation, the patient was able to continue daily life with assistance.

Traumatic brain injury (TBI) following CPA is associated with extremely unfavorable outcomes. Very few patients with acute subdural hematomas presenting with accidental hypothermia and CPA have been reported to recover. No suitable strategies have been clearly established for the rewarming performed following accidental hypothermia in patients with TBI. Our experience with this patient suggests that therapeutic hypothermia might improve the outcome in some patients with severe brain injury. It also appears that the method used for rewarming might play an important role in the therapy for TBI with accidental hypothermia.

Keywords: Traumatic brain injury; hypothermia; cardiac arrest.

Introduction

The neurological outcomes of traumatic brain injury (TBI) are extremely poor in patients resuscitated from cardiopulmonary arrest (CPA). In fact, TBI with hypoxia/hypotension is reported to be one of the most common causes of secondary brain damage. Several studies have shown at least a doubling of mortality in brain-injured patients with hypoxia and hypotension [9, 11, 14]. In this study we describe a patient who received intensive treatment and ultimately survived after presenting with CPA involving accidental hypothermia in the aftermath of a traumatic brain injury.

The optimal strategy for rewarming from accidental hypothermia in TBI patients remains unclear. Our patient with accidental hypothermia (31.5 °C) underwent active core rewarming from to 33 to 34 °C, followed by therapeutic hypothermia with slow rewarming. This experience may prove useful when considering the need for warming and the method by which it is applied. Specifically, the present case suggests that rewarming is important in patients with TBI involving accidental hypothermia and that therapeutic hypothermia has the potential to improve outcome in selected patients with severe TBI.

Case history

A 57-year-old man was discovered in front of a restaurant in coma with a gasping respiration pattern. By the time emergency medical services reached the scene, the patient's body temperature had fallen to 31.5 °C. On arrival at our emergency room, bladder temperature was 31.8 °C and no spontaneous respirations were present. Carotid pulse was absent. Electrocardiogram indicated pulseless electrical activity. Active core rewarming to 33 °C was performed using a warming blanket and intravenous infusion of warm crystalloid.

Cardiopulmonary resuscitation (CPR) had been administered with an intravenous injection of epinephrine (adrenaline, 1 mg) and atropine (1 mg). Return of spontaneous circulation was finally observed 25 minutes after the onset of CPA.

A brain computed tomography (CT) revealed an acute subdural hematoma with effacement of the basal cisterns (Fig. 1A). The cause of the brain injury was unclear, as no witnesses had been present at the scene and the patient had been comatose at hospital admission. An emergency craniotomy was performed after

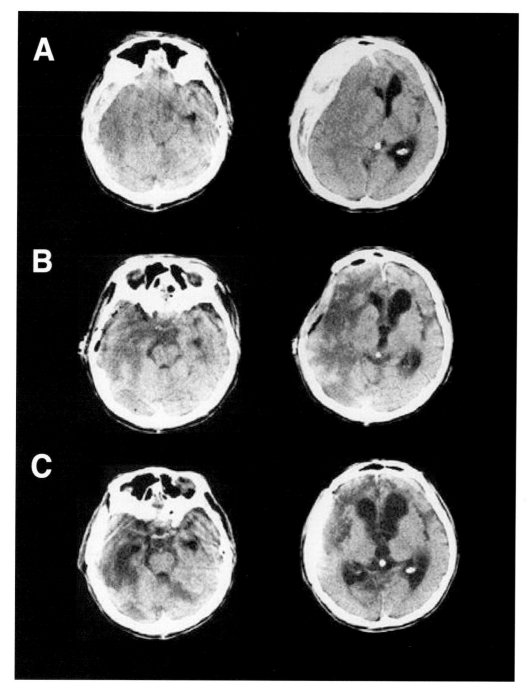

Fig. 1. Sequential changes in brain CT after admission. (A) Brain CT revealed an acute subdural hematoma on admission. (B) No brain swelling was observed on CT 4 days after admission. (C) Ventricular dilatations were observed at 6 weeks after admission

rapid rewarming to 33 °C, with bladder temperature maintained at 33 to 34 °C during the surgery. No brain swelling or episodes of intraoperative hypotension were observed. The pupils measured 3.5 mm after surgery, but both were unreactive to light. Therapeutic hypothermia was induced using a water-circulating blanket to confer brain protection. The patient received sedation by intravenous administration of midazolam (0.1 mg/kg body weight/h initially) and buprenorphine (0.05 mg/h), with dose adjustments as needed for the management of mechanical ventilation. Paralysis was induced by a continuous infusion of pancuronium (0.05 mg/kg body weight) to prevent shivering.

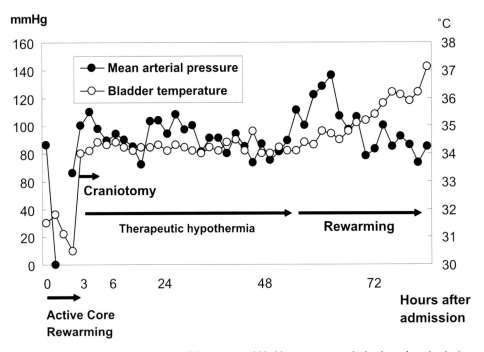

Fig. 2. Sequential changes in mean arterial pressure and bladder temperature during hypothermia. Active core rewarming to 34 °C was done using a warming blanket and intravenous infusion of warm crystalloid. Therapeutic hypothermia was continued for 3 days with gradual rewarming of the patient to 37.0 °C at a rate of 1 °C/day. No hypotensive episodes were observed after craniotomy

The patient was cooled (34.0 °C) continuously for 2 days and then gradually rewarmed to 37.0 °C at a rate no faster than 0.1 °C/h and 1.0 °C/d (Fig. 2). No evidence of intracranial hypertension was observed on brain CT 4 days after admission (Fig. 1B).

The patient was able to respond to verbal requests and eat meals by himself after a rehabilitation program, but he remained aphasic for 6 weeks after onset. He was discharged to a satellite hospital with a rating of severe disability on the Glasgow Outcome Scale. CT at the time of discharge showed atrophy of the cerebral cortex and ventricular dilation (Fig. 1C). The patient currently lives in a partially-dependent state at the satellite hospital.

Discussion

The CPA suffered by this patient might have been wholly or partially attributable to the following conditions: 1) increased intracranial pressure caused by brain contusion/hematoma, resulting in brainstem compression, brain herniation, and subsequent respiratory arrest; 2) secondary brain damage caused by anoxia-hypoxia and hypotension; 3) hypothermia. Even when return of spontaneous circulation is observed after successful CPR in patients with TBI, only a few patients survive over the long term and those who do have a poor prognosis for daily life. Isolated brain injuries with hypotension are associated with increased mortality [8].

The recovery of our patient after hypoxia and severe hypotension was somewhat surprising. In view of this outcome, this case suggests that therapeutic strategies after TBI with accidental hypothermia should be considered further. Recent clinical trials of therapeutic hypothermia suggest that this treatment can improve outcomes after resuscitation from CPA due to cardiogenic origins [1, 2, 13]. Although a clinical study of TBI failed to identify significant improvements in outcomes with therapeutic hypothermia [3], the present report suggests that such treatment might be beneficial in selected patients with TBI.

Two methods are applied for rewarming from accidental hypothermia (30 °C to <34 °C) [4]: passive rewarming and active external rewarming (truncal areas only). To avoid the risks of dysrhythmia and coagulopathy [15] that arise when the core temperature drops below 33 °C, the latter method, active external rewarming, is generally preferred [4]. Dysrhythmia and coagulopathy were not observed in the case reported here. A comparison between normothermic (37 to 38 °C) and hypothermic (32 to 33 °C) groups by Clifton et al. revealed no difference in the incidence of delayed traumatic intracerebral hemorrhage due to coagulopathy [3, 12]. Standard rewarming rates from accidental hypothermia range from 1 to 3 °C/h for cardiac arrest patients [5, 6]. Suitable rewarming rates from accidental hypothermia with TBI remain controversial, however. Each rewarming method has advantages and disadvantages, and no controlled studies have been performed to compare them in humans. Clifton et al. [3] reported that brain-injured patients with hypothermia on admission should not be rewarmed. They based this advice on a subgroup analysis revealing a significant improvement in the outcome of brain-injured patients with hypothermia on admission. They did not, however, provide any data or indicate the methods used for rewarming. The present case suggests that the method of rewarming might play an important role in improving outcomes in TBI with accidental hypothermia. Posttraumatic hypothermia followed by rapid rewarming offered no beneficial effects for neuronal outcome [10]. Our group previously reported [7] that marked changes in alternation of vascular resistance at rewarming and active core rewarming to normothermia may lead to elevations in cerebral blood volume and intracranial pressure, which in turn may adversely affect final outcomes. For these reasons, we have determined that active core rewarming to normothermia at a rate of 1 to 3 °C/h might worsen the patient outcome. After active core rewarming to 33 to 34 °C, therapeutic hypothermia with slow rewarming could be effective for the treatment of patients with severe brain injuries accompanied by accidental hypothermia < 33 °C.

Conclusion

We report the case of a patient with severe brain injury who survived CPA with accidental hypothermia. Although the neurological outcome was poor, the survival of our patient suggests that therapeutic hypothermia might have the potential to improve survival and functional outcome in TBI patients with accidental hypothermia who experience hypoxia and hypotension. This case may provide information useful for planning rewarming treatment in TBI patients with accidental hypothermia.

References

1. Bernard SA, Buist M (2003) Induced hypothermia in critical care medicine: a review. Crit Care Med 31: 2041–2051
2. Bernard SA, Gray TW, Buist MD, Jones BM, Silvester W, Gutteridge G, Smith K (2002) Treatment of comatose survivors of out-of-hospital cardiac arrest with induced hypothermia. N Engl J Med 346: 557–563
3. Clifton GL, Miller ER, Choi SC, Levin HS, McCauley S, Smith KR Jr, Muizelaar JP, Wagner FC Jr, Marion DW, Luerssen TG, Chesnut RM, Schwartz M (2001) Lack of effect of induction of hypothermia after acute brain injury. N Engl J Med 344: 556–563
4. Cummins RO, Field JM, Hazinski MF (2003) Special resuscitation situations part 1: hypothermia. In: Cummins RO (ed) ACLS for experienced providers – the reference textbook. American Heart Association, Texas, pp 83–95
5. Giesbrecht GG, Schroeder M, Bristow GK (1994) Treatment of mild immersion hypothermia by forced-air warming. Aviat Space Environ Med 65: 803–808
6. Greif R, Rajek A, Laciny S, Bastanmehr H, Sessler DI (2000) Resistive heating is more effective than metallic-foil insulation in an experimental model of accidental hypothermia: A randomized controlled trial. Ann Emerg Med 35: 337–345
7. Kinoshita K, Hayashi N, Sakurai A, Utagawa A, Moriya T (2003) Importance of hemodynamics management in patients with severe head injury and during hypothermia. Acta Neurochir [Suppl] 86: 373–376
8. Mahoney EJ, Biffl WL, Harrington DT, Cioffi WG (2003) Isolated brain injury as a cause of hypotension in the blunt trauma patient. J Trauma 55: 1065–1069
9. Marshall LF, Becker DP, Bowers SA, Cayard C, Eisenberg H, Gross CR, Grossman RG, Jane JA, Kunitz SC, Rimel R, Tabaddor K, Warren J (1983) The National Traumatic Coma Data Bank. Part 1: Design, purpose, goals, and results. J Neurosurg 59: 276–284
10. Matsushita Y, Bramlett HM, Alonso O, Dietrich WD (2001) Posttraumatic hypothermia is neuroprotective in a model of traumatic brain injury complicated by a secondary hypoxic insult. Crit Care Med 29: 2060–2066
11. Miller JD, Becker DP (1982) Secondary insults to the injured brain. J R Coll Surg Edinb 27: 292–298
12. Resnick DK, Marion DW, Darby JM (1994) The effect of hypothermia on the incidence of delayed traumatic intracerebral hemorrhage. Neurosurgery 34: 252–255; discussion 255–256
13. The Hypothermia After Cardiac Arrest Study Group (2002) Mild therapeutic hypothermia to improve the neurologic outcome after cardiac arrest. N Engl J Med 346: 549–556
14. Wald SL, Shackford SR, Fenwick J (1993) The effect of secondary insults on mortality and long-term disability after severe head injury in a rural region without a trauma system. J Trauma 34: 377–381; discussion 381–382
15. Wolberg AS, Meng ZH, Monroe DM 3rd, Hoffman M (2004) A systematic evaluation of the effect of temperature on coagulation enzyme activity and platelet function. J Trauma 56: 1221–1228

Correspondence: Kosaku Kinoshita, Department of Emergency and Critical Care Medicine, Nihon University School of Medicine, 30-1 Oyaguchi Kamimachi, Itabashi ku, Tokyo 173-8610, Japan. e-mail: kosaku@med.nihon-u.ac.jp

Clinical characteristics of postoperative contralateral intracranial hematoma after traumatic brain injury

M. Furukawa, K. Kinoshita, T. Ebihara, A. Sakurai, A. Noda, Y. Kitahata, A. Utagawa, T. Moriya, K. Okuno, and K. Tanjoh

Department of Emergency and Critical Care Medicine, Nihon University School of Medicine, Tokyo, Japan

Summary

Objectives. To investigate the clinical characteristics of contralateral intracranial hematoma (ICH) after traumatic brain injury.

Methods. The subjects included 149 patients with traumatic ICH treated by hematoma evacuation. The patients were retrospectively divided into a bilateral ICH (B-ICH) group and unilateral ICH (U-ICH) group after craniotomy using brain CT scans for comparison of the following parameters: complicated expanded brain bulk from the cranial window, hypotension during craniotomy, and outcome.

Results. Post-craniotomy brain CT scans revealed U-ICH in 106 patients and B-ICH in 43 patients. Average Glasgow Coma Scale on arrival did not differ between the groups, but a higher proportion of patients in the B-ICH group deteriorated after admission ($p = 0.02$). The B-ICH patients also exhibited a significantly higher rate of expanded brain bulk from the cranial window ($p < 0.05$). No significant difference was observed between the groups with hypotension during craniotomy. The B-ICH group exhibited a lower rate of favorable outcome ($p < 0.05$) and higher mortality ($p < 0.05$).

Conclusion. The B-ICH patients had a worse outcome than the U-ICH patients. Contralateral ICH was difficult to forecast based on pre- and intraoperative clinical conditions. Subdural hematoma or contusional ICH was frequently observed as a contralateral ICH.

Keywords: Traumatic brain injury; intracranial hematoma; hypotension, clinical characteristics.

Introduction

Traumatic intracranial hematoma is a severe insult with poor outcome. It often has mass effect and requires surgical management urgently. Outbreak or enhancement of a contralateral intracranial hematoma can develop after decompressive craniotomy. When this happens, the patient is sent back to the operating room after postoperative brain computed tomography (CT) scan for contralateral surgical intervention. A spreading contralateral intracranial hematoma might be preventable if detected early enough. Few articles, however, have described the clinical characteristics of bilateral traumatic intracranial hematoma. In this study we sought to clarify these characteristics and determine whether the development of a contralateral hematoma expansion could be predicted from the clinical findings before and during craniotomy.

Materials and methods

The subjects included 149 patients who arrived with evidence of traumatic intracranial hematoma on brain CT and underwent emergency craniotomy and hematoma evacuation in the emergency center of a university hospital between 1993 and 1999. Patients with multiple injuries or hypotension before craniotomy (systolic blood pressure < 90 mmHg) were excluded.

Brain CT scans were recorded on arrival and just after operation. Based on these CT findings, the patients were retrospectively divided into 2 groups: those with confirmed hematoma on the non-operative side (bilateral intracranial hematoma: B-ICH) and those with no hematoma on the non-operative side (unilateral intracranial hematoma: U-ICH). The group with a confirmed presence of hematoma on the non-operative side was further divided into 2 subgroups based on timing of the detection of the hematoma using brain CT on the non-operative side on arrival (pre-B-ICH) or after the craniotomy (post-B-ICH). Four parameters were studied in each of these groups: Glasgow coma scale (GCS) on arrival, complicated expanded brain bulk from the cranial window, hypotension during craniotomy, and outcome.

Hypotension during craniotomy was defined as systolic arterial blood pressure of <90 mmHg during emergency craniotomy for 10 minutes or more, in spite of steps taken to correct hypotension.

Deterioration into a coma state was assessed in patients whose GCS score fell from a 9 or more to an 8 or less within 24 hours due to neurological deterioration.

Outcome was evaluated on the Glasgow Outcome Scale.

Statistical analysis was performed using StatView 5.0 (SAS Institute Inc., Cary, NC). Student *t*-test was used to compare clinical data between groups. Data are expressed as mean ± SD. A p-value of less than 0.05 was considered significant.

Table 1. *Clinical characteristics of ICH patients after TBI.*

	U-ICH	B-ICH		p
		Pre-B-ICH	Post-B-ICH	
Patients	106	43		
		15 (34.9%)	28 (65.1%)	
Age (range)	52.1 ± 19.8 (6–90)	51.7 ± 19.2 (19–86)		NS
GCS ≤ 8 on arrival (n)	76.4% (81)	83.7% (36)		NS
Average GCS on arrival	6.7	5.7		NS
Deteriorated after admission (n)	36.0% (9/25)	85.7% (6/7)		p = 0.02
Expanded brain bulk from the cranial window (n)	10.4% (11)	23.3% (10)		p < 0.05
Intraoperative hypotension (n)	12.3% (13)	16.3% (7)		NS
Favorable outcome (n)	42.5% (45)	23.3% (10)		p < 0.05
Mortality (n)	42.5% (45)	65.1% (28)		p < 0.05

B-ICH Bilateral intracranial hematoma; *GCS* Glasgow Coma Scale; *ICH* intracranial hematoma; *NS* not significant; *TBI* traumatic brain injury; *U-ICH* unilateral intracranial hematoma.

Results

The results of our investigation are detailed in Table 1. Brain CT scans revealed traumatic U-ICH in 106 patients and B-ICH in 43 patients. Fifteen of the contralateral hematomas in the B-ICH group were detected on arrival (pre-B-ICH), while the other 28 were detected after craniotomy (post-B-ICH).

Patients in the B-ICH and U-ICH groups ranged in age from 19 to 86 years and 6 to 90 years, respectively, with no significant difference in age between groups.

The average GCS on arrival did not differ between the groups. A higher proportion of patients in the B-ICH group deteriorated into a comatose state before the craniotomy.

No difference was observed between the groups in hypotensive episodes during the craniotomy. The B-ICH patients exhibited a higher rate of expanded brain bulk from the cranial window. The cause of the expanded brain bulk in the U-ICH group was associated with brain swelling (vascular engorgement) after treatment for hypotension using massive fluid infusion or vasopressors. There was no evidence of contralateral hematoma in the brain CT scans after the craniotomy in the U-ICH group. The main cause of the expanded brain bulk in the B-ICH group was impossible to discern due to the combined contralateral hematoma with brain swelling after treatment for hypotension.

Postoperative brain CT scans identified acute epidural hematoma in 5 patients, acute subdural hematoma in 12, and contusional intracerebral hematoma on the contralateral side in 11.

The B-ICH group exhibited a lower rate of favorable outcome (good recovery or moderate disability on Glasgow Outcome Scale) and higher mortality.

Discussion

The GCS on admission did not differ between the 2 groups investigated in this study. Moreover, the contralateral hematoma was only detected after the craniotomy in 65% of the patients in the B-ICH group. These results suggest that clinical parameters such as GCS or an initial brain CT on admission may not be predictive of B-ICH before craniotomy.

Deterioration into a comatose state after traumatic brain injury has recently been estimated to occur in about 10.5% to 25.1% brain-injured patients [6, 7, 10]. Intracranial hematoma has been cited as the most frequent cause of the deterioration, and mass lesions requiring surgical management are present in 75% to 81% of patients [9, 10, 13]. Our study indicated a very high (85.7%) incidence of deterioration into a comatose state in B-ICH patients, versus a much lower incidence (36.0%) in patients with U-ICH. Patients with traumatic B-ICH after craniotomy had a significantly high mortality (65.1%) in the present study. In earlier studies, however, favorable outcomes (good recovery and moderate disability) were reported in 56% of patients with acute epidural hematoma, 10% to 30% of patients with acute subdural hematoma, and 9% of patients with cerebral hemorrhage [2, 3, 5, 11, 12, 14]. Further studies will be needed to clarify the clinical characteristics and the operative strategies for B-ICH in order to improve outcome.

Intraoperative hypotension has often developed during decompressive craniotomy for severe traumatic brain injury. Kinoshita *et al.* [4] reported that the expanded brain bulk from the cranial window was not always caused by expansion of contralateral hematoma, and that the total infusion volumes treated for intraoperative hypotension during craniotomy were associated with intraoperative acute brain swelling. This finding formed the basis for our decision to exclude from the present study patients who had hypotension before craniotomy. Two recent papers [1, 8] describe intraoperative brain swelling as a sign suggestive of contralateral hematoma development. In mentioning "intraoperative brain swelling," they refer to expanded brain bulk from the cranial window. This may not be entirely appropriate, however, as it can be difficult to estimate whether the expanded brain bulk occurs due to an expansion of contralateral hematoma or due to swelling of brain substance during craniotomy [4]. Given the differences in pathophysiology and clinical strategies for treatment, our group thought that brain swelling (vascular engorgement) after treatment for hypotension should be clearly distinguished from expanded brain bulk caused by a contralateral hematoma, a condition with compression from the opposite side. This was our rationale for coining the term "expanded brain bulk from the cranial window."

Expanded brain bulk from the cranial window during craniotomy was significantly more frequent in the B-ICH group than in the U-ICH group, but the main cause of the condition in the former could not be confirmed. It appeared that the expanded brain bulk from the cranial window did not always indicate development or expansion of a contralateral hematoma. Confirmation by imaging such as brain CT might be necessary when brain bulk expands in association with treatment for unstable hemodynamics during craniotomy. We conclude that the occurrence of a contralateral hematoma could not be confirmed solely on the basis of intraoperative findings.

Subdural hematoma or contusional intracerebral hematoma was frequently observed as a contralateral intracranial hematoma on postoperative brain CT. Subdural hematoma or contusional intracerebral hematoma is often accompanied by damage of the brain substance. These results suggest that B-ICH might lead to extensive damage of the brain and ultimately cause poor outcome.

Conclusions

The B-ICH patients had a worse outcome than the U-ICH patients. Subdural hematoma or contusional ICH was frequently observed as a contralateral ICH. Development of contralateral hematoma could not be predicted from pre- and intraoperative findings.

References

1. Cohen JE, Rajz G, Itshayek E, Umansky F (2004) Bilateral acute epidural hematoma after evacuation of acute subdural hematoma: brain shift and the dynamics of extraaxial collections. Neurol Res 26: 763–766
2. Haselsberger K, Pucher R, Auer LM (1988) Prognosis after acute subdural or epidural haemorrhage. Acta Neurochir (Wien) 90: 111–116
3. Hatashita S, Koga N, Hosaka Y, Takagi S (1993) Acute subdural hematoma: severity of injury, surgical intervention, and mortality. Neurol Med Chir (Tokyo) 33: 13–18
4. Kinoshita K, Kushi H, Sakurai A, Utagawa A, Saito T, Moriya T, Hayashi N (2004) Risk factors for intraoperative hypotension in traumatic intracranial hematoma. Resuscitation 60: 151–155
5. Koc RK, Akdemir H, Oktem IS, Meral M, Menku A (1997) Acute subdural hematoma: outcome and outcome prediction. Neurosurg Rev 20: 239–244
6. Lobato RD, Rivas JJ, Gomez PA, Castaneda M, Canizal JM, Sarabia R, Cabrera A, Munoz MJ (1991) Head-injured patients who talk and deteriorate into coma. Analysis of 211 cases studied with computerized tomography. J Neurosurg 75: 256–261
7. Marshall LF, Toole BM, Bowers SA (1983) The National Traumatic Coma Data Bank. Part 2: Patients who talk and deteriorate: implications for treatment. J Neurosurg 59: 285–288
8. Matsuno A, Katayama H, Wada H, Morikawa K, Tanaka K, Tanaka H, Murakami M, Fuke N, Nagashima T (2003) Significance of consecutive bilateral surgeries for patients with acute subdural hematoma who develop contralateral acute epi- or subdural hematoma. Surg Neurol 60: 23–30; discussion 30
9. Rockswold GL, Pheley PJ (1993) Patients who talk and deteriorate. Ann Emerg Med 22: 1004–1007
10. Rockswold GL, Leonard PR, Nagib MG (1987) Analysis of management in thirty-three closed head injury patients who "talked and deteriorated". Neurosurgery 21: 51–55
11. Stening WA, Berry G, Dan NG, Kwok B, Mandryk JA, Ring I, Sewell M, Simpson DA (1986) Experience with acute subdural haematomas in New South Wales. Aust N Z J Surg 56: 549–556
12. Stone JL, Lowe RJ, Jonasson O, Baker RJ, Barrett J, Oldershaw JB, Crowell RM, Stein RJ (1986) Acute subdural hematoma: direct admission to a trauma center yields improved results. J Trauma 26: 445–450
13. Tan JE, Ng I, Lim J, Wong HB, Yeo TT (2004) Patients who talk and deteriorate: a new look at an old problem. Ann Acad Med Singapore 33: 489–493
14. Wilberger JE Jr, Harris M, Diamond DL (1991) Acute subdural hematoma: morbidity, mortality, and operative timing. J Neurosurg 74: 212–218

Correspondence: Makoto Furukawa, Department of Emergency and Critical Care Medicine, Nihon University School of Medicine, 30-1 Oyaguchi Kamimachi Itabashi-ku, Tokyo 173-8610, Japan. e-mail: makotof@med.nihon-u.ac.jp

Human Intracranial Hemorrhage

Diagnostic impact of the spectrum of ischemic cerebral blood flow thresholds in sedated subarachnoid hemorrhage patients

C. Compagnone[1], F. Tagliaferri[1], E. Fainardi[2], A. Tanfani[1], R. Pascarella[3], M. Ravaldini[1], L. Targa[1], and A. Chieregato[1]

[1] Neurorianimazione, Ospedale M. Bufalini, Cesena, Italy
[2] Neuroradiologia, Arcispedale S. Anna, Ferrara, Italy

Summary

Objective. Ischemia is the main cause of secondary damage in subarachnoid hemorrhage (SAH). Cerebral blood flow (CBF) measurement is useful to detect critical values. We analyzed the diagnostic impact of CBF ischemic thresholds to predict a new low attenuation area on computed tomography (CT) due to failure of large vessel perfusion.

Methods. We analyzed 48 xenon CT (Xe-CT) studies from 10 patients with SAH. CBF measurements were obtained by means of Xe-CT and cortical regions of interest (ROIs). The ROIs which appeared in a hypoattenuation area were recorded. Cortical CBF was tested for specificity and sensitivity as a predictor of hypoattenuation by means of a receiver operating characteristic curve.

Results. Mean age was 58 (SD ± 12.4) years. The median Fisher score and Hunt & Hess scale were 2 and 3, respectively. The area under the receiver operating characteristic curve was 0.912 (CI 0.896 to 0.926). The cut-off value for best accuracy was 6 mL/100 g/min, with a likelihood ratio of 37.

Conclusion. The present study suggests a threshold of 6 mL/100 g/min as a predictor of a new low attenuation area. However, each clinician should choose the most useful threshold according to pre-test probability and the cost/effectiveness ratio of the applied therapies.

Keywords: Subarachnoid hemorrhage; ischemia; xenon CT.

Introduction

Ischemia is a frequent complication after aneurysmal subarachnoid hemorrhage (SAH) caused by vasospasm [19], thromboembolic events related to embolization [17], inadvertent intraoperative vascular occlusion [6], or cerebral herniation [7]. Computed tomography (CT) is considered to be a sensitive and practical imaging method for detecting ischemic lesions after SAH, which appear as well-demarcated parenchymal regions of low attenuation relative to adjacent normal brain tissue [16]. However, hypoattenuation of ischemic tissue, as depicted by CT scans, becomes fully apparent only 24 hours after the onset of symptoms [11].

On the other hand, xenon-enhanced CT (Xe-CT) technology has proven to be a powerful tool to demonstrate the presence of cerebral blood flow (CBF) disturbances in patients with ischemic stroke, SAH, and head injury [3, 8, 14]. However, there is no consensus about ischemic thresholds, which are often derived from animal models. Failure of electrical activity and neurological dysfunctions occurs at 18–20 mL/100 g/min [1, 13] in awake subjects, but when the patient is comatose from disease and/or sedation [4, 9], an ischemic threshold CBF level might be as low as 10 mL/100 g/min.

The purpose of this study was to analyze the diagnostic impact of a spectrum of CBF ischemic thresholds to predict a new low attenuation area on the CT scan, indicative of ischemia.

Materials and methods

Patient selection

From January 2001 to December 2003, our neurointensive care unit admitted 173 patients with aneurysmal SAH. Of these, 85 patients underwent Xe-CT studies when physiological parameters were stable. We selected for analysis 10 patients with more than 4 Xe-CT studies who did not have any low attenuation areas on the initial Xe-CT. A total of 48 Xe-CT studies were analyzed. The physiological data from Xe-CT studies were recorded. The amount of traumatic SAH on initial CT scans was determined using Fisher's classification [5]. Clinical severity at admission was graded according to the Hunt and Hess scale [10]. The outcome at 6 months post-injury was measured using a Glasgow Outcome Score [12].

Treatment and management

Decisions concerning aneurysm treatment were made based on a combination of factors (aneurysm location, size, and shape, patient age, Hunt and Hess scale, Fisher score, presence of hematoma, and accessibility of resources). Patient monitoring involved transcranial Doppler, intracranial pressure, and cerebral perfusion pressure measurements. Clinical assessment was performed only in patients who were lightly sedated, with good Hunt-Hess scores, and no intracranial hypertension. When there was a suspicion of vasospasm, a Xe-CT was performed. If a low CBF was found, hypertensive hypervolemic therapy and increased sedation, up to electroencephalogram burst suppression, were instituted. Serial Xe-CTs were performed over the next few days.

Stable Xe-CT CBF studies

CBF measurements were performed using a CT scanner (Picker 5000) equipped for Xe-CT CBF imaging (Xe/TC system-2TM, DDP, Inc., Houston, TX). Cortical regions of interest (ROIs) were measured by dedicated software (Xe-CT System Version 1.0, 1998, Diversified Diagnostic Products Inc., Houston, TX).

An independent researcher (FT) evaluated the ROI density coefficients in the following Xe-CT studies. The corresponding ROIs, which evolved in a hypoattenuation area due to impairment of arterial perfusion (peri-herniation or postsurgical vascular distortion, occlusion, or vasospasm), were recorded and matched to their cortical regional CBF (rCBF) values from the previous Xe-CT study.

Statistical analysis

The presence of a new low attenuation area in combination with a CBF value obtained from the same ROI found in the previous Xe-CT study prompted a receiver operating characteristic curve analysis. The area under the curve was measured and a cut-off value obtained using the statistical approach involving the best accuracy (MedCalc 7.5, Frank Schoonijans, Belgium). Diagnostic test results were also evaluated using methods considered to be more relevant for the "diagnostic impact" (the likelihood ratio [LR]) of diagnostic thresholds [18]. The latter was categorized according to "Rule In" (above 10 of LR), "Indeterminate High" (above 1 of LR), "Indeterminate" (equal to 1 of LR), "Indeterminate Low" (below 1 of LR), and "Rule Out" (above 0.3 of LR).

Results

The demographic and clinical parameters of our patients are reported in Table 1. The physiological data on CBF studies were: intracranial pressure 18.4 mmHg (SD ± 5.6), cerebral perfusion pressure 80.2 mmHg (SD ± 13.5), and paCO2 39.9 mmHg (SD ± 4.8). The median time elapsing between consecutive studies was 3 days. We analyzed 1663 ROIs (mean area 400.2; SD ± 89.8 mm^2; mean rCBF 27.9 mL/100 g/min; SD ± 13.7); 110 (7.3%) of them developed low attenuation. The principal cause of a new low attenuation area was vasospasm (62%). The area under the receiver operating characteristic curve was 0.912 (CI 0.896 to 0.926). The threshold of 6 mL/100 g/min had a sensitivity of 37%, a specificity of 99%, and a LR of 37 (Table 2).

Discussion

In the present study, the threshold of 6 mL/100 g/min was more representative of low attenuation than the traditionally accepted threshold of 18 mL/100 g/min [13]. Physiologic and therapeutic reasons could explain our results. Brain tissue at risk for irreversible damage is classically described as the ischemic penumbra due to a critical CBF. The enlargement of permanent damage and ischemia in the penumbra might be dependent on 3 variables: the level of residual blood flow (collateral blood flow), the duration of ischemia, and the individual susceptibility of neurons [9]. The treatment of SAH with suspicion of ischemia is consistent with the increasing cerebral perfusion pressure, reduction of blood viscosity, and a reduction of cerebral metabolic rate for oxygen, by means of metabolic suppression (deep sedation and/or burst suppression). The latter strategy could interfere with the relationship between the CBF threshold and ischemia, and rCBF reduction due to metabolic suppression could be improperly classified as ischemia. In fact, CBF and cerebral metabolic rate for oxygen are physiologically coupled [15] and the deep sedation/burst suppression can reduce metabolic rate and reduce CBF thresholds for ischemia [2].

To verify the final impact of our diagnostic tests, we must consider 2 other factors: the pre-test probability and potential adverse events. The pre-test probability is the patient's probability of having ischemia based on clinical experience, statistical prevalence, practice databases, and the accuracy of diagnostic tests. With a higher pre-test probability, it is possible to accept a range of CBF with a low LR. Conversely, when the chance to predict *a priori* that the fate of the tissue is incomplete, thresholds having elevated LR may be more useful. However, highly diagnostic LR may detect biological phenomena without any possibility of treatment benefit.

Our results, from both the statistical approach as well as the "diagnostic impact," suggest a threshold of 6 mL/100 g/min. Nevertheless, each clinician should choose the most useful threshold according to the pre-test probability and the cost/effectiveness ratio of the therapies applied. In our setting, the ischemic threshold was lower (6 mL/100 g/min) than convention (18 mL/100 g/min) [13]. This finding might be related to our treatment protocol (deep sedation, hypertension, hemodilution, and hypervolemia) [16]. Future studies will be performed to confirm this supposition.

Table 1. *Patient demographics and clinical parameters.*

		Mean (SD)	Median	n (%)
Patients				10
Age, y		58 (12.4)		
Sex, male/female				4/6
Fisher Score			2	
Hunt & Hess Scale			3	
Clipping/embolization				6/4
Normal pupillary reactivity				10 (100)
Bleeding Site	Anterior communicating artery			5 (50)
	Posterior communicating artery			4 (40)
	Anterior cerebral artery			1 (10)
Maximum TIL	Standard[#]			0 (0)
	Reinforced[##]			5 (50)
	Extreme[###]			5 (50)
Therapy	ICP monitoring			9 (90)
	CSF drainage			9 (90)
	Sedation/analgesia			10 (100)
	Burst suppression			4 (40)
	Catecholamines (Mean arterial pressure improvement)			2 (20)
	Catecholamines (arterial hypertension)			6 (60)
GOS at 6 months	Good recovery or moderate disability			4 (40)
	Severe disability			2 (20)
	Persistent vegetative state or death			2 (20)
	Missing			2 (20)
Xenon Studies				48
New Low Attenuation Area				8 (80)
Causes of Low Attenuation	Postsurgical			2 (25)
	Vasospasm			5 (62.5)
	Vascular distortion			1 (12.5)
Time elapsed between basal Xe-CT and Xe-CT having new attenuation area (days)			3 (2)	
Frequency distribution of xenon studies per patients	Patients with 4 studies			6 (60)
	Patients with 5 studies			1 (10)
	Patients with 6 studies			2 (20)
	Patients with 7 studies			1 (10)

[#] Standard: Sodium infusion, sedation, scheduled cerebrospinal fluid drainage, mannitol, or mild hyperventilation. [##] Reinforced: Moderate hyperventilation, arterial pressure improvement, propofol and benzodiazepine, indometacin, evacuation of contusion or hematoma. [###] Extreme: Deep hyperventilation, arterial hypertension, muscle paralysis, external decompression, internal decompression, barbiturates or hypothermia. *CSF* Cerebrospinal fluid, *GOS* Glasgow Outcome Score, *ICP* intracranial pressure, *TIL* therapeutic intervention level, *Xe-CT* Xenon-enhanced computed tomography.

Table 2. *Suggested thresholds according to classes of likelihood ratio values [22]*

CBF mL/100 g/min	Hypoattenuation present (sensibility) n (%)	Hypoattenuation absent (1-specificity) n (%)	Likelihood ratio	Diagnostic impact
<6	41 (37%)	18 (1%)	37	Rule In
≥6 to <15	41 (37%)	218 (14%)	2.6	Indeterminate High
≥15 to <20	12 (11%)	196 (13%)	1.21	Indeterminate
≥20 to <25	10 (9%)	222 (14%)	0.84	Indeterminate Low
>25	6 (5%)	898 (57%)	0.09	Rule Out
	Total: 110 (100%)	Total: 1552 (100%)		

CBF Cerebral blood flow.

References

1. Astrup J, Siesjo BK, Symon L (1981) Thresholds in cerebral ischemia – the ischemic penumbra. Stroke 12: 723–725
2. Bergsneider M, Hovda DA, Lee SM, Kelly DF, McArthur DL, Vespa PM, Lee JH, Huang SC, Martin NA, Phelps ME, Becker DP (2000) Dissociation of cerebral glucose metabolism and level of conciousness during the period of metabolic depression following human traumatic brain injury. J Neurotrauma 17: 389–401
3. Chieregato A, Fainardi E, Tanfani A, Martino C, Pransani V, Cocciolo F, Targa L, Servadei F (2003) Mixed dishomogeneous hemorrhagic brain contusions. Mapping of cerebral blood flow. Acta Neurochir [Suppl] 86: 333–337
4. Diringer MN, Videen TO, Yundt K, Zazulia AR, Aiyagari V, Dacey RG Jr, Grubb RL, Powers WJ (2002) Regional cerebrovascular and metabolic effects of hyperventilation after severe traumatic brain injury. J Neurosurg 96: 103–108
5. Fisher CM, Kistler JP, Davis JM (1980) Relation of cerebral vasospasm to subarachnoid hemorrhage visualized by computerized tomographic scanning. Neurosurgery 6: 1–9
6. Fridriksson S, Saveland H, Jakobsson KE, Edner G, Zygmunt S, Brandt L, Hillman J (2002) Intraoperative complications in aneurysm surgery: a prospective national study. J Neurosurg 96: 515–522
7. Graham DI, Adams JH, Doyle D (1978) Ischaemic brain damage in fatal non-missile head injuries. J Neurol Sci 39: 213–234
8. Hayashi T, Suzuki A, Hatazawa J, Kanno I, Shirane R, Yoshimoto T, Yasui N (2000) Cerebral circulation and metabolism in the acute stage of subarachnoid hemorrhage. J Neurosurg 93: 1014–1018
9. Heiss WD (2000) Ischemic penumbra: evidence from functional imaging in man. J Cereb Blood Flow Metab 20: 1276–1293
10. Hunt WE, Hess RM (1968) Surgical risk as related to time of intervention in the repair of intracranial aneurysms. J Neurosurg 28: 14–20
11. Inoue Y, Takemoto K, Miyamoto T, Yoshikawa N, Taniguchi S, Saiwai S, Nishimura Y, Komatsu T (1980) Sequential computed tomography scans in acute cerebral infarction. Radiology 135: 655–662
12. Jennett B, Bond M (1975) Assessment of outcome after severe brain damage. Lancet 1: 480–484
13. Jones TH, Morawetz RB, Crowell RM, Marcoux FW, FitzGibbon SJ, DeGirolami U, Ojemann RG (1981) Thresholds of focal cerebral ischemia in awake monkeys. J Neurosurg 54: 773–782
14. Kaufmann AM, Firlik AD, Fukui MB, Wechsler LR, Jungries CA, Yonas H (1999) Ischemic core and penumbra in human stroke. Stroke 30: 93–99
15. Lebrun-Grandie P, Baron JC, Soussaline F, Loch'h C, Sastre J, Bousser MG (1983) Coupling between regional blood flow and oxygen utilization in the normal human brain. A study with positron tomography and oxygen 15. Arch Neurol 40: 230–236
16. Mayberg MR, Batjer HH, Dacey R, Diringer M, Haley EC, Heros RC, Sternau LL, Torner J, Adams HP Jr, Feinberg W, Thies W (1994) Guidelines for the management of aneurismal subarachnoid hemorrhage. A statement for healthcare professionals from a special writing group of the Stroke Council, American Heart Association. Circulation 90: 2592–2605
17. Qureshi AI, Luft AR, Sharma M, Guterman LR, Hopkins LN (2000) Prevention and treatment of thromboembolic and ischemic complications associated with endovascular procedures: part II – clinical aspects and recommendations. Neurosurgery 46: 1360–1375; discussion 1375–1376
18. Sackett DL, Straus SE, Richardson WS, Rosemberg W, Haynes RB (2000) Diagnosis and screening. In: Sackett DL (ed) Evidence-based medicine: how to practice and teach EBM, 2nd edn. Churchill Livingstone, Edinburgh, pp 67–93
19. Weir B, Macdonald RL, Stoodley M (1999) Etiology of cerebral vasospasm. Acta Neurochir [Suppl] 72: 27–46

Correspondence: Arturo Chieregato, Servizio di Anestesia e Rianimazione, Ospedale M. Bufalini, Viale Ghirotti 286 – 47023 Cesena, Italy. e-mail: achiere@ausl-cesena.emr.it

Pharmacological brain cooling with indomethacin in acute hemorrhagic stroke: antiinflammatory cytokines and antioxidative effects

K. Dohi[1], H. Jimbo[2], Y. Ikeda[4], S. Fujita[2], H. Ohtaki[3], S. Shioda[3], T. Abe[2], and T. Aruga[1]

[1] Department of Critical Care and Emergency Medicine, Showa University School of Medicine, Tokyo, Japan
[2] Department of Neurosurgery, Showa University School of Medicine, Tokyo, Japan
[3] Department of Anatomy, Showa University School of Medicine, Tokyo, Japan
[4] Department of Neurosurgery, Tokyo Medical University Hachioji Medical Center

Summary

We evaluated the effects of a novel pharmacological brain cooling (PBC) method with indomethacin (IND), a nonselective cyclooxygenase inhibitor, without the use of cooling blankets in patients with hemorrhagic stroke. Forty-six patients with hemorrhagic stroke (subarachnoid hemorrhage; n = 35, intracerebral hemorrhage; n = 11) were enrolled in this study.

Brain temperature was measured directly with a temperature sensor. Patients were cooled by administering transrectal IND (100 mg) and a modified nasopharyngeal cooling method (positive selective brain cooling) initially. Brain temperature was controlled with IND 6 mg/kg/day for 14 days. Cerebrospinal fluid concentrations of interleukin-1β (CSF IL-1β) and serum bilirubin levels were measured at 1, 2, 4, and 7 days. The incidence of complicating symptomatic vasospasm after subarachnoid hemorrhage was lower than in non-PBC patients. CSF IL-1β and serum bilirubin levels were suppressed in treated patients.

IND has several beneficial effects on damaged brain tissues (anticytokine, free radical scavenger, antiprostaglandin effects, etc.) and prevents initial and secondary brain damage. PBC treatment for hemorrhagic stroke in patients appears to yield favorable results by acting as an antiinflammatory cytokine and reducing oxidative stress.

Keywords: Cyclooxygenase inhibitor; interleukin-1β; stroke; hypothermia; free radical scavenger.

Introduction

Neurons are extremely vulnerable to hyperthermia compared to other cell types [17]. Brain temperature elevates during the early phase of severe brain damage caused by cerebral vascular accidents or severe head injury [3]. Hyperthermia is believed to occur after brain damage due to dysfunction of the selective brain cooling mechanisms [5], destruction of the hypothalamus, the abnormal release of catecholamines, and the production of endogenous pyrogens such as inflammatory cytokines [8, 14, 21]. Elevation in cerebrospinal fluid concentrations of interleukin-1β (CSF IL-1β) induces hyperthermia [20]. Hyperthermia and increased concentrations of inflammatory cytokines caused by primary brain damage induce secondary brain damage [11, 17].

Neuroprotective effects of hypothermia have been described [2]. Clinically, even mild brain hypothermia (33 °C to 35 °C) achieved by surface cooling is neuroprotective [10, 16]; however, the clinical use of brain hypothermia therapy has been limited, and imprecise temperature control may cause systemic complications.

Indomethacin (IND) is a strong cyclooxygenase (COX) inhibitor widely used as a nonsteroidal antiinflammatory drug. Recently, the COX inhibitors, including IND, have been shown to have not only antipyretic and antiinflammatory effects, but also other pharmacological effects (Table 1) including protection against neuronal cell death [7, 9].

Heme oxygenases, the rate-limiting enzymes in heme degeneration, catalyze the cleavage of the heme ring to form ferrous irons, carbon monoxide, and biliverdin. Biliverdin is rapidly metabolized to bilirubin, which is known to have powerful antioxidant properties [4, 18, 19]. Serum bilirubin concentration has recently become a marker of oxidative stress in patients with brain damage [4].

In the present study, we administered a pharmacological brain cooling (PBC) method with IND to patients with cerebrovascular diseases. We also measured CSF IL-1β and serum bilirubin concentrations in these patients to evaluate the combined effect of PBC and IND. This study was approved by the Ethics Committee of Showa University.

Materials and methods

Patients

The study group included 46 patients with intracerebral hemorrhage (ICH) (n = 11; Glasgow Coma Scale, mean = 5.9) and subarachnoid hemorrhage (SAH) (n = 35; World Federation of Neurosurgical Societies Grade 2, n = 5; Grade 3, n = 3; Grade 4, n = 6; Grade 5, n = 21) who were admitted to Showa University Hospital from January 1997 to March 2000. The mean ages of the patients in the ICH and SAH groups were 58.6 and 68.2 years. The patients in the SAH group underwent early operation. Thirteen patients with SAH were not treated with the PBC method, serving as controls, and were compared with the ICH group. The ICH control group (Glasgow Coma Scale, mean = 6.9) consisted of 30 patients and were compared with the SAH group (SAH control group: World Federation of Neurosurgical Societies Grade 1, n = 1; Grade 2, n = 7; Grade 3, n = 3; Grade 4, n = 4; Grade 5, n = 15). The mean ages of patients in the ICH and SAH control groups were 68.3 and 66.3 years.

PBC method

The PBC method was introduced at the early postoperative stage in the SAH group, and as early as possible after admission in the ICH Group.

Induction of brain cooling

Brain temperature was measured directly with a ventriculostomy catheter (intracranial pressure and brain temperature sensor; 4HMT, Integra NeuroSciences, Hampshire, UK). While vital signs were monitored, transrectal IND (100 mg) was administered to the patients. Positive selective brain cooling was also performed for rapid introduction of brain cooling [5, 12]. A balloon catheter was inserted to direct chilled air (8 to 12 L/min) into each side of the nasal cavity. The chilled air was exhaled only through the oral cavity.

Control of brain temperature

Brain temperature was maintained at 36.5 to 37.5 °C by administering IND suppositories (6 mg/kg/day) and regulating the room temperature without the use of cooling blankets. Additional IND suppositories were administered (maximum of 600 mg/day) to maintain temperature control.

Rewarming

Rewarming was controlled to within 1 °C/day by decreasing the IND dose and regulating the room temperature.

Measurement of CSF IL-1β concentrations

CSF IL-1β concentrations in SAH patients were measured on days 1, 2, 4, and 7 by enzyme-linked immunoabsorbent assay (PBC group, n = 6; control group, n = 6).

Measurement of serum bilirubin

On days 1, 2, 4, and 7, 5 ml aliquots of blood were collected in the morning. Bilirubin concentrations in the serum were measured for 4 consecutive days using an enzyme assay (PBC group, n = 46; control group, n = 13).

Statistical analysis

Data were evaluated using SigmaStat software (Systat Software Inc., Point Richmond, CA) using analysis of variance followed by Tukey's test or by Student t-test and are expressed as the mean ± SEM.

Results

Clinical course

The mean dose of IND was 267.3 ± 121.6 mg/day. The outcomes 3 months after admission were evaluated using the Glasgow Outcome Scale. In the ICH PBC group, 5 patients were scored as "moderate disability" (MD), 3 had "severe disability" (SD), 3 were classified as "vegetative state" (VS), and none were "dead" (D). In the control group, there were 2 MD, 3 SD, 6 VS, and 2 D. In the SAH PBC group, 11 patients were scored as "good recovery" (GR), 3 MD, 5 SD, 2 VS, and 14 D. In the control group, 5 were GR, 6 MD, 3 SD, 2 VS, and 14 D. The incidence of symptomatic vasospasm after SAH was suppressed ($p < 0.01$): 17.1% in the PBC group and 46.7% in the non-PBC group. In the ICH patients with hemiparesis on admission, weakness was significantly improved in the PBC group (81.8%) compared to the non-PBC group (33.3%). Anal bleeding due to the insertion of suppositories was observed in 2 patients. Common complications associated with brain hypothermia, such as severe infections including pneumonia, were observed in only 2 patients.

CSF IL-1β concentrations

In the PBC group, CSF IL-1β concentrations were suppressed significantly 1 and 2 days after attack (Fig. 1a).

Serum bilirubin concentrations

In the PBC group, serum bilirubin concentrations were suppressed significantly 1 and 2 days after attack (Fig. 1b).

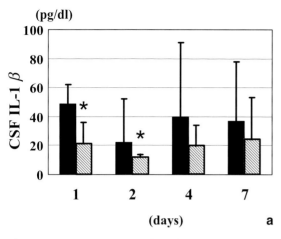
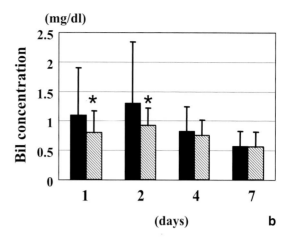

Fig. 1. (a) CSF IL-1β concentrations after hemorrhagic stroke. In the PBC group, CSF IL-1β concentrations were suppressed compared to the non-PBC group. Each value is mean ± SEM for 6 patients per group. (*p < 0.05) (b) Serum bilirubin concentrations after hemorrhagic stroke. In the PBC group, plasma bilirubin concentrations were suppressed compared to the non-PBC group. Each value is mean ± SEM. (*p < 0.05)
▨ PBC (+); ■ PBC (−)

Table 1. *Effects of pharmacological brain cooling with indomethacin.*

Desired Effects	Complications
• Control of intracranial pressure	• Gastrointestinal ulcer
• Inhibition of neurotoxic factors	• Degeneration of coagulation system
• Control of brain hyperthermia	
• Prevention of brain edema	• Anal laceration by suppository
• Maintenance of blood-brain barrier	• Renal failure
• Prevention of endothelial cell damage	• Shock (drug allergy)
	• Asthma

Discussion

Effects of PBC (Table 1)

COX is an important enzyme that metabolizes arachidonic acid found in unsaturated fatty acids in prostaglandin. There are 3 isozymes: COX-1 (constitutional type), COX-2 (inducible type), and COX-3, which represents a new COX family [1]. IND is one of the strong nonselective COX inhibitors.

COX inhibitors, including IND, induce various systemic and local pharmacological effects, and are used commonly as nonsteroidal antiinflammatory drugs and to treat various diseases. COX inhibitors directly prevent delayed neuronal cell death after ischemia and spinal injury [9]. We used IND in the present study since it has already been used in patients with brain tumors and head injury for its beneficial effect in cases of elevated intracranial pressure [6]. Recent advances in molecular biology have shown that various biologically active substances, including COX, have complex roles in brain injury. Inflammatory cytokines and free radicals are induced in the early phases of brain injury and promote primary and secondary tissue damage. These substances also play a role in the systemic inflammatory response syndrome [13, 15]. Inflammatory cytokine levels and COX are also associated with vasospasm after SAH [13, 15]. IND acts directly as a free radical scavenger [7].

We have demonstrated that IND suppresses production of CSF IL-1β and reduces oxidative stress. These findings indicate that PBC improves outcomes and minimizes neurological deficits by decreasing the incidence of cerebral vasospasm through its effect on antiinflammatory cytokines and free radical scavengers as well as suppression of hyperthermia. In particular, the decreased occurrence of symptomatic vasospasm in the PBC group is highly significant, since vasospasm is a serious problem associated with conventional brain hypothermia. Our results demonstrate that PBC induces effects different from those of conventional brain cooling achieved by the use of cooling blankets.

IND is a nonselective COX inhibitor that causes adverse events such as gastrointestinal ulcer. Safer and more effective COX-2 inhibitors should be evaluated in the PBC method. We have started using a COX-2 inhibitor clinically, though in a small number of patients thus far. Further research to evaluate the different effects of COX-1, COX-2, and COX-3 on neurons, endothelial cells, as well as the entire body, is required.

Conclusion

We describe a novel pharmacological method for brain cooling using IND. PBC induces many effects including suppression of inflammatory cytokines and reduction in oxidative stress. PBC is a safe alternative to conventional brain hypothermia management, which often results in systemic complications.

Acknowledgments

This research was partially supported by the Ministry of Education, Science, Sports and Culture, Grant-in-Aid for Scientific Research (C), 16591815, 2004.

References

1. Chandrasekharan NV, Dai H, Roos KL, Evanson NK, Tomsik J, Elton TS, Simmons DL (2002) COX-3, a cyclooxygenase-1 variant inhibited by acetaminophen and other analgesic/antipyretic drugs: cloning, structure, and expression. Proc Natl Acad Sci USA 99: 13926–13931
2. Clifton GL, Jiang JY, Lyeth BG, Jenkins LW, Hamm RJ, Hayes RL (1991) Marked protection by moderate hypothermia after experimental traumatic brain injury. J Cereb Blood Flow Metab 11: 114–121
3. Dohi K, Jimbo H, Ikeda Y (2000) Usefulness of continuous brain temperature monitoring after SAH. Jpn J Crit Care Med 12: 103–104
4. Dohi K, Mochizuki Y, Satoh K, Jimbo H, Hayashi M, Toyoda I, Ikeda Y, Abe T, Aruga T (2003) Transient elevation of serum bilirubin (a heme oxygenase-1 metabolite) level in hemorrhagic stroke: bilirubin is a marker of oxidant stress. Acta Neurochir [Suppl] 86: 247–249
5. Dohi K, Jimbo H, Abe T, Aruga T (2005) A novel and simple selective brain cooling method by nasopharyngeal cooling (positive selective brain cooling method). Acta Neurochir [Suppl] [in press]
6. Harrigan MR, Tuteja S, Neudeck BL (1997) Indomethacin in the management of elevated intracranial pressure: a review. J Neurotrauma 14: 637–650
7. Ikeda Y, Matsumoto K, Dohi K, Jimbo H, Sasaki K, Satoh K (2001) Direct superoxide scavenging activity of nonsteroidal anti-inflammatory drugs: determination by electron spin resonance using the spin trap method. Headache 41: 138–141
8. Imaizumi Y, Mizushima H, Matsumoto H, Dohi K, Matsumoto Ki, Ohtaki H, Funahashi H, Matsunaga S, Horai R, Asano M, Iwakura Y, Shioda S (2001) Increased expression of interleukin-1β in mouse hippocampus after global cerebral ischemia. Acta Histochem Cytochem 34: 357–362
9. Kondo F, Kondo Y, Gomez-Vargas M, Ogawa N (1998) Indomethacin inhibits delayed DNA fragmentation of hippocampal CA1 pyramidal neurons after transient forebrain ischemia in gerbils. Brain Res 791: 352–356
10. Marion DW, Obrist WD, Carlier PM, Penrod LE, Darby JM (1993) The use of moderate therapeutic hypothermia for patients with severe head injuries: a preliminary report. J Neurosurg 79: 354–362
11. Mizushima H, Zhou CJ, Dohi K, Horai R, Asano M, Iwakura Y, Hirabayashi T, Arata S, Nakajo S, Takaki A, Ohtaki H, Shioda S (2002) Reduced postischemic apoptosis in the hippocampus of mice deficient in interleukin-1. J Comp Neurol 448: 203–216
12. Nagasaka T, Brinnel H, Hales JR, Ogawa T (1998) Selective brain cooling in hyperthermia: the mechanisms and medical implications. Med Hypotheses 50: 203–211
13. Nakagawa K, Hirai K, Aoyagi M, Yamamoto K, Hirakawa K, Katayama Y (2000) Bloody cerebrospinal fluid from patients with subarachnoid hemorrhage alters intracellular calcium regulation in cultured human vascular endothelial cells. Neurol Res 22: 588–596
14. Ohtaki H, Yin L, Nakamachi T, Dohi K, Kudo Y, Makino R, Shioda S (2004) Expression of tumor necrosis factor alpha in nerve fibers and oligodendrocytes after transient focal ischemia in mice. Neurosci Lett 368: 162–166
15. Osuka K, Suzuki Y, Watanabe Y, Takayasu M, Yoshida J (1998) Inducible cyclooxygenase expression in canine basilar artery after experimental subarachnoid hemorrhage. Stroke 29: 1219–1222
16. Shiozaki T, Sugimoto H, Taneda M, Yoshida H, Iwai A, Yoshioka T, Sugimoto T (1993) Effect of mild hypothermia on uncontrollable intracranial hypertension after severe head injury. J Neurosurg 79: 363–368
17. Sminia P, Haveman J, Ongerboer de Visser BW (1989) What is a safe heat dose which can be applied to normal brain tissue? Int J Hyperthermia 5: 115–117
18. Stocker R, Yamamoto Y, McDonagh AF, Glazer AN, Ames BN (1987) Bilirubin is an antioxidant of possible physiological importance. Science 235: 1043–1046
19. Tomaro ML, Batlle AM (2002) Bilirubin: its role in cytoprotection against oxidative stress. Int J Biochem Cell Biol 34: 216–220
20. Watanabe T, Morimoto A, Murakami N (1991) ACTH response in rats during biphasic fever induced by interleukin-1. Am J Physiol 261: R1104–1108
21. Yin L, Ohtaki H, Nakamachi T, Dohi K, Iwai Y, Funahashi H, Makino R, Shioda S (2003) Expression of tumor necrosis factor alpha (TNFalpha) following transient cerebral ischemia. Acta Neurochir [Suppl] 86: 93–96

Correspondence: Kenji Dohi, Department of Critical Care and Emergency Medicine, Showa University School of Medicine, 1-5-8 Hatanodai, Shinagawa-ku, Tokyo, Japan. e-mail: kdop@med.showa-u.ac.jp

The significance of crossovers after randomization in the STICH trial

K. S. M. Prasad, B. A. Gregson, P. S. Bhattathiri, P. Mitchell, and **A. D. Mendelow (on behalf of the STICH investigators)**

Regional Neurosciences Centre, Newcastle General Hospital, Newcastle Upon Tyne, UK

Summary

Introduction. Of all forms of stroke, spontaneous intracerebral haemorrhage (ICH) causes the highest morbidity and mortality. The Surgical Trial in Intracerebral Haemorrhage (STICH) found no difference in outcomes between patients randomized to surgical or conservative treatment.

Patients and methods. Of 530 patients randomized to initial conservative treatment, 140 crossed over to surgery. This study examines the variables associated with crossover.

Results. Dominant features of the crossover group were: male, ($p = 0.04$), right-sided clot ($p = 0.03$), lobar clot ($p = 0.003$), clot volume (median 64 mL for crossovers vs. 38 mL for others, $p < 0.00001$), midline shift (median 6 mm for crossovers vs. 3 mm for others, $p < 0.00001$), superficial clot (median 1.3 mm for crossovers vs. 11.5 mm for others, $p < 0.00001$), and randomization within 12 hours of ictus ($p < 0.0005$). Thalamic location ($p = 0.002$) was under-represented. Intraventricular haemorrhage, hydrocephalus, and focal deficits were not associated with crossover. Craniotomy was the method of evacuation in 85% of crossover patients.

Conclusions. Crossover to surgery was more likely when ICH had these features: Right side, lobar location, superficial, large volume, big shift, and early randomization. Crossovers formed a worse prognostic group compared to non-crossovers. Surgery did not affect trial results, which were analyzed by intention-to-treat.

Keywords: Crossover; intracerebral haemorrhage; surgery.

Introduction

Spontaneous intracerebral haemorrhage has the highest mortality of all forms of stroke, and accounts for 20% of all stroke-related neurological deficits [5]. The Surgical Trial in Intracerebral Haemorrhage (STICH) reported a neutral trial finding with no difference in outcomes at 6 months between those randomized to early surgery and those receiving initial conservative treatment. An unexpected feature of this trial was the crossover rate. Crossovers to surgery in prospective randomized controlled trials (PRCT) cause problems in comparing groups as well as affecting the power of the trial. We explored the features and impact of crossovers in STICH.

Materials and methods

STICH had a parallel group design in which a total of 1033 patients from 83 centers in 27 countries were recruited over an 8-year period [5]. The patients were randomized to early surgery (haematoma evacuation) or initial conservative treatment, but crossovers were permitted for clinical and ethical reasons. The extended 8-point Glasgow Outcome Score (GOS) at 6 months was used as the primary outcome measure. Analysis was by intention-to-treat. We analyzed STICH randomization and outcome data and the pre-randomization CT scans coded using a specified protocol [3].

Results

The reasons for crossover to surgery from the initial conservative treatment group were rebleeding (n = 17), neurological deterioration (n = 82), clinical deterioration (n = 20), no improvement with conservative treatment (n = 4), raised intracranial pressure (n = 3), edema (n = 5), altered consciousness (n = 2), coma (n = 1), aneurysm (n = 1), not waking after external ventricular drain (n = 1), family request (n = 1), and reason not recorded (n = 3).

Of the 1033 patients who were recruited, 503 were randomized to early surgery while 530 drew initial conservative treatment. Of these, 496 were assessable in the surgery group while 529 were assessable in the initial conservative treatment group at 2 weeks. Complete outcome data at 6 months was available for 468 patients among those randomized to early surgery while 497 were assessable in the initial conservative treatment group. Thirty-one patients (6%) allocated to surgery did not have the procedure while 140 (26%)

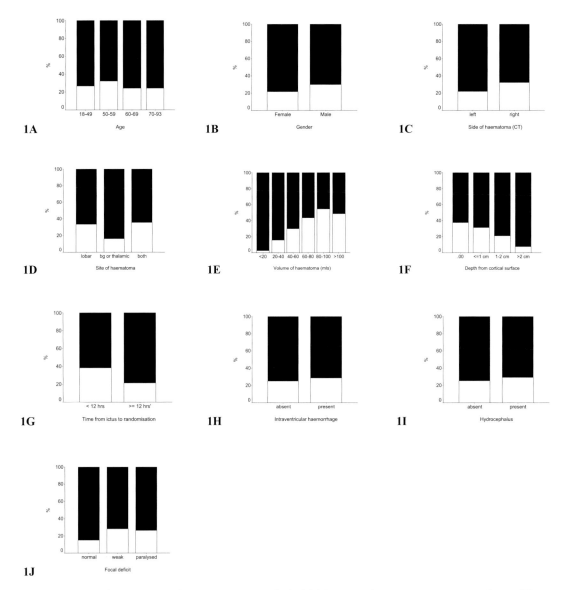

Fig. 1. Graph showing percentage of crossovers to surgery in the initial conservative treatment group according to different ages (1A), gender (1B), side of haematoma (1C), site of haematoma (1D), volume of haematoma (1E), haematomas at different depths from cortical surface (1F), ictus to randomization time intervals (1G), presence or absence of IVH (1H), presence or absence of hydrocephalus (1I), and presence or absence of focal deficits (1J). Had surgery ■ NO, □ YES

from the conservative treatment group crossed over to surgery.

Patients randomized to surgery but did not have it were a very small group and are not considered in further detail here. Crossovers from initial conservative treatment to surgery are included in all further analyses in this paper.

Crossover rates for patients with a Glasgow Coma Scale (GCS) score of 15 was 14.5% while for patients with a GCS score less than 10 was 34.9%. The drop in GCS between randomization and surgery for the crossovers averaged 3 points. Age did not have any particular association with crossovers and frequencies were matched in the different quartiles (Fig. 1A). Males were more common in the crossovers ($p = 0.04$; Fig. 1B). Right-sided clots were more likely than left-sided ones in the crossover group ($p = 0.03$; Fig. 1C). The site of the haematoma had a positive association with crossovers with lobar clots ($p = 0.003$, Fig. 1D) while thalamic haematomas were infrequent ($p = 0.002$). As expected, larger haematomas tended to crossover more often (Fig. 1E). Seventy percent of the crossovers had a clot less than 10 mm from the surface ($p < 0.0005$; Fig. 1F). Of those randomized with-

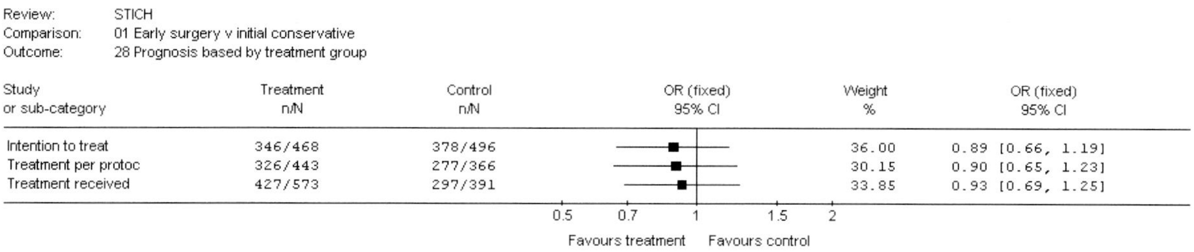

Fig. 2. Forest plot and odds ratio for outcome (alternate methods of analysis)

in 12 hours after the ictus, 38.4% crossed over, while the crossover rate for those randomized later was 21.4% (p < 0.0005, Fig. 1G). There were no differences in crossover rates in subjects with or without intraventricular haemorrhage (IVH) (Fig. 1H), hydrocephalus (Fig. 1I), or focal deficits (Fig. 1J). The median clot volume for those crossing over was 64 mL (IQR 44 to 85) compared to 38 mL (IQR 21 to 58) for those who did not (p < 0.00001). Midline shift also had a predictable association. The median shift in crossovers was 6 mm (IQR 4 to 9), while it was 3 mm (IQR 1 to 6) for the others (p < 0.00001). Haematomas closer to the surface were more likely to be crossovers than deeper ones, with median depth from the cortical surface for crossovers being 1.3 mm (IQR 0 to 10.9) and 11.5 mm (IQR 0 to 18.2) in those that did not (p < 0.00001). Of the 140 patients who were crossovers to surgery, 119 (85%) had craniotomy as the method of haematoma evacuation. Favorable outcome was experienced in 22% of the crossover group and 24% of those who continued to have conservative treatment (p = 0.7).

STICH used the intention-to-treat analysis primarily, but analyses using methods such as "treatment per protocol" or "treatment received" did not change the basic conclusions (Fig. 2).

Discussion

PRCT are always vulnerable to dropouts, protocol violations, and crossovers. Crossovers occur due to evolution of the clinical problem, thus removing the equipoise that existed at randomization or due to selection/efficacy bias. Sometimes crossovers are used deliberately, as in crossover trials to use within-patient comparisons rather than between-patient comparisons [4], but that methodology is impossible in surgical trials.

Another constraint for PRCTs involving surgical options is that true blinding of treatment is not possible. In various coronary bypass grafting trials, crossover rates to surgery have been in the range of 25% to 38% over periods up to 5 years [6, 9]. In STICH, crossover from initial conservative treatment to surgical treatment was 26%. An important requirement of PRCTs is determination of sample size. If crossovers are not factored in, the trial can suffer from loss of power [8]. When statistical power is lost, a truly effective treatment may be erroneously considered to be no better than control [9]. Often sample size determination is based on anticipated type I and II error rates and other assumptions such as Cox's proportional hazards [8]. However, STICH used published data and information from a prospective pilot study to include a safety margin of 25% to offset effects on sample size of protocol violations and crossovers [5]. The target sample size of 800 was exceeded even when protocol violations and crossovers were excluded.

Crossovers are significant confounding elements when analysis of outcome is by initial assigned treatment (intention-to-treat). Attributing success to the assigned treatment, but ignoring crossovers, would falsely associate outcome to the treatment modality from which the patient had crossed over. Similarly, if success was simply tagged to the treatment received after crossover, it would miss failures of the first assigned treatment [2].

For these reasons the classic intention-to-treat analysis has been doubted for its lack of sensitivity in detecting the 'true' effect of a treatment. This has been compared with other types of analyses, such as: 1) exclusion of crossovers from analysis; 2) transferring crossovers to the new treatment group from the time of treatment change; 3) censoring crossovers from the time of the crossover; 4) counting crossovers retrospectively from the date of randomization, but in the new group. After comparison of all these methods, the validity of the intention-to-treat analysis has been reaffirmed [6]. There is extensive support in the literature that data occurring after crossover has little bearing on the comparison of 2 treatments [7]. STICH primarily used the intention-to-treat analysis, but analyses using methods such as "treatment per protocol" and "treatment received" did not change the basic conclusions (Fig. 2).

Features of the crossovers in STICH displayed a predictable pattern. Disease evolution and loss of clinical equipoise were likely reasons for crossover when randomization was undertaken early (<12 hours) [5]. Crossover took place with an average drop in GCS by 3 points. Larger clots, lobar location, nondominant hemisphere involvement, superficial haematomas, and greater midline shift had higher associations with crossovers. These are in keeping with conventional surgical practice. Established focal neurological deficits did not spur crossovers. Not surprisingly, patients with IVH and hydrocephalus were not associated with higher crossover rates, probably related to the poor prognosis associated with such cases [1].

Conclusions

There was a greater rate of crossovers to surgery in STICH in men, and in patients with right-sided clots, larger haematomas, lobar location, superficial presence, increased midline shift, and in those randomized early (<12 hours). Thalamic bleeds had a significantly lower crossover rate. IVH, hydrocephalus, and focal deficits had no association with crossovers. Of the crossovers from the initial conservative treatment group, 85% underwent craniotomy for haematoma evacuation.

Crossover rates in STICH were not higher than expected. Crossovers formed a significantly worse prognostic group than non-crossovers. Ineffective surgery did not affect trial results analyzed by intention-to-treat. Our results suggest that surgeons in the STICH trial were more likely to operate (despite initial randomization) in patients with large, superficial right-sided haematomas with a greater midline shift.

References

1. Arene NU, Fernandes H, Wilson S, Wooldridge T, Grivas AG, Deogaonkar M, Mendelow AD (1998) Intraventricular hemorrhage from spontaneous intracerebral hemorrhage and aneurysmal rupture. In: von Wild KRH (ed) Pathophysiological principles and controversies in neurointensive care: minimizing mortality in head injured patients by "the Lund concept"? W. Zuckschwerdt Verlag, Munich
2. Barlow W, Azen S (1990) The effect of therapeutic treatment crossovers on the power of clinical trials. The Silicone Study Group. Control Clin Trials 11: 314–326
3. Bhattathiri PS, Gregson B, Prasad KS, Mitchell P, Soh C, Mitra D, Gholkar A, Mendelow AD (2003) Reliability assessment of computerized tomography scanning measurements in intracerebral hematoma. Neurosurg Focus 15: E6
4. Cleophas TJM (1995) A simple analysis of crossover studies with one-group interaction. Int J Clin Pharmacol Ther 33: 322–327
5. Mendelow AD, Gregson BA, Fernandes HM, Murray GD, Teasdale GM, Hope DT, Karimi A, Shaw MD, Barer DH; STICH investigators (2005) Early surgery versus initial conservative treatment in patients with spontaneous supratentorial intracerebral haematomas in the International Surgical Trial in Intracerebral Haemorrhage (STICH): a randomised trial. Lancet 365: 387–397
6. Peduzzi P, Detre K, Wittes J, Holford T (1991) Intent-to-treat analysis and the problem of crossovers. An example from the Veterans Administration coronary bypass surgery study. J Thorac Cardiovasc Surg 101: 481–487
7. Pocock SJ (1983) Clinical trials: a practical approach. Wiley, New York, pp 114
8. Porcher R, Levy V, Chevret S (2002) Sample size correction for treatment crossovers in randomized clinical trials with a survival endpoint. Control Clin Trials 23: 650–661
9. Weinstein GS, Levin B (1989) Effect of crossover on the statistical power of randomized studies. Ann Thorac Surg 48: 490–495

Correspondence: A. D. Mendelow, Department of Neurosurgery, Newcastle General Hospital, Newcastle Upon Tyne NE4 6BE, UK. e-mail: a.d.mendelow@ncl.co.uk

Intraventricular hemorrhage and hydrocephalus after spontaneous intracerebral hemorrhage: results from the STICH trial

P. S. Bhattathiri, B. Gregson, K. S. M. Prasad, and A. D. Mendelow (on behalf of the S.T.I.C.H. investigators)

Regional Neurosciences Centre, Newcastle General Hospital, Newcastle Upon Tyne, UK

Summary

Introduction. Intraventricular hemorrhage (IVH), either independent of or as an extension of intracranial bleed, is thought to carry a grave prognosis. Although the effect of IVH on outcome in patients with subarachnoid hemorrhage has been extensively reviewed in the literature, reports of spontaneous intracerebral hemorrhage (ICH) in similar situations have been infrequent. The association of hydrocephalus in such situations and its influence on outcome is also uncertain.

Patients and methods. As a sub-analysis of data obtained through the international Surgical Trial in Intracerebral Hemorrhage (STICH), the impact of IVH, with or without the presence of hydrocephalus, on outcome in patients with spontaneous ICH was analyzed. CT scans of randomized patients were examined for IVH and/or hydrocephalus. Other characteristics of hematoma were evaluated to see if they influenced outcome, as defined by the STICH protocol [9].

Results. Favorable outcomes were more frequent when IVH was absent (31.4% vs. 15.1%; $p < 0.00001$). The presence of hydrocephalus lowered the likelihood of favorable outcome still further to 11.5% ($p = 0.031$). In patients with IVH, early surgical intervention had a more favorable outcome (17.8%) compared to initial conservative management (12.4%) ($p = 0.141$).

Conclusion. The presence of IVH and hydrocephalus are independent predictors of poor outcome in spontaneous ICH. Early surgery is of some benefit in those with IVH.

Keywords: Intracerebral hemorrhage; intraventricular hemorrhage; hydrocephalus; outcome.

Introduction

Primary intracerebral hemorrhage (ICH) is a common and devastating disorder that often results in long-term disability and a socio-economic burden to society. Clinical and radiological findings following ICH not only help in treatment planning and prognostication, but also aid rehabilitation specialists to develop treatment goals, anticipate long-term patient care needs, and educate and train caregivers [2].

The recently concluded international Surgical Trial in Intracerebral Hemorrhage (STICH) [9] gave insight into the outcome of ICH with early surgical intervention when compared to initial conservative treatment. Intraventricular hemorrhage (IVH), either independently or as an extension of an intracranial bleed, is thought to carry a grave prognosis [10, 14]. Various treatment options to address the intraventricular blood load has been of particular interest recently [3, 5, 7, 11]. Although the effect of IVH on outcome in patients with subarachnoid hemorrhage has been reviewed in the literature extensively [4, 8, 12], reports of spontaneous ICH in similar situations have been reported less frequently. The association of hydrocephalus in such situations and its influence on outcome is also uncertain, although it has been identified as one of the major factors determining mortality in primary IVH [14].

Materials and methods

As detailed in the STICH report [9], a parallel group trial design was used to compare outcomes of patients having ICH treated with early surgery versus initial conservative treatment. In addition to clinical data, pre-randomization computerized tomography (CT) scans were collected from centers all over the world either as hard copies or in electronic format. Scans were re-evaluated for the presence of IVH and/or hydrocephalus, as well as information on site, volume, midline shift, depth, and other special characteristics, if any. At 6-month follow-up, the 8-point Glasgow Outcome Score was used as the primary outcome measure. Analysis was by intention-to-treat. As a sub-analysis of data obtained through STICH, the impact of IVH and the presence of hydrocephalus on outcome were analyzed. Patient factors and hematoma characteristics were also evaluated to see if they had any bearing on outcome in this subset of patients.

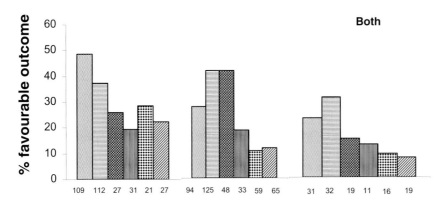

Fig. 1. Comparison of favorable outcome (%) in patients with no intraventricular hemorrhage (*no ivh*), intraventricular hemorrhage (*ivh*), and associated hydrocephalus (*ivh + hcp*) with respect to treatment; early surgery (*ES*) and initial conservative treatment (*ICT*). Total numbers in each group are shown on the horizontal axis. ▨ no ivh ES, ▤ no ivh ICT, ▩ ivh ES, ▥ ivh ICT, ▦ ivh + hcp ES, ▧ ivh + hcp ICT

Results

Patients (n = 1033) from 83 centers in 27 countries were randomized to early surgery (n = 503) or initial conservative treatment (n = 530) [9]. Of 468 patients randomized to early surgery, 122 (26%) had a favorable outcome compared with 118 (24%) of 496 randomized to initial conservative treatment (odds ratio 0.89 [95% CI, 0.66–1.19]; p = 0.414); absolute benefit 2% (−3.2 to 7.7); relative benefit 10% (−13 to 33). Pre-randomization CT scans were collected and analyzed for 960 (93%) patients. Of these, 950 had adequate images to provide for analysis and 902 had follow-up. Scans were reviewed by the Research Registrars after establishing inter-observer and intra-observer variability [1].

Of the 902 patients with follow-up, 42% (377) had IVH of whom 55% (208) went on to develop hydrocephalus. Mean age was 60.9 years (± 13.6) when IVH was present and 62.9 years (± 12.2) when hydrocephalus was present, compared to mean age of 60.7 years (± 12.4) when neither were present. Median Glasgow Coma Scale score changed from 12 when IVH was absent to 11 when IVH was present, or 10 when hydrocephalus was present as well (p < 0.00001, Kruskal-Wallis test). Fifty-eight percent of patients ≥ 50 years of age developed hydrocephalus following IVH compared to only 41% of those <50 years of age (p = 0.007).

Outcome

Favorable outcomes were more frequent (31%) when IVH was absent compared to when IVH was present (15%; p < 0.00001), with or without associated hydrocephalus. The presence of hydrocephalus lowered the likelihood of favorable outcome still further to 12% in comparison to the presence of IVH alone (20%, p = 0.031). In patients with IVH, early surgical intervention had more favorable outcomes (18%) when compared to initial conservative management (12%) (p = 0.141). Age did not have any significant effect on outcome in these patients.

As one would expect, deeper hematomas were associated with greater IVH (20%; p < 0.00001) and hydrocephalus (29%) than lobar ones (17% and 15%). Outcome was different between lobar and deep hematomas when hydrocephalus was present (p = 0.001).

There was marginal benefit for the early surgical group over the conservative group when the ICH was lobar in location, whereas there was a slight, although not statistically significant, detrimental effect of surgery when the ICH was deep with no IVH (Fig. 1).

The average midline shift for having a favorable outcome was significantly different in both IVH (p = 0.008) and hydrocephalus groups (p = 0.008), whereas there was no significant difference (p = 0.059) when neither was present (Table 1). Increasing volume of the ICH also correlated with unfavorable outcome

Table 1. *Percentage of favorable outcomes in 902 patients after spontaneous intracerebral hematoma.*

	No IVH	IVH only	IVH + Hydrocephalus
Midline shift (cm)			
0	36% (111)	33% (21)	25% (28)
≤1	31% (371)	19% (124)	11% (141)
≥1	12% (17)	9% (23)	3% (38)
Hematoma volume (mL)			
≤40	36% (245)	32% (54)	13% (67)
40–80	29% (179)	19% (67)	13% (75)
>80	16% (43)	6% (35)	5% (55)
Side of hematoma			
Left	30% (249)	24% (92)	11% (116)
Right	34% (239)	16% (71)	11% (88)
Primary site			
Lobar	42% (233)	21% (61)	24% (51)
Deep	23% (268)	17% (105)	6% (155)

IVH Intraventricular hemorrhage.
Number of patients in parentheses.

with IVH ($p = 0.003$) and without IVH ($p = 0.003$) and did not have any significant correlation in presence of hydrocephalus ($p = 0.167$). Left-sided ICH with IVH had a tendency for better outcome than right-sided ones, although it was not statistically significant.

To summarize, factors that most affected outcome were midline shift, volume, and deep location.

Discussion

Spontaneous or secondary IVH is a marker of poor prognosis for hemorrhagic stroke. It can cause hydrocephalus requiring ventricular shunt placement, resulting in permanent neurological deficit or death [6, 14]. Severe IVH caused by extension from subarachnoid hemorrhage or ICH leads to hydrocephalus and often poor outcome, although no randomized controlled trials have been conducted thus far [13]. Most studies regarding treatment and outcome following IVH have been conducted in neonates.

STICH was designed to look at outcome measures in adolescent or adult ICH patients receiving early surgical intervention versus initial conservative management. Although no overall benefit was shown from early surgery when compared with initial conservative treatment [9], there may be subsets showing some benefit of one over the other.

A systematic review compared conservative treatment, extraventricular drainage, and extraventricular drainage combined with fibrinolysis [13]. The poor outcome rate was 90% for conservative treatment, 89% for extraventricular drainage, and 34% for extraventricular drainage with fibrinolytic agents.

Conclusion

This analysis of the STICH data shows the detrimental effect of the presence of IVH and consequent hydrocephalus on outcome in patients with spontaneous ICH. Early surgery might be of some benefit in those with IVH. Further studies are necessary to elucidate this problem and evaluate various treatment options.

References

1. Bhattathiri PS, Gregson B, Prasad KS, Mitchell P, Soh C, Mitra D, Gholkar A, Mendelow AD (2003) Reliability assessment of computerized tomography scanning measurements in intracerebral hematoma. Neurosurg Focus 15: E6
2. Diamond P, Gale S, Stewart K (2003) Primary intracerebral haemorrhage – clinical and radiologic predictors of survival and functional outcome. Disabil Rehabil 25: 689–698
3. Findlay JM, Jacka MJ (2004) Cohort study of intraventricular thrombolysis with recombinant tissue plasminogen activator for aneurysmal intraventricular hemorrhage. Neurosurgery 55: 532–538
4. Franz G, Brenneis C, Kampfl A, Pfausler B, Poewe W, Schmutzhard E (2001) Prognostic value of intraventricular blood in perimesencephalic nonaneurysmal subarachnoid hemorrhage. J Comput Assist Tomogr 25: 742–746
5. Kumar R, Pathak A, Mathuriya SN, Khandelwal N (2003) Intraventricular sodium nitroprusside therapy: a future promise for refractory subarachnoid hemorrhage-induced vasospasm. Neurol India 51: 197–202
6. Lapointe M, Haines S (2002) Fibrinolytic therapy for intraventricular hemorrhage in adults. Cochrane Database Syst Rev: CD003692
7. Mayfrank L, Kim Y, Kissler J, Delsing P, Gilsbach JM, Schroder JM, Weis J (2000) Morphological changes following experimental intraventricular haemorrhage and intraventricular fibrinolytic treatment with recombinant tissue plasminogen activator. Acta Neuropathol (Berl) 100: 561–567
8. Mayfrank L, Hutter BO, Kohorst Y, Kreitschmann-Andermahr I, Rohde V, Thron A, Gilsbach JM (2001) Influence of intraventricular hemorrhage on outcome after rupture of intracranial aneurysm. Neurosurg Rev 24: 185–191
9. Mendelow AD, Gregson BA, Fernandes HM, Murray GD, Teasdale GM, Hope DT, Karimi A, Shaw MD, Barer DH; STICH investigatos (2005) Early surgery versus initial conservative treatment in patients with spontaneous supratentorial intracerebral haematomas in the International Surgical Trial in Intracerebral Haemorrhage (STICH): a randomised trial. Lancet 365: 387–397
10. Naff NJ (1999) Intraventricular Hemorrhage in Adults. Curr Treat Options Neurol 1: 173–178
11. Naff NJ, Hanley DF, Keyl PM, Tuhrim S, Kraut M, Bederson

J, Bullock R, Mayer SA, Schmutzhard E (2004) Intraventricular thrombolysis speeds blood clot resolution: results of a pilot, prospective, randomized, double-blind, controlled trial. Neurosurgery 54: 577–584

12. Nakagawa T, Suga S, Mayanagi K, Akaji K, Inamasu J, Kawase T; Keio SAH Cooperative Study Group (2005) Predicting the overall management outcome in patients with a subarachnoid hemorrhage accompanied by a massive intracerebral or full-packed intraventricular hemorrhage: a 15-year retrospective study. Surg Neurol 63: 329–335

13. Nieuwkamp DJ, de Gans K, Rinkel GJ, Algra A (2000) Treatment and outcome of severe intraventricular extension in patients with subarachnoid or intracerebral hemorrhage: a systematic review of the literature. J Neurol 247: 117–121

14. Passero S, Ulivelli M, Reale F (2002) Primary intraventricular haemorrhage in adults. Acta Neurol Scand 105: 115–119

Correspondence: A. D. Mendelow, Department of Neurosurgery, Newcastle General Hospital, Westgate Road, Newcastle Upon Tyne NE4 6BE, UK. e-mail: Parameswaran.Bhattathiri@nuth.nhs.uk

Changes in coagulative and fibrinolytic activities in patients with intracranial hemorrhage

T. Ebihara, K. Kinoshita, A. Utagawa, A. Sakurai, M. Furukawa, Y. Kitahata, Y. Tominaga, N. Chiba, T. Moriya, K. Nagao, and K. Tanjoh

Department of Emergency and Critical Care Medicine, Nihon University School of Medicine, Tokyo, Japan

Summary

Objective. To investigate whether any changes occur in the coagulative/fibrinolytic cascade in patients with subarachnoid hemorrhage (SAH) or hypertensive intracerebral hemorrhage (HICH).

Design and methods. Subjects included 143 patients with intracranial hemorrhage (SAH, n = 50; HICH, n = 82; ROSC-SAH [return of spontaneous circulation after cardiopulmonary arrest due to SAH], n = 11). Coagulative and fibrinolytic factors were measured in blood samples taken on admission.

Results. The prothrombin fragment 1+2 level was significantly higher ($p < 0.005$) in SAH patients than in HICH patients. The fibrinolytic factors (plasmin alpha 2-plasmin inhibitor complex, D-dimer, or fibrinogen degradation products) in SAH and ROSC-SAH were both significantly higher than those in HICH, but the significance of difference was stronger in the case of ROSC-SAH ($p < 0.05$).

Discussion. Both coagulative and fibrinolytic activities were altered after the onset of SAH. These results demonstrate that the coagulative/fibrinolytic cascade might be activated via different mechanisms in different types of stroke. It remains unclear, however, whether a significant alteration of the fibrinolytic cascade in patients with ROSC-SAH might be a nonspecific phenomenon attributable to the reperfusion after collapse.

Keywords: Coagulation; fibrinolysis; subarachnoid hemorrhage; cardiopulmonary arrest; return of spontaneous circulation; intracerebral hemorrhage; PTF1+2; plasmin inhibitor complex; D-dimer; fibrinogen degradation products.

Introduction

A recent study [20] reported that mechanisms secondary to subarachnoid hemorrhage (SAH) could contribute to the development and progression of extracerebral organ dysfunction by promoting systemic inflammation. About 77% of documented extracerebral organ system failures have been reported to occur in conjunction with systemic inflammatory response syndrome [5]. Moreover, the inflammatory and procoagulant host responses have been found to be closely related [4]. Inflammatory cytokines such as tumor necrosis factor alpha, interleukin-1beta, and interleukin-6 are capable of activating coagulation and inhibiting fibrinolysis, while the procoagulant thrombin is capable of stimulating multiple inflammatory pathways [1, 4].

The various relationships among the activated coagulative/fibrinolytic cascade, patient prognosis, and the incidence of complications have recently been described in stroke patients [6, 19]. It remains unclear, however, whether the mechanisms activating these cascade changes differ among various types of stroke patients, such as SAH and hypertensive intracerebral hemorrhage (HICH). Knowledge is also scanty on coagulative/fibrinolytic activities in patients who experience return of spontaneous circulation (ROSC) from cardiopulmonary arrest (CPA) due to SAH, one of the most common causes of sudden death [17]. The objective of our study was to examine the coagulative/fibrinolytic cascade and its mechanism of activation in stroke patients.

Materials and methods

SAH and HICH patients scoring 8 or less on the Glasgow Coma Scale (GCS) and patients experiencing ROSC from SAH-induced CPA (ROSC-SAH) were studied at a university hospital. ROSC-SAH patients whose spontaneous circulation returned in an ambulance or in the hospital were included. All ROSC-SAH patients were diagnosed as SAH by computed tomography. Coagulation and fibrinolytic activities were retrospectively compared between the patient groups.

We investigated 143 intracranial hemorrhage patients divided into 3 groups: SAH (n = 50), ROSC-SAH (n = 11), and HICH (n = 82). Peripheral blood samples were drawn immediately after hospitaliza-

Table 1. *Patient characteristics at admission*

Variable	SAH	ROSC-SAH	HICH	p Value
No. of patients	50	11	82	NS
Age, yrs	60.3 ± 15.0	60.1 ± 12.0	59.9 ± 15.6	NS
Male gender, %	46.0	27.3	75.6	<0.0005
GCS	4.9 ± 1.8	3.0 ± 0.0	5.0 ± 1.8	<0.005[a], <0.001[b]
Body temperature, centigrade	36.3 ± 1.2	35.3 ± 0.7	36.4 ± 1.2	NS
Blood glucose level, mg/dL	221.3 ± 74.3	311.8 ± 85.6	182.0 ± 49.2	<0.005[a], <0.0001[b], <0.05[c]

[a] Denotes significance in the comparison between SAH and ROSC-SAH, [b] denotes significance in the comparison between ROSC-SAH and HICH, [c] denotes the significance in the comparison between SAH and HICH. *SAH* Subarachnoid hemorrhage, *ROSC-SAH* return of spontaneous circulation after cardiopulmonary arrest due to subarachnoid hemorrhage, *HICH* hypertensive intracerebral hemorrhage, *GCS* Glasgow Coma Scale, *NS* not significant.

tion in patients with SAH and HICH. The samples from the ROSC-SAH group were drawn after the ROSC.

The modulating factors of coagulation cascade were measured using the following methods: Antithrombin III was measured by the chromogenic synthesized substrate method; prothrombin fragment 1+2 (PTF1+2) was assayed by enzyme-linked immunosorbent assay; protein C was measured by latex photometric immunoassay; activated protein C was evaluated using the activated prothrombin time method; thrombomodulin was assayed by enzyme immunoassay.

The modulating factors of the fibrinolytic cascade were measured using the following methods: Tissue plasminogen activator plasminogen activator inhibitor-1 complex (tPA-PAI1) was assayed by the enzyme immunoassay; plasmin alpha-2 plasmin inhibitor complex (PIC), D-dimer, and fibrinogen degradation products were assayed using latex photometric immunoassay.

StatView for Windows version 5.0 (SAS Institute, Cary, NC) was used for statistical analysis. Proportions were compared using the Fisher exact test. Continuous variables were compared using the unpaired Student t test or Mann-Whitney's U test, and presented as mean ± SD. Two-tailed p values of <0.05 were used to indicate statistical significance.

Results

Table 1 shows patient characteristics at admission. The proportion of males was significantly lower in the ROSC-SAH group than in the other groups ($p < 0.0005$). All patients in the ROSC-SAH scored 3 points on the GCS, and the GCS score for the group as a whole was significantly lower than the scores in the other groups ($p < 0.005$, ROSC-SAH vs. SAH; $p < 0.001$, ROSC-SAH vs. HICH). Blood glucose levels were higher than the reference value (70–109 mg/dL) in all groups, and remarkably elevated in the ROSC-SAH group.

Comparison of coagulative/fibrinolytic activities between SAH and HICH patients

The antithrombin III level was significantly lower in the HICH group than in the SAH group ($p < 0.05$), but the change was within reference values. Changes in protein C, activated protein C, and thrombomodulin levels were comparable in these 2 groups. The PTF1+2 level in the SAH group was significantly higher than that in the HICH group ($p < 0.005$). Changes in tPA-PAI1 levels were not significant in either group, but the levels were remarkably higher than the reference value in both groups. The levels of PIC, D-dimer, and fibrinogen degradation products exceeded the reference values in both groups and were significantly higher in the SAH group than in the HICH group ($p < 0.001$, $p < 0.01$, $p < 0.005$, respectively) (Table 2).

Comparison of coagulative/fibrinolytic activities between SAH and ROSC-SAH

While the antithrombin III level in the ROSC-SAH patients was below the reference value, it was not significantly different from that of the SAH patients. Changes in PTF1+2, protein C, activated protein C, and thrombomodulin levels were not significant either. Changes in tPA-PAI1 levels were not significant in either group, but they were above the reference value in both groups. The levels of PIC, D-dimer, and fibrinogen degradation products exceeded the reference value in both groups, but the only difference between the groups was a significantly higher PIC in the ROSC-SAH patients ($p < 0.05$) (Table 2).

Discussion

Medical complications such as systemic inflammatory response syndrome, pneumonia, sepsis, and multiple organ dysfunction syndrome are among the leading causes of late morbidity and mortality in

Table 2. *Changes in coagulative/fibrinolytic activities in patients with SAH, ROSC-SAH, and HICH*

Variable	SAH	ROSC-SAH	HICH	p Value
AT (79–121%)	93.9 ± 17.8	70.5 ± 0.7	82.2 ± 19.1	<0.05[c]
PTF1+2 (0.4–1.4 nmol/L)	7.8 ± 7.1	2.7 ± 1.5	3.5 ± 3.0	<0.005[c]
Protein C (70–150%)	95.8 ± 26.2	80.3 ± 29.0	88.6 ± 26.3	NS
APC (64–146 %)	79.1 ± 28.9	60.0 ± 3.6	86.5 ± 27.8	NS
TM (1.8–4.1 FU/mL)	9.6 ± 21.4	3.3 ± 1.0	4.8 ± 3.8	NS
tPA-PAI1 (<15 ng/mL)	21.0 ± 16.0	20.8 ± 9.0	26.3 ± 45.1	NS
PIC (<0.8 µg/mL)	9.1 ± 11.9	25.4 ± 30.9	2.8 ± 3.5	<0.05[a], <0.001[c]
D-dimer (<1.0 µg/mL)	22.5 ± 34.6	41.8 ± 53.4	7.5 ± 17.4	<0.01[c]
FDP (<4 µg/mL)	6.2 ± 9.2	9.2 ± 10.4	2.0 ± 3.3	<0.005[c]

Numerical values in parentheses denote the reference values of the coagulation/fibrinolysis factor.
[a] Denotes significance in the comparison between SAH and ROSC-SAH, [c] denotes significance in the comparison between SAH with HICH.
SAH Subarachnoid hemorrhage, *ROSC-SAH* return of spontaneous circulation after cardiopulmonary arrest due to subarachnoid hemorrhage, *HICH* hypertensive intracerebral hemorrhage, *AT* antithrombin, *PTF1+2* prothrombin fragment 1+2, *APC* activated protein C, *TM* thrombomodulin, *tPA-PAI1* tissue-plasminogen activator plasminogen activator inhibitor-1 complex, *PIC* plasmin alpha2-plasmin inhibitor complex, *FDP* fibrinogen degradation products, *NS* not significant.

aneurysmal SAH. Le Roux *et al.* [8] found a significant association between multiple medical complications with unfavorable outcome in both Hunt-Hess grades 1 to 3 [8] and Hunt-Hess grades 4 and 5 [9]. The mechanisms leading to the development of these medical complications after SAH are still poorly understood, however.

Disseminated intravascular coagulation is characterized by the widespread activation of coagulation, which results in intravascular fibrin formation and, ultimately, thrombotic occlusion of small and midsize vessels [2, 13, 14]. Intravascular coagulation can also compromise the blood supply to organs, and when combined with hemodynamic and metabolic derangements, it can contribute to multiple organ dysfunction syndrome [10]. Early activation of both coagulative and fibrinolytic systems following SAH has also been demonstrated [10]. Moreover, SAH patients who suffered CPA (ROSC-SAH group) had higher plasma levels of PIC, a marker of increased fibrinolytic activity, than SAH patients without CPA.

Several reports have described this altered coagulation/fibrinolytic cascade activity in SAH patients [6, 15, 16, 19], and altered coagulation activity has been found to correlate with poor outcome [15]. Other authors have concluded that altered fibrinolytic activity is linked with neurological deficits and a higher incidence of complications [6, 15, 16]. These results suggest that activation of both coagulation and fibrinolysis in SAH patients without CPA may be strongly associated with clinical severity, correlated with the development of multiple organ dysfunction syndrome and worsened outcome. It remains unclear, however, whether a significant alteration of fibrinolytic cascade in ROSC-SAH might be a nonspecific phenomenon attributable to reperfusion after collapse.

As is true for almost all systemic inflammatory responses, the derangement of coagulation and fibrinolysis in disseminated intravascular coagulation is mediated by several proinflammatory cytokines [11, 18]. Recent reports have demonstrated that the efficacy of activated protein C in sepsis might be rooted in the protein's ability to modulate both coagulation and inflammation [12]. In vitro experiments have shown that activated protein C can inhibit neutrophil binding to selectins, potentially blocking tight leukocyte adhesion. Protein C also inhibits tumor necrosis factor-alpha secretion from monocytes and other cell lines by interfering with nuclear factor-kappaB nuclear translocation, and it has been shown to prevent organ damage in experimental models of sepsis [3]. The level of activated protein C in our study tended to descend below the reference value in patients resuscitated from cardiopulmonary arrest following SAH, although the difference failed to reach a significant level among the SAH, ROSC-SAH, and HICH groups.

We also observed an elevation of plasma tPA-PAI1 above the reference value in all 3 groups. Johansson *et al.* [7] recently reported that the tPA/PAI-1 complex, a novel fibrinolytic marker, is independently associated with the development of a first-ever stroke, especially hemorrhagic stroke. They proposed that elevated tPA-PAI-1 complex levels could reflect a more severe form of endothelial dysfunction caused by an advanced form of atherosclerotic disease. If our results can be taken to support their hypothesis, the disturbances in fibrinolysis might precede a cerebrovascular event and the coagulative/fibrinolytic cascade might be activated via different mechanisms in different types of stroke. Ikeda et al. [6] reported that elevated levels of cerebrospinal tPA-PAI-1 complex are associated with neurological outcome and the occurrence of vasospasm in severe SAH patients. Noting its role in inducing a more advanced form of atherosclerotic disease, we proposed that this early alteration of the coagulative/fibrinolytic cascade with endothelial dysfunction might affect the incidence of vasospasm after SAH. Nonetheless, we have yet to elucidate the role of endogenous activated protein C, tPA-PAI1, or PIC in coordinating the innate immune response in endothelium-based inflammation.

Our study has some limitations. The gravely ill condition of the patients following stroke (GCS 8 or less) or ROSC from CPA limits the applicability of our results. These findings might simply indicate a poor clinical condition such as severe ischemia inapplicable to stroke patients of a more favorable grade. The lack of detailed data on medical complications, on the incidence of vasospasm, and on overall outcomes following stroke in the limited number of patients included in this study calls for further investigations to elucidate the importance of the data, as well as the status of other types of diseases and more favorable grades of stroke.

Conclusions

Coagulative and fibrinolytic activities were both altered after SAH. The activation of fibrinolytic cascade was higher in patients with ROSC-SAH than in those with SAH. These results demonstrate that the coagulative/fibrinolytic cascades might be activated via different mechanisms in different types of stroke. It remains unclear, however, whether the significant alteration of the fibrinolytic cascade in patients with ROSC-SAH might be a nonspecific phenomenon attributable to the reperfusion after collapse.

References

1. Bernard GR, Vincent JL, Laterre PF, LaRosa SP, Dhainaut JF, Lopez-Rodriguez A, Steingrub JS, Garber GE, Helterbrand JD, Ely EW, Fisher CJ Jr, Recombinant Human Protein C Worldwide Evaluation in Severe Sepsis (PROWESS) Study Group (2001) Efficacy and safety of recombinant human activated protein C for severe sepsis. N Engl J Med 344: 699–709
2. Bone RC (1992) Modulators of coagulation. A critical appraisal of their role in sepsis. Arch Intern Med 152: 1381–1389
3. Esmon CT (2001) Protein C anticoagulant pathway and its role in controlling microvascular thrombosis and inflammation. Crit Care Med 29 [Suppl] 7: 48–51; discussion 51–52
4. Esmon CT, Taylor FB Jr, Snow TR (1991) Inflammation and coagulation: linked processes potentially regulated through a common pathway mediated by protein C. Thromb Haemost 66: 160–165
5. Gruber A, Reinprecht A, Illievich UM, Fitzgerald R, Dietrich W, Czech T, Richling B (1999) Extracerebral organ dysfunction and neurologic outcome after aneurysmal subarachnoid hemorrhage. Crit Care Med 27: 505–514
6. Ikeda K, Asakura H, Futami K, Yamashita J (1997) Coagulative and fibrinolytic activation in cerebrospinal fluid and plasma after subarachnoid hemorrhage. Neurosurgery 41: 344–349; discussion 349–350
7. Johansson L, Jansson JH, Boman K, Nilsson TK, Stegmayr B, Hallmans G (2000) Tissue plasminogen activator, plasminogen activator inhibitor-1, and tissue plasminogen activator/plasminogen activator inhibitor-1 complex as risk factors for the development of a first stroke. Stroke 31: 26–32
8. Le Roux PD, Elliott JP, Downey L, Newell DW, Grady MS, Mayberg MR, Eskridge JM, Winn HR (1995) Improved outcome after rupture of anterior circulation aneurysms: a retrospective 10-year review of 224 good-grade patients. J Neurosurg 83: 394–402
9. Le Roux PD, Elliott JP, Newell DW, Grady MS, Winn HR (1996) Predicting outcome in poor-grade patients with subarachnoid hemorrhage: a retrospective review of 159 aggressively managed cases. J Neurosurg 85: 39–49
10. Levi M, Ten Cate H (1999) Disseminated intravascular coagulation. N Engl J Med 341: 586–592
11. Levi M, van der Poll T, ten Cate H, van Deventer SJ (1997) The cytokine-mediated imbalance between coagulant and anticoagulant mechanisms in sepsis and endotoxaemia. Eur J Clin Invest 27: 3–9
12. Levi M, Choi G, Schoots I, Schultz M, van der Poll T (2004) Beyond sepsis: activated protein C and ischemia-reperfusion injury. Crit Care Med 32 [Suppl] 5: 309–312
13. Marder VJ, Feinstein DI, Francis CW, Colman RW (1994) Consumptive thrombohemorrhagic disorders. In: Colman RW, Hirsh J, Marder VJ, Salzman EW (eds) Hemostasis and Thrombosis: Basic Principles and Clinical Practice, 3rd edn. J.B. Lippincott, Philadelphia, pp 1023–1063
14. Muller-Berghaus G, ten Cate H, Levi MM (1998) Disseminated intravascular coagulation. In: Verstraete M, Fuster V, Topol EJ (eds) Cardiovascular Thrombosis: Thrombocardiology and Thromboneurology, 2nd edn. Lippincott-Raven, Philadelphia, pp 114–120, 781–801
15. Nina P, Schisano G, Chiappetta F, Luisa Papa M, Maddaloni E, Brunori A, Capasso F, Corpetti MG, Demurtas F (2001) A study of blood coagulation and fibrinolytic system in spontaneous subarachnoid hemorrhage. Correlation with Hunt-Hess grade and outcome. Surg Neurol 55: 197–203
16. Peltonen S, Juvela S, Kaste M, Lassila R (1997) Hemostasis and fibrinolysis activation after subarachnoid hemorrhage. J Neurosurg 87: 207–214
17. Shapiro S (1996) Management of subarachnoid hemorrhage patients who presented with respiratory arrest resuscitated with bystander CPR. Stroke 27: 1780–1782
18. van der Poll T, Buller HR, ten Cate H, Wortel CH, Bauer KA, van Deventer SJ, Hack CE, Sauerwein HP, Rosenberg RD, ten

Cate JW (1990) Activation of coagulation after administration of tumor necrosis factor to normal subjects. N Engl J Med 322: 1622–1627
19. Vergouwen MD, Frijns CJ, Roos YB, Rinkel GJ, Baas F, Vermeulen M (2004) Plasminogen activator inhibitor-1 4G allele in the 4G/5G promoter polymorphism increases the occurrence of cerebral ischemia after aneurismal subarachnoid hemorrhage. Stroke 35: 1280–1283
20. Yoshimoto Y, Tanaka Y, Hoya K (2001) Acute systemic inflammatory response syndrome in subarachnoid hemorrhage. Stroke 32: 1989–1993

Correspondence: Takayuki Ebihara, Department of Emergency and Critical Care Medicine, Nihon University School of Medicine, 30-1, Oyaguchi Kamimachi, Itabashi-ku, Tokyo 173-8610, Japan. e-mail: ebi@med.nihon-u.ac.jp

The effect of hematoma removal for reducing the development of brain edema in cases of putaminal hemorrhage

M. Okuda, R. Suzuki, M. Moriya, M. Fujimoto, C. W. Chang, and T. Fujimoto

Department of Neurosurgery, Showa University Fujigaoka Hospital, Yokohama, Japan

Summary

Introduction. Surgical intervention in putaminal hemorrhage has been a controversial issue. The aim of this research is to evaluate the benefits of surgery for reducing the development of brain edema.

Materials and methods. Sixteen cases of putaminal hemorrhage were examined. Eight patients were treated conservatively (C group), and the other 8 patients were treated surgically (S group). Head CT scans were performed on the day of onset (day 0) in C group or performed just after surgery (day 0) in S group, and performed again once per period on days 1–7, 8–14, and 15–21. The volume of the mass including hematoma and edema (H + E) was measured using CT scans and the $(H + E)/H_0$ ratios were calculated (H_0; hematoma volume on day 0). The $(H + E)/H_0$ ratios for each period were compared statistically between the 2 groups using a *t*-test.

Results. The mean values of $(H + E)/H_0$ ratios at each period were 2.19, 2.63, 2.53 in C group, and 1.29, 1.29, 0.66 in S group. The values in S group were significantly lower as compared with C group in every period ($p < 0.01$, <0.05, <0.01).

Conclusions. Hematoma volume reduction by surgery reduced the development of brain edema.

Keywords: Brain edema; intracerebral hemorrhage; hematoma removal.

Introduction

Surgical treatment for intracerebral hemorrhage (ICH), including hematoma removal by craniotomy and stereotactic hematoma evacuation, has long been controversial. Since Cushing [1] reported the first operatively treated case of ICH with localizing symptoms in 1903, many medical researchers have studied the effectiveness of surgery for ICH.

In 1961, McKissock *et al.* [8] reported the first randomized trial in which 180 patients were included. They reported that surgically treated cases demonstrated worse prognoses than conservatively treated cases. Since then, there has been a tendency not to perform surgical treatment for ICH in Western countries. However, the pre-microneurosurgical and pre-computed tomography (CT) era during which that investigation was conducted should be taken into account.

Recent meta-analyses of published trials of craniotomy for ICH by Fernandes *et al.* [2] were also found to be inconclusive, indicating that more information is needed from randomized studies to determine the role of surgery in ICH.

On the other hand, in Japan, a large-scale retrospective study by Kanaya *et al.* [5] in 1980, in which 7010 patients were included, has revealed an improvement in mortality and morbidity rates in the cases of grade 3 (stupor), 4a (semicoma without herniation signs), 4b (semicoma with herniation signs), and 5 (deep coma) following hematoma evacuation by craniotomy. Since then, there has been a tendency to perform surgical treatment for ICH in Japan [6].

In a report from 2004, Hattori *et al.* [3] reported that stereotactic hematoma evacuation is clearly of value in selected patients with spontaneous putaminal hemorrhage whose eyes are closed, but will open in response to strong stimuli on admission.

Until now there has been no consensus that surgical treatment produces a better outcome than conservative treatment, because a number of mechanisms appear to be involved in ICH, especially edema formation around a hematoma.

In our hospital, on the basis of past literature and from our own experience, we usually perform surgical treatment for ICH when the hematoma volume is more than about 40 cm^3, the patient is suited to a neurological grading of 3 (stupor) or 4 (semicoma), and no other conditions such as aging, anti-coagulant use, or chronic renal failure are an obstacle.

The object of this paper is to evaluate the benefits of surgery for reducing the development of brain edema by measuring the volume of the mass including the hematoma and surrounding edema in a quantitative manner and analyzing the data statistically.

Materials and methods

Sixteen cases of putaminal hemorrhage were selected according to the volume of the hematoma. Patients with hematoma volume of 20 cm^3 or less at the time of onset and who had been treated conservatively and patients whose hematoma volume had been reduced to 20 cm^3 or less as a result of surgery were included in the present study.

The patients were hospitalized during a 5-year period between 1999 and 2003. All patients were admitted to our hospital within 24 hours after onset, and the diagnosis of idiopathic putaminal hemorrhage was based on CT findings. We excluded patients with a past history of cerebrovascular diseases such as ICH or cerebral infarction, intraventricular extension, underlying aneurysm or vascular malformation, anticoagulant use, trauma, or death before 28 days after onset.

Eight patients were treated conservatively (C group) and 8 patients were treated surgically (S group). Both groups received administration of osmotic antidiuretics and other intensive medical treatments such as antihypertensives.

Surgery consisted of hematoma removal by craniotomy or stereotactic hematoma evacuation, the former for 3 cases and the latter for 5 cases. Stereotactic hematoma evacuations were performed using the Komai CT-guided stereotaxic system, without administration of argatroban or tissue plasminogen activator (tPA).

There were 5 men and 3 women ranging in age from 41 to 82 years (mean 57.6 ± 14.8 years old) in the C group, and 4 men and 4 women from 33 to 77 years (mean 55.9 ± 14.9 years old) in the S group.

In the C group, the mean volume of the hematoma was 20 cm^3 or less at the time of onset (mean 13.4 ± 4.8 cm^3). Three cases occurred in the right hemisphere, and the other 5 cases in the left.

In the S group, surgery was performed within 72 hours after the onset; the mean volume of the hematoma before surgery was 46.1 ± 9.7 cm^3 and it was reduced to 20 cm^3 or less as the result of surgery (mean 10.4 ± 5.4 cm^3). Three cases occurred in the right hemisphere, and the other 5 cases in the left.

There were no significant differences between these 2 groups with regard to age, sex, hematoma volume (compared the day of onset in C group and the day after surgery in S group), side of the hematoma, and history of hypertension or diabetes. Hematoma enlargement was not observed in either group.

We defined the day of onset as day 0 in the C group, and the day just after surgery as day 0 in the S group. Head CT scans were performed on the day of onset (day 0) in C group or just after surgery (day 0) in S group, and repeated on periods of days 1–7, 8–14, and 15–21, respectively, one scan per period.

First, we calculated the volume of hematoma on day 0 (H_0). We measured the area of the hematoma in each CT slice using NIH Image software, multiplied the area and the thickness of each slice, and summed them all. The product is H_0.

Next, we calculated hematoma volume (H) plus perihematomal edema volume (E) (H + E) in each CT film since day 1. We defined (E) as a lower density area than the corresponding area in the contralateral hemisphere [4]. We measured the area of mass including (H) and (E), multiplied the area and the thickness of each CT slice, and summed them all. The product was (H + E).

Finally, we determined the $(H + E)/H_0$ ratio, which indicated the increased volume compared to the hematoma volume on day 0, in each CT film in every period. The $(H + E)/H_0$ ratios were compared statistically between the C group and the S group in every period by performing a t-test. Differences were regarded as significant when the probability value was less than 0.05.

We evaluated the neurological status of each patient on day 21 using the Modified Rankin Scale.

Results

Figure 1 shows representative cases from the C group (upper row) and the S group (lower row), respectively. The volume of edema peaked on days 8–14 in both groups.

The mean values of $(H + E)/H_0$ ratios at each period (days 1–7, 8–14, and 15–21) were 2.19, 2.63, 2.53 in the C group, and 1.29, 1.29, 0.66 in the S group. The mean values of the $(H + E)/H_0$ ratios in the S group were significantly lower compared with the C group in every period ($p < 0.01$, <0.05, <0.01) (Fig. 2).

The results of our study indicate the benefits of surgery for reducing the development of brain edema in patients with putaminal hemorrhage. However, there was an apparent inconsistency in that the S group showed a worse prognosis than the C group with regard to neurological status on day 21 as evaluated by the Modified Rankin Scale. The scores indicated 2.5 as the mean value in the C group, and 3.75 in the S group ($p < 0.05$).

Discussion

In this study, by measuring volume of the hematoma and edema in a quantitative manner and analyzing the data statistically, we found that surgical treatment was considered to reduce the development of brain edema with putaminal hemorrhage. Considering the correlation between brain ICH-induced edema and surgical treatment, we review the current knowledge of how the brain tissue is injured in ICH.

Brain injury due to ICH consists of 2 steps. Primary brain damage is due to hematoma formation. Hematoma itself causes mechanical and immediate destruction of the neural structure. Secondary brain damage is brought about by edema formation and ischemia around a hematoma. Edema formation in the perihematomal area is considered to play an important role in brain damage.

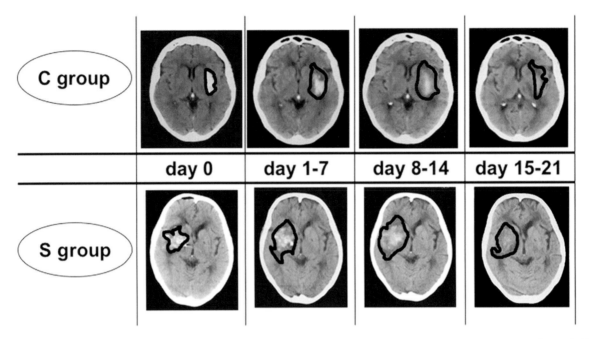

Fig. 1. The upper row shows the representative course of a conservatively treated case, and the lower row shows that of a surgically treated case. The volume of edema peaked on days 8–14 in both cases (*C group*; conservatively treated group, *S group*; surgically treated group)

The pathomechanism for the development of brain edema associated with ICH seems to be quite different from that of cold lesion edema, edema induced by ischemia, and other sorts of edema. Suzuki *et al*. [9] demonstrated that the ICH-induced edema peaked between 10 to 20 days after the onset and prolonged clinical deterioration.

A number of mechanisms appear to be involved in edema formation after ICH and have not yet been fully understood. Xi *et al*. [10] reported that at least 3 phases of edema are involved in ICH. These include a very early phase involving hydrostatic pressure and clot retraction, a second phase involving the activation of the coagulation cascade and thrombin production, and a third phase involving red blood cell lysis and hemoglobin-induced neuronal toxicity. That is, toxic components of blood seem to be closely correlated with edema formation.

Recent research has identified thrombin as a key mediator in the development of edema in animal models, prompting further studies on the administration of a thrombin inhibitor such as argatroban. Kitaoka *et al*. [7] reported in 2002 that intracerebral injection or the systemic administration of high-dose argatroban in the acute phase of ICH significantly reduced edema and suggested that argatroban may be an effective therapy for ICH-induced edema. Clot removal after tPA treatment or infusion of tPA directly into the hematoma have also been actively tested.

Hematoma removal may be the effective cure because it reduced brain edema volume and blood-brain barrier disruption and improved cerebral tissue pressure, and because it results in the removal of toxic components such as thrombin at the same time [10].

We emphasize that, whether medicines such as argatroban and tPA are used or not, surgical reduction of the fluid component at the site of the hematoma would be effective for reducing the development of brain edema because thrombin existing in fluid components around the clot would be reduced. In our study, edema formation was reduced after surgical intervention compared with the conservatively treated cases, although the hematoma volume just after surgery remained at the same level between the 2 groups. We did not use either argatroban or tPA.

Although 2 of 8 cases were treated with ventricular drainage for hydrocephalus and this influence cannot be ignored, the result of our current study supports the possibility that surgical treatment itself can prevent brain edema by reducing toxic components of blood such as thrombin in ICH.

Regarding the worse prognosis for the S group as evaluated by the Modified Rankin Scale in the current study, the result appeared to be affected by the neuro-

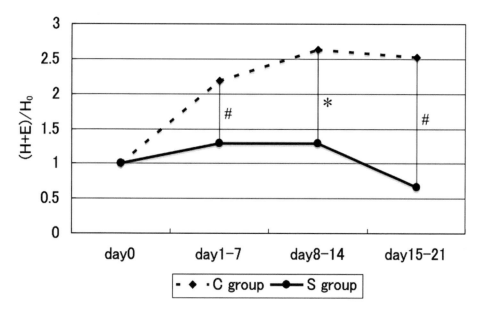

Fig. 2. The broken line shows the chronological change of $(H + E)/H_0$ ratios in the conservatively treated group, and the solid line shows that in the surgically treated group. The mean values of the $(H + E)/H_0$ ratios in S group are significantly lower compared with C group in every period (# = p < 0.01; * = p < 0.05) (C group; conservatively treated group, S group; surgically treated group)

logical status at the time of onset between the 2 groups. The mean score on the Glasgow Coma Scale in our C group was 13.75 at the time of onset, and 10.25 in our S group; it was significantly lower in the S group (p < 0.05).

We intend to enhance the quality of research with a more secure protocol, many more cases, analysis of the fluid components gained by surgery, and by seeking to promote our understanding of the connection between edema formation in ICH and surgical treatment.

References

1. Cushing H (1903) The blood-pressure reaction of acute cerebral compression, illustrated by cases of intracerebral hemorrhage. Am J Med Sci 125: 1017–1044
2. Fernandes HM, Gregson B, Siddique S, Mendelow AD (2000) Surgery in intracerebral hemorrhage. The uncertainty continues. Stroke 31: 2511–2516
3. Hattori N, Katayama Y, Maya Y, Gatherer A (2004) Impact of stereotactic hematoma evacuation on activities of daily living during the chronic period following spontaneous putaminal hemorrhage: a randomized study. J Neurosurg 101: 417–420
4. Inaji M, Tomita H, Tone O, Tamaki M, Suzuki R, Ohno K (2003) Chronological changes of perihematomal edema of human intracerebral hematoma. Acta Neurochir [Suppl] 86: 445–448
5. Kanaya H, Yukawa H, Itoh Z, Kutsuzawa H, Kagawa M, Kanno T, Kuwabara T, Mizukami M, Araki G, Irito T (1980) Grading and the indications for treatment in ICH of the basal ganglia (cooperative study in Japan). Spontaneous intracerebral haematomas, pp 268–274
6. Kanaya H, Saiki I, Ohuchi T, Kamata K, Endo H, Mizukami M, Kagawa M, Kaneko M, Ito Z (1983) Hypertensive intracerebral hemorrhage in Japan: update on surgical treatment. In: Mizukami M (ed) Hypertensive intracerebral hemorrhage. Raven Press, New York, pp 147–163
7. Kitaoka T, Hua Y, Xi G, Hoff JT, Keep RF (2002) Delayed argatroban treatment reduces edema in a rat model of intracerebral hemorrhage. Stroke 33: 3012–3018
8. McKissock W, Richardson A, Taylor J (1961) Primary intracerebral hemorrhage: a controlled trial of surgical and conservative treatment in 180 unselected cases. Lancet 278: 221–226
9. Suzuki R, Ohno K, Hiratsuka H, Inaba Y (1985) Chronological changes in brain edema in hypertensive intracerebral hemorrhage observed by CT and xenon-enhanced CT. In: Inaba Y, Klatzo I, Spatz M (eds) Brain edema: Proceedings of the 6th International Symposium, November 7–10, 1984, in Tokyo. Springer, Berlin Heidelberg New York Tokyo, pp 612–620
10. Xi G, Keep RF, Hoff JT (2002) Pathophysiology of brain edema formation. Neurosurg Clin N Am 13: 371–383

Correspondence: R. Suzuki, Department of Neurosurgery, Showa University Fujigaoka Hospital, 1-30 Fujigaoka, Aoba-ku, Yokohama, Japan. e-mail: ryuta@med.showa-u.ac.jp, or muneo.okuda@nifty.com

Spontaneous intracerebral hemorrhage in humans: hematoma enlargement, clot lysis, and brain edema

G. Wu[1], G. Xi[2], and F. Huang[1]

[1] Department of Neurosurgery, Huashan Hospital, Fudan University, Shanghai, China
[2] Department of Neurosurgery, University of Michigan Medical School, Ann Arbor, Michigan, USA

Summary

Early hematoma enlargement and delayed clot lysis contribute to brain injury after intracerebral hemorrhage (ICH). We investigated hematoma growth, clot lysis, and brain edema formation in patients with spontaneous ICH.

A total of 17 spontaneous ICH patients who received regular medication were chosen for this study. All patients had their first CT scan within 5 hours of onset of symptoms (day 0). The patients then underwent second, third, and fourth CT scans at 1, 3, and 10 days later. Hematoma size and absolute and relative brain edema volumes were measured. Hematoma enlargement was defined as a >33% increase in volume. Relative brain edema volume = absolute brain edema volume/hematoma size. Hematoma enlargement occurred in 4 of the 17 ICH patients (24%) within the first 24 hours. The hematoma sizes were reduced significantly at day 10 ($p < 0.05$) because of clot lysis. However, both absolute and relative brain edema increased gradually with time ($p < 0.01$).

These results suggest that delayed brain edema following ICH may result from hematoma lysis. This study also shows that early hematoma enlargement occurs in Chinese patients with ICH. Reducing early hematoma growth and limiting clot lysis-induced brain toxicity could be potential therapies for ICH.

Keywords: Intracerebral hemorrhage; brain edema; hematoma growth; hematoma lysis; computed tomography.

Introduction

Spontaneous intracerebral hemorrhage (ICH) is a common and often fatal stroke subtype lacking effective management [6]. ICH is estimated to account for 10–15% of all strokes in the United States [10]. The incidence of ICH is more common in China [17]. Brain edema contributes to brain damage after ICH [14]. Experimental and clinical investigations have demonstrated that early hematoma enlargement and delayed clot lysis contribute to ICH-induced brain injury [15]. The natural history and pathogenesis of hematoma and perihematomal edema in human ICH have not been well-studied. We investigated hematoma growth, clot lysis, and brain edema formation in Chinese ICH patients.

Materials and methods

Study design

In-patients who experienced spontaneous ICH (n = 17) in the Huashan Hospital at Fudan University between 2003 and 2004 were studied. All hematomas were located in the supratentorial area. Patients received regular medication. Exclusion parameters for this study included: 1) traumatic hemorrhage with initial or subsequent intraventricular extension, subsequent subarachnoid hemorrhage, or underlying aneurysm or vascular malformation, 2) death, or 3) undergoing surgical treatment within 10 days. All patients received non-contrast brain computed tomography (CT) scans within 6 hours of ICH onset. After the first CT scan, patients underwent second, third, and fourth CT scans at 1, 3, and 10 days after ICH onset. Hematoma enlargement was defined as a >33% increase in volume [2].

Measurement of hematoma and perihematomal edema volumes

All CT pictures were converted from CT machine to personal computer using the accessory software of the CT machine (e-Film), and NIH Image J 1.29 software (National Institutes of Health, USA) was used to analyze hematoma and edema volume.

We reset the calibration of the image according to the scale on the CT slices. Then the range of hematoma and perihematomal edema were marked out (for hematoma, the grey value was >130, and for the edema area, the grey value was 55–90). The area of hematoma and perihematomal edema of each CT slice was measured by NIH Image J. The above steps were repeated 3 times. Relative brain edema volume = absolute brain edema volume/hematoma size.

Statistical analysis

All data in this study are presented as mean ± SD. Data were analyzed with ANOVA using the Scheffe F test. Significance levels were measured at $p < 0.05$.

Results

A total of 17 patients were chosen for this study. Demographic and clinical features of the study patients are summarized in Table 1. All patients received non-contrast brain CT within 6 hours of ICH onset, and mean time was 2.6 hours after symptom onset. Ten patients had a history of hypertension (3–20 years) and 5 patients had a history of diabetes mellitus (4–13 years).

Hematoma enlargement occurred in 4 of the 17 ICH patients (24%) within the first 24 hours. Average hematoma sizes of all 17 patients were 20.9 ± 18.8, 23.8 ± 16.5, 21.2 ± 14.8, and 12.4 ± 10.3 cm^3 at days 0, 1, 3, and 10, respectively. The hematoma sizes were reduced significantly at day 10 ($p < 0.05$) because of clot lysis (Table 2, Fig. 1). However, both absolute and relative brain edema volumes increased gradually with time (absolute edema: 46.4 ± 30.1 at day 10 vs. 20.4 ± 13.2 cm^3 at day 0, $p < 0.05$; relative edema: 6.1 ± 6.5 at day 10 vs. 1.3 ± 0.8 at day 0, $p < 0.05$; Table 2). Figure 1 shows serial CT scans in 2 ICH patients.

Discussion

Our study demonstrates that early hematoma enlargement contributes to brain injury in Chinese ICH patients. Hematoma enlargement occurred in 4 of 17 ICH patients (24%) within the first 24 hours.

Several recent investigations evaluated the rate of hematoma enlargement after initial presentation [1, 2, 4, 7]. Early enlargement of the hematoma after the ictus is associated with midline shift and accelerates neurological deterioration [1, 16]. The precise mechanisms of hematoma growth are not known, but most hematoma enlargement occurs within the first 24 hours [2, 7]. Broderick et al. [1] recognized that early hematoma growth is associated with early neurological deterioration. An ongoing clinical trial focuses on early treatment with activated Factor VIIa aimed at preventing hematoma enlargement and reducing ICH-induced brain injury [8].

We also found that hematoma size decreases and perihematomal brain edema increases during the first 10 days after ICH in humans, suggesting delayed brain edema following ICH may result from hematoma lysis. Other studies have demonstrated that perihematomal edema peaks several days after ICH [3, 11]. In rats, brain edema peak occurs on the third or fourth day after experimental ICH [13]. This delayed brain edema may be related to erythrocyte lysis, because infusion of packed erythrocytes causes edema after about 3 days but not earlier when the erythrocytes remain intact [13]. A clinical study of brain edema after ICH also indicates that delayed edema is related to significant midline shift after ICH [16].

Delayed brain edema in ICH patients may be due to erythrocyte lysis and iron toxicity. Our previous studies demonstrated that an intracerebral infusion of hemoglobin and its degradation products, hemin, iron, and bilirubin, cause the formation of brain edema within 24 hours. Hemoglobin itself induces heme oxygenase-1 up-regulation in the brain, and heme oxygenase inhibition by tin-protoporphyrin reduces hemoglobin-induced brain edema. In addition, an intraperitoneal injection of a large dose of deferoxamine, an iron chelator, attenuates brain edema induced by hemoglobin. These results indicate that hemoglobin

Table 1. *Clinical data after admission.*

Gender (n)	Male (10), female (7)
Age, y	59 ± 13
Systolic pressure, mmHg	176 ± 26
Diastolic pressure, mmHg	105 ± 13
Blood glucose, mM	9 ± 3
Glasgow Coma Scale	11 ± 3
Hematoma location, n (%)	
– Right basal ganglia	6 (36)
– Left basal ganglia	4 (24)
– Right temporal lobe	2 (12)
– Left temporal lobe	3 (18)
– Right parieto-occipital	1 (6)
– Right frontal lobe	1 (6)

Table 2. *Volumes of hematoma and perihematomal brain edema.*

Volume	Day 0	Day 1	Day 3	Day 10
Volume of hematoma (cm^3)	20.9 ± 18.8	23.8 ± 16.5	21.2 ± 14.8	12.4 ± 10.2#
Absolute edema volume (cm^3)	20.4 ± 13.2	30.7 ± 15.4#	42.6 ± 23.8	46.4 ± 30.1
Relative edema volume	1.3 ± 0.8	1.7 ± 1.1	2.5 ± 1.7	6.1 ± 6.5

# $p < 0.05$ vs. day 0.

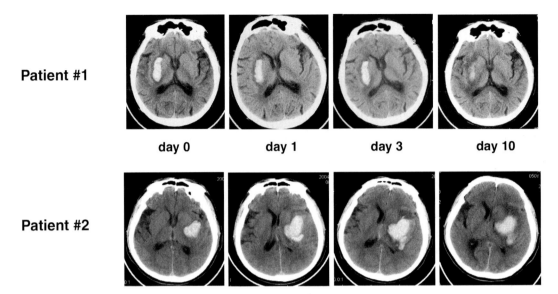

Fig. 1. Serial CT scans of 2 ICH patients

causes brain injury by itself and through its degradation products [5]. Also, investigations have demonstrated that iron overload occurs in the brain after ICH, and iron chelation with deferoxamine reduces perihematomal edema [9, 12].

In conclusion, early hematoma enlargement occurred in our study population of Chinese ICH patients, and delayed perihematomal edema development was associated with erythrocyte lysis. Reducing early hematoma growth and limiting clot lysis-induced brain toxicity could be potential therapies for ICH.

References

1. Broderick JP, Brott TG, Tomsick T, Barsan W, Spilker J (1990) Ultra-early evaluation of intracerebral hemorrhage. J Neurosurg 72: 195–199
2. Brott T, Broderick J, Kothari R, Barsan W, Tomsick T, Sauerbeck L, Spilker J, Duldner J, Khoury J (1997) Early hemorrhage growth in patients with intracerebral hemorrhage. Stroke 28: 1–5
3. Enzmann DR, Britt RH, Lyons BE, Buxton JL, Wilson DA (1981) Natural history of experimental intracerebral hemorrhage: sonography, computed tomography and neuropathology. AJNR Am J Neuroradiol 2: 517–526
4. Fujii Y, Tanaka R, Takeuchi S, Koike T, Minakawa T, Sasaki O (1994) Hematoma enlargement in spontaneous intracerebral hemorrhage. J Neurosurg 80: 51–57
5. Huang FP, Xi G, Keep RF, Hua Y, Nemoianu A, Hoff JT (2002) Brain edema after experimental intracerebral hemorrhage: role of hemoglobin degradation products. J Neurosurg 96: 287–293
6. Kase CS, Caplan LR (1994) Intracerebral Hemorrhage. Butterworth-Heinemann, Boston
7. Kazui S, Naritomi H, Yamamoto H, Sawada T, Yamaguchi T (1996) Enlargement of spontaneous intracerebral hemorrhage. Incidence and time course. Stroke 27: 1783–1787
8. Mayer SA, Brun NC, Begtrup K, Broderick J, Davis S, Diringer MN, Skolnick BE, Steiner T; Recombinant Activated Factor VII Intracerebral Hemorrhage Trial Investigators (2005) Recombinant activated factor VII for acute intracerebral hemorrhage. N Eng J Med 352: 777–785
9. Nakamura T, Keep R, Hua Y, Schallert T, Hoff JT, Xi G (2004) Deferoxamine-induced attenuation of brain edema and neurological deficits in a rat model of intracerebral hemorrhage. J Neurosurg 100: 672–678
10. Qureshi AI, Tuhrim S, Broderick JP, Batjer HH, Hondo H, Hanley DF (2001) Spontaneous intracerebral hemorrhage. N Engl J Med 344: 1450–1460
11. Tomita H, Ito U, Ohno K, Hirakawa K (1994) Chronological changes in brain edema induced by experimental intracerebral hematoma in cats. Acta Neurochir Suppl (Wien) 60: 558–560
12. Wu J, Hua Y, Keep RF, Nakamura T, Hoff JT, Xi G (2003) Iron and iron-handling proteins in the brain after intracerebral hemorrhage. Stroke 34: 2964–2969
13. Xi G, Keep RF, Hoff JT (1998) Erythrocytes and delayed brain edema formation following intracerebral hemorrhage in rats. J Neurosurg 89: 991–996
14. Xi G, Keep RF, Hoff JT (2002) Pathophysiology of brain edema formation. Neurosurg Clin N Am 13: 371–383
15. Xi G, Fewel ME, Hua Y, Thompson BG, Hoff J, Keep R (2004) Intracerebral hemorrhage: pathophysiology and therapy. Neurocrit Care 1: 5–18
16. Zazulia AR, Diringer MN, Derdeyn CP, Powers WJ (1999) Progression of mass effect after intracerebral hemorrhage. Stroke 30: 1167–1173
17. Zhang LF, Yang L, Hong Z, Yuan GG, Zhou BF, Zhao LC, Huang YN, Chen J, Wu YF; Collaborative Group of China Multicenter Study of Cardiovascular Epidemiology (2003) Proportion of different subtypes of stroke in China. Stroke 34: 2091–2096

Correspondence: Fengping Huang, Department of Neurosurgery, Huashan Hospital, Fudan University, 12 Wulumuqi Zhong Road, Shanghai 200040 China. e-mail: fengpinghuang@hotmail.com

Evaluation of acute perihematomal regional apparent diffusion coefficient abnormalities by diffusion-weighted imaging

E. Fainardi[1], M. Borrelli[1], A. Saletti[1], R. Schivalocchi[2], M. Russo[3], C. Azzini[4], M. Cavallo[2], S. Ceruti[1], R. Tamarozzi[1], and A. Chieregato[5]

[1] Neuroradiology Unit, Department of Neuroscience, Arcispedale S. Anna, Ferrara, Italy
[2] Neurosurgery Unit, Department of Neuroscience, Arcispedale S. Anna, Ferrara, Italy
[3] Neurology Unit, Department of Neuroscience, Arcispedale S. Anna, Ferrara, Italy
[4] Neurology Unit, Ospedale S. Bortolo, Vicenza, Italy
[5] Neurocritical Care Unit, Ospedale M. Bufalini, Cesena, Italy

Summary

In this study, we investigated 40 patients (18 male, 22 female; mean age = 64.5 ± 11.0; GCS = 9 to 14) with acute supratentorial spontaneous intracerebral hemorrhage (SICH) at admission by using a 1-tesla magnetic resonance imaging (MRI) unit equipped for single-shot echo-planar spin-echo isotropic diffusion-weighted imaging (DWI) sequences. All DWI studies were obtained within 48 hours after symptom onset. Regional apparent diffusion coefficient (rADC) values were measured in 3 different regions of interest (ROIs) drawn freehand on the T2-weighted images at b 0 s/mm^2 on every section in which hematoma was visible: 1) the perihematomal hyperintense area; 2) 1 cm of normal appearing brain tissue surrounding the perilesional hyperintense rim; 3) an area mirroring the region including the clot and perihematomal hyperintense area placed in the contralateral hemisphere. rADC mean values were higher in perihematomal hyperintense and in contralateral than in normal appearing areas ($p < 0.001$), with increased rADC mean levels in all regions examined.

Our findings show that rADC values indicative of vasogenic edema were present in the perihematomal area and in normal appearing brain tissue located both ipsilateral and contralateral to the hematoma, with lower levels in non-injured areas located in the T2 hyperintense rim around the clot.

Keywords: Intracerebral hemorrhage; apparent diffusion coefficient; diffusion-weighted imaging.

Introduction

Neither surgical nor medical treatment has been shown to improve clinical outcome after spontaneous intracerebral hemorrhage (SICH) [13]. Several investigations have recently focused on the mechanisms underlying edema formation in brain tissue surrounding the hematoma, widely considered to be a major cause of secondary damage contributing to delayed deterioration following SICH [6]. Although the development of perihematomal edema has been clearly demonstrated in animal models [10, 18, 19], the precise nature of this edema reaction remains to be elucidated [6, 13]. Studies performed with diffusion-weighted imaging (DWI) in humans may distinguish between cytotoxic and vasogenic edema, showing decreased apparent diffusion coefficient (ADC) values in cytotoxic edema and increased ADC values in vasogenic edema [14]. Perihematomal edema can be cytotoxic [8] or vasogenic [12, 15] in hyperacute phases, vasogenic [2, 3, 12] or combined cytotoxic and vasogenic [4] in acute phases, and vasogenic in subacute [2, 3, 12] stages of hematoma temporal evolution. ADC levels measured in the whole lesion (hemorrhagic core plus perihematomal area) are low in hyperacute, elevated in acute, decreased in early subacute, and slightly increased in late subacute hematomas [7]. Similar chronological fluctuations in ADC values are seen in the perihematomal area [4]. Since cellular components are irreversibly injured in cytotoxic and preserved in vasogenic edema [1], it is important to clarify which type of edema occurs early in the perihemorrhagic area. In the present study, we sought to evaluate perilesional ADC changes in patients with acute SICH.

Materials and methods

Patient selection

Forty patients with acute supratentorial SICH (18 male and 22 female; mean age = 64.5 ± 11.0) who underwent DWI studies within

48 hours after symptom onset were included in the study. Time of ictus was considered as the last time the patient was known to be neurologically normal. Patients with infratentorial hemorrhage, hematoma related to tumor, trauma, coagulopathy, aneurysms, vascular malformations, hemorrhagic transformation of brain infarction, intraventricular extension of their hemorrhage, and patients who had undergone surgical hematoma evacuation were excluded. Disease severity was scored in all patients at entry using the Glasgow Coma Scale (GCS) [16]. Hematoma location was classified as either basal ganglia or lobar. Hematoma volume was calculated using the formula $A \times B \times C/2$ [9]. Informed consent was obtained from each patient or from relatives before DWI studies were performed.

DWI studies

DWI examinations were performed on a 1-tesla magnetic resonance imaging (MRI) unit (Signa Horizon LX™, GE Medical System, Milwaukee, WI) equipped for isotropic DWI single-shot echo-planar spin-echo sequences. Axial images covering the whole brain were obtained by single-shot echo-planar spin-echo sequences (TR = 10000 ms; TE = 109 ms; slice thickness = 5 mm; interslice spacing = 0; matrix size = 96 × 96; FOV = 28 cm; NEX = 1; b-value = 1000 s/mm²). ADC maps were generated for each patient using an imaging workstation (Advantage Windows; GE Medical System, Milwaukee, WI) supplied with a dedicated software package (Functool; GE Medical System, Milwaukee, WI). Regional ADC (rADC) values were measured in 3 different regions of interest (ROIs) drawn freehand on the T2-weighted images at b 0 s/mm² on every section in which the hematoma was visible: 1) the perihematomal hyperintense area; 2) 1 cm of normal appearing brain tissue surrounding the perilesional hyperintense rim; 3) an area mirroring the region showing the clot and perihematomal hyperintense area in the contralateral hemisphere (Fig. 1). rADC values within the hemorrhagic core were not evaluated due to the presence of susceptibility artifacts. rADC levels were expressed in s/mm² [11]. Because cerebrospinal fluid (CSF) contamination was not excluded from rADC measurements, rADC values lower than 80×10^{-5} and higher than 92×10^{-5} s/mm² were considered to be suggestive of cytotoxic and vasogenic edema, respectively [17].

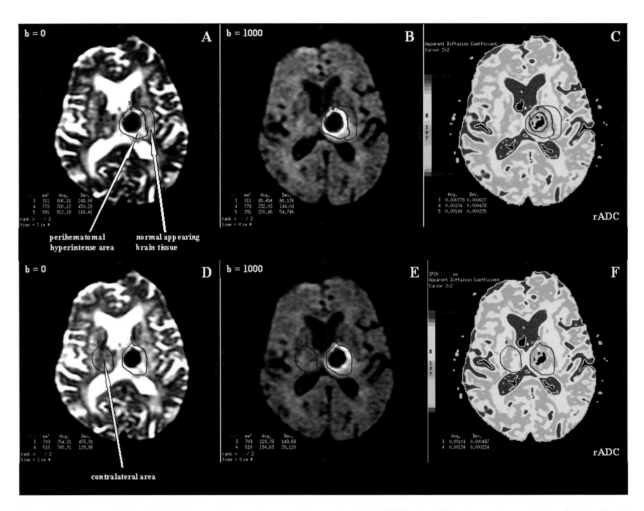

Fig. 1. DWI images obtained within 48 hours after symptom onset in a patient with SICH in the left thalamus. *Images A, B, and C* show hand-drawn ROIs placed around the perihematomal hyperintense area and normal appearing brain tissue surrounding the perilesional hyperintense rim in T2-weighted sequences at b = 0 s/mm² (A), in isotropic DWI sequences at b = 1000 s/mm² (B), and in ADC maps (C). *Image D, E, and F* depict hand-drawn ROIs including the clot and perihematomal hyperintense area, and a mirroring area located in the contralateral hemisphere in T2-weighted sequences at b = 0 s/mm² (D), in isotropic DWI sequences at b = 1000 s/mm² (E), and in ADC maps (F)

Statistical analysis

Mann-Whitney U test was used to compare mean values among the various groups. The Spearman rank correlation coefficient test was used to identify possible relationships among the different variables. Statistical significance was set at $p < 0.05$.

Results

Overall, SICH was observed within basal ganglia regions in 10 patients and within lobar regions in the remaining 30 patients. GCS scores ranged between 9 and 14. Mean hematoma volume was 17.2 ± 12 (range = 2–48.3). The mean time from symptom onset and DWI studies was 39.9 ± 4.4 hours (range = 29.2 to 47 hours; median = 40.2 hours). As illustrated in Fig. 2, rADC mean values were higher in perihematomal hyperintense tissue and contralateral parenchyma than in normal appearing areas ($p < 0.001$). There was no difference between perihematomal hyperintense and contralateral areas. The analysis of absolute values revealed that rADC mean levels were increased in perihematomal hyperintense ($111.7 \pm 38.2 \times 10^{-5}$ s/mm^2), in normal appearing ($93.8 \pm 11.4 \times 10^{-5}$ s/mm^2) and in contralateral ($104.9 \pm 20 \times 10^{-5}$ s/mm^2) areas. No definite correlations were observed between perihematomal rADC mean levels and hematoma volume.

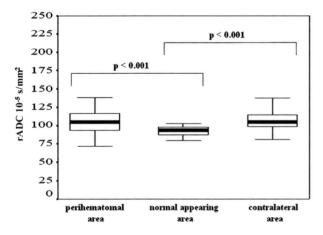

Fig. 2. Regional ADC *(rADC)* mean values expressed in 10^{-5} s/mm^2 in perihematomal hyperintense area, in normal appearing tissue surrounding the perilesional hyperintense rim, and in a mirroring contralateral area including the clot and perihematomal hyperintense area from 40 patients with SICH. rADC mean values were higher in perihematomal hyperintense and contralateral areas than in normal appearing areas ($p < 0.001$). The boundaries of the box represent the 25th to 75th quartile. The line within the box indicates the median. The whiskers above and below the box correspond to highest and lowest values, excluding outliers

Discussion

We have investigated rADC values in the region peripheral to the hematoma to verify whether perihematomal edema is cytotoxic or vasogenic in the acute phase. In accordance with previous studies [2, 3, 12], perihemorrhagic rADC values were elevated, suggesting ongoing accumulation of excess fluid in the extracellular space consistent with vasogenic edema. These findings are in contrast with those obtained by other investigators who have shown that cytotoxic and vasogenic edema can coexist in the perihematomal zone when a different topographical approach is used [4]. A possible explanation for acute edema formation surrounding the clot in the extracellular compartment comes from experimental studies documenting that vasogenic edema development around an intracerebral hemorrhage during the acute stage is mainly mediated by the clotting of blood with liberation of the remaining serum proteins, especially thrombin, into surrounding brain parenchyma [10] rather than by the toxic effect of hemoglobin degradation products derived from erythrocyte lysis [19]. In agreement with recent observations [7, 15], an inverse correlation between perihematomal rADC levels and hematoma size described previously [3] was not found. Our results argue against a major role for mechanical compression of small perilesional blood vessels related to hematoma mass effect in early perihemorrhagic edema development [6].

Interestingly, rADC values indicative of vasogenic edema were detected in non-lesioned regions located around the perihemorrhagic hyperintense rim and in the contralateral hemisphere. Concordance between our data and those obtained in other reports [3, 7, 15] is of particular relevance since they confirm that normal appearing brain tissue is not always normal when a focal hemorrhagic lesion is present. Global elevation of hydrostatic pressure resulting in abnormal water shift from blood vessels to the extracellular space could account for the occurrence of the increase in whole-brain ADC levels [7]. Alternatively, an intense and diffuse inflammatory response due to remote toxic effects of clotting proteins and red blood cell lysis components may contribute to the vasogenic edema observed in non-affected brain areas adjacent and distant to the SICH [10]. As seen in Fig. 1, the high rADC values obtained in normal appearing ROIs, manually outlined ipsilateral and contralateral to the hematoma, could be ascribed to inclusion of small and partially confluent focal hyperintensity on T2-weighted imaging. In fact, it has been demonstrated that ADC values are elevated in regions of leukoaraiosis as a consequence of axonal loss replaced by interstitial water [5].

Regional ADC levels were greater in contralateral than in ipsilateral normal appearing regions. Because the CSF signal was not excluded from rADC measurements, this apparent paradox was probably attributable to an effacement of the contiguous CSF spaces with elevated ADC values determined by the hematoma mass effect.

In conclusion, rADC values reflecting vasogenic edema were found in the perihematomal area and in normal appearing brain tissue located both ipsilateral and contralateral to the hematoma, with less pronounced values in the non-injured area located around the T2 hyperintense rim of the clot. Edema formation in the perilesional T2 high signal area did not seem to relate to the hematoma size. These findings suggest that an acute SICH is associated with both local and global edematous brain responses, indicating that DWI with analysis of rADC values represents a powerful tool for the evaluation of early edema development occurring around an acute hemorrhagic focal lesion.

References

1. Ayata C, Ropper AH (2002) Ischaemic brain oedema. J Clin Neurosci 9: 113–124
2. Carhuapoma JR, Wang P, Beauchamp NJ, Keyl PM, Hanley DF, Barker PB (2000) Diffusion-weighted MRI and proton MR spectroscopy imaging in the study of secondary neuronal injury after intracerebral hemorrhage. Stroke 31: 726–732
3. Carhuapoma JR, Barker PB, Hanley DF, Wang P, Beauchamp NJ (2002) Human brain hemorrhage: quantification of perihematoma edema by use of diffusion-weighted MR imaging. AJNR Am J Neuroradiol 23: 1322–1326
4. Forbes KP, Pipe JG, Heiserman JE (2003) Diffusion-weighted imaging provides support for secondary neuronal damage from intraparenchymal hematoma. Neuroradiology 45: 363–367
5. Helenius J, Soinne L, Salonen O, Kaste M, Tatlisumak T (2002) Leukoaraiosis, ischemic stroke, and normal white matter on diffusion-weighted MRI. Stroke 33: 45–50
6. Hoff JT, Xi G (2003) Brain edema from intracerebral hemorrhage. Acta Neurochir [Suppl] 86: 11–15
7. Kamal AK, Dyke JP, Katz JM, Liberato B, Filippi CG, Zimmerman RD, Uluğ AM (2003) Temporal evolution of diffusion after spontaneous supratentorial intracranial hemorrhage. AJNR Am J Neuroradiol 24: 895–901
8. Kidwell CS, Saver JL, Mattiello J, Warach S, Liebeskind DS, Starkman S, Vespa PM, Villablanca JP, Martin NA, Frazee J, Alger JR (2001) Diffusion-perfusion MR evaluation of perihematomal injury in hyperacute intracerebral hemorrhage. Neurology 57: 1611–1617
9. Kothari RU, Brott T, Broderick JP, Barsan WG, Sauerbeck LR, Zuccarello M, Khoury J (1996) The ABCs of measuring intracerebral hemorrhage volumes. Stroke 27: 1304–1305
10. Lee KR, Colon GP, Betz AL, Keep RF, Kim S, Hoff JT (1996) Edema from intracerebral hemorrhage: the role of thrombin. J Neurosurg 84: 91–96
11. Maldjian JA, Listerud J, Moonis G, Siddiqi F (2001) Computing diffusion rates in T2-dark hematomas and areas of low T2 signal. AJNR Am J Neuroradiol 22: 112–118
12. Morita N, Harada M, Yoneda K, Nishitani H, Uno M (2002) A characteristic feature of acute haematomas in the brain on echo-planar diffusion-weighted imaging. Neuroradiology 44: 907–911
13. Qureshi AI, Tuhrim S, Broderick JP, Batjer HH, Hondo H, Hanley DF (2001) Spontaneous intracerebral hemorrage. N Engl J Med 344: 1450–1460
14. Schaefer PW, Buonanno FS, Gonzalez RG, Schwamm LH (1997) Diffusion-weighted imaging discriminates between cytotoxic and vasogenic edema in a patient with eclampsia. Stroke 28: 1082–1085
15. Schellinger PD, Fiebach JB, Hoffmann K, Becker K, Orakcioglu B, Kollmar R, Jüttler E, Schramm P, Schwab S, Sartor K, Hacke W (2003) Stroke MRI in intracerebral hemorrhage. Is there a perihemorrhagic penumbra? Stroke 34: 1674–1680
16. Teasdale G, Jennett B (1974) Assessment of coma and impaired consciousness. A practical scale. Lancet 2: 81–84
17. Ulug AZ, Beauchamp N, Bryan RN, van Zijl PCM (1997) Absolute quantitation of diffusion constants in human stroke. Stroke 28: 483–490
18. Wagner KR, Xi G, Hua Y, Kleinholtz M, de Courten-Myers GM, Myers RE, Broderick JP, Brott TG (1996) Lobar intracerebral hemorrhage model in pigs: rapid edema development in perihematomal white matter. Stroke 27: 490–497
19. Xi G, Hua Y, Keep RF, Hoff JT (1998) Erythrocytes and delayed edema formation following intracerebral hemorrhage in rats. J Neurosurg 89: 991–996

Correspondence: Enrico Fainardi, Unità Operativa di Neuroradiologia, Dipartimento di Neuroscienze, Arcispedale S. Anna, Corso della Giovecca 203, 44100 Ferrara, Italy. e-mail: henryfai@tin.it

Reperfusion of low attenuation areas complicating subarachnoid hemorrhage

F. Tagliaferri[1], C. Compagnone[1], E. Fainardi[2], A. Tanfani[1], R. Pascarella[3], F. Sarpieri[1], L. Targa[1], and A. Chieregato[1]

[1] Neurorianimazione, Ospedale M. Bufalini, Cesena, Italy
[2] Neuroradiologia, Arcispedale S. Anna, Ferrara, Italy

Summary

Hypoattenuation areas shown on brain CT scans after subarachnoid hemorrhage (SAH) are believed to be associated with persistent ischemia. The aim of this study was to evaluate regional cerebral blood flow (rCBF) in hypoattenuation areas and its evolution over time by means of Xenon CT (Xe-CT).

We enrolled 16 patients with SAH who developed a hypoattenuation area in the middle cerebral artery territory. Patients were studied at time zero (the first Xe-CT), within 24 to 96 hours, and 96 hours after the initial Xe-CT.

We analyzed 19 hypoattenuation areas caused by vascular distortion, vasospasm, or post-surgical embolization in 48 Xe-CT studies. Areas of hypoattenuation were divided in 2 groups according to initial rCBF. In the first group (n = 15), rCBF was initially above 6 mL/100 gr/min but only 2 were still ischemic (rCBF < 18 mL/100 gr/min) 96 hours after the first Xe-CT, while 7 (58%) were hyperemic. Conversely, in the second group with severe ischemia (rCBF < 6 mL/100 gr/min; n = 4) mean rCBF increased (p = 0.08) but still remained below the ischemic threshold.

In severely ischemic lesions, rCBF reperfusion occurs but is probably marginally relevant. Conversely, in lesions not initially severely ischemic, residual CBF gradually improved and frequently became hyperemic. The functional recovery of these zones remains to be evaluated.

Keywords: Subarachnoid hemorrhage; ischemia; cerebral blood flow; Xenon-CT.

Introduction

The most common complication of aneurysmal subarachnoid hemorrhage (SAH) is cerebral ischemia [11]. Low attenuation zones due to poor perfusion in a major vessel territory after SAH are usually associated with persistent ischemia. Hypodense lesions consistent with cerebral infarctions (40 to 60%) are common on follow-up computerized tomography (CT) scans among survivors [6]. However, spontaneous reperfusion can follow ischemia. Many reports describe ischemic events after SAH, but none explain the evolution over time of the regional cerebral blood flow (rCBF) following ischemia. The knowledge of rCBF values in low-density areas related to SAH may be of potential interest in outcome prediction as well as in treatment planning. It is reasonable to believe that patients with low-density areas associated with normal rCBF levels have better outcomes. The practice of regional monitoring of hypodense edematous areas on CT could benefit from rCBF measurements, which might explain brain tissue oxygen or microdialysis regional data and improve the physiological background that sustains specific therapies [12]. The aim of our study was to evaluate CBF in low attenuation areas over time by repeated rCBF measures with Xenon CT (Xe-CT).

Materials and methods

From June 2000 to January 2003, 169 patients with aneurysmal SAH were treated in the Intensive Care Unit at Bufalini Hospital, Cesena-Italy. Sixteen of them developed new hypoattenuation areas greater than 1 cm^2 in the middle cerebral artery (MCA) territory and were studied with at least 2 Xe-CTs. Parenchymal hypoattenuation on CT was defined as a visually well-recognized cerebral region of abnormally increased radiolucency relative to other parts of the same structure or to its contralateral counterpart covering the same vascular territory [13]. There were 3 causes of major vessel distortion: distortion due to elevated intracranial pressure (ICP), vasospasm, and post-clipping or post-embolization procedures. The site of the bleeding aneurysm was assessed by angiography. Clinical and radiological severity on admission was graded according to Hunt and Hess [2] and Fisher [1] scores. ICP, arterial pressure, body temperature, and end-tidal CO_2 were continuously monitored. Patients were also monitored with a Swan-Ganz or a PICCO catheter as needed. Additional monitoring included transcranial Doppler and electroencephalography.

A staircase treatment protocol to maintain ICP below 20 mmHg was applied, consisting of sedation and analgesia to a level reducing noxious stimulation known to increase ICP, intermittent cerebrospinal fluid drainage to control ICP and to wash out bloody cerebrospinal fluid, control of serum sodium, and normocapnia. In cases with refractory elevated ICP, benzodiazepine and fentanyl treatment was combined with propofol or barbiturate to induce burst suppression, if necessary. Cerebral perfusion pressure (CPP) was maintained above 70 mmHg with crystalloid input and with norepinephrine and dobutamine if needed. Patients with critical rCBF and/or suspected vasospasm were treated with further elevation of CPP up to 90 mmHg.

Once a focal lesion was found in the MCA territory and rCBF measured, the treatment applied was consistent with ICP level, the extent of the area of the lesion, the cardiovascular reserve status, and the risk of medical complications. In patients with elevated ICP and large low-density areas that had ischemic or hyperemic rCBF values, treatment consisted of external decompression, CPP levels no higher than 70 mmHg, and deep sedation frequently induced with diazepam, propofol, and fentanyl. Conversely, patients with borderline rCBF values in low-density areas were treated with prolonged CPP elevation (90 mmHg).

Outcome was evaluated 1 year later using the Glasgow Outcome Score (GOS) [3]. We combined good recovery and moderate disability into one group (good), and severe disability, persistent vegetative state, or dead into a second group (poor).

CBF studies

When a hypoattenuation area was detected by CT, CBF was measured using a CT scanner (Picker 5000) equipped for Xe-CT CBF imaging (Xe/TC system-2, DDP, Inc., Houston, TX). Regions of interest larger than 1 cm² were drawn freehand around the low-density area on the CT scan. The rCBF mean values of each region of interest were expressed in mL/100 gr/min. Two different groups were defined according to the rCBF of the lesions seen on the first Xe-CT. The first group had severely ischemic lesions (CBF $\leq$ 6 mL/100 g/min) [7] on the first Xe-CT, while the second group had less severe ischemic lesions on the first Xe-CT (CBF > 6 mL/100 g/min). Moderate ischemia was defined as rCBF > 6 and $\leq$18 mL/100 gr/min [4]. Relative hyperemia was defined as CBF > 33.9 and $\leq$55.3 mL/100 g/min, while absolute hyperemia was CBF above 55.3 mL/100 g/min [10]. The distribution of the studies were categorized in 3 time periods: 1) first Xe-CT with a hypoattenuation zone, 2) 24 to 96 hours, and 3) 96 hours after the first Xe-CT.

The time intervals between both the onset of SAH and surgical or endovascular procedures and Xe-CT CBF measurements were recorded.

Statistical analysis

Comparisons of mean rCBF among different time groups were carried out by ANOVA. Comparison of mean rCBF between the group with severely ischemic lesions on the first Xe-CT and the group with less severe ischemic lesions was done using the unpaired 2-tailed *t* test. Comparison of percentage of hyperemic and ischemic lesions and outcome between groups was done by means of Chi-square (Data Desk v 6.1.1, by Data Description, Inc., Ithaca, NY, USA). A value of p < 0.05 was considered significant.

Results

Patient characteristics are reported in Table 1. The median Hunt and Hess score was 3 (interquartile range; IQR 2). The median Fisher score was 3 (IQR 1). The 16 patients developed 19 hypoattenuation le-

Table 1. *Patient characteristics.*

Patients		n	16
Age (years)		Mean (SD)	52 (12)
Sex (%)		Female	62
Fisher Score	I	n (%)	1 (6)
	II	n (%)	1 (6)
	III	n (%)	6 (38)
	IV	n (%)	8 (50)
Hunt & Hess Score	I	n (%)	4 (25)
	II	n (%)	3 (19)
	III	n (%)	1 (6)
	IV	n (%)	8 (50)
	V	n (%)	0 (0)
Clipping/Embolization		n/n	12/4
Bleeding Site	AcoA	n (%)	5 (31)
	MCA	n (%)	7 (44)
	PcoA	n (%)	2 (12.5)
	ICA	n (%)	2 (12.5)
Outcome at 12 months	Good recovery or moderate disability	n (%)	7 (44)
	Severe disability	n (%)	5 (31)
	Persistent vegetative state or death	n (%)	3 (19)
	Missing	n (%)	1 (6)

AcoA Anterior communicating artery; *MCA* middle cerebral artery; *PcoA* posterior communicating artery; *ICA* internal carotid artery.

sions in the MCA territory on plain CT scans. The first Xe-CT of a hypoattenuated lesion was performed at a median time of 3 (IQR 5) days after bleeding. The median time between aneurysm exclusion and the first Xe-CT was 1.5 (IQR 4) days. The 19 lesions were analyzed by 48 Xe-CT studies with a median of 3 Xe-CT studies per lesion (low attenuation lesion/Xe-CT study). Therefore, a total of 52 low attenuation lesion/Xe-CT studies were subjected to statistical analysis. The mean physiological parameters recorded during the 48 Xe-CT studies were ICP = 21 $\pm$ 6 mmHg, CPP = 74 $\pm$ 13 mmHg, and PaCO$_2$ = 39 $\pm$ 5 mmHg. The initially less severe ischemic lesions were due to vascular distortion (n = 1), vasospasm (n = 3), or post-surgical embolization (n = 8). In the first group with severely ischemic lesions on the first Xe-CT, the mean rCBF was higher in subsequent Xe-CTs (p = 0.08), failing to reach a mean rCBF above 18 mL/100 gr/min. In the group of less severe ischemic lesions on the first Xe-CT, the mean rCBF rose (p = 0.24) with an increased frequency hyperemic lesions (p = 0.053) (Table 2). This behavior was consistent with relative hyperemia. Conversely, absolute hyperemia peaked at 24 to 96 hours after the first Xe-CT and declined thereafter.

Poor outcome was found at 12 months in 3 (75%) patients with severely ischemic lesions on the first Xe-

Table 2. *Data from 48 Xe-CT studies comparing lesions according to the threshold of severe ischemia.*

		First CT	24–96 hrs	>96 hrs	p-value
Initially severe ischemic low-density lesions (n = 4)	mL/100 gr/min mean ± SD (n)	3 ± 1 (4)*	4.3 ± 2.6 (4)#	14 ± 12 (2)°	0.08
	% of relative hyperemic lesions (n)	0	0	0	1
	% of absolute hyperemic lesions (n)	0	0	0	
	% of lesions with rCBF < 18 mL/100 gr/min (n)	100 (4)	100 (4)	50 (1)	
Initially less severe ischemic low-density lesions (n = 15)	mL/100 gr/min mean ± SD (n)	22 ± 19 (15)*	34 ± 30 (15)#	35 ± 17 (12)°	0.24
	% of relative hyperemic lesions (n)	6.6 (1)	13.3 (2)	40 (6)	0.053
	% of absolute hyperemic lesions (n)	6.6 (1)	26.6 (4)	6.6 (1)	
	% of lesions with rCBF < 18 mL/100 gr/min (n)	60 (9)	40 (6)	5 (2)	

* $p < 0.03$; # $p < 0.03$; ° $p < 0.05$.

CT and 6 (50%) patients with less severe ischemic lesions ($p = 0.68$).

Discussion

This study suggests that hypoattenuated lesions on plain CT scans usually associated with current ischemia do not always correspond with reduced rCBF below the ischemic threshold and that rCBF recovery is consistent in most lesions. However, only in the initially less severe ischemic lesions will stable and satisfactory final rCBF levels return.

Reperfusion or recovery of rCBF may be due to the reversal of an initial fall in rCBF in the MCA territory, suggested by the pathological hyperemia which appeared early and disappeared after 96 hours. Similar behavior has already been described in patients with acute stroke [9]. Conversely, relative hyperemia tends to consolidate later, suggesting that focal derangement of microcirculation due to post-ischemic vasodilatation subsides and normal circulation recovers. Whether reperfusion in low-density areas contributes to cellular viability and better neurological outcome remains to be evaluated.

Since cerebral ischemia is generally described as the most frequent complication after SAH [11] and is associated with poor outcome [5], the value and evolution of rCBF within hypodensity areas may help with management decisions [8].

References

1. Fisher CM, Kistler JP, Davis JM (1980) Relation of cerebral vasospasm to subarachnoid hemorrhage visualized by computerized tomographic scanning. Neurosurgery 6: 1–9
2. Hunt WE, Hess RM (1968) Surgical risk as related to time of intervention in the repair of intracranial aneurysms. J Neurosurg 28: 14–20
3. Jennett B, Bond M (1975) Assessment of outcome after severe brain damage. Lancet 1: 480–484
4. Jones TH, Morawetz RB, Crowell RM, Marcoux FW, FitzGibbon SJ, DeGirolami U, Ojemann RG (1981) Thresholds of focal cerebral ischemia in awake monkeys. J Neurosurg 54: 773–782
5. Juvela S (1995) Aspirin and delayed cerebral ischemia after aneurysmal subarachnoid hemorrhage. J Neurosurg 82: 945–952
6. Juvela S, Siironen J, Varis J, Poussa K, Porras M (2005) Risk factors for ischemic lesions following aneurysmal subarachnoid hemorrhage. J Neurosurg 102: 194–201
7. Kaufmann AM, Firlik AD, Fukui MB, Wechsler LR, Jungries CA, Yonas H (1999) Ischemic core and penumbra in human stroke. Stroke 30: 93–99
8. Latchaw RE, Yonas H, Hunter GJ, Yuh WT, Ueda T, Sorensen AG, Sunshine JL, Biller J, Wechsler L, Higashida R, Hademenos G (2003) Guidelines and recommendations for perfusion imaging in cerebral ischemia: a scientific statement for healthcare professionals by the writing group on perfusion imaging, from the Council on Cardiovascular Radiology of the American Heart Association. Stroke 34: 1084–1104
9. Marchal G, Young AR, Baron JC (1999) Early postischemic hyperperfusion: pathophysiologic insights from positron emission tomography. J Cereb Blood Flow Metab 19: 467–482
10. Obrist WD, Langfitt TW, Jaggi JL, Cruz J, Gennarelli TA (1984) Cerebral blood flow and metabolism in comatose patients with acute head injury. Relationship to intracranial hypertension. J Neurosurg 61: 241–253
11. Roos YB, de Haan RJ, Beenen LF, Groen RJ, Albrecht KW, Vermeulen M (2000) Complications and outcome in patients with aneurysmal subarachnoid haemorrhage: a prospective hospital based cohort study in the Netherlands. J Neurol Neurosurg Psychiatry 68: 337–341
12. Stocchetti N, Chieregato A, De Marchi M, Croci M, Benti R, Grimoldi N (1998) High cerebral perfusion pressure improves low values of local brain tissue O2 tension (PtiO2) in focal lesions. Acta Neurochir [Suppl] 71: 162–165
13. Wardlaw JM, Sellar R (1994) A simple practical classification of cerebral infarcts on CT and its interobserver reliability. AJNR Am J Neuroradiol 15: 1933–1939

Correspondence: Arturo Chieregato, Servizio di Anestesia e Rianimazione, Ospedale M. Bufalini, Viale Ghirotti 286, 47023 Cesena, Italy. e-mail: achiere@ausl-cesena.emr.it

Human Cerebral Ischemia

Stroke in the young: relationship of active cocaine use with stroke mechanism and outcome

A. Nanda[1], P. Vannemreddy[1], B. Willis[1], and R. Kelley[2]

[1] Department of Neurosurgery, Louisiana State University Health Sciences Center, Shreveport, LA, USA
[2] Department of Neurology, Louisiana State University Health Sciences Center, Shreveport, LA, USA

Summary

Background. Cocaine and other vasoactive substances are known causes of cerebrovascular disease. Ictus during drug intake adversely affects outcome.

Materials and methods. A retrospective review revealed 42 patients with cocaine abuse and stroke. Aneurysmal bleed occurred in 15 patients; the rest had stroke. The outcome of stroke because of cocaine intoxication was analyzed.

Results. Mean age for stroke was 38 (± 8.5 SD) years; males outnumbered females (20:7) similar to the pattern seen in subarachnoid hemorrhage (SAH) following aneurysm rupture. Nine had intracerebral hematomas, 6 had SAH with intracerebral hemorrhage (ICH)/infarct, 1 had transverse myelopathy. Transient ischemic attack was identified in 4. Carotid occlusion was found in 2, and slow-flow in the vertebrobasilar system in 1. Fifteen were known hypertensives.

Cocaine was the principal substance in all patients; 7 used other substances including marijuana and heroin. Three patients had HIV, 3 had hepatitis, 2 had syphilis, and 1 had tuberculosis. Urinalysis was positive for cocaine metabolites in 15; 2 had late analysis. Nine had ICH or SAH with poor neurological status at admission and died. Cocaine intoxication correlated with fatal cerebrovascular accident (CVA) ($p < 0.001$) and poor Glasgow Outcome Score (GOS) ($p < 0.001$).

Conclusion. Stroke and cocaine use correlated with fatal CVA and poor outcome. Prompt diagnostic intervention may reveal the incidence of CNS injury with cocaine abuse.

Keywords: Cocaine; stroke; aneurysm; subarachnoid hemorrhage.

Introduction

The Substance Abuse and Mental Health Administration reported that nearly 2.5 million Americans admitted occasional and 600 000 admitted frequent cocaine use in 1995 [43]. These statistics show that a large number of individuals are exposing themselves to potentially adverse health consequences associated with cocaine use, the most commonly documented system being cardiovascular [5]. Hospital admissions for major cerebrovascular abnormalities associated with cocaine abuse have been less than 3% [13]. However, recent literature illustrates that the incidence of cocaine-related cerebrovascular disease is rapidly increasing [6]. The likelihood of developing a stroke in cocaine users may be as much as 14 times greater than that in age-matched non-cocaine using people [36]; 25% to 60% of these strokes are attributable to cerebral ischemia [7, 24]. The etiology of cocaine-induced brain ischemia involves vasospasm, platelet aggregation, pathological changes in the cerebral vasculature, and impaired cellular oxygenation [18, 27]. Rupture of intracranial aneurysms and arteriovenous malformations have been detected in nearly half of the patients with hemorrhagic strokes due to cocaine abuse. A temporal association exists between the onset of hemorrhagic and ischemic strokes and cocaine administration, the majority developing within 1 hour [29, 32].

Materials and methods

A computer search of ICD-9-CM (International Classification of Diseases, World Health Organization, 9th Revision, Clinical Modification) was utilized for retrieval of data on cocaine abuse and cerebrovascular disease. Over a 7-year period, 42 patients admitted to the hospital with this diagnosis were identified. Our review included patient demographics, comorbid factors like hypertension, nature of stroke, urinalysis findings, outcome measured by Glasgow Outcome Score (GOS) and angiographic findings. Data were coded and entered in to multivariate analysis using SigmaStat (version 3.0, Systat Software Inc., Chicago, IL).

Results

There are 2 categories of patients in the study; one with aneurysmal subarachnoid hemorrhage and the other group with hemorrhagic and non-hemorrhagic

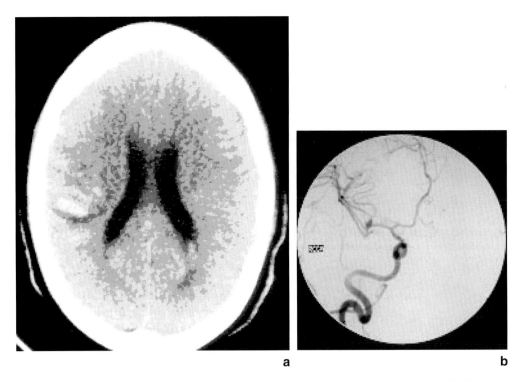

Fig. 1. (a) Plain CT scan of brain showing intracerebral and subarachnoid hemorrhage on right side. (b) Cerebral angiogram showing internal carotid injection on right side with saccular aneurysm arising from middle cerebral artery bifurcation

cerebrovascular incidents including transient ischemic attack (TIA), infarcts, intracerebral hemorrhage (ICH), and subarachnoid hemorrhage (SAH).

Aneurysmal SAH

Among 42 patients, aneurysmal SAH was seen in 15 patients with a mean age of 38.4 years (± 11.2 SD). Five were males and 10 were females. Ten patients experienced ictus during intake of the drug, while others were affected a few hours after intake. Most aneurysms were located on proximal sites and all except one were in the anterior circulation.

In comparison with the mean size of aneurysms in our own data base, aneurysms associated with cocaine were significantly smaller and tended to be multiple. The mean size of the aneurysms was 8 mm, almost 50% of them 5 mm or smaller in size, a difference from the average aneurysm size in our database ($p < 0.05$).

Nearly 60% of the patients presented with a good clinical grade (Hunt and Hess classification). One patient, in grade V, died before angiography and was found to have a proximal internal carotid artery aneurysm. No patient had evidence of vasospasm in our study. Most aneurysms were small and proximal in location; one was on the basilar artery.

All patients underwent clipping of the ruptured aneurysms. A good outcome was achieved in 80%, comparable with our overall outcome in aneurysm patients.

Case illustration

(Case 1) GM, a 26-year-old female, had severe headaches and altered sensorium following nasal insufflation of cocaine. Computed tomography (CT) scan of the brain showed an ICH with extension into the sylvian fissure (Fig. 1a). An emergency angiogram revealed a multi-lobulated aneurysm at the bifurcation of the middle cerebral artery on the side of the ICH (Fig. 1b). Clipping of the aneurysm achieved a good result.

Cerebrovascular events

There were 20 men and 7 women with a mean age of 38 years (± 8.5 SD). Nine patients had an ICH, 6 had SAH with ICH/infarct, 13 had infarcts, and 4 had

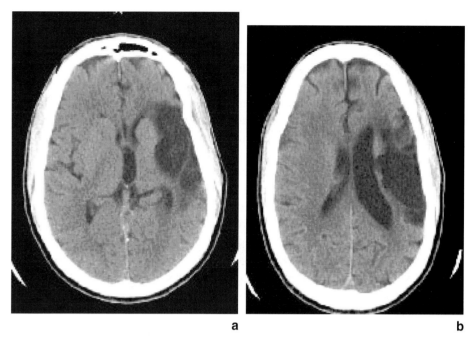

Fig. 2. (a, b) Delayed MR scan of brain showing infarct in middle cerebral artery territory in a young man who presented with sudden onset right-sided extremity weakness during cocaine abuse

TIA. Fifteen had hypertension and were getting treatment for it. All had cocaine as the principal abuse substance, while 7 used other stimulants including marijuana and heroin. At the time of admission, urinalysis revealed positive cocaine metabolites in 15 while 2 had late analysis. Other patients arrived in an elective fashion. Nine had other diseases including tuberculosis (n = 1), syphilis (n = 2), hepatitis (n = 3), and HIV (n = 3).

Nine patients admitted emergently died. All of them had ICH and/or SAH with poor neurological status at the time of arrival (33.33%). Other reasons for admission of patients with cocaine abuse included TIA and small non-hemorrhagic infarcts.

Statistical analysis

Forward stepwise regression analysis for cerebrovascular events revealed that a positive urinalysis was a strong predictor of severe stroke (hemorrhagic with ICH or SAH; $p < 0.001$) while other comorbid factors like hypertension, cigarette smoking, alcohol, age, and gender had no correlation with stroke severity. Hypertension, as a comorbid factor, had no relationship with the nature of the stroke.

Stepwise regression model for outcome (GOS) showed that a positive urinalysis for cocaine metabolites and stroke severity had a correlation with poor outcome ($p < 0.001$ and $p < 0.01$, respectively). Comorbid factors, including age and hypertension, did not reveal any association with outcome.

Case illustrations

(Case 2) SP, a 33-year-old man, was admitted with right-sided weakness, global aphasia, and altered sensorium following usage of cocaine and possibly other agents that the patient was unable to name. Cocaine metabolites were found in the urine. CT was negative for hemorrhage. Echocardiography was negative for thrombus. Magnetic resonance imaging (MRI) studies were positive for an occluded middle cerebral artery on left side (Fig. 2a–b).

(Case 3) SB, a 22-year-old female presented with sudden onset of right-sided limb weakness during intake of cocaine. She was admitted to the hospital. A CT scan did not reveal significant abnormality. Her urinalysis was positive for cocaine metabolites. She recovered completely from weakness over the next few hours. A follow-up MRI of the brain was normal. The neurological event this patient had was secondary to spasm of cerebral vessels during cocaine intake.

Discussion

Cocaine is a highly lipid soluble alkaloid that reaches a 5:1 brain-plasma ratio with a half-life of approximately 1 hour [17]. It potentiates the effects of monoamines by blocking reuptake of norepinephrine, thus increasing the bio-availability of catecholamines. This effect, combined with increased sensitivity of catecholamine receptors, leads to sympathetic hyperactivity and transient hypertension [11, 21, 23]. Hypertension is a significant risk factor for cerebrovascular accident and aneurysmal formation as well as rupture [25, 38, 40].

Cocaine-induced hemodynamic changes play a vital role in the formation and rupture of intracranial aneurysms. The transcranial Doppler ultrasound study by Van de Bor et al. [47] on infants exposed to cocaine in utero strongly supports the hypothesis that aneurysmal rupture is a direct consequence of the hemodynamic effects of cocaine; the transient but repeated bouts of hypertension are transmitted, almost unchanged, to the saccular aneurysms [9] and result in exceedingly high intra-aneurysmal wall tension that increases the risk of rupture. Because of the relative nondistendability of the aneurysmal sac, which is deficient in functional elastin and collagen, the intraluminal stress within the aneurysm can be nearly 10 times that of cerebral arteries at a particular pressure [3, 34]. Frequent bursts of hypertension can induce an injury to the arterial wall, the formation and enlargement of aneurysms, and even subsequent rupture [1, 25, 37, 41]. In the current study, rupture of the aneurysms occurred during intake of cocaine in the majority of the patients.

Cocaine and its metabolites have been shown to be potent cerebral vasoconstrictors in animal models [26]. In almost 80% of long term cocaine users, a focal perfusion defect develops, which is a subtle form of cerebrovascular dysfunction, and may be secondary to cocaine-induced cerebral vasoconstriction [42, 48]. In human volunteers, a similar phenomenon was demonstrated by Kaufmann et al. [18]. Their findings also suggested a dose-related cumulative residual effect in which repeated cocaine exposures produce delayed and/or prolonged vasoconstriction. In both formation and growth of an aneurysm, narrowing of vessels has a direct influence: the nozzle effect of a jet of blood emerging through a constricted area results in turbulence that induces damage to the vessel wall [14, 46]. However, the duration of cocaine-induced vasoconstriction is unclear. The active metabolites, benzylecgonine and ecgonine, are known to remain in the brain long after exposure to the drug and the defective areas of cerebral blood flow may remain 10 days after intake [50]. The smaller size aneurysms at the time of rupture and the young age of patients might be indicators of these vasoactive properties of cocaine and resultant damage to the cerebral vasculature [29].

A recent study by Su et al. [44] showed that cerebral vascular smooth muscle can undergo rapid apoptosis in response to cocaine in a concentration-dependent manner. Cocaine-induced apoptosis plays a major role in brain-microvascular damage, cerebral vascular toxicity, and stroke.

Our study has shown that strokes tend to be hemorrhagic and present in poor neurological status in people who take cocaine actively. Most of them in poor clinical grade succumb to the intracranial pathology. The manner in which cocaine induces cerebrovascular disease is multifactorial.

Nassogne et al. [30] showed that exposure of fetal mouse brain co-cultures to cocaine does not affect the viability of glial cells, but selectively inhibits neurite outgrowth, followed by loss of neurons through an unknown mechanism. The major metabolites of cocaine appeared to have no detectable effects on neurons, indicating that apoptosis could be due to cocaine itself [31]. Inappropriate neuronal apoptosis in cocaine-exposed fetal brain could perturb the neurodevelopment program and contribute to quantitative neuronal defects. Similar laboratory data suggest that cocaine induces apoptosis in the smooth muscle of the cerebral vasculature predisposing the brain to development of multiple cerebrovascular diseases. Kaufmann et al. [18], using human volunteers, administered cocaine in small doses and suggested that low cocaine doses are sufficient to induce cerebrovascular dysfunction. Their data also suggested a dose-response relationship between cocaine and vasoconstriction. In a small dose (0.2 or 0.4 mg/kg), cocaine when administered intravenously to human volunteers induced vasoconstriction on magnetic resonance angiography. A similar phenomenon may be responsible in one of our patients who presented with vascular insufficiency syndrome but had normal MRI brain 1 year later. Kaufmann et al. [18] confirmed the findings of Su et al. [44] that cocaine may have cumulative effect in producing cerebrovascular dysfunction in addition to its acute vasoconstrictive effect.

According to Levine et al. [24], 0.1% of in-patient

admissions had a cerebrovascular event temporally related to cocaine use. Petitti *et al.* [36] also reported that cocaine and/or amphetamine use was a major independent risk factor for stroke in young women. The occurrence of hemorrhagic strokes exceeds that of cerebral infarctions [19], although significant numbers also present with ischemic episodes [24]. Our study also showed a large number of patients with hemorrhage (ICH, SAH, or both) from aneurysmal or non-aneurysmal causes. There is a cause and effect relationship between a cerebrovascular event and cocaine as shown by our study, where the majority of aneurysms bled during drug abuse. Similar clinical studies have also shown the same phenomenon [12, 19, 24]. Apart from spasm, other possible mechanisms for increased stroke risk include emboli from drug impurities [8], paradoxical fat embolism [2], infectious emboli from heart [16], acute severe increase in blood pressure [35], cardiomyopathy induced emboli [29], hypoxia during drug overdose [51], and allergic reactions to the drug or its additives [8]. Cocaine also enhances the response of platelets to arachidonic acid in vitro, thus promoting thrombus formation [45]. Chronic cocaine use increases platelet levels, enhances adenosine diphosphate platelet activation, and augments sporadic release of platelet-bound alpha granules. These activities may be mediated by cocaine-induced increases in monoamine levels, particularly of 5-HT [20].

About 70% of events occurring with intranasal or intravenous cocaine use are caused by hemorrhage. Hemorrhages may be intracerebral, intraventricular, or subarachnoid. In the majority of studies, most strokes occurred during the first hour of drug use and blood pressure may not be elevated at all times or at the time or presentation. Sometimes hypertension is labeled as the cause of hemorrhage.

Although vasculitis was often mentioned as a possible cause, no cases of vasculitis have been found in several large postmortem series [22, 28, 33]. Nolte *et al.* [33] found almost 60% of the fatal intracranial hemorrhages were associated with cocaine use. This is similar to our finding where all fatalities occurred from ICH. Vascular changes described histologically in cocaine abuse include abnormal internal elastic lamina infolding and tunica media disruption in cerebral infarction [20], arteriolar and periarteriolar fibrosis in the nasal mucosa of cocaine snorters [4], elastic disruption and subendothelial edema in submucosal intestinal arterioles in a case of intestinal ischemia [10], and coronary artery intimal hyperplasia with platelet thrombi in ischemic heart disease [39]. Vascular changes may appear as spasm on angiography and the beaded appearance of vessels sometimes goes on to vasculitis. Large arteries also may go into spasm and produce infarction. In some instances, platelet activation and subsequent thrombus formation can induce vascular occlusion.

The biological half-life of cocaine in blood is approximately 1 hour, with less than 5% of cocaine appearing unchanged in urine. Most urine excretion of cocaine and its metabolites occurs within the first 24 hours after administration regardless of route. The duration of detection of urinary cocaine metabolites depends upon 2 factors: the amount of cocaine absorbed or injected and the sensitivity of the drug assay used. Smoking cocaine provides the fastest route of entry into the cerebral circulation (6 to 8 seconds) and the intravenous route takes twice as much time; nasal insufflation produces peak levels in 30 to 60 minutes [15, 49].

The true incidence of cocaine-induced cerebrovascular events, including aneurysmal rupture, remains obscure. The pharmacokinetics and pathophysiological mechanisms involved in cocaine-induced CNS damage are also unclear. From our data and other publications, it is evident that most strokes and deaths due to hemorrhage in the young have a relationship with cocaine use. Prompt diagnostics for the evaluation of cocaine metabolites in body fluids might help us understand the true nature of the cocaine-related CNS injury and further management strategies.

References

1. Austin GM, Schievink W, Williams R (1989) Controlled pressure-volume factors in the enlargement of intracranial aneurysms. Neurosurgery 24: 722–730
2. Bitar S, Gomez CR (1993) Stroke following injection of a melted suppository. Stroke 24: 741–743
3. Burton A (1954) Relation of structure to function of tissues of wall of blood vessels. Physiol Rev 34: 619–622
4. Chow JM, Robertson AL, Stein RJ (1990) Vascular changes in the nasal submucosa of chronic cocaine addicts. J Forensic Sci 30: 740–745
5. Cregler LL, Mark H (1986) Medical complications of cocaine abuse. N Engl J Med 315: 1495–1500
6. Daras M (1996) Neurologic complications of cocaine. In: Majewska MD (ed) Neurotoxicity and Neuropathology Associated with Cocaine Abuse (NIH Publication No. 4019). U.S. Department of Health and Human Services, Washington, DC, pp 43–65
7. Daras M, Tuchman AJ, Koppel BS, Samkoff LM, Weitzner I, Marc J (1994) Neurovascular complications of cocaine. Acta Neurol Scand 90: 124–129

8. Delaney P, Estes M (1980) Intracranial hemorrhage with amphetamine abuse. Neurology 30: 1125–1128
9. Ferguson GG (1972) Direct measurement of mean and pulsatile blood pressure at operation in human intracranial saccular aneurysms. J Neurosurg 36: 560–563
10. Garfia A, Valverde JL, Borondo JC, Candenas I, Lucena J (1990) Vascular lesions in intestinal ischemia induced by cocaine-alcohol abuse: report of a fatal case due to overdose. J Forensic Sci 35: 740–745
11. Gawin FH (1991) Cocaine addiction: psychology and neurophysiology. Science 251: 1580–1586
12. Heye N, Hankey GJ (1996) Amphetamine-associated stroke. Cerebrovasc Dis 6: 149–155
13. Jacob IG, Roszler MH, Kelly JK, Klein MA, Kling GA (1989) Cocaine abuse: neurovascular complications. Radiology 170: 223–227
14. Jain K (1963) Mechanism of rupture of intracranial saccular aneurysms. Surgery 54: 347–350
15. Jatlow PI (1987) Drug of abuse profile: cocaine. Clin Chem 33: 66B–71B
16. Kaku DA, Lowenstein DH (1990) Emergence of recreational drug abuse as a major risk factor for stroke in young adults. Ann Intern Med 113: 821–827
17. Karch SB (1991) Introduction to the forensic pathology of cocaine. Am J Forensic Med Pathol 12: 126–131
18. Kaufman MJ, Levin JM, Ross MH, Lange N, Rose SL, Kukes TJ, Mendelson JH, Lukas SE, Cohen BM, Renshaw PF (1998) Cocaine-induced cerebral vasoconstriction detected in humans with magnetic resonance angiography. JAMA 279: 376–380
19. Klonoff DC, Andrews BT, Obana WG (1989) Stroke associated with cocaine use. Arch Neurol 46: 989–993
20. Konzen JP, Levine SR, Garcia JH (1995) Vasospasm and thrombus formation as possible mechanisms of stroke related to alkaloidal cocaine. Stroke 26: 1114–1118
21. Koob GF, Bloom FE (1988) Cellular and molecular mechanisms of drug dependence. Science 242: 715–723
22. Krendel DA, Ditter SM, Frankel MR, Ross WK (1990) Biopsy-proven cerebral vasculitis associated with cocaine abuse. Neurology 40: 1092–1094
23. Kuhar MJ, Ritz MC, Boja JW (1991) The dopamine hypothesis of the reinforcing properties of cocaine. Trends Neurosci 14: 299–302
24. Levine SR, Brust JC, Futrell N, Ho KL, Blake D, Millikan CH, Brass LM, Fayad P, Schultz LR, Selwa JF, Welch KM (1990) Cerebrovascular complications of the use of the "crack" form of alkaloidal cocaine. N Engl J Med 323: 699–704
25. Locksley HB (1966) Natural history of subarachnoid hemorrhage, intracranial aneurysms and arteriovenous malformations: Based on 6368 cases in the cooperative study. J Neurosurg 25: 219–239
26. Madden JA, Konkol RJ, Keller PA, Alvarez TA (1995) Cocaine and benzoylecgonine constrict cerebral arteries by different mechanisms. Life Sci 56: 679–686
27. Martinez N, Diez-Tejedor E, Frank A (1996) Vasospasm/thrombus in cerebral ischemia related to cocaine abuse. Stroke 27: 147–148 [letter]
28. Morrow PL, McQuillen JB (1993) Cerebral vasculitis associated with cocaine abuse. J Forensic Sci 38: 732–738
29. Nanda A, Vannemreddy P, Polin RS, Willis BK (2000) Intracranial aneurysms and cocaine abuse: analysis of prognostic indicators. Neurosurgery 46: 1063–1069
30. Nassogne MC, Evrard P, Courtoy PJ (1995) Selective neuronal toxicity of cocaine in embryonic mouse brain cocultures. Proc Natl Acad Sci USA 92: 11029–11033
31. Nassogne M, Louahed J, Evrard P, Courtoy PJ (1997) Cocaine induces apoptosis in cortical neurons of fetal mice. J Neurochem 68: 2442–2450
32. Neiman J, Haapaniemi HM, Hillbom M (2000) Neurological complications of drug abuse: pathophysiological mechanisms. Eur J Neurol 7: 595–672
33. Nolte KB, Brass LM, Fletterick CF (1996) Intracranial hemorrhage associated with cocaine abuse: a prospective autopsy study. Neurology 46: 1291–1296
34. Nystrom S (1963) Development of intracranial aneurysms as revealed by electron microscopy. J Neurosurg 20: 324–337
35. Perez Ja Jr, Arsura EL, Strategos S (1999) Metamphetamine-related stroke: four cases. J Emerg Med 17: 469–471
36. Petitti DB, Sidney S, Quesenberry C, Bernstein A (1998) Stroke and cocaine or amphetamine use. Epidemiology 9: 596–600
37. Quigley MR, Heiferman K, Kwaan HC, Vidovich D, Nora P, Cerullo LJ (1987) Bursting pressure of experimental aneurysms. J Neurosurg 67: 288–290
38. Sacco RL, Wolf PA, Bharucha NE, Meeks SL, Kannel WB, Charette LJ, McNamara PM, Palmer EP, D'Agostino R (1984) Subarachnoid and intracerebral hemorrhage: natural history, prognosis, and precursive factors in the Framingham Study. Neurology 34: 847–854
39. Simpson RLW, Edwards WD (1986) Pathogenesis of cocaine induced ischemic heart disease. Arch Pathol Lab Med 110: 479–484
40. Stehbens WE (1962) Hypertension and cerebral aneurysms. Med J Aust 2: 8–10
41. Stehbens WE (1989) Etiology of intracranial berry aneurysms. J Neurosurg 70: 823–831
42. Strickland TL, Mena I, Villanueva-Meyer J, Miller BL, Cummings J, Mehringer CM, Satz P, Myers H (1993) Cerebral perfusion and neuropsychological consequences of chronic cocaine use. J Neuropsychiatry Clin Neurosci 5: 419–427
43. Substance Abuse and Mental Health Services Administration (1996) Preliminary estimates from the 1995 national household survey on drug abuse (Advanced Report No. 18). U.S. Department of Health and Human Services, Washington, DC
44. Su J, Li J, Li W, Altura BT, Altura BM (2003) Cocaine induces apoptosis in cerebral vascular muscle cells: potential roles in strokes and brain damage. Eur J Phamcol 482: 61–66
45. Togna G, Tempesta E, Togna AR, Dolci N, Cebo B, Caprino L (1985) Platelet responsiveness and biosynthesis of thromboxane and prostacycline in response to in vitro cocaine treatment. Hemostasis 15: 100–107
46. Tolstedt G, Bell J (1963) Production of experimental aneurysms in the canine aorta. Angiology 14: 459–464
47. Van de Bor M, Walther FJ, Sims ME (1990) Increased cerebral blood flow velocity in infants of mothers who abuse cocaine. Pediatrics 85: 733–736
48. Volkow ND, Mullani N, Gould KL, Adler S, Krajewski K (1988) Cerebral blood flow in chronic cocaine users: a study with positron emission tomography. Br J Psychiatry 152: 641–648
49. Warner EA (1993) Cocaine abuse. Ann Intern Med 119: 226–235
50. Weiss RD, Gawin FH (1988) Protracted elimination of cocaine metabolites in long-term high-dose cocaine abusers. Am J Med 85: 879–880
51. Zuckermann GB, Ruiz DC, Keller IA, Brooks J (1996) Neurologic complications following intranasal administration of heroin in an adolescent. Ann Pharmacother 30: 778–781

Correspondence: Anil Nanda, Department of Neurosurgery, LSU Health Sciences Center, Shrevport, LA 71130, USA. e-mail: pvanne@lsu.hsc.edu

Brain oxygen metabolism may relate to the temperature gradient between the jugular vein and pulmonary artery after cardiopulmonary resuscitation

A. Sakurai, K. Kinoshita, K. Inada, M. Furukawa, T. Ebihara, T. Moriya, A. Utagawa, Y. Kitahata, K. Okuno, and K. Tanjoh

Department of Emergency and Critical Care Medicine, Nihon University School of Medicine, Tokyo, Japan

Summary

Objective. A gradient between the jugular vein temperature and core body temperature has been reported in animal and clinical studies; however, the pathophysiological meaning of this phenomenon remains unclear. This study was conducted to identify the temperature gradient between the jugular vein and pulmonary artery in comatose patients after cardiopulmonary resuscitation.

Materials and methods. The temperatures of the jugular vein and pulmonary artery were measured in 19 patients at 6 and 24 hours after cardiopulmonary resuscitation. Jugular venous blood saturation (SjO_2; %) was also measured concomitantly. The patients were divided into 2 groups: high SjO_2 ($SjO_2 > 75\%$: H-group; n = 10) and normal SjO_2 ($SjO_2 \leq 75\%$: N-group; n = 9). The temperature gradient was calculated by subtracting the temperature of the pulmonary artery from that of the jugular vein (jugular – pulmonary = $dT\,°C$). Statistical significance was defined as $p < 0.05$.

Results. dT was significantly lower in the H-group than in the N-group at 6 hours (0.120 ± 0.011: mean $\pm$ SD vs. 0.389 ± 0.036: $p = 0.0012$) and 24 hours (0.090 ± 0.005 vs. 0.256 ± 0.030: $p = 0.0136$) after cardiopulmonary resuscitation.

Conclusion. The temperature gradient between the jugular vein and pulmonary artery was significantly lower in patients with high SjO_2 after cardiopulmonary resuscitation. This temperature gradient may be reflected in brain oxygen metabolism.

Keywords: Jugular vein temperature; pulmonary artery temperature; SjO_2; $CMRO_2$.

Introduction

Fever is a frequent and important problem in patients with neurological injury [4]. Among brain-injured patients with hypothermia, regional differences in brain temperature appear to increase [5, 9, 10]. In temperature measurements taken immediately after neurological injury, local brain temperatures are often significantly higher than core body temperatures [3, 4, 9, 10]. This differential between local brain and core body temperatures ranges from 0.1 to 2.0 °C [4, 9, 10], and it has been clinically and experimentally demonstrated to increase even further when the core temperature climbs above 38 °C [1, 3, 5]. The pathophysiological meaning of this phenomenon remains unclear, however.

Clinical studies [6, 8] have reported fluctuations in this temperature differential between brain and core body in patients with traumatic brain injury or stroke. The authors of these studies have suggested that the temperature gradient in critical brain conditions might be reflected in brain blood flow and oxygen metabolism. Few studies, however, have reported the temperature gradient between the central nervous system and core body temperatures in comatose patients after resuscitation from cardiopulmonary arrest (CPA). The objective of our study was to identify whether this temperature gradient in comatose patients resuscitated from CPA reflects brain blood flow and oxygen metabolism.

Materials and methods

The temperatures of the jugular vein and pulmonary artery were measured in 19 patients at 6 and 24 hours after CPA. Patients with traumatic brain injury or cerebrovascular accident were excluded by computed tomography (CT). The jugular venous blood saturation (SjO_2; %) was measured concomitantly using a jugular vein catheter (Opticath P540-H, Abbott Laboratories, Chicago, IL) inserted into the right jugular bulb and depicted on skull radiographs to confirm appropriate placement. All SjO_2 values were obtained by re-calibrated data for blood samples drawn through the catheter. A pulmonary artery catheter (OptiQ, Abbott Laboratories, Chicago, IL) was inserted into the pulmonary artery. Appropriate placement was confirmed by pressure waves and a chest radiograph.

All patients were managed by controlled ventilation (maintaining $PaCO_2$ at 35 to 45 mmHg), sedation, and neuromuscular blocking agents. Systemic blood pressure was kept above 90 mmHg.

A Pittsburgh cerebral performance category of 1 (good recovery) or 2 (moderate disability) was defined as a favorable outcome. The other 3 categories – 3 (severe disability), 4 (vegetative state), and 5 (death) – were defined as unfavorable outcomes.

The patients were retrospectively divided into 2 groups based on the SjO_2 at 6 hours after resuscitation: normal SjO_2 ($SjO_2 \leq 75\%$: N-group; n = 9) and high SjO_2 ($SjO_2 > 75\%$: H-group; n = 10). The temperature gradient was calculated by subtracting the temperature of the pulmonary artery from that of the jugular vein (jugular – pulmonary = dT °C). Non-paired t-test and Fisher exact test were employed for the comparisons between the 2 groups. Statistical significance was defined as $p < 0.05$.

Results

The profiles of the 2 groups are summarized in Table 1. There were no significant differences between the groups in gender, age, rate of cardiac etiology with CPA, rate of witnessed CPA, or outcome.

Brain CT scans revealed no brain swelling, basal cistern, or evidence of intracranial hypertension.

Figure 1 compares the temperature gradients (dT) between the N-group and H-group at 6 hours (A) and 24 hours (B) after resuscitation. dT was significantly lower in the H-group than in the N-group at both 6 hours (0.120 ± 0.011: mean $\pm$ SD vs. 0.389 ± 0.036: $p = 0.0012$) and 24 hours (0.090 ± 0.005 vs. 0.256 ± 0.030: $p = 0.0136$) after cardiopulmonary resuscitation.

Discussion

Fluctuations in the gradient between brain and core body temperatures are frequently observed in brain-injured patients [3, 5, 6]. Brain temperature seems to be determined by 3 major factors: 1) the production of local heat by metabolic processes in the brain; 2) the rate of local cerebral blood flow; and 3) the arterial blood temperature [1]. We have reported that temperature gradient differences between brain and bladder have a significant inverse correlation with SjO_2 at cerebral perfusion pressure (CPP) > 50 mmHg in patients with traumatic brain injury and stroke [8]. A higher SjO_2 indicates brain hyperemia or blood contamination from the external carotid artery after brain death [2]. Hyperemia is defined as a cerebral blood flow in excess of the metabolic demand. In this study we also found that the temperature gradient between the jugular vein and pulmonary artery was significantly lower

Table 1. *Patient characteristics*

	N-group	H-group	p-value
Gender (male/female)	6/3	5/5	0.6499
Age (mean $\pm$ SD)	56.7 $\pm$ 15.0	55.2 $\pm$ 15.5	0.8364
Cause of CPA (cardiac/non-cardiac)	3/6	2/8	0.6285
CPA witnessed/non-witnessed	7/2	6/4	0.6285
Outcome (favorable*/unfavorable**)	3/6	3/7	0.9999

N-group Normal SjO_2 group; *H-group* high SjO_2 (>75%) group; *CPA* cardiopulmonary arrest.
* Favorable outcome defined as Pittsburgh cerebral performance category 1 (good recovery) or 2 (moderate disability), ** unfavorable outcomes defined as Pittsburgh cerebral performance category 3 (severe disability), 4 (vegetative state), or 5 (death)

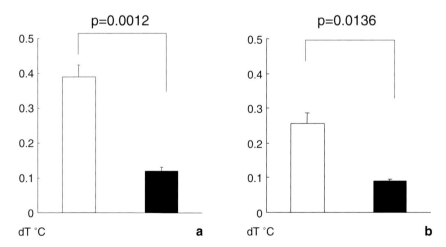

Fig. 1. Comparison of temperature gradient between the N-group (open bars) and H-group (closed bars) at 6 hours (a) and 24 hours (b) after resuscitation. dT = temperature of the pulmonary artery subtracted from the temperature of the jugular vein

in a high SjO$_2$ group than in the normal group, and it remained so for 24 hours. These results corroborated those from our previous study [8]. We surmise that a long-lasting elevation of SjO$_2$ for at least 24 hours after resuscitation from CPA might indicate a lower brain oxygen metabolism rather than hyperperfusion following anoxia-ischemia. This would be consistent with earlier results from ischemia-reperfusion models demonstrating postischemic hyperperfusion lasting only a few minutes [7]. Taken together, these findings have two implications: first, that brain blood flow and oxygen metabolism might reflect fluctuations in the temperature gradient between the jugular vein and pulmonary artery of comatose patients after cardiopulmonary resuscitation; and second, that a reduced gradient might indicate a lower oxygen metabolism at SjO$_2$ > 75%.

Our decision to measure the jugular vein temperature instead of the brain tissue temperature in this study was based on the report by Ao indicating that jugular vein temperature reflects brain temperature [1]. A major limitation of our study was the inability to monitor intracranial pressure (ICP) or CPP. While CT brain scans revealed no evidence of increased ICP, further studies will be needed to elucidate the importance of temperature gradient data together with data on ICP and CPP in comatose patients after resuscitation from CPA.

Conclusion

This study showed that the temperature gradient between the jugular vein and pulmonary artery was significantly lower in a high-SjO$_2$ group (SjO$_2$ > 75%) than in a normal group (SjO$_2$ ≤ 75%) among comatose patients after cardiopulmonary resuscitation. This temperature gradient may also reflect brain blood flow and oxygen metabolism.

References

1. Ao H, Moon JK, Tanimoto H, Sakanashi Y, Terasaki H (2000) Jugular vein temperature reflects brain temperature during hypothermia. Resuscitation 45: 111–118
2. Gupta AK, Matta BF (2000) Cerebral oximetry. In: Matta BF, Menon DK, Turner JM (eds) Textbook of neuroanaesthesia and critical care. Greenwich Medical Media Ltd, London, pp 133–146
3. Hayashi N, Hirayama T, Udagawa A, Daimon W, Ohata M (1994) Systemic management of cerebral edema based on a new concept in severe head injury patients. Acta Neurochir [Suppl] 60: 541–543
4. Kilpatrick MM, Lowry DW, Firlik AD, Yonas H, Marion DW (2000) Hyperthermia in the neurosurgical intensive care unit. Neurosurgery 47: 850–855; discussion 855–856
5. Mellergard P, Nordstrom CH (1991) Intracerebral temperature in neurosurgical patients. Neurosurgery 28: 709–713
6. Rumana CS, Gopinath SP, Uzura M, Valadka AB, Robertson CS (1998) Brain temperature exceeds systemic temperature in head-injured patients. Crit Care Med 26: 562–567
7. Safar P (1993) Cerebral resuscitation after cardiac arrest: research initiatives and future directions. Ann Emerg Med 22: 324–349
8. Sakurai A, Kinoshita K, Atsumi T, Moriya T, Utagawa A, Hayashi N (2003) Relation between brain oxygen metabolism and temperature gradient between brain and bladder. Acta Neurochir [Suppl] 86: 251–253
9. Schwab S, Spranger M, Aschoff A, Steiner T, Hacke W (1997) Brain temperature monitoring and modulation in patients with severe MCA infarction. Neurology 48: 762–767
10. Verlooy J, Heytens L, Veeckmans G, Selosse P (1995) Intracerebral temperature monitoring in severely head injured patients. Acta Neurochir (Wien) 134: 76–78

Correspondence: Atsushi Sakurai, Department of Emergency and Critical Care Medicine, Nihon University School of Medicine, 30-1 Oyaguchi-Kamimachi, Itabashi-ku, Tokyo, 173-8610, Japan. e-mail: sakurai@med.nihon-u.ac.jp

Imaging/Monitoring

Intracranial pressure monitoring: modeling cerebrovascular pressure transmission

M. L. Daley[1], C. W. Leffler[2], M. Czosnyka[3], and J. D. Pickard[3]

[1] Department of Electrical and Computer Engineering, University of Memphis, Memphis, Tennessee, USA
[2] Department of Physiology, University of Tennessee Health Science Center, Memphis, Tennessee, USA
[3] Academic Neurosurgical Unit, Addenbrooke's Hospital, Cambridge, UK

Summary

Objectives. To examine changes in cerebrovascular pressure transmission derived from arterial blood pressure (ABP) and intracranial pressure (ICP) recordings by autoregressive moving average modeling technique.

Methods. Digitized ICP and ABP recordings were obtained from patients with brain injury. Two groups were defined: Group A with 4 patients who demonstrated plateau waves, and Group B with 4 intracranial hypertensive, hypoperfused patients. For each 16.5 s interval, mean values of ICP, ABP, cerebral perfusion pressure (CPP), and corresponding highest modal frequency (HMF) of cerebrovascular pressure transmission were computed.

Results. Mean values of CPP and HMF of 56.2 mmHg and 2.0 Hz for Group A were significantly higher (p < 0.005) than corresponding mean values of 31.9 mmHg and 0.744 Hz for Group B. The mean value of the slope of the regression line between HMF and CPP for group A of −0.034 Hz/mmHg was significantly different (p < 0.025) than the mean value of 0.0077 Hz/mmHg for Group B. Computations of HMF, pressure reactivity, and correlation pressure reactivity index on continuous pressure recordings are illustrated.

Conclusions. Values of HMF of cerebrovascular pressure transmission are inversely related to CPP when pressure regulation is thought to be intact, and directly related when regulation is likely lost.

Keywords: Plateau waves; cerebrovascular pressure transmission; highest modal frequency.

Introduction

Protocols for the intensive care management of patients with traumatic brain injury such as controlled cerebral perfusion pressure (CPP) [8], controlled intracranial pressure (ICP) [6], or the Lund therapy [1] are substantially different. However, a recent retrospective study comparing the CPP- and ICP-oriented protocols indicates better outcomes occur when the former protocol is applied during active pressure regulation of cerebral blood flow and the latter protocol is more appropriate when pressure regulation of cerebral blood flow is impaired [5]. Given the lack of clinically practical methods to continuously evaluate pressure regulation, trauma centers generally have no alternative but to follow the approach that one protocol fits all.

Indirect methods designed to assess regulation of cerebral blood flow continuously based on analysis of the readily-available arterial blood pressure (ABP) and ICP recordings have been proposed [3, 5]. A recent laboratory study has shown that numerical analysis system identification techniques applied to recordings of ABP and ICP can be used to derive a mathematical model of cerebrovascular pressure transmission [4]. Furthermore, during pressure regulation of cerebral blood flow, the highest modal frequency (HMF) of cerebrovascular pressure transmission is inversely related to CPP [4]. In contrast, following brain injury and during loss of pressure regulation, the HMF varies directly with CPP [4].

In our current study we used this same mathematical technique to examine clinical pressure recordings obtained from 2 groups of patients. One group consisted of patients with plateau waves. For this group it was assumed that partial pressure regulation was intact before and following a wave. The other group consisted of 4 patients with severe intracranial hypertension and hypoperfusion. All patients in this latter group were assumed to have impaired pressure regulation.

The purpose of this study was to determine the relationship between HMF and CPP for each group to test the hypothesis that during active pressure regulation, the relationship between HMF and CPP is an inverse one, whereas during loss of pressure regulation the HMF varies directly with CPP.

Materials and methods

Patients

Measurement of ICP and ABP is standard monitoring in the management of severely head injured young adult patients in Addenbrooke's Hospital, Cambridge, United Kingdom. Each set of recordings was labeled in such a manner that the patients studied here could not be identified by the Memphis investigators. The Institutional Review Board at The University of Memphis approved the data analysis protocol. Two groups were defined: Group A consisting of 4 patients who demonstrated a plateau wave, and Group B consisting of 4 patients with intracranial hypertension and hypoperfusion. Pressure recordings over a 6.6 hour period were obtained from 1 additional patient.

Pressure recording analysis

Monitoring included invasive ABP from the radial or dorsalis pedis artery. ICP was monitored using an intraparenchymal probe (Camino ICP transducer, Integra LifeSciences Corp., Plainsboro, NY, or Codman ICP MicroSensors, Codman & Shurtleff, Inc., Raynham, MA). ICP and ABP recordings were sampled at 30 Hz. For each 16.5 s interval, mean values of ICP, ABP, CPP, and the corresponding HMF of cerebrovascular pressure transmission were computed. The length of the pressure recordings obtained from each patient in Groups A and B varied from 16.2 minutes to 117 minutes. Because the aim of this study was to compare the results of the mathematical modeling technique derived from pressure recordings obtained during both active pressure regulation and loss of pressure regulation, analysis of pressure recordings from Group A was done on segments before and following a wave. To reconstruct the sampled recordings to a more accurate description of the actual pressure recordings, a moving average filter was applied to every 2 sample values of the sampled recordings.

Numerical methods

The details of the numerical analysis have been described previously. In brief, the mathematical structure of the model used for identification modeling of cerebrovascular pressure transmission by the autoregressive moving average (ARMAX) is based on a Windkessel model of ICP dynamics that has been successfully used to interpret bedside tests of cerebrovascular autoregulation [2] and is of the form:

$$Y((n+3)T) + a_2 Y((n+2)T) + a_1 Y((n+1)T) + a_0 Y(nT)$$
$$= b_1 U(nT) + b_2 U((n+1)T) + b_3 U((n+2)T). \quad (1)$$

Specifically, in the above equation, $Y(nT)$ and $U(nT)$ represent ICP and ABP, respectively, and T represents the sampling epoch of 33 ms. For each 500 paired samples of pressure values representing 16.5 s segments, the autoregressive moving average (ARMAX) numerical technique was applied using MATLAB System Identification Toolbox software (The MathWorks, Inc., Natick, MA) to obtain the minimum least square error set of constants a_0, a_1, a_2 and b_1, b_2, b_3 for equation (1) above. These constants are used to derive the equivalent continuous-time differential equation description of cerebrovascular pressure transmission of the form:

$$d^3 Y(t)/dt^3 + aa_2 d^2 Y(t)/dt^2 + aa_1 dY(t)/dt + aa_0 Y(t)$$
$$= bb_1 U(t) + bb_2 dU(t)/dt + bb_3 d^2 U(t)/dt^2. \quad (2)$$

The eigenvalues of this differential equation are the modal radian frequencies of the cerebrovascular pressure transmission and are the roots of the polynomial equation:

$$z^3 + aa_2 z^2 + aa_1 z^1 + aa_0 = 0. \quad (3)$$

The highest modal radian frequency is defined as the eigenvalue with the greatest absolute value and is converted to modal frequency by division by 2*pi.

The ABP recording obtained from either the radial artery or dorsalis pedis artery does not have the same time relationship to the ICP recording as does an ABP recording obtained from one of the major arteries entering the intracranial subarachnoid space. To account for the time difference between recordings, the computation is done on 10 paired recordings. These paired recordings are obtained by shifting the start of the 16.5 s ICP recording in steps of 33 ms over a range of ± 165 ms relative to the same 16.5 s ABP recording. As a result, 10 sets of constants each derived by the ARMAX technique were determined, and the best fit set which produced minimum least square error was selected to describe the cerebrovascular pressure transmission between ABP and ICP and the corresponding value of HMF.

For the patient monitored continuously over a 6.6 hour period, a continuous computation of the correlation pressure reactivity index (PrX) was computed as the correlation of 40 consecutive 5 s averaged samples of ICP and ABP. In addition, the pressure reactivity index (PRI) defined as the slope of the regression line of the relationship between mean ICP and mean ABP was determined from 11 consecutive samples of mean ICP and mean ABP with each value derived over a 16.5 s interval. The slope parameter of the HMF/CPP was similarly determined by 11 consecutive samples of mean HMF and mean CPP with each value derived over a 16.5 s interval.

Results

Examples of ICP and ABP recordings from patients in Groups A and B and the corresponding changes in the HMF of cerebrovascular pressure transmission are shown (Fig. 1). During the 26-minute monitoring period, patient A demonstrated an apparent plateau wave (Fig. 1a). The corresponding relationship between the HMF and CPP is also shown (Fig. 1c). Both the computed correlation coefficient and slope of the regression line values for this relationship were negative at -0.29 and -0.061 Hz/mmHg, respectively. During the 130-minute monitoring period, patient B demonstrated a marked increase in ICP and an increase in ABP. The corresponding relationship between the HMF and CPP is also shown (Fig. 1f). The computed correlation coefficient and slope of the regression line values for this were positive at 0.831 and 0.0151 Hz/mmHg, respectively. Grand mean values ($\pm$SD) of ABP, ICP, CPP, HMF, and the slope of the regression line and correlation coefficient of the relationship between HMF and CPP of patients and with the degree of significance between the corresponding means for Groups A and B are given in Table 1.

To illustrate the use of indirect continuous methods to assess pressure regulation of cerebral blood flow, paired recordings obtained over a 6.6 hour period were evaluated using the slope parameter of regression lines of the relationships between HMF and CPP, the PRI, and the PrX. During the monitoring period, ICP, ABP, and CPP remained stable with mean values ($\pm$SD) of 21.2 ($\pm$4.8) mmHg, 92.9 ($\pm$22.7) mmHg, and 71.6 ($\pm$21.8) mmHg; however, brief artifacts, likely due to instrumentation or patient manipulation, can be noted (Fig. 2).

Discussion

In this study, the ARMAX modeling technique was used on clinical recordings obtained from 2 distinct patient groups. Group A consisted of patients who demonstrated plateau waves. Because the onset and termination of a plateau wave are thought to relate to active vasodilation and vasoconstriction [7], the recording intervals just prior and following the wave were used for analysis and assumed to be obtained during active pressure regulation of cerebral blood flow. For this group, values of HMF were found to decrease with increasing CPP. In contrast, the severely intracranial hypertensive, hypoperfused patients in Group B with mean values ($\pm$SD) of ICP and CPP of 58.3 ($\pm$12.6) mmHg and 31.9 ($\pm$16.1) mmHg, respectively, were assumed to have impaired pressure regulation. These patients demonstrated a direct relationship between HMF and CPP (Table 1). These findings of the use of the numerical modeling technique of cerebrovascular pressure transmission support the conclusion of a recent laboratory study that during active pressure regulation of cerebral blood flow the relationship between HMF and CPP is an inverse one; during impaired pressure regulation a direct relationship exists between HMF and CPP [4].

This modeling technique is based on a functional physiological compatible parameter model of a dynamic equilibrium between cerebrospinal fluid volume

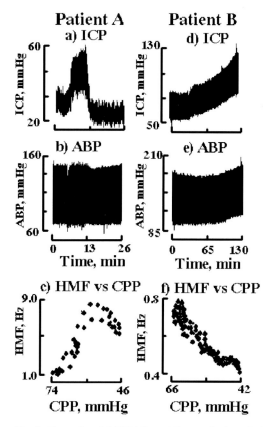

Fig. 1. Examples of ARMAX modeling method applied to clinical pressure recordings. (a) ICP recording from patient with assumed active pressure regulation in Group A. (b) ABP recording corresponding to (a). (c) Derived values of HMF from recordings in (a and b) plotted against corresponding values of CPP. Values of HMF decrease with increasing CPP. (d) ICP recording from patient with assumed impaired pressure regulation in Group B. (e) ABP recording corresponding to (d). (f) Derived values of HMF from recordings in (d and e) plotted against corresponding values of CPP. Values of HMF increase with increasing CPP

Table 1. Grand mean ($\pm$SD) of ABP, ICP, CPP, HMF, regression line slope, and correlation coefficient between HMF and CPP for patient groups A and B

Patient Group	ABP (mmHg)	ICP (mmHg)	CPP (mmHg)	HMF (Hz)	Slope Hz/mmHg	Correlation coefficient
A (n = 4)	94.4 ($\pm$3.4)	29.6 ($\pm$4.9)	56.2 ($\pm$6.5)	2.0 ($\pm$0.20)	−0.034 ($\pm$0.02)	−0.35 ($\pm$0.24)
p <[1]	n.s.	.01	.05	.005	.025	.05
B (n = 4)	86.2 ($\pm$29.9)	58.3 ($\pm$12.6)	31.9 ($\pm$16.1)	0.74 ($\pm$0.22)	0.0077 ($\pm$0.008)	0.24 ($\pm$0.40)

[1] Level of significance between grand means was determined using Student *t*-test. *ABP* arterial blood pressure, *CPP* cerebral perfusion pressure, *HMF* highest modal frequency, *ICP* intracerebral hemorrhage.

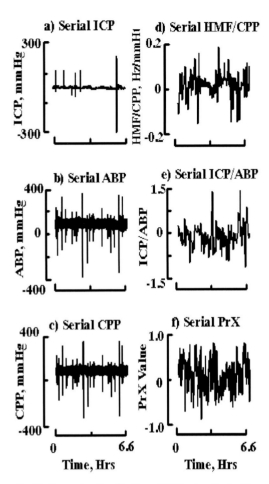

Fig. 2. Example of application of indirect methods of assessment of pressure regulation of cerebral blood flow on continuous 6.6 hour recording. (a) Continuous recording of ICP with a mean value ($\pm$SD) of 21.2 ($\pm$4.8) mmHg. Marked pressure peaks are apparent artifacts and not physiological. (b) Corresponding continuous recording of ABP with a mean value ($\pm$SD) of 92.9 ($\pm$22.7) mmHg. Marked pressure peaks are apparent artifacts and not physiological. (c) Computed recording of CPP derived as the difference between ABP and ICP with mean value of 71.6 ($\pm$21.8) mmHg. (d) Serial HMF/CPP. Serial values of the slope of the regression line of the relationship between HMF and CPP computed from 11 consecutive paired values of HMF and CPP representing 182 minute intervals are plotted. These slope parameter values were greater than zero 60% of the time. (e) Serial ICP/ABP. Serial values of the slope of regression line of the relationship between ICP and ABP computed from 11 consecutive paired values of HMF and CPP are plotted. These slope parameter values were positive 49% of the time. (f) Serial PrX. Serial values of the correlation index computed on 40 consecutive 5-minute averages of ICP and ABP recordings are plotted and found to be positive 68% of the time

and cerebral blood volume [2]. Given this model, the analytical description of cerebrovascular pressure transmission is expressed as the third order differential equation (see equation 3). As a result, it is possible to interpret changes in modes of cerebrovascular pressure transmission, particularly the HMF, in terms of changes in the resistance and compliance of the arterial-arteriolar bed, cerebral venous compliance, and intracranial compliance. During pressure regulation of cerebral blood flow, increases in CPP result in active vasoconstriction that causes an increase in resistance of the arteriolar bed, cerebral venous compliance, and intracranial compliance. When these parameters of the model are increased, HMF decreases systematically. Thus, the inverse relationship between HMF and CPP demonstrated by the patients in Group A and by the model of cerebrovascular pressure transmission is consistent with active vasoconstriction and pressure regulation. Conversely, during impaired pressure regulation, increases in CPP result in a passive vasodilation that causes decreased resistance of the arteriolar bed, cerebral venous compliance, and intracranial compliance. When these parameters of the model are decreased, HMF increases systematically. Thus, the direct relationship between HMF and CPP demonstrated by the patients in Group A and by the model is passive vasodilation and impaired regulation. Because monitoring the HMF of cerebrovascular pressure transmissions may link directly to physiological properties of the cerebrovasculature, it can be used to support pure black-box indirect assessment such as the PRI and PrX indices. As shown in Figure 2, the values of HMF, PRI, and PrX were positive during the 6.6 hour monitoring period 60%, 49%, and 68% of the time, indicating that during the monitoring period the patient had impaired pressure regulation of cerebral blood flow.

In conclusion, the findings of this study support the premise that when the HMF varies indirectly with CPP, pressure regulation of cerebral blood flow is intact; whereas, when the HMF varies directly with CPP, pressure regulation is impaired. Because the HMF is a mode of cerebrovascular pressure transmission based on a physiological compatible model of ICP dynamics, changes in this parameter can be related to changes in arterial-arteriolar resistance, vascular compliances, and intracranial compliance. Future studies will be designed to examine the changes of the other two modes of the model during active and impaired pressure regulation.

Acknowledgments

This project was partially supported by the UK Government Technology Foresight Initiative and the Medical Research Council (Grant No. G9439390 ID 65883) (MC and JDP). This research was also supported in part by the National Heart, Lung, and Blood Institute, National Institutes of Health, USA.

References

1. Asgeirsson B, Grande PO, Nordstrom CH (1994) A new therapy of post-trauma brain oedema based on haemodynamic principles for brain volume regulation. Intensive Care Med 20: 260–267
2. Czosnyka M, Piechnik S, Richards HK, Kirkpatrick P, Smielewski P, Pickard JD (1997) Contribution of mathematical modeling to the interpretation of bedside tests of cerebrovascular autoregulation. J Neurol Neurosurg Psychiatry 63: 721–731
3. Czosnyka M, Smielewski P, Kirkpatrick P, Laing RJ, Menon D, Pickard JD (1997) Continuous assessment of the cerebral vasomotor reactivity in head injury. Neurosurgery 41: 11–17; discussion 17–19
4. Daley ML, Pourcyrous M, Timmons SD, Leffler CW (2004) Assessment of cerebrovascular autoregulation: changes of highest modal frequency of cerebrovascular pressure transmission with cerebral perfusion pressure. Stroke 35: 1952–1956
5. Howells T, Elf K, Jones PA, Ronne-Engstrom E, Piper I, Nilsson P, Andrews P, Enblad P (2005) Pressure reactivity as a guide in the treatment of cerebral perfusion pressure in patients with brain trauma. J Neurosurg 102: 311–317
6. Juul N, Morris GF, Marshall SB, Marshall LF (2000) Intracranial hypertension and cerebral perfusion pressure: influence on neurological deterioration and outcome in severe head injury. The Executive Committee of the International Selfotel Trial. J Neurosurg 92: 1–6
7. Rosner MJ, Becker DP (1984) Origin and evolution of plateau waves. Experimental observations and a theoretical model. J Neurosurg 60: 312–324
8. Rosner MJ, Rosner SD, Johnson AH (1995) Cerebral perfusion pressure: management protocol and clinical results. J Neurosurg 83: 949–962

Correspondence: Michael L. Daley, Department of Electrical and Computer Engineering, The University of Memphis, Engineering Science Building, Rm. 208B, Memphis, TN 38152-3180, USA. e-mail: mdaley@memphis.edu

Use of ICM+ software for on-line analysis of intracranial and arterial pressures in head-injured patients

K. Guendling[1,2], P. Smielewski[1], M. Czosnyka[1], P. Lewis[3], J. Nortje[4], I. Timofeev[1], P. J. Hutchinson[1], and J. D. Pickard[1]

[1] Academic Department of Clinical Neuroscience, Addenbrooke's Hospital, Cambridge, UK
[2] Department of Neurosurgery, University of Giessen, Giessen, Germany
[3] Prince Alfred Hospital, Melbourne, Australia
[4] Department of Anesthesiology, Addenbrooke's Hospital, Cambridge, UK

Summary

Objective. To summarize our experience from the first 2 years of use of the ICM+ software in our Neurocritical Care Unit (NCCU).

Materials and methods. Ninety-five head-injured patients (74 males, 21 females), average age 36 years, were managed in the NCCU. Intracranial pressure (ICP) was monitored using Codman intraparenchymal probes and arterial blood pressure (ABP) was measured from the radial artery. Signals were monitored by ICM+ software calculating mean values of ICP, ABP, cerebral perfusion pressure (CPP) and various indices describing pressure reactivity, compensation and vascular waveforms of ICP (pulse amplitude, respiratory, and slow waves), etc.

Results. Mean ICP was 17 mmHg, mean CPP was 73 mmHg. Seven patients showed permanent disturbance of cerebral autoregulation (mean pressure reactivity index above 0.3). Pressure reactivity index demonstrated significant U-shape relationship with CPP, suggesting loss of pressure reactivity at too low (CPP < 55 mmHg) and too high CPPs (CPP > 95 mmHg). Mean ICP was inversely correlated with respiratory rate (R = 0.46; p < 0.0001; reciprocal model).

Conclusion. The new version of ICM+ software proved to be useful clinically in the NCCU. It allows continuous monitoring of pressure reactivity and exploratory analysis of factors implicating intracranial hypertension.

Keywords: Head injury; intracranial pressure; cerebral perfusion pressure; autoregulation; computer monitoring.

Introduction

Major improvements in outcome after traumatic brain injury (TBI) over the past 20 years have been achieved, not only because of new drugs and therapies, but also through the identification and monitoring of secondary brain insults [12]. Within the first few days after injury, vital mechanisms to match the blood supply of the brain to its energy demands are often impaired [7, 9]. This leaves the brain vulnerable to relatively minor insults such as transient falls in arterial blood pressure (ABP) or arterial oxygen saturation with profound effects on outcome. The new generation of brain sensors (intracranial pressure [ICP], tissue oxygen, pH, temperature, intracerebral microdialysis, transcranial Doppler, laser Doppler, near-infrared spectroscopy) enables on-line acquisition of vital information on cerebral metabolism and blood supply and function, enabling clinicians to recognize secondary insults and optimize management [10, 11]. However, to process and analyze on-line the large amount of data generated continuously by bedside monitors in order to facilitate decision-making is a complex task. The problems of data filtration, integration, appropriate analysis and prognostic interpretation await a satisfactory solution, which has been only partly addressed in previous studies [4, 8, 13].

The first specialized computer-based systems for neurointensive care were introduced in the early 1970's. Initially, these systems were aimed at monitoring ICP and ABP, allowing calculation of CPP and a basic analysis of the pulsatile ICP waveform. In contrast, contemporary systems are sophisticated, multichannel, digital trend recorders with built-in options for complex signal processing [13].

The intensive care multimodality monitoring system adopted in our Cambridge Neurosurgical Unit is based on software for the standard IBM-compatible personal computer, equipped with a digital to analogue converter and RS232 serial interface. The first version of the software was introduced into clinical practice in Poland, Denmark, and the United Kingdom in the middle 1980's and has been extended into a system for multimodal neuro-intensive care monitoring (ICM) and waveform analysis ICP [4] used in Cambridge, UK, and other centers in Europe and the United States. Most data has been derived from head-injured [5, 14] and hydrocephalus patients [1, 6]. However, the same or similar techniques are being increasingly applied to those suffering from severe stroke, subarachnoid hemorrhage, cerebral infections, encephalopathy, liver failure, idiopathic intracranial hypertension.

Over past 2 years, new software called ICM+, has been introduced into clinical practice. We summarize our clinical experience from use of the software in the management of patients after TBI.

Materials and methods

Following head injuries of various etiology, 95 patients (74 males, 21 females) with an average age of 36 years were managed in the Neurocritical Care Unit (NCCU) between 2003 and 2005. Median admission Glasgow Coma Scale (GCS) score was 6 (range 3 to 14), with 23% of patients having a GCS of 9 or greater, but deteriorating later. ICP was monitored using Codman intraparenchymal probes and ABP was measured from a peripheral artery. Signals were captured by personal computers running ICM+ software, calculating mean values of ICP, ABP, cerebral perfusion pressure (CPP) and various indices describing pressure reactivity, compensation, and vascular waveforms of ICP.

The software reads analog signals through the analog-to-digital converter (Data Translation 9800 USB box) with the sampling frequency of 50 Hz. Data were processed (processing is fully programmable) and, for TBI patients, configuration was set to calculate average values of the following variables every minute:

ABP	mean arterial blood pressure
aABP	pulse amplitude of arterial blood pressure
AMP	pulse amplitude of ICP waveform
CPP	mean cerebral perfusion pressure
HR	mean heart rate
ICP	mean intracranial pressure
PRx	pressure reactivity index
RAC	index describing moving correlation between pulse waveform of ICP and mean CPP
RAP	index describing pressure-volume compensatory reserve
Resp	amplitude of respiratory waveform
RespRate	mean respiratory rate
Slow	slow waves of ICP (equivalent time periods from 20 seconds to 3 minutes)

Particular care was paid to organization of the front page display, which shows time trends of ABP, ICP, CPP, and pressure reactivity. For clarity, the pressure reactivity index (PRx) is also displayed at the bottom of the screen as risk-graph, converting information about reactivity to colors: green = good; red = impaired (Fig. 1). Optimal CPP [14] was also calculated on-line (Fig. 2).

Results

Artifact-free time of signal recording, which provided good quality output suitable for data analysis, was around 92% of the time our patients spent in the NCCU.

Averaged values of monitored variables and their standard deviations are given in Table 1. PRx plotted as a function of CPP showed a significant relationship (ANOVA: $p < 0.023$) indicating loss of cerebral pressure-reactivity for low CPP (CPP < 55 mmHg) and for high CPPs (CPP > 95 mmHg). This averaged trend (Fig. 2b) emphasizes the validity of individual methodology for tracing optimal CPP (Fig. 2a) using finite-time (3 to 6 hours) plots of PRx versus CPP.

In total, 10% of patients had permanently impaired cerebral autoregulation (PRx > 0.2). The software was helpful in analyzing ICP respiratory waves. Respiratory amplitude was not associated with any other parameters characterizing intracranial hypertension or cerebral perfusion. But, respiratory rate demonstrated a strong inverse relationship to ICP ($R = 0.46$; $p < 0.0001$) indicating that lower mean ICP is present with higher frequency ventilation (20 to 25 cycles per minute).

Discussion

In the established environment of a clinical neuroscience department, enormous quantities of data can be captured, from which information regarding cerebral autoregulation, cerebrospinal compensatory reserve, oxygenation, metabolite production and function can be obtained. Recognition of changing cerebrovascular hemodynamics and oxygenation demands can result from reliable monitoring techniques, as well as sophisticated and time-consuming signal analysis provided by dedicated computer support [4, 8, 13].

The flexibility of such systems allows wide-range signal analysis, which can generate data chaos. Thus, the modern user must decide which parameters should be considered and how the data should be interpreted. This information must be presented in a manner that is comprehensible to medical and nursing staff. Although personal computers with designated software are portable, they have yet to gain widespread clinical acceptance as an intensive care tool. They are seen as stand-alone instruments requiring specialized skills for their operation, and occupying precious space. In contrast, a commercial hardware system with a cus-

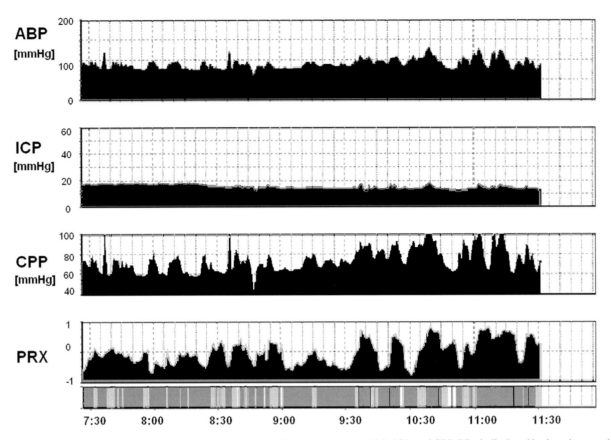

Fig. 1. Example of the 'front page' of the software. Standard display shows mean ABP, ICP, and CPP. PRx is displayed both as time trend and color graph (here reduced to grey-scale). White indicates good autoregulation and black indicates poor pressure-reactivity. In this example, ICP was quite stable but PRx deteriorated in the second half of the recording, probably due to elevated CPP

tomized console can be more user-friendly, but less flexible and more expensive.

PRx – global index of cerebrovascular reactivity

Useful ICP-derived variable is the PRx, based on assessing the response of ICP to spontaneous fluctuations in ABP [5]. Using computational methods, PRx is determined by calculating the correlation coefficient between 40 consecutive, time-averaged data points (over 6- to 10-second periods) of ICP and ABP. A positive PRx signifies a positive gradient of the regression line between the slow components of ABP and ICP, which has been shown to be associated with passive behavior of a non-reactive vascular bed. A negative value of PRx reflects normal reactive cerebral vessels, as ABP waves provoke inversely correlated waves in ICP.

Earlier work has shown a correlation between PRx and outcome [14]. Our results show that the PRx/CPP plot replicates a U-shape curve often seen in individual cases: pressure reactivity is disturbed by both too low and too high CPPs.

Optimization of CPP

Many attempts have been made to find an optimal value for CPP; however, there is no method available currently that is accurate enough to be clinically useful.

In a group of retrospectively-evaluated patients, the greater the distance between current and "optimal" CPP, the worse the outcome [14]. This potentially useful parameter attempts to refine CPP-directed therapy. Both, too low CPP (indicating ischemia) and too high CPP (indicating hyperemia) are detrimental. Hence, it has been suggested that CPP should be optimized online to maintain cerebral perfusion in the most favorable state (Fig. 2).

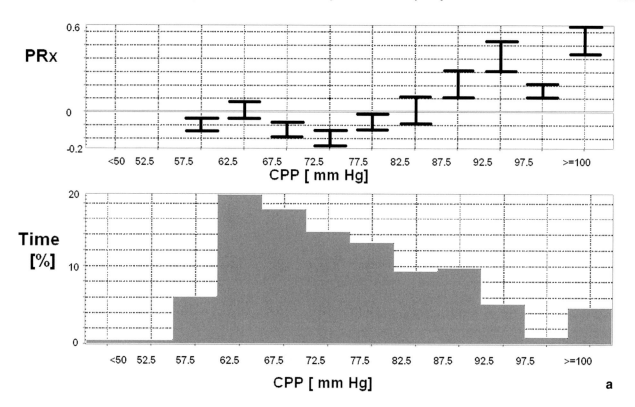

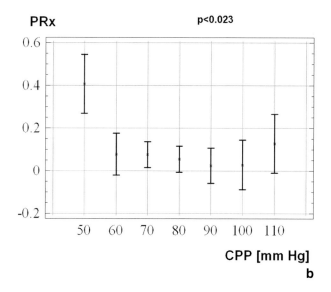

Fig. 2. (a) Individual plot of PRx versus CPP (upper panel) and histogram of CPP, used to analyze on-line optimal CPP. This is a value of CPP corresponding to the lowest PRx (i.e., the best PRx) updated on-line using past 3-hour trends of PRx and CPP. (b) Averaged plot of PRx versus CPP in a group of 95 patients. The plot replicates U-shape often seen in individual cases: PRx is disturbed by too low and too high CPP (ANOVA; $p < 0.023$)

Relationship between ICP and respiratory rate

Head trauma is a significant cause of death and disability, especially in young males, and is associated with raised ICP. Raised ICP is defined as pressure greater than 20 mmHg, and appears most commonly in about 50 to 75% of patients with serve head injury who remain comatose after resuscitation [3, 11]. In the past, raised ICP has been found to be associated with a poorer outcome from injury. Higher ICP, particularly higher peak ICP levels, correlate with mortality and morbidity [2, 12].

It is difficult to establish a universal "normal value" for ICP because it depends on age, body position, and

Table 1. *Mean values and standard deviations of the monitored variables.*

Variables	Units	Average	Standard deviation
ABP (arterial blood pressure)	mmHg	90.7	17.1
aABP (pulse amplitude of ABP)	mmHg	20.1	6.2
AMP (pulse amplitude of ICP waveform)	mmHg	4.78	7.98
CPP (cerebral perfusion pressure)	mmHg	73.4	20.8
HR (heart rate)	Beats/minute	72.8	17.5
ICP (intracranial pressure)	mmHg	17.3	20.3
PRx (pressure reactivity index)		0.0558	0.1639
RAC (index describing moving correlation between pulse waveform ICP and mean CPP)		1.11	9.22
RAP (index describing pressure-volume compensator reserve)		0.148	0.174
Resp (amplitude of respiratory waveform)	mmHg	0.515	0.450
RespRate (respiratory rate)	Cycles/minute	14.44	3.39
Slow (slow waves of ICP)	mmHg	5.97	7.28

clinical conditions. Our results show an inverse relationship between respiratory rate and mean ICP, i.e., a high respiratory rate signifies low ICP.

However, several questions need to be addressed to determine the mechanistic and clinical significance of this relationship. While the association of respiratory rate and ICP is intriguing, these data do not provide a complete explanation for the underlying pathophysiology, because higher frequency ventilation may control intracranial hypertension. And, higher respiratory rate leads to a lower arterial CO_2 with vasoconstriction of the cerebral blood vessels, causing ICP to fall.

Optional diagnostic tools

Program monitors gather input variables in the pre-programmed manner and analyzes them according to the programmed configuration, saving the output data in 2 separate files. First file contains time trends of the analyzed signals, and all calculations as well as comments and remarks introduced during monitoring. The second file contains raw data (input signals defined by the user). This file may be viewed and analyzed using various spectral analysis methods, or processed directly on-line.

Software also enables off-line data analysis, including files from the old ICM software, text files, and cerebrovascular laboratory software known also as BioSAn, i.e. 'text files'.

The software enables modification of existing methods of brain monitoring and development of new algorithms through extensive programming of signal analysis. It aids the integration of physiological monitoring with clinical observations.

In addition to continuous assessment provided by the time trends of indices, it is sometimes necessary to introduce external excitation to the measured system and to quantify its response. Examples are an increase in the ventilator rate to induce a change in arterial CO_2 content, brief compression of the common carotid artery to induce a momentary drop in CPP, or controlled infusion of saline into the cerebrospinal fluid space in order to challenge the compensatory reserve. Such an intervention provides an opportunity for more accurate assessment of the queried system characteristics than analysis of spontaneous fluctuations originating from it. On-line tools available to assess these diagnostic tests help gain additional insight into the developing pathology as well as allow for cross-calibration of continuous time trends.

Conclusion

ICM+ software proved to be useful in the management of patients after TBI. The study revealed that respiratory rate and mean ICP were inversely related. Therefore, higher frequency ventilation may be helpful to control intracranial hypertension.

Acknowledgments

This project was supported by the UK Government Technology Foresight Initiative, and the Medical Research Council (Grant No G9439390 ID 65883).

The authors are indebted to all the team participating in data collection: Mrs. Pippa Al-Rawi, Mrs. Helen Seeley, Mrs. Carole Turner, Mrs. Colette O'Kane, Mrs. Shirley Love, Mrs. Diana Simpson, Dr. Eric Schmidt, Dr. Stefan Piechnik, Dr. Andreas Raabe, Mr. Eric Guazzo, Dr. David Menon, Dr. Arun Gupta, Mr. Peter Kirkpatrick, Mr. Rupert Kett-White, Mr. Pwawanjit Minhas, Mr. Rodney Laing, and all nursing and research staff of NCCU.

P. J. Hutchinson is supported by an Academy of Medical Sciences PPP Foundation Senior Surgical Scientist Fellowship. M. Czosnyka is on unpaid leave from Warsaw University of Technology, Poland.

ICM+ software is licensed by University of Cambridge, Cambridge Enterprise (www.nerosurg.cam.ac.uk/icmplus). PS and MC have a financial interest in a fraction of the licensing fee.

References

1. Bech-Azeddine R, Gjerris F, Waldemar G, Czosnyka M, Juhler M (2005) Intraventricular or lumbar infusion test in adult communicating hydrocephalus? Practical consequences and clinical outcome of shunt operation. Acta Neurochir (Wien) 147: 1027–1036
2. Chambers IR, Treadwell L, Mendelow AD (2001) Determination of threshold levels of cerebral perfusion pressure and intracranial pressure in severe head injury by using receiver-operating characteristic curves: an observational study in 291 patients. J Neurosurg 94: 412–416
3. Chesnut RM, Marshall LF, Klauber MR, Blunt BA, Baldwin N, Eisenberg HM, Jane JA, Marmarou A, Foulkes MA (1993) The role of secondary brain injury in determining outcome from severe head injury. J Trauma 34: 216–222
4. Czosnyka M, Whitehouse H, Smielewski P, Kirkpatrick P, Guazzo EP, Pickard JD (1994) Computer supported multimodal bed-side monitoring in neuro intensive care. Int J Clin Monit Comput 11: 223–232
5. Czosnyka M, Smielewski P, Piechnik S, Steiner LA, Pickard JD (2001) Cerebral autoregulation following head injury. J Neurosurg 95: 756–763
6. Czosnyka M, Czosnyka Z, Momjian S, Pickard JD (2004) Cerebrospinal fluid dynamics. Physiol Meas 25: R51–R76
7. Ginsberg MD, Zhao W, Alonso OF, Loor-Estades JY, Dietrich WD, Busto R (1997) Uncoupling of local cerebral glucose metabolism and blood flow after acute fluid-percussion injury in rats. Am J Physiol 272: H2859–H2868
8. Howells T, Elf K, Jones PA, Ronne-Engstrom E, Piper I, Nilsson P, Andrews P, Enblad P (2005) Pressure reactivity as a guide in the treatment of cerebral perfusion pressure in patients with brain trauma. J Neurosurg 102: 311–317
9. Kelly DF, Martin NA, Kordestani R, Counelis G, Hovda DA, Bergsneider M, McBride DQ, Shalmon E, Herman D, Becker DP (1997) Cerebral blood flow as a predictor of outcome following traumatic brain injury. J Neurosurg 86: 633–641
10. Kett-White R, Hutchinson PJ, Czosnyka M, Boniface S, Pickard JD, Kirkpatrick PJ (2002) Multi-modal monitoring of acute brain injury. Adv Tech Stand Neurosurg 27: 87–134
11. Kirkpatrick PJ, Czosnyka M, Pickard JD (1996) Multimodal monitoring in neuro-intensive care. J Neurol Neurosurg Psychiatry 80: 131–139
12. Miller JD, Becker DP, Ward JD, Sullivan HG, Adams WE, Rosner MJ (1977) Significance of intracranial hypertension in severe head injury. J Neurosurg 47: 503–516
13. Smielewski P, Czosnyka M, Zabolotny W, Kirkpatrick P, Richards HK, Pickard JD (1997) A computing system for the clinical and experimental investigation of cerebrovascular reactivity. Int J Clin Monit Comput 14: 185–198
14. Steiner LA, Czosnyka M, Piechnik SK, Smielewski P, Chatfield D, Menon DK, Pickard JD (2002) Continuous monitoring of cerebrovascular pressure reactivity allows determination of optimal cerebral perfusion pressure in patients with traumatic brain injury. Crit Care Med 30: 733–738

Correspondence: P. Smielewski, Academic Neurosurgery, Box 167, Addenbrooke's Hospital, Cambridge CB2 2QQ, UK. e-mail: ps10011@medschl.cam.ac.uk

Monitoring and interpretation of intracranial pressure after head injury

M. Czosnyka, P. J. Hutchinson, M. Balestreri, M. Hiler, P. Smielewski, and J. D. Pickard

Department of Clinical Neurosciences, Neurosurgical Unit, University of Cambridge, Addenbrooke's Hospital, Cambridge, UK

Summary

Objective. To investigate the relationships between long-term computer-assisted monitoring of intracranial pressure (ICP) and indices derived from its waveform versus outcome, age, and sex.

Materials and methods. From 1992 to 2002, 429 sedated and ventilated head-injured patients were continuously monitored. ICP and arterial blood pressure (ABP) were recorded directly and stored in bedside computers. Additional calculated variables included: 1) Cerebral perfusion pressure (CPP) = ABP – ICP; 2) a PRx calculated as a moving correlation coefficient between slow waves (of periods from 20 seconds to 3 minutes) of ICP and ABP.

Results. Fatal outcome was associated with higher ICP ($p < 0.000002$), worse PRx ($p < 0.0006$), and lower CPP ($p < 0.001$). None of these parameters differentiated severely disabled patients from patients with a favorable outcome. Higher average ICP, lower CPP, worse outcome, and worse pressure reactivity were observed in females than in males (age-matched). Worse outcome, lower mean ICP, worse PRx, and higher CPP were significantly associated with the older age of patients.

Conclusion. High ICP and low PRx are strongly associated with fatal outcome. There is a considerable heterogeneity amongst patients; optimization of care depends upon observing the time-trends for the individual patient.

Keywords: Head injury; intracranial pressure; pressure reactivity; outcome.

Introduction

Several novel methods of brain monitoring, including brain tissue oxygenation, cerebral microdialysis, and cerebral blood flow, have recently been studied and compared to more traditional techniques. However, intracranial pressure (ICP) and mean arterial blood pressure (ABP) are still gold standards in neurocritical care monitoring. Although there is no class-1 evidence that monitoring of ICP has the potential to improve outcome [7], there is consensus that without this modality the management of severely head injured patients is far from optimal. Cerebral perfusion pressure (CPP) and ICP have become therapeutic targets to prevent potentially life-threatening cerebral hypoperfusion. Therefore, several protocols for the management of acutely head-injured patients are based on a CPP-oriented therapy [9], an ICP-oriented management [6], or a mixture of both [8].

Beginning in September 1991, bedside computer-supported systems were used in our Neurosurgical Critical Care Annexe until 1993, and then in our Neurosciences Critical Care Unit (NCCU) from 1994 onward [3]. Its purpose has been to continuously monitor physiological parameters such as ICP, ABP, and CPP, and pressure-derived indices describing the state of brain homeostasis. The resultant large dataset has been used to examine our 10-year experience with head-injury monitoring. Some particular aspects are summarized, such as the relationship between Glasgow Coma Score (GCS) and outcome, influence of gender [1] and age [4] on outcome, and usefulness of ICP waveform analysis in order to predict complications associated with intracranial hypertension [2].

Patients and methods

This retrospective analysis is based on 492 head-injured patients admitted to the Annexe and NCCU between January 1992 and December 2001. Only patients with invasive monitoring of ICP and ABP over a period greater than 12 hours and connected to a bedside computerized system were included in the study. It is important to emphasize that the studied group is not representative of all admissions to the unit. Patients who were admitted and discharged promptly or died soon after admission are not included in the analysis.

Patients were sedated, mechanically ventilated, and paralyzed in order to maintain ICP below 25 mmHg. Systemic hypotension was treated with fluids and vasoactive drugs. CPP was kept above 60 to 70 mmHg to avoid secondary ischemic insults. Episodes of intracranial hypertension were treated with mild hyperventilation ($PaCO2 > 4.0$ kPa), moderate hypothermia, and boluses of mannitol and thiopentone. An external ventricular drain was inserted when feasible, depending on the size of ventricles on computed tomography scan. In 1997, a standard ICP/CPP-oriented protocol for head injury was introduced, consisting of more aggressive management of intracranial hypertension and stricter control of ABP, aiming to minimizing the detrimental effects of hypoperfusion on brain tissue [8].

Data were analyzed retrospectively as part of the standard clinical audit and no additional intervention was associated with the bedside computer data capture of the monitored variables.

Monitoring and data analysis

ICP was monitored by an intraparenchymal probe (Camino ICP transducer in 12 patients [Integra Neurosciences, Plainsboro, NJ] or Codman ICP MicroSensors in 446 patients [Codman & Shurtleff Inc., Raynham, MA]) or through a ventricular drain and an external pressure transducer (34 cases; Baxter Healthcare Corp., Round Lake, IL) prior to 1994. ABP was monitored invasively. Signals were sampled from the analogue output of the monitors at 30 Hz, digitized (12 bits analogue-to-digital converter), analyzed as 6-second averages, and subsequently converted to 1-minute averages.

From each of the 40 samples of 6-second mean ICP and ABP, a moving correlation was calculated (pressure-reactivity index [PRx]). By averaging such a moving index over a minimum half an hour interval, pressure reactivity could be assessed. A positive correlation between ABP and ICP revealed a passive, non-reactive cerebrovascular bed. A negative correlation is specific for a reactive bed (Fig. 1A). A positive value of PRx has been previously demonstrated to be a strong predictor of fatal outcome following head injury [2, 10].

Apart from the calculation of global averaged values of ICP, ABP, CPP, and PRx, patients were classified as having gross intracranial hypertension if their mean ICP was above 35 mmHg for at least 4 hours continuously. Post-resuscitation GCS was used for analysis. The Glasgow Outcome Score (GOS) was determined at 6 months, either by follow-up clinic or by questionnaire.

Results

Of 492 patients included in the computer-supported monitoring, 429 were suitable for analysis, with adequate quality of the continuous recording of ICP and ABP and reliable outcome follow-up. Mean age was 34 ($\pm$16.7 years) and median admission GCS was 6 (range 3 to 15; 20% of patients had an initial GCS above 8); 21% of the patients were female. Overall, 28% of patients had a good outcome, 21% were moderately disabled, 22% severely disabled, 2% remained in a persistent vegetative state, and 27% died at 6 months.

Impact of ICP, CPP, and pressure reactivity on outcome

Outcome rates were distributed unevenly along the observed range of ICP. Mortality showed a clear breakpoint, increasing from 17% to 47% when averaged ICP increased above 20 mmHg ($p < 0.0001$; the

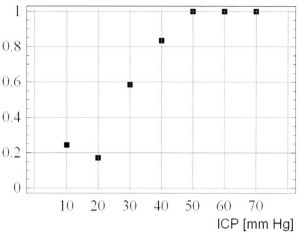

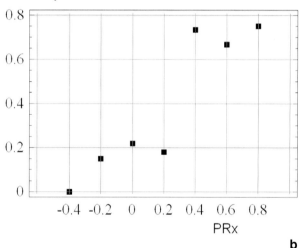

Fig. 1. (a) Mortality rate expressed as a function of intracranial pressure (*ICP*). (b) Mortality rate expressed as a function of pressure-reactivity (*PRx*)

exact threshold of ICP that minimized the p value of difference in mortality rate was 23 mmHg) (Fig. 1). This was mirrored by a decrease in good/moderate outcome rate. Severe disability rate did not show any remarkable changes dependent on ICP.

The mortality rate indicated a threshold rise from 20% to 70% when pressure reactivity deteriorated (averaged PRx increased above 0.3; $p < 0.01$) (Fig. 1). The relationship between CPP and the mortality rate revealed 2 areas where mortality rate increased. For CPP below 55 mmHg mortality was 81%, while above 55 mmHg it was only 23% ($p < 0.0001$). For

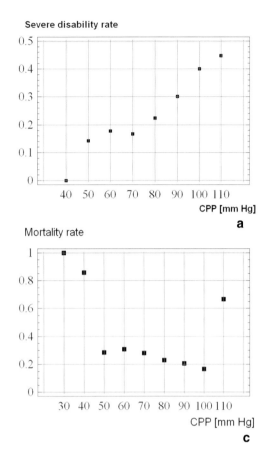
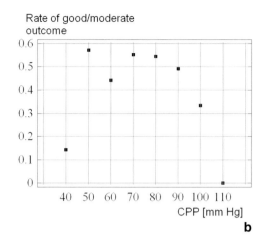

Fig. 2. Outcomes as a function of cerebral perfusion pressure (*CPP*). (a) Mortality/persistent vegetative state. (b) Good/moderate. (c) Severe disability

CPP above 95 mmHg mortality was 50%, while below was 20%, although this difference was not significant.

However, a CPP greater than 95 mmHg had a detrimental effect on good and moderate outcome. For CPP above 95 mmHg the rate of good and moderate outcome was 28%, while below it was 50% ($p < 0.033$).

The severe disability rate showed a tendency to steadily increase with CPP ($r = 0.87$; $p = 0.02$), suggesting either that a greater CPP does not help achieve a favorable outcome, or that these patients had more severe injury requiring greater intensity of treatment (Fig. 2).

Gender-related differences

ICP was significantly greater (median 15.9 mmHg, range 3.6 mmHg to 79 mmHg in males versus median 17.5 mmHg, range 5.7 mmHg to 106 mmHg in females, $p = 0.036$) and CPP significantly lower in females (males median: 76.7 mmHg vs. females 73.6 mmHg; $p = 0.007$). PRx was worse in females than males (females: 0.1 vs. males 0.04; $p = 0.022$).

ABP was not different between the 2 groups. The mortality rate in females was significantly higher (females 35%, males 24%; $p = 0.029$) and rate of favorable outcome was lower (females 40%, males 25%; $p = 0.047$). The median GOS was worse in females (severe disability) than in males (moderate disability).

The rate of gross intracranial hypertension (averaged ICP above 35 mmHg) was greater in females (11%) than males (4.1%; $p = 0.012$). Averaged ICP was compared between both sexes in different outcome groups. Significant differences were only found in patients who died or persisted in a vegetative state: in females, mean ICP was 21.5 (range 9.4 to 106) mmHg in females versus 19.3 (range 3.6 to 75) mmHg in males; $p = 0.036$ (Mann-Whitney). In other outcome groups, ICP was distributed uniformly and reached a median of 16.2 mmHg both in females and males.

Influence of age

Elderly people had a worse outcome after brain trauma; the relationship between GOS and age was significant and positive ($R = 0.301$; $p < 0.0001$). The initial GCS assessment correlated with age (R = 0.14; p < 0.01), indicating that there was a tendency to score better in elderly patients.

Monitored variables appeared to be associated with age: mean ICP had a weak tendency to decrease with age (R = −0.14; p < 0.01), ABP tended to increase with age although insignificantly (p = 0.18), and CPP increased with age (R = 0.19; p = 0.0004). Pressure reactivity indicated worsening of cerebrovascular control with age (PRx: R = 0.24; p = 0.003).

Discussion

Raised ICP determines outcome in terms of life and death. Above 20 to 25 mmHg, the mortality rate increases dramatically. An even more spectacular threshold can be demonstrated when PRx is considered. There is a sudden increase in mortality rate from 20% to 70% when averaged PRx increases above 0.3. Low CPP with aggressive CPP-oriented therapy is seldom an issue. The lower values of CPP observed in patients who died were mainly related to the higher values of ICP, as no significant difference was found between mean values of ABP in the different outcome groups.

The finding that a high CPP (above 95 mmHg) may reduce the rate of favorable outcome was surprising. This may suggest that an excessive increase in ABP to improve brain perfusion may be detrimental. Low ICP and CPP within the range of 50 to 90 mmHg are clinical findings that seem to justify a lower CPP target [6]. Our study is based on material gathered over a long time interval and has not been influenced by one consistent protocol. Our policy has changed a few times over the period, finishing with a mixed CPP- and ICP-oriented protocol [8].

In this group of traumatic brain injury (TBI) patients, females had a significantly greater rate of fatal outcome than males [1]. This was associated with a higher incidence of gross intracranial hypertension in females compared to males. One practical implication may be that aggressive treatment of intracranial hypertension should be administered more readily in young females than in males, as refractory intracranial hypertension is more likely to develop and lead to fatal outcome in females. Indeed, in patients with only moderately elevated global ICP (<25 mmHg), the gender-related difference in the mortality rate becomes non-significant and decreases further when this critical threshold is lowered. If ICP remains normal in TBI patients, the likelihood of a good outcome is comparable for females and males. However, if ICP rises, it appears to be more difficult to control in young females than males. In spite of the postulated neuroprotective role of estrogen in experimental studies, the susceptibility to brain swelling reported in the recently published clinical audit [5] is very likely to be an important factor.

The initial GCS in elderly patients admitted to Addenbrooke's NCCU indicates that the primary injury is usually slightly less severe. Also the post-injury course seems to be more favorable in the elderly from the point of view of brain protection against secondary insults: ICP seems to be lower and CPP higher in elderly patients. What, then, makes outcome worse if these traditionally outcome-linked factors are more favorable? It may be that critical thresholds may become less favorable in elderly patients. However, such an analysis would be impossible using our material. The only variables which deteriorated with age in our data were vascular pressure reactivity and autoregulation [4]. This association may suggest that worsening of the indices of blood-flow regulation with age is responsible for the worse outcomes. Indeed, in all our previous studies, when these indices were considered they were strong and independent predictors of outcome after TBI.

Conclusion

In addition to low CPP and high ICP, cerebral PRx should become a target in post-injury intensive care. Vasopressors should be used with moderation. Young women should be treated more aggressively when they develop intracranial hypertension (possibly including surgical decompression), as their mortality rate associated with raised ICP is greater than in males.

Acknowledgments

The authors are in debt to all team members participating in data collection: Dr. M. Hiler, Mrs. Pippa Al-Rawi, Mrs. Helen Seeley, Mrs. Carole Turner, Mrs. Colette O'Kane, Mrs. Shirley Love, Mrs. Diana Simpson, Dr. Eric Schmidt, Dr. Stefan Piechnik, Dr. P. Smielewski, Dr. Andreas Raabe, Mr. Eric Guazzo, Dr. David Menon, Dr. Arun Gupta, Mr. Peter Kirkpatrick, Mr. Rupert Kett-White, Mr. Pwawanjit Minhas, Mr. Rodney Laing, and all the nursing and research staff of NCCU.

P.J.H. is supported by an Academy of Medical Sciences Health Foundation Senior Surgical Scientist Fellowship.

M. Czosnyka is on unpaid leave from Warsaw University of Technology, Poland.

This project was supported by the U.K. Government Technology Foresight Initiative, and the Medical Research Council (Grant No G9439390 ID 65883).

References

1. Balestreri M, Steiner LA, Czosnyka M (2003) Sex-related differences and traumatic brain injury. J Neurosurg 99: 616–617
2. Balestreri M, Czosnyka M, Steiner LA, Schmidt E, Smielewski P, Matta B, Pickard JD (2004) Intracranial hypertension: what additional information can be derived from ICP waveform after head injury? Acta Neurochir (Wien) 146: 131–141
3. Czosnyka M, Whitehouse H, Smielewski P, Kirkpatrick P, Guazzo EP, Pickard JD (1994) Computer supported multimodal bed-side monitoring for neuro intensive care. Int J Clin Monit Comput 11: 223–232
4. Czosnyka M, Balestreri M, Steiner L, Smielewski P, Hutchinson PJ, Matta B, Pickard JD (2005) Age, intracranial pressure, autoregulation, and outcome after brain trauma. J Neurosurg 102: 450–454
5. Farin A, Deutsch R, Biegon A, Marshall LF (2003) Sex-related differences in patients with severe head injury: greater susceptibility to brain swelling in female patients 50 years of age and younger. J Neurosurg 98: 32–36
6. Grande PO (2000) Pathophysiology of brain insult. Therapeutic implications with the Lund Concept. Schweiz Med Wochenschr 130: 1538–1543
7. Marmarou A (1992) Increased intracranial pressure in head injury and influence of blood volume. J Neurotrauma 9 [Suppl] 1: S327–S332
8. Menon DK (1999) Cerebral protection in severe brain injury: physiological determinants of outcome and their optimisation. Br Med Bull 55: 226–258
9. Rosner MJ, Rosner SD, Johnson AH (1995) Cerebral perfusion pressure: management protocol and clinical results. J Neurosurg 83: 949–962
10. Steiner LA, Czosnyka M, Piechnik SK, Smielewski P, Chatfield D, Menon DK, Pickard JD (2002) Continuous monitoring of cerebrovascular pressure reactivity allows determination of optimal cerebral perfusion pressure in patients with traumatic brain injury. Crit Care Med 30: 733–738

Correspondence: Marek Czosnyka, Neurosurgical Unit, Addenbrooke's Hospital, Box 167, Cambridge, UK. e-mail: Mc141@medschl.cam.ac.uk

Experimental Traumatic Brain Injury

The temporal profile of edema formation differs between male and female rats following diffuse traumatic brain injury

C. A. O'Connor[1], I. Cernak[2], and R. Vink[1]

[1] Department of Pathology, University of Adelaide, Adelaide SA, Australia
[2] Department of Neuroscience, Georgetown University, Washington, DC, USA

Summary

Although female hormones are known to influence edema formation following traumatic brain injury (TBI), no studies have actually compared the temporal profile of edema formation in both male and female rats following diffuse TBI. In this study, male, female, and female ovariectomized rats were injured using the 2 m impact acceleration model of diffuse TBI. The temporal profile of brain water content was assessed over 1 week post-trauma. Male animals demonstrated increased ($p < 0.05$) edema at 5 hours, 24 hours, 3 days, 4 days, and 5 days after TBI with a peak at 5 hours post-injury. This time point was associated with increased blood-brain barrier (BBB) permeability. In contrast, intact females showed increased levels of edema ($p < 0.05$) at 5 hours, 24 hours, 3 days, and 4 days post-TBI, with a peak at 24 hours. No BBB opening was present in intact females at 5 hours. Female animals demonstrated more edema than male animals at 24 hours, but less at 5 hours, 3 days, and 5 days. Ovariectomy produced an edema profile that was similar to that observed in males. The temporal profile of edema formation after TBI seems to depend on endogenous hormone levels, a difference which may have an influence on clinical management.

Keywords: Edema; gender; progesterone; estrogen; neurotrauma; brain swelling.

Introduction

Cerebral edema is a serious consequence of traumatic brain injury (TBI), resulting in increased intracranial pressure and possibly death [12]. In young victims of trauma, it has been reported that brain edema may be associated with up to 50% of all deaths [8]. Currently, there is no effective treatment in clinical practice, with interventions such as mannitol, corticosteroids, hyperthermia, barbiturates, and drainage of cerebrospinal fluid having either limited success or being completely ineffective [12].

A number of experimental TBI studies have demonstrated that female gonadal hormones, particularly progesterone, may significantly attenuate post-traumatic edema formation. For example, normal cycling female rats develop less edema than males 24 hours after TBI [15]. Females at the high estrogen stage of their cycle (proestrus) demonstrate 50% of the edema which males developed, whereas females that are high in progesterone (pseudo-pregnant) show virtually no evidence of edema.

In contrast to the experimental TBI literature, the effects of female gonadal hormones on edema and brain swelling in human head injury are less clear. A recent report examining this issue has shown that females under 50 years of age have worse edema and brain swelling after TBI than males [7], suggesting that the female gonadal hormones have no beneficial effect on edema formation.

Although the reasons for these differences are unclear, it may be that the experimental models used to date have generally been focal in nature, while clinical injury tends to be more often diffuse. Alternatively, differences may be related to the time points chosen to measure edema, particularly in temporal edema profiles between the genders. Accordingly, the current study compares the temporal profile of edema formation in male, female, and female ovariectomized rats in a diffuse model of TBI.

Materials and methods

Induction of injury

Male (n = 35; 5/group; 380–450 g), intact female (n = 35; 5/group; 310–400 g), and ovariectomized female (n = 20; 5/group; 310–400 g) Sprague-Dawley, out-bred rats were injured using the impact acceleration model of diffuse TBI [13] as previously described [14]. Briefly, under halothane anesthesia, the skull was exposed by a midline incision and a stainless steel disc (10 mm in diameter and

3 mm in depth) was fixed rigidly with polyacrylamide adhesive to the animal's skull centrally between lambda and bregma. The rats were subsequently placed on a 10 cm foam bed and subjected to brain injury induced by dropping a 450 g brass weight a distance of 2 m onto the stainless steel disc. During all surgical procedures and in the immediate recovery period, rectal temperature was maintained at 37 °C by use of a thermostatically controlled heating pad. Subsequent to injury, all wounds were sutured, anesthesia was terminated and, when stable, animals were returned to their cages.

Ovariectomy

A subgroup of female rats (n = 30) was surgically ovariectomized at 7 weeks of age. Briefly, anesthesia was induced using halothane (3% induction followed by 1% maintenance) and a bilateral ovariectomy performed by ligation and dissection of the ovaries. Rectal temperature in all animals was maintained at 37 °C with a thermostatically controlled heating pad. Animals were allowed to recover for approximately 9 weeks, after which TBI was induced. In female animals that were not ovariectomized, the stage of estrus was determined using vaginal smears [11]. At the time of trauma, 66% of the intact females were in diestrus, 17% in estrus, and 17% in metestrus.

Determination of edema

At pre-selected time points (5 hours to 5 days), animals were re-anesthetized with halothane and decapitated. All sham animals (n = 5/group) were decapitated 30 minutes after surgery. Brains were rapidly removed from the skull, the olfactory bulbs and cerebellum discarded, and the cortex and subcortex separated. The cortex and subcortex of each rat was placed separately into pre-weighed and labeled glass vials with quick-fit lids to prevent evaporation. After weighing for wet water content, the vials (glass lids removed) were then placed in an oven for 72 hours at 100 °C. Vials and brain segments were then re-weighed to obtain dry weight content. Edema in each brain sample was calculated using the wet/dry method formula [6], where %water = [(wet wt − dry wt)/wet weight] × 100.

Determination of blood-brain barrier permeability

Evans blue (EB) is a serum albumin tracer. Extravasation of EB at 5 hours after TBI (n = 5/group) was used for determination of blood-brain barrier (BBB) permeability. Briefly, at 4 hours after injury, 2% EB was injected intravenously at a dose of 2 ml/kg. Animals were then re-anesthetized at 5 hours with halothane and perfused using saline to remove intravascular EB dye. Animals were then decapitated, the brains removed and homogenized in phosphate buffered saline. Trichloroacetic acid was then added to precipitate protein, and the samples were cooled and centrifuged. The resulting supernatant was measured for absorbance of EB using a spectrophotometer.

Data analysis

Data are shown as mean ± SEM. Statistical significance was determined by analysis of variance, followed by a Tukey HSD test to determine specific differences between groups. A p value of 0.05 was considered significant.

Results

The temporal profile for edema development in the cortex of male, female, and ovariectomized female animals is shown in Fig. 1.

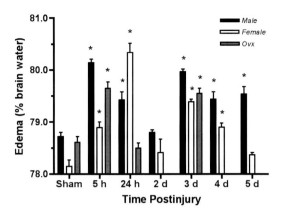

Fig. 1. Temporal changes in brain water content in male, female, and female ovariectomized (*Ovx*) rats following severe diffuse traumatic brain injury (* = p < 0.05 versus sham value)

The results in the cortex were similar to results obtained in the subcortices (results not shown). Prior to injury, brain water content was similar amongst all 3 groups. After injury, male animals had significant edema development at 5 hours, 24 hours, 3 days, 4 days, and 5 days when compared to sham values. The water content was not different from sham values at 2 days post-trauma. These results were consistent with the biphasic edema profile previously demonstrated after TBI using diffusion weighted magnetic resonance imaging [2, 4]. The maximum level of edema was recorded 5 hours post-injury.

In intact females, increases in brain water content after trauma were noted at 5 hours, 24 hours, 3 days, and 4 days. However, in contrast to males, the largest increase occurred at the 24-hour time point. There was also a more rapid decline in edema over time to the extent that by 5 days after injury, brain water content had returned to normal. A significant gender by time interaction was observed ($F(6, 56) = 15.434$; $p < 0.01$) in the cortex of both groups, confirming that the temporal profile for edema formation was different between the groups. Tukey post hoc comparisons illustrated that females had less edema after injury than male animals at 5 hours, 3 days, and 5 days. In contrast, female animals demonstrated more edema than male animals at the 24-hour time point. There were no differences observed at the 48-hour point.

Ovariectomized female animals also demonstrated increases in cortical water content after injury (Fig. 1). Tukey post hoc comparisons demonstrated that the differences occurred at 5 hours and 3 days with no edema present at 24 hours. This edema profile in the

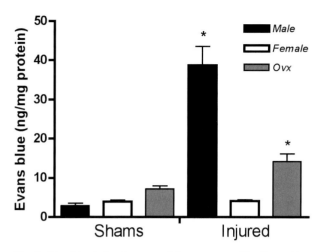

Fig. 2. Blood-brain barrier permeability as assessed by Evans blue penetration at 5 hours following traumatic brain injury in male, female, and female ovariectomized (Ovx) rats (* = p < 0.05 versus sham value)

ovariectomized female animals was, in fact, very similar to that observed in male animals.

BBB permeability at 5 hours after TBI is shown in Fig. 2. Male rats demonstrated a marked increase in BBB permeability after TBI, consistent with previously published reports [1, 3, 18]. In contrast, intact female animals demonstrated no increase in BBB permeability, whereas ovariectomized female rats did show an increase in BBB permeability, although the degree of EB extravasation was quantitatively far less than that observed in the males.

Discussion

A number of studies have now reported effects of female gonadal hormones in central nervous system injury, including ischemic injury [5, 10], controlled cortical impact injury [9, 15, 16], and bilateral medial frontal cortex injury [17, 19]. These models are generally focal in nature, with a number of them also involving a profound ischemic component. The degree of edema in these forms of injury is high, with increases in water content from 6–10% in male animals having been recorded [16, 19]. Clinical trauma involves a significant degree of diffuse axonal injury, and rodent models of diffuse TBI have shown small increases in brain water content of less than 3% [18]. Accordingly, it was of interest to determine if gender differences in edema exist in diffuse models of TBI.

Our results indicate that the profile for edema formation in females after diffuse TBI is considerably different from that noted in males. Female animals displayed an increase in edema at 5 hours, but this was much less than that observed in the males. Unlike the males, the early edema in females was not followed by a decline in water content, but by an increase at 24 hours that was greater than that observed in the males. Thereafter, female brain water levels were less than that observed in males at every time point, falling to non-significant levels by day 5. Most intriguing is our finding that ovariectomy conferred a male edema profile on ovariectomized animals, with a maximum at 5 hours and significant edema at later time points. The presence of endogenous levels of hormones not only attenuated the edema development after trauma, but also altered the temporal profile of edema formation.

A significant increase in EB extravasation occurred in males 5 hours after trauma but not in intact females. The fact that males have greater edema formation at this time point is therefore consistent with a larger potential contribution of vasogenic edema in male animals in the period immediately after TBI. Ovariectomy in female animals resulted in more EB extravasation than that observed in intact female animals, although this increased BBB permeability was less in quantitative terms than that observed in males. Moreover, the higher degree of BBB permeability was associated with increased edema formation in ovariectomized females at this time point, suggesting greater involvement of vasogenic processes in edema formation. However, when compared to males, the degree of early edema was almost identical in both groups, suggesting that more processes are involved in the early phase of edema after TBI than simply increased BBB permeability.

In conclusion, we have demonstrated that male and female animals have different temporal profiles of edema formation and variable degrees of BBB permeability after diffuse TBI. These differences may, in part, be accounted for by the effects of female gonadal hormones, although it has yet to be determined which hormones are involved.

References

1. Barzo P, Marmarou A, Fatouros P, Corwin F, Dunbar J (1996) Magnetic resonance imaging-monitored acute blood-brain barrier changes in experimental traumatic brain injury. J Neurosurg 85: 1113–1121
2. Barzo P, Marmarou A, Fatouros P, Hayasaki K, Corwin F (1997) Contribution of vasogenic and cellular edema to traumatic brain swelling measured by diffusion-weighted imaging. J Neurosurg 87: 900–907

3. Beaumont A, Marmarou A, Hayasaki K, Barzo P, Fatouros P, Corwin F, Marmarou C, Dunbar J (2000) The permissive nature of blood brain barrier (BBB) opening in edema formation following traumatic brain injury. Acta Neurochir [Suppl] 76: 125–129
4. Cernak I, Vink R, Zapple DN, Cruz MI, Ahmed F, Chang T, Fricke ST, Faden AI (2004) The pathobiology of moderate diffuse traumatic brain injury as identified using a new experimental model of injury in rats. Neurobiol Dis 17: 29–43
5. Chen J, Chopp M, Li Y (1999) Neuroprotective effects of progesterone after transient middle cerebral artery occlusion in rat. J Neurol Sci 171: 24–30
6. Elliot KAC, Jasper H (1949) Measurement of experimentally induced brain swelling and shrinkage. Am J Pathol 157: 122–129
7. Farin A, Deutsch R, Biegon A, Marshall LF (2003) Sex-related differences in patients with severe head injury: greater susceptibility to brain swelling in female patients 50 years of age and younger. J Neurosurg 98: 32–36
8. Feickert HJ, Drommer S, Heyer R (1999) Severe head injury in children: impact of risk factors on outcome. J Trauma 47: 33–38
9. Galani R, Hoffman SW, Stein DG (2001) Effects of the duration of progesterone treatment on the resolution of cerebral edema induced by cortical contusions in rats. Restor Neurol Neurosci 18: 1–6
10. Kumon Y, Kim SC, Tompkins P, Stevens A, Sakaki S, Loftus CM (2000) Neuroprotective effect of postischemic administration of progesterone in spontaneously hypertensive rats with focal cerebral ischemia. J Neurosurg 92: 848–852
11. Maeda K, Ohkura S, Tsukarmura T (2000) Physiology of reproduction. In: Krinke G (ed), The Laboratory Rat. Academic Press, London
12. Marmarou A (1994) Traumatic brain edema: an overview. Acta Neurochir Suppl (Wien) 60: 421–424
13. Marmarou A, Foda MA, van den Brink W, Campbell J, Kita H, Demetriadou K (1994) A new model of diffuse brain injury in rats. Part I: Pathophysiology and biomechanics. J Neurosurg 80: 291–300
14. O'Connor CA, Cernak I, Vink R (2003) Interaction between anesthesia, gender, and functional outcome task following diffuse traumatic brain injury in rats. J Neurotrauma 20: 533–541
15. Roof RL, Duvdevani R, Stein DG (1993) Gender influences outcome of brain injury: progesterone plays a protective role. Brain Res 607: 333–336
16. Roof RL, Duvdevani R, Heyburn JW, Stein DG (1996) Progesterone rapidly decreases brain edema: treatment delayed up to 24 hours is still effective. Exp Neurol 138: 246–251
17. Shear DA, Galani R, Hoffman SW, Stein DG (2002) Progesterone protects against necrotic damage and behavioral abnormalities caused by traumatic brain injury. Exp Neurol 178: 59–67
18. Vink R, Young A, Bennett CJ, Hu X, Connor CO, Cernak I, Nimmo AJ (2003) Neuropeptide release influences brain edema formation after diffuse traumatic brain injury. Acta Neurochir [Suppl] 86: 257–260
19. Wright DW, Bauer ME, Hoffman SW, Stein DG (2001) Serum progesterone levels correlate with decreased cerebral edema after traumatic brain injury in male rats. J Neurotrauma 18: 901–909

Correspondence: Robert Vink, Department of Pathology, University of Adelaide, Adelaide, SA, Australia. e-mail: Robert.Vink@adelaide.edu.au

The effect of intravenous fluid replacement on the response to mannitol in experimental cerebral edema: an analysis of intracranial pressure, serum osmolality, serum electrolytes, and brain water content

H. E. James

University of Florida Jacksonville and Wolfson Children's Hospital, Jacksonville, FL, and the University of California San Diego, San Diego, CA, USA

Summary

Albino rabbits that had undergone a cryogenic insult over the left parieto-occipital cortex were analyzed for serum osmolality, serum electrolytes, brain water content, and intracranial pressure (ICP) following either a baseline infusion of intravenous (IV) fluid (45 mL total) for 3 hours or above-maintenance isotonic saline (73.5 ± 12 mL or 90.5 ± 1.5 mL) and mannitol therapy. The subgroups were compared amongst themselves and to sham-operated controls. Serum osmolality was elevated in the higher-dose mannitol subgroup compared with maintenance IV fluids subgroup (1 g/kg/h vs 1 g/kg/3 h; $p < 0.05$), accompanied by an insignificant reduction of serum sodium. A significant reduction in brain water in the injured left hemisphere was seen following high-dose mannitol in the subgroup that received less IV (maintenance) fluids than the group that received above-maintenance IV fluids ($p < 0.025$). No reduction in brain water was seen in the subgroup that received above-maintenance IV fluids (non-treated groups). Reduction of ICP was not found in the lower mannitol dose group. We conclude that the ability of mannitol to reduce cerebral edema is related to the total amount of IV fluid replacement. This implies that the amount of IV crystalloid fluid that is administered to patients with cerebral edema and raised ICP requiring mannitol for control needs to be carefully monitored.

Keywords: Edema; mannitol; osmolality; intracranial pressure.

Introduction

The use of hypertonic agents in clinical practice for the control of intracranial pressure (ICP) and brain edema has been extensively documented [9–11, 15, 18, 23]. Concerns relating to the clinical use of hypertonic agents are: hypertonic dehydration [1, 5, 8, 19, 25], rebound elevation of ICP [13, 26, 27], acute changes in serum osmolality and sodium, and loss of ICP-reducing effects following repeated administrations of the agents [11, 21, 22]. The influence of the volume and type of intravenous (IV) fluid administration on serum, cerebrospinal fluid, and brain interstitial fluid osmolality, and the response of ICP to hypertonic therapy has been debated [6, 14, 16, 17, 20].

The purpose of our study was to determine the effects of hypertonic mannitol administration in normal rabbits and in rabbits with brain edema secondary to a cryogenic insult in the presence of isotonic fluid administration in varying amounts over a fixed time period.

Materials and methods

Albino rabbits weighing 2.5 to 3.0 kg were anesthetized with halothane (2%), placed in a stereotactic head-holding device, and their scalp incised under local anesthesia (bupivacaine 1%). The calvarium was exposed and, with a 12.5 mm diameter circular trephine, an opening was created in the skull at the left parieto-occipital region. A stainless steel probe then was immersed and equilibrated in liquid nitrogen and applied to the intact dura for 90 seconds to create a cold injury [3]. The bone remnant was sutured to the skull, the scalp incision closed, and topical antibiotic ointment (bacitracin) applied to the suture line. All animals received 1 mL Evans blue solution intravenously to document the extent of blood-brain barrier breakdown. The rabbits then received 50 mg/kg of ampicillin intraperitoneally. Another group of animals received a similar operation but without a cryogenic insult, constituting the sham-operated (control) group.

Twenty-four hours following the operation at the time of maximal brain edema [7, 10], the surviving animals underwent the experimental trials. They were re-anesthetized with halothane (2%), intubated with an endotracheal tube and continuously ventilated with a mixture of oxygen (50% O_2, nitrous oxide 50%) and halothane (0.5%). Arterial and central venous (femoral) lines were placed with local anesthesia (bupivacaine 1%). Blood gases were frequently determined in order to maintain a $PaCO_2$ in the 37 to 43 torr range (Harvard Apparatus, Harvard Instruments, Framingham, MA). The animals were then placed in the stereotactic head-holding device and an 18-gauge needle was inserted into the cisterna magna. The scalp incision was opened and the calvarium exposed. Two platinum electrodes were placed in the skull on each side of the midline for continuous

Table 1. *Experimental subgroups*

Group 1	Group 2	Group 3
	Mannitol (1 g/kg/3 h = 0.33 g/kg/h)	Mannitol (1 g/kg/h)
Sham-operated (controls)	*sham-operated (controls)*	*sham-operated (controls)*
a) Maintenance fluids	a) maintenance fluids	a) maintenance fluids
b) High-volume fluids	b) high-volume fluids	b) high-volume fluids
Cryogenic insult (edema)	*cryogenic insult (edema)*	*cryogenic insult (edema)*
a) Maintenance fluids	a) maintenance fluids	a) maintenance fluids
b) High-volume fluids	b) high-volume fluids	b) high-volume fluids

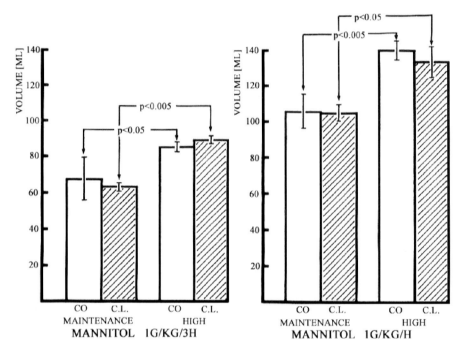

Fig. 1. Amount of IV fluids administered to the various groups. Note: Group 2 (A) received mannitol at 0.33 g/kg/h, while Group 3 (B) received mannitol at 1.0 g/kg/h. *CO* Controls; *CL* cold lesion; *Maintenance* maintenance fluid groups; *High* above-maintenance fluid groups

EEG recording. The cisterna magna needle, and the arterial and venous lines were connected to transducers for continuous display of ICP, systolic arterial pressure, and central venous pressure on a multi-channel polygraph (Hewlett-Packard, Model 7758B System, Palo Alto, CA).

The animals were divided into 3 groups. Group 1 consisted of the animals without mannitol therapy. Group 2 received 0.33 g/kg/h of 20% mannitol IV infusion over 3 hours. Group 3 received an IV infusion of 20% mannitol at 1 g/kg/h for 3 hours (high-dose mannitol). Each group was in turn subdivided into the following subgroups: sham-operated (controls), cryogenic insult, maintenance IV fluids, and above-maintenance IV fluids (vide infra). The various groups and subgroups are shown in Table 1.

IV infusions were initiated following stabilization of the $PaCO_2$. One subgroup received 4 to 8 mL/h and the second subgroup received 15 mL/h of isotonic saline solution. The amounts of fluids have been summated for all subgroups (Fig. 1). Group 1 had 15 mL/h baseline isotonic saline and had no mannitol. Group 2 had 62.5 ± 2.4 mL/h isotonic saline infusion for the maintenance subgroup, and 90.5 ± 1.5 mL/h for the above-maintenance subgroup. For Group 3, the maintenance subgroup had 47.7 ± 7.7 mL/h and the above-maintenance subgroup had 73.5 ± 12.0 mL/h. Both Groups 2 and 3 had mannitol therapy. Blood samples for serum osmolality, sodium, and potassium were taken prior to mannitol infusion at 60, 120, and 180 minutes after mannitol infusion was initiated. Upon completion of 3 hours of mannitol therapy, the animals were killed by IV air embolization. Rapid craniectomy permitted prompt removal of the brain, which was placed in cold kerosene for gravimetry studies. Samples were taken in front and behind the areas of hemorrhage and necrosis resulting from the cold lesion, where Evans blue extravasation indicated a disturbance of the blood-brain barrier. An equivalent area from the opposite hemisphere was sampled. In sham-operated animals, homologous samples were taken from the parieto-occipital regions of each hemisphere [12].

Authorization for this investigation was obtained from the Animal Investigation Committee of the University of California San Diego.

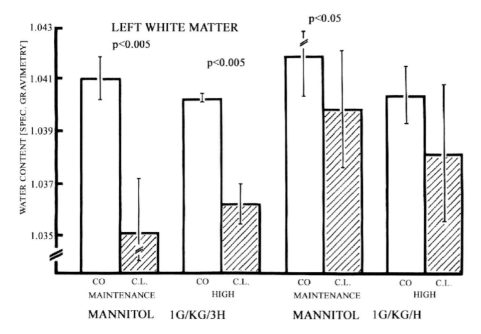

Fig. 2. Specific gravimetry values of the left hemisphere white matter samples (mean and SD) of the various groups. Note: Group 2 (A) received mannitol at 0.33 g/kg/h, while Group 3 (B) received mannitol at 1.0 g/kg/h. *CO* Controls; *CL* cold lesion; *Maintenance* maintenance fluids; *High* above-maintenance fluids

Results

Systolic arterial pressure ranged from 85 to 120 torr in all groups. Central venous pressure failed to change in all groups. There was no statistical difference in serum osmolality in any of the subgroups except that receiving maintenance IV fluids ($p < 0.05$). Serum osmolality rose 120 minutes into treatment in all subgroups that progressed through the 3 hours of the experimental trial.

The pre-treatment serum sodium for Group 2 with mannitol was 133.8 ± 7.4 mEq/L, and at 180 minutes it was 134.3 ± 3.8 mEq/L in the control subgroup with maintenance fluids. A similar change during the mannitol infusion was seen for the cold lesion subgroups.

In Group 1, the sodium had a decreasing trend during the infusion, but this was not significant. Serum potassium was not altered significantly in any of the groups studied.

There was no significant difference in brain water data between subgroups. A similar lack of significance was seen between Group 3 subgroups following mannitol therapy. There was no statistical difference between the sham-operated subgroups of low-dose mannitol and high-dose mannitol (Fig. 2). Likewise, there was no difference between equivalent subgroups that received maintenance or above-maintenance IV fluids.

Comparison of ICP showed no difference between any of the study groups.

Discussion

Our previous work determined that large volumes of isotonic IV fluids given in a short time do not lead to an increase in brain water content or ICP in the mechanically ventilated animal [12]. However, the effectiveness of mannitol in the same setting may be different.

In previous studies of albino rabbits with cytotoxic brain edema, the administration of a bolus of mannitol produced a prompt reduction of ICP. When ICP reached its nadir, analysis of gray matter water content demonstrated a significant reduction compared to untreated controls [10]. In a subsequent study, similar bolus administrations produced a prompt reduction in ICP; however, with this form of cryogenic edema there was no significant reduction of brain water [10]. In both studies, animals received no IV fluid replacement other than the amount necessary to maintain patent monitoring systems. The simultaneous administration of mannitol and furosemide in those experiments did not further reduce ICP or brain water content [10, 19].

In our current study, animals were given isotonic IV fluid loads that were in excess of their resting re-

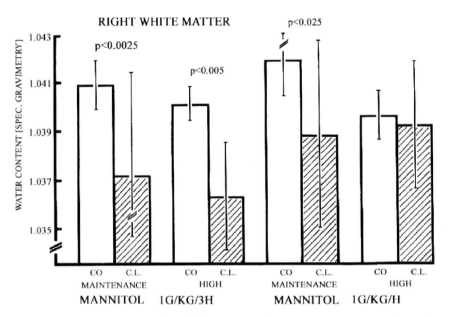

Fig. 3. Specific gravimetry values of the right hemisphere white matter samples (mean and SD) of the various groups. Note: The cold lesion was on the left side. *CO* Controls; *CL* cold lesion; *Maintenance* maintenance fluids; *High* above-maintenance fluids

quirements. In the subgroups that received the lower mannitol dose (1 g/kg over 3 hours), serum osmolality trended upward in some animals, however, with the higher dose (1 g/kg/h for 3 hours), osmolality rose significantly as previously reported [2, 9, 11, 24, 28]. A reduction in serum sodium was noted in the higher mannitol dose group, but the change was not significant.

Earlier, we found no cause and effect between administered fluid volume and resultant brain edema [12] with or without a cold lesion (Figs. 2 and 3). However, brain water fell significantly in the white matter of the left hemisphere (the lesion side) in animals that had maintenance IV fluid. The high-dose mannitol subgroup also had less edema than the low-dose subgroup on the lesion side (Fig. 2). This was not the case in the subgroup that had high-volume IV fluids (Fig. 2).

These findings demonstrate that isotonic fluid given rapidly in large volumes may not increase brain edema, but it may alter intracranial dynamics caused by changes in the movement of sodium and water, which may then lead to impaired effectiveness of mannitol in extracting brain water and reducing ICP. Alteration of cerebral circulation and changes in brain blood volume may also account for the lack of reduction in ICP [4]. Further investigations into the effects of hypertonic agents and IV replacement fluids are needed.

Acknowledgments

This study has been supported in part by the Foundation for Pediatric and Laser Neurosurgery, Inc.

The author would like to acknowledge Sylvia Schneider, who performed the technical work of this study, and Jennifer Santarone who prepared the manuscript.

References

1. Aviram A, Pfau A, Czaczkes JW, Ullmann TD (1967) Hyperosmolality with hyponatremia, caused by inappropriate administration of mannitol. Am J Med 42: 648–650
2. Becker DP, Vries JK (1972) The alleviation of increased intracranial pressure by the chronic administration of osmotic agents. In: Brock M, Dietz H (eds) Intracranial Pressure. Springer-Verlag, New York, pp 310–315
3. Beks JW, ter Weeme CA (1967) The influence of urea and mannitol on increased intraventricular pressure in cold-induced cerebral edema. Acta Neurochir (Wien) 16: 97–107
4. Bruce DA, Langfitt TW, Miller JD, Schutz H, Vapalahti MP, Stanek A, Goldberg HI (1973) Regional cerebral blood flow, intracranial pressure, and brain metabolism in comatose patients. J Neurosurg 38: 131–144
5. Buhrley LE, Reed DJ (1972) The effect of furosemide on sodium-22 uptake into the cerebrospinal fluid and brain. Exp Brain Res 14: 503–510
6. Goluboff B, Shenkin HA, Haft H (1964) The effect of mannitol and urea on cerebral hemodynamics and cerebrospinal fluid pressure. Neurology 14: 891–898
7. Herrmann HD, Neuenfeldt D (1972) Development and regression of a disturbance of the blood-brain barrier and of edema in tissue surrounding a circumscribed cold lesion. Exp Neurol 34: 115–120
8. Javid M, Anderson J (1959) The effect of urea on cerebrospinal

fluid pressure in monkeys before and after bilateral nephrectomy. J Lab Clin Med 53: 484–489
9. James HE, Langfitt TW, Kumar VS, Ghostine SY (1977) Treatment of intracranial hypertension: analysis of 105 consecutive, continuous recordings of intracranial pressure. Acta Neurochir (Wien) 36: 189–200
10. James HE, Bruce DA, Welsh F (1978) Cytotoxic edema produced by 6-aminonicotinamide and its response to therapy. Neurosurgery 3: 196–200
11. James HE (1980) Methodology for the control of intracranial pressure with hypertonic mannitol. Acta Neurochir (Wien) 51: 161–172
12. James HE, Schneider S (1993) Effects of acute isotonic saline administration on serum osmolality, serum electrolytes, brain water content and intracranial pressure. Acta Neurochir [Suppl] 57: 89–93
13. Javid M, Gilboe D, Cesario T (1964) The rebound phenomenon and hypertonic solutions. J Neurosurg 21: 1059–1066
14. Langfitt TW (1961) Possible mechanisms of action of hypertonic urea in reducing intracranial pressure. Neurology 11: 196–209
15. Leech P, Miller JD (1974) Intracranial volume-pressure relationships during experimental brain compression in primates. 3. Effects of mannitol and hyperventilation. J Neurol Neurosurg Psychiatry 37: 1105–1111
16. Marshall LF, Smith RW, Rauscher LA, Shapiro HM (1978) Mannitol dose requirements in brain-injured patients. J Neurosurg 48: 169–172
17. Miller JD, Leech P (1975) Effects of mannitol and steroid therapy on intracranial volume-pressure relationships in patients. J Neurosurg 42: 274–281
18. Pappius HM, Dayes LA (1965) Hypertonic urea. Its effect on the distribution of water and electrolytes in normal and edematous brain tissue. Arch Neurol 13: 395–402
19. Pollay M, Fullenwider C, Roberts PA, Stevens FA (1983) Effect of mannitol and furosemide on blood-brain osmotic gradient and intracranial pressure. J Neurosurg 59: 945–950
20. Prill A, Volles E, Dahlmann W (1972) Stabilization and disturbance of osmoregulation in the cerebrospinal fluid. In: Brock M, Dietz H (eds) Intracranial pressure. Springer Berlin Heidelberg New York Tokyo, pp 303–308
21. Szewczykowski J, Sliwka S, Kunicki A, Korsack-Sliwka J, Dziduszko J, Dytko P, Augustyniak B (1975) Computer-assisted analysis of intraventricular pressure after mannitol administration. J Neurosurg 43: 136–141
22. Shenkin HA, Goluboff B, Haft H (1962) The use of mannitol for the reduction of intracranial pressure in intracranial surgery. J Neurosurg 19: 897–901
23. Shenkin HA, Bexier HS, Bouzarth WF (1976) Restricted fluid intake. Rational management of the neurosurgical patient. J Neurosurg 45: 432–436
24. Silber SJ, Thompson N (1972) Mannitol induced central nervous system toxicity in renal failure. Invest Urol 9: 310–312
25. Stuart FP, Torres E, Fletcher D, Moore FD (1970) Effects of single, repeated and massive mannitol infusion in the dog: structural and functional changes to the kidney and brain. Ann Surg 172: 190–204
26. Troupp H, Valtonen S, Vapalahti M (1971) Intraventricular pressure after administration of dehydrating agents to severely brain injured patients: is there a rebound phenomenon? Acta Neurochir (Wien) 24: 89–95
27. Wise BL, Chater N (1962) The value of hypertonic mannitol solution in decreasing brain mass and lowering cerebrospinal fluid pressure. J Neurosurg 19: 1038–1043
28. Wise BL (1963) Effects of infusion of hypertonic mannitol on electrolyte balance and on osmolarity of serum and cerebrospinal fluid. J Neurosurg 20: 961–996

Correspondence: Hector E. James, Lucy Gooding Pediatric Neurosurgery Center, University of Florida HSC/Jacksonville, 836 Prudential Drive, Pavilion Building, Suite 1005, Jacksonville, FL 32207, USA. e-mail: FNLPNS@aol.com

Matrix metalloproteinase-9 is associated with blood-brain barrier opening and brain edema formation after cortical contusion in rats

Y. Shigemori, Y. Katayama, T. Mori, T. Maeda, and T. Kawamata

Department of Neurological Surgery, Nihon University School of Medicine, Tokyo, Japan

Summary

Matrix metalloproteinases (MMPs) are associated with blood-brain opening and may be involved in the pathophysiology of acute brain injury. Previous research demonstrated that knockout mice deficient in MMP-9 subjected to transient focal cerebral ischemia had reduced blood-brain barrier (BBB) disruption and attenuated cerebral infarction.

In this study, we examined MMP-9 up-regulation, BBB disruption, and brain edema formation after cortical impact injury in rats. Cortical contusion was induced by controlled cortical impact. Animals were sacrificed at intervals after injury. MMP up-regulation was assessed by gelatin zymography, and BBB integrity was evaluated using Evans blue dye with a spectrophotometric assay. Brain water content was measured by comparing wet and dry weights of each hemisphere as an indicator of brain edema.

Zymograms showed elevated MMP-9 as early as at 3 hours after injury, reaching a maximum at 18 hours. Peak levels of BBB disruption occurred 6 hours after injury. Brain edema became progressively more severe, peaking 24 hours after injury. Compared to control group, treatment with MMP-inhibitor GM6001 significantly reduced BBB disruption 6 hours and brain water content ($85.9 \pm 0.5\%$ vs. $82.6 \pm 0.3\%$; $p < 0.05$) 24 hours after injury. These findings suggest that MMP-9 may contribute to BBB disturbance and subsequent brain edema after traumatic brain injury.

Keywords: Matrix metalloproteinase; traumatic brain injury; blood-brain barrier.

Introduction

Recent studies have suggested that matrix metalloproteinases (MMPs) increase after brain injury such as in ischemia, trauma, and neurodegenerative disorders. These proteins might play a critical role in degradation of the extracellular matrix, disruption of the blood-brain barrier (BBB), facilitation of leukocyte infiltration, and inflammatory responses following brain injury. The MMP family comprises endopeptidases with zinc-binding catalytic regions [11]. Asahi *et al.* have demonstrated that knockout mice deficient in MMP-9 subjected to transient focal cerebral ischemia had reduced BBB disruption, and attenuated edema and infarction [2].

In the present study, we examined the time course of MMP-9 up-regulation, BBB disruption, and brain edema formation after cortical impact injury in rats.

Materials and methods

Animal model

Male Wistar rats (200–250 g) were anesthetized with a gas mixture of 66% nitrous oxide, 33% oxygen, and 1% halothane. The head of each animal was secured in a stereotaxic apparatus, and the rectal temperature maintained at 37.0–38.0 °C using a heat pad and lamp. An 8 mm diameter craniotomy, centered 3 mm caudal to the bregma and 3 mm lateral to the midline, was performed on the left side of the parietal cranium. Cortical contusion was induced with a controlled cortical impact device, as described in detail elsewhere [8]. A 5 mm diameter injury tip, 6 m/sec impact velocity, and 3 mm penetration depth were employed for the injury induction. These parameters were chosen to provide a moderate level of cortical contusion.

Gelatin zymography

Gelatin zymography was used to assess the MMP-9 levels, as described previously [10]. At 1 hour, 3 hours, 6 hours, 18 hours, and 24 hours after trauma, animals were killed with an overdose of pentobarbital sodium (100 mg/kg, i.p.) and perfused transcardially with chilled (4 °C) phosphate-buffered saline (PBS), pH 7.4. The brains were rapidly removed, and damaged brain tissue within the traumatized hemisphere was homogenized in lysis buffer, which contained protease inhibitors on ice. Following centrifugation, the supernatant was collected, and the total protein concentrations were determined using the Bradford assay (Bio-Rad Laboratories, Hercules, CA). Equal amounts (50 μg) of total protein extracts were prepared. Protein samples were loaded and separated on 10% Tris-Glycine gel with 0.1% gelatin as substrate. After the separation by electrophoresis, the gel was re-natured and then incubated with developing buffer at 37 °C for 24 hours. Following development, the gel was stained with 0.5% Coomassie Blue R-250 for 30 minutes and then de-stained

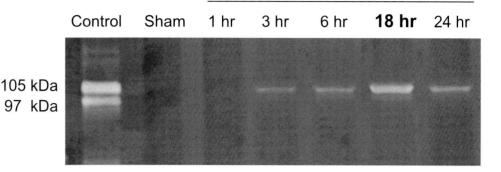

Fig. 1. Representative gelatin zymogram showing elevation of MMP-9 as early as 3 hours after injury, reaching a maximum at 18 hours after injury. The control lane was loaded with murine MMP-9 standards

appropriately. Purified proteins of murine MMP-9 were purchased from Chemicon International, Inc. (Temecula, CA). The zymogram gels were quantified by standard densitometry techniques.

Assessment of BBB permeability

The degree of BBB disruption was evaluated quantitatively by fluorescent detection of extravasated Evans blue dye [9, 14]. After trauma, 2% Evans blue in PBS was injected intravenously (4 ml/kg) and allowed to circulate for at least 1 hour. The animals were anesthetized deeply, and perfused transcardially with PBS through the left ventricle until a colorless perfusion fluid was obtained from the right atrium. The brains were removed and separated into hemispheres. The materials were stored at $-70\,°C$ until use. The sample weight of each hemisphere was measured. Following homogenization of the materials with 1.2 ml of 50% trichloroacetic acid in distilled water for 40 seconds, centrifugation was carried out for 20 minutes at 10,000 rpm. The supernatant was extracted and diluted 1:4 in 100% ethanol. A fluorescent plate reader (KC4) was used at an excitation wavelength of 620 nm and an emission wavelength of 680 nm.

Assessment of brain edema

Brain edema was evaluated on the basis of brain water content. The brain water content was quantitated by the wet-dry weight method, as described previously [8]. Briefly, animals were killed by decapitation under deep pentobarbital anesthesia, the brain was removed, and each hemisphere was gently blotted with tissue paper to remove the small quantities of adsorbent cerebrospinal fluid. The tissue samples were rapidly weighed with a basic precision scale, and then dried to constant weight in a vacuum oven at $105\,°C$ for 24 hours to obtain the dry weight. The percent water content of each tissue sample was calculated according to the following equation: $\%H_2O = [(Wet\ Weight - Dry\ Weight)/Wet\ Weight] \times 100$.

Statistical analysis

Data are expressed as the mean ± SEM. The data were analyzed statistically among the groups, employing the unpaired Student t-test or repeated measures analysis of variance. P values of less than 0.05 were considered significant.

Results

MMP up-regulation

Traumatic brain injury (TBI) induced via controlled cortical impact resulted in a significant elevation of MMP-9 protein levels (Fig. 1). MMP-9 was primarily expressed as the higher molecular weight zymogen (105 kDa). No clear evidence of the lower molecular weight cleaved (activated) form of MMP-9 was noted. MMP-9 was not detected in sham-operated control brains subjected to craniotomy alone. At 3 hours after injury, the MMP-9 level was elevated and reached a maximum at 18 hours after injury. Elevated MMP-9 levels persisted until 24 hours after trauma. The control lane was loaded with murine MMP-9 standards.

BBB permeability

Extravasated Evans blue dye was observed in the injured cortex and ipsilateral hippocampus. Fluorescence measurements confirmed that BBB disruption increased within 3 hours after injury induction. The peak level of BBB disruption was observed in the traumatized hemisphere 6 hours after injury and then gradually decreased (Fig. 2).

Brain edema

Increased brain edema corresponds to increased brain water content. Brain edema was first noted at 6 hours after injury, became progressively more severe over time, and peaked 24 hours after injury (Fig. 3).

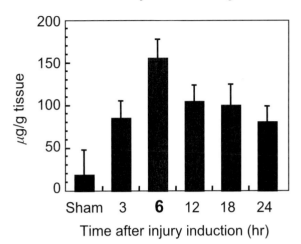

Fig. 2. Time course of BBB disruption after cortical contusion examined on the basis of Evans blue extravasation. Peak level of BBB disruption was observed 6 hours after injury. (*BBB* blood-brain barrier)

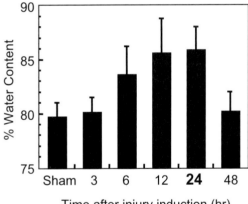

Fig. 3. Time course of brain edema after cortical contusion evaluated from measurements of water content by the wet-dry weight method. Brain edema became progressively more severe over time and peaked 24 hours after injury

The contralateral hemisphere did not exhibit significant changes in brain water content (data not shown).

MMP inhibitor reduced BBB disruption and edema after TBI

Compared to the untreated control group, treatment with GM6001, an MMP inhibitor, significantly reduced the BBB disruption at 6 hours and the brain water content at 24 hours after injury (85.9 ± 0.5% vs. 82.6 ± 0.3%; $p < 0.05$) (Fig. 4). GM6001 (100 mg/kg) was injected intraperitoneally immediately after injury, as described previously [7].

Discussion

The progression of secondary brain damage following trauma involves a multifactorial cascade of pathophysiology including excitotoxicity [5], oxidative stress [6], inflammation [1], and abnormal apoptosis [3]. Recent studies have suggested that MMPs increase after cerebral ischemia [2], TBI [16], and neurodegenerative disorders [4]. These proteins might play a critical role in degradation of the extracellular matrix, disruption of the BBB [13], and a facilitation of leukocyte infiltration [12]. MMPs are also induced by cytokines which are produced in association with inflammation [15]. In cerebral ischemia, emphasis has been placed on the fact that MMP-9 can digest matrix proteins present in the vascular basal lamina including collagen, fibronectin, and laminin. Damage to the vascular integrity would then lead to a disrupted BBB function and increased vasogenic edema [13]. Since edema can play a critical role in TBI, this specific pathway may be involved as well. Inhibition of such cascades could thus exert a therapeutic effect on TBI.

In the present study, we focused on MMP-9 upregulation in cerebral contusion, which leads to BBB disruption and subsequent edema formation. The level of MMP-9 increased and BBB disruption and subsequent edema formation after TBI became progressively more severe over time. GM6001, an MMP inhibitor, significantly reduced BBB disruption at 6 hours and the brain water content at 24 hours after injury. The results of our study imply a mechanistic connection between increased levels of MMP-9 protein and brain tissue damage. We acknowledge that the time lag for each individual result reflects regulatory mechanisms. Our findings indicate that MMP-9 contributes to BBB disturbance and subsequent brain edema after TBI, although the precise mechanisms involved remain to be elucidated.

In conclusion, the present study demonstrates that MMP-9 contributes to BBB disturbance and subsequent brain edema after TBI. Further investigations into the deleterious activities of MMP-9 could reveal new therapeutic targets in TBI.

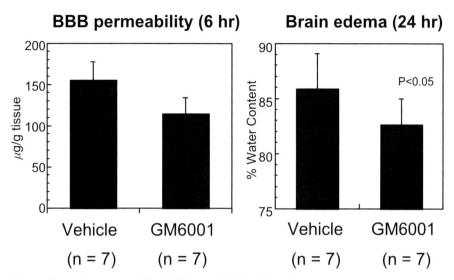

Fig. 4. Effects of treatment with GM6001, an MMP inhibitor. Compared to untreated control group, treatment with GM6001 significantly reduced BBB disruption at 6 hours and the brain water content at 24 hours after injury. GM6001 (100 mg/kg) was injected intraperitoneally immediately after injury

References

1. Arvin B, Neville LF, Barone FC, Feuerstein GZ (1996) The role of inflammation and cytokines in brain injury. Neurosci Biobehav Rev 20: 445–452
2. Asahi M, Wang X, Mori T, Sumii T, Jung JC, Moskowitz MA, Fini ME, Lo EH (2001) Effects of matrix metalloproteinase-9 gene knock-out on the proteolysis of blood-brain barrier and white matter components after cerebral ischemia. J Neurosci 21: 7724–7732
3. Colicos MA, Dash PK (1996) Apoptotic morphology of dentate gyrus granule cells following experimental cortical impact injury in rats: possible role in spatial memory deficits. Brain Res 739: 120–131
4. Fiotti N, Zivadinov R, Altamura N, Nasuelli D, Bratina A, Tommasi MA, Bosco A, Locatelli L, Grop A, Cazzato G, Guamieri G, Giansante C, Zorzon M (2004) MMP-9 microsatellite polymorphism and multiple sclerosis. J Neuroimmunol 152: 147–153
5. Katayama Y, Maeda T, Koshinaga M, Kawamata T, Tsubokawa T (1995) Role of excitatory amino acid-mediated ionic fluxes in traumatic brain injury. Brain Pathol 5: 427–435
6. Kawamata T, Katayama Y, Maeda T, Mori T, Aoyama N, Kikuchi T, Uwahodo Y (1997) Antioxidant, OPC-14117, attenuates edema formation and behavioral deficits following cortical contusion in rats. Acta Neurochir [Suppl] 70: 191–193
7. Keogh B, Sheahan BJ, Atkins GJ, Mills KH (2003) Inhibition of matrix metalloproteinases ameliorates blood-brain barrier disruption and neuropathological lesions caused by avirulent Semliki Forest virus infection. Vet Immunol Immunopathol 94: 185–190
8. Mori T, Katayama Y, Kawamata T, et al (1996) Progressive edema formation in contused brain: Role of tissue osmolality and ion concentrations. Adv Neurotrauma Res 8: 15–18
9. Mori T, Wang X, Kline AE, Siao CJ, Dixon CE, Tsirka SE, Lo EH (2001) Reduced cortical injury and edema in tissue plasminogen activator knockout mice after brain trauma. Neuroreport 12: 4117–4120
10. Mori T, Wang X, Aoki T, Lo EH (2002) Downregulation of matrix metalloproteinase-9 and attenuation of edema via inhibition of ERK mitogen activated protein kinase in traumatic brain injury. J Neurotrauma 19: 1411–1419
11. Nagase H, Woessner JF Jr (1999) Matrix metalloproteinases. J Biol Chem 274: 21491–21494
12. Romanic AM, White RF, Arleth AJ, Ohlstein EH, Barone FC (1998) Matrix metalloproteinase expression increases after cerebral focal ischemia in rats: inhibition of matrix metalloproteinase-9 reduces infarct size. Stroke 29: 1020–1030
13. Rosenberg GA, Navratil M, Barone F, Feuerstein G (1996) Proteolytic cascade enzymes increase in focal cerebral ischemia in rat. J Cereb Blood Flow Metab 16: 360–366
14. Uyama O, Okamura N, Yanase M, Narita M, Kawabata K, Sugita M (1988) Quantitative evaluation of vascular permeability in the gerbil brain after transient ischemia using Evans blue fluorescence. J Cereb Blood Flow Metab 8: 282–284
15. Vecil GG, Larsen PH, Corley SM, Herx LM, Besson A, Goodyer CG, Yong VW (2000) Interleukin-1 is a key regulator of matrix metalloproteinase-9 expression in human neurons in culture and following mouse brain trauma in vivo. J Neurosci Res 61: 212–224
16. Wang X, Jung J, Asahi M, Chwang W, Russo L, Moskowitz MA, Dixon CE, Fini ME, Lo EH (2000) Effects of matrix metalloproteinase-9 gene knock-out on morphological and motor outcomes after traumatic brain injury. J Neurosci 20: 7037–7042

Correspondence: Yutaka Shigemori, Department of Neurological Surgery, Nihon University School of Medicine, 30-1 Oyaguchi-Kamimachi, Itabashi-ku, Tokyo 173-8610, Japan. e-mail: yutaka@lares.dti.ne.jp

Delayed precursor cell marker response in hippocampus following cold injury-induced brain edema

T. Nakamura[1,2], O. Miyamoto[2], S. Yamashita[1], R. F. Keep[3], T. Itano[2], and S. Nagao[1]

[1] Department of Neurological Surgery, Kagawa University Faculty of Medicine, Kagawa, Japan
[2] Department of Neurobiology, Kagawa University Faculty of Medicine, Kagawa, Japan
[3] Department of Neurosurgery, University of Michigan Medical School, Ann Arbor, Michigan

Summary

The purpose of this study was to examine the possibility of neuronal remodeling and repair after cold injury-induced brain edema using immunoassays of nestin, 3CB2, and TUC-4. Male ddN strain mice were subjected to cold-induced cortical injury. Animals were divided into the following 6 groups: 1) 1-day after injury, 2) 1-week after injury, 3) 2-weeks after injury, 4) 1-month after injury, 5) sham, and 6) normal controls. Brain water content measurement, Western blot analysis, histological examination, and neurobehavioral examination were performed.

Brain water content was significantly increased in the ipsilateral cortex at 1-day after injury. At 1-day and 1-week after injury, immunoreactivity of nestin, 3CB2, and TUC-4 were absent. Nestin was expressed in 3CB2-positive astrocytes at 1-month after injury, and nestin expression with TUC-4 was present in the hippocampal cell layer. Neurobehavioral function of the 1-month after injury group was significantly improved compared with function 1-day after injury. These results suggest that delayed precursor cell marker expression in glia and neuron-like cells might be part of adaptation to the injury. Although brain injury causes brain edema and neuronal death, there is the possibility of remodeling.

Keywords: Nestin; 3CB2; TUC-4; remodeling; behavior; cold-induced brain injury; mice.

Introduction

Pathological events in traumatic brain injury may be divided into primary and secondary brain injury [10]. Primary brain injury occurs at the time of the insult, whereas secondary injury follows during the recovery period. Furthermore, reparative processes may occur following damage to the tissue. Eventually, the reaction results in the formation of gliotic scar [8]. However, there are few data on long-term reparative processes after traumatic brain injury.

In the present study, we investigated evidence of nestin, 3CB2, and TUC-4 immunoreactivity, as potential precursor cell markers, in a cold-induced brain injury model. In addition, we examined whether the appearance of cells positive for these markers might coincide temporally with a recovery in behavioral deficit.

Materials and methods

Animals and cold-induced cortical injury

ddN strain mice (25–30 g body weight), inbred in our laboratory (Nakamura *et al.*, 1999), were used for this study. Animal protocols were approved by the Kagawa University Animal Committee. Mice were given free access to food and water prior to experiment. Animals were anesthetized with sodium pentobarbital (30 mg/kg i.p.), with supplemental doses given as necessary, and then placed in a stereotactic frame (Narishige Instruments, Tokyo, Japan). Rectal temperature was maintained at 37 °C using a feedback-controlled heating pad (CMA, Stockholm, Sweden) during the operation. The scalp was incised at the midline and the skull exposed. A cold-induced brain injury was generated by application of a metal probe (3 mm in diameter) cooled with liquid nitrogen to the exposed right parietal bone for 20 seconds. The animals were then divided into 6 experimental groups: 1) 1-day after injury, 2) 1-week after injury, 3) 2-week after injury, 4) 1-month after injury, 5) sham control, and 6) normal control.

Water content

Animals were re-anesthetized with sodium pentobarbital (50 mg/kg i.p.) and decapitated. Brains were removed, and a coronal brain slice (approximately 2 mm thick at the level of injury) was cut with a blade. The brain slice was divided into 2 hemispheres along the midline, and each hemisphere was dissected into the cortex and the basal ganglia. A total of 3 samples from each brain were obtained: the ipsilateral and contralateral cortex and the cerebellum, which served as a control. Brain samples were immediately weighed on an electric analytical balance to obtain the wet weight. Brain samples were then dried at 100 °C for 24 hours to obtain the dry weight. Brain water content was determined as: (Wet Weight − Dry Weight)/Wet Weight.

Western blot analysis

Animals were anesthetized with sodium pentobarbital (50 mg/kg i.p.) before undergoing intracardiac perfusion with 0.1 mol phosphate-buffered saline (PBS; pH 7.4). The brains were removed and the ipsilateral and contralateral hippocampi separated. Western blot analysis was performed. Briefly, 25 µg of protein from each sample was separated by sodium dodecyl sulfate polyacrylamide gel electrophoresis and transferred to a Hybond-C pure nitrocellulose membrane (Amersham, Piscataway, NJ). The membranes were blocked in Carnation nonfat milk. Membranes were probed with a 1:1000 dilution of the primary antibodies, monoclonal mouse anti-nestin or anti-3CB2 (Developmental Studies Hybridoma Bank, University of Iowa, Iowa City, IA) followed by a 1:1500 dilution of the secondary antibody (peroxidase-conjugated goat anti-mouse; Bio-Rad Laboratories, Hercules, CA). The antigen-antibody complexes were visualized with a chemiluminescence system (Amersham) and exposed to film. The relative densities of bands were analyzed with NIH Image (National Institutes of Health, Washington, DC).

Histological examination

For histological examination, animals were sacrificed under deep anesthesia using sodium pentobarbital (50 mg/kg i.p.). The brains were transcardially perfused with 4% phosphate-buffered paraformaldehyde after flushing with 0.1 mol PBS. The brains were removed and placed in fixative overnight. Adjacent coronal paraffin sections (10 µm thick) were taken. Sections of the CA1 area (including the dorsal hippocampal area) around 2.0 mm posterior to bregma were obtained.

For evaluation of neuro-degeneration following cold-induced brain injury, the terminal deoxynucleotidyl transferase-mediated dUTP nick end-labeling technique was performed on adjacent brain sections to detect DNA double-strand breaks. ApopTag Peroxidase Kits (Serologicals Corp., Temecula, CA) were used in this study. In this method, 0.05 mol PBS was used as solution for dilution and washing. The sections were made permeable with 1% Triton X-100 for 30 minutes and endogenous peroxidases quenched with 2% H_2O_2 for 20 minutes. After washing with PBS, the sections were incubated in a moist chamber at 37 °C for 60 minutes with TdT enzyme. After anti-digoxigenin peroxidase had been applied to the sections for 30 minutes at room temperature, peroxidase was detected with DAB. The labeling target of this method was the new 3′-OH DNA ends generated by DNA fragmentation. The omission of the terminal deoxynucleotidyl transferase was used as the negative control. Sections were counterstained with hematoxylin.

Double labeling was performed. Briefly, sections were incubated overnight at 4 °C with monoclonal mouse anti-nestin antibody and polyclonal rabbit anti-glial fibrillary acidic protein (GFAP) antibody diluted, respectively, to 1:100 and 1:200 in 1% skim milk solution in 0.01 mol PBS. After washing with 0.01 mol PBS, the reactions for nestin and GFAP were visualized after incubation for 2 hours at room temperature with anti-mouse immunoglobulin (IgG) conjugated with fluorescein isothiocyanate (Vector Laboratories, Burlingame, CA) and anti-rabbit IgG conjugated with Texas red avidin D (Vector). The secondary antibodies were both diluted 1:40 in 1% skim milk solution in 0.01 mol PBS. Finally, the sections were rinsed in 0.01 mol PBS and visualized using a confocal laser-scanning microscope (LSM-GB200; Olympus, Tokyo, Japan). Other sections were used for double-labeling of nestin and 3CB2. Monoclonal mouse anti-3CB2 antibody was diluted to 1:100 and anti-mouse immunoglobulin M conjugated with Texas red avidin D (1:40) (Vector) was used as the second antibody. Additionally, some sections were used for double-labeling of nestin and immature neuronal marker TUC-4. Polyoclonal rabbit anti-TUC-4 antibody was diluted to 1:100 and anti-rabbit IgG conjugated with Texas red avidin D (1:40) (Vector) was used as the second antibody.

Behavioral test

The radial maze and test procedure used in the present study was similar to that previously reported by Nakamura *et al.* [11]. The central part of the maze was 22 cm in diameter. The arms (25 cm long, 6 cm high, 6 cm wide) were made of transparent Plexiglas for the spatial memory task. Arms were baited with small food pellets. The maze was always oriented in the same direction in space. Twelve hours prior to training, the body weight was verified to be at 80% of the pre-test body weight. Animals (n = 10) were trained on 4 consecutive days, with 1 trial per day. An error was noted if the animal entered an arm previously visited. The number of errors was recorded. The neurological test was scored by an investigator blind to the treatment group.

Statistical analysis

All data are presented as mean ± SD. Data from water content and Western blot analysis was analyzed using Student *t*-test. Neuro-behavioral data were used by 2-way analysis of variance, followed by Scheffe's post hoc test.

Results

Water content

The brain water content was significantly increased in the ipsilateral cortex in ddN mice 1-day after injury compared with the sham controls (80.9 ± 0.9% vs 77.9 ± 0.4%; $p < 0.01$).

TUNEL staining

We observed staining of some clustered cells in the core area of the cold-induced lesion at 1-day after injury. There was expansion of lesions stained by TUNEL-positive cells to include positive CA1 pyramidal cells in the ipsilateral hippocampus at 1-week after injury. The TUNEL-positive cells were already resolved at 1-month after injury. Migration changes were seen in the ipsilateral hippocampal stratum radiatum area at 1-month after injury by counter staining.

Western blot analysis

By Western blot analysis, there was no increase in nestin, 3CB2, and TUC-4 protein levels in the ipsilateral hippocampus from 1-day to 2-weeks after injury. However, there was strong nestin (791 ± 144% of contralateral; $p < 0.01$), 3CB2 (524 ± 215% of contralateral; $p < 0.05$), and TUC-4 (356 ± 127% of contralateral; $p < 0.05$) expression at 1-month after injury (Fig. 1).

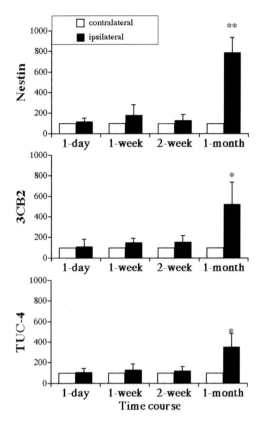

Fig. 1. Western blot analysis of the time course of nestin, 3CB2, and TUC-4 expression in the contralateral and ipsilateral hippocampus 1-day, 1-week, 2-weeks, and 1-month after injury (n = 3 in each group). Values are mean ± SD; *p < 0.05 and **p < 0.01 compared with the contralateral hippocampus

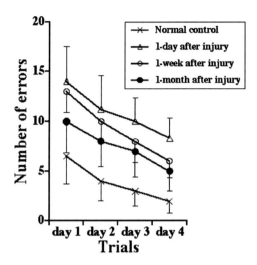

Fig. 2. Number of performance errors for 4 experimental groups in 4 days of trials using the 8-arm radial maze. Neurobehavioral performance deteriorated immediately after injury, (1-day injury group vs control), but then there was a significant improvement over 1-month. Values are mean ± SD, n = 10. (See text for a description of statistics)

Immunohistochemistry

Nestin, 3CB2, and TUC-4 were absent from astrocytes and neurons in sham and normal control groups. After induction of brain injury, nestin, 3CB2, and TUC-4 immunoreactivity was also absent at 1-day, 1- and 2-weeks after injury, similar to the Western blot findings. Nestin expression was observed in the ipsilateral hippocampal stratum pyramidale and radiatum area at 1-month after injury. 3CB2 was expressed in astrocytes and TUC-4 was expressed in neuron-like cells in the ipsilateral hippocampus at 1-month.

Interestingly, nestin-positive neuron-like cells were observed in the ipsilateral hippocampal CA1 cell layer at 1-month after injury. However, there was no immunoreactivity for 3CB2 in the ipsilateral CA1 cell band. The number of nestin-positive cells in the CA1 band was significantly increased at 1-month after injury (18.7 ± 12.7 cells/mm; p < 0.05) compared with normal controls, and 1-day and 1-week injury groups. Nestin-positive neuron-like cells could be observed in each animal 1-month after injury.

Nestin appeared to be localized to both glial cells and neurons in the ipsilateral hippocampus at 1-month after injury, as assessed using double-labeling for nestin and GFAP, nestin and 3CB2, or nestin and TUC-4. Immunoreactivity for both nestin and GFAP was present in the ipsilateral hippocampus CA1 pyramidal cell layer and in the hippocampal striatum radiatum. Interestingly, nestin-positive neuron-like cells that were not GFAP immunoreactive were observed in the ipsilateral hippocampus CA1 cell layer. The nestin-positive neuron-like cells co-expressed TUC-4.

Behavioral test

Figure 2 shows the number of errors made in the 8-arm radial maze. The analysis revealed significant major effects of the groups ($F_{4,108} = 18.0$; $p < 0.01$) and trials ($F_{3,108} = 23.4$; $p < 0.01$). No interaction between the groups and trials was observed ($F_{12,108} = 0.5$; $p > 0.05$). The number of errors in the 1-day, 1-week, and 1-month after injury groups was significantly increased in comparison with those in the sham control group ($p < 0.01$). The number of errors decreased in the 1-month after injury group, a significant difference when compared with the 1-day after injury group ($p < 0.05$). These results indicate that there were marked neurological deficits after injury, with progressive recovery of function over 1-month.

Discussion

Degeneration and remodeling following brain injury

Neuronal cell death after traumatic brain injury can be divided into 2 categories; acute cell death due to primary injury, and delayed cell death as a result of secondary injury. In the present study, there was expansion of lesions stained by TUNEL-positive cells to include positive CA1 pyramidal cells in the hippocampus at 1-week after injury. Thus, delayed cell death appeared to continue for at least 1-week after cold-induced brain injury.

The possibility of neurogenesis exists following acute and delayed cell death in traumatic brain injury. Several studies have indicated some remodeling changes following brain injury [4, 16, 18]. Suzuki and Choi [16] demonstrated repair and reconstruction of the cortical plate following brain injury to the neonatal rat. Dash et al. [4] showed that brain injury increased the production of new granule neurons in dentate gyrus. Yang et al. [18] demonstrated that an increase in reactive astrocytes at the immediate site of cerebral cortical injury in rat was related to tissue remodeling. However, the precise mechanisms involved in remodeling following brain injury are not fully understood. In the present study, dynamic histological changes were seen by immunohistochemical examination.

Precursor cell marker expression in hippocampus following cortical injury

Adult brain tissue consists of neurons and glia that are generated by precursor cells from the embryonic ventricular zone. During postnatal development of the central nervous system, intermediate filament proteins are subjected to a remodeling process [17]. Nestin is a distinct neurofilament protein expressed transiently in immediate precursors to neurons and glia [1] and nestin expression is regarded as correlating with progenitor cells in the central nervous system [3].

In general, glial cells are generated after neurons during development. Radial glia are an exception to this rule, however, being generated before neurogenesis and neuronal migration [14]. Radial glia are mitotically active throughout neurogenesis [9]. Noctor et al. [13] showed that neurons migrate along clonally-related radial glia, and that proliferative radial glia generate neurons. In the present study, the appearance of 3CB2-positive radial glial cells had a similar time course to that of nestin-positive cells. This is the first report of 3CB2 expression following brain injury in an in vivo model. Moreover, nestin immunoreactive cells in the hippocampal CA1 cell layer were GFAP-negative at 1-month after injury and they were neuron-like in appearance. These cells co-expressed the immature neuron marker TUC-4. Whether these cells will develop into mature neurons requires further study. The present and other studies suggest that nestin might play an important role in neuronal remodeling.

Behavioral function after brain injury

Traumatic brain injury can cause neurobehavioral impairment clinically [7] and experimentally [2]. The radial maze has been used to study hippocampal function in ischemic models [6, 12]. In the present study, hippocampal damage could be detected histologically 1-week after injury. Memory function in all of the brain injury groups (1-day, 1-week, and 1-month after injury groups) was significantly decreased compared to sham controls. However, the number of errors was significantly decreased in the 1-month after injury group compared to the 1-day after injury group. This result suggests improvement of memory function long-term after experimental brain injury. The improvement in neurobehavioral function after brain injury may reflect the remodeling changes in the damaged hippocampus.

Conclusions

The present study suggests that delayed nestin expression in glia and neuron-like cells might be a part of adaptation to the injury and recovery of function. Progress in the regenerative treatment using neurotrophic factors [5] and neural stem cells [15] has been remarkable.

Treatment strategies for traumatic brain injury have been directed at attenuating secondary or delayed injury. Our data shows that although experimental brain injury causes neuronal death, there is the possibility of remodeling. As damaged brain tissue has the potential for neurogenesis, it is important to utilize this potential in treatment of traumatic brain injury.

Acknowledgments

This study was supported by a Grant-in-Aid for Scientific Research from the Japanese Society for the Promotion of Science. The authors thank Drs. Testuro Negi, Shin-ichi Yamagami and Kazunori Sumitani (Kagawa University Faculty of Medicine) for their excellent technical assistance.

References

1. Clarke SR, Shetty AK, Bradley JL, Turner DA (1994) Reactive astrocytes express the embryonic intermediate neurofilament nestin. Neuroreport 5: 1885–1888
2. Colicos MA, Dixon CE, Dash PK (1996) Delayed, selective neuronal death following experimental cortical impact injury in rats: possible role in memory deficits. Brain Res 739: 111–119
3. Dahlstrand J, Lardelli M, Lendahl U (1995) Nestin mRNA expression correlates with the central nervous system progenitor cell state in many, but not all, regions of the developing central nervous system. Brain Res Dev Brain Res 84: 109–129
4. Dash PK, Mach SA, Moore AN (2001) Enhanced neurogenesis in the rodent hippocampus following traumatic brain injury. J Neurosci Res 63: 313–319
5. Galvin KA, Oorschot DE (2003) Continuous low-dose treatment with brain-derived neurotrophic factor or neurotrophin-3 protects striatal medium spiny neurons from mild neonatal hypoxia/ischemia: a stereological study. Neuroscience 118: 1023–1032
6. Katoh A, Ishibashi C, Shiomi T, Takahara Y, Eigyo M (1992) Ischemia-induced irreversible deficit of memory function in gerbils. Brain Res 577: 57–63
7. Kunishio K, Matsumoto Y, Kawada S, Miyoshi Y, Matsuhisa T, Moriyama E, Norikane H, Tanaka R (1993) Neuropsychological outcome and social recovery of head-injured patients. Neurol Med Chir (Tokyo) 33: 824–829
8. Malhotra SK, Shnitka TK, Elbrink J (1990) Reactive astrocytes-a review. Cytobios 61: 133–160
9. Misson JP, Edwards MA, Yamamoto M, Caviness VS Jr (1988) Mitotic cycling of radial glial cells of the fetal murine cerebral wall: a combined autoradiographic and immunohistochemical study. Brain Res 466: 183–190
10. Nakamura T, Miyamoto O, Kawai N (1999) Therapeutic window of hypothermia after cold-induced injury in mice. In: Chiu, WT (ed) International conference on recent advances in neurotraumatology. Monduzzi Editore, Bologna, Italy, pp 187–191
11. Nakamura T, Miyamoto O, Yamagami S, Toyoshima T, Negi T, Itano T, Nagao S (1999) The chronic cell death with DNA fragmentation after post-ischaemic hypothermia in the gerbil hippocampus. Acta Neurochir (Wien) 141: 407–412; discussion 412–413
12. Nakamura T, Miyamoto O, Kawai N, Negi T, Itano T, Nagao S (2001) Long-term activation of the glutamatergic system associated with N-methyl-D-aspartate receptors after postischemic hypothermia in gerbils. Neurosurgery 49: 706–713; discussion 713–714
13. Noctor SC, Flint AC, Weissman TA, Dammerman RS, Kriegstein AR (2001) Neurons derived from radial glial cells establish radial units in neocortex. Nature 409: 714–720
14. Rakic P (1972) Model of cell migration to the superficial layers of fetal monkey neocortex. J Comp Neurol 145: 61–83
15. Riess P, Zhang C, Saatman KE, Laurer HL, Longhi LG, Raghupathi R, Lenzlinger PM, Lifshitz J, Boockvar J, Neugebauer E, Snyder EY, McIntosh TK (2002) Transplanted neural stem cells survive, differentiate, and improve neurological motor function after experimental traumatic brain injury. Neurosurgery 51: 1043–1052; discussion 1052–1054
16. Suzuki M, Choi BH (1991) Repair and reconstruction of the cortical plate following closed cryogenic injury to the neonatal rat cerebrum. Acta Neuropathol (Berl) 82: 93–101
17. Wei LC, Shi M, Chen LW, Cao R, Zhang P, Chan YS (2002) Nestin-containing cells express glial fibrillary acidic protein in the proliferative regions of central nervous system of postnatal developing and adult mice. Brain Res Dev Brain Res 139: 9–17
18. Yang HY, Lieska N, Kriho V, Wu CM, Pappas GD (1997) A subpopulation of reactive astrocytes at the immediate site of cerebral cortical injury. Exp Neurol 146: 199–205

Correspondence: Takehiro Nakamura, Department of Neurological Surgery, Kagawa University Faculty of Medicine, 1750-1 Ikenobe, Miki-cho, Kita-gun, Kagawa 761-0793 Japan. e-mail: nakamura@umich.edu

Granulocyte colony-stimulating factor does not affect contusion size, brain edema or cerebrospinal fluid glutamate concentrations in rats following controlled cortical impact

O. W. Sakowitz[1], C. Schardt[1], M. Neher[1], J. F. Stover[2], A. W. Unterberg[1], and K. L. Kiening[1]

[1] Department of Neurosurgery, University of Heidelberg, Heidelberg, Germany
[2] Department of Surgical Intensive Care, University Hospital, Zurich, Switzerland

Summary

Introduction. Granulocyte colony-stimulating factor (G-CSF) is an established treatment in the neutropenic host. Usage in head-injured patients at risk for infection may aggravate brain damage. In contrast, evidence of G-CSF neuroprotective effects has been reported in rodent models of focal cerebral ischemia. We investigated effects of G-CSF in acute focal traumatic brain injury (TBI) in rats.

Methods. Thirty-six male Sprague-Dawley rats were anesthetized with 1.2% to 2.0% isoflurane and subjected to controlled cortical impact injury (CCII). Thirty minutes following CCII, either vehicle or G-CSF was administered intravenously. Animals were sacrificed 24 hours following CCII. Glutamate concentrations were determined in cisternal cerebrospinal fluid (CSF). Brain edema was assessed gravimetrically. Contusion size was estimated by 2,3,5-triphenyltetrazolium chloride staining and volumetric analysis.

Results. Dose-dependent leukocytosis was induced by infusion of G-CSF. Physiological variables were unaffected. Water content of the traumatized hemisphere and CSF glutamate concentrations were unchanged by treatment. Contusion volume was similar in all groups.

Conclusions. A single injection of G-CSF did not influence cortical contusion volume, brain edema, or glutamate concentrations in CSF determined 24 hours following CCII in rats. G-CSF, administered 30 minutes following experimental TBI, failed to exert neuroprotective effects.

Keywords: Cranio-cervical trauma; glutamate; growth factor; neuroprotection; traumatic brain injury.

Introduction

Multifold neuroprotective treatment strategies have been explored to counter the dreaded sequelae of traumatic brain injury (TBI). While it is generally accepted that the main target is to avoid or ameliorate secondary injuries, the primary injuries are often not amenable to treatment. Yet, less severely injured neural tissue could potentially be resuscitated by the same pathways that stimulate cell division, growth, and survival. As such, cytokine growth factors have increasingly gained attention in recent years [1, 11, 18].

Cytokine granulocyte colony-stimulating factor (G-CSF) is a 19.6 kD 175 amino acid polypeptide secreted by monocytes, macrophages, and neutrophils after cell activation. Its role in hematopoiesis is stimulation of the neutrophil/granulocyte cell lineage [8].

Recent experimental studies suggest that G-CSF, as a growth factor, could be neuroprotective in models of glutamate toxicity, focal ischemia, and intracerebral hemorrhage, and improve neurological outcome [6, 13–16]. Inducible expression of G-CSF and its respective receptor has been demonstrated in neural tissues [10, 11, 14, 18]. The underlying mechanisms of neuroprotection are not fully understood, but anti-apoptotic signaling [14], hematopoietic stem-cell liberation [5, 16, 20], and neoangiogenesis [4, 16] may play a role.

The current study tested whether or not G-CSF administered soon after controlled cortical impact injury (CCII) reduces brain edema formation and contusion volumes 24 hours after injury.

Methods

Animals

For the present study, a total of 36 male Sprague-Dawley rats (347 ± 52 g) (Charles River, Germany) were used. Animals were accustomed to the laboratory for approximately 24 hours before the study was performed. The experimental protocol was approved by the committee for animal research in Karlsruhe, Germany (35-9185.81/G165/03).

Table 1. *Physiological variables determined before trauma, before and after infusion, and at 24 hours following trauma (N = 6 per treatment group)**

	Granulocyte Colony-Stimulating Factor			
	Vehicle	10 µg/kg	100 µg/kg	p
Mean arterial pressure (mmHg)				
Before CCII/infusion	72 ± 5	73 ± 3	73 ± 6	NS
After CCII/infusion	78 ± 5	79 ± 8	79 ± 8	NS
Heart rate (bpm)				
Before CCII/infusion	348 ± 37	356 ± 43	343 ± 18	NS
After CCII/infusion	362 ± 42	379 ± 39	369 ± 43	NS
Core temperature (°C)				
Before CCII/infusion	36.6 ± 1.4	36.6 ± 0.9	36.8 ± 0.5	NS
After CCII/infusion	36.4 ± 0.3	36.4 ± 0.3	36.3 ± 0.3	NS
pH				
Before CCII/infusion	7.38 ± 0.05	7.4 ± 0.03	7.41 ± 0.05	NS
After CCII/infusion	7.37 ± 0.02	7.36 ± 0.04	7.38 ± 0.04	NS
24 h after CCII	7.46 ± 0.04	7.45 ± 0.04	7.47 ± 0.04	NS
pCO_2 (mmHg)				
Before CCII/infusion	42.8 ± 8.4	43.0 ± 2.5	41.4 ± 5.3	NS
After CCII/infusion	43.0 ± 4.3	43.7 ± 2.9	39.9 ± 6.5	NS
24 h after CCII	36.2 ± 5.7	36.9 ± 4.5	34.8 ± 6.2	NS
pO_2 (mmHg)				
Before CCII/infusion	145.5 ± 12.9	150.5 ± 13.9	146.8 ± 7.1	NS
After CCII/infusion	155.9 ± 4.4	149.9 ± 11.9	151.5 ± 21.2	NS
24 h after CCII	150.4 ± 28.8	124.9 ± 24.7	142.7 ± 24.8	NS
Glucose (g/dL)				
After CCII/infusion	225 ± 40	207 ± 35	223 ± 51	NS
24 h after CCII	197 ± 23	183 ± 26	175 ± 20	NS
Hemoglobin (g/dL)				
Before CCII/infusion	14.7 ± 1.2	15.1 ± 1.3	14.4 ± 1.3	NS
After CCII/infusion	14.8 ± 1.4	14.6 ± 0.6	13.5 ± 1.8	NS
24 h after CCII	12.3 ± 1.1	11.9 ± 1.5	10.8 ± 2.0	NS
Leukocytes (nL)				
Before CCII/infusion	8.6 ± 2.5	9.2 ± 2.2	9.0 ± 1.7	NS
24 h after CCII	8.3 ± 2.9	9.2 ± 3.9	12.3 ± 6.0**	<0.05

* Values are mean ± SD, ** $p < 0.05$ compared to vehicle-treated animals. *NS* Not significant.

Experimental procedures

After induction of anesthesia with isoflurane (5%, 3 minutes), animals were weighed and positioned. All animals breathed spontaneously under inhalational anesthesia (isoflurane: 1.0% to 2.5%; fraction of inspirational oxygen 30%). The caudal artery was catheterized to allow continuous recording of mean arterial blood pressure and serial blood sampling to determine arterial blood gases. The right femoral vein was cannulated with care not to injure the femoral nerve. Lines were locked with normal saline, or perfused at a rate of 0.5 mL/h, respectively. Samples for differential blood cell counts were obtained at the beginning and at the end of all experiments. Rectal temperature was measured during anesthesia and maintained between 37° and 38 °C using a homeothermic heating pad.

Brain trauma was induced with the CCII device using a 5 mm bolt which was pneumatically driven at a velocity of 7 m/sec^{-1} (100 p.s.i.), a penetration depth of 1 mm, and a contact time of 0.15 second, as described previously [17].

At 30 minutes after CCII, either G-CSF at standard (10 µg/kg) or high dose (100 µg/kg) was infused intravenously over 30 minutes in a randomized and blinded fashion. Control animals received iso-volume amounts of vehicle only.

Following infusion, catheters were removed and the animals returned to their cages with free access to food and water for 24 hours.

Outcome variables

At 24 hours following CCII, animals were deeply anaesthetized, exsanguinated, and cerebrospinal fluid (CSF) was carefully aspirated from the cerebromedullary cistern by puncture with a 27-gauge needle. Paired samples of CSF and heparinized arterial plasma were deproteinized using centrifugal filtration through 10.000 kD cut-off polyethersulfone membranes (Vivaspin 500, Vivascience, Hannover, Germany). Glucose, lactate, and glutamate were determined in protein-free samples using the CMA600 enzyme-kinetic analyzer (CMA, Solna, Sweden). Brains were removed quickly, either for gravimetric analysis of hemispheric water content (n = 18) or staining of 6 2-mm sections in a 2% solution of 2,3,5-triphenyltetrazolium chloride (n = 18). Stained slices were immersed in 4% paraformaldehyde and scanned for determination of contusion volumes using

ImageJ 1.6 software (National Institutes of Health, Bethesda, MD, USA).

Statistical analysis

Statistical analysis was performed using SPSS 13.0 software (SPSS Inc., Chicago, IL, USA). Data averaged among groups are reported as mean ± SD. For testing of averages and differences between groups, analysis of variance (ANOVA) was employed. Post-hoc *t* tests for paired comparisons were adjusted according to the Bonferroni method. A p-value of less than 0.05 was considered significant.

Results

Physiological variables, i.e., blood pressure, heart rate, rectal temperature, arterial blood gases, hemoglobin, or serum glucose concentrations, remained in the physiological range at all times. No significant differences were seen among groups. High-dose G-CSF treatment resulted in moderate leukocytosis (12 ± 6 nL. vs. 8 ± 3) in vehicle controls (Table 1).

Metabolic parameters were determined in CSF obtained from the cerebromedullary cistern postmortem. Lactate/glucose ratios were calculated as an index of anaerobic neurometabolism. No significant differences in CSF glucose, lactate, or glutamate were found in comparison to vehicle-treated animals (Table 2).

Hemispheric water content was increased in the left (traumatized) hemisphere, but unaffected by either dose of G-CSF (Fig. 1). Contusion volumes varied within treatment groups, but distributions were not altered when compared to control animals (29 ± 21, 24 ± 23, and 33 ± 31 mm³, respectively) (Fig. 2).

Discussion

Lack of efficacy in our study is in contradiction with multiple experimental studies of ischemia models [6, 14, 16]. Only in one other experimental TBI study was G-CSF tested with similar results, despite continued treatment for 7 days following trauma [15].

Ischemia and TBI have many pathophysiological features in common. Still they are different disease entities, and mechanisms involved in ischemia can have a lesser importance in TBI [2]. One can only speculate that transient vessel occlusion and reperfusion *per se* provides a better ground for cytokine growth factor treatment. Likewise, the primary injury caused by mechanical disruption of the cortex in our model might have been too advanced for salvage by G-CSF treatment.

At present, neither dose-response curves, nor optimal dosing regimens have been established for neuro-

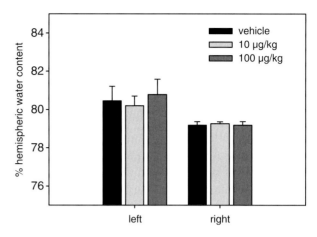

Fig. 1. Hemispheric water content derived from wet/dry weights and according to treatment 24 hours following focal contusion of the left parietal cortex. Error-bars indicate mean ± SD. N = 6 per treatment group

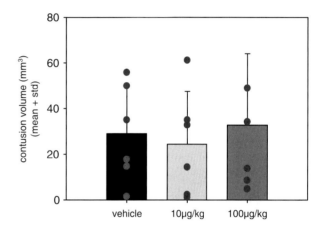

Fig. 2. Contusion volumes determined by computerized volumetric analysis of 2,3,5-triphenyltetrazolium chloride-stained brain slices 24 hours following injury. Error-bars indicate mean ± SD. N = 6 per treatment group

Table 2. *Concentrations of metabolites glucose, lactate, and glutamate determined in cisternal cerebrospinal fluid at 24 hours following trauma (N = 6 per treatment group)*

	Granulocyte Colony-Stimulating Factor			
	Vehicle	10 µg/kg	100 µg/kg	p
Glucose (mmol)	3.1 ± 0.1	3.2 ± 0.3	3.5 ± 0.6	NS
Lactate (mmol)	6.5 ± 0.8	6.7 ± 0.9	6.4 ± 1.0	NS
Glucose/lactate ratio	2.1 ± 0.3	2.0 ± 0.2	1.8 ± 0.5	NS
Glutamate (µmol)	13.7 ± 12.3	12.4 ± 5.8	12.7 ± 6.9	NS

* Values are mean ± SD. Lactate/glucose ratios calculated individually. *NS* Not significant.

protective G-CSF treatment. Recombinant G-CSF is FDA-approved for treatment of the neutropenic host. The standard dose is 10 μg/kg/day. In mice, rats, hamsters, dogs, and monkeys, doses of >1 μg/kg/day of G-CSF stimulate neutrophil production in bone marrow. Untoward effects are rare, with the only irreversible effects seen in monkeys at doses exceeding 1 mg/kg/day when hyperleukocytosis and cerebral leukostasis caused intracerebral hemorrhage and infarction [9]. Clinical studies in patients with TBI have indicated its safety, though no clear benefit in prevention of nosocomial infections were demonstrated [7].

The dosages used in our study can be regarded as safe. The neuroprotective effects found by others were observed with intravenous or subcutaneous application of similar doses. Single doses as well as repeated injections have been shown to convey neuroprotection [13, 14, 16].

Although intrathecal synthesis of G-CSF and its receptor is known to occur, blood-brain barrier (BBB) penetration of extracerebral G-CSF has not been proven. However, cytokine growth factors with similar protein characteristics (e.g., granulocyte-macrophage colony-stimulating factor, erythropoietin), do penetrate [3, 12]. In a pretreatment study of experimental brain injury in rats, G-CSF increased BBB damage, but development of cerebral edema and leukocyte-infiltration remained unchanged. Treatment-related BBB disruption is less likely to occur as a direct effect of G-CSF compared to the secondary effects of systemic leukocytosis [19]. With a single application of G-CSF leukocytosis is mild, however, as shown in this study.

In summary, a potential neuroprotective effect of G-CSF in TBI cannot be ruled out. Further studies are needed, where optimal physiological, histopathological, and functional outcome parameters are followed for times exceeding our time-window of observation. Additionally, details of drug delivery and dose-response curves need to be established.

Conclusion

In a rodent model of focal cortical contusion, a single injection of G-CSF at 10 μg/kg or 100 μg/kg did not influence cortical contusion volume, brain edema, or glutamate concentrations in CSF determined 24 hours following CCII. Overall, G-CSF administered 30 minutes following experimental TBI failed to exert neuroprotective effects.

Acknowledgments

We thank Prof. S. Schwab (Dept. of Neurology, Heidelberg) for his help with our study. Test substance and randomization, provided by Dr. A. Schneider and coworkers (Axaron Bioscience AG, Heidelberg), is gratefully acknowledged. Laboratory space was generously allocated by Prof. W. Kuschinsky (Dept. of Physiology, Heidelberg).

References

1. Berezovskaya O, Maysinger D, Fedoroff S (1996) Colony stimulating factor-1 potentiates neuronal survival in cerebral cortex ischemic lesion. Acta Neuropathol (Berl) 92: 479–486
2. Bramlett HM, Dietrich WD (2004) Pathophysiology of cerebral ischemia and brain trauma: similarities and differences. J Cereb Blood Flow Metab 24: 133–150
3. Brines ML, Ghezzi P, Keenan S, Agnello D, de Lanerolle NC, Cerami C, Itri LM, Cerami A (2000) Erythropoietin crosses the blood-brain barrier to protect against experimental brain injury. Proc Natl Acad Sci USA 97: 10526–10531
4. Chen J, Zhang ZG, Li Y, Wang L, Xu YX, Gautam SC, Lu M, Zhu Z, Chopp M (2003) Intravenous administration of human bone marrow stromal cells induces angiogenesis in the ischemic boundary zone after stroke in rats. Circ Res 92: 692–699
5. Chopp M, Li Y (2002) Treatment of neural injury with marrow stromal cells. Lancet Neurol 1: 92–100
6. Gibson CL, Bath PM, Murphy SP (2005) G-CSF reduces infarct volume and improves functional outcome after transient focal cerebral ischemia in mice. J Cereb Blood Flow Metab 25: 431–439
7. Heard SO, Fink MP, Gamelli RL, Solomkin JS, Joshi M, Trask AL, Fabian TC, Hudson LD, Gerold KB, Logan ED (1998) Effect of prophylactic administration of recombinant human granulocyte colony-stimulating factor (filgrastim) on the frequency of nosocomial infections in patients with acute traumatic brain injury or cerebral hemorrhage. The Filgrastim Study Group. Crit Care Med 26: 748–754
8. Hill CP, Osslund TD, Eisenberg D (1993) The structure of granulocyte-colony-stimulating factor and its relationship to other growth factors. Proc Natl Acad Sci USA 90: 5167–5171
9. Keller P, Smalling R (1993) Granulocyte colony stimulating factor: animal studies for risk assessment. Int Rev Exp Pathol 34 Pt A: 173–188
10. Kleinschnitz C, Schroeter M, Jander S, Stoll G (2004) Induction of granulocyte colony-stimulating factor mRNA by focal cerebral ischemia and cortical spreading depression. Brain Res Mol Brain Res 131: 73–78
11. Malipiero UV, Frei K, Fontana A (1990) Production of hemopoietic colony-stimulating factors by astrocytes. J Immunol 144: 3816–3821
12. McLay RN, Kimura M, Banks WA, Kastin AJ (1997) Granulocyte-macrophage colony-stimulating factor crosses the blood-brain and blood-spinal cord barriers. Brain 120: 2083–2091
13. Park HK, Chu K, Lee ST, Jung KH, Kim EH, Lee KB, Song YM, Jeong SW, Kim M, Roh JK (2005) Granulocyte colony-stimulating factor induces sensorimotor recovery in intracerebral hemorrhage. Brain Res 1041: 125–131
14. Schabitz WR, Kollmar R, Schwaninger M, Juettler E, Bardutzky J, Scholzke MN, Sommer C, Schwab S (2003) Neuroprotective effect of granulocyte colony-stimulating factor after focal cerebral ischemia. Stroke 34: 745–751

15. Sheibani N, Grabowski EF, Schoenfeld DA, Whalen MJ (2004) Effect of granulocyte colony-stimulating factor on functional and histopathologic outcome after traumatic brain injury in mice. Crit Care Med 32: 2274–2278
16. Shyu WC, Lin SZ, Yang HI, Tzeng YS, Pang CY, Yen PS, Li H (2004) Functional recovery of stroke rats induced by granulocyte colony-stimulating factor-stimulated stem cells. Circulation 110: 1847–1854
17. Stover JF, Unterberg AW (2000) Increased cerebrospinal fluid glutamate and taurine concentrations are associated with traumatic brain edema formation in rats. Brain Res 875: 51–55
18. Wesselingh SL, Gough NM, Finlay-Jones JJ, McDonald PJ (1990) Detection of cytokine mRNA in astrocyte cultures using the polymerase chain reaction. Lymphokine Res 9: 177–185
19. Whalen MJ, Carlos TM, Wisniewski SR, Clark RS, Mellick JA, Marion DW, Kochanek PM (2000) Effect of neutropenia and granulocyte colony stimulating factor-induced neutrophilia on blood-brain barrier permeability and brain edema after traumatic brain injury in rats. Crit Care Med 28: 3710–3717
20. Willing AE, Vendrame M, Mallery J, Cassady CJ, Davis CD, Sanchez-Ramos J, Sanberg PR (2003) Mobilized peripheral blood cells administered intravenously produce functional recovery in stroke. Cell Transplant 12: 449–454

Correspondence: Oliver W. Sakowitz, Department of Neurosurgery, University of Heidelberg, Im Neuenheimer Feld 400, 69120 Heidelberg, Germany. e-mail: oliver.sakowitz@med.uni-heidelberg.de

Unilateral spatial neglect and memory deficit associated with abnormal β-amyloid precursor protein accumulation after lateral fluid percussion injury in Mongolian gerbils

S. Li[1], T. Kuroiwa[2], N. Katsumata[2], S. Ishibashi[3], L. Sun[3], S. Endo[4], and K. Ohno[1]

[1] Department of Neurosurgery, Graduate School of Medicine, Tokyo Medical and Dental University, Tokyo, Japan
[2] Department of Neuropathology, Graduate School of Medicine, Tokyo Medical and Dental University, Tokyo, Japan
[3] Department of Neurology, Graduate School of Medicine, Tokyo Medical and Dental University, Tokyo, Japan
[4] Department of Animal Research Center, Graduate School of Medicine, Tokyo Medical and Dental University, Tokyo, Japan

Summary

The purpose of this study was to investigate cognitive/memory dysfunctions and the pathological process contributing to such dysfunction following moderate lateral fluid percussion injury (LFPI) in Mongolian gerbils.

Mongolian gerbils were subjected to moderate LFPI (1.3–1.6 atm). During 7 days post-trauma, spatial cognitive and memory dysfunctions were evaluated by T-maze test (TMT). At 6 hours, 24 hours, and 7 days post-injury, animals were sacrificed and the brains were prepared for Kluver-Barrera staining and immunostaining of β-amyloid precursor protein (APP).

In LFPI animals, the spontaneous alternation rate in the TMT remained below the random alternation rate (<50%) on all post-injury test days. These animals also showed a transient tendency to choose only the right arm (ipsilateral to the injury) in the TMT at 6 hours and 24 hours after injury. Significant accumulation of APP was found widespread in the ipsilateral hemisphere including directly injured cortex, subcortical white matter, and hippocampal formation at 6 hours and 24 hours post-injury, while on day 7, the increased immunoreactivity of APP subsided.

These results suggest that the widespread axonal degeneration of the white matter might contribute to the unilateral spatial neglect and memory deficit in the acute stage after LFPI.

Keywords: Diffuse axonal injury; fluid percussion injury; gerbil; spatial neglect.

Introduction

Cognitive dysfunctions including memory impairment are common neurological sequelae of traumatic brain injury (TBI), and their underlying mechanisms are poorly understood [9]. Smith *et al.* [11] concluded that posttraumatic spatial learning and memory dysfunctions primarily result from the selective vulnerability of the hippocampus. However, it is becoming evident that hippocampal neuronal death may not account for the entire spectrum of cognitive deficiency [4, 10]. Evidence has shown that white matter injury may also be a major determinant of clinical outcome. The studies of TBI in both human and in a non-human primate model of experimental injury suggest a direct link between the extent of diffuse axonal injury and the ensuing patient morbidity [1, 2].

Lateral fluid percussion injury (LFPI) replicates several clinically relevant features of human TBI and has been widely used in investigations of the biomechanical responses, neurological syndromes, and pathology observed in human closed-head injury [7]. In the present study, we induced LFPI in Mongolian gerbils, an excellent animal model for assessing behavior after experimental cerebral ischemia. During 7 days post-trauma, cognitive/working memory deficit was evaluated by T-maze test (TMT). Immunohistochemistry of β-amyloid precursor protein (APP) was performed to detect axonal injury, and the correlation between behavioral dysfunction and the abnormal accumulation of APP was examined.

Materials and methods

The animal experiments were approved by the Animal Care and Use Committee of Tokyo Medical and Dental University. Thirty male Mongolian gerbils ranging in age from 6 to 8 months and in weight from 65 to 80 g were used for this experiment. The animals were housed in groups of 3 and maintained on a 12-hour light/dark cycle with unlimited access to food and water.

Surgery and fluid percussion injury

Thirty animals were randomly divided into 2 groups of 15: a sham-operated group (SHAM) and an LFPI group (LFPI). LFPI was induced as previously described [8]. Briefly, each animal was anesthetized with ketamine hydrochloride (50 mg/kg, i.m.), supplemented as necessary. All wounds were infiltrated with 2.0% lidocaine hydrochloride during surgical preparation and throughout the experiment. The animals were allowed to breathe spontaneously throughout all surgical procedures. Each animal was placed in a stereotaxic frame and a round craniotomy (3.5 mm in diameter) was made on the right parietal cortex with center coordinates midway between the bregma and lambda and 2.5 mm lateral to the midline. LFPI of moderate severity (1.3–1.6 atm) was induced. SHAM animals received anesthesia and underwent all of the surgical procedures except delivery of the LFPI.

T-maze test

Animals performed the TMT at pre-injury, 6 hours, 24 hours, and 3, 5, and 7 days post-injury, respectively. Each animal was allowed to alternate between the left and right goal arms of a T-shaped maze (60 (stem) × 25 (arm) × 10 (width) cm) throughout a 15-trial continuous alternation session. The animal's behavior was traced with a video-tracking system (Bio Research Center, Nagoya, Japan). The spontaneous alternation rate (SAR) was calculated as the ratio of alternating choices to total number of choices (50%, random choice; 100%, alternation at every trial; 0%, no alternation). We also calculated the percentage of choices of the goal arm ipsilateral to the injured hemisphere (right-biased rate).

Histology and APP immunohistochemistry

Animals were anesthetized and perfused with 4% paraformaldehyde at 6 hours (LFPI-6 h, n = 3; SHAM-6 h, n = 3), 24 hours (LFPI-24 h, n = 3; SHAM-24 h, n = 3), and 7 days (LFPI-7 days, n = 9; SHAM-7 days, n = 9). The brain was cut into 6 serial coronal sections every 2.0 mm from the level of the anterior pole of the caudate nucleus to the posterior pole. The sections were embedded in paraffin. Each coronal section was sliced to a thickness of 4 μm. Kluver-Barrera staining and immunostaining of APP were performed. For APP immunostaining, primary antibody (clone 22C11; dilution 1:500; Boehringer Mannheim, Mannheim, Germany) was applied and slides were placed in the refrigerator overnight. Further rinsing was done with phosphate-buffered saline and secondary antibody applied. Avidin-Biotin (Vector Laboratories, Burlingame, CA) complex was used for antibody detection along with 3,3′-diaminobenzidine tetrahydrochloride to increase staining intensity. Counterstaining was done with hematoxylin and tissue was then dehydrated and cover-slipped.

Results

The animals tended to choose the left and right arms with equal frequency in the TMT before LFPI. In the LFPI group, the SAR remained below the random alternation rate (<50%) on all post-injury test days (Fig. 1B).

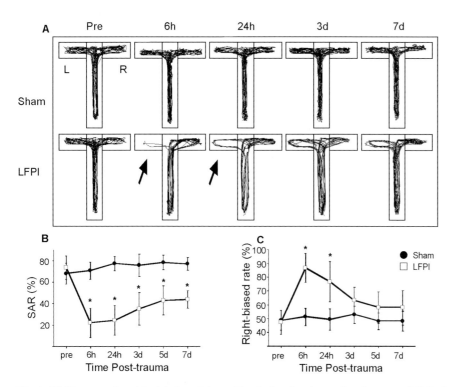

Fig. 1. TMT was employed to detect spatial cognitive dysfunction (arrows) and memory deficit after LFPI (A). In the LFPI group, the SAR remained below the random alternation rate (<50%) on all post-injury test days (B). A transient tendency to choose only the right arm (ipsilateral to the injury) was also found in TMT at 6 hours and 24 hours after the injury (C). Data are presented as mean ± SD, n = 18. *p < 0.05

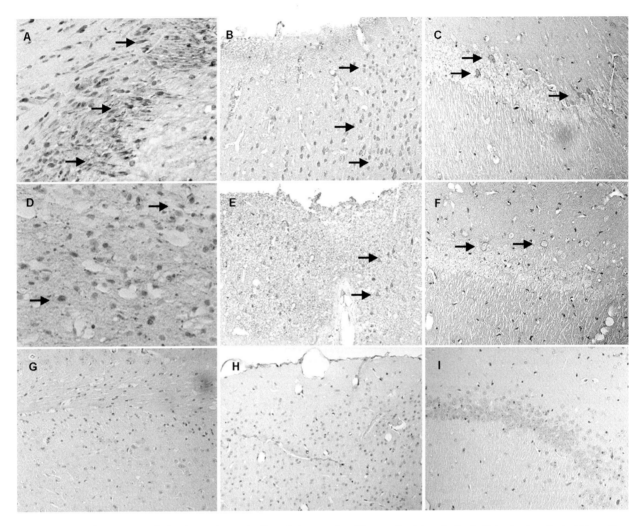

Fig. 2. Abnormal APP accumulation (arrows) was significantly increased at 6 hours after injury in the ipsilateral subcortical white matter, cortex, and hippocampus (A–C). At day 7, the APP accumulation in these regions subsided but could still be found (D–F) when compared with the sham group (G–I).

The LFPI animals also showed a transient tendency to choose only the right arm (ipsilateral to the injury) in the TMT after the injury (Fig. 1A, C). From day 3, no significant biased alternation was found in this group. Animals in the SHAM group showed no biases in the TMTs at any point during the testing period.

On day 7 after injury, Kluver-Barrera staining revealed cortical necrosis at the impact site, widespread subcortical white matter rarefaction, and neuronal loss in the ipsilateral CA3 region and dentate gyrus. Significant widespread accumulation of APP was found in the ipsilateral hemisphere including the directly injured cortex, subcortical white matter, and hippocampal formation at 6 hours and 24 hours postinjury. On day 7, the increased immunoreactivity of APP subsided but could still be found in these regions (Fig. 2).

Discussion

In the present study, gerbils were used to evaluate posttraumatic spatial cognitive and memory deficits using a T-maze, a reliable and easily-operated tool for cognitive evaluation after brain injury [3]. The T-maze spontaneous alternation task has been reported as being effective for testing exploratory behavior and working memory. In animals with disturbed cognition/memory caused by hippocampal injury, the SAR is known to decrease to approximately 50%, the level of random choice [3]. In the present study, the SAR was significantly lowered at 6 hours and 24 hours in the LFPI animals (below 25% on average). On following test days it returned to the level of approximately the random alternation rate (50%), indicative of a significant and prolonged memory impairment caused by LFPI. However, the very low SAR at 6 hours and 24 hours cannot be explained by posttraumatic memory deficit alone, but also involves unilateral spatial neglect. Analysis of right-biased alternation in the TMT also indicated an acute visual neglect of the side contralateral to the injury (Fig. 1C). Interestingly, we found the temporal profile of the immunoreactivity of APP to be very similar to that of unilateral neglect. APP has been found to be a useful marker for axonal damage. The interruption of axonal transport due to traumatic or ischemic insults has been shown to result in the accumulation of APP, indicative of dysfunction and perhaps eventual discontinuity of the axon [14].

Taken together, findings from the present study strongly suggest the involvement of white matter damage in the patho-mechanisms of acute unilateral spatial neglect following LFPI. In clinical cases, unilateral spatial neglect is usually associated with lesions in the parieto-occipital lobes of the non-dominant hemisphere, and its mechanism is still unclear [5]. Vallar [13] suggested that unilateral spatial neglect is a multifarious disorder, frequently associated with extensive subcortical lesions involving the thalamus, the basal ganglia, and the subcortical white matter. A recent clinical study of lesion anatomy indicated that patients with unilateral neglect and a spatial working memory deficit were most likely to have damage to parietal white matter [6]. Therefore, we concluded the acute unilateral spatial neglect found at 6 hours and 24 hours was associated with the acute axonal injury reflected in the extensively and significantly increased APP immunoreactivity at the same time points.

The abnormal APP accumulation found in our present study showed a diffuse pattern in the acute phase after the LFPI, indicating that LFPI in gerbils is a useful animal model for investigation of diffuse axonal injury. Some studies have indicated that diffuse axonal injury may mediate significant memory and learning deficits [12]. Theoretically, the interruption of a sufficient number of axons could disrupt the flow of information in the brain and produce neurological deficits. Further study should be performed to clarify the effect of axonal damage in different brain regions on functional deficits following TBI of various severities. A combination of TMT and immunohistochemical markers should be useful in such a study.

References

1. Adams JH, Doyle D, Ford I, Gennarelli TA, Graham DI, McLellan DR (1989) Diffuse axonal injury in head injury: definition, diagnosis and grading. Histopathology 15: 49–59
2. Gennarelli TA, Thibault LE, Adams JH, Graham DI, Thompson CJ, Marcincin RP (1982) Diffuse axonal injury and traumatic coma in the primate. Ann Neurol 12: 564–574
3. Gerlai R (1998) A new continuous alternation task in T-maze detects hippocampal dysfunction in mice. A strain comparison and lesion study. Behav Brain Res 95: 91–101
4. Lyeth BG, Jenkins LW, Hamm RJ, Dixon CE, Phillips LL, Clifton GL, Young HF, Hayes RL (1990) Prolonged memory impairment in the absence of hippocampal cell death following traumatic brain injury in the rat. Brain Res 526: 249–258
5. Maeshima S, Truman G, Smith DS, Dohi N, Nakai K, Itakura T, Komai N (1997) Is unilateral spatial neglect a single phenomenon? A comparative study between exploratory-motor and visual-counting tests. J Neurol 244: 412–417
6. Malhotra P, Jager HR, Parton A, Greenwood R, Playford ED, Brown MM, Driver J, Husain M (2005) Spatial working memory capacity in unilateral neglect. Brain 128: 424–435
7. McIntosh TK, Vink R, Noble L, Yamakami I, Fernyak S, Soares H, Faden AL (1989) Traumatic brain injury in the rat: characterization of a lateral fluid-percussion model. Neuroscience 28: 233–244
8. Qian L, Ohno K, Maehara T, Tominaga B, Hirakawa K, Kuroiwa T, Takakuda K, Miyairi H (1996) Changes in ICBF, morphology and related parameters by fluid percussion injury. Acta Neurochir (Wien) 138: 90–98
9. Sanders MJ, Sick TJ, Perez-Pinzon MA, Dietrich WD, Green EJ (2000) Chronic failure in the maintenance of long-term potentiation following fluid percussion injury in the rat. Brain Res 861: 69–76
10. Scheff SW, Baldwin SA, Brown RW, Kraemer PJ (1997) Morris water maze deficits in rats following traumatic brain injury: lateral controlled cortical impact. J Neurotrauma 14: 615–627
11. Smith DH, Okiyama K, Thomas MJ, Claussen B, McIntosh TK (1991) Evaluation of memory dysfunction following experimental brain injury using the Morris water maze. J Neurotrauma 8: 259–269
12. Uzzell BP, Dolinskas CA, Wiser RF, Langfitt TW (1987) Influence of lesions detected by computed tomography on outcome and neuropsychological recovery after severe head injury. Neurosurgery 20: 396–402
13. Vallar G (2001) Extrapersonal visual unilateral spatial neglect and its neuroanatomy. Neuroimage 14: S52–S58
14. Yam PS, Takasago T, Dewar D, Graham DI, McCulloch J (1997) Amyloid precursor protein accumulates in white matter at the margin of a focal ischaemic lesion. Brain Res 760: 150–157

Correspondence: Toshihiko Kuroiwa, Department of Neuropathology, Tokyo Medical and Dental University, Yushima 1-5-45, Bunkyo-Ku, Tokyo 113-8510, Japan. e-mail: t.kuroiwa.npat@mri.tmd.ac.jp

Alteration of gap junction proteins (connexins) following lateral fluid percussion injury in rats

A. Ohsumi, H. Nawashiro, N. Otani, H. Ooigawa, T. Toyooka, A. Yano, N. Nomura, and K. Shima

Department of Neurosurgery, National Defense Medical College, Saitama, Japan

Summary

Gap junctions are intercellular channels that mediate the cytoplasmic exchange of small hydrophilic molecules and are formed by a family of integral membrane proteins called connexins (Cxs). Cx43 is expressed predominantly in astrocytes, while Cx36 is expressed in neurons. In this study, we show alteration of Cx43 and Cx36 in the hippocampus after traumatic brain injury in rats.

Adult male Sprague-Dawley rats were subjected to lateral fluid percussion injury of moderate severity. Brain coronal sections were used for immunohistochemistry with Cx43 and Cx36 antibodies. Cx43 immunoreactivity was increased in reactive astrocytes in the damaged hippocampus 24 hours after injury, and persisted for 72 hours. On the other hand, Cx36 immunoreactivity increased in CA3 neurons 1 hour after injury, and decreased later. These results indicate that gap junctions might participate in the pathophysiological process after traumatic brain injury.

Keywords: Connexin; traumatic brain injury; gap junction.

Introduction

Gap junctions are intercellular channels that mediate the direct cytoplasmic exchange of small hydrophilic molecules and ions [1] and are considered to be critical for tissue homeostasis [4]. Gap junction proteins, connexins (Cx), are expressed by various cells in the central nervous system [2]. Cx43 is expressed predominantly in astrocytes, while Cx36 is expressed in neurons. Gap junctions may be neuroprotective [9] or harmful [8] under ischemic conditions. Our study focuses on the expression and distribution of Cx after traumatic brain injury (TBI) in vivo.

Experimental studies suggest that selective vulnerability is observed in CA3 pyramidal neurons, dentate hilar neurons, and cortical neurons [3, 10] after fluid percussion brain injury (FPI) in rats. In addition, TBI causes the early loss of astrocytes [12], and induces reactive astrocytes [5] in the hippocampus after FPI. Reactive astrocytes are the most prominent response forms after injury. A recent study suggested that the existence of GFAP-positive astrocytes was more extensive in Cx43+/+ than in Cx43+/− mice and reactive astrocytes may reduce neuronal apoptosis under ischemia by regulating extracellular conditions through their gap junctions [9]. In the present study, we show the alteration of Cx43 and Cx36 immunoreactivity in the rat brain after TBI.

Materials and methods

Animal experimental procedures

Eighty-eight adult male Sprague-Dawley rats weighing 300 to 400 g were used. The rats were anesthetized with sodium pentobarbital (50 mg/kg i.p.) and fixed in a stereotaxic flame. A 4.8 mm craniectomy was made over the right parietal bone (centered at 4.0 mm posterior and 3.0 mm lateral to the bregma). A plastic Luer-loc was fixed over the craniectomy site with dental cement. The next day the rats were anesthetized with isoflurane in a 2:1 mixture of nitrous oxide and oxygen, and intubated. The rats were then subjected to moderate severity FPI (2.6–2.8 atm, 12 ms) using a Dragonfly device (Dragonfly Research and Development, Inc., Ridgeley, WV). Rectal temperatures and blood pressures were monitored continuously. Arterial blood samples were analyzed intermittently. Sham control animals were subjected to the same procedures except for percussion injury. The Animal Care and Use Committee of the National Defense Medical College approved all animal procedures.

Tissue preparation

For immunohistochemical analysis, rats were sacrificed at 5, 30, 60 minutes and 6, 24, 72 hours after FPI (n = 5 per each time point). They were perfused transcardially with 4% buffered paraformaldehyde. The brains were removed and embedded in paraffin, and then serial coronal sections (5 μm thick) were prepared.

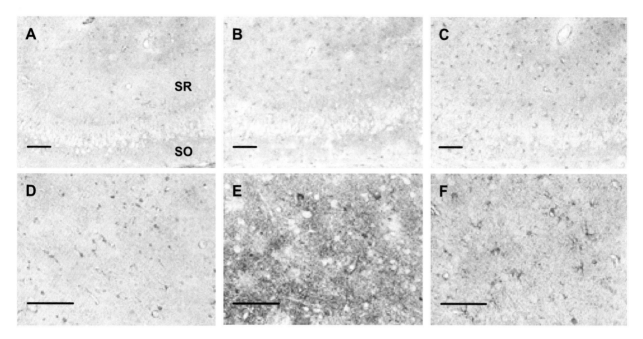

Fig. 1. Time course of Cx43 immunoreactivity in the ipsilateral hippocampal CA3 subfield (A, B, C) and cortex (D, E, F) (A and D, sham; B and E, 24 hours after injury; C and F, 72 hours after injury). The immunoreactivity for Cx43 was enhanced in astrocytes 72 hours after injury. In cortex, Cx43 immunoreactivity was also induced 72 hours after injury. SO, stratum oriens; SR, stratum radiatum. Scale bars, 100 μm

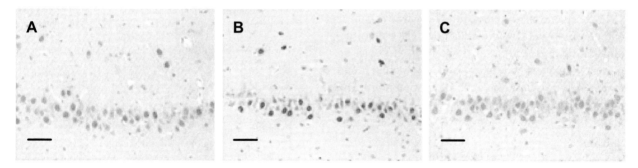

Fig. 2. Time course of immunoreactivity for Cx36 in the ipsilateral CA3 region (A, sham; B, 1 hour after injury; C, 72 hours after injury). Cx36 immunoreactivity increased in pyramidal neurons at 1 hour after injury (B), then gradually decreased to control levels (C). Scale bars, 50 μm

Immunohistochemistry

To examine the distribution of the immunoreactive cells, we performed immunohistochemistry using anti-Cx43 and -Cx36 antibodies. Histofine MAX-PO (MULTI) (Nichirei Biosciences Inc., Tokyo, Japan) was used for immunostaining by monoclonal anti-Cx43 antibody (C8093; Sigma-Aldrich Co., St. Louis, MO) and polyclonal anti-Cx36 antibody (H-130; Santa Cruz Biotechnology, Inc., Santa Cruz, CA). For evaluation of morphologic change, adjacent sections were counterstained with hematoxylin and eosin.

Results

The alterations of Cx43 immunoreactivity in the ipsilateral CA3 subfield are shown in Fig. 1. Immunoreactivity for Cx43 was reduced in astrocytes up to 6 hours after injury, recovered by 24 hours, and increased at 72 hours. The immunoreactivity for Cx43 in the contralateral region was similar to sham controls (data not shown). Cx43 immunoreactivity in the ipsilateral cortex was induced 72 hours after injury compared to sham controls.

Immunoreactivity for Cx36 in the ipsilateral CA3 region was observed in pyramidal neuronal layers (Fig. 2). Cx36 immunoreactivity was induced in neurons 1 hour after injury, then gradually decreased to control levels. Immunoreactivity for Cx36 in the CA1 subfield was not detected after injury.

Discussion

Selective vulnerability of neurons was recognized after TBI in the hippocampal CA3 subfield and dentate hilus [9]. The mechanisms of posttraumatic selective vulnerability are not fully understood. On the other hand, astrocytes have a particularly important role in the uptake of glutamate by the injured brain demonstrated by large influxes of extracellular glutamate [7]. In addition, reactive astrocytes induce the expression of neurotrophins and nerve growth factors. We speculate that the reactive astrocytes play an important role in the development of astrogliosis after injury.

Astrocytic gap junctions propagate intercellular signal transduction, which exacerbates cell injury induced by calcium overload, oxidative stress, metabolic inhibition [8]. In contrast, gap junctions could also be reasoned to be neuroprotective because astroglial cells remove potassium or glutamate efficiently. As a result, neurons are not subjected to large depolarizations with the consequent excitotoxicity [6]. Conversely, global ischemia induces selective up-regulation of Cx32 and Cx36 in the vulnerable CA1 before the onset of neuronal death [11].

The mechanisms of posttraumatic selective vulnerability are also not fully understood. The role of gap junction communication after injury in vivo has not been investigated. The present study shows that immunoreactivity for Cx43 was reduced up to 6 hours after injury; however, immunoreactivity for Cx43 was induced in astrocytes 72 hours after injury. The induction of Cx36 immunoreactivity was also observed in CA3 pyramidal neurons early after injury, whereas the alteration of Cx36 immunoreactivity was not recognized in the hippocampal CA1 subfield. These findings suggest that astrocytic injury occurred early after TBI, then reactive astrocytes appeared later. The neuronal gap junctions were activated before neuronal damage developed.

In conclusion, Cx43 might contribute to neuronal regeneration at a late stage after TBI, and Cx36 might play a role in the pathophysiological process early after injury. Therefore, gap junction communication may be important molecular targets for clarifying the mechanisms of cell injury in the central nervous system after TBI.

References

1. Bennett MV, Barrio LC, Bargiello TA, Spray DC, Hertzberg E, Saez JC (1991) Gap junctions: new tools, new answers, new questions. Neuron 6: 305–320
2. Bruzzone R, Ressot C (1997) Connexins, gap junctions and cell-cell signaling in the nervous system. Eur J Neurosci 9: 1–6
3. Cortez SC, McIntosh TK, Noble LJ (1989) Experimental fluid percussion brain injury: vascular disruption and neuronal and glial alterations. Brain Res 482: 271–282
4. Dermietzel R (1998) Gap junction wiring: a 'new' principle in cell-to-cell communication in the nervous system? Brain Res Brain Res Rev 26: 176–183
5. Dietrich WD, Alonso O, Halley M (1994) Early microvascular and neuronal consequences of traumatic brain injury: a light and electron microscopic study in rats. J Neurotrauma 11: 289–301
6. Hansson E, Muyderman H, Leonova J, Allansson L, Sinclair J, Blomstrand F, Tholin T, Nilsson M, Ronnback L (2000) Astroglia and glutamate in physiology and pathology: aspects on glutamate transport, glutamate-induced cell swelling and gap-junction communication. Neurochem Int 37: 317–329
7. Katayama Y, Becker DP, Tamura T, Ikezaki K (1990) Early cellular swelling in experimental traumatic brain injury: a phenomenon mediated by excitatory amino acids. Acta Neurochir Suppl (Wien) 51: 271–273
8. Lin JH, Weigel H, Cotrina ML, Liu S, Bueno E, Hansen AJ, Hansen TW, Goldman S, Nedergaard M (1998) Gap-junction-mediated propagation and amplification of cell injury. Nat Neurosci 1: 494–500
9. Nakase T, Fushiki S, Naus CC (2003) Astrocytic gap junctions composed of connexin 43 reduce apoptotic neuronal damage in cerebral ischemia. Stroke 34: 1987–1993
10. Nawashiro H, Shima K, Chigasaki H (1995) Selective vulnerability of hippocampal CA3 neurons to hypoxia after mild concussion in the rat. Neurol Res 17: 455–460
11. Oguro K, Jover T, Tanaka H, Lin Y, Kojima T, Oguro N, Grooms SY, Bennett MV, Zukin RS (2001) Global ischemia-induced increases in the gap junctional proteins connexin 32 (Cx32) and cx36 in hippocampus and enhanced vulnerability of Cx32 knock-out mice. J Neurosci 21: 7534–7542
12. Zhao X, Ahram A, Berman RF, Muizelaar JP, Lyeth BG (2003) Early loss of astrocytes after experimental traumatic brain injury. Glia 44: 140–152

Correspondence: Atsushi Ohsumi, Department of Neurosurgery, National Defense Medical College, Namiki 3-2, Tokorozawa, Saitama, Japan. e-mail: atsuoh@ndmc.ac.jp

Zinc protoporphyrin IX attenuates closed head injury-induced edema formation, blood-brain barrier disruption, and serotonin levels in the rat

P. Vannemreddy[1], A. K. Ray[2,3], R. Patnaik[2,3], S. Patnaik[4], S. Mohanty[5], and H. S. Sharma[1,2]

[1] Laboratory of Neuroanatomy, Department of Medical Cell Biology, Biomedical Center, Uppsala University, Uppsala, Sweden
[2] Department of Surgical Sciences, Anesthesiology and Intensive Care, University Hospital, Uppsala University, Uppsala, Sweden
[3] Department of Biomedical Engineering, Institute of Technology, Banaras Hindu University, Varanasi, India
[4] Department of Pharmacy, Institute of Technology, Banaras Hindu University, Varanasi, India
[5] Department of Neurosurgery, Institute of Medical Sciences, Banaras Hindu University, Varanasi, India

Summary

The role of heme oxygenase (HO) in closed head injury (CHI) was examined using a potent HO and guanylyl cyclase inhibitor, zinc protoporphyrin (Zn-PP) in the rat. Blood-brain barrier (BBB) permeability to Evans blue and radioiodine, edema formation, and plasma and brain levels of serotonin were measured in control, CHI, and Zn-PP-treated CHI rats. CHI was produced by an impact of 0.224 N on the right parietal bone by dropping 114.6 g weight from a height of 20 cm in anesthetized rats. This concussive injury resulted in edema formation and brain swelling 5 hours after insult that was most pronounced in the contralateral hemisphere. The whole brain was edematous and remained in a semi-fluid state. Microvascular permeability disturbances to protein tracers were prominent in both cerebral hemispheres and the underlying cerebral structures. Plasma and brain serotonin showed pronounced increases and correlated with edema formation. Pretreatment with Zn-PP (10 mg/kg, i.p) 30 minutes before or after CHI attenuated edema formation, brain swelling, plasma and brain serotonin levels, and microvascular permeability at 5 hours. Brain edema, BBB permeability, and serotonin levels were not attenuated when the compound was administered 60 minutes post-CHI suggesting that HO is involved in cellular and molecular mechanisms of edema formation and BBB breakdown early after CHI.

Keywords: Closed head injury; edema; heme oxygenase; zinc protoporphyrin; blood-brain barrier; serotonin.

Introduction

Closed head injury (CHI) results in instant death in many victims [6, 15, 16, 19]. In the United States, CHI accounts for at least 2000 admissions to hospital per million population [2, 15, 16]. About 400,000 new cases are added each year, and many patients have long-term disabilities [6, 15]. Swelling of the brain in a closed cranial compartment is largely responsible for instant deaths [10]. Clinical cases may show diffuse injury with brain shift, mass lesions, or brain stem injury that are responsible for high mortality rates [18, 19]. Diffuse injuries with brain swelling may leave patients in a persistent vegetative state [2, 7, 9]. Unfortunately, there are few proven therapies available now. Efforts are needed to understand the molecular mechanisms of early pathophysiological events and to explore the therapeutic potentials of neuroprotective agents in order to minimize edema formation and cell death.

It is likely that CHI-induced micro-hemorrhage, oxidative stress, and generation of free radicals contribute to blood-brain barrier (BBB) breakdown and vasogenic brain edema formation [4, 5, 40]. Extravasation of blood and blood degradation products in the brain parenchyma are potential sources of free radical generation and may have key roles in the induction of brain swelling [36]. Hemoglobin is metabolized by the enzyme heme oxygenase (HO) after lysis of red blood cells, releasing iron, carbon monoxide (CO), and biliverdin [14]. CO is a free radical gas, similar to nitric oxide (NO), which can induce profound cell and tissue injury [37]. Since CO, like NO, is a molecule with very short half-life (<5 seconds) [12, 37, 38], its involvement in cell and tissue injury is largely based on studies using its synthesizing enzyme, HO.

The role of HO in CHI is not well understood. We examined HO using zinc protoporphyrin IX (Zn-PP), a potent HO and guanylyl cyclase inhibitor compound [3, 17, 20, 39] in a rat model [5]. Since blockade of serotonin synthesis appears to be neuroprotective in this CHI model [5], plasma and brain levels of the amine were also measured in animals treated with Zn-PP.

Materials and methods

Animals

Experiments were carried out on 48 young male rats (250–300 g) housed at 21 ± 1 °C room temperature, on a 12-hour light, 12-hour dark schedule. Food and tap water were supplied ad libitum before the experiments.

Anesthesia

All experiments were carried out under urethane anesthesia (1.5 g/kg, i.p.). This dose was sufficient to induce a grade IV anesthesia for more than 12 hours [5]. Urethane is a long-lasting irreversible anesthetic that acts mainly at the cerebral cortical level [24]. Thus, arterial blood pressure, heart rate, and respiration were stable throughout the experimental period [22].

A new model of CHI

We developed a new animal model of CHI that is easily reproduced and induces severe brain edema in the rat [5]. The model involves an impact of 0.224 N on the right parietal skull bone during anesthesia (Fig. 1), achieved by dropping a 114.6 g weight from a height of 20 cm through a guide-tube [5]. The animals were allowed to survive 1 hour, 2 hours, and 5 hours after injury. The biomechanical forces generated by this impact diffusely penetrate to the underlying brain tissues to induce a powerful concussive brain injury. A few animals (<5%) had minor skull fracture and were not included in this study. Untraumatized urethane-anesthetized rats were used as controls. These experiments were approved by the Ethics Committee of Uppsala University, and Banaras Hindu University.

Treatment with HO inhibitor, Zn-PP

Zn-PP (Tocris Bioscience, Avonmouth, UK) was administered (10 mg/kg, i.p) in a group of rats 30 minutes before CHI [11, 17, 21, 26, 28, 39]. In other groups of animals, Zn-PP was given either 30 or 60 min after brain trauma. The animals were allowed to survive 5 hours after CHI [5].

BBB permeability

BBB permeability in the cerebral cortex of both hemispheres was measured using Evans blue albumin (2%, 3 mL/kg, i.v.) and $^{[131]}$Iodine [1, 25, 33, 34]. These tracers were allowed to circulate for 5 minutes, and the intravascular tracer was washed out by transcardiac perfusion with 0.9% saline [22]. About 1 mL of arterial blood was withdrawn via heart puncture for whole blood radioactivity before perfusion [26, 35]. BBB permeability was expressed by percentage increase in the radioactivity in brain over the whole blood concentration [29, 31].

Brain water content

Brain water content in the right and left cerebral cortices was determined using the difference in sample wet and dry weights [30, 32]. The right cerebral cortex and the left cerebral cortex were dissected out, weighed immediately, and placed in an oven maintained at 90 °C for 72 hours or until the dry weight of the samples became constant [33]. The percentage of volume swelling was calculated from changes in the brain water content [24, 33].

Measurement of serotonin in plasma and brain

Plasma and brain serotonin were measured using a fluorometric assay [5, 22, 25]. About 1 mL of whole blood was collected after cardiac puncture. Plasma was obtained by centrifugation. Plasma (0.5 mL) and brain samples were diluted to 4 mL in 0.4 N ice-cooled perchloric acid and were centrifuged at 4 °C to separate out proteins [22]. The extraction of serotonin from 1 mL aliquots from plasma or brain was performed using butanol in a salt-saturated and alkaline medium (pH 10) and purified using n-heptane (Extra Pure, Merck). The fluoropores were developed by incubating the samples with ninhydrin at 75 °C for 30 minutes and measured in duplicate samples at room temperature using a spectrophotofluorometer at excitation 385 nm and emission 490 nm wave lengths (Aminco-Bowman, USA) [5, 28].

Statistical analysis

ANOVA following Dunnet's test for multiple group comparisons with one control group was applied to evaluate the statistical significance of the data obtained. A p-value less than 0.05 was considered significant.

Results

Effect of Zn-PP on brain edema in CHI

Five hours after CHI, the whole brain was considerably edematous, softened, and remained in a semi-fluid state. A marked increase in brain water content and brain swelling was observed that was more pronounced in the contralateral than the ipsilateral cerebral hemisphere (Table 1). Pretreatment with Zn-PP markedly attenuated brain swelling 5 hours after CHI compared to the untreated traumatized group (Table

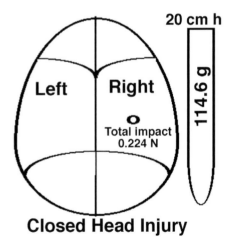

Fig. 1. Closed head injury model in rat. Under anesthesia, a 114.6 g weight (non-piercing) was dropped from a height of 20 cm through a guide-tube on a predetermined location on the right parietal bone. The skull was firmly held to avoid displacement during impact. This procedure generated an impact of 0.224 N on the surface of skull. A few animals (<5%) showed minor skull fracture and were not included in this study [5]

Table 1. *Effects of HO inhibitor Zn-PP on BBB permeability, brain edema formation, and plasma and brain serotonin levels following CHI in rats.*

Groups	n	Brain water content %		BBB permeability [131]Iodine %		Brain serotonin (µg/g)		Plasma serotonin (µg/mL)
		Right injured	Left intact	Right injured	Left intact	Right injured	Left intact	
Control	6	78.34 ± 0.23	78.03 ± 0.27	0.38 ± 0.06	0.40 ± 0.08	0.68 ± 0.24	0.70 ± 0.32	0.34 ± 0.08
5 h after CHI	8	81.34 ± 0.89**	82.47 ± 0.67**a	1.78 ± 0.56**	2.14 ± 0.32**a	1.89 ± 0.45**	2.23 ± 0.18**a	0.68 ± 0.11**
Zn-PP# + CHI	6	79.32 ± 0.22*b	80.16 ± 0.21*b	0.67 ± 0.23*b	0.76 ± 0.44*b	0.93 ± 0.28*b	1.16 ± 0.21*b	0.48 ± 0.11*b

BBB Blood-brain barrier; *CHI* closed head injury; *HO* heme oxygenase; *Zn-PP* zinc protoporphyrin.
\# = 10 mg/kg, i.p. 30 min after CHI; a = $p < 0.05$ from injured half; b = $p < 0.05$ from CHI; ** = $p < 0.01$ from control.

1). This effect on edema was also evident when the HO inhibitor was administered 30 minutes after CHI (Table 1). However, no significant reduction in brain water content or volume swelling was noted when Zn-PP was given 1 hour after trauma (results not shown).

Effect of Zn-PP on BBB permeability in CHI

Microvascular permeability disturbances to Evans blue and radioiodine tracers were prominent at 5 hours in both cerebral hemispheres as well as in underlying cerebral structures (Table 1). However, the extravasation of protein tracers was higher in the contralateral hemisphere compared to the side ipsilateral to injury. Pretreatment with Zn-PP (10 mg/kg, i.p) 30 minutes before or 30 minutes after CHI significantly attenuated the enhanced BBB permeability to protein tracers seen at 5 hours (Table 1). However, Zn-PP was ineffective when administered 60 minutes after CHI (results not shown).

Effect of Zn-PP on plasma and brain serotonin levels in CHI

There were pronounced increases in serotonin in both traumatized and contralateral hemispheres 5 hours after CHI. This increase in the contralateral hemisphere was higher than the injured cortex (Table 1). The plasma serotonin also increased significantly from the control group.

Pretreatment with Zn-PP markedly attenuated increased plasma and brain serotonin levels 5 hours after CHI. The increase in serotonin levels was diminished by the HO-inhibitor when given 30 minutes after CHI (Table 1). No changes in plasma or brain serotonin were seen when Zn-PP was administered 60 minutes after CHI (results not shown).

Discussion

Treatment with an HO and guanylyl cyclase inhibitor compound, Zn-PP, within 30 minutes of CHI markedly attenuates BBB disruption and brain edema formation. This new observation suggests that CO participates in the early phase of the pathophysiology after CHI, a concept consistent with the fact that administration of Zn-PP 1 hour after CHI did not reduce brain edema formation and/or BBB breakdown.

The increase in brain edema in the contralateral hemisphere suggests that the model can be used to study contre coup mechanisms in the brain. Physical forces following impact on the intact skull will be transmitted to the opposite hemisphere causing massive damage compared to the injured side [5]. Our observations further show a close parallelism between serotonin levels, BBB dysfunction, and brain edema formation in CHI.

Increased levels of serotonin in plasma and brain closely correspond to BBB breakdown. Furthermore, Zn-PP administered either 30 minutes before or after CHI was able to reduce plasma and brain serotonin levels effectively, together with brain edema formation and BBB breakdown. On the other hand, serotonin levels, BBB disruption, and brain edema formation did not fall when Zn-PP was administered 1 hour after CHI, indicating an interaction between HO and serotonin in CHI.

Up-regulation of HO-1 and HO-2 occurs following various types of centra nervous system insult [1, 8, 22, 23, 25–27, 30–34]. Previous reports from our laboratory showed up-regulation of HO-2 5 hours after focal spinal cord injury (SCI) in the rat, which closely corresponds to cell and tissue injury [25, 26, 30–32]. Inhibition of HO-2 expression in the cord caused by either topical application of neurotrophins [31, 32] or by pretreatment with the serotonin synthesis inhibitor, p-chlorophenylalanine [25], markedly attenuated edema formation, microvascular permeability disturbances, and cell injury [26]. Up-regulation of HO is associated with increased production of CO that may contribute to cell and tissue injury similar to NO [24, 26, 28, 29, 33, 38]. These observations suggest that an up-regulation of HO and subsequent generation of CO appears to be an instrumental factor causing cell and tissue injury in CHI. To confirm this hypothesis further, studies on HO expression in CHI are needed, and are currently being investigated in our laboratory.

Brain injury is a complex event that includes physical destruction of microvessels, alterations in local and global microcirculation, as well as permeability changes in vessel walls leading to leakage of plasma constituents into the brain microenvironment [19, 24]. Early events following focal brain trauma are influenced by a number of compounds which are released or become activated in and around the primary lesion [24, 26]. These chemical mediators of the inflammatory response include biogenic amines, arachidonic acid derivatives, free radicals, histamine, and bradykinin [21–24, 26, 28, 35].

Various neurochemicals interact in vivo influencing cell and molecular functions in synergy, and play important roles in BBB disturbances, edema formation, and cell injury [1, 22, 26, 28]. And, inhibitors of serotonin, prostaglandin, histamine, and NO synthesis before injury attenuate microvascular permeability disturbances and edema formation [24, 28]. Furthermore, blockade of serotonin synthesis prior to SCI attenuates trauma-induced HO-2 up-regulation in the spinal cord [33], indicating an interaction between serotonin and HO in SCI, and suggesting that serotonin somehow influences trauma-induced HO expression. In the present study, HO inhibition attenuated CHI-induced increases in brain and plasma serotonin levels, supporting the concept that there are interactions among various endogenous substances that are released or affected during secondary injury cascades.

Serotonin is a powerful neurochemical involved in BBB disruption and edema formation [22, 25]. A focal incision into the brain or spinal cord induces profound increases in plasma and tissue serotonin levels [5]. Elevated levels of tissue and blood serotonin therefore influence microvascular permeability disturbances and edema formation following CHI. However, the probable mechanism(s) by which serotonin influences HO production or vice versa is still unclear from this study. Micro-hemorrhages are a known inducer of HO-1 expression [13, 14]. Since platelets are also very rich in serotonin content, the possibility exists that extravasation of blood components into the cerebral compartment somehow contributes to HO expression and vice versa [38]. An increased level of brain serotonin following CHI may also result from breakdown of the BBB and from direct release of amines from central monoaminergic neurons following trauma [22, 24].

That Zn-PP attenuates edema formation and BBB breakdown supports the involvement of HO in CHI pathophysiology. Other studies on the effects of HO inhibitors on cell injury and edema formation following injury in vivo and in vitro are in line with this hypothesis [3, 11–14, 17, 20, 39]. The mechanisms by which HO inhibitors confer neuroprotection are not yet clear. It appears that HO inhibitors exert anti-inflammatory effects, since Zn-PP reduces infarct size and edema following cerebral ischemia [11, 13]. The use of Zn-PP in the present study does not indicate that HO-1 or HO-2 expression is related to cell injury in CHI. Previous reports from our laboratory and others suggest that HO-2 expression is injurious to the cell and that HO-1 up-regulation has some beneficial effects [26]. It would be important to examine HO-1 and HO-2 expression in CHI in Zn-PP treated rats to clarify this point.

Conclusion

Our observations indicate that CHI induces profound brain swelling within a short period (5 hours). This swelling appears to be caused by leakage of plasma proteins through a disrupted BBB. Early intervention with HO-inhibitor Zn-PP significantly attenuated brain edema formation, BBB leakage, and elevated circulation and brain serotonin levels, suggesting that HO and serotonin are working in synergy during the early phase of CHI pathophysiology.

Acknowledgments

This study was supported by grants from the Swedish Medical Research Council, No. 2710; the University Grants Commission, New Delhi, India; the Indian Medical Research Council, New Delhi, India. The technical assistance of Mrs. Kärstin Flink, and the secretarial assistance of Mrs. Aruna Sharma is acknowledged with thanks.

References

1. Alm P, Sharma HS, Sjöquist PO, Westman J (2000) A new antioxidant compound H-290/51 attenuates nitric oxide synthase

and heme oxygenase expression following hyperthermic brain injury. An experimental study using immunohistochemistry in the rat. Amino Acids 19: 383–394
2. Bergemalm PO, Lyxell B (2005) Appearances are deceptive? Long-term cognitive and central auditory sequelae from closed head injury. Int J Audiol 44: 39–49
3. Blumenthal SB, Kiemer AK, Tiegs G, Seyfried S, Holtje M, Brandt B, Holtje HD, Zahler S, Vollmar AM (2005) Metalloporphyrins inactivate caspase-3 and -8. FASEB J 19: 1272–1279
4. Chieregato A, Fainardi E, Morselli-Labate AM, Antonelli V, Compagnone C, Targa L, Kraus J, Servadei F (2005) Factors associated with neurological outcome and lesion progression in traumatic subarachnoid hemorrhage patients. Neurosurgery 56: 671–680
5. Dey PK, Sharma HS (1984) Influence of ambient temperature and drug treatments on brain oedema induced by impact injury on skull in rats. Indian J Physiol Pharmacol 28: 177–186
6. Farin A, Marshall LF (2004) Lessons from epidemiologic studies in clinical trials of traumatic brain injury. Acta Neurochir [Suppl] 89: 101–107
7. Fortin S, Godbout L, Braun CM (2003) Cognitive structure of executive deficits in frontally lesioned head trauma patients performing activities of daily living. Cortex 39: 273–291
8. Gordh T, Sharma HS, Azizi M, Alm P, Westman J (2000) Spinal nerve lesion induces upregulation of constitutive isoform of heme oxygenase in the spinal cord. An immunohistochemical investigation in the rat. Amino Acids 19: 373–381
9. Heitger MH, Macaskill MR, Jones RD, Anderson TJ (2005) The impact of mild closed head injury on involuntary saccadic adaptation: evidence for the preservation of implicit motor learning. Brain Inj 19: 109–117
10. Hlatky R, Valadka AB, Goodman JC, Robertson CS (2004) Evolution of brain tissue injury after evacuation of acute traumatic subdural hematomas. Neurosurgery 55: 1318–1323
11. Kadoya C, Domino EF, Yang GY, Stern JD, Betz AL (1995) Preischemic but not postischemic zinc protoporphyrin treatment reduces infarct size and edema accumulation after temporary focal cerebral ischemia in rats. Stroke 26: 1035–1038
12. Leffler CW, Balabanova L, Fedinec AL, Parfenova H (2005) Nitric oxide increases carbon monoxide production by piglet cerebral microvessels. Am J Physiol Heart Circ Physiol [Epub ahead of print]
13. Levere RD, Escalante B, Schwartzman ML, Abraham NG (1989) Role of heme oxygenase in heme-mediated inhibition of rat brain Na+-K+-ATPase: protection by tin-protoporphyrin. Neurochem Res 14: 861–864
14. Maines MD (1988) Heme oxygenase: function, multiplicity, regulatory mechanisms, and clinical applications. FASEB J 2: 2557–2568
15. Marshall RS (2004) The functional relevance of cerebral hemodynamics: why blood flow matters to the injured and recovering brain. Curr Opin Neurol 17: 705–709
16. Nolan S (2005) Traumatic brain injury: a review. Crit Care Nurs Q 28: 188–194
17. Panizzon KL, Dwyer BE, Nishimura RN, Wallis RA (1996) Neuroprotection against CA1 injury with metalloporphyrins. Neuroreport 7: 662–666
18. Ratcliff G, Colantonio A, Escobar M, Chase S, Vernich L (2005) Long-term survival following traumatic brain injury. Disabil Rehabil 27: 305–314
19. Schmidt OI, Heyde CE, Ertel W, Stahel PF (2005) Closed head injury – an inflammatory disease? Brain Res Brain Res Rev 48: 388–399
20. Serfass L, Burstyn JN (1998) Effect of heme oxygenase inhibitors on soluble guanylyl cyclase activity. Arch Biochem Biophys 359: 8–16
21. Sharma HS, Westman J (1998) Brain Functions in Hot Environment. Elsevier, Amsterdam, pp 1–516
22. Sharma HS (1999) Pathophysiology of blood-brain barrier, brain edema and cell injury following hyperthermia: New role of heat shock protein, nitric oxide and carbon monoxide. An experimental study in the rat using light and electron microscopy. Acta Universitatis Upsaliensis 830: 1–94
23. Sharma HS (2000) Degeneration and regeneration in the CNS. New roles of heat shock proteins, nitric oxide and carbon monoxide. Amino Acids 19: 335–337
24. Sharma HS (2005) Pathophysiology of blood-spinal cord barrier in traumatic injury and repair. Curr Pharm Des 11: 1353–1389
25. Sharma HS, Westman J (2003) Depletion of endogenous serotonin synthesis with p-CPA, attenuates upregulation of constitutive isoform of heme oxygenase-2 expression, edema formation and cell injury following a focal trauma to the rat spinal cord. Acta Neurochir Suppl 86: 389–394
26. Sharma HS, Westman J (2004) The heat shock proteins and heme oxygenase response in central nervous system injuries. In: Sharma HS, Westman J (eds) Blood-spinal cord and brain barriers in health and disease. Elsevier Academic Press, San Diego, pp 329–360
27. Sharma HS, Alm P, Westman J (1997) Upregulation of hemeoxygenase-II in the rat spinal cord following heat stress. In: Nielsen-Johanssen B, Nielsen R (eds) Thermal physiology 1997. The August Krogh Institute, Copenhagen, pp 135–138
28. Sharma HS, Alm P, Westman J (1998) Nitric oxide and carbon monoxide in the brain pathology of heat stress. Prog Brain Res 115: 297–333
29. Sharma HS, Drieu K, Alm P, Westman J (1999) Upregulation of neuronal nitric oxide synthase, edema and cell injury following heat stress are reduced by pretreatment with EGB-761 in the rat. J Therm Biol 24: 439–446
30. Sharma HS, Alm P, Sjöquist PO, Westman J (2000) A new antioxidant compound H-290/51 attenuates upregulation of constitutive isoform of heme oxygenase (HO-2) following trauma to the rat spinal cord. Acta Neurochir [Suppl] 76: 153–157
31. Sharma HS, Nyberg F, Gordh T, Alm P, Westman J (2000) Neurotrophic factors influence upregulation of constitutive isoform of heme oxygenase and cellular stress response in the spinal cord following trauma. An experimental study using immunohistochemistry in the rat. Amino Acids 19: 351–361
32. Sharma HS, Westman J, Gordh T, Alm P (2000) Topical application of brain derived neurotrophic factor influences upregulation of constitutive isoform of heme oxygenase in the spinal cord following trauma. An experimental study using immunohistochemistry in the rat. Acta Neurochir [Suppl] 76: 365–369
33. Sharma HS, Drieu K, Westman J (2003) Antioxidant compounds EGB-761 and BN-52021 attenuate brain edema formation and hemeoxygenase expression following hyperthermic brain injury in the rat. Acta Neurochir [Suppl] 86: 313–319
34. Sharma HS, Sjöquist PO, Alm P (2003) A new antioxidant compound H-290/51 attenuates spinal cord injury induced expression of constitutive and inducible isoforms of nitric oxide synthase and edema formation in the rat. Acta Neurochir [Suppl] 86: 415–420
35. Sharma HS, Badgaiyan RD, Mohanty S, Alm P, Wiklund L (2005) Neuroprotective effects of nitric oxide synthase inhibitors in spinal cord injury induced pathophysiology and motor functions. An experimental study in the rat. Ann NY Acad Sci 1053 [in press]
36. Tavazzi B, Signoretti S, Lazzarino G, Amorini AM, Delfini R, Cimatti M, Marmarou A, Vagnozzi R (2005) Cerebral oxidative

stress and depression of energy metabolism correlate with severity of diffuse brain injury in rats. Neurosurgery 56: 582–589
37. Verma A, Hirsch DJ, Glatt CE, Ronnett GV, Snyder SH (1993) Carbon monoxide: a putative neural messenger. Science 259: 381–384
38. Vreman HJ, Stevenson DK (1988) Heme oxygenase activity as measured by carbon monoxide production. Anal Biochem 168: 31–38
39. Wagner KR, Hua Y, de Courten-Myers GM, Broderick JP, Nishimura RN, Lu SY, Dwyer BE (2000) Tin-mesoporphyrin, a potent heme oxygenase inhibitor, for treatment of intracerebral hemorrhage: in vivo and in vitro studies. Cell Mol Biol (Noisy-le-grand) 46: 597–608
40. Zohar O, Schreiber S, Getslev V, Schwartz JP, Mullins PG, Pick CG (2003) Closed-head minimal traumatic brain injury produces long-term cognitive deficits in mice. Neuroscience 118: 949–955

Correspondence: Hari Shanker Sharma, Department of Surgical Sciences, Anesthesiology & Intensive Care Medicine, University Hospital, SE-75185 Uppsala, Sweden. e-mail: Sharma@surgsci.uu.se

A novel neuroprotective compound FR901459 with dual inhibition of calcineurin and cyclophilins

H. Uchino[1,3], S. Morota[4], T. Takahashi[1], Y. Ikeda[2], Y. Kudo[4], N. Ishii[2], B. K. Siesjö[5], and F. Shibasaki[3]

[1] Department of Anesthesiology, Hachioji Medical Center, Tokyo Medical University, Tokyo, Japan
[2] Department of Neurosurgery, Hachioji Medical Center, Tokyo Medical University, Tokyo, Japan
[3] Department of Molecular and Cell Physiology, Tokyo Metropolitan Institute of Medical Science, Tokyo, Japan
[4] Laboratory of Cellular Neurobiology, Tokyo University of Pharmacy and Life Science, Tokyo, Japan
[5] Arendala 702 22478 Lund, Sweden

Summary

Brain ischemia leads to severe damage in the form of delayed neuronal cell death. In our study, we show that the marked neuroprotection of the new immunosuppressant FR901495 in forebrain ischemia is due not only to inhibition of calcineurin, but also to protection against mitochondrial damage caused by mitochondrial permeability transition pore formation through cyclophilin D, one of the prolyl *cis/trans* isomerase family members. These findings shed light on the clinical application and development of new drugs for the treatment of ischemic damage in the brain as well as in the heart and liver.

Keywords: Ischemic brain damage; calcineurin; cyclophilin D; FR901495; mitochondrial permeability transition.

Introduction

Brain stroke caused by global or focal ischemia presents major medical, social, and economic problems in every country. Transient ischemia, which occurs in patients suffering from transient cardiac arrest, leads to severe brain damage selectively affecting vulnerable regions such as the CA1 sector of the hippocampus, medium-sized neurons in the caudate-putamen, and neocortical neurons in layers 3–5 [15]. A neuroprotective effect of immunosuppressants on ischemic brain damage was first reported for tacrolimus (FK506) in rats with focal ischemia [20]. Pretreatment with cyclosporin A (CsA) in rats was also found to reduce forebrain ischemic damage, provided that measures were taken to allow penetration of the drug through the blood-brain barrier [23, 24]. Although no direct comparison has been made, the effect of FK506 in this model of ischemia seems less pronounced [8].

Brain ischemia is one of the principal causes of acute organ failure that may involve regulatory events acting at the level of mitochondrial permeability transition (MPT) [3]. The MPT pore has been reported to be formed from a complex of the voltage-dependent anion channel (VDAC), adenine nucleotide translocase (ANT), and cyclophilin D (CyPD), a prolyl *cis/trans* isomerase, at contact sites between the mitochondrial outer and inner membranes [6]. In vitro, under pseudo-pathological conditions of oxidative stress, relatively high Ca^{2+}, and low ATP, the VDAC-ANT-CyPD complex can recruit a number of other proteins, including Bax and Bad, and the complex is involved in cell death [7, 9, 12]. The apoptotic pathway is amplified by the release of apoptogenic proteins from the mitochondrial inner membrane space, including cytochrome c, apoptosis-inducing factor, and some proteases [12]. The release of these proteins, particularly cytochrome c, triggers the assembly of a complex activating caspase 9 and caspase 3 [4, 29]. Studies of cultured cells have shown that transient deprivation of oxygen and/or nutrients can lead to delayed assembly of MPT and cell death [14].

CsA and FK506 are specific inhibitors of immunophilines, i.e., prolyl *cis/trans* isomerases, and of calcineurin, a serine/threonine phosphatase 2B that is abundant in the central nervous system [16]. Recently, calcineurin was reported to induce cell death in an activity-dependent manner [21] partly through dephosphorylation of a proapoptotic protein, Bad, one of the Bcl-2 family members [26]. Bad can be phosphorylated on serine 136 by AKT/PKB/RAC or on serine 112 by a mitochondria-anchored cAMP-dependent protein kinase A [5]. The sub-cellular localization of the apoptosis-inducing protein Bad is regulated by these kinases and phosphatases. Only non-phosphorylated Bad is capable of interacting with and antagonizing anti-apoptotic Bcl-2 or Bcl-XL on the outer membrane of mitochondria [27]. Bad phosphorylation results in the redistribution of this protein to the cytosol, where it may interact with 14-3-3 protein [28]. In contrast, the calcium/calmodulin-dependent phosphatase, calcineurin, dephosphorylates Bad, restoring its ability to bind antiapoptotic Bcl-2 family members and trigger a mitochondrial permeability transition. On the basis of the neuroprotective mechanism of this drug, we found a novel natural compound, FR901459, with 4 substitutions of amino acids in comparison with CsA. FR901459, a novel immunosuppressant, has been isolated from the fermentation broth of Stachybotrys chartarum No. 19392.

Given this background, we suspected that FR901459 acts as a neuroprotectant by inhibiting calcineurin and/or cyclophilins such as CypD, which in turn regulate MPT pore formation.

Materials and methods

Rat model of transient forebrain ischemia

Animal experiments were performed in accordance with our institutional guidelines for animal research. Briefly, prior to ischemia, Wistar rats (aged 8 to 10 weeks) with a body weight of 300 to 350 g were fasted overnight but allowed water ad libitum. Under anesthesia with 3.5% isoflurane in 70% N_2O and 30% O_2 through an intubation tube and respirator, venous and arterial catheters were inserted, and ligatures were loosely placed around each common carotid artery. After the surgical procedures had been completed, 50 IU heparin was given intravenously and ischemia was induced by bilateral common carotid artery occlusion and exsanguination to a blood pressure of 50 mmHg. After 10 minutes of ischemia, cerebral circulation was restored by removal of the carotid ligatures and reinfusion of the shed blood. Blood pressure and temperature were continuously recorded, and arterial PO_2, PCO_2, and pH were controlled before and after ischemia. Stereotactic insertion of a needle (Hamilton syringe, outer diameter 450 μm) into the hippocampus on one side was performed under anesthesia 1 week before ischemia in animals designed for CsA and FR901459 treatment, but not FK506 treatment. CsA, FR901459, and FK506 were administrated intravenously at doses of 5, 10 mg/mL or 10, 20, 30 mg/kg or 0.5, 1, 2 mg/mL respectively, 3 days before ischemia, followed by intraperitoneal administration of the same dose once a day.

Assay of swelling of isolated mitochondria

We isolated non-synaptic brain mitochondria from 300 g male Wistar rats (Kyudo Laboratories Inc., Japan) as described previously [19]. Mitochondrial swelling was estimated from changes of light scattering (at 90° to the incident light beam) at 540 nm (for both excitation and emission wavelengths) measured in mitochondrial suspensions (0.5 mg protein in 2 mL). Different concentrations of each drug (CsA 50 to 1000 nmol, FK506 1000 nmol, FR901495 10 to 500 nmol) were added before $CaCl_2$ administration. Experiments were performed in a water-jacketed cuvette holder at 37 °C using a PerkinElmer LS-50B fluorescence spectrometer (Wellesley, MA). Each experiment was terminated by the addition of alamethicin (40 mg/mg protein) to induce maximal swelling of the whole mitochondrial population.

Histological analysis

For hematoxylin and eosin staining, the brains were perfused with cold 0.1 mol phosphate buffer (pH 7.4) and fixed with 4% paraformaldehyde in 0.1 mol phosphate buffer. The brains were removed, post-fixed overnight in the same fixative solution, and embedded in paraffin. Sections (8 μm thick) of the brain were prepared and stained with hematoxylin and eosin. Histopathological outcome in the CA1 sector of the hippocampus was examined after 7 days of recovery following 10 minutes of ischemia in vehicle-injected animals and those post-treated with CsA or FR901459 or FK506. Quantification of brain damage after 1 week of recovery was performed using light microscopy at a magnification of ×400 by direct visual counting of acidophilic (necrotic) neurons. All necrotic and surviving hippocampal CA1 neurons were counted in 1 coronal section at the level of bregma −3.8 mm, and the percentage of necrotic neurons was calculated. All sections were evaluated in a blinded manner.

Results

Comparison of neuroprotective effects of FR901459, CsA, and FK506 in rat forebrain ischemia

The structure of FR901459 is almost the same as that of CsA, with only a difference in 4 amino acid residues (Fig. 1a). Previous reports have shown potent neuroprotection of CsA and FK506 with or without needle penetration (NP). We have already confirmed no effect of NP, and examined the brain damage 7 days after initiation of reperfusion in animals treated with FR901459, CsA, and FK506. In vehicle-treated animals with NP, mean damage to the CA1 sector was 93.7 ± 4.8% (Fig. 1b, column 1). Pretreatment with 5 mg/kg CsA with NP showed weak suppression of CA1 neuronal damage (58.5 ± 7.5%, $p < 0.01$ vs. vehicle, Fig. 1b, column 2). Nearly total suppression of cell damage (3.7 ± 3.8%, $p < 0.001$ vs. vehicle, Fig. 1b, column 3) was observed in animals pretreated with NP + 10 mg/kg CsA.

In contrast, pretreatment with 0.5 mg/kg FK506 without NP showed no protective effect (Fig. 1b, column 4). Pretreatment with 1 and 2 mg/kg FK506 without NP showed similar suppression of cell damage (63.5 ± 5.8%, $p < 0.01$ vs. vehicle, Fig. 1b, column 5

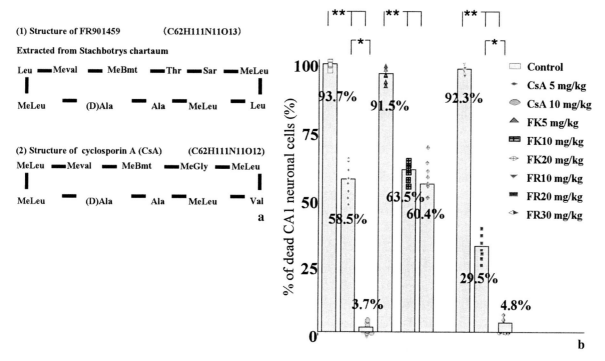

Fig. 1. Characteristics of FR901459 and histopathological analysis of each immunosuppressant. (a) Structures of FR901459 and CsA: FR901459 was isolated from fermentation broth of Stachybotrys chartarum No. 19392, and found to be a member of the cyclosporin family. It is structurally distinct from any other cyclosporins discovered thus far, in that Leu is present at position 5 instead of Val. (b) Comparison of effect of FR901495 with each immunosuppressant: Histopathological outcome in CA1 region of hippocampus after 10 minutes of forebrain ischemia followed by 7 days of recovery. Data are from vehicle-treated animals (column 1: n = 8), and animals pre-treated with 5 or 10 mg/kg CsA (columns 2, 3: n = 8, respectively), 0.5, 1, or 2 mg/kg FK506 (columns 4, 5, 6: n = 8, respectively), or 10, 20, or 30 mg/kg FR901495 (columns 7, 8, 9: n = 8, respectively). (** $p < 0.001$, * $p < 0.01$; one way ANOVA with post-hoc Scheffé test)

and $60.4 \pm 9.8\%$, $p < 0.01$ vs. vehicle, Fig. 1b, column 6). Mean damage in 10 mg/kg FR901459-treated animals with NP was $92.3 \pm 2.5\%$, showing no protective effect (Fig. 1b, column 7). FR901459 20 mg/kg showed a strong protective effect ($29.5 \pm 5.6\%$, $p < 0.01$ vs. vehicle, Fig. 1b, column 8). Almost the same effect was exerted with 30 mg/kg FR901459 ($4.8 \pm 2.7\%$, $p < 0.001$ vs. vehicle, Fig. 1b, column 9). These results suggest that suppression of both calcineurin and CyPD is essential for an immunosuppressant to have a neuroprotective effect.

Comparison of stabilizing effects of FR901495, CsA, and FK506 on mitochondria

We examined the effects of FR901459, CsA, and FK506 on mitochondrial function, focusing especially on MPT pore regulation. We analyzed the effects on Ca^{2+} (40 nmol/mg)-induced swelling in isolated brain mitochondria. Figure 2a shows that CsA dose-dependently reduced the rate of Ca^{2+}-induced mitochondrial swelling, whereas FK506 had no effect (Fig. 2b). FR901459 also showed a dose-dependent reduction in the rate of Ca^{2+}-induced mitochondrial swelling; this inhibitory effect was 10 times as strong as that of CsA (Fig. 2c).

Discussion

The present results have an important bearing on the molecular effects of immunosuppressants CsA, FK506, and FR901459. FR901459, a novel immunosuppressant, has been isolated from the fermentation broth of Stachybotrys chartarum No. 19392 [19]. The molecular formula of FR901459 was determined as $C_{62}H_{111}N_{11}O_{13}$. FR901459 was found to be a member of the cyclosporin family. However, it is structurally distinct from any other cyclosporins discovered thus far, in that Leu is present at position 5 instead of Val. FR901459 was capable of prolonging the survival time of skin allografts in rats with one-third the potency of CsA.

Based on data published in the literature [1, 21, 22], we assumed that the common effects of the 3 drugs

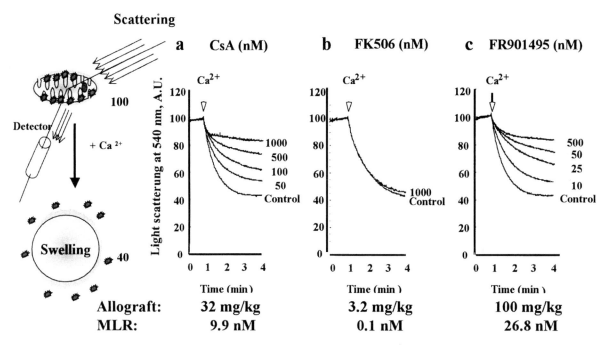

Fig. 2. Comparison of drugs to prevent mitochondrial swelling. (a) Effect of CsA on Ca^{2+}-induced swelling of isolated brain mitochondria: Mitochondrial swelling was induced by addition of 40 nmol Ca^{2+}/mg protein, and estimated from the changes of light scattering (at 90° to the incident light beam) at 540 nm (for both excitation and emission wavelengths) measured at 37 °C in mitochondrial suspensions (0.5 mg protein in 2 mL) using the same buffer as for the mitochondrial Ca^{2+} experiments previously described [25]. The curves show Ca^{2+}-induced swelling with 0 (control), 50, 100, 500, and 1000 nmol CsA. The rate of swelling represents the initial slope of the recorded curves, which is normalized to the curve without CsA (control). (b) Effect of FK506 on Ca^{2+}-induced swelling: The inhibitory effect on mitochondrial swelling was studied in animals given 0 (control) and 1000 nmol FK506. Neither 100 nmol nor 10,000 nmol FK506 had any inhibitory effect (data not shown). (c) Ca^{2+}-induced swelling with 0 (control), 10, 25, 50, and 500 nmol FR901495: The rate of swelling represents the initial slope of the recorded curves, which is normalized to the curve without FR901495 (control). Control represents 100% and the other values are expressed as percent of the control rate

are related to their inhibition of calcineurin, a serine/threonine phosphatase, while the additional effects of CsA and FR901459, suggested by previous reports, are due to blocking of the MPT pore. It has already been suggested that the death pathway activated by calcineurin involves dephosphorylation of Bad, a proapoptotic member of the Bcl-2 family of proteins, similar to the in vitro and in vivo results reported previously [26]. FK506 0.5 mg/kg and FR901459 10 mg/kg did not show neuroprotective effects. There are several speculations about this result. One is that calcineurin inhibition alone is not enough to rescue neuronal cells. The other is that an additional effect of CsA and FR901459 is related to other mechanisms than activation of calcineurin. We need to investigate the importance of the inhibition of cyclophilins, especially CyPD that has the capacity to open the MPT pore. Apart from preventing calcineurin activation, CsA and FR901459 act by blocking the assembly of an MPT pore by combining with CyPD [10]. This would explain the additional protection offered by CsA and FR901459, and the differential effects of the 3 immunosuppressants in terms of mitochondrial swelling. A tentative suggestion, which requires verification, is that CsA and FR901459 must remain bound to CyPD to prevent mitochondrial failure at a time when other factors, such as mitochondrial calcium accumulation [17] and down-regulation of anti-apoptotic members or up-regulation of proapoptotic members of the Bcl-2 family [9], tend to promote mitochondrial membrane depolarization. However, a low dose of CsA and FR901459 did not act to protect neuronal cells, even though FR901459 inhibited CyPD 10 times as strongly as did CsA. This means that CyPD inhibition alone is not enough to prevent ischemic neuronal cell death.

From our results, it is suggested that both calcineurin and immunophilin (CyPD) inhibition are necessary to induce neuroprotection.

Obviously, the clinical usefulness of CsA and FR901459 for neuroprotection is limited by the requirement to enhance blood-brain barrier penetration by producing a brain lesion, by intracarotid administration, by an increase in the plasma concentration to levels that may cause toxic side effects, or by some carrier that is able to deliver the drugs into the brain.

After ischemic reperfusion, calcineurin and CyPD play an important role to induce cell death [2, 18]. Further investigation is needed to analyze the relationship between calcineurin and immunophilin, which may regulate apoptosis and necrosis.

In summary, we postulate that the mechanisms for the marked neuroprotective effects of FR901459 are due to the inhibition of a calcium-dependent phosphatase, calcineurin, as well as blockade of the MPT pore, most likely through inhibition of CyPD. Elucidation of these mechanisms will greatly contribute to the development of new treatments for ischemic injury not only in the brain, but also other organs such as the heart [11] and liver [13].

References

1. Asai A, Qiu J, Narita Y, Chi S, Saito N, Shinoura N, Hamada H, Kuchino Y, Kirino T (1999) High level calcineurin activity predisposes neuronal cells to apoptosis. J Biol Chem 274: 34450–34458
2. Baines CP, Kaiser RA, Purcell NH, Blair NS, Osinska H, Hambleton MA, Brunskill EW, Sayen MR, Gottlieb RA, Dorn GW, Robbins J, Molkentin JD (2005) Loss of cyclophilin D reveals a critical role for mitochondrial permeability transition in cell death. Nature 434: 658–662
3. Bernardi P, Scorrano L, Colonna R, Petronilli V, Di Lisa F (1999) Mitochondria and cell death. Mechanistic aspects and methodological issues. Eur J Biochem 264: 687–701
4. Chen J, Nagayama T, Jin K, Stetler RA, Zhu RL, Graham SH, Simon RP (1998) Induction of caspase-3-like protease may mediate delayed neuronal death in the hippocampus after transient cerebral ischemia. J Neurosci 18: 4914–4928
5. Coghlan VM, Perrino BA, Howard M, Langeberg LK, Hicks JB, Gallatin WM, Scott JD (1995) Association of protein kinase A and protein phosphatase 2B with a common anchoring protein. Science 267: 108–111
6. Crompton M (1999) The mitochondrial permeability transition pore and its role in cell death. Biochem J 341: 233–249
7. Datta SR, Dudek H, Tao X, Masters S, Fu H, Gotoh Y, Greenberg ME (1997) Akt phosphorylation of BAD couples survival signals to the cell-intrinsic death machinery. Cell 91: 231–241
8. Drake M, Friberg H, Boris-Moller F, Sakata K, Wieloch T (1996) The immunosuppressant FK506 ameliorates ischaemic damage in the rat brain. Acta Physiol Scand 158: 155–159
9. Ferrer I, Lopez E, Blanco R, Rivera R, Ballabriga J, Pozas E, Marti E (1998) Bcl-2, Bax, and Bcl-x expression in the CA1 area of the hippocampus following transient forebrain ischemia in the adult gerbil. Exp Brain Res 121: 167–173
10. Friberg H, Connern C, Halestrap AP, Wieloch T (1999) Differences in the activation of the mitochondrial permeability transition among brain regions in the rat correlate with selective vulnerability. J Neurochem 72: 2488–2497
11. Green DR, Reed JC (1998) Mitochondria and apoptosis. Science 281: 1309–1312
12. Griffiths EJ, Halestrap AP (1993) Protection by cyclosporin A of ischemia/reperfusion-induced damage in isolated rat hearts. J Mol Cell Cardiol 25: 1461–1469
13. Hayashi T, Nagasue N, Kohno H, Chang YC, Nakamura T (1991) Beneficial effect of cyclosporine pretreatment in canine liver ischemia. Enzymatic and electronmicroscopic studies. Transplantation 52: 116–121
14. Khaspekov L, Friberg H, Halestrap A, Viktorov I, Wieloch T (1999) Cyclosporin A and its nonimmunosuppressive analogue N-Me-Val-4-cyclosporin A mitigate glucose/oxygen deprivation-induced damage to rat cultured hippocampal neurons. Eur J Neurosci 11: 3194–3198
15. Kirino T (1982) Delayed neuronal death in the gerbil hippocampus following ischemia. Brain Res 239: 57–69
16. Liu J, Farmer JD Jr, Lane WS, Friedman J, Weissman I, Schreiber SL (1991) Calcineurin is a common target of cyclophilin-cyclosporin A and FKBP-FK506 complexes. Cell 66: 807–815
17. Martins E, Inamura K, Themner K, Malmqvist KG, Siesjö BK (1988) Accumulation of calcium and loss of potassium in the hippocampus following transient cerebral ischemia: a proton microprobe study. J Cereb Blood Flow Metab 8: 531–538
18. Nakagawa T, Shimizu S, Watanabe T, Yamaguchi O, Otsu K, Yamagata H, Inohara H, Kubo T, Tsujimoto Y (2005) Cyclophilin D-dependent mitochondrial permeability transition regulates some necrotic but not apoptotic cell death. Nature 434: 652–658
19. Sakamoto K, Tsujii E, Miyauchi M, Nakanishi T, Yamashita M, Shigematsu N, Tada T, Izumi S, Okuhara M (1993) FR901459, a novel immunosuppressant isolated from Stachybotrys chartarum No. 19392. Taxonomy of the producing organism, fermentation, isolation, physico-chemical properties and biological activities. J Antibiot (Tokyo) 46: 1788–1798
20. Sharkey J, Butcher SP (1994) Immunophilins mediate the neuroprotective effects of FK506 in focal cerebral ischemia. Nature 371: 336–339
21. Shibasaki F, McKeon F (1995) Calcineurin functions in Ca (2+)-activated cell death in mammalian cells. J Cell Biol 131: 735–743
22. Sugano N, Ito K, Murai S (1999) Cyclosporin A inhibits H_2O_2-induced apoptosis of human fibroblasts. FEBS Lett 447: 274–276
23. Uchino H, Elmer E, Uchino K, Lindvall O, Siesjö BK (1995) Cyclosporin A dramatically ameliorates CA1 hippocampal damage following transient forebrain ischemia in the rat. Acta Physiol Scand 155: 469–471
24. Uchino H, Elmer E, Uchino K, Li PA, He QP, Smith ML, Siesjö BK (1998) Amelioration by cyclosporin A of brain damage in transient forebrain ischemia in the rat. Brain Res 812: 216–226
25. Uchino H, Minamikawa-Tachino R, Kristian T, Perkins G, Narazaki M, Siesjo BK, Shibasaki F (2002) Differential neuroprotection by cyclosporin A and FK506 following ischemia corresponds with differing abilities to inhibit calcineurin and the mitochondrial permeability transition. Neurobiol Dis 10: 219–233
26. Wang HG, Pathan N, Ethell IM, Krajewski S, Yamaguchi Y, Shibasaki F, McKeon F, Bobo T, Franke TF, Reed JC (1999) Ca2+-induced apoptosis through calcineurin dephosphorylation of BAD. Science 284: 339–343
27. Yang E, Zha J, Jockel J, Boise LH, Thompson CB, Korsmeyer SJ (1995) Bad, a heterodimeric partner for Bcl-XL and Bcl-2, displaces Bax and promotes cell death. Cell 80: 285–291

28. Zha J, Harada H, Yang E, Jockel J, Korsmeyer SJ (1996) Serine phosphorylation of death agonist BAD in response to survival factor results in binding to 14-3-3 not BCL-X (L). Cell 87: 619–628
29. Zou H, Henzel WJ, Liu X, Lutschg A, Wang X (1997) Apaf-1, a human protein homologous to C. elegans CED-4, participates in cytochrome c-dependent activation of caspase-3. Cell 90: 405–413

Correspondence: H. Uchino, Department of Anesthesiology, Hachioji Medical Center, Tokyo Medical University, 1163 Tate-machi, Hachioji, Tokyo 193-0998, Japan. e-mail: h-uchi@tokyo-med.ac.jp

Search for novel gene markers of traumatic brain injury by time differential microarray analysis

Y. Ishikawa[1], H. Uchino[3], S. Morota[3,5], C. Li[2], T. Takahashi[3], Y. Ikeda[4], N. Ishii[3], and F. Shibasaki[2]

[1] Gene Analysis Center, Tokyo Metropolitan Institute of Medical Science, Tokyo, Japan
[2] Department of Molecular Cell Physiology, Tokyo Metropolitan Institute of Medical Science, Tokyo, Japan
[3] Department of Anesthesiology, Tokyo Medical University Hachioji Medical Center, Hachioji, Japan
[4] Department of Neurosurgery, Tokyo Medical University Hachioji Medical Center, Hachioji, Japan
[5] Laboratory of Cellar Neurobiology, School of Life Sciences, Tokyo University of Pharmacy and Life Science, Hachioji, Japan

Summary

Neuronal and glial cell death caused by axonal injury sometimes contributes to whole brain pathology after traumatic brain injury (TBI). We show that neuroprotection by 2 types of immunosuppressants, cyclosporin A (CsA) and tacrolimus (FK506), in a cryogenic brain injury model results from inhibition of calcineurin and protection from mitochondrial damage caused by formation of a mitochondrial permeability transition pore induced by cyclophilin D (CyPD), one of the prolyl cis/trans isomerase family members. We evaluated why CsA is neuroprotective by microarray analysis of gene expression in the cryogenic brain injury rat model. Analyses of expression patterns demonstrated that expression of over 14,000 genes changed between the groups with and without CsA treatment, and about 350 genes among them were extracted showing a significant difference. We learned that the differential expression of several gene targets showed specific patterns in a time-dependent manner. These results may help elucidate the mechanisms of neuronal cell death after TBI and the neuroprotective effects of CsA after TBI.

Keywords: Traumatic brain injury; gene expression; cold injury; immunosuppression.

Introduction

The characteristics of neuronal degeneration after traumatic brain injury (TBI) are biphasic, consisting of the primary mechanical insult and progressive secondary injury [5]. Mitochondrial dysfunction is one of the important consequences of TBI. However, the pathogenesis of TBI, particularly the differential expression of gene targets, remains unclear.

Recent reports have shown that cyclosporin A (CsA) protects against TBI by prevention of mitochondrial permeability transition during exposure to high levels of calcium or oxidative stress [6], suggesting that in vivo pharmacological preservation of mitochondrial function may provide an avenue for treatment of acute injury and edema.

We previously reported that treatment of TBI with CsA dramatically reduced forebrain ischemic damage in rats, and suggested an inhibitory role for both serine/threonine phosphatase 2B calcineurin and prolyl cis/trans isomerase cyclophilin D in CsA neuroprotection [3]. However, real targets for TBI are still unknown.

Severe head injury is commonly encountered in Japan, and contributes to the high rate of morbidity and mortality as well as social and economic problems. Effective neuroprotective agents are needed for brain-injured humans. CsA has shown promise as a neuroprotectant and is now in a clinical trial for the treatment of head injury in the United States.

Our study evaluates the mechanism of CsA using microarray analysis of gene expression in a cryogenic brain injury rat model.

Materials and methods

Surgical procedures

Fifty male Wistar rats, weighing 250 to 300 g each, were anesthetized with pentobarbital (50 mg/kg i.p.). A midline scalp incision and a craniectomy were performed in the right parietal region. A cortical cryogenic injury was produced by placing a metal probe cooled with dry ice on the intact dura of the right parietal region at the level of bregma -3.8 mm. The skin was closed with sutures. The rats were sacrificed at 1 hour (n = 5), 6 hours (n = 5), 12 hours (n = 5), and 24 hours (n = 5) after lesion production.

Histpathology of brain damage by cryogenic injury

The brains were perfused with cold 0.1 mol phosphate buffer (pH 7.4) and fixed with 4% paraformaldehyde in 0.1 mol phosphate buffer. The brains were removed, post-fixed overnight in the same fixative solution, and embedded in paraffin. Sections (8 μm thick) of the brain were prepared and stained with hematoxylin and eosin. Histopathological outcome in the CA1 sector of hippocampus was assessed 7 days after recovery. Vehicle-injected animals and those treated with CsA or FK506 following 10 minutes of ischemia were compared. Quantification of brain damage after 1 week of recovery was performed using light microscopy at a magnification of 40× by direct visual assessment of the damaged area (2 mm) in the neocortex and the damaged area (2 mm) at the level of bregma -3.8 mm. All sections were evaluated in a blinded manner (Fig. 1).

Measurement of brain water content

Brain water content was determined by the wet-dry weight method. Animals were killed by decapitation. The brain was removed immediately and each hemisphere was weighed to obtain the wet weight. The tissue was then dried in a 100 °C oven for 24 hours and re-weighed to obtain the dry weight. Brain water content, expressed as a percentage of the wet weight, was calculated (wet weight − dry weight/wet weight × 100).

RNA isolation, cRNA synthesis, and labeling

At the indicated time point of recirculation, each animal was decapitated under anesthesia. The brain was quickly removed and kept in cold phosphate-buffered saline. Using curved forceps inserted along the hippocampal fissure, the dentate gyrus was unfolded from the hippocampus, exposing the CA1 region. Total RNA was isolated and purified by RNeasy Mini Kit (Qiagen Inc., Valencia, CA). Quality and quantity of the preparations were assessed by A260/280 absorbance. Purified RNA was precipitated with ethanol and stored at -80 °C for later use.

Synthesized cRNA and cyanine-3 CTP or cyanine-5 CTP labeling was used for amplification and labeling of total RNA using a Low RNA Input Linear Amp Kit (Agilent Technologies, Tokyo, Japan) according to the manufacturer's instructions. Labeled cRNA was purified by RNeasy Mini Kit (Qiagen), and stored at -80 °C after ethanol was precipitated for later use.

Array hybridization, washing, and scanning

Fluorescent linear amplified cRNA was hybridized to Rat Oligo Microarray (Agilent Technologies), according to the manufacturer's instructions. This microarray slide contained about 20,000 features including controls of 60-mer oligo probes. Labeled cRNA was blended with each appropriate TBI-treated sample or control sample from non-TBI-treated CA1, and optimized by reduced-input labeling protocols. After hybridization at 60 °C for 17 to 18 hours, microarrays were washed and the signals were scanned on a GenePix 4000B scanner (Amersham Biosciences, Tokyo, Japan).

Data processing and statistical analysis

Expression data were extracted from scanned microarray images using GenePix Pro 4.0 software (Amersham Pharmacia Biotech, Piscataway, NJ). Dye-normalized and background-subtracted intensity and ratio data in Excel (Microsoft Corp., Redmond, WA) were compared between control and test group to extract distinctive data, or analyzed by Gene Spring 6.2 software (Agilent Technologies) to make a gene tree cluster and analysis of variance (ANOVA). Principal components analysis (PCA) is a decomposition technique with set expression patterns known as principal components. To elucidate groups of changed genes, we employed k-means clustering. All data were assembled in 6 clusters. We collated the expression pattern of the selected gene and clustered patterns by the Statistical Group Comparison algorithm (within Gene Spring) to find similarly expressed genes. ANOVA was performed to find the difference in each gene expression (as the *t*-test p value). We tested 1-way ANOVA for a statistical group comparison by parametric test and selected the gene p value as <0.05.

Results and discussion

In this study, we examined TBI in the cryogenic brain injury rat model for protective effects of CsA. CsA is known to protect against delayed neuronal cell death by its ability to prevent mitochondrial permeability transition during exposure to high levels of calcium or oxidative stress.

The results of histopathological comparisons of TBI injured regions, with or without the immunosuppressants CsA or FK506, indicate a dramatic reduction of the TBI damaged area (Fig. 1a–d). Both immunosuppressants caused a significant change in wet weight of the whole brain (Fig. 1e; $p < 0.05$).

The area of damaged brain after TBI was about half that of control when CsA or FK508 was given. The potent neuroprotective effect of CsA resulted from inhibition of serine/threonine phosphatase 2B calcineurin and prolyl cis/trans isomerase cyclophilin D, which were expressed specifically in the mitochondrial matrix and regulated the mitochondria permeability transition (MPT) pore. The mechanism of neuronal cell death that accompanies TBI resembles ischemic neuronal cell death.

Analysis of the gene expressional pattern demonstrated that over 14,000 genes changed between the groups with or without CsA treatment (Fig. 2a). Further analysis disclosed that the differential expression of several gene targets showed specific patterns in a time-dependent manner. These results may help elucidate the mechanisms of neuronal cell death after TBI and the CsA effect on TBI.

Expressional data were extracted from scanned microarray images. For clear visualization of up- and down-regulated or time-dependent expression of genes, we performed clustering analysis and classified them by expression patterns (Fig. 2b). All genes were analyzed by the PCA method. These linear pattern combinations show the 8 most significant patterns seen in all expressed genes (Fig. 2c). We used these

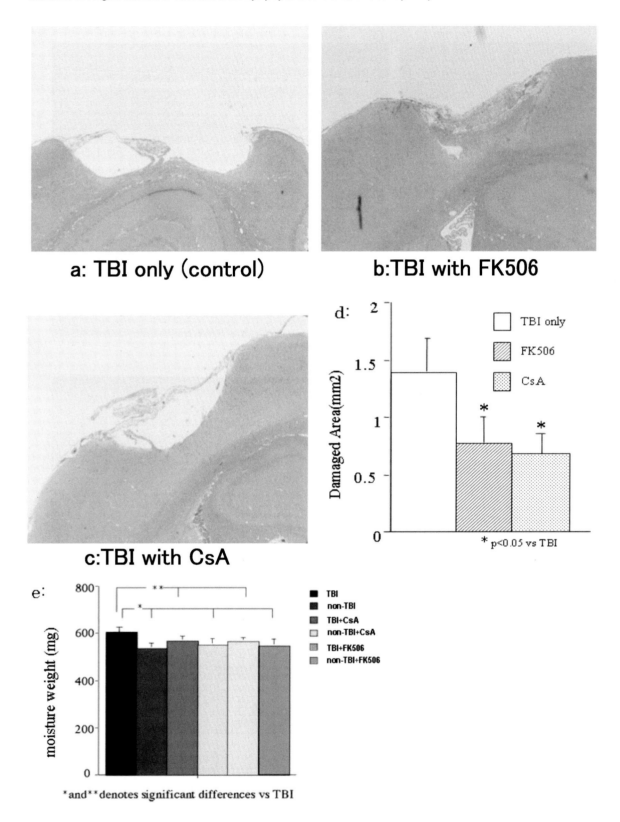

Fig. 1. Effect of FK506 and CsA in TBI. (a–c) Histopathological comparison of TBI region with or without immunosuppressants (40×). (d) Result of measurement of damaged area. Immunosuppressants reduce damaged area significantly. (e) Moisture weight of whole brain with TBI and immunosuppressants. Significant differences observed between were using one-way ANOVA with post hoc Scheffe's test ($*p < 0.01$; $**p < 0.001$)

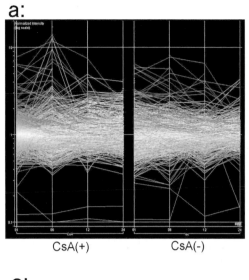
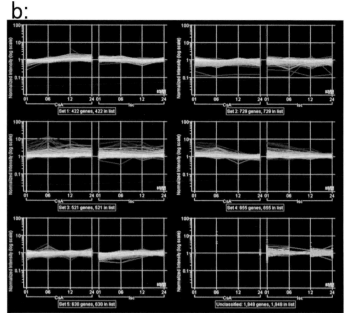
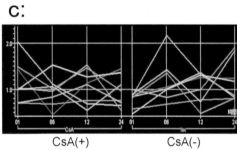
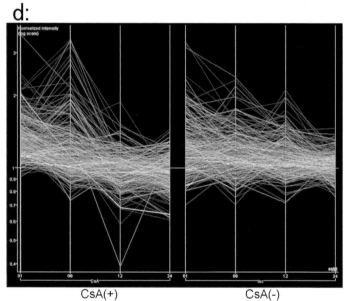

Fig. 2. Results of microarray analysis. (a) mRNA expression pattern (left: CsA[+], right: CsA[−]). X-axis indicates time series and Y-axis is expression ratio. (b) Cluster analysis of expression pattern of mRNA. (c) Result of PCA analysis. These linear combinations show the 8 most significant patterns in all expressed genes. (d) Gene expression pattern of cluster 3 genes. This was an isolated cluster of up-regulated genes without CsA. Green line was a significant expressional pattern derived from PCA analysis. (e) Significant expressional genes

e: significant gene expression

Unknown	181	ApoE related	1
Transcriptional factor	14	Tumor related factor	1
Receptor / channel	45	FKBP related	1
Proteome	7	Cytochrome related	2
Living construction	3	Heat shock Protein	2
Protein / enzymes	66	Ribosomal proteins	25
Cyclophilin D	2	PKC related	1
Narcotic receptor	1	Promoter	1
Ubiquichine	1	Mitochondrial	1
G-protein related	2	Binding proteins	2
Apoptosis related	1	Cyclin related	1
protease	1	calcitonin	1
Cathepsin	1	**total**	363

results as a reference, divided according to similarity of the expression patterns (Fig. 2d). We listed up to 363 distinctive genes (Fig. 2e). These genes fall into many groups; for example, receptor and some anion homeostasis-related genes, transcriptional factors, ribosomal proteins, cyclophilin D, etc. Our results are similar to a previous study in which the listed genes are often identified when cells are damaged by ischemia. The changes of expression of the listed genes with CsA treatment suggest that the effect of this drug is anti-ischemic.

About half of the extracted genes are unknown. The role of an individual gene is not clear yet; however, the function of these unknown genes might be expected by the expression pattern similar to CsA sensitivity.

Based on published data [1, 3, 4], we assumed that the common effects of the 2 drugs are related to inhibition of calcineurin, a serine/threonine phosphatase, while the additional effect of CsA, as suggested by previous reports, is due to its effect in blocking the MPT pore. It has already been suggested that the death pathway activated by calcineurin involves dephosphorylation of Bad, a proapoptotic member of the Bcl-2 family of proteins, similar to in vitro and in vivo results reported previously [7].

Conclusion

CsA is more effective than FK506 in TBI according to histopathological analysis. Inhibition of the increase in brain water content is particularly interesting and could contribute to a reduction in brain damage [2]. There are 2 speculations about these results. One is that calcineurin inhibition alone is not enough to rescue neuronal cells. The other is that the additional effect of CsA is related to mechanisms other than activation of calcineurin. We considered the possibility that this partly represents inhibition of cyclophilins, especially a CyPD that has the capacity to block the MPT pore.

We plan to identify other genes to clarify our statistical analysis and to find new targets that will allow us to understand the mechanisms of ischemia and the neuroprotective effect of CsA.

Reference

1. Asai A, Qiu J, Narita Y, Chi S, Saito N, Shinoura N, Hamada H, Kuchino Y, Kirino T (1999) High level calcineurin activity predisposes neuronal cells to apoptosis. J Biol Chem 274: 34450–34458
2. Sharkey J, Butcher SP (1994) Immunophilins mediate the neuroprotective effects of FK506 in focal cerebral ischemia. Nature 371: 336–339
3. Shibasaki F, Hallin U, Uchino H (2002) Calcineurin as a multifunctional regulator. J Biochem (Tokyo) 131: 1–15
4. Sugano N, Ito K, Murai S (1999) Cyclosporin A inhibits H_2O_2-induced apoptosis of human fibroblasts. FEBS Lett 447: 274–276
5. Sullivan PG, Rabchevsky AG, Waldmeier PC, Springer JE (2005) Mitochondrial permeability transition in CNS trauma: cause or effect of neuronal cell death? J Neurosci Res 79: 231–239
6. Uchino H, Ishii N, Shibasaki F (2003) Calcineurin and cyclophilin D are differential targets of neuroprotection by immunosuppressants CsA and FK506 in ischemic brain damage. Acta Neurochir Suppl 86: 105–111
7. Wang HG, Pathan N, Ethell IM, Krajewski S, Yamaguchi Y, Shibasaki F, McKeon F, Bobo T, Franke TF, Reed JC (1999) Ca2+-induced apoptosis through calcineurin dephosphorylation of BAD. Science 284: 339–343

Correspondence: Hiroyuki Uchino, Dept of Anesthesiology, Hachioji Medical Center, Tokyo Medical University, 1163 Tate-machi, Hachioji, Tokyo 193-0998, Japan. e-mail: h-uchi@tokyo-med.ac.jp

Diffusion tensor feature in vasogenic brain edema in cats

F. Y. Zhao[1], T. Kuroiwa[2], N. Miyasakai[3], F. Tanabe[3], T. Nagaoka[4], H. Akimoto[4], K. Ohno[4], and A. Tamura[5]

[1] Bio-Organic and Natural Products Research Laboratory, McLean Hospital, Harvard Medical School, Boston, MA, USA
[2] Department of Neuropathology, Tokyo Medical and Dental University, Tokyo, Japan
[3] Department of Gynecology and Obstetrics, Tokyo Medical and Dental University, Tokyo, Japan
[4] Department of Neurosurgery, Tokyo Medical and Dental University, Tokyo, Japan
[5] Department of Neurosurgery, Teikyo University, Japan

Summary

We investigated the correlation between the changes in diffusion tensor magnetic resonance imaging, regional water content, and tissue ultrastructure after vasogenic brain edema induced by cortical cold lesioning. In this cat model, E3 in the white matter was dominantly increased while fractional anisotropy (FA) was significantly decreased 8 hours after cortical cold lesioning. This finding indicates that water diffusion in the cortical white matter mainly increased perpendicularly rather than parallel to the direction of the nerve fibers. Additionally, in the area where edema is mild or moderate (tissues with water content of 65% to 75%), FA in the chronic phase was significantly lower than that in the acute phase. Histological examination demonstrated disordered arrangement of nerve fibers, highly dissociated neuronal fibers due to extracellular accumulation of protein rich-fluid, and enlarged interfiber spaces in the acute phase.

Keywords: Diffusion tensor MR imaging; vasogenic brain edema; cortical cold lesioning; ultrastructure.

Introduction

Many disease processes in white matter show regional vulnerability such as multiple sclerosis, leukoaraiosis, etc. Diffusion weighted imaging possesses unique sensitivity at the imaging voxel scale to map microscopic tissue structure characteristics. Its clinical use has been expanding rapidly. The diffusion tensor and its eigen values may be used to express the degree of diffusion anisotropy present in the tissue. Therefore, it has been widely used to evaluate white matter diseases such as stroke, multiple sclerosis, schizophrenia, vascular dementia, leukoaraiosis, trauma, and hypertension [1–3, 8]. However, little is known about diffusion tensor changes in vasogenic brain edema. The purpose of this study was to examine the correlation between changes in diffusion tensor imaging and regional tissue water content and tissue ultrastructure changes in vasogenic edema.

Materials and methods

The experiments were performed according to a protocol approved by the Committee on Animal Research of the Tokyo Medical and Dental University. Six adult cats weighing 4 ± 0.5 kg were used in the experiments. Vasogenic edema was induced in the white matter of cats by cortical cold lesioning as previously described [4, 5]. Briefly, after initial anesthesia with ketamine (30 mg/kg), a 15-mm left parietal craniotomy (12 mm anterior to the auditory meatus) was made with a dental drill. The cortical cold lesion was made by applying a $-40\,^\circ$C cooled metal plate to the dura covering the suprasylvian gyrus for 90 seconds. A 2% solution of Evans blue dye was injected intravenously soon after the operation. The cat was then intubated and artificially ventilated under anesthesia (1.5% isoflurane). Catheters were placed in the right femoral artery to allow monitoring of blood pressure and blood gas levels and in the right femoral vein to allow injection of drugs and contrast agent. Body temperature was maintained at 37 $^\circ$C by a feedback-controlled water jacket.

The MR images were acquired using a 4.7-T experimental system, which has a 330-mm horizontal bore magnet and a 65 mT/m maximum gradient capability (Unity INOVA, Varian, Inc., Palo Alto, CA). Diffusion tensor magnetic resonance (MR) imaging was performed using a multisection, spin-echo sequence, and the diffusion gradients were applied in turn along the 6 non-colinear directions. All analyses were performed with the aid of a workstation-based image analysis system (Sparc 10; Sun Microsystems, Mountain View, CA). Trace (D) maps were generated as described previously [6, 7]. Voxel by voxel, fractional anisotropy (FA) were calculated by using the following algorithms:

$$\text{FA} = \frac{\sqrt{(\lambda_1 - \lambda_2)^2 + (\lambda_2 - \lambda_3)^2 + (\lambda_1 - \lambda_3)^2}}{\sqrt{2}\sqrt{\lambda_1^2 + \lambda_2^2 + \lambda_3^2}}$$

where λn = the eigen values describing a diffusion tensor. For each section, these values were composed into FA maps. After MR imaging, the cats were sacrificed. Regions of interest (ROI) in the subcortical white matter of the left suprasylvian gyrus and the deep white matter (left semioval center under the suprasylvian gyrus) were drawn from T2-weighted images obtained before lesioning, and Trace (D) and FA values corresponding to each of these ROI were determined. After the final MR image was obtained, 4 cats were sacrificed under pentobarbital anesthesia (50 mg/kg) and the brains were sliced coronally corresponding to the MR imaging. For measurement of water content, tissue samples were taken from the sites corresponding to the ROI in the white matter and were determined as previously described [6]. The mirror-image coronal block was then immersion fixed with buffered formalin and stained with hematoxylin and eosin, Klüver-Barrera Luxol fast blue, and Bodian silver immunoglobulin stain for light microscopic examination. For electron microscopic examination, the remaining 2 cats were perfused transcardially with a 3% paraformaldehyde buffered solution and 1% glutaraldehyde after induction of anesthesia. Their brains were cut into coronal sections, and a block corresponding to the ROI of the FA map was chosen and sampled.

Results are expressed as the mean ± SD. Changes in systemic parameters, ADC values, and water content were analyzed using a one-way analysis of variance and Scheffé F test. The relationships between the FA or Trace (D) and water content were assessed using linear regression analysis and unpaired Student t test. Differences at probability values of less than 0.05 were considered to be statistically significant.

Results

Eight hours after cold lesioning, Trace, E1, E2, and E3 in the cortical white matter were increased 190%, 180%, 180%, and 280% from baseline values (just after the cold lesioning), respectively (Fig. 1). Correspondingly, FA was significantly decreased (46%). A linear correlation ($Y = 5.251e^{-5} \times X - 0.003$; $R = 0.82$) was observed between Trace and regional tissue water content in both acute and chronic phases after cold le-

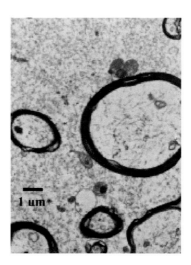

Fig. 2. Electron micrograph of deep white matter on the lesion side, demonstrating highly dissociated neuronal fibers with enlarged interfiber and extracellular spaces. The coronal section of dendrites appeared pale and nonhomogeneous. Intermicrotuble and interneurofilament spaces of the dendrites were enlarged compared to contralateral side

sioning. The relationship between FA and regional water content during the acute phase showed an all-or-nothing response, while that during the chronic phase showed a linear correlation. Histological examination demonstrated microvacuolation, disordered arrangement of nerve fibers, highly dissociated neuronal fibers due to extracellular accumulation of protein-rich fluid, and enlarged interfiber spaces (Fig. 2).

Discussion

As the diffusion of water molecules move within tissues, they encounter various restrictions and obstruction (for example, myelin in nerve fibers). Therefore, instead of observing "free" diffusion of water, we more often observe restricted diffusion. The motion of free diffusion (such as in cerebrospinal fluid, $\lambda 1 \approx \lambda 2 \approx \lambda 3$) can be visualized as a sphere, while the motion of restricted diffusion (such as in white matter, $\lambda 1 > [\lambda 2 \approx \lambda 3]$) can be visualized as an ellipsoid. In the latter, it is defined as the $\lambda 1$ (E1) dominant diffusion direction. We originally assumed that water diffusion would mainly be along the direction parallel to the nerve fiber in vasogenic brain edema. However, E3 had a greater increase (280% increase from baseline value) than E1 (180% increase) in cortical white matter 8 hours after cold lesioning. This indicates that water diffusion in the cortical white matter mainly increases perpendicularly rather than in a direction parallel to

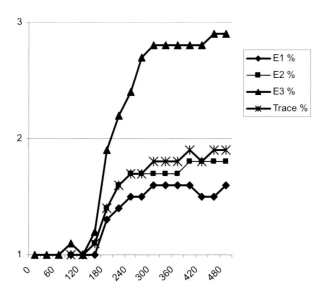

Fig. 1. Time course of diffusion tensor in acute phases. Trace, E1, E2, and E3 in the cortical white matter 8 hours after cold lesioning were increased 190%, 180%, 180%, and 280% from baseline values (just after the cold lesioning), respectively

the nerve fibers in the acute phase of vasogenic brain edema. In areas where edema is mild or moderate (tissues with water content of 65% to 75%), FA in the chronic phase is significantly lower than that in the acute phase. This finding indicates that even after the resolution of brain edema, FA still remains relatively low and is probably associated with the persistence of histological changes in the cortical white matter after cold lesioning. Further investigation is required, however.

References

1. Choi SH, Na DL, Chung CS, Lee KH, Na DG, Adair JC (2000) Diffusion-weighted MRI in vascular dementia. Neurology 54: 83–89
2. Jones DK, Lythgoe D, Horsfield MA, Simmons A, Williams SC, Markus HS (1999) Characterization of white matter damage in ischemic leukoaraiosis with diffusion tensor MRI. Stroke 30: 393–397
3. Kubicki M, Westin C-F, Maier SE, Frumin M, Nestor PG, Salisbury DF, Kikinis R, Jolesz FA, McCarley RW, Shenton ME (2002) Uncinate fasciculus findings in schizophrenia: a magnetic resonance diffusion tensor imaging study. Am J Psychiatry 159: 813–820
4. Kuroiwa T, Cahn R, Juhler M, Goping G, Campbell G, Klatzo I (1985) Role of extracellular proteins in the dynamics of vasogenic brain edema. Acta Neuropathol (Berl) 66: 3–11
5. Kuroiwa T, Nagaoka T, Ueki M, Yamada I, Miyasaka N, Akimoto H (1998) Different apparent diffusion coefficient: water content correlations of gray and white matter during early ischemia. Stroke 29: 859–865
6. Kuroiwa T, Nagaoka T, Ueki M, Yamada I, Miyasaka N, Akimoto H, Ichinose S, Okeda R, Hirakawa K (1999) Correlations between the apparent diffusion coefficient, water content, and ultrastructure after induction of vasogenic brain edema in cats. J Neurosurg 90: 499–503
7. Kuroiwa T, Nagaoka T, Miyasaka N, Akimoto H, Zhao F, Yamada I, Ueki M, Ichinose S (2000) Time course of trace of diffusion tensor [Trace(D)] and histology in brain edema. Acta Neurochir [Suppl] 76: 191–194
8. Werring DJ, Clark CA, Barker GJ, Thompson AJ, Miller DH (1999) Diffusion tensor imaging of lesions and normal-appearing white matter in multiple sclerosis. Neurology 52: 1626–1632

Correspondence: Fengyu Zhao, McLean Hospital/Harvard Medical School, 115 Mill St., Belmont, MA02478, USA. e-mail: zhaofy@mclean.harvard.edu

Bolus tracer delivery measured by MRI confirms edema without blood-brain barrier permeability in diffuse traumatic brain injury

A. Beaumont[1], P. Fatouros[3], T. Gennarelli[1], F. Corwin[3], and A. Marmarou[2]

[1] Dept of Neurosurgery, Medical College of Wisconsin, Milwaukee, WI, USA
[2] Dept of Neurosurgery, Medical College of Virginia, Richmond, VA, USA
[3] Dept of Radiology, Medical College of Virginia, Richmond, VA, USA

Summary

Introduction. Previous studies have shown that edema formation after diffuse traumatic brain injury (TBI) with secondary insult is cytotoxic and not vasogenic. This assumption is based on observations of reduced apparent diffusion coefficient (ADC) and lack of significant accumulation of intravascular tracer in brain tissue. However, ADC reduction does not exclude vasogenic edema, and intravascular tracer can only accumulate when it reaches the tissue and is not perfusion limited. This study aims to confirm tissue delivery of intravascular tracer and lack of BBB opening during a phase of rapid brain swelling after diffuse TBI.

Methods. Rats were exposed to either TBI using the impact acceleration model combined with 30 minutes of hypoxia and hypotension, or sham injury. At 2 or 4 hours after injury, ADC and tissue water content were assessed using MRI. Gd-DTPA was given followed by a combination of rapid T2 imaging (60 seconds) and T1 imaging (30 minutes). Signal intensity changes were analyzed to determine a bolus effect (dynamic susceptibility contrast) and longer term tissue accumulation of Gd-DTPA.

Results. Mean increase in cortical water content on the left was 0.8% at 2 hours, 2.1% at 4 hours; on the right it was 0.5% at 2 hours and 1.7% at 4 hours ($p < 0.05$). Mean ADC reduction over 4 hours was 0.04×10^{-3} mm^2/s on the left and 0.06×10^{-3} mm^2/s on the right. Kinetic analysis of signal intensity changes after Gd-DTPA showed no significant difference in inward transfer coefficient (BBB permeability) between sham injury and 2 or 4 hours post-injury. T2 imaging showed consistent tissue delivery of a bolus of Gd-DTPA to the tissue at 2 and 4 hours post-injury, comparable to sham animals.

Conclusions. Progressive cerebral edema formation after diffuse TBI occurred during ADC reduction and without continued BBB permeability. Tissue delivery of Gd-DTPA was confirmed, verifying that lack of tracer accumulation is due to an intact BBB and not to limited perfusion.

Keywords: Magnetic resonance imaging; brain edema; blood-brain barrier; traumatic brain injury.

Introduction

The role of blood-brain barrier (BBB) damage in posttraumatic brain swelling is not well understood. Recent studies have highlighted the importance of a cellular swelling process in edema formation after injury, as assessed by measurement of the apparent diffusion coefficient (ADC) of water [1, 2]. Other studies have suggested a permissive role for BBB damage [3]. Experimental studies of diffuse traumatic brain injury (TBI), in comparison to focal injury, have not demonstrated more than transient opening of the BBB. There are several methodological limitations in studies of the BBB. Of special relevance to TBI is the problem of flow-limited diffusion; if intravascular tracer cannot reach the tissue because of low blood flow, then BBB damage may be underestimated or undetected.

Magnetic resonance imaging (MRI) techniques are useful for assessing BBB damage after injury because it also provides other information such as ADC and degree of tissue water content concurrently over multiple time-points. However, visual assessment of signal intensity changes with intravascular administration of gadolinium-diethylenetriamine pentaacetic acid (Gd-DTPA) may underestimate tissue accumulation. Therefore, numerical assessment of signal intensity changes is recommended. In addition, bolus delivery of Gd-DTPA to tissue has been shown to generate a rapid and transient drop in signal intensity on T2-weighted images [7], the so-called dynamic susceptibility contrast phenomenon.

The aim of this study was to evaluate BBB damage in experimental diffuse TBI using MRI with Gd-DTPA, and to demonstrate tracer delivery to the tissue definitively in order to rule out underestimation of BBB damage due to flow-limited diffusion. A second goal of the study was to demonstrate progressive brain swelling in the absence of prolonged BBB opening, thereby confirming that posttraumatic brain swelling is cellular and not vasogenic in origin.

Materials and methods

In this study, experimental diffuse TBI was combined with a secondary insult of hypoxia and hypotension. BBB damage was assessed by serial measures of tissue T1-weighted contrast change after intravascular administration of Gd-DTPA. Tissue delivery of tracer was demonstrated using T2 imaging immediately after bolus administration. Tissue water content was assessed from the calculation of absolute values of T1 for the tissue. The ADC was measured at the same time as tissue water content.

All animals received humane care in compliance with the Guide for the Care & Use of Laboratory Animals (National Research Council, National Academy Press, Washington D.C., 1996). Adult male Sprague-Dawley rats (350 to 380 g; n = 16) were exposed to the impact acceleration model of diffuse brain injury [5] using a weight of 450 g over 2 m. Injury was combined with hypoxia and hypotension. Sham-injured animals (n = 6) underwent all procedures except for the impact of the weight or the imposition of secondary insult. Secondary insults of hypoxia and hypotension were applied by reduction of FiO_2 to 12%, resulting in arterial PO_2 levels of 30 to 40 mmHg, and arterial blood pressures of 30 to 40 mmHg. Secondary insults were initiated immediately after injury and maintained for 30 minutes.

At either 2 hours (n = 9) or 4 hours (n = 7) after injury, animals were placed in a 2.35 T, 40 cm bore magnet (Biospec, Bruker Instruments, Billerica, MA). Initial measures of tissue ADC and water content were made, followed by baseline T1 and T2 images according to methods described below. All images were obtained from a 3 mm thick slice, positioned 7.5 mm caudal to the anterior pole of the cerebrum. Following baseline imaging, each animal was infused with an intravenous bolus of 0.2 mmol/kg Gd-DTPA (Omniscan, Nycomed, Wayne, PA). Sixty seconds of rapid T2 imaging were performed for description of the signal intensity change due to dynamic susceptibility, followed by 30 minutes of T1 imaging to assess longer-term Gd-DTPA accumulation assessed by changes in T1 signal intensity.

Changes in T1 or T2 signal intensity were used to calculate Gd-DTPA concentration in the bolus and in the tissue, respectively, for regions of interest. Conversions to concentrations were based on previously defined calibration curves derived from known standards. The time course of Gd-DTPA accumulation acquired from the T1 images was then subjected to kinetic analysis using a previously described kinetic model [6] in order to derive features of BBB permeability. Concentration profiles were fitted to a derived equation using a non-linear least squares algorithm (Levenberg-Marquardt).

Mean arterial blood pressure was assessed continuously from the time of injury. BBB permeability parameters were compared with measured changes in tissue water and ADC values. Statistical significance was assessed using ANOVA, with appropriate post hoc tests (Fisher least significant difference, Newmann-Keuls) and p-values less than 0.05 were considered significant.

MRI measurements

At the time of assessment, animals were placed in a 2.35 T, 40 cm bore magnet (Bruker Instruments) equipped with a 12 cm inner diameter actively shielded gradient insert. RF excitation and reception were performed using a 4.5 cm helmet coil. In order to minimize any macroscopic motion artifacts, the rat's head was rigidly supported with a specially designed stereotactic device, including both ear and mouth supports mounted inside a Plexiglas cylinder. For evaluation of Gd-DTPA accumulation secondary to BBB damage, serial T1 images were obtained from a 3 mm thick slice, 7.5 mm caudal to the frontal pole. Imaging parameters used were TR = 700 ms, TE = 22 ms, FOV 4 cm^2, with a 64 × 64 matrix.

To demonstrate tissue delivery of Gd-DTPA, T2 imaging was used as described above. An example of a T2 MRI sequence with sufficient TR is the gradient echo method, acquired with TR/TE values of 27/20 ms, respectively, using a matrix size of 32 × 32 pixels and a 4 cm^2 FOV. These parameters generated an image acquisition time of 1000 msec. Images were repeated sequentially at maximum speed in order to provide TR of 1 second.

ADC measurements were performed using a 2-dimensional spin echo imaging technique (diffusion weighted imaging) appropriately modified to include diffusion-sensitizing gradients along the readout (horizontal) direction with a duration of 4 ms and a gradient separation of 20 ms. Each dataset consisted of a single coronal slice (3 mm thick) positioned 7.5 mm caudal to the frontal pole imaged with a 64 × 64 matrix using a TR/TE of 1500/33 ms and FOV 4 cm^2. Diffusion weighing factors, or *b* values, of 10, 340, 670, and 1000 s/mm^2 were used (maximum gradient strength of 23 G/cm). Pure ADC maps were calculated for each slice from the diffusion-weighted images using a pixel-by-pixel 3-parameter least squares fit to the magnitude image data. The effect of the frequency encoding gradients was included in the ADC calculations.

The concept of utilizing MRI for measuring brain water is based on laboratory and clinical studies directed toward noninvasive monitoring of brain edema formation and resolution [4]. Briefly, pure T1 maps are generated and then converted to water maps by means of the following equation:

$$\frac{1}{W} = 0.907 + \frac{0.407}{T1}$$

T1 is the measured T1 value of the tissue expressed in seconds and W is the tissue water content measured in gm H_2O/gm tissue.

Results

Figures 1a–c show profiles of tissue tracer concentration over time for 1 minute after tracer injection in sham animals, 2 and 4 hours after injury. There are no appreciable differences between the profiles; therefore, comparable quantities of tracer are delivered to the tissue in the injured animals compared with sham animals. Figures 1d–f show the profile of tracer concentration change in the tissue over 30 minutes after injection, based on T1W imaging. Figure 1f is taken from muscle, and represents the accumulation of Gd-DTPA in a tissue without a blood-brain barrier. Gd-DTPA follows a characteristic wash-in and wash-out profile. In contrast, there is minimal Gd-DTPA accumulation over 30 minutes in either the left (Fig. 1d) or right (Fig. 1e) cortex of 2-hour and 4-hour injured animals, consistent with either an impermeable BBB or severe flow-limited diffusion.

Application of the concentration-time data to a pre-

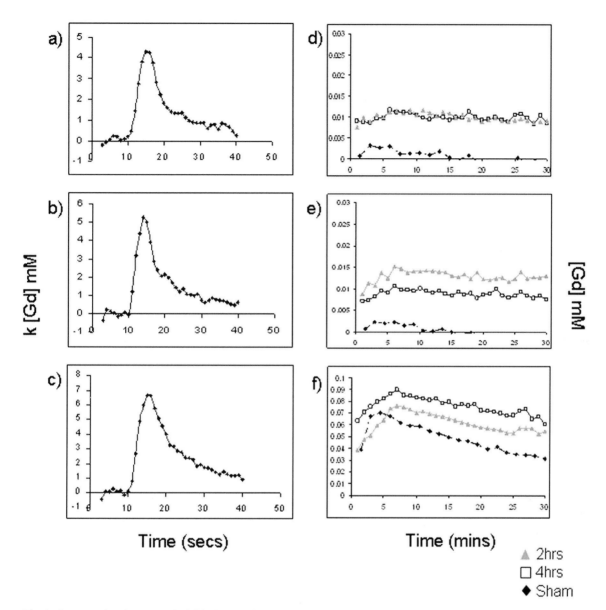

Fig. 1. Concentration time curves for initial bolus delivery of Gd-DTPA to the tissue (a–c) derived from signal intensity change due to dynamic susceptibility. Concentration time curves for Gd-DTPA accumulation over 30 minutes after infusion (d–f) showing longer term accumulation of tracer derived from T1W signal intensity changes under sham conditions, 2 and 4 hours after injury, (d) left cortex, (e) right cortex, (f) paraspinal muscle

viously published kinetic model [6] was performed. The derived parameter D_1, according to this model, is the product of the extracellular volume (V_e), peak plasma tracer concentration (A_0), and K_1, the inward transfer coefficient. Because V_e and A_0 are considered to be relatively constant, changes in D_1 are thought to reflect changes in K_1 closely. Table 1 shows the mean calculated D_1 values (uM/min) for each group in the left and right cortex. They are not different and do not appear to be influenced by trauma with secondary insult or time after trauma. Table 1 also shows the tissue water content and ADC values obtained in sham animals and at 2 and 4 hours after injury. At 2 hours after injury, tissue water content rose from 79.5% to 80.3% in the left hemisphere, and from 79.3% to 79.8% in the right hemisphere. Tissue water content continued to rise by 4 hours after injury to 81.6% and 81.0% in the left and right hemispheres, respectively. The rise in tissue water content was accompanied by a steady decline in ADC from 0.66 ($\times 10^{-3}$ mm^2/sec) to 0.62 in the left hemisphere and from 0.70 to 0.64 in the right hemisphere.

Table 1. *Comparison between mean values (±SD) for D1 (uM/min), tissue water content (%) and ADC (×10⁻³ mm²/sec) values in each experimental group.*

	Left			Right		
	Sham	2 hours	4 hours	Sham	2 hours	4 hours
D1	1.6 ± 1.0	2.0 ± 1.0	1.0 ± 0.2	2.0 ± 2.0	2.0 ± 2.0	1.0 ± 0.3
Tissue Water	79.5 ± 0.7	80.3 ± 1.4	81.6 ± 1.9	79.3 ± 0.9	79.9 ± 1.2	81.0 ± 2.5
ADC	0.66 ± 0.03	0.64 ± 0.09	0.62 ± 0.17	0.70 ± 0.06	0.65 ± 0.09	0.64 ± 0.16

D1 is the derived parameter from kinetic analysis of the gadolinium-diethylenetriamine pentaacetic acid concentration time-curves and is closely related to the inward transfer coefficient; D1 represents blood-brain barrier permeability. Tissue water is derived from tissue T1 data and is expressed as a percentage. ADC is the apparent diffusion coefficient of water. Reduction of apparent diffusion coefficient is associated with intracellular water accumulation.

Conclusions

The contribution of cytotoxic and vasogenic edema to posttraumatic cerebral swelling is not entirely clear. Recent studies have highlighted the importance of cytotoxic swelling [1, 2], and possibly a permissive role of BBB damage [3]. Although BBB damage has been well documented experimentally in models of focal contusion, nothing more than a transient opening has been demonstrated in experimental models of diffuse injury. In order to be sure that measures of BBB damage are not flawed, it is necessary to demonstrate delivery of tracer to the tissue.

Using rapid T2 imaging, this study confirms that an intravascular bolus of Gd-DTPA reaches the brain in rodents with diffuse TBI. The study also demonstrates that there is minimal accumulation of Gd-DTPA over 30 minutes of circulation time, consistent with an intact BBB. This was confirmed by application of a kinetic model that demonstrated no difference in BBB permeability between sham animals and animals 2 hours and 4 hours after injury. Despite the lack of BBB permeability, the injured brains showed progressive accumulation of water associated with ADC reduction up to 4 hours after injury.

For the first time, this study demonstrates progressive edema formation associated with ADC reduction and without BBB permeability in the context of confirmed tracer delivery, with all measures obtained at the same time in the same experimental subject. Our data confirms the dominant role of cellular swelling after TBI. However, further studies will be needed to characterize better the nature of edema associated with diffuse TBI.

References

1. Albensi BC, Knoblach SM, Chew BG, O'Reilly MP, Faden AI, Pekar JJ (2000) Diffusion and high resolution MRI of traumatic brain injury in rats: time course and correlation with histology. Exp Neurol 162: 61–72
2. Barzo P, Marmarou A, Fatouros P, Hayasaki K, Corwin F (1997) Contribution of vasogenic and cellular edema to traumatic brain swelling measured by diffusion-weighted imaging. J Neurosurg 87: 900–907
3. Beaumont A, Marmarou A, Hayasaki K, Barzo P, Fatouros P, Corwin F, Marmarou C, Dunbar J (2000) The permissive nature of blood brain barrier (BBB) opening in edema formation following traumatic brain injury. Acta Neurochir [Suppl] 76: 125–129
4. Marmarou A, Fatouros P, Ward J, Appley A, Young H (1990) In vivo measurement of brain water by MRI. Acta Neurochir [Suppl] 51: 123–124
5. Marmarou A, Foda MA, van den Brink W, Campbell J, Kita H, Demetriadou K (1994) A new model of diffuse brain injury in rats. Part I: Pathophysiology and biomechanics. J Neurosurg 80: 291–300
6. Su MY, Jao JC, Nalcioglu O (1994) Measurement of vascular volume fraction and blood-tissue permeability constants with a pharmacokinetic model: studies in rat muscle tumors with dynamic Gd-DTPA enhanced MRI. Magn Reson Med 32: 714–724
7. Villringer A, Rosen BR, Belliveau JW, Ackerman JL, Lauffer RB, Buxton RB, Chao YS, Wedeen VJ, Brady TJ (1988) Dynamic imaging with lanthanide chelates in normal brain: contrast due to magnetic susceptibility effects. Magn Reson Med 6: 164–174

Correspondence: Andrew Beaumont, Department of Neurosurgery, Medical College of Wisconsin, 9200 W Wisconsin Ave., Milwaukee, WI 53226, USA. e-mail: abeaumont@neuroscience.mcw.edu

Experimental Intracranial Hemorrhage

Delayed profound local brain hypothermia markedly reduces interleukin-1β gene expression and vasogenic edema development in a porcine model of intracerebral hemorrhage

K. R. Wagner[1,4], S. Beiler[1,4], C. Beiler[1,4], J. Kirkman[5], K. Casey[5], T. Robinson[5], D. Larnard[5], G. M. de Courten-Myers[2], M. J. Linke[4], and M. Zuccarello[3,4]

[1] Department of Neurology, University of Cincinnati College of Medicine, Cincinnati, OH, USA
[2] Department of Pathology and Laboratory Medicine, University of Cincinnati College of Medicine, Cincinnati, OH, USA
[3] Department of Neurosurgery, University of Cincinnati College of Medicine, Cincinnati, OH, USA
[4] Medical Research Service, Department of Veterans Affairs Medical Center, Cincinnati, OH, USA
[5] Seacoast Technologies, Inc., Portsmouth, NH, USA

Summary

White matter (lobar) intracerebral hemorrhage (ICH) can cause edema-related deaths and life-long morbidity. In our porcine model, ICH induces oxidative stress, acute interstitial and delayed vasogenic edema, and up-regulates interleukin-1β (IL-1β), a proinflammatory cytokine-linked to blood-brain barrier (BBB) opening. In brain injury models, hypothermia reduces inflammatory cytokine production and protects the BBB. Clinically, however, hypothermia for stroke treatment using surface and systemic approaches can be challenging. We tested the hypothesis that an alternative approach, i.e., local brain cooling using the ChillerPad System, would reduce IL-1β gene expression and vasogenic edema development even if initiated several hours after ICH.

We infused autologous whole blood (3.0 mL) into the frontal hemispheric white matter of 20 kg pentobarbital-anesthetized pigs. At 3 hours post-ICH, we performed a craniotomy for epidural placement of the ChillerPad. Chilled saline was then circulated through the pad for 12 hours to induce profound local hypothermia (14 °C brain surface temperature). We froze brains in situ at 16 hours after ICH induction, sampled perihematomal white matter, extracted RNA, and performed real-time RT-PCR.

Local brain cooling markedly reduced both IL-1β RNA levels and vasogenic edema. These robust results support the potential for local brain cooling to protect the BBB and reduce injury after ICH.

Keywords: Intracerebral hemorrhage; white matter; hypothermia; cytokines.

Introduction

Intracerebral hemorrhage (ICH) is the stroke subtype with the highest mortality and morbidity [10, 30]. Almost half of ICH patients will die, often within the first 48 hours, while at most 20% of ICH survivors will return to normal lives [6, 21]. Intracerebral bleeds not only occur spontaneously, but can also follow thrombolytic treatment for ischemic stroke and myocardial infarction [13, 29]. Lobar ICH (white matter hemorrhage) causes edema-related deaths at twice the rate versus other locations [23]. White matter ICH can also damage fiber tracts resulting in permanent neurological deficits [12, 19].

We have developed a large animal (porcine) lobar ICH model to study the pathophysiology, pathochemistry, and treatment of this disease, including surgical clot removal [2, 39]. In this animal model as in human ICH, both early perihematomal edema and delayed vasogenic edema following blood-brain barrier (BBB) opening are prominent [9, 39]. White matter is more vulnerable to vasogenic edema development than is gray [20]. Thus, our model is useful for studying edema-induced injury following ICH. Neuropathologically, astrogliosis, demyelination, and cystic necrosis in perihematomal white matter, all features of human ICH-induced brain damage, develop in this model [37]. This model has been used to examine ICH-induced pathophysiologic and pathochemical events in white matter [2, 38–40, 44, 49].

We have observed that oxidative stress (protein carbonyl formation) and up-regulation of heme oxygenase-1 gene expression develops rapidly in perihematomal white matter following ICH [42]. Protein carbonyl formation also develops after plasma infusions into the frontal white matter, indicating that plasma components alone can induce oxidative stress [46]. This may relate to an interaction between holo-transferrin and thrombin leading to intracellular iron deposition [27]. Early oxidative stress occurs in a gray matter ICH model [48] and damages DNA [26]. We demonstrated that the transcription factor, nuclear factor kappaB (NF-κB), which is responsive to oxidative stress [15, 17], is activated in perihematomal white matter after ICH [43, 45]. This transcription factor is a primary mediator for the rapid and coordinated induction of central nervous system genes which respond to injury and other pathological stimuli in brain, including the proinflammatory cytokine, interleukin (IL)-1β [1, 3, 5, 32, 47].

A renewed interest has emerged in the last several years in treating stroke by surface or systemic hypothermia techniques [16, 38]. Mild to moderate hypothermia protects the BBB and reduces brain injury in animal stroke models [7, 8]. However, clinically, systemic hypothermia for stroke treatment is challenging with prolonged times to target temperatures, difficult sedation protocols, and adverse events including shivering and pneumonia.

We describe an alterative approach, i.e., local profound brain cooling using a new device and technology, the ChillerPad system (Seacoast Technologies, Inc., Portsmouth, NH) in a porcine ICH model. Previously we demonstrated the effectiveness of the epidurally-placed device to reduce vasogenic edema development following ICH in this model [38]. We tested the hypothesis that local profound brain cooling using the ChillerPad system could reduce IL-1β gene expression even if initiated several hours after inducing ICH.

Materials and methods

Animal surgical preparations and intracerebral blood infusions

The animal protocol for this study was approved by the Institutional Animal Care and Use Committee of The Cincinnati Veterans Affairs Medical Center. Pigs were Yorkshire mixed breed and were obtained from a local farm (Yeazal Farms, Eaton, OH). They received food and water ad libitum. We previously described our surgical and intracerebral blood infusion methodologies in detail [39–41]. Pigs (~15 kg) were initially anesthetized with ketamine, 25 to 30 mg/kg, i.m., followed by intravenous pentobarbital (35 mg/kg). Anesthesia was maintained by continuous pentobarbital infusions (10 mg/kg/hr). We intubated pigs by mouth, mechanically ventilated them (air plus 0.5 liters/min oxygen), and catheterized femoral vessels (arteries to record blood pressure, measure arterial respiratory gases and pH; veins to infuse saline, pentobarbital). We monitored and controlled core temperatures ($38.5 \pm 0.5\,°C$). We infused arterial blood (3.0 mL) through a 20-gauge Teflon catheter implanted into the frontal hemispheric white matter. We infused Evans blue (i.v., 1 mL/kg of 2% wt/vol solution) at 1.5 hours following ICH to test BBB permeability.

Local brain hypothermia using the ChillerPad system

The ChillerPad (Seacoast Technology, Inc.) is a 2-cm device comprised of coiled polyester tubing encased in a polyurethane skin containing integral thermocouples. Chilled saline is circulated through the pad to produce a temperature of $14\,°C \pm 1\,°C$ on its contact surface. Beginning at 3 hours post-ICH to enable epidural placement of the ChillerPad, a circular craniotomy (~2 cm diameter) was made with a trephine centered approximately 2 cm laterally from the sagittal ridge and 1 cm anterior to the coronal suture. The cooling pad was placed on the dura and held in place with either the bone flap or a gauze pad sutured in place. Hypothermia was conducted for 12 hours by circulating chilled saline through the device. The brain was then allowed to rewarm for 30 minutes prior to removing the pad and initiating in situ brain freezing. The brains of control normothermic ICH animals were not cooled and were frozen in situ at 16 hours post-ICH.

Brain freezing, coronal sections, photography, hematoma and edema volume quantitation

Brains were frozen in situ using liquid nitrogen 12 hours following onset of local brain cooling (16 hours following hematoma induction) as previously described [39–41, 49]. Following decapitation, heads were stored at $-70\,°C$ until sectioned. Coronal sections (5 mm) were cut through the frozen heads using a band saw. Both sides of coronal sections containing the hematoma and/or edema were photographed with a digital camera. The hematoma and the visible perihematomal edema volumes were determined by importing the color images into Image Tool, a freeware image analysis system (University of Texas) for computer-assisted morphometry [41]. Areas obtained by the Image Tool software were identical to those obtained with NIH Image. Areas from both sides of the slices were then averaged, volumes calculated from the slice thickness, and then summed for total hematoma and edema volumes.

RNA isolation, reverse transcription and real-time polymerase chain reaction (PCR)

White matter tissue (10 mg) adjacent to the hematomas was sampled from the frozen coronal sections in a glove box at $-15\,°C$. White matter from the same gyrus or location in the corresponding contralateral hemisphere served as the control. Total RNA was isolated from white matter samples using Trizol reagent (GibcoBRL/Life Technologies, Carlsbad, CA) and cDNAs were synthesized from RNA using reverse transcriptase and random hexamer primers (Promega, Madison, WI) [42, 45]. Real-time PCR was conducted using BioRad's iCycler instrument, and cDNA templates plus sense and antisense LUX fluorogenic primers (Invitrogen, Carlsbad, CA) designed for porcine IL-1β and GAPDH (housekeeping gene). Relative quantitation for IL-1β was obtained by threshold cycle analyses normalized against GAPDH.

Statistical analysis

Hematoma and edema volumes and IL-1 differences were compared between the 2 groups by *t* test (Statgraphics, Manugistics, Inc., Rockville, MD). Differences were considered significant at $p < 0.05$.

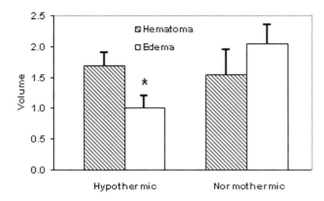

Fig. 1. Brains from animals with local hypothermia frozen in situ 16 hours following ICH and 12 hours after initiation of local brain cooling have markedly reduced (51%) perihematomal Evans blue staining, indicating reduced vasogenic edema development. Hematoma volumes were similar in both groups. Volume measurements (*mL*) are mean ± SEM, N = 3 in each group. *p = 0.027 versus edema volume in normothermic animals

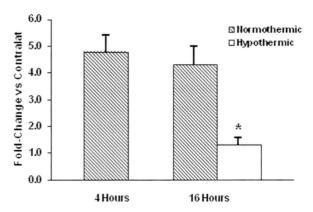

Fig. 2. IL-1β cytokine gene expression measured by real-time PCR was up-regulated in edematous white matter adjacent to the hematoma at 4 and 16 hours following normothermic ICH. In contrast, 12 hours of local brain cooling significantly reduced up-regulated IL-1β gene expression to contralateral control levels. Values are mean ± SEM, N = 3 in each group. *p < 0.025

Results

Coronal brain sections from normothermic lobar ICH pigs frozen in situ at 16 hours demonstrated the typical location of the induced hematoma with clearly visible perihematomal edematous white matter regions. This white matter tissue had a marked increase in water content (>10%) and was blue-stained after opening of the BBB to Evans blue containing albumin, leading to development of vasogenic edema [39–41]. In contrast, brains from animals with local hypothermia frozen in situ 16 hours after ICH and 12 hours after initiation of local brain cooling had markedly reduced (51%) perihematomal Evans blue staining, indicating reduced vasogenic edema development ($p < 0.03$ edema volume in hypothermic versus normothermic animals) (Fig. 1).

Following normothermic ICH, there was a 4.2-fold up-regulation in expression of the proinflammatory IL-1β cytokine gene measured by real-time reverse transcriptase PCR in edematous white matter adjacent to the hematoma at 16 hours (Fig. 2). In contrast, 12 hours of local brain cooling significantly reduced up-regulated IL-1β gene expression essentially to control levels.

Discussion

In this report, we describe our studies on local (focal) brain cooling using the ChillerPad system (Seacoast Technologies, Inc.) and expression of the proinflammatory cytokine gene, IL-1β, in perihematomal white matter in porcine lobar ICH. Specifically, our findings show that local hypothermia significantly reduced the level of this gene expression, even if delayed until 3 to 4 hours following ICH. We previously observed that local (focal) brain cooling reduced vasogenic edema [38]. Based on results from the literature [1, 18], the reduction in IL-1β expression produced by local hypothermia may be responsible for reduced BBB injury.

Whole-body hypothermia has long been studied both experimentally and clinically for treating brain injuries, including stroke and trauma [16]. However, because of its related systemic problems, including increased risk of infection and various systemic complications, its use has never been widely instituted. Interest in hypothermia was recently renewed when reports demonstrated its effectiveness in treating out-of-hospital cardiac arrest [4, 36]. Studies of surface cooling blankets and endovascular cooling catheters to induce hypothermia in stroke patients have been recently reported [16, 24]. However, whole-body hypothermia continues to be challenging because of the amount of cooling time necessary to achieve target temperatures (up to several hours), shivering, and adverse events. Since stroke patients are often elderly and may have additional medical problems, the general use of whole-body hypothermia for stroke treatment appears unlikely. In contrast, local or focal brain cooling is a potential approach that may be useful for treating ICH patients, especially to extend the window for surgical treatment. This may also be a useful approach to prevent ischemic injury in aneurysm surgery [11, 38].

We have reported that oxidative stress and NF-κB activation rapidly occur (within the first hours) in white matter adjacent to the hematoma following ICH [42, 45]. Since NF-κB is a redox-sensitive transcription factor activated in gray matter after ICH [17], this ultra-early oxidative stress may participate in activating the IL-1 gene expression that we observed in the present study. In this regard, LY341122, an antioxidant that is neuroprotective, prevented activation of NF-κB following cerebral ischemia [34]. NF-κB is a well-described transcription factor for target genes including the proinflammatory cytokines [15]. Our previous findings demonstrated rapidly-activated NF-κB in perihematomal white matter as previously described in gray matter.

Microglia may be the cells responsible for the rapid generation of reactive oxygen species following ICH since they are known to secrete potentially toxic molecules including oxygen and nitrogen free radicals [14]. Microglial activation is a hallmark of various brain disorders [14]. Interestingly, a recent report using 2-photon microscopy demonstrates that these cells are highly active, even in their resting state, continually surveying their microenvironment with extremely motile processes and protrusions that can rapidly undergo responses [28]. Indeed, upon BBB disruption, immediate and focal activation occur, with these cells switching their behavior from patrolling to shielding the injured site.

Relevant to our findings following ICH, the plasma proteins that enter the brain parenchyma following ICH activate microglia [14, 25, 31]. In vitro, serum addition stimulates superoxide production by microglial cultures [33]. Up-regulation of HO-1 expression has been demonstrated immunocytochemically in morphologically appearing microglia following intracerebral infusion of lysed blood [31]. We previously reported early induction of HO-1 in perihematomal white matter following ICH [42].

Our previous findings demonstrated early (1 to 2 hour) up-regulation of IL-1β mRNA in perihematomal white matter [43, 45]. These findings are confirmed by the real-time PCR data presented in this report. The proinflammatory role of IL-1β has been well-described in brain tissue and is considered to be an important mediator of diverse forms of acute neurodegeneration and chronic neurological conditions [1, 32], as well as inducing BBB opening and edema [18]. The significance of IL-1β expression is supported by recent findings that overexpression of the IL-1 receptor antagonist, IL-1ra, reduces thrombin-induced edema [22]. This early up-regulation of IL-1β gene expression appears to be continued and expanded after ICH, based on our recent DNA microarray findings in a rat ICH model [35]. Elevated expression of IL-1β and related genes including IL-1β converting enzyme (ICE, caspase-1) and a proinflammatory cytokine, IL-18, that is activated by ICE, were present at 24 hours. In this regard, we have also demonstrated immunostaining with a porcine monoclonal antibody for IL-1β protein in white matter glial cells adjacent to the hematoma at 24 hours post-ICH. Some of these cells appear to have microglial morphology while others appear to be astrocytes. Additional double-labeling studies are required to identify the specific cell types. Temporal studies in progress using recently available porcine gene arrays from Affymetrix (Santa Clara, CA) will identify both the acute and subacute changes in these cytokine pathways in perihematomal white and gray matter after ICH.

In summary, these results from several different brain injury models point toward a component(s) in the plasma itself that induces white matter edema and oxidative stress leading to activation of downstream NF-κB-dependent target gene expression including proinflammatory cytokines [44]. Although proinflammatory cytokines also appear to participate in tissue recovery and repair, in the acute setting these molecules are generally associated with central nervous system tissue injury [5]. Because these important mediators of increased BBB permeability, leukocyte infiltration, and secondary lesion expansion and cell death are rapidly up-regulated after ICH, ultra-early treatment aimed at these processes may provide protection until surgical intervention with clot removal can be initiated.

Acknowledgments

This study was supported by National Institutes of Health grant R01-NS30652 and funds from the Office of Research and Development, Medical Research Service, Department of Veterans Affairs.

References

1. Allan SM, Rothwell NJ (2003) Inflammation in central nervous system injury. Philos Trans R Soc Lond B Biol Sci 358: 1669–1677
2. Andaluz N, Zuccarello M, Wagner KR (2002) Experimental animal models of intracerebral hemorrhage. Neurosurg Clin N Am 13: 385–393

3. Barone FC, Feuerstein GZ (1999) Inflammatory mediators and stroke: new opportunities for novel therapeutics. J Cereb Blood Flow Metab 19: 819–834
4. Bernard SA, Gray TW, Buist MD, Jones BM, Silvester W, Gutteridge G, Smith K (2002) Treatment of comatose survivors of out-of-hospital cardiac arrest with induced hypothermia. N Engl J Med 346: 557–563
5. Bethea JR (2000) Spinal cord injury-induced inflammation: a dual-edged sword. Prog Brain Res 128: 33–42
6. Broderick JP, Brott TG, Duldner JE, Tomsick T, Huster G (1993) Volume of intracerebral hemorrhage. A powerful and easy-to-use predictor of 30-day mortality. Stroke 24: 987–993
7. Dietrich WD, Kuluz JW (2003) New research in the field of stroke: therapeutic hypothermia after cardiac arrest. Stroke 34: 1051–1053
8. Dietrich WD, Busto R, Halley M, Valdes I (1990) The importance of brain temperature in alterations of the blood-brain barrier following cerebral ischemia. J Neuropathol Exp Neurol 49: 486–497
9. Dul K, Drayer BP (1994) CT and MR imaging of intracerebral hemorrhage. In: Kase CS, Caplan LR (eds) Intracerebral Hemorrhage. Butterworth-Heinemann, Boston, pp 73–98
10. Foulkes MA, Wolf PA, Price TR, Mohr JP, Hier DB (1988) The Stroke Data Bank: design, methods, and baseline characteristics. Stroke 19: 547–554
11. Frietsch T, Kirsch JR (2004) Strategies of neuroprotection for intracranial aneurysms. Best Pract Res Clin Anaesthesiol 18: 595–630
12. Fukui K, Iguchi I, Kito A, Watanabe Y, Sugita K (1994) Extent of pontine pyramidal tract Wallerian degeneration and outcome after supratentorial hemorrhagic stroke. Stroke 25: 1207–1210
13. Gebel JM, Brott TG, Sila CA, Tomsick TA, Jauch E, Salisbury S, Khoury J, Miller R, Pancioli A, Duldner JE, Topol EJ, Broderick JP (2000) Decreased perihematomal edema in thrombolysis-related intracerebral hemorrhage compared with spontaneous intracerebral hemorrhage. Stroke 31: 596–600
14. Gonzalez-Scarano F, Baltuch G (1999) Microglia as mediators of inflammatory and degenerative diseases. Annu Rev Neurosci 22: 219–240
15. Grilli M, Memo M (1999) Possible role of NF-kappaB and p53 in the glutamate-induced pro-apoptotic neuronal pathway. Cell Death Differ 6: 22–27
16. Hammer MD, Krieger DW (2003) Hypothermia for acute ischemic stroke: not just another neuroprotectant. Neurologist 9: 280–289
17. Hickenbottom SL, Grotta JC, Strong R, Denner LA, Aronowski J (1999) Nuclear factor-kappaB and cell death after experimental intracerebral hemorrhage in rats. Stroke 30: 2472–2477
18. Holmin S, Mathiesen T (2000) Intracerebral administration of interleukin-1beta and induction of inflammation, apoptosis, and vasogenic edema. J Neurosurg 92: 108–120
19. Kazui S, Kuriyama Y, Sawada T, Imakita S (1994) Very early demonstration of secondary pyramidal tract degeneration by computed tomography. Stroke 25: 2287–2289
20. Kimelberg HK (1995) Current concepts of brain edema. Review of laboratory investigations. J Neurosurg 83: 1051–1059
21. Lisk DR, Pasteur W, Rhoades H, Putnam RD, Grotta JC (1994) Early presentation of hemispheric intracerebral hemorrhage: prediction of outcome and guidelines for treatment allocation. Neurology 44: 133–139
22. Masada T, Hua Y, Xi G, Yang GY, Hoff JT, Keep RF (2001) Attenuation of intracerebral hemorrhage and thrombin-induced brain edema by overexpression of interleukin-1 receptor antagonist. J Neurosurg 95: 680–686
23. Mayer SA, Sacco RL, Shi T, Mohr JP (1994) Neurologic deterioration in noncomatose patients with supratentorial intracerebral hemorrhage. Neurology 44: 1379–1384
24. Mayer SA, Kowalski RG, Presciutti M, Ostapkovich ND, McGann E, Fitzsimmons BF, Yavagal DR, Du YE, Naidech AM, Janjua NA, Claassen J, Kreiter KT, Parra A, Commichau C (2004) Clinical trial of a novel surface cooling system for fever control in neurocritical care patients. Crit Care Med 32: 2508–2515
25. Murakami K, Kawase M, Kondo T, Chan PH (1998) Cellular accumulation of extravasated serum protein and DNA fragmentation following vasogenic edema. J Neurotrauma 15: 825–835
26. Nakamura T, Keep RF, Hua Y, Hoff JT, Xi G (2005) Oxidative DNA injury after experimental intracerebral hemorrhage. Brain Res 1039: 30–36
27. Nakamura T, Xi G, Park JW, Hua Y, Hoff JT, Keep RF (2005) Holo-transferrin and thrombin can interact to cause brain damage. Stroke 36: 348–352
28. Nimmerjahn A, Kirchhoff F, Helmchen F (2005) Resting microglial cells are highly dynamic surveillants of brain parenchyma in vivo. Science 308: 1314–1318
29. NINDS rt-PA Stroke Study Group (1995) Tissue plasminogen activator for acute ischemic stroke. The National Institute of Neurological Disorders and Stroke rt-PA Stroke Study Group. N Engl J Med 333: 1581–1587
30. Qureshi AI, Tuhrim S, Broderick JP, Batjer HH, Hondo H, Hanley DF (2001) Spontaneous intracerebral hemorrhage. N Engl J Med 344: 1450–1460
31. Richmon JD, Fukuda K, Maida N, Sato M, Bergeron M, Sharp FR, Panter SS, Noble LJ (1998) Induction of heme oxygenase-1 after hyperosmotic opening of the blood-brain barrier. Brain Res 780: 108–118
32. Rothwell NJ (1999) Annual review prize lecture cytokines – killers in the brain? J Physiol 514 (Pt 1): 3–17
33. Si QS, Nakamura Y, Kataoka K (1997) Albumin enhances superoxide production in cultured microglia. Glia 21: 413–418
34. Stephenson D, Yin T, Smalstig EB, Hsu MA, Panetta J, Little S, Clemens J (2000) Transcription factor nuclear factor-kappa B is activated in neurons after focal cerebral ischemia. J Cereb Blood Flow Metab 20: 592–603
35. Tang Y, Lu A, Aronow BJ, Wagner KR, Sharp FR (2002) Genomic responses of the brain to ischemic stroke, intracerebral haemorrhage, kainate seizures, hypoglycemia, and hypoxia. Eur J Neurosci 15: 1937–1952
36. The Hypothermia After Cardiac Arrest Study Group (2002) Mild therapeutic hypothermia to improve the neurologic outcome after cardiac arrest. N Engl J Med 346: 549–556
37. Wagner KR, Broderick JP (2001) Hemorrhagic stroke: pathophysiological mechanisms and neuroprotective treatments. In: Lo EH, Marwah J (eds) Neuroprotection. Prominent Press, Scottsdale, pp 471–508
38. Wagner KR, Zuccarello M (2005) Local brain hypothermia for neuroprotection in stroke treatment and aneurysm repair. Neurol Res 27: 238–245
39. Wagner KR, Xi G, Hua Y, Kleinholz M, de Courten-Myers GM, Myers RE, Broderick JP, Brott TG (1996) Lobar intracerebral hemorrhage model in pigs: rapid edema development in perihematomal white matter. Stroke 27: 490–497
40. Wagner KR, Xi G, Hua Y, Kleinholz M, de Courten-Myers GM, Myers RE (1998) Early metabolic alterations in edematous perihematomal brain regions following experimental intracerebral hemorrhage. J Neurosurg 88: 1058–1065
41. Wagner KR, Xi G, Hua Y, Zuccarello M, de Courten-Myers GM, Broderick JP, Brott TG (1999) Ultra-early clot aspiration after lysis with tissue plasminogen activator in a porcine model

of intracerebral hemorrhage: edema reduction and blood-brain barrier protection. J Neurosurg 90: 491–498
42. Wagner KR, Packard BA, Hall CL, Smulian AG, Linke MJ, de Courten-Myers GM, Packard LM, Hall NC (2002) Protein oxidation and heme oxygenase-1 induction in porcine white matter following intracerebral infusions of whole blood or plasma. Dev Neurosci 24: 154–160
43. Wagner KR, Dean C, Beiler S, Knight J, Packard BA, Hall CL, de Courten-Myers GM (2003) Rapid activation of pro-inflammatory signaling cascades in perihematomal brain regions in a porcine white matter intracerebral hemorrhage model. J Cereb Blood Flow Metab 23 [Suppl] 1: 277
44. Wagner KR, Sharp FR, Ardizzone TD, Lu A, Clark JF (2003) Heme and iron metabolism: role in cerebral hemorrhage. J Cereb Blood Flow Metab 23: 629–652
45. Wagner KR, Beiler S, Dean C, Hall CL, Knight J, Packard BA, Smulian AG, Linke MJ, de Courten-Myers GM (2004) NFκB activation and pro-inflammatory cytokine gene upregulation in white matter following porcine intracerebral hemorrhage. In: Krieglstein J, Klumpp S (eds) Pharmacology of cerebral ischemia 2004. Medpharm Scientific Publishers, Stuttgart, pp 185–194
46. Wagner KR, Dean C, Beiler S, Bryan DW, Packard BA, Smulian AG, Linke MJ, de Courten-Myers GM (2005) Plasma infusions into porcine cerebral white matter induce early edema, oxidative stress, pro-inflammatory cytokine gene expression and DNA fragmentation: Implications for white matter injury with increased blood-brain barrier permeability. Curr Neurovasc Res 2: 149–155
47. Wang CX, Shuaib A (2002) Involvement of inflammatory cytokines in central nervous system injury. Prog Neurobiol 67: 161–172
48. Wu J, Hua Y, Keep RF, Nakamura T, Hoff JT, Xi G (2003) Iron and iron-handling proteins in the brain after intracerebral hemorrhage. Stroke 34: 2964–2969
49. Xi G, Wagner KR, Keep RF, Hua Y, de Courten-Myers GM, Broderick JP, Brott TG, Hoff JT, Muizelaar JP (1998) Role of blood clot formation on early edema development after experimental intracerebral hemorrhage. Stroke 29: 2580–2586

Correspondence: Kenneth R. Wagner, Research Service (151), Department of Veterans Affairs Medical Center, 3200 Vine Street, Cincinnati, OH 45220, USA. e-mail: wagnerkr@email.uc.edu

Alterations in intracerebral hemorrhage-induced brain injury in the iron deficient rat

J. Shao[1], G. Xi[2], Y. Hua[2], T. Schallert[2,3], and B. T. Felt[4]

[1] Department of Pediatrics, Children's Hospital, Zhejiang University, China
[2] Department of Neurosurgery, University of Michigan Medical School, Ann Arbor, MI, USA
[3] Department of Psychology, University of Texas, Austin, TX, USA
[4] Center for Human Growth and Development, University of Michigan, Ann Arbor, MI, USA

Summary

Background. Iron contributes to brain edema and cellular toxicity after intracerebral hemorrhage (ICH). Knowledge regarding ICH in the context of iron deficiency anemia (IDA), a common nutritional disorder, is limited.

Objective. To determine the effect of IDA on brain and behavioral outcome after ICH in rats.

Methods. Six-week-old male rats (n = 75) were randomized to non-IDA or IDA groups. After 1 month of iron sufficient or deficient diets, 100 μl autologous blood was infused into the right basal ganglia (BG). Brains were assessed for iron concentration, regional water content, BG transferrin, and transferrin receptor concentrations after ICH. Recovery of upper extremity sensorimotor function was assessed. Brain and behavioral variables were compared by diet group. Significance was set at $p < 0.05$.

Results. Whole brain iron was decreased and water content was increased for IDA rats in injured cortex and BG at day 3 ($p < 0.05$) compared with non-IDA rats. Transferrin and transferrin receptor content were increased in injured BG for IDA compared to non-IDA in the first week after ICH ($p < 0.05$). IDA rats had greater left vibrissae-stimulated forelimb-placing deficits and forelimb-use asymmetry than non-IDA after ICH ($p < 0.05$).

Conclusions. Brain iron status may be an important determinant of injury severity and recovery after ICH.

Keywords: Intracerebral hemorrhage; iron deficiency anemia; edema; transferrin; behavior.

Introduction

Intracerebral hemorrhage (ICH), a common stroke subtype, is often associated with neurologic deficits in surviving patients. Development of a hematoma within the brain parenchyma triggers a series of events including edema formation and consequent space-occupying effects, and tissue destruction [11]. The role of hemoglobin degradation products such as iron, have been investigated in non-anemic ICH models. Recent studies suggest that iron and oxidative stress contribute to edema formation after ICH [10, 15, 20]. In addition, the iron-trafficking proteins transferrin (Tf), transferrin receptor (TfR), and ferritin, are thought to play a role by modulating iron-induced cellular toxicity, lipid peroxidation, and free radical formation [4, 12].

Iron deficiency anemia (IDA) is a common nutritional disorder that affects approximately 25% of young children worldwide [19]. This problem occurs when dietary iron intake is insufficient for growth and maintenance needs or excessive iron losses occur, such as with chronic illness conditions, injury, or repeated medical procedures. In addition to hemoglobin production, iron is important for numerous processes including neurotransmitter synthesis, oxidative metabolism, and myelination [1]. IDA reduces whole brain iron concentration in animal models. However, the brain does not respond in a homogeneous fashion; iron content varies by brain region in response to IDA [2, 5, 6]. Similarly, while whole brain Tf content is increased with IDA, there is a heterogeneous response by brain region [5, 8]. The iron-trafficking proteins Tf and TfR are thought to regulate regional brain iron content in response to local needs.

One might hypothesize that brain edema formation and behavioral outcome after ICH would be less severe in the context of IDA due to less available iron secondary to less hemoglobin, number of red cells, and serum iron. However, in a first study of ICH in the context of IDA, we found that brain edema and iron-trafficking proteins were increased and functional outcome was worse [18]. In this paper we review these findings and extend brain and behavioral outcome measures after ICH in rats with or without anemia.

Materials and methods

Animals and experimental groups

Six-week-old male Sprague-Dawley rats (Harlan Sprague Dawley, Indianapolis, IN) were randomly assigned to receive 1 of 2 diets (Harlan Teklad, Madison, WI) that varied in iron content but were sufficient for protein and other micronutrients. The iron sufficient diet contained 40 mg/kg iron (non-IDA group, n = 32). The iron deficient diet contained 3 to 6 mg/kg iron (IDA group, n = 43). Rats were fed ad-lib on their respective diets until brain assessment. ICH surgery occurred after 1 month on the diets. All animals were pair-housed in a temperature-controlled (25 °C) room with a 12:12 light:dark cycle. Approval for this protocol was provided by the University of Michigan Committee on Use and Care of Animals.

ICH surgery

After weighing, rats were anesthetized with pentobarbital (40 mg/kg, i.p.) then placed on a feedback-controlled warming pad. Rectal temperatures were maintained at $36.5 \pm 0.5\,°C$. Blood for arterial gases (PO_2, PCO_2), pH, hematocrit, serum iron, and glucose were obtained by femoral artery catheterization. Blood pressure was monitored. A right basal ganglia (BG) hematoma was created by slowly infusing 100 μl autologous whole blood (from a rat within the same diet group) through a 26-gauge needle using a micro-infusion pump (10 μL/min; Harvard Apparatus, Holliston, MA) after creating a 1 mm burr hole on the right coronal suture, 3.5 mm lateral to midline. Sham-operated IDA rats received needle insertion into the right BG only (n = 6). Serum iron was determined by standard methods [3].

Brain assessments

At day 3 after ICH surgery, injured and non-injured BG and cortex and the non-injured cerebellum (control) were measured for water content in 5 to 6 animals in each diet group, as previously described [18]. Regional samples were weighed to 0.0001 g before and after drying for 24-hours at 100 °C (gravity oven, Blue M Electric, Watertown, WI). Brain water content was expressed as (wet weight − dry weight)/wet weight.

At days 1, 3, 7, and 28 after ICH surgery, Tf and TfR Western blot analyses were performed for injured and non-injured BG and cortex of 3 to 4 animals per group per day, as previously described [18]. Briefly, brain regions were sonicated in 0.5 mL of Western sample buffer. Protein/samples (50 μg; Bio-Rad [Hercules, CA] protein assay kits for protein measurement) were run on 7.5% sodium dodecyl sulfate-polyacrylamide gel electrophoresis then transferred to a Hybond-C pure nitrocellulose membrane (Amersham Biosciences, Piscataway, NJ). Membranes were probed with 1:2500 polyclonal rabbit anti-human Tf (DakoCytomation, Carpenteria, CA) or 1:500 monoclonal mouse anti-human TfR (Zymed Laboratories, San Francisco, CA), then second antibody (1:2500 dilution, Bio-Rad) was applied. Antigen-antibody complexes were visualized with a chemiluminescence system (Amersham) and exposed to photosensitive film. Relative complex densities were analyzed using NIH image software (NIH Image, Version 1.61).

Total non-heme brain iron concentration was determined for IDA and non-IDA rats (n = 6) using spectrophotometric methods [6].

Behavioral assessments

Upper extremity function was assessed at days 3, 7–14, and 21–28 after ICH [7, 9, 17]. Vibrissae-stimulated forelimb placement was measured by 10 trials at each forelimb and expressed as percent (right [injured BG side] or left [non-injured BG side]/total trials). Forelimb use was videotaped after placing rats, singly, in a 20 cm × 30 cm plastic cylinder until 20 forelimb placements were observed. Forelimb-use asymmetry was calculated as (right upper extremity + $\frac{1}{2}$both)/Total (right + left + both) × 100. The corner test consisted of counting the observed direction (right or left) of turns with rearing to exit a 30° corner at 8 trials per day. For the final assessment, the sticker test, adhesive patches were placed at the radial aspect of each upper extremity and the number of first touches and sticker removals were counted and averaged for three, 3-minute trials each day.

Statistical analysis

Hematology, behavior, and brain variables were compared by diet group or day after ICH by Student t-test and Mann-Whitney U test. A heterogenous variance model was used to control for the effect of gel for Tf and TfR measurements on comparisons by diet group. Diet group, day after ICH, and their interaction were assessed using Proc Mix. Significance was set at $p < 0.05$.

Results

Physiological measures, hematology, brain iron, and regional water content are reported in Table 1. Body weight, hematocrit, serum iron, brain iron, and blood pressure were significantly reduced for IDA as compared to non-IDA rats. Brain water content was significantly increased in injured BG and cortex for rats in the IDA group as compared to non-IDA animals.

Table 1. *Physiologic measures by diet group.*

	IDA (n = 43)	non-IDA (n = 32)
Body weight (g)	242.8 ± 53.2*	275.2 ± 60.7
Hematocrit (%)	19.8 ± 4.6**	42.9 ± 3.3
Serum iron (μg/L)	4.83 ± 2.3**	30.32 ± 9.6
Total brain iron	7.19 ± 0.65**	12.19 ± 1.90
ICH surgery		
– pH	7.41 ± 0.03*	7.42 ± 0.03
– PaO_2 (mmHg)	91.7 ± 12.6	86.7 ± 9.7
– $PaCO_2$ (mmHg)	43.7 ± 8.5	42.4 ± 4.7
– Base excess	2.1 ± 2.8	3.2 ± 2.4
– Glucose (mg/dL)	111.1 ± 17.8	110.7 ± 18.8
– Mean arterial blood pressure	104.3 ± 24.4*	117.2 ± 27.6
Water content day 3 (%)		
– Right (injured) basal ganglia	81.82 ± 0.41*	80.61 ± 1.03
– Left (non-injured) basal ganglia	77.93 ± 0.42	77.72 ± 0.21
– Right (injured) cortex	80.58 ± 0.47*	79.76 ± 0.49

Student t-test: * $p < 0.05$; ** $p < 0.001$.
Values are mean ± SD.

There were no significant differences by diet group for non-injured BG, cortex, or cerebellum (control) (Table 1).

As previously reported, there were no significant differences in Tf or TfR content in non-injured BG for IDA versus non-IDA rats [18]. In the injured BG, Tf content was significantly greater for IDA than non-IDA rats at day 1 (IDA: 6819 ± 336; non-IDA: 5100 ± 180) and at day 7 (IDA: 6431 ± 436; non-IDA: 4297 ± 702) (both $p < 0.05$). There was a significant effect in diet group and day after ICH, and a significant diet group by day interaction for Tf content in the injured BG ($F = 3.70$, $p < 0.03$). TfR content was also significantly increased in IDA compared to non-IDA rats in the injured BG at day 1 (IDA: 7639 ± 824; non-IDA: 4001 ± 210) and at day 3 (IDA: 6810 ± 750; non-IDA: 4741 ± 294) (both $p < 0.05$). Over time, there was significant effect for diet and day, but no diet group by day interaction for TfR content in the injured BG.

IDA sham rats did not show impairments in the 2 upper extremity assessments performed for this group: vibrissae-stimulated forelimb-placing and the forelimb asymmetry test [18]. The IDA-ICH rats had significantly fewer left vibrissae-stimulated forelimb placements in the first 2 weeks after ICH and a trend for impaired placing after 1 month of recovery [18]. The IDA-ICH rats had greater forelimb asymmetry compared to non-IDA ICH rats during the first week and continuing for 1 month after ICH [18]. Rats from each diet group did not differ significantly for performance on the corner test on any day. Sticker test performance was also relatively preserved for IDA rats. However, IDA rats had a trend for fewer left upper extremity touches during the second week after ICH (Fig. 1).

Discussion

This study demonstrates that IDA worsens aspects of brain injury and behavioral recovery after ICH as compared to the non-IDA condition in the rat. Rats with IDA had greater brain edema at day 3 and significantly greater Tf and TfR protein expression in the injured BG after ICH as compared to those without anemia. In addition, the IDA rats had greater and longer-lasting forelimb function deficits after ICH. The mechanisms that underlie the findings of worse brain and behavioral outcome in the context of IDA are not yet clear. However, the observed differences in

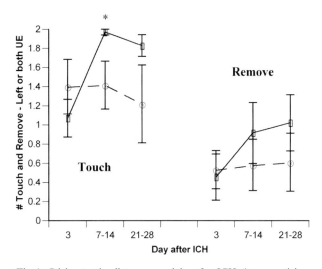

Fig. 1. Sticker test by diet group and day after ICH. Average sticker touch and removal at left or both upper extremities (*UE*) is compared by IDA and non-IDA diet groups after ICH. *$F = 4.878$, $p < 0.06$. ——— non-IDA; - - - IDA

edema formation and brain iron handling suggest potential pathways for the poorer functional outcome in IDA rats after ICH.

A previous investigation demonstrated that the degree of brain edema and its resolution correlates directly with behavioral outcome after ICH [9]. In the non-anemic rat, brain edema resolved and behavioral recovery stabilized within 2 weeks after ICH. However, in the present study, IDA rats showed greater deficits and more asymmetry of upper extremity function on 3 of 4 sensorimotor behavioral tests than non-IDA rats after ICH. On one, the forelimb asymmetry test, IDA rats continued to show impairments as compared to the non-IDA group for at least 1 month after surgery. These behavioral findings may suggest that IDA rats have greater initial and/or longer-lasting edema after ICH and/or altered brain iron trafficking leading to a greater risk for damage related to cellular toxicity effects.

Brain edema due to ICH exacerbates brain injury due to space-occupying effects and direct tissue destruction. The mechanisms of edema formation have been investigated and several key factors have been identified [13, 22, 23]. Thrombin contributes to the early phase of brain edema development within the first several hours after ICH [13, 22]. Hua and colleagues [9] demonstrated that thrombin worsens brain edema and behavioral outcome and that hirudin, a thrombin inhibitor, improves these measures after

ICH in non-anemic rats. Since the IDA rats in the present study had increased serum volume (about 33%), the absolute amount of thrombin delivered in the IDA infusate could have been greater. If true, one might expect that rats with IDA would have greater early edema formation. To clarify this potential mechanism, it would be useful to assess brain edema earlier in the course after ICH, as well as to measure the concentration of prothrombin in IDA sera and investigate whether thrombin inhibitors reduce edema formation in the IDA rats.

Since delayed edema formation, which occurs about 3 days after ICH, involves erythrocyte lysis and cellular toxicity due to hemoglobin and its degradation products, one might expect that delayed edema formation would play less of a role in ICH brain injury with IDA [21]. The IDA rats in the present study had lower hematocrit, serum iron, and brain iron, suggesting less available iron to support this mechanism. However, it may not be the amount of iron, but rather the management of iron after ICH that is key to the risk of delayed edema formation as well as cellular toxicity. Brain non-heme iron has been demonstrated to increase 3-fold in the peri-hematomal zone after ICH in non-IDA rats [21]. Local iron overload may lead to lipid peroxidation and the formation of free radicals. In addition, oxidative brain injury after ICH has been demonstrated and antioxidants block toxicity to neurons related to excess hemoglobin and iron [20]. The study by Wu and colleagues [21] suggests that after ICH, iron concentrations are greater in the injured versus the non-injured hemisphere. It would be interesting to explore this after ICH with IDA.

We hypothesize that the greater expression of Tf and TfR in the perihematomal area observed in the IDA rats in the present study may be a maladaptive or, an adaptive but insufficient response to ICH in the context of IDA. Since baseline Tf and TfR protein levels in the non-injured BG did not differ as a consequence of IDA in this study, the increased levels after ICH may suggest that it is the response of these proteins to injury that is different in the iron deficient brain. Tf and TfR are important to iron transport and in the non-anemic rat are increased in the injury area after ICH [21]. It has been hypothesized that the increase of iron-trafficking proteins is related to clearance of locally high iron concentrations in the injured areas of the brain. In the context of IDA, whole brain Tf concentration is increased but there is considerable regional variability for Tf and TfR expression [5, 8]. Tf has also been demonstrated in oligodendrocytes, and TfR appears to be localized to neuron-like cells in the peri-hematomal area after ICH [14, 21]. Future studies might explore whether greater oxidative injury occurs as a consequence of local iron overload, and the response of iron trafficking proteins after ICH in the context of IDA.

In conclusion, IDA rats have greater brain edema at day 3, increased iron-regulatory protein response during at least the first week, and greater impairment of sensorimotor function and recovery after ICH than non-IDA rats. Our results suggest that iron status may be an important determinant of the severity of brain injury with ICH and subsequent recovery of function. Poorer outcome in the context of IDA has been demonstrated in another brain injury model, that of hypoxia ischemia in the developing rat [16]. The appropriate management of brain injury may depend on the iron status of the individual patient. Given the common occurrence of IDA worldwide, further studies are needed to understand the mechanisms unique to ICH with IDA.

References

1. Baynes RD, Bothwell TH (1990) Iron deficiency. Annu Rev Nutr 10: 133–148
2. Beard JL, Connor JR, Jones BC (1993) Iron in the brain. Nutr Rev 51: 157–170
3. Brittenham G (1979) Spectrophotometric plasma iron determination from fingerpuncture specimens. Clin Chim Acta 91: 203–211
4. Chiueh CC (2001) Iron overload, oxidative stress, and axonal dystrophy in brain disorders. Pediatr Neurol 25: 138–147
5. Erikson K, Pinero DJ, Connor JR, Beard JL (1997) Regional brain iron, ferritin and transferrin concentrations during iron deficiency and iron repletion in developing rats. J Nutr 127: 2030–2038
6. Felt BT, Lozoff B (1996) Brain iron and behavior of rats are not normalized by treatment of iron deficiency anemia during early development. J Nutr 126: 693–701
7. Felt BT, Schallert T, Shao J, Liu Y, Li X, Barks JD (2002) Early appearance of functional deficits after neonatal excitotoxic and hypoxic-ischemic injury: fragile recovery after development and role of the NMDA receptor. Dev Neurosci 24: 418–425
8. Han J, Day JR, Connor JR, Beard JL (2003) Gene expression of transferrin and transferrin receptor in brains of control vs iron-deficient rats. Nutr Neurosci 6: 1–10
9. Hua Y, Schallert T, Keep RF, Wu J, Hoff JT, Xi G (2002) Behavioral tests after intracerebral hemorrhage in the rat. Stroke 33: 2478–2484
10. Huang FP, Xi G, Keep RF, Hua Y, Nemoianu A, Hoff JT (2002) Brain edema after experimental intracerebral hemorrhage: role of hemoglobin degradation products. J Neurosurg 96: 287–293
11. Kase CS, Caplan LR (1994) Intracerebral Hemorrhage. Butterworth-Heinemann, Boston

12. Koeppen AH, Dickson AC, McEvoy JA (1995) The cellular reactions to experimental intracerebral hemorrhage. J Neurol Sci 134 Suppl: 102–112
13. Lee KR, Colon GP, Betz AL, Keep RF, Kim S, Hoff JT (1996) Edema from intracerebral hemorrhage: the role of thrombin. J Neurosurg 84: 91–96
14. Masada T, Hua Y, Xi G, Yang GY, Hoff JT, Keep RF (2001) Attenuation of intracerebral hemorrhage and thrombin-induced brain edema by overexpression of interleukin-1 receptor antagonist. J Neurosurg 95: 680–686
15. Nakamura T, Keep RF, Hua Y, Schallert T, Hoff JT, Xi G (2004) Deferoxamine-induced attenuation of brain edema and neurological deficits in a rat model of intracerebral hemorrhage. J Neurosurg 100: 672–678
16. Rao R, de Ungria M, Sullivan D, Wu P, Wobken JD, Nelson CA, Georgieff MK (1999) Perinatal brain iron deficiency increases the vulnerability of rat hippocampus to hypoxic ischemic insult. J Nutr 129: 199–206
17. Schallert T, Woodlee MT (2003) Brain-dependent movements and cerebral-spinal connections: key targets of cellular and behavioral enrichment in CNS injury models. J Rehabil Res Dev 40 (4 Suppl 1): 9–17
18. Shao J, Xi G, Hua Y, Schallert T, Felt B (2005) Intracerebral hemorrhage in the iron deficient rat. Stroke 36: 660–664
19. Stoltzfus RJ (2001) Defining iron-deficiency anemia in public health terms: A time for reflection. J Nutr 131: 565S–567S
20. Wu J, Hua Y, Keep RF, Schallert T, Hoff JT, Xi G (2002) Oxidative brain injury from extravasated erythrocytes after intracerebral hemorrhage. Brain Res 953: 45–52
21. Wu J, Hua Y, Keep RF, Nakamura T, Hoff JT, Xi G (2003) Iron and iron-handling proteins in the brain after intracerebral hemorrhage. Stroke 34: 2964–2969
22. Xi G, Keep RF, Hoff JT (1998) Erythrocytes and delayed brain edema formation following intracerebral hemorrhage in rats. J Neurosurg 89: 991–996
23. Xi G, Wagner KR, Keep RF, Hua Y, de Courten-Myers GM, Broderick JP, Brott TG, Hoff JT, Muizelaar JP (1998) Role of blood clot formation on early edema development after experimental intracerebral hemorrhage. Stroke 29: 2580–2586

Correspondence: Barbara True Felt, Center for Human Growth and Development, University of Michigan, 1000SW NIB, 300 North Ingalls Street, Ann Arbor, MI 48109-0406, USA. e-mail: truefelt@umich.edu

Neuroprotective effect of hyperbaric oxygen in a rat model of subarachnoid hemorrhage

R. P. Ostrowski[1], A. R. T. Colohan[2], and J. H. Zhang[1,2]

[1] Department of Physiology, Department of Surgery, Loma Linda University, Loma Linda, CA, USA
[2] Division of Neurosurgery, Department of Surgery, Loma Linda University, Loma Linda, CA, USA

Summary

Acute brain ischemia after subarachnoid hemorrhage (SAH) induces oxidative stress in brain tissues. Up-regulated NADPH oxidase (NOX), a major enzymatic source of superoxide anion in the brain, may contribute to early brain injury after SAH. We evaluated the effects of hyperbaric oxygen (HBO) on protein expression of gp91^{phox} catalytic subunit of NOX, lipid peroxidation as a marker of oxidative stress, and on neurological and neuropathological outcomes after SAH.

Twenty-nine male Sprague-Dawley rats (300 to 350 g) were randomly allocated to control (sham operation), SAH (endovascular perforation), and SAH treated with HBO groups (2.8 ATA for 2 hours, at 1 hour after SAH). Cerebral blood flow was measured using laser Doppler flowmetry. Rats were sacrificed after 24 hours and brain tissues collected for histology (Nissl staining and gp91^{phox} immunohistochemistry) and biochemistry. Mortality and neurological scores were evaluated.

Neuronal injury associated with enhanced gp91^{phox} immunostaining was observed in the cerebral cortex after SAH. The lipid peroxidation product, malondialdehyde, accumulated in the ipsilateral cerebral cortex. HBO treatment reduced expression of NOX, diminished lipid peroxidation, and reduced neuronal damage. HBO caused a drop in mortality and ameliorated functional deficits.

HBO-induced neuroprotection after SAH may involve down-regulation of NOX and a subsequent reduction in oxidative stress.

Keywords: Subarachnoid hemorrhage; lipid peroxidation; NADPH oxidase; hyperbaric oxygen.

Introduction

Oxidative stress is a major cause of brain damage after subarachnoid hemorrhage (SAH). Free radicals triggered by SAH induce lipid peroxidation with subsequent apoptosis and vasospasm [28, 30]. Nicotinamide adenine dinucleotide phosphate (NADPH) oxidase (NOX) that generates superoxide is a major enzymatic producer of free radicals in the brain. Neuronal expression of phagocytic-type NOX was shown in the rodent brain [37, 45]. The active form of NOX is a multi-component system including cytoplasmic subunits: regulatory p40^{phox}, p47^{phox}, activator p67^{phox}, and membrane-bound flavocytochrome b$_{558}$ consisting of catalytic gp91^{phox} and p22^{phox} subunits [42]. Up-regulation of cerebral vascular NOX shows its peak between 12 and 24 hours after SAH in a single-injection rat model, contributing to the impairment of autoregulatory vasodilation [38]. No data on NOX expression in cerebral tissues after SAH are currently available. Until recently it was believed that oxidative stress after SAH rests with increase in mitochondrial leak of superoxide or with auto-oxidation of hemoglobin, both of which initiate free radical cascades underlying development of vasospasm [1, 24, 26, 27]. In addition, oxidative stress has been studied predominately in blood injection models that are suitable for investigating vasospasm rather than the acute sequelae of SAH [35].

The phagocytic phenotype of NOX is expressed in neurons [37, 45] of the brain and, by producing excess free radicals, may contribute to oxidative brain damage and apoptosis in cerebral tissues after experimental stroke [21]. We found that hyperbaric oxygen (HBO) in a perforation model of SAH reduced apoptosis and necrosis-like cell death, and blood-brain barrier rupture through inhibition of hypoxia-inducible factor 1α (HIF-1α) and its target genes [33]. Since it has been postulated that free radicals stabilize HIF-1α upon hypoxia [4], HIF-1α inhibition by HBO may suggest that suppression of NOX mediates HBO-induced effect [11]. Consistently, an inhibition of NOX by diphenyleneiodonium is capable of abolishing HIF-1α accumulation and the hypoxic induction of its target genes [10]. HBO can suppress other potentially detrimental enzymes such as COX-2, the activation of which may theoretically lead through the NOX pathway [34, 43]. Even though HBO itself may induce oxidative stress, we postulate that it may have a prolonged inhibitory effect on pro-oxidant mechanisms triggered by SAH as opposed to transient increase in free radicals upon hyperbaric oxygenation [19]. The aim was to study effects of HBO treatment on 1) NOX expression in ipsilateral cerebral cortex, 2) lipid peroxidation, 3) morphology of neuronal injury, and 4) mortality and neurological score in a rat model of SAH.

Materials and methods

SAH animal model

Twenty nine male Sprague-Dawley rats (300 to 350 g) were randomly assigned to the following groups: control (sham operation), SAH without treatment, and SAH treated with HBO. Rats were transorally intubated and mechanically ventilated throughout the operation. Rectal temperature was maintained at 37 °C. SAH was induced according to the method described by Bederson et al. in Schwartz's modification [2, 36] under ketamine (100 mg/kg, i.p.) and xylazine (10 mg/kg, i.p.) anesthesia. The left femoral artery was cannulated for blood pressure recording and withdrawal of blood samples. Blood glucose, hematocrit, and blood gas were periodically measured (1610 pH/blood gas analyzer). Neurological status was evaluated using the modified Garcia scoring system, in which normal animals score 18 points and lower scores indicate brain injury [9, 20]. Animals were sacrificed after 24 hours to study lipid peroxidation, morphology of brain injury (Nissl staining), and cortical expression of gp91^{phox} by immunohistochemistry. All experimental procedures complied with the Guide for the Care and Use of Laboratory Animals (National Institutes of Health publication no. 85-23) and were approved by the Animal Care and Use Committee at Loma Linda University.

Cerebral blood flow measurements

Cerebral blood flow (CBF) in the contralateral cerebral cortex was measured by laser Doppler flowmeter (Periflux System 5000, Perimed, Jarfalla, Sweden) through skull bones that were thinned to translucency with a microdrill. A small, straight laser probe was attached to the skull with acrylic glue at the site localized 5 mm lateral and 1 mm posterior to bregma. Recording was started at the beginning of surgery and continued during and 1 hour after SAH induction. At this time point, rats were either transferred to the HBO chamber or returned to their cages.

HBO treatment

HBO (100% oxygen at 2.8 ATA and 2 hours duration) was applied 1 hour after SAH. Animals were placed in a small research hyperbaric chamber (Sechrist Industries, Anaheim, CA) equipped with carbonate crystals to prevent CO_2 accumulation.

Histology and immunohistochemistry

Anesthetized rats were perfused transcardially with 200 mL of ice-cold 0.1 mol phosphate-buffered saline (PBS) followed by 400 mL of 10% buffered formalin (n = 3). Brains were collected, post-fixed in the same fixative overnight and kept in 30% sucrose until they sank. Coronal tissue sections 10 μm thick were cut on a cryostat (CM3050S, Leica Microsystems AG, Wetzlar, Germany). For Nissl staining, sections were embedded in 0.1% cresyl violet for 5 minutes, dehydrated in Flex tissue specimen system (Richard-Allan Scientific, Kalamazoo, MI), cleared in xylenes, coverslipped with Permount, and observed under a light microscope (Olympus BX51, Melville, NY).

For immunohistochemical analysis of tissue gp91^{phox} expression, sections were hydrated with 0.01 mol PBS and treated briefly with 3% hydrogen peroxide. Sections were incubated overnight at 4 °C with goat polyclonal anti-gp91^{phox} antibody (Santa Cruz Biotechnology, Santa Cruz, CA) diluted 1:100, followed by a secondary antibody and 3,3'-diaminobenzidine tetrahydrochloride (DAB) staining using an ABC kit (Santa Cruz Biotechnology). Immunostaining with the omission of the primary antibody served as a negative control.

Lipid peroxidation assay

Twenty-four hours after SAH under deep anesthesia (n = 5), rat brains were perfused with cold 0.1 mol PBS to remove blood. The level of malondialdehyde (MDA) was measured using a LPO-586 kit (Oxis Research, Portland, OR). Left cerebral cortices were homogenized in phosphate buffer (pH 7.4), with 0.5 mol butylated hydroxytoluene in acetonitrile. The homogenates were centrifuged at 3000 g for 10 minutes at 4 °C. Protein concentration was measured by means of a detergent-compatible protein assay (BioRad Laboratories, Hercules, CA). Equal amounts of proteins in each sample reacted with a chromogenic reagent at 45 °C for 60 minutes. The samples were centrifuged at 15000 g for 10 minutes at 4 °C, and supernatants measured spectrophotometrically at 586 nm. The level of MDA was calculated in pmol/mg protein based on the standard curve [37].

Statistical analysis

Data are expressed as mean ± SEM. One-way ANOVA and post hoc Holm-Sidak tests were applied to verify statistical significance of differences between means. A p-value of <0.05 was considered significant.

Results

Mortality and neurological score

Mortality was 30.33% after SAH and 11.11% in rats treated with HBO after SAH. No animal died in the control group. Animals in the SAH without treatment group presented low neurological scores associated with severe functional impairment (6.13 ± 0.61; n = 8). HBO treatment reduced SAH-induced neurological deficits, resulting in greater scores (13.13 ± 0.90; n = 8) that were still significantly decreased compared to values in sham operated rats (17.63 ± 0.18; n = 8).

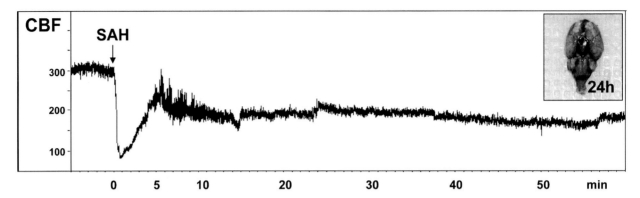

Fig. 1. Representative trace of cortical microflow (laser Doppler flowmetry; in arbitrary units) shows acute impairment of cortical CBF after SAH followed by a tendency toward recovery and a secondary hypoperfusion that is maintained 60 minutes after bleeding. Inset shows the corresponding extent of hemorrhage that clearly involves the circle of Willis

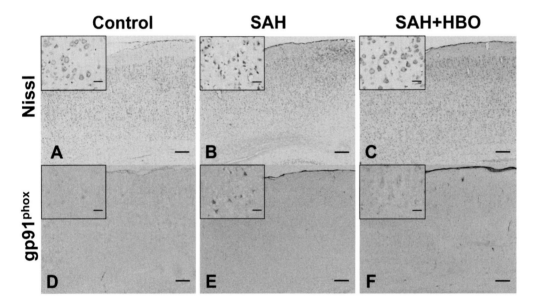

Fig. 2. HBO reduces neuronal injury and gp91^{phox} expression in cerebral tissues. The left panels show samples from control rats that had normal appearance of neurons and low gp91^{phox} expression in the cerebral cortex. The middle panel is a sample of ipsilateral cerebral cortex 24 hours after SAH. Features of cell injury and strong gp91^{phox} expression in the cortical neurons are shown. The right panel is from rats treated with HBO and shows that among well-preserved neurons, only a few gp91^{phox}-positive cells are present. Scale bars in low magnification photographs represent 200 μm, and 30 μm in insets

Cerebral blood flow

The lowest level of CBF, $20.3 \pm 2.9\%$ (no treatment group; $n = 8$) and $17.7 \pm 2.2\%$ (with HBO treatment group; $n = 8$) of baseline, contralaterally to the perforated artery, was found 1 minute after induction of SAH (Fig. 1). This immediate ischemic flow was followed by a trend toward recovery of CBF, which at 1 hour reached $65.9 \pm 6.7\%$ and $62.0 \pm 9.5\%$ of baseline in rats without treatment and in rats assigned to HBO groups, respectively. No significant differences in CBF were found between untreated rats and rats assigned to HBO treatment within 1 hour from initial bleeding. Similarly, physiological variables did not show any differences between groups.

Nissl staining

Extensive damage of cortical neurons was observed 24 hours after SAH without treatment. Darkened and shrunken perikarya and condensed nuclei were dominant features of injured neurons (Fig. 2B). However, well-preserved neurons, with only slightly darkened cytoplasm, were noted in animals treated with HBO (Fig. 2C).

NOX immunohistochemistry

Noticeable but weak gp91^{phox} immunolabeling was present in control cortex (Fig. 2D). SAH caused a tremendous increase in gp91^{phox} immunoreactivity in the cerebral cortex that was ipsilateral to the perforated artery (Fig. 2E). Deposits of DAB stain were most dense in cells overlying the cortex. Immunostained cells showed distinct neuronal morphology. HBO treatment resulted in diminution of gp91^{phox} immunoreactivity within neurons with a residual staining at the cell periphery (Fig. 2F).

Lipid peroxidation in three experimental groups

Twenty-four hours after SAH, brain content of MDA was 8.02 ± 1.25 pmol/mg protein in the control group, 48.27 ± 4.40 pmol/mg protein in the SAH without treatment group ($p < 0.05$ vs. control), but was reduced more than 3-fold in SAH rats treated with HBO (13.99 ± 2.06 pmol/mg protein, $p < 0.05$ vs. SAH) (n = 4).

Discussion

Acute ischemia after SAH usually has a severe global impact, but is more profound ipsilateral to the perforated artery [13, 16]. The reduction in CBF to ~20% of pre-SAH levels, observed by us contralaterally, indicates severe SAH. Indeed, large amounts of blood and macroscopic features of brain edema were observed upon collection of brain samples (Fig. 1). The level of CBF impairment after SAH was similar in rats assigned to no treatment and HBO treatment groups. This confirms that HBO-induced differences in studied parameters were not related to an uneven impact of the initial bleeding.

The main finding of this study is that HBO-induced neuroprotection involves down-regulation of neuronal NOX and diminution of lipid peroxidation after SAH. SAH increases free radicals, presumably through a mitochondrial leak and their excess formation by enzymatic sources with a dominant role of gp91^{phox}-containing NOX in brain tissues. Free radicals may contribute to cell necrosis through damaging effects on lipid bilayers within the cell [23]. It has also been reported that a NOX-dependent burst of reactive oxygen species may trigger neuronal apoptosis [39]. Additionally, the increasing trend of enzymatic lipid peroxidation after SAH is associated with a reduction in antioxidant enzymatic activities [7].

Oxidative stress has been reported clinically and in animals after SAH [8, 40]. However, negative results of lipid peroxide assay have been reported in a single-injection SAH model [25]. In our perforation model that closely depicts acute intracranial phenomena after SAH, a remarkable increase in lipid peroxidation and gp91^{phox} expression, a catalytic subunit of NOX that is responsible for a formation of superoxide anion [42]. This also occurred in the cerebral cortex, which has been reported by Noda *et al.* [32] to be the most oxidative stress-prone structure of the brain regions studied.

The fact that hyperbaric oxygen can reduce oxidative stress in the hemorrhagic brain has not been recognized previously. However, the induction of endogenous antioxidants has been reported after global ischemia treated with HBO. This was assumed to be the mechanism underlying improved morphology of neurons [31]. Here we tested a somewhat provocative hypothesis proposing inhibition of a pro-oxidant component of oxidative stress after SAH. Oxygenation at high baricity is a broadly studied model of oxidative stress [12, 15]. There have been inconsistent results from different research teams regarding HBO treatment-induced free radical production. Mink and Dutka [29] showed enhanced free radical formation but no increase in lipid peroxidation products (2.8 atm, 75 minutes) and Elayan *et al.* [6] reported negative results of hydroxyl radical detection (3 ATA, 2 hours). In contrast, others published results showing enhancement of both free radical generation and lipid breakdown after HBO [5, 32, 44]. If this is the case, the delayed decrease in lipid peroxidation found after HBO in our study would constitute a rebound phenomenon that could involve primarily a transient increase in free radicals observed with hyperbaric oxygenation [32]. Down-regulated NOX expression could underlie HBO effect in these settings. Most studies have shown good correlation between free radical production and expression of NOX subunits [17, 22]. It is not clear, however, when the pro-oxidant cellular status triggered by HBO may undergo transition into suppression of oxidative stress. To address this issue, markers of oxidative stress should be investigated at time points that are closer to the HBO treatment. Additionally, more remote outcomes should be investigated to find out whether repeated HBO is beneficial for the brain. It has been reported that multiple HBO treatments may enhance lipid peroxidation in human blood [3].

It would be interesting to study whether HBO treatment also down-regulates other types of NOX with respect to their role in free radical production. This would include NOX4 in neurons, as well as NOX1 and NOX4 in cerebral vessels [22, 41]. Recent observations have shown that both NOX inhibition, with resultant decrease in superoxide anion production and hyperbaric oxygenation, attenuate vasospasm after SAH. This may further suggest an inhibitory effect of HBO on NOX signaling [18, 46].

In the present study, the lowering effect of HBO on NOX expression was robust in the ipsilateral cerebral cortex where it may be coupled with inhibition of HIF-1α and its target genes including vascular endothelial growth factor and proapoptotic BNIP3 [33]. It is known that negative HIF-1α mutants are protected against hypoxic injuries in vivo through inhibition of its target genes mediating apoptotic pathways [14]. In a previous study, we found that HBO suppressed HIF-1α in the same experimental setting. Therefore, it cannot be excluded that HBO reduces HIF-1α accumulation through suppression of NOX. Further studies are needed to evaluate the role of NOX isoforms in SAH-induced hypoxic signaling, subsequent brain injury, and mechanisms of hyperbaric oxygen treatment.

References

1. Arai T, Takeyama N, Tanaka T (1999) Glutathione monoethyl ester and inhibition of the oxyhemoglobin-induced increase in cytosolic calcium in cultured smooth-muscle cells. J Neurosurg 90: 527–532
2. Bederson JB, Germano IM, Guarino L (1995) Cortical blood flow and cerebral perfusion pressure in a new noncraniotomy model of subarachnoid hemorrhage in the rat. Stroke 26: 1086–1091
3. Benedetti S, Lamorgese A, Piersantelli M, Pagliarani S, Benvenuti F, Canestrari F (2004) Oxidative stress and antioxidant status in patients undergoing prolonged exposure to hyperbaric oxygen. Clin Biochem 37: 312–317
4. Chandel NS, McClintock DS, Feliciano CE, Wood TM, Melendez JA, Rodriguez AM, Schumacker PT (2000) Reactive oxygen species generated at mitochondrial complex III stabilize hypoxia-inducible factor-1alpha during hypoxia: a mechanism of O2 sensing. J Biol Chem 275: 25130–25138
5. Chavko M, Harabin AL (1996) Regional lipid peroxidation and protein oxidation in rat brain after hyperbaric oxygen exposure. Free Radic Biol Med 20: 973–978
6. Elayan IM, Axley MJ, Prasad PV, Ahlers ST, Auker CR (2000) Effect of hyperbaric oxygen treatment on nitric oxide and oxygen free radicals in rat brain. J Neurophysiol 83: 2022–2029
7. Gaetani P, Lombardi D (1992) Brain damage following subarachnoid hemorrhage: the imbalance between anti-oxidant systems and lipid peroxidative processes. J Neurosurg Sci 36: 1–10
8. Gaetani P, Pasqualin A, Baena R, Borasio E, Marzatico F (1998) Oxidative stress in the human brain after subarachnoid hemorrhage. J Neurosurg 89: 748–754
9. Garcia JH, Wagner S, Liu KF, Hu XJ (1995) Neurological deficit and extent of neuronal necrosis attributable to middle cerebral artery occlusion in rats. Statistical validation. Stroke 26: 627–634
10. Gleadle JM, Ebert BL, Ratcliffe PJ (1995) Diphenylene iodonium inhibits the induction of erythropoietin and other mammalian genes by hypoxia. Implications for the mechanism of oxygen sensing. Eur J Biochem 234: 92–99
11. Goyal P, Weissmann N, Grimminger F, Hegel C, Bader L, Rose F, Fink L, Ghofrani HA, Schermuly RT, Schmidt HH, Seeger W, Hanze J (2004) Upregulation of NAD(P)H oxidase 1 in hypoxia activates hypoxia-inducible factor 1 via increase in reactive oxygen species. Free Radic Biol Med 36: 1279–1288
12. Groger M, Speit G, Radermacher P, Muth CM (2005) Interaction of hyperbaric oxygen, nitric oxide, and heme oxygenase on DNA strand breaks in vivo. Mutat Res 572: 167–172
13. Grote E, Hassler W (1988) The critical first minutes after subarachnoid hemorrhage. Neurosurgery 22: 654–661
14. Helton R, Cui J, Scheel JR, Ellison JA, Ames C, Gibson C, Blouw B, Ouyang L, Dragatsis I, Zeitlin S, Johnson RS, Lipton SA, Barlow C (2005) Brain-specific knock-out of hypoxia-inducible factor-1alpha reduces rather than increases hypoxic-ischemic damage. J Neurosci 25: 4099–4107
15. Ito T, Yufu K, Mori A, Packer L (1996) Oxidative stress alters arginine metabolism in rat brain: effect of sub-convulsive hyperbaric oxygen exposure. Neurochem Int 29: 187–195
16. Jarus-Dziedzic K, Czernicki Z, Kozniewska E (2003) Acute decrease of cerebrocortical microflow and lack of carbon dioxide reactivity following subarachnoid haemorrhage in the rat. Acta Neurochir [Suppl] 86: 473–476
17. Kim SH, Won SJ, Sohn S, Kwon HJ, Lee JY, Park JH, Gwag BJ (2002) Brain-derived neurotrophic factor can act as a pronecrotic factor through transcriptional and translational activation of NADPH oxidase. J Cell Biol 159: 821–831
18. Kocaogullar Y, Ustun ME, Avci E, Karabacakoglu A, Fossett D (2003) The role of hyperbaric oxygen in the management of subarachnoid hemorrhage. Intensive Care Med
19. Kot J, Sicko Z, Wozniak M (2003) Oxidative stress during oxygen tolerance test. Int Marit Health 54: 117–126
20. Kusaka G, Ishikawa M, Nanda A, Granger DN, Zhang JH (2004) Signaling pathways for early brain injury after subarachnoid hemorrhage. J Cereb Blood Flow Metab 24: 916–925
21. Kusaka I, Kusaka G, Zhou C, Ishikawa M, Nanda A, Granger DN, Zhang JH, Tang J (2004) Role of AT1 receptors and NAD(P)H oxidase in diabetes-aggravated ischemic brain injury. Am J Physiol Heart Circ Physiol 286: H2442–H2451
22. Lassegue B, Clempus RE (2003) Vascular NAD(P)H oxidases: specific features, expression, and regulation. Am J Physiol Regul Integr Comp Physiol 285: R277–R297
23. Lipton P (1999) Ischemic cell death in brain neurons. Physiol Rev 79: 1431–1568
24. Macdonald RL, Weir BK (1994) Cerebral vasospasm and free radicals. Free Radic Biol Med 16: 633–643
25. Marzatico F, Gaetani P, Baena R, Silvani V, Fulle I, Lombardi D, Ferlenga P, Benzi G (1989) Experimental subarachnoid hemorrhage. Lipid peroxidation and Na+,K(+)-ATPase in different rat brain areas. Mol Chem Neuropathol 11: 99–107
26. Marzatico F, Gaetani P, Cafe C, Spanu G, Baena R (1993) Antioxidant enzymatic activities after experimental subarachnoid hemorrhage in rats. Acta Neurol Scand 87: 62–66

27. Matz PG, Copin JC, Chan PH (2000) Cell death after exposure to subarachnoid hemolysate correlates inversely with expression of CuZn-superoxide dismutase. Stroke 31: 2450–2459
28. Matz PG, Fujimura M, Lewen A, Morita-Fujimura Y, Chan PH (2001) Increased cytochrome c-mediated DNA fragmentation and cell death in manganese-superoxide dismutase-deficient mice after exposure to subarachnoid hemolysate. Stroke 32: 506–515
29. Mink RB, Dutka AJ (1995) Hyperbaric oxygen after global cerebral ischemia in rabbits does not promote brain lipid peroxidation. Crit Care Med 23: 1398–1404
30. Mori T, Nagata K, Town T, Tan J, Matsui T, Asano T (2001) Intracisternal increase of superoxide anion production in a canine subarachnoid hemorrhage model. Stroke 32: 636–642
31. Mrsic-Pelcic J, Pelcic G, Vitezic D, Antoncic I, Filipovic T, Simonic A, Zupan G (2004) Hyperbaric oxygen treatment: the influence on the hippocampal superoxide dismutase and Na+,K+-ATPase activities in global cerebral ischemia-exposed rats. Neurochem Int 44: 585–594
32. Noda Y, McGeer PL, McGeer EG (1983) Lipid peroxide distribution in brain and the effect of hyperbaric oxygen. J Neurochem 40: 1329–1332
33. Ostrowski RP, Colohan AR, Zhang JH (2005) Mechanisms of hyperbaric oxygen-induced neuroprotection in a rat model of subarachnoid hemorrhage. J Cereb Blood Flow Metab 25: 554–571
34. Peng T, Lu X, Feng Q (2005) NADH oxidase signaling induces cyclooxygenase-2 expression during lipopolysaccharide stimulation in cardiomyocytes. FASEB J 19: 293–295
35. Prunell GF, Mathiesen T, Diemer NH, Svendgaard NA (2003) Experimental subarachnoid hemorrhage: subarachnoid blood volume, mortality rate, neuronal death, cerebral blood flow, and perfusion pressure in three different rat models. Neurosurgery 52: 165–175
36. Schwartz AY, Masago A, Sehba FA, Bederson JB (2000) Experimental models of subarachnoid hemorrhage in the rat: a refinement of the endovascular filament model. J Neurosci Meth 96: 161–167
37. Serrano F, Kolluri NS, Wientjes FB, Card JP, Klann E (2003) NADPH oxidase immunoreactivity in the mouse brain. Brain Res 988: 193–198
38. Shin HK, Lee JH, Kim KY, Kim CD, Lee WS, Rhim BY, Hong KW (2002) Impairment of autoregulatory vasodilation by NAD(P)H oxidase-dependent superoxide generation during acute stage of subarachnoid hemorrhage in rat pial artery. J Cereb Blood Flow Metab 22: 869–877
39. Tammariello SP, Quinn MT, Estus S (2000) NADPH oxidase contributes directly to oxidative stress and apoptosis in nerve growth factor-deprived sympathetic neurons. J Neurosci 20: RC53
40. Turner CP, Panter SS, Sharp FR (1999) Anti-oxidants prevent focal rat brain injury as assessed by induction of heat shock proteins (HSP70, HO-1/HSP32, HSP47) following subarachnoid injections of lysed blood. Brain Res Mol Brain Res 65: 87–102
41. Vallet P, Charnay Y, Steger K, Ogier-Denis E, Kovari E, Herrmann F, Michel JP, Szanto I (2005) Neuronal expression of the NADPH oxidase NOX4, and its regulation in mouse experimental brain ischemia. Neuroscience 132: 233–238
42. Vignais PV (2002) The superoxide-generating NADPH oxidase: structural aspects and activation mechanism. Cell Mol Life Sci 59: 1428–1459
43. Yin W, Badr AE, Mychaskiw G, Zhang JH (2002) Down regulation of COX-2 is involved in hyperbaric oxygen treatment in a rat transient focal cerebral ischemia model. Brain Res 926: 165–171
44. Zaleska MM, Floyd RA (1985) Regional lipid peroxidation in rat brain in vitro: possible role of endogenous iron. Neurochem Res 10: 397–410
45. Zhang X, Dong F, Ren J, Driscoll MJ, Culver B (2005) High dietary fat induces NADPH oxidase-associated oxidative stress and inflammation in rat cerebral cortex. Exp Neurol 191: 318–325
46. Zheng JS, Zhan RY, Zheng SS, Zhou YQ, Tong Y, Wan S (2005) Inhibition of NADPH oxidase attenuates vasospasm after experimental subarachnoid hemorrhage in rats. Stroke 36: 1059–1064

Correspondence: Robert P. Ostrowski, Department of Physiology, Loma Linda University, Risley Hall, Room 219, Loma Linda, CA 92350, USA. e-mail: rostro2104@yahoo.com

Iron-induced oxidative brain injury after experimental intracerebral hemorrhage

T. Nakamura[1,2], R. F. Keep[1], Y. Hua[1], S. Nagao[2], J. T. Hoff[1], and G. Xi[1]

[1] Department of Neurosurgery, University of Michigan, Ann Arbor, Michigan, USA
[2] Department of Neurological Surgery, Kagawa University Faculty of Medicine, Kagawa, Japan

Summary

We investigated the occurrence of DNA damage in brain after intracerebral hemorrhage (ICH) and the role of iron in such injury.

Male Sprague-Dawley rats received an infusion of 100 μL autologous whole blood or 30 μL FeCl$_2$ into the right basal ganglia and were sacrificed 1, 3, or 7 days later. 8-hydroxyl-2′-deoxyguanosine (8-OHdG) was analyzed by immunohistochemistry, while the number of apurinic/apyrimidinic abasic sites (AP sites) was also quantified. 8-OHdG and AP sites are two hallmarks of DNA oxidation. DNA damage was also examined using PANT and TUNEL labeling. Dinitrophenyl (DNP) was measured by Western blot to compare the time course of protein oxidative damage to that of DNA. DNA repair APE/Ref-1 and Ku-proteins were also measured by Western blot. Bipyridine, a ferrous iron chelator, was used to examine the role of iron in ICH-induced oxidative brain injury.

An increase in 8-OHdG, AP sites, and DNP levels, and a decrease in APE/Ref-1 and Ku levels were observed. Abundant PANT-positive cells were also observed in the perihematomal area 3 days after ICH. Bipyridine attenuated ICH-induced changes in PANT and DNP. These results suggest that iron-induced oxidation causes DNA damage in brain after ICH and that iron is a therapeutic target for ICH.

Keywords: Intracerebral hemorrhage; iron; oxidative DNA injury; 8-OHdG; AP sites; DNP; PANT; APE/Ref-1; Ku-proteins; brain edema.

Introduction

Intracerebral hemorrhage (ICH) is a common and often fatal subtype of stroke. Iron is one of the hemoglobin degradation products and iron overload in the brain can cause free radical formation and oxidative damage such as lipid peroxidation after ICH [12]. There are several potential targets for oxidative damage following ICH.

We hypothesized that iron-induced oxidative DNA damage occurs after ICH and that it contributes to ICH-induced brain injury. Formation of the DNA modification 8-hydroxyl-2′-deoxyguanosine (8-OHdG) and apurinic/apyrimidinic (AP) sites are 2 oxidative DNA injury markers [5, 7]. This study examines the effect of ICH and intracerebral infusion of iron on these 2 parameters. Whether DNA damage might result in single and double strand breaks was examined using PANT and TUNEL staining, respectively [12]. The time course of DNA oxidative damage was compared to that in proteins using an anti-dinitrophenyl (DNP) antibody which can be used to detect protein oxidation [8]. AP endonuclease (APE)/Ref-1 and Ku-proteins are multifunctional proteins associated with DNA repair but which are decreased following cerebral ischemia [6, 7]. In addition, we examined the effect of bipyridine, a ferrous iron chelator, on ICH-induced oxidative brain injury.

Materials and methods

Animal preparation and experimental groups

Animal protocols were approved by the University of Michigan Committee on the Use and Care of Animals. Male Sprague-Dawley rats, each weighing 300 to 400 g, were used for all experiments. The animals were anesthetized with pentobarbital (40 mg/kg i.p.) and the right femoral artery was catheterized to sample blood for intracerebral infusion. The rats were positioned in a stereotaxic frame and a 26-gauge needle was inserted stereotaxically into the right basal ganglia (coordinates: 0.2 mm anterior, 5.5 mm ventral, and 3.5 mm lateral to the bregma). Autologous whole blood (100 μL) or FeCl$_2$ (10 mM, 30 μL) was infused at a rate of 10 μL/min with the use of a microinfusion pump.

This study was performed in 3 parts. Part 1 evaluated the time course of iron accumulation, oxidation, and DNA injury after ICH. Iron accumulation around hematoma was measured by Perl's iron staining. 8-OHdG were investigated by immunohistochemistry (n = 3 each time point). The number of AP sites was measured quantitatively (n = 3–6 each time point). DNP, APE/Ref-1, and Ku-proteins were investigated by Western blot analysis (n = 3 each time point). TUNEL and PANT staining investigated the time course of DNA damage (n = 3 each time point).

Part 2 examined the effect of iron on oxidation and DNA damage. In this part, rats received an intracaudate injection of FeCl$_2$ or a needle insertion (n = 3 each time point). The rats were sacrificed 24 hours later.

Part 3 investigated the effect of bipyridine (2,2′-dipyridyl) on DNP and PANT staining. Animals were immediately treated with either bipyridine (25 mg/kg in 1 mL saline i.p. per 12 hours) or vehicle (1 mL saline i.p. each time) after ICH.

Iron staining (Perl's) and immunohistochemistry (8-OHdG)

For detection of ferric iron, a modified Perl's staining was performed [9]. In immunostaining, the avidin-biotin complex technique was used. The primary antibody was mouse anti-8-OHdG monoclonal antibody (10 μg/mL) purchased from Oxis International Inc. (Portland, OR). The second antibody was anti-mouse IgG antibody (1:150) (Vector Laboratories, Burlingame, CA).

Detection of AP sites in DNA

DNA extraction was performed using a DNA isolation kit produced by Dojindo Molecular Technologies Inc. (Gaithersburg, MD). The aldehyde reactive probe (ARP) labeling and quantification of AP sites were performed by the AP sites assay kit (Dojindo). The ARP-labeled DNA was quantified using a 96-well microplate, similar to an enzyme-linked immunoabsorbent assay study. The wells were subjected to optical density measurement at 630 nm. ARP assays were performed in triplicate and the means were calculated. The data, expressed as the number of AP sites per 100,000 nucleotides, were calculated based on the linear calibration curve generated for each experiment using ARP-DNA standard solutions.

Western blot analysis (DNP, APE/Ref-1, and Ku-proteins)

Briefly, 50 μg proteins for each were separated by sodium dodecyl sulfate polyacrylamide gel electrophoresis and transferred to a Hybond-C pure nitrocellulose membrane (Amersham, Piscataway, NJ). Membranes were probed with a 1:1000 dilution of the primary antibody and a 1:1500 dilution of the second antibody (BIO-RAD Laboratories, Hercules, CA). The antigen-antibody complexes were visualized with a chemiluminescence system (Amersham) and exposed to film. The relative densities of bands were analyzed with NIH Image (Version 1.61, National Institutes of Health, USA).

Detection of DNA single- and double-strand breaks by PANT and TUNEL staining

PANT and TUNEL staining were performed on adjacent brain sections to detect DNA single- and double-strand breaks according to the method described by Wu *et al.* [12].

Statistical analysis

All data in this study are presented as mean ± SD. Data were analyzed using analysis of variance, followed by Scheffe's post hoc test. Significance levels were measured at $p < 0.05$.

Results

Time course of iron accumulation, oxidation DNA injury after ICH

Table 1 shows the summary of changes in some parameters associated with iron accumulation, oxidation

Table 1. *Summary of changes in iron accumulation, oxidation, and DNA injury after ICH.*

	1 day	3 days	7 days
Iron	+	+++	++
8-OHdG	+	++	+
AP sites	++	+++	+
DNP	++	+++	++
APE/Ref-1	−	−−−	−
Ku-proteins	−	−−−	(±)
TUNEL staining	(±)	(±)	(±)
PANT staining	(±)	+++	(±)

+ Weak positive; ++ moderate positive; +++ strong positive; − slight decrease; −− moderate decrease; −−− extreme decrease; (±) neutral.
ICH Intracerebral hemorrhage; *8-OHdG* 8-hydroxyl-2′-deoxyguanosine; *AP* apurinic/apyrimidinic; *DNP* dinitrophenyl; *APE* apurinic/apyrimidinic endonuclease.

DNA injury following ICH. An increase in 8-OHdG, AP sites, and DNP levels, and a decrease of APE/Ref-1 and Ku levels were observed in the ipsilateral basal ganglia, especially 3 days after ICH. Abundant PANT-positive cells were also observed in the perihematomal area 3 days after ICH.

Influence of iron on oxidation and DNA damage

DNP protein levels in the ipsilateral basal ganglia after ferrous iron injection were increased compared with the sham ipsilateral and the Fe^{++} injection-contralateral basal ganglia 24 hours after Fe^{++} injection ($p < 0.01$, Fig. 1A). PANT-positive cells were also detected in the ipsilateral basal ganglia 24 hours after Fe^{++} injection. There were no PANT-positive cells in the contralateral basal ganglia or in the sham-ipsilateral basal ganglia (Fig. 1B).

Effect of bipyridine, a ferrous iron chelator, on ICH

Bipyridine treatment given immediately after ICH reduced DNP protein levels in the ipsilateral basal ganglia compared to vehicle-treated animals (3 days post-ICH, $p < 0.01$; Fig. 2A). Similarly, while PANT-positive cells were detected in the perihematomal area in vehicle-treated rats at 3 days after ICH (Fig. 2B-b), with bipyridine treatment given immediately after ICH, there were no PANT-positive cells detected in the ipsilateral basal ganglia (Fig. 2B-c).

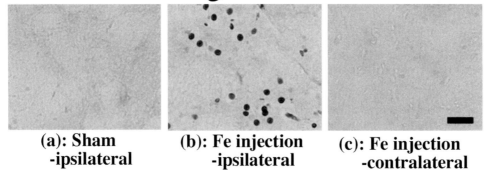

Fig. 1. (A) Western blot analysis showing the DNP concentration in the sham-ipsilateral (lanes 1–3), the Fe^{++} injection-ipsilateral (lanes 4–6), and the Fe^{++} injection-contralateral (lanes 7–9) basal ganglia 24 hours. Equal amounts of protein (50 μg) were used. (B) PANT staining in the sham-ipsilateral (a), Fe^{++} injection-ipsilateral (b), and Fe^{++} injection-contralateral (c) basal ganglia 24 hours after ICH. Bar = 20 μm

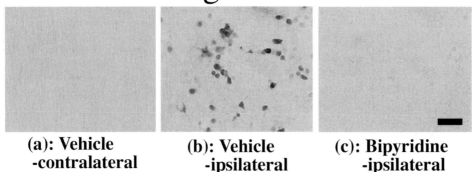

Fig. 2. The effect of bipyridine treatment (25 mg/kg i.p. given immediately after ICH) on DNP expression following ICH. (A) Western blot analysis showing DNP concentration in the vehicle contralateral (lanes 1–3), vehicle ipsilateral (lanes 4–6), and bipyridine treatment-ipsilateral (lanes 7–9) basal ganglia 3 days after ICH. Equal amounts of protein (50 μg) were used. (B) PANT staining in the vehicle-contralateral (a), vehicle-ipsilateral (b), and bipyridine treatment-ipsilateral (c) 3 days after ICH. Bar = 20 μm

Discussion

Iron accumulation and oxidative DNA damage after ICH

Although iron is essential for normal brain function, iron overload can cause brain injury [2]. After ICH, iron concentrations in the brain can reach very high levels. In the present study, iron-positive cells were found in the perihematomal area as early as the first day, detected by Perl's staining. DNA is vulnerable to oxidative stress, and 8-OHdG and AP sites are sensitive markers of such DNA injury [5]. Normally, APE, a DNA repair enzyme, repairs AP sites in DNA [7]. Ku-proteins are also DNA repair proteins [6]. APE and Ku is constitutively expressed in the non-injured brain, but can decrease due to oxidative DNA damage [6, 7]. DNA damage can result from at least 2 pathways: endonuclease-mediated DNA fragmentation or oxidative injury [3]. In the present study, we could not find obvious double-strand breaks by the TUNEL method, but did detect single-strand breaks by PANT staining. We also found abundant positive cells around the hematoma 3 days after ICH. As with the 8-OHdG immunoreactivity and AP sites results, this may reflect oxidative damage. The oxidative proteins are analyzed for carbonyl content by immunoblot with anti-DNP antibody and a specific band could be detected at 50 kDa [10]. The DNP protein levels and the number of AP sites and 8-OHdG immunoreactivity all peaked at 3 days after ICH in this study. These results suggest that brain oxidative damage peaks about 3 days after ICH.

Influence of iron on brain oxidation and DNA damage

It is known that iron can react with lipid hydroperoxides to produce free radicals, which contribute to neuronal damage during ischemia/reperfusion [11]. In vitro, exposure to $FeSO_4$ results in lipid peroxidation in neurons and an increase in apoptotic cell death [13]. The present study shows that infusion of $FeCl_2$ into the caudate induces DNP expression and PANT-positive cells 24 hours later, suggesting the Fe^{2+} might contribute to ICH-induced oxidative stress and DNA damage, a hypothesis supported by our findings on the effects of bipyridine.

Effect of iron chelation on ICH

Bipyridine is a small molecular weight (MW 220) ferrous iron chelator. It is hydrophobic, so that at physiological pH it partitions into cell membranes and binds iron as it passes through this lipid environment [1]. In a previous study, we found that deferoxamine, a ferric iron chelator, attenuates brain edema and neurological deficits in a rat ICH model [9]. These results suggest that both ferrous and ferric iron chelators such as bipyridine and deferoxamine could be useful for the treatment of brain edema following ICH. Because of their permeability, both bipyridine and deferoxamine are capable of chelating intracellular iron, although deferoxamine also chelates extracellular iron [4].

Conclusion

The present study suggests that iron-induced oxidation causes DNA damage in the brain following ICH. Oxidative stress and iron chelation are potential therapeutic targets for ICH.

References

1. Bridges KR, Cudkowicz A (1984) Effect of iron chelators on the transferrin receptor in K562 cells. J Biol Chem 259: 12970–12977
2. Connor JR, Menzies SL, Burdo JR, Boyer PJ (2001) Iron and iron management proteins in neurobiology. Pediatr Neurol 25: 118–129
3. Graham SH, Chen J (2001) Programmed cell death in cerebral ischemia. J Cereb Blood Flow Metab 21: 99–109
4. Ikeda Y, Ikeda K, Long DM (1989) Comparative study of different iron-chelating agents in cold-induced brain edema. Neurosurgery 24: 820–824
5. Kasai H, Crain PF, Kuchino Y, Nishimura S, Ootsuyama A, Tanooka H (1986) Formation of 8-hydroxyguanine moiety in cellular DNA by agents producing oxygen radicals and evidence for its repair. Carcinogenesis 7: 1849–1851
6. Kim GW, Noshita N, Sugawara T, Chan PH (2001) Early decrease in DNA repair proteins, Ku70 and Ku86, and subsequent DNA fragmentation after transient focal cerebral ischemia in mice. Stroke 32: 1401–1407
7. Lewen A, Sugawara T, Gasche Y, Fujimura M, Chan PH (2001) Oxidative cellular damage and the reduction of APE/Ref-1 expression after experimental traumatic brain injury. Neurobiol Dis 8: 380–390
8. Nagasawa T, Hatayama T, Watanabe Y, Tanaka M, Niisato Y, Kitts DD (1997) Free radical-mediated effects on skeletal muscle protein in rats treated with Fe-nitrilotriacetate. Biochem Biophys Res Commun 231: 37–41
9. Nakamura T, Keep RF, Hua Y, Schallert T, Hoff JT, Xi G (2004) Deferoxamine-induced attenuation of brain edema and neurological deficits in a rat model of intracerebral hemorrhage. J Neurosurg 100: 672–678

10. Shacter E, Williams JA, Lim M, Levine RL (1994) Differential susceptibility of plasma proteins to oxidative modification: examination by Western blot immunoassay. Free Radic Biol Med 17: 429–437
11. Siesjo BK, Agardh CD, Bengtsson F (1989) Free radicals and brain damage. Cerebrovasc Brain Metab Rev 1: 165–211
12. Wu J, Hua Y, Keep RF, Schallert T, Hoff JT, Xi G (2002) Oxidative brain injury from extravasated erythrocytes after intracerebral hemorrhage. Brain Res 953: 45–52
13. Zhang Z, Wei T, Hou J, Li G, Yu S, Xin W (2003) Iron-induced oxidative damage and apoptosis in cerebellar granule cells: attenuation by tetramethylpyrazine and ferulic acid. Eur J Pharmacol 467: 41–47

Correspondence: Guohua Xi, Department of Neurosurgery, University of Michigan, 5550 Kresge I, Ann Arbor, MI 48109-0532, USA. e-mail: guohuaxi@umich.edu

Deferoxamine reduces CSF free iron levels following intracerebral hemorrhage

S. Wan[1,2], Y. Hua[1], R. F. Keep[1], J. T. Hoff[1], and G. Xi[1]

[1] Department of Neurosurgery, University of Michigan Medical School, Ann Arbor, Michigan, USA
[2] Department of Neurosurgery, First Affiliated Hospital, Zhejiang University, China

Summary

Iron overload occurs in brain after intracerebral hemorrhage (ICH). Deferoxamine, an iron chelator, attenuates perihematomal edema and oxidative stress in brain after ICH. We investigated the effects of deferoxamine on cerebrospinal fluid (CSF) free iron and brain total iron following ICH.

Rats received an infusion of 100-μL autologous whole blood into the right basal ganglia, then were treated with either deferoxamine (100 mg/kg, i.p., administered 2 hours after ICH and then at 12-hour intervals for up to 7 days) or vehicle. The rats were killed at different time points from 1 to 28 days for measurement of free and total iron. Behavioral tests were also performed. Free iron levels in normal rat CSF were very low (1.1 ± 0.4 μmol). After ICH, CSF free iron levels were increased at all time points. Levels of brain total iron were also increased after ICH ($p < 0.05$). Deferoxamine given 2 hours after ICH reduced free iron in CSF at all time points. Deferoxamine also reduced ICH-induced neurological deficits ($p < 0.05$), but did not reduce total brain iron.

In conclusion, CSF free iron levels increase after ICH and do not clear for at least 28 days. Deferoxamine reduces free iron levels and improves functional outcome in the rat, indicating that it may be a potential therapeutic agent for ICH patients.

Keywords: Cerebral hemorrhage; iron; behavior; deferoxamine.

Introduction

Experimental studies have demonstrated that iron overload occurs after intracerebral hemorrhage (ICH) and contributes to ICH-induced brain injury [25]. Our previous study showed that non-heme iron increases about 3-fold after ICH in a rat model [22]. The major source of iron accumulation in the brain is hemoglobin after erythrocyte lysis [22, 23]. However, a recent study found that iron bound to transferrin in the plasma also results in brain injury after ICH [13]. Deferoxamine, an iron chelator, attenuates acute perihematomal brain edema and oxidative stress [11].

Free iron can cause free radical formation and oxidative brain damage. The natural history of free iron accumulation following ICH is still not clear. We investigated the time course of free and total iron in the brain after ICH. The effects of deferoxamine on free iron in cerebrospinal fluid (CSF), total iron in the brain, and behavioral outcomes following ICH were also examined.

Materials and methods

Animal preparation and intracerebral infusion

The University of Michigan Committee on the Use and Care of Animals approved the protocols for these studies. Male Sprague-Dawley rats each weighing 300 to 400 g (Charles River Laboratories, Wilmington, MA) were used. Aseptic techniques were utilized in all surgical procedures. Animals were anesthetized with pentobarbital (50 mg/kg, i.p.). The right femoral artery was catheterized for continuous blood pressure monitoring and blood sampling. Blood was obtained from the catheter for analysis of pH, PaO_2, $PaCO_2$, hematocrit, and glucose, and as the source for the intracerebral blood infusion. Body temperature was maintained at 37.5 °C using a feedback-controlled heating pad. Animals were positioned in a stereotactic frame (David Kopf Instruments, Tujunga, CA) and a cranial burr hole (1 mm) was drilled on the right coronal suture 4.0 mm lateral to the midline. Autologous blood was withdrawn from the right femoral artery and infused (100 μL) immediately into the right caudate nucleus through a 26-gauge needle at a rate of 10 μL per minute using a microinfusion pump (Harvard Apparatus Inc., Holliston, MA). Coordinates were 0.2 mm anterior and 3.5 mm lateral to the bregma with a depth of 5.5 mm. After intracerebral infusion, the needle was removed and the skin incision closed with suture.

Experimental groups

Rats were divided to 2 groups. All rats had an ICH. In the first group, rats received deferoxamine treatment (100 mg/kg, i.p., 2 hours after ICH and at 12-hour intervals thereafter). The second group received the same amount of vehicle. The rats (6 to 9 rats/group/time point) were then killed at 1, 3, 7, 14, or 28 days later for total brain tissue iron and CSF free iron determination. All animals underwent behavioral testing until sacrificed.

Free iron determination

The rats were anesthetized with pentobarbital. CSF was obtained by puncture of the cisterna magna 1, 3, 7, 14, and 28 days after ICH and stored at $-80\,°C$ before determination. Free iron in CSF was determined according to the method described by Nilsson et al. [14].

Total brain tissue iron determination

Rats were killed at 1, 3, 7, 14, and 28 days after ICH. Brains were perfused with saline before decapitation and then removed. A coronal slice ~ 4 mm thick around the injection needle tract was cut, divided into ipsilateral and contralateral sides, and weighed. The brain was then homogenized with 2 mL 0.1 mol phosphate-buffered saline and stored at $-80\,°C$ before determination. Total brain tissue iron ($\mu g/g$ tissue weight) was determined according to the method described by Fish [1].

Behavioral tests

All animals were tested before and after surgery and scored by investigators who were blinded to both neurological and treatment conditions. Three behavioral assessments were used: forelimb-placing, forelimb-use asymmetry, and corner-turn tests [4].

(A) Forelimb-placing test

Forelimb placing was scored using a vibrissae-elicited forelimb-placing test. Independent testing of each forelimb was induced by brushing the vibrissae ipsilateral to that forelimb on the edge of a tabletop once per trial for 10 trials. Intact animals placed the forelimb quickly onto the countertop. Percent of successful placing responses were determined. A previous study showed a reduction in successful responses in the forelimb contralateral to the site of injection after ICH [4].

(B) Forelimb-use asymmetry test

Forelimb use during explorative activity was analyzed by videotaping rats in a transparent cylinder for 3 to 10 minutes depending on the degree of activity during the trial. Behavior was quantified first by determining the occasions when the non-impaired ipsilateral (I) forelimb was used as a percentage of total number of limb-use observations on the cylinder wall. Second, the occasions when the impaired forelimb contralateral (C) to the blood-injection site were used as a percentage of total number of limb-use observations on the wall. Third, the occasions when both (B) forelimbs were used simultaneously as a percentage of total number of limb-use observations on the wall. A single overall limb-use asymmetry score was calculated as: Limb use asymmetry score = $[I/(I + C + B)] - [C/(I + C + B)]$.

(C) Corner-turn test

Each rat was allowed to proceed into a 30° corner. To exit the corner, the rat could turn either left or right. The direction was recorded. The test was repeated 10 to 15 times, with at least 30 seconds between each trial, and the percentage of right turns calculated. Only turns involving full rearing along either wall were included. The rats were not picked up immediately after each turn so they did not develop an aversion for turning around.

Statistical analysis

Student t test and Mann-Whitney U test were used to compare brain iron and behavioral data. Values are mean $\pm$ SD. Statistical significance was set at $p < 0.05$.

Results

All physiological variables were measured immediately before intracerebral infusion. Mean arterial blood pressure, blood pH, PaO_2, $PaCO_2$, hematocrit, and blood glucose level were controlled within normal ranges (mean arterial blood pressure, 70 to 100 mmHg; blood pH, 7.40 to 7.50; PaO_2, 80 to 120 mmHg; $PaCO_2$, 35 to 45 mmHg; hematocrit, 38 to 43%; blood glucose level, 80 to 130 mg/dL).

Deferoxamine reduces free iron levels in CSF following ICH

Free iron levels in the normal CSF were very low in the rat (1.1 ± 0.4 μmol). After ICH, free iron levels in CSF were increased at the first day (8.5 ± 1.3 μmol) and peaked at the third day (14.2 ± 5.0 μmol). CSF free iron remained at high levels for at least 28 days (6.2 ± 1.1 μmol). Deferoxamine treatment initiated 2 hours after ICH reduced free iron in CSF at all time points (e.g., day 3: 6.7 ± 2.0 μmol versus 14.2 ± 5.0 μmol in the vehicle-treated group, $p < 0.05$).

Deferoxamine fails to reduce total brain tissue iron levels in the ipsilateral hemisphere following ICH

The levels of total brain tissue iron also increased in the ipsilateral hemisphere after ICH (e.g., day 1: 264 ± 55 μg/g versus 87 ± 13 μg/g in the contralateral side, $p < 0.01$), and remained elevated for at least 4 weeks (255 ± 61 μg/g versus 85 ± 17 μg/g in the contralateral side, $p < 0.01$). Deferoxamine treatment initiated 2 hours after ICH did not reduce total brain tissue iron in the ipsilateral hemisphere following ICH at all time points (e.g., day 1: 257 ± 41 μg/g versus 264 ± 55 μg/g in the vehicle-treated group, $p > 0.05$; day 3: 227 ± 41 μg/g versus 243 ± 46 μg/g in the vehicle-treated group, $p > 0.05$).

Deferoxamine treatment ameliorates neurological deficits after ICH

Deferoxamine treatment reduced ICH-induced neurological deficits in rats. Corner-turn scores were improved at all time points in the deferoxamine-treated group compared with the vehicle group (e.g., day 1: $88.4 \pm 3.0\%$ versus $97.7 \pm 1.2\%$, $p < 0.05$; day 7: $69.6 \pm 20.0\%$ versus $83.9 \pm 16.8\%$, $p < 0.05$).

Forelimb-placing scores were also improved in the deferoxamine-treated group compared with the vehicle group (e.g., day 3: 73.3 ± 6.5% versus 46.7 ± 7.8%, p < 0.01; day 7: 87.1 ± 3.1% versus 52.4 ± 6.7%, p < 0.05). There was also an improvement in ICH-induced forelimb-use asymmetry associated with deferoxamine therapy (e.g., day 1: 26.8 ± 3.2% versus 47.2 ± 3.4%, p < 0.05; day 7: 10.0 ± 3.6% compared with 31.0 ± 4.3%, p < 0.05).

Discussion

The present study shows that free iron levels in CSF increase on the first day, peak on the third day, and remain high for at least 28 days after ICH. Systemic deferoxamine administration reduces free iron contents in CSF and improves functional outcomes after ICH in rats. However, deferoxamine has little effect on brain total iron after ICH.

Free iron levels in CSF increase as early as the first day after ICH. The sources of free iron are either the hematoma itself or circulating blood in the presence of a disrupted blood-brain barrier. Before erythrocyte lysis and hemoglobin breakdown, iron may come from serum of the clot and/or through a leaky blood-brain barrier. Our recent studies have shown that iron-positive cells (Perls' staining) that appear to be neurons are located around the hematoma 24 hours after ICH and that deferoxamine reduces acute brain edema when given 6 hours after ICH [11, 22]. After erythrocyte lysis, hemoglobin releases iron; free iron concentrations in the brain reach even higher levels. Neither CSF free iron nor brain total iron returns to normal levels within 28 days. Thus, there is the potential for long-term iron-mediated damage following ICH.

Iron-induced brain damage may result from oxidative stress. Oxidative brain injury plays an important role in ICH [12]. Iron can stimulate the formation of free radicals leading to neuronal damage. It is known that ferrous (Fe^{2+}) and ferric (Fe^{3+}) iron react with lipid hydroperoxides to produce free radicals [19]. Intracerebral injection of iron causes lipid peroxidation, brain edema, neuronal damage, and focal epileptiform paroxysmal discharges [3, 5, 6, 11, 20, 21].

Systemic administration of deferoxamine reduces free iron levels in CSF and improves functional outcomes after ICH. Our previous studies demonstrated that deferoxamine attenuates hematoma- and hemoglobin-induced brain edema [6, 11]. These results suggest that deferoxamine may be a therapeutic agent for ICH patients. Deferoxamine, an iron chelator, is an FDA-approved drug for the treatment of acute iron intoxication and for chronic iron overload due to transfusion-dependent anemias. Deferoxamine can rapidly penetrate the blood-brain barrier and accumulate in the brain tissue at a significant concentration after systemic administration [8, 15]. Deferoxamine chelates iron by forming a stable complex that prevents iron from entering into further chemical reactions. It readily chelates iron from ferritin and hemosiderin but not readily from transferrin. Deferoxamine binds ferric iron and prevents the formation of hydroxyl radical via the Fenton/Haber-Weiss reaction. Furthermore, deferoxamine reduces hemoglobin-induced brain Na^+/K^+ ATPase inhibition and neuronal toxicity [2, 17, 18]. Favorable effects of iron chelator therapy have been reported in various cerebral ischemia models [7, 10].

Although deferoxamine is an iron chelator, it can have other effects. Thus, it can act as a direct free radical scavenger [7, 10] and it can induce ischemic tolerance in the brain [16]. The latter has been demonstrated in vivo and in vitro and may be related to a deferoxamine induction of hypoxia-inducible transcription factor 1 binding to DNA [16].

It is well-known that thrombin formation and iron are 2 major factors causing brain injury after ICH [9, 24, 25]. Indeed, iron and thrombin can interact to cause brain damage [13]. Nakamura *et al.* [13] found that intracerebral co-administration of holo-transferrin (holo-Tf; iron-loaded transferrin) with thrombin causes brain edema, oxidative damage, and DNA fragmentation. These effects were not found in rats treated with holo-Tf alone or apo-transferrin, a non-iron-loaded transferrin, with thrombin. Thus, the presence of holo-Tf and thrombin during the formation of an intracerebral hematoma may participate in ICH-induced brain injury. Reducing brain free iron levels by deferoxamine may result in less thrombin-related brain injury after ICH.

To determine whether or not deferoxamine enhances the clearance of iron from the brain after ICH, brain total iron levels in the ipsilateral and contralateral hemisphere with or without deferoxamine were also determined. Brain total iron levels were increased after ICH, but deferoxamine did not reduce total iron levels in the ipsilateral hemisphere, suggesting that deferoxamine cannot enhance iron export after ICH, at least in this model.

In summary, CSF free iron levels are increased at

the first day after ICH and remain high for at least 4 weeks. Deferoxamine reduces free iron levels, oxidative brain injury, edema, and neurological deficits, suggesting that it may be a potential therapeutic agent for ICH patients.

Acknowledgments

This study was supported by grants NS-017760, NS-039866, and NS-047245 from the National Institutes of Health, and Scientist Development Grant 0435354Z from American Heart Association.

References

1. Fish WW (1988) Rapid colorimetric micromethod for the quantitation of complexed iron in biological samples. Methods Enzymol 158: 357–364
2. Guo Y, Regan RF (2001) Delayed therapy of hemoglobin neurotoxicity. Acad Emerg Med 8: 510
3. Hammond EJ, Ramsay RE, Villarreal HJ, Wilder BJ (1980) Effects of intracortical injection of blood and blood components on the electrocorticogram. Epilepsia 21: 3–14
4. Hua Y, Schallert T, Keep RF, Wu J, Hoff JT, Xi G (2002) Behavioral tests after intracerebral hemorrhage in the rat. Stroke 33: 2478–2484
5. Hua Y, Keep RF, Hoff JT, Xi G (2003) Thrombin preconditioning attenuates brain edema induced by erythrocytes and iron. J Cereb Blood Flow Metab 23: 1448–1454
6. Huang F, Xi G, Keep RF, Hua Y, Nemoianu A, Hoff JT (2002) Brain edema after experimental intracerebral hemorrhage: role of hemoglobin degradation products. J Neurosurg 96: 287–293
7. Hurn PD, Koehler RC, Blizzard KK, Traystman RJ (1995) Deferoxamine reduces early metabolic failure associated with severe cerebral ischemic acidosis in dogs. Stroke 26: 688–694
8. Keberle H (1964) The biochemistry of desferrioxamine and its relation to iron metabolism. Ann NY Acad Sci 119: 758–768
9. Lee KR, Colon GP, Betz AL, Keep RF, Kim S, Hoff JT (1996) Edema from intracerebral hemorrhage: the role of thrombin. J Neurosurg 84: 91–96
10. Liachenko S, Tang P, Xu Y (2003) Deferoxamine improves early postresuscitation reperfusion after prolonged cardiac arrest in rats. J Cereb Blood Flow Metab 23: 574–581
11. Nakamura T, Keep R, Hua Y, Schallert T, Hoff J, Xi G (2004) Deferoxamine-induced attenuation of brain edema and neurological deficits in a rat model of intracerebral hemorrhage. J Neurosurg 100: 672–678
12. Nakamura T, Keep RF, Hua Y, Hoff JT, Xi G (2005) Oxidative DNA injury after experimental intracerebral hemorrhage. Brain Res 1039: 30–36
13. Nakamura T, Xi G, Park JW, Hua Y, Hoff JT, Keep RF (2005) Holo-transferrin and thrombin can interact to cause brain damage. Stroke 36: 348–352
14. Nilsson UA, Bassen M, Savman K, Kjellmer I (2002) A simple and rapid method for the determination of "free" iron in biological fluids. Free Radic Res 36: 677–684
15. Palmer C, Roberts RL, Bero C (1994) Deferoxamine posttreatment reduces ischemic brain injury in neonatal rats. Stroke 25: 1039–1045
16. Prass K, Ruscher K, Karsch M, Isaev N, Megow D, Priller J, Scharff A, Dimagl U, Meisel A (2002) Desferrioxamine induces delayed tolerance against cerebral ischemia in vivo and in vitro. J Cereb Blood Flow Metab 22: 520–525
17. Regan RF, Panter SS (1993) Neurotoxicity of hemoglobin in cortical cell culture. Neurosci Lett 153: 219–222
18. Sadrzadeh SM, Anderson DK, Panter SS, Hallaway PE, Eaton JW (1987) Hemoglobin potentiates central nervous system damage. J Clin Invest 79: 662–664
19. Siesjo BK, Agardh CD, Bengtsson F (1989) Free radicals and brain damage. Cerebrovasc Brain Metab Rev 1: 165–211
20. Willmore LJ, Sypert GW, Munson JV, Hurd RW (1978) Chronic focal epileptiform discharges induced by injection of iron into rat and cat cortex. Science 200: 1501–1503
21. Willmore LJ, Rubin JJ (1982) Formation of malonaldehyde and focal brain edema induced by subpial injection of $FeCl_2$ into rat isocortex. Brain Res 246: 113–119
22. Wu J, Hua Y, Keep RF, Nakamura T, Hoff JT, Xi G (2003) Iron and iron-handling proteins in the brain after intracerebral hemorrhage. Stroke 34: 2964–2969
23. Xi G, Keep RF, Hoff JT (1998) Erythrocytes and delayed brain edema formation following intracerebral hemorrhage in rats. J Neurosurg 89: 991–996
24. Xi G, Reiser G, Keep RF (2003) The role of thrombin and thrombin receptors in ischemic, hemorrhagic and traumatic brain injury: deleterious or protective? J Neurochem 84: 3–9
25. Xi G, Fewel ME, Hua Y, Thompson BG, Hoff J, Keep R (2004) Intracerebral hemorrhage: pathophysiology and therapy. Neurocrit Care 1: 5–18

Correspondence: Guohua Xi, Department of Neurosurgery, University of Michigan Medical School, R5550 Kresge I Bldg., Ann Arbor, MI 48109-0532, USA. e-mail: guohuaxi@umich.edu

Up-regulation of brain ceruloplasmin in thrombin preconditioning

S. Yang[1,3], Y. Hua[1], T. Nakamura[1], R. F. Keep[1,2], and G. Xi[1]

[1] Department of Neurosurgery, University of Michigan Medical School, Ann Arbor, Michigan, USA
[2] Department of Physiology, University of Michigan Medical School, Ann Arbor, Michigan, USA
[3] Zhejiang University School of Medicine, Hangzhou, China

Summary

Pretreatment with low-dose thrombin attenuates brain edema induced by iron or intracerebral hemorrhage (ICH). Ceruloplasmin is involved in iron metabolism by oxidizing ferrous iron to ferric iron. The present study examines whether thrombin modulates brain ceruloplasmin levels and whether exogenous ceruloplasmin reduces brain edema induced by ferrous iron in vivo.

In the first set of experiments, rats received intracerebral infusion of saline or 1 U thrombin into the right basal ganglia. Rats were killed 1, 3, or 7 days later for Western blot analysis and RT-PCR analysis. In the second set of experiments, rats received either ferric iron, ferrous iron, or ferrous iron plus ceruloplasmin, then were killed 24 hours later for brain edema measurement. We found that ceruloplasmin protein levels in the ipsilateral basal ganglia increased on the first day after thrombin stimulation and peaked at day 3. Brain ceruloplasmin levels were higher after thrombin infusion than after saline injection. RT-PCR showed that brain ceruloplasmin mRNA levels were also up-regulated after thrombin injection ($p < 0.05$). We also found ipsilateral brain edema after intracerebral infusion of ferrous iron but not ferric iron at 24 hours. Co-injection of ferrous iron with ceruloplasmin reduced ferrous iron-induced brain edema ($p < 0.05$). Our results demonstrate that thrombin increases brain ceruloplasmin levels and exogenous ceruloplasmin reduces ferrous iron-induced brain edema, suggesting that ceruloplasmin up-regulation may contribute to thrombin-induced brain tolerance to ICH by limiting the injury caused by ferrous iron released from the hematoma.

Keywords: Brain edema; ceruloplasmin; iron; preconditioning; thrombin.

Introduction

High concentrations of thrombin cause brain edema and cell death, but thrombin in low concentrations is neuroprotective [6, 13, 16]. Thus, we found that prior treatment with a low dose of thrombin attenuates the brain edema induced by intracerebral hemorrhage (ICH) or iron [3, 17], and significantly reduces infarct size in a rat model of middle cerebral artery occlusion [8]. We termed this phenomena thrombin preconditioning (TPC), or thrombin-induced brain tolerance.

Iron overload occurs in the brain after ICH and contributes to ICH-induced brain injury [9, 14, 18]. Although TPC can reduce iron-induced brain edema, the mechanisms of thrombin-induced tolerance to iron are not clear. Thrombin may induce brain protection through up-regulating brain iron-handling proteins. Ceruloplasmin, which can be synthesized in the brain, is involved in iron metabolism by oxidizing ferrous iron to ferric iron [11].

The present study examines whether thrombin can modulate brain ceruloplasmin levels and whether exogenous ceruloplasmin can reduce brain edema induced by ferrous iron in vivo.

Materials and methods

Animal preparation and intracerebral infusion

The University of Michigan Committee on the Use and Care of Animals approved the protocols for these animal studies. Male Sprague-Dawley rats each weighing 300 to 400 g (Charles River Laboratories, Wilmington, MA) were used in our study. Aseptic precautions were utilized in all surgical procedures. Animals were anesthetized with pentobarbital (50 mg/kg, i.p.). The right femoral artery was catheterized for continuous blood pressure monitoring and blood sampling. Blood was obtained from the catheter for analysis of pH, PaO_2, $PaCO_2$, hematocrit and glucose. Body temperature was maintained at 37.5 °C using a feedback-controlled heating pad. The animals were positioned in a stereotactic frame (David Kopf Instruments, Tujunga, CA) and a cranial burr hole (1 mm) was drilled on the right coronal suture 4.0 mm lateral to the midline. Thrombin, iron, or saline was infused into the right caudate nucleus through a 26-gauge needle at a rate of 10 µL per minute using a microinfusion pump (Harvard Apparatus Inc., Holliston, MA). After intracerebral infusion, the needle was removed and the skin incision closed with suture.

Experimental groups

There were 2 sets of experiments in this study. In the first set, pentobarbital anesthetized rats received an intracerebral infusion of saline or 1 U thrombin into the right basal ganglia. Rats were killed 1, 3, or 7 days later for Western blot analysis and reverse transcription polymerase chain reaction (RT-PCR) analysis. In the second set, rats received 50 µL of either 0.2 mmol ferric iron, 0.2 mmol ferrous iron, or 0.2 mmol ferrous iron plus 10 µmol ceruloplasmin, then were killed 24 hours later for brain edema measurement.

Brain water and ion content measurement

Rats were killed by decapitation under deep pentobarbital anesthesia (60 mg/kg i.p.). The brains were removed immediately and a coronal brain slice (approximately 4 mm thick) 3 mm from the frontal pole was cut with a blade. The brain samples were then divided into cortex or basal ganglia (ipsilateral or contralateral). A total of 5 samples from each brain were obtained: the ipsilateral cortex and basal ganglia, the contralateral cortex and basal ganglia, and cerebellum. Brain water content was measured by the wet/dry weight method and sodium ion content was measured by flame photometry [15].

Western blot analysis

Western blot analysis was performed on days 1, 3, and 7 after thrombin infusion using previously described methods [16]. In brief, protein concentration was determined by Bio-Rad protein assay kit. Samples (50 µg protein) were run on a polyacrylamide gel and then transferred to pure nitrocellulose membrane (Amersham Biosciences, Piscataway, NJ). For ceruloplasmin measurements, membranes were probed with 1:2500 dilution of rabbit anti-human ceruloplasmin antibody (Dako Cytomation, Carpinteria, CA), followed by a 1:2500 dilution of the secondary antibody (peroxidase-conjugated goat anti-rabbit antibody, Vector Laboratories, Burlingame, CA). The antigen-antibody complexes were visualized with a chemiluminescence system (Amersham) and exposed to photosensitive film. The relative densities were analyzed using NIH Image software, version 1.62 (National Institutes of Health, Bethesda, MD).

Reverse transcription (RT) and polymerase chain reaction (PCR)

Rats were re-anesthetized with pentobarbital (60 mg/kg, i.p.) and killed by decapitation. The brains were removed and a 3-mm thick coronal brain slice was cut with a blade approximately 4 mm from the frontal pole. The ipsilateral and the contralateral basal ganglia were sampled for RT-PCR [2].

PCR was performed with 15 µL of the reverse transcriptase reaction mixture (PerkinElmer, Wellesley, MA) containing 25 mmol $MgCl_2$, dNTP, 10× PCR buffer II and AmpliTaq DNA polymerase in a final volume of 50 µL. The rat GPI-anchored ceruloplasmin primers (NIH GenBank database) corresponded to nucleotides 2928 to 2951 (sense primer, 5'-GTA TGT GAT GGC TAT GGG CAA TGA-3') and 3355 to 3376 (antisense primer, 5'-CCT GGA TGG AAC TGG TGA TGG A-3'). Rat GAPDH primers (5'-CTCAGTGTAGCCCAGGATGC-3', 5'-ACCACCATGGA-GAAGGCTGG-3') were used to amplify GAPDH mRNA, a housekeeping gene used as a control. Amplification was performed in a DNA thermal cycler (MJ Research, Waltham, MA). Samples were subjected to 30 cycles (94 °C, 1 minute; 58 °C, 1.5 minutes; and 72 °C, 2 minutes). PCR production was analyzed by the use of electrophoresis on a 1% agarose gel. Gels were visualized with ethidium bromide staining and ultraviolet transillumination. Photographs were taken with black and white film (Polaroid Corp., Waltham, MA) and analyzed using NIH image 1.62.

Statistical analysis

All data are presented as mean ± SD. Data were analyzed with ANOVA using the Scheffe post hoc test or Student t test. Significance levels were measured at $p < 0.05$.

Results

Mean arterial blood pressure, blood pH, PaO_2, $PaCO_2$, hematocrit, and blood glucose were controlled within normal ranges.

To test whether thrombin can up-regulate brain ceruloplasmin, mRNA and protein levels were determined by RT-PCR and Western blot. RT-PCR analysis followed by scanning densitometry of bands revealed that brain ceruloplasmin mRNA levels in the ipsilateral basal ganglia were markedly up-regulated on the first day after thrombin injection (from undetectable to 1847 ± 740 pixels, $p < 0.05$). Ceruloplasmin protein levels in the ipsilateral basal ganglia increased on day 1, peaked at day 3, and were still higher than normal at day 7 after thrombin infusion. Compared with saline control, brain ceruloplasmin levels in the ipsilateral basal ganglia were higher after thrombin infusion (day 3: 8833 ± 1250 vs. 3863 ± 961 pixels in the saline control group, $p < 0.01$) (Fig. 1).

Brain edema developed in the ipsilateral basal ganglia 24 hours after intracerebral infusion of 50 µL 0.2 mmol ferrous iron ($80.6 \pm 0.9\%$ vs. $78.2 \pm 0.6\%$ in the contralateral side; n = 7) but not 50 µL 0.2 mmol ferric iron ($78.4 \pm 0.4\%$ vs. $78.1 \pm 0.6\%$ in the contralateral side, n = 7). Co-injection of ferrous iron with ceruloplasmin reduced ferrous iron-induced brain edema (79.0 ± 1.2 vs. $80.9 \pm$ p < 0.05), sodium ion accumulation (213 ± 22 vs. 294 ± 80 µEq/g dry wt, $p < 0.05$) and potassium loss (393 ± 22 vs. 354 ± 22 µEq/g dry wt, $p < 0.05$) in the ipsilateral basal ganglia.

Discussion

Our results demonstrate that TPC increases brain ceruloplasmin levels and that exogenous ceruloplasmin reduces ferrous iron-induced brain edema, suggesting that up-regulation of ceruloplasmin may contribute to thrombin-induced brain tolerance to ICH by limiting the injury caused by ferrous iron released from the hematoma.

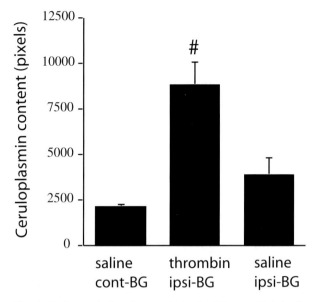

Fig. 1. Brain ceruloplasmin protein levels (*Western blot*) in the ipsilateral basal ganglia (*ipsi-BG*) and contralateral basal ganglia (*cont-BG*) 3 days after saline and thrombin infusion. Values are mean ± SD. #p < 0.01 vs. other groups

induced brain tolerance against iron-induced brain injury because they may represent naturally-occurring pathways that can be manipulated pharmacologically to limit brain injury associated with iron overload. It should be noted that iron overload in the brain is associated with many neurodegenerative disorders [12].

Although the mechanisms of thrombin-induced brain tolerance against iron are still not well understood, our present study indicates that ceruloplasmin may be involved. The brain can produce ceruloplasmin [7] and we found that thrombin up-regulates ceruloplasmin in the rat brain at both mRNA and protein levels. Ceruloplasmin is an α2-glycoprotein that oxidizes toxic ferrous iron to less-toxic ferric iron. Our study demonstrates that ceruloplasmin can reduce ferrous iron-induced brain edema. In addition, ceruloplasmin is involved in iron efflux from cultured astrocytes [5]. An important question is whether thrombin-induced ceruloplasmin up-regulation enhances iron efflux from brain after ICH. Ceruloplasmin also possesses antioxidant functions and prevents free radical injury [10, 11]. Mice lacking ceruloplasmin have iron deposition in areas of the brain and have increased lipid peroxidation [10], while mice lacking both ceruloplasmin and its homolog hephaestin have increased retinal iron and retinal neurodegeneration [1].

Iron accumulates in the brain after ICH and an iron chelator, deferoxamine, reduces ICH-and hemoglobin-induced brain edema, suggesting that iron plays an important role in perihematomal brain edema formation [4, 9, 14]. We demonstrated that TPC reduces brain edema induced by intracerebral infusion of iron or lysed erythrocytes [3]. It is important to understand the mechanisms behind the thrombin-

In addition to ceruloplasmin, other iron handling proteins including transferrin and transferrin receptor may contribute to thrombin-induced brain tolerance to iron. We have shown that brain transferrin and

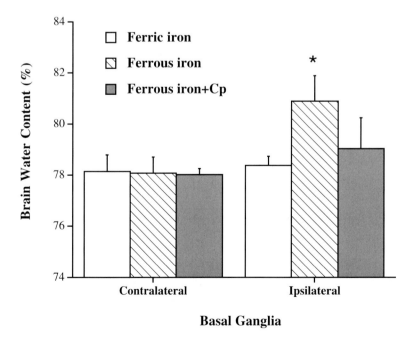

Fig. 2. Brain water content in the basal ganglia 24 hours after intracerebral injection of 50 μL of 0.2 mmol ferric iron, 0.2 mmol ferrous iron, or 0.2 mmol ferrous iron plus 10 μmol ceruloplasmin (*Cp*). Values are mean ± SD. *p < 0.05 vs. other groups

transferrin receptor levels are increased after intracerebral thrombin infusion [3]. Many studies implicate transferrin and transferrin receptor in the transport of iron from blood to brain across the blood-brain barrier, but a recent investigation indicates that there also is rapid efflux of transferrin from brain to blood, suggesting that transferrin may help iron clearance [19].

In summary, thrombin increases brain ceruloplasmin levels and induces brain tolerance to iron. Manipulating brain ceruloplasmin content may limit iron-mediated brain injury.

Acknowledgments

This study was supported by grants NS-017760, NS-039866 and NS-047245 from the National Institutes of Health and the Scientist Development Grant 0435354Z from American Heart Association.

References

1. Hahn P, Qian Y, Dentchev T, Chen L, Beard J, Harris ZL, Dunaief JL (2004) Disruption of ceruloplasmin and hephaestin in mice causes retinal iron overload and retinal degeneration with features of age-related macular degeneration. Proc Natl Acad Sci USA 101: 13850–13855
2. Hua Y, Xi G, Keep RF, Wu J, Jiang Y, Hoff JT (2002) Plasminogen activator inhibitor-1 induction after experimental intracerebral hemorrhage. J Cereb Blood Flow Metab 22: 55–61
3. Hua Y, Keep RF, Hoff JT, Xi G (2003) Thrombin preconditioning attenuates brain edema induced by erythrocytes and iron. J Cereb Blood Flow Metab 23: 1448–1454
4. Huang F, Xi G, Keep RF, Hua Y, Nemoianu A, Hoff JT (2002) Brain edema after experimental intracerebral hemorrhage: role of hemoglobin degradation products. J Neurosurg 96: 287–293
5. Jeong SY, David S (2003) Glycosylphosphatidylinositol-anchored ceruloplasmin is required for iron efflux from cells in the central nervous system. J Biol Chem 278: 27144–27148
6. Jiang Y, Wu J, Hua Y, Keep RF, Xiang J, Hoff JT, Xi G (2002) Thrombin-receptor activation and thrombin-induced brain tolerance. J Cereb Blood Flow Metab 22: 404–410
7. Klomp LWJ, Farhangrazi ZS, Dugan LL, Gitlin JD (1996) Ceruloplasmin gene expression in the murine central nervous system. J Clin Invest 98: 207–215
8. Masada T, Xi G, Hua Y, Keep RF (2000) The effects of thrombin preconditioning on focal cerebral ischemia in rats. Brain Res 867: 173–179
9. Nakamura T, Keep R, Hua Y, Schallert T, Hoff J, Xi G (2004) Deferoxamine-induced attenuation of brain edema and neurological deficits in a rat model of intracerebral hemorrhage. J Neurosurg 100: 672–678
10. Patel BN, Dunn RJ, Jeong SY, Zhu Q, Julien JP, David S (2002) Ceruloplasmin regulates iron levels in the CNS and prevents free radical injury. J Neurosci 22: 6578–6586
11. Qian ZM, Wang Q (1998) Expression of iron transport proteins and excessive iron accumulation in the brain in neurodegenerative disorders. Brain Res Rev 27: 257–267
12. Thompson KJ, Shoham S, Connor JR (2001) Iron and neurodegenerative disorders. Brain Res Bull 55: 155–164
13. Vaughan PJ, Pike CJ, Cotman CW, Cunningham DD (1995) Thrombin receptor activation protects neurons and astrocytes from cell death produced by environmental insults. J Neurosci 15: 5389–5401
14. Wu J, Hua Y, Keep RF, Nakamura T, Hoff JT, Xi G (2003) Iron and iron-handling proteins in the brain after intracerebral hemorrhage. Stroke 34: 2964–2969
15. Xi G, Keep RF, Hoff JT (1998) Erythrocytes and delayed brain edema formation following intracerebral hemorrhage in rats. J Neurosurg 89: 991–996
16. Xi G, Keep RF, Hua Y, Xiang JM, Hoff JT (1999) Attenuation of thrombin-induced brain edema by cerebral thrombin preconditioning. Stroke 30: 1247–1255
17. Xi G, Hua Y, Keep RF, Hoff JT (2000) Induction of colligin may attenuate brain edema following intracerebral hemorrhage. Acta Neurochir Suppl 76: 501–505
18. Xi G, Fewel ME, Hua Y, Thompson BG, Hoff J, Keep R (2004) Intracerebral hemorrhage: pathophysiology and therapy. Neurocrit Care 1: 5–18
19. Zhang Y, Pardridge WM (2001) Rapid transferrin efflux from brain to blood across the blood-brain barrier. J Neurochem 76: 1597–1600

Correspondence: Guohua Xi, Department of Neurosurgery, University of Michigan Medical School, R5550 Kresge I Bldg., Ann Arbor, MI 48109-0532, USA. e-mail: guohuaxi@umich.edu

Hydrocephalus in a rat model of intraventricular hemorrhage

K. R. Lodhia[1], P. Shakui[1], and R. F. Keep[1,2]

[1] Department of Neurosurgery, University of Michigan Medical School, Ann Arbor, MI, USA
[2] Department of Physiology, University of Michigan Medical School, Ann Arbor, MI, USA

Summary

The aims of the current study were 1) to establish an adult rat model of intraventricular hemorrhage (IVH) and post-hemorrhagic ventricular dilatation, and 2) to examine the role of alterations in cerebrospinal fluid (CSF) drainage and parenchymal injury in that dilatation.

Rats underwent infusion of 200 μl of autologous blood over 15 minutes. The rats were used to measure hematoma mass, ventricular dilatation, and cortical mantle volume (with T2 imaging), resistance to CSF absorption, and brain edema (as a marker of brain injury). IVH resulted in ventricular dilatation peaking at day 2 but persisting for at least 8 weeks. Although there was an increased resistance to CSF absorption at 3 days, it returned to normal at day 7. Long-term ventricular dilatation was not associated with an alteration in cortical mantle volume, although there was evidence of cortical damage (edema). It is possible that initial ventricular distension (due to the hematoma and the impaired CSF drainage) in combination with periventricular white matter damage results in structural changes that prevent total recoil once the hematoma has resolved and CSF drainage is normalized, leading to long-term hydrocephalus.

Keywords: Intraventricular hemorrhage; hydrocephalus; CSF absorption.

Introduction

In adults, parenchymal cerebral hematomas close to the ventricular system often extend into the ventricles. Indeed, it has been estimated that about 40% of intracerebral hemorrhage cases have an intraventricular component [1]. Such intraventricular hemorrhage (IVH) is a predictor of poor outcome following intracerebral hemorrhage [2]. IVH can also result from intraventricular vascular malformations, aneurysms, or tumors, as well the insertion or removal of ventricular catheters [3]. In some cases of IVH there is long-term post-hemorrhagic ventricular dilatation. Such dilatation is also a predictor of poor outcome [6, 9]. However, the underlying cause of the dilatation is still uncertain.

The aims of the current study were, therefore, to establish an adult rat model of IVH with post-hemorrhagic ventricular dilatation. This was then used to examine the effects of IVH on cerebrospinal fluid (CSF) drainage and parenchymal injury in order to examine the potential role of altered CSF dynamics and hydrocephalus *ex vacuo* in post-hemorrhagic ventricular dilatation.

Methods

In all the experiments in this report, adult male Sprague Dawley rats (300 to 400 g) were anesthetized with pentobarbital (50 mg/kg; i.p.) and a femoral artery cannula was inserted. The rats were then placed in a stereotactic frame and 200 μl of arterial blood or CSF was infused over 15 minutes into the right lateral ventricle, 1.7 mm lateral to the sagittal suture, 0.6 mm posterior to the coronal suture and at a depth of 4.5 mm. The animals were then allowed to recover from anesthesia.

In preliminary experiments, other potential IVH models were tested. Blood was infused at a faster rate (200 μl/4 min) with or without prior infusion of thrombin (1 unit) or activated Factor VII (2 units). These experiments did not produce a well-defined intraventricular hematoma and blood gathered in the cisterna magna and on the base of the brain. These models were, therefore, not employed.

Utilizing the 200 μl/15 min infusion regimen, the following measurements were made:

1) Hematoma size was assessed 1 hour, 3 days, and 1 week after IVH. Animals were re-anesthetized and sacrificed at each time point. The intraventricular clot was dissected out and weighed.

2) Ventricular dilatation was assessed by T2 imaging using a 7T small animal MRI scanner with serial thin cut slices (Fig. 1). Six rats with IVH and 6 rats with CSF infusion were followed up to 8 weeks by performing serial brain MRI scans to assess ventricular volume.

Cortical mantle volume was measured in 3 normal (no operation), 3 IVH, and 3 CSF-infused rats. All volume measurements were calculated from serial coronal T2 images using MRI volumetric analy-

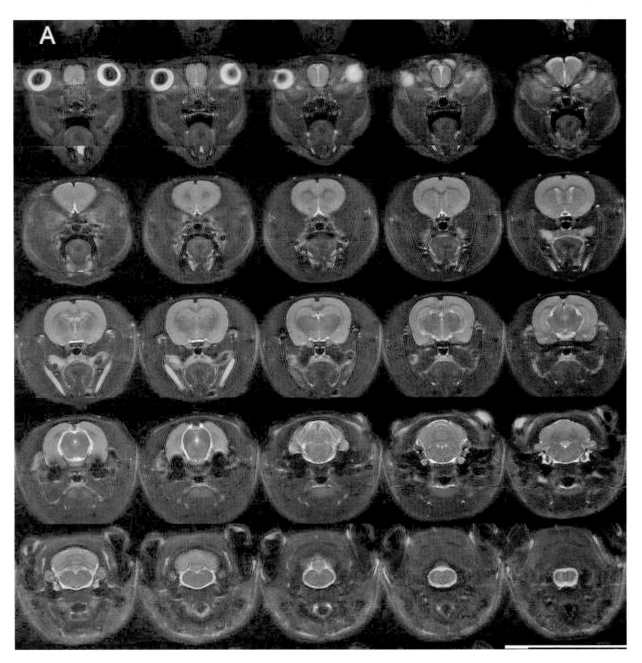

Fig. 1. Serial magnetic resonance imaging of a control rat (A), and a rat 2 days after IVH (B). Note the marked ventricular dilatation in (B). The degree of dilatation was greater ipsilateral (right) to the site of blood injection

sis software. Measurements were taken rostral-to-caudal from the anterior commissure to the splenium of the corpus callosum, and superior-to-inferior by measuring from 1.5 mm lateral to the mid-sagittal plane around the periphery to the mid-equatorial line in the coronal plane. Full thickness of the mantle was measured all the way to the periventricular surfaces in all sections.

3) Resistance to CSF absorption was measured at day 3 and day 7 in IVH and CSF-infused controls (n = 4 per group). Resistance was measured by the method of Jones and Gratton [4]. CSF was infused into the right lateral ventricle at 0, 2, 5, 10, 15, and 20 µl/min while CSF pressure was monitored. CSF plateau pressures were plotted against CSF infusion rate and the resistance to absorption calculated as the slope of this relationship using linear regression analysis.

4) Brain edema was measured at 3 days after IVH (n = 5) or CSF (n = 4) infusion. The ipsi- and contralateral anterior and posterior cortices were dissected and weighed before and after drying at 100 °C. The water content per g dry weight was then calculated.

Values presented in the text are given as means ± SE. Comparisons were made by t-test or analysis of variance with a Newman-Keuls post-hoc test for multiple comparisons.

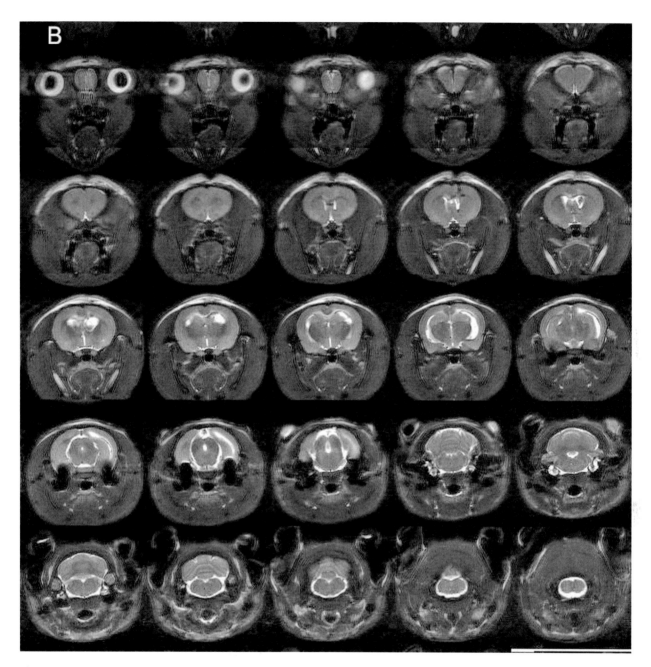

Fig. 1 (continued)

Results

Infusing 200 μl of autologous blood over 15 minutes produced a sizeable hematoma within the ventricular system. One hour after infusion, the extractable hematoma volume was 58 ± 4 mg. By 3 days, hematoma volume was 15 ± 2 mg and by 1 week, there was almost complete hematoma resolution (1 ± 1 mg).

By MRI, the lateral ventricular volume after IVH peaked on day 2 (62.9 ± 9.6 mm^3 vs. 7.7 ± 0.2 mm^3 in controls; n = 6 per group, $p < 0.01$). This post-hemorrhagic ventricular dilatation (Fig. 1) was usually bilateral, and there was still dilatation after 8 weeks although to a reduced degree (28.6 ± 10.4 mm^3 vs. 6.7 ± 0.2 mm^3 in controls; $p < 0.05$). Cortical mantle volume at 8 weeks was similar in normal, IVH- and CSF-infused sham rats (126 ± 1, 124 ± 4, and 127 ± 3 mm^3, respectively).

The IVH caused an increase in the resistance to CSF absorption at 3 days (16.0 ± 4.7 vs. 5.3 ± 0.5 mm H_2O/min/µL in controls; n = 4 per group; p < 0.05). By 7 days, however, the resistance in the IVH group had dropped (p < 0.05 vs. day 3) to a level not unlike that of controls (6.3 ± 0.7 vs. 5.3 ± 0.4 H_2O/min/µL; n = 4).

IVH also induced brain edema at day 3 in the posterior cortex ipsilateral (4.09 ± 0.04 vs. 3.76 ± 0.03 g water/g dry weight in controls; n = 5 and 4 per group, p < 0.001) and contralateral (4.00 ± 0.05 vs. 3.68 ± 0.04 g water/g dry weight in controls; p < 0.01) to the site of injection. There was also ipsi- and contralateral anterior cortex edema after IVH, although this was less than found in the posterior cortex.

Discussion

The slow infusion of blood employed in this model produced a reliable intraventricular hematoma that resolved over the first week. This rate of resolution is faster than is found in human IVH [10] or in porcine models of IVH [5], where clot resolution occurs over weeks. This is probably the result of differences in absolute hematoma volumes in the different species.

The IVH model used in this study resulted in ventricular dilatation, peaking at day 2. There was some resolution of the hydrocephalus by 8 weeks, but this was still incomplete. In a porcine model, Pang et al. [8] also reported long-term hydrocephalus after intraventricular blood infusion. However, Mayfrank et al. [5], in another pig model, found that the ventricular dilatation resolved within 6 weeks. The reason for this difference is uncertain.

In rat, early dilatation appears to be due to the presence of the hematoma and impairment in the CSF flow pathways (a 3-fold increase in the resistance to absorption). However, dilatation remained even when CSF outflow dynamics returned to normal and the intraventricular hematoma was resolved. Mayfrank et al. [5], in a porcine IVH model, also found that the resistance to CSF absorption was elevated early after IVH but returned to normal by day 7, even though the pigs continued to have ventricular dilatation at that time point.

Given that the long-term post-hemorrhagic hydrocephalus in the rat model does not appear to be caused by altered CSF absorption, it might represent *ex vacuo* changes secondary to brain parenchymal atrophy.

IVH did induce damage to the cortex as evinced by brain edema and periventricular white matter damage (data not shown). However, our cortical mantle volume measurements failed to support the hypothesis that the post-hemorrhagic dilatation represented *ex vacuo* hydrocephalus. No brain atrophy was found in the IVH group at 8 weeks. It is possible that the initial ventricular distension (due to the hematoma and the impaired CSF drainage) in combination with periventricular white matter damage results in structural changes that prevent total brain recoil once the hematoma has been resolved and CSF drainage is normalized, leading to long-term hydrocephalus.

There are ongoing trials of fibrinolytics to assess their potential for removing intraventricular blood and, thus, diminishing mass and neurotoxic effects of the IVH and improving CSF drainage [7]. There are concerns, however, that fibrinolytics, such as tissue plasminogen activator (tPA), may have adverse parenchymal effects. The current rat model should provide a setting in which to study the effects of tPA treatment or other therapeutic interventions.

Acknowledgments

K.L. was supported by a training grant from the National Institutes of Health (T32 NS07222).

References

1. Broderick JP, Brott T, Tomsick T, Miller R, Huster G (1993) Intracerebral hemorrhage more than twice as common as subarachnoid hemorrhage. J Neurosurg 78: 188–191
2. Daverat P, Castel JP, Dartigues JF, Orgogozo JM (1991) Death and functional outcome after spontaneous intracerebral hemorrhage. A prospective study of 166 cases using multivariate analysis. Stroke 22: 1–6
3. Engelhard HH, Andrews CO, Slavin KV, Charbel FT (2003) Current management of intraventricular hemorrhage. Surg Neurol 60: 15–22
4. Jones HC, Gratton JA (1989) The effect of cerebrospinal fluid pressure on dural venous pressure in young rats. J Neurosurg 71: 119–123
5. Mayfrank L, Kissler J, Raoofi R, Delsing P, Weis J, Kuker W, Gilsbach JM (1997) Ventricular dilatatation in experimental intraventricular hemorrhage in pigs. Characterization of cerebrospinal fluid dynamics and the effects of fibrinolytic treatment. Stroke 28: 141–148
6. Mohr G, Ferguson G, Khan M, Malloy D, Watts R, Benoit B, Weir B (1983) Intraventricular hemorrhage from ruptured aneurysm. A retrospective analysis of 91 cases. J Neurosurg 58: 482–487
7. Naff NJ, Hanley DF, Keyl PM, Tuhrim S, Kraut M, Bederson J, Bullock R, Mayer SA, Schmutzhard E (2004) Intraventricular

thrombolysis speeds clot resolution: results of a pilot, prospective, randomized, double-blind, controlled trial. Neurosurgery 54: 577–584
8. Pang D, Sclabassi RJ, Horton JA (1986) Lysis of intraventricular blood clot with urokinase in a canine model: part 3. Effects of intraventricular urokinase on clot lysis and posthemorrhagic hydrocephalus. Neurosurgery 19: 553–572
9. Shapiro AS, Campbell RL, Scully T (1994) Hemorrhagic dilation of the fourth ventricle: an ominous predictor. J Neurosurg 80: 805–809
10. Yamamoto Y, Waga S (1982) Persistent intraventricular hematoma following ruptured aneurysm. Surg Neurol 17: 301–303

Correspondence: Keith R. Lodhia, Department of Neurosurgery, University of Michigan Medical School, R5550 Kresge I Bldg., Ann Arbor, MI 48109-0532, USA. e-mail: klodhia@umich.edu

Early hemostatic therapy using recombinant factor VIIa in a collagenase-induced intracerebral hemorrhage model in rats

N. Kawai, T. Nakamura, and S. Nagao

Department of Neurological Surgery, Kagawa University School of Medicine, Kagawa, Japan

Summary

Neurological deterioration during the first day after intracerebral hemorrhage (ICH) is associated with early hematoma growth in 18 to 38% of patients. While clinical studies continue to evaluate efficacy of activated recombinant factor VII (rFVIIa) for reducing frequency of early hematoma growth, there have been no studies investigating the effect of rFVIIa on early hematoma growth. We used a collagenase-induced ICH model in the rat to evaluate the effects of rFVIIa on early hematoma growth.

Two hours after injection of 0.14 U of type IV bacterial collagenase in 10 μL of saline into the basal ganglia, a small amount of blood collected in the striatum. The ICH gradually increased in size, extending posteriorly to the thalamus by 24 hours after injection. Intravenous administration of rFVIIa immediately after collagenase injection decreased average hematoma volume at 24 hours compared with vehicle-treated group (168.1 ± 13.4 mm^3 vs. 118.3 ± 23.0 mm^3, $p < 0.01$). There was also a decrease in total hemoglobin content in rats treated with rFVIIa compared with vehicle-treated rats (optical density at 550 nm: 0.87 ± 0.08 vs. 0.71 ± 0.09, $p < 0.05$). There was no difference in cortical brain water content overlying the hematoma between the rFVIIa- and vehicle-treated groups (81.4 ± 0.7% vs. 81.7 ± 0.4%). Our study indicates that treatment with rFVIIa may be useful in reducing the frequency of early hematoma growth in ICH patients.

Keywords: Factor VII; brain edema; hematoma; intracerebral hemorrhage.

Introduction

Intracerebral hemorrhage (ICH) represents approximately 15% of stroke cases in the United States and 20% to 30% in Asian populations [30]. The 30-day mortality rate is 35% to 50%, and most survivors are neurologically disabled, with only 20% of patients becoming fully independent at 6 months [3]. In contrast to the successful therapeutic advances for ischemic stroke and subarachnoid hemorrhage, there remains a lack of effective treatment for ICH. Neurological deterioration frequently begins during the first day and is attributed to the development of brain edema and mass effect surrounding the hematoma [16]. Deterioration during the first day after bleeding is also strongly associated with early hematoma growth, and hematoma volume is an important predictor of 30-day mortality [2, 4, 10, 11, 17]. Studies involving patients scanned within 3 hours of the onset of ICH have shown that early hematoma growth, documented by subsequent CT scans, occurs in 18% to 38% of patients [4, 9, 12, 17].

The mechanism by which hematomas enlarge is unclear. Bleeding has been thought to be completed within minutes of onset and hematoma growth has been assumed to be the result of rebleeding from the initial site of arterial or arteriolar rupture [14]. Although this may be true in many cases, several lines of recent evidence have suggested that hematoma growth is due to bleeding into the congested and damaged tissue in the regions around the hematoma [24, 25]. As hematoma growth is a dynamic process in acute ICH, intervention with ultra-early hemostatic therapy could minimize, and possibly even prevent, early hematoma growth. The rapid action of recombinant activated factor VII (rFVIIa) at the local bleeding site, coupled with its low risk of systemic adverse effects, makes this agent a potentially valuable hemostatic treatment during the high-risk stage of ICH [8].

One of the key prerequisites for identifying an effective therapy for ICH is the development of an animal model that accurately mimics the dynamic processes involved in human ICH. Infusion of bacterial collagenase into the striatum disrupts vasculature and causes bleeding in the surrounding brain tissue [29]. Magnetic resonance imaging and histopathological examination have shown that hematoma enlargement occurs for up to 4 hours after the collagenase injection in this model of ICH [6]. We slightly modified the procedure reported by Rosenberg et al. [29] by increasing the amount of solution and decreasing the concentration of collagenase in the solution. This allowed slower hematoma development in animals, and enabled evaluation of the effects of rFVIIa on early hematoma enlargement. Although no experimental studies investigating the effect of rFVIIa on ICH have been published, a clinical study is currently underway to evaluate its efficacy in reducing the frequency of early hematoma growth in patients with acute ICH [8, 23, 24].

Materials and methods

The Ethics Committees for Animal Experiments at Kagawa University approved the experimental protocols used in this study. Animals were allowed free access to food and water before the experiment.

Experimental model

A total of 66 male Sprague-Dawley rats were used in this study. The rats were anesthetized with intraperitoneal sodium pentobarbital (60 mg/kg). A polyethylene (PE-50) catheter was introduced into the femoral artery to obtain blood samples for analysis of blood gases, blood pH, hematocrit, and blood glucose concentration. Rectal temperature was maintained at 37.5 °C during the surgery using a feedback-controlled heating system.

The rat was positioned in a stereotactic frame and the scalp was incised along the midline. Using sterile technique, a 1-mm burr hole was opened in the skull near the left coronal suture 3-mm lateral to the midline. A blunt 26-gauge needle was inserted into the left caudate putamen (striatum) under stereotactic guidance (coordinates: 0.2 mm anterior, 5.5 mm ventral, and 3 mm lateral to the bregma). Solutions containing 0.14 U of bacterial collagenase (Type IV, Sigma-Aldrich, St. Louis, MO) in 10 µL of saline were infused into the brain over a period of 10 minutes using a microinfusion pump (Eicom EPS-26, Kyoto, Japan). The stereotactic needle was removed 5 minutes after completion of infusion. The burr hole was sealed with cyanoacrylate glue and the incision was closed with sutures. The rats were placed in a warm box and allowed to recover from the anesthesia and given free access to food and water.

Experimental protocols

This study was divided into 3 parts. The first part examined hematoma growth after the collagenase infusion. Animals received an infusion of 10 µL of collagenase solution (0.14 U) into the left basal ganglia. Hematoma volume was evaluated at 2, 4, 6, 12, and 24 hours after collagenase infusion (n = 5 for each time point). Control animals received 10 µL of saline into the basal ganglia and were sacrificed at 24 hours (n = 5).

The second part of the study investigated whether general administration of rFVIIa reduces hematoma enlargement. Treated-animals received an intravenous injection of 120 µg/kg of rFVIIa (NovoSeven, Novo Nordisk, Denmark) in 200 µL water immediately after collagenase infusion. Control animals received vehicle (200 µL of water) immediately after collagenase infusion. Recombinant human VIIa was reported to inhibit the bleeding tendency induced by warfarin treatment in rats [7]. The dose of rFVIIa used in this study (120 µg/kg) was chosen based on a previously determined range used in an experimental study (50 and 250 µg/kg) [7] and in a randomized clinical trial for patients with ICH (10- to 120-µg/kg bolus dose) [8]. At 24 hours after collagenase infusion, animals were sacrificed for measurements of hematoma volume (n = 6 in each group) or brain hemoglobin content (n = 6 in each group).

The third part of the study evaluated the effect of general administration of rFVIIa on brain edema formation 24 hours after collagenase infusion. Treated-animals (n = 6) received an intravenous injection of 120 µg/kg of rFVIIa in 200 µL water immediately after collagenase infusion. Control animals (n = 6) received vehicle (200 µL of water) immediately after collagenase infusion.

Analytical methods

Morphometric measurement of hematoma volume

Animals were sacrificed by decapitation under deep sodium pentobarbital anesthesia (100 mg/kg), and the brains were rapidly removed and sectioned coronally at 2-mm intervals. After taking photographs using a digital camera, the hemorrhage area for each slice was measured with the use of a computerized image analysis system (ImageJ, version 1.32, National Institutes of Health, Bethesda, MD). Total hematoma volume was calculated by summing the clot area in each section and multiplying the distance (2 mm) between sections.

Spectrophotometric hemoglobin assay

A modified spectrophotometric assay was used to determine blood volume (hemoglobin) in the brain after ICH [5]. The ipsilateral cerebral hemisphere was collected from each animal. Distilled water (3 mL) was added to each hemisphere, followed by homogenization for 30 seconds, sonication on ice with an ultrasonicator for 1 minute, and centrifugation at 13 000 rpm for 30 minutes. The hemoglobin containing supernatant was collected, and 400 µL of Drabkin's solution was added to a 100-µL aliquot. Fifteen minutes later, the optical density of the solution at 550-nm wavelength was measured to assess hemoglobin content.

Measurement of brain water content

After sacrificing the rats by decapitation, the brains were rapidly removed and 2 coronal slices of 4-mm thickness were cut 4 mm from the frontal pole. The brain slices were divided along the midline and the cortex was separated from the basal ganglia bilaterally. Tissue samples were immediately weighed on an electronic analytical balance to the nearest 0.1 mg to obtain the wet weight (WW). Tissue samples were then dried in an oven at 110 °C for 24 hours and weighed again to obtain the dry weight (DW). The formula $(WW - DW)/WW \times 100$ was used to calculate the brain water content and expressed as a percentage of WW.

Data analysis

All data in this study are presented as mean ± SD. Data were analyzed using Student t test or analysis of variance (ANOVA) with Fisher PLSD test using StatView version 5.0 (SAS Institute, Chicago, IL). A 2-tailed probability value of less than 0.05 was used to indicate a significant difference.

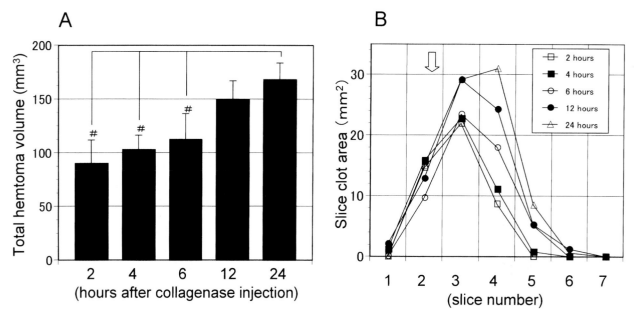

Fig. 1. Time course of hematoma growth after intracerebral infusion of type IV bacterial collagenase solution into the striatum. The average hematoma volume at 24 hours after collagenase injection was significantly larger compared with the values at 2, 4, and 6 hours after the collagenase injection (A) (#$p < 0.001$). The hematoma mainly increased in size in slices posterior to the collagenase injection site (arrow) (B)

Results

There were no significant differences in the values of arterial blood pressure, blood gas tension, blood pH, hematocrit level, body temperature, or blood glucose concentration between the 2 groups.

Hematoma enlargement

Two hours after collagenase injection, a small amount of blood had already collected in the striatum extending along with the external capsule, and the average hematoma volume was 90.5 ± 21.3 mm^3 (Fig. 1A). At 4 hours after collagenase injection, the amount of blood that had collected in the striatum had become larger and roughly spherical in shape; the average hematoma volume was 103.3 ± 12.9 mm^3 (Fig. 1A). The clot occupied almost the entire area of the striatum, extending to the ventricular wall medially and corpus callosum superiorly. The amount of blood gradually increased in size and extended to the thalamus posteriorly by 24 hours after collagenase injection (Fig. 1B); the average hematoma volume was 112.7 ± 23.5 mm^3 at 6 hours, 149.8 ± 17.0 mm^3 at 12 hours, and 168.4 ± 14.9 mm^3 at 24 hours (Fig. 1A). When comparison was made with the value at 2 hours after collagenase injection, the average hematoma volume was larger at 12 hours ($p < 0.001$) and 24 hours ($p < 0.001$). When comparison was made with the value at 4 hours, the average hematoma volume was also larger at 12 hours ($p < 0.01$) and at 24 hours ($p < 0.001$). When comparison was made with the value at 6 hours, the average hematoma volume was larger only at 24 hours ($p < 0.001$).

Effects of rFVIIa on hematoma enlargement

Blood collection in the basal ganglia was assessed morphometrically 24 hours after collagenase injection in rats receiving either rFVIIa or vehicle. Intravenous administration of rFVIIa immediately after collagenase injection significantly decreased the average hematoma volume compared with vehicle-treated rats (168.1 ± 13.4 mm^3 vs. 118.3 ± 23.0 mm^3, $p < 0.01$) (Fig. 2A). Brain hemoglobin content was also used to assess hematoma mass. Again, there was a decrease in hemispheric total hemoglobin content in rats treated with rFVIIa compared with vehicle-treated rats (OD at 550 nm: 0.87 ± 0.08 vs. 0.71 ± 0.09, $p < 0.05$) (Fig. 2B). Since the hemoglobin concentration in the blood was the same in the rFVIIa and vehicle groups, this indicates that the clot mass was significantly smaller in the rFVIIa-treated rats.

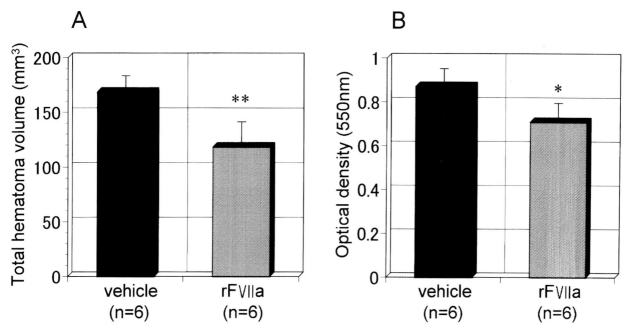

Fig. 2. Hematoma volume (A) and brain hemoglobin (optical density) (B) 24 hours after intracerebral infusion of collagenase solution. Immediately after collagenase injection, rats received an intravenous infusion of rFVIIa (120 µg/kg) or vehicle (water). Values are mean ± SD. **$p < 0.01$ and *$p < 0.05$ compared with vehicle

Effects of rFVIIa on brain water content

Brain water content in the cortex overlying the hematoma was determined 24 hours after collagenase injection using the drying/weighing method in rats receiving either rFVIIa or vehicle. In spite of reduced hematoma volume by administration of rFVIIa, there were no differences in ipsilateral cortical water content between rFVIIa-treated and vehicle-treated groups (anterior: 81.4 ± 0.7% vs. 81.7 ± 0.4%, posterior: 81.7 ± 0.6% vs. 81.7 ± 0.5%).

Discussion

Early hematoma growth after ICH

Neurological deterioration during the first day after ICH is strongly associated with early hematoma growth, and the volume of the hematoma is an important predictor of 30-day mortality [2, 4, 10, 11, 17]. The time course for progression of ICH in humans is controversial. Herbstein and Schaumburg [14] injected ^{51}Cr-labeled erythrocytes into 11 patients with hypertensive ICH between 1 to 2 and 4 to 5 hours after onset. Postmortem examination revealed no significant radioactivity in the primary hematoma, suggesting that bleeding had ceased within at least 2 to 5 hours after onset [14]. On the other hand, cerebral angiography performed 1.5 to 7 hours after onset showed extravasation of the contrast medium from perforating arteries in 7 patients with ICH [27]. An association between early hematoma growth and irregular clot morphology, which is presumably the result of multifocal bleeding, has also been reported [18]. Initial bleeding has been thought to be completed quickly in many cases as a result of clotting and tamponade by surrounding brain tissue. Hematoma growth has been assumed to result from rebleeding from the initial site of arterial or arteriolar rupture. Although this may be true in many cases, there have been several case reports in which active bleeding seemed to last more than 6 hours after onset of ICH. Recent studies suggest that early hematoma growth may also result from secondary bleeding into perilesional tissue in the periphery of the initial clot [24, 25]. Simultaneous CT and single-photon emission CT studies have demonstrated instances in which ICH growth results from the addition of discrete hemorrhages within the no-flow zone around the existing clot [24, 25]. These clinical data suggest that early hematoma growth is a dynamic process in acute ICH and in some cases may result from bleeding into a "penumbra" of damaged and congested brain tissue immediately surrounding a hematoma.

Collagenase-induced ICH model in rats

In the collagenase-induced ICH model, histopathological studies have shown that erythrocytes appear around blood vessels at the needle puncture site within the first hour and there is an extensive hematoma 4 hours after collagenase infusion with tissue disruption by extravasated erythrocytes [29]. Enlargement of the hematoma was observed for up to 4 hours after collagenase injection in this model of ICH using magnetic resonance imaging and examining comparable histological sections [6]. We have slightly modified the procedure reported by Rosenberg *et al.* [29] by increasing the amount of solution and decreasing the concentration of collagenase in the solution. In our model, there was already a small amount of hemorrhage in the striatum spreading along the external capsule 2 hours after collagenase injection. The relatively large amount of injection solution used in this study readily spread through the white matter. By 4 hours, the extravasated blood in the striatum was almost contiguous with the external capsule and became roughly spherical, occupying almost the entire area of the striatum, extending to the ventricular wall medially and corpus callosum superiorly. ICH gradually increased in size and extended to the thalamus posteriorly by 24 hours after collagenase injection. Our modification has the benefit of producing a slow-growing ICH of uniform shape and reproducible size in the basal ganglia. In this model, the effects might be caused by tissue compression and "infusion edema" because of the relatively large amount of solution injection. However, rats that received 10 μL of vehicle (saline) injection did not exhibit an increase in brain water content in the basal ganglia and overlying cortex (data not shown).

Effects of rFVIIa on hematoma enlargement

Because hematoma growth is a dynamic process during an acute ICH, intervention with ultra-early hemostatic therapy could minimize and possibly even prevent early hematoma growth. FVIIa is an important natural initiator of hemostasis exerting its primary effects locally in regions of endothelial disruption and vascular injury [13]. Factor VII forms a complex with exposed tissue factor at local bleeding sites, activating the hemostatic cascade locally to form a hemostatic plug. A pharmacological dose of rFVIIa amplifies this process. Long-term clinical use for the treatment of hemophilic patients with inhibitors, rFVIIa has been associated with low risk of systemic coagulation and thromboembolic complications and has shown good clinical results for treating intracranial hemorrhage [1, 22, 31]. Clinical trials also indicate that rFVIIa promotes hemostasis in neurosurgical patients with normal coagulation activity [15, 28]. Clinical studies are currently ongoing to evaluate its efficacy for reducing the frequency of early hematoma growth in patients with acute ICH [8, 23, 24]. Arrest of early hematoma growth might reduce the frequency of neurological deterioration by preventing early worsening related to hematoma growth, as well as late deterioration from perihematomal edema and mass effect.

The ideal animal model has not been established for early hematoma growth and prevention of hematoma enlargement using pharmacological intervention. While our model does not faithfully mimic the complex and dynamic nature of human ICH, it does resemble human ICH dynamics and enables evaluation of the effect of rFVIIa on early hematoma growth. As a clotted hematoma forms, plasma rich in thrombin quickly seeps into surrounding tissue [32]. Thrombin causes blood-brain barrier disruption and enhances brain edema formation in rats [20, 21]. In spite of reduced hematoma volume by rFVIIa administration, there was no reduction in the brain water content in the overlying cortex between the rFVIIa-treated and vehicle-treated groups in our experiment. A pharmacological dose of rFVIIa amplifies thrombin formation at the clot-brain interface and may exacerbate brain edema formation in rFVIIa-treated animals. However, results from phase II trials have shown that there was no dose-related effect of rFVIIa on edema-to-ICH volume ratio and even a high dose of rFVIIa (160 μg/kg) did not exacerbate brain edema formation around ICH [26]. A recent study has shown that the thrombin inhibitor argatroban can reduce ICH-induced edema formation in rats [19]. Further studies are necessary to evaluate the effect of combined treatment using rFVIIa and argatroban on ICH growth and perihematomal brain edema formation in rats. While we administered the rFVIIa immediately after collagenase injection into the brain, studies of delayed treatments of rFVIIa on hematoma growth should also be performed before endorsing the clinical use of rFVIIa in patients with acute ICH.

In conclusion, the present experimental study indicates that treatment with rFVIIa may be useful in reducing the frequency of early hematoma growth in patients with ICH. Our experimental results further justify the ongoing clinical trial of rFVIIa on human ICH.

References

1. Arkin S, Cooper HA, Hutter JJ, Miller S, Schmidt ML, Seibel NL, Shapiro A, Warrier I (1998) Activated recombinant human coagulation factor VII therapy for intracranial hemorrhage in patients with hemophilia A or B with inhibitors: results of the NovoSeven emergency-use program. Hemostasis 28: 93–98
2. Broderick JP, Brott TG, Duldner JE, Tomsick T, Huster G (1993) Volume of intracerebral hemorrhage. A powerful and easy-to-use predictor of 30-day mortality. Stroke 24: 987–993
3. Broderick JP, Adama HP Jr, Barsan W, Feinberg W, Feldmann E, Grotta J, Kase C, Krieger D, Mayberg M, Tilley B, Zabramski JM, Zuccarello M (1999) Guidelines for the management of spontaneous intracerebral hemorrhage: a statement for healthcare professionals from a special writing group of the Stroke Council, American Heart Association. Stroke 30: 905–915
4. Brott T, Broderick J, Kothari R, Barsen W, Tomsick T, Sauerbeck L, Spilker J, Duldner J, Khoury J (1997) Early hemorrhage growth in patients with intracerebral hemorrhage. Stroke 28: 1–5
5. Choudhri TF, Hoh BL, Solomon RA, Connolly ES Jr, Pinsky DJ (1997) Use of a spectrophotometric hemoglobin assay to objectively quantify intracerebral hemorrhage in mice. Stroke 28: 2296–2302
6. Del Bigio MR, Yan HJ, Buist R, Peeling J (1996) Experimental intracerebral hemorrhage in rats. Stroke 27: 2312–2320
7. Diness V, Lund-Hansen T, Hedner U (1990) Effect of recombinant human FVIIa on warfarin-induced bleeding in rats. Thromb Res 59: 921–929
8. Erhardtsen E (2002) Ongoing NovoSeven trials. Intensive Care Med 28: S248–S255
9. Fujii Y, Tanaka R (1998) Predictors of hematoma growth? Stroke 29: 2442–2443
10. Fujii Y, Tanaka R, Takeuchi S, Koike T, Minakawa T, Sasaki O (1994) Hematoma enlargement in spontaneous intracerebral hemorrhage. J Neurosurg 80: 51–57
11. Fujii Y, Takeuchi S, Sasaki O, Minakawa T, Tanaka R (1998) Multivariate analysis of predictors of hematoma enlargement in spontaneous intracerebral hemorrhage. Stroke 29: 1160–1166
12. Fujitsu K, Muramoto M, Ikeda Y, Inada Y, Kim I, Kuwabara T (1990) Indications for surgical treatment of putaminal hemorrhage. Comparative study based on serial CT and time-course analysis. J Neurosurg 73: 518–525
13. Hedner U (1998) Recombinant activated factor VII as a universal haemostatic agent. Blood Coagul Fibrinolysis 9 Suppl 1: S147–S152
14. Herbstein DJ, Schaumberg HH (1974) Hypertensive intracerebral hematoma. An investigation of the initial hemorrhage and rebleeding using Cr 51-labeled erythrocytes. Arch Neurol 30: 412–414
15. Karadimov D, Binev K, Nachkov Y, Platikanov V (2003) Use of activated recombinant factor VII (NovoSeven) during neurosurgery. J Neurosurg Anesthesiol 15: 330–332
16. Kase C, Mohr JP (1986) General features of intracerebral hemorrhage. In: Barnett HJM (ed) Stroke: Pathophysiology, Diagnosis, and Management. Churchill Livingstone, New York, pp 497–523
17. Kazui S, Naritomi H, Yamamoto H, Sawada T, Yamaguchi T (1996) Enlargement of spontaneous intracerebral hemorrhage. Incidence and time course. Stroke 27: 1783–1787
18. Kazui S, Minematsu K, Yamamoto H, Sawada T, Yamaguchi T (1997) Predisposing factors to enlargement of spontaneous intracerebral hematoma. Stroke 28: 2370–2375
19. Kitaoka T, Hua Y, Xi G, Hoff JT, Keep RF (2002) Delayed argatroban treatment reduces edema in a rat model of intracerebral hemorrhage. Stroke 33: 3012–3018
20. Lee KR, Colon GP, Betz AL, Keep RF, Kim S, Hoff JT (1996) Edema from intracerebral hemorrhage: the role of thrombin. J Neurosurg 84: 91–96
21. Lee KR, Kawai N, Kim S, Sagher O, Hoff JT (1997) Mechanisms of edema formation after intracerebral hemorrhage: effects of thrombin on cerebral blood flow, blood-brain barrier permeability, and cell survival in a rat model. J Neurosurg 86: 272–278
22. Lin J, Hanigan WC, Tarantino M, Wang J (2003) The use of recombinant activated factor VII to reverse warfarin-induced anticoagulation in patients with hemorrhage in the central nervous system: preliminary findings. J Neurosurg 98: 737–740
23. Mayer SA (2002) Intracerebral hemorrhage: natural history and rationale of ultra-early hemostatic therapy. Intensive Care Med 28: S235–S240
24. Mayer SA (2003) Ultra-early hemostatic therapy for intracerebral hemorrhage. Stroke 34: 224–229
25. Mayer SA, Lignelli A, Fink ME, Kessler DB, Thomas CE, Swarup R, Van Heertum RL (1998) Perilesional blood flow and edema formation in acute intracerebral hemorrhage: a SPECT study. Stroke 29: 1791–1798
26. Mayer SA, Brun NC, Broderick J, Davis S, Diringer MN, Skolnick BE, Steiner T for the Europian/AustralAsia NovoSeven ICH Trial Investigators (2005) Safety and feasibility of recombinant factor VIIa for acute intracerebral hemorrhage. Stroke 36: 74–79
27. Mizukami M, Araki G, Mihara H, Tomita T, Fujinaga R (1972) Arteriographically visualized extravasation in hypertensive intracerebral hemorrhage. Stroke 3: 527–537
28. Pickard JD, Kirkpatrick PJ, Melson T, Andreasen RB, Gelling L, Fryer T, Matthews J, Minhas P, Hutchinson PJA, Menon D, Downey SP, Kendall I, Clark J, Carpenter TA, Williams E, Persson L (2000) Potential role of NovoSeven in the prevention of bleeding following aneurismal subarachnoid haemorrhage. Blood Coagul Fibrinolysis 11 Suppl 1: S117–S120
29. Rosenberg GA, Mun-Bryce S, Wesley M, Kornfeld M (1990) Collagenase-induced intracerebral hemorrhage in rats. Stroke 21: 801–807
30. Sacco RL, Mayer SA (1994) Epidemiology of intracerebral hemorrhage. In: Feldmann E (ed) Intracerebral Hemorrhage. Futura Publishing, Armonk, NY, pp 3–23
31. Scmidt ML, Gamerman S, Smith HE, Scott JP, DiMichele DM (1994) Recombinant activated factor VII (rFVIIa) therapy for intracranial hemorrhage in hemophilia A patients with inhibitors. Am J Hematol 47: 36–40
32. Wagner KR, Xi G, Hua Y, Kleinholz M, de Courten-Myers GM, Myers RE, Broderick JP, Brott TG (1996) Lobar intracerebral hemorrhage model in pigs: rapid edema development in perihematomal white matter. Stroke 27: 490–497

Correspondence: Nobuyuki Kawai, Department of Neurological Surgery, Kagawa University School of Medicine, 1750-1 Miki-cho, Kita-gun, Kagawa 761-0793, Japan. e-mail: nobu@kms.ac.jp

Effects of endogenous and exogenous estrogen on intracerebral hemorrhage-induced brain damage in rats

T. Nakamura[1,2], G. Xi[1], R. F. Keep[1], M. Wang[3], S. Nagao[2], J. T. Hoff[1], and Y. Hua[1]

[1] Department of Neurosurgery, University of Michigan Medical School, Ann Arbor, Michigan, USA
[2] Department of Neurological Surgery, Kagawa University Faculty of Medicine, Kagawa, Japan
[3] Department of Neurology, University of Michigan Medical School, Ann Arbor, Michigan, USA

Summary

The present study examined differences in intracerebral hemorrhage (ICH)-induced brain injury in male and female rats, whether delayed administration of 17β-estradiol can reduce ICH-induced brain damage, and whether these effects are estrogen receptor (ER)-dependent.

Male and female Sprague-Dawley rats received an infusion of 100-μL autologous whole blood into the right basal ganglia. The effects of 17β-estradiol (5 mg/kg, i.p.) on ICH-induced brain injury were examined by measuring brain edema and neurological deficits 24 hours later. Heme oxygenase-1 (HO-1) was investigated by immuno-analysis. Brain edema was significantly less in female compared to male rats. The ER antagonist ICI182,780 exacerbated ICH-induced brain edema in female but not in male rats, suggesting that ER activation during ICH is protective in female rats. Administration of 17β-estradiol to male (but not female) rats significantly reduced brain edema, neurological deficits, and ICH-induced increases in brain HO-1 levels when given 2 hours after ICH. This study showed that female rats have less ICH-induced injury than male rats. ER is involved in limiting ICH-induced injury in female rats. ICH-injury in male rats can be reduced by 17β-estradiol. Since 17β-estradiol treatment was effective in male rats, it could be a potential therapeutic agent for ICH.

Keywords: Cerebral hemorrhage; brain edema; estrogen receptor.

Introduction

Evidence suggests that estrogen is both a natural neuroprotectant and a potential therapeutic agent for cerebrovascular disease. Several mechanisms may contribute to the effect of estrogen on brain injury. Estrogen is known to be vasoactive in the cerebral circulation under normal and ischemic conditions [5]. Harder and Coulson [3] showed that estrogen receptors (ERs) are present in vascular smooth muscle cells. There are, however, direct effects of estrogen on neurons, as evidenced by the fact that estrogen protects neuronal cultures from a variety of stresses including oxidative stress and excitotoxicity [2]. Estrogen exerts protective effects in ischemic stroke through ER activation [10]. There is also evidence that estrogens can protect via ER-independent mechanisms [4].

Many patients with an intracerebral hemorrhage (ICH) deteriorate progressively because of secondary edema formation [6]. We have shown that pretreatment with 17β-estradiol attenuates brain edema after ICH in male mice [8]. Whether delayed treatment would be beneficial after ICH is unknown, however, there is a paucity of data on the underlying neuroprotective mechanisms of estrogen in ICH and, in particular, whether those mechanisms involve ER activation. The present study examined whether there are differences in ICH-induced brain injury in male and female rats, whether delayed administration of 17β-estradiol can reduce ICH-induced brain damage, and whether these effects are ER-dependent.

Materials and methods

Animal preparation and intracerebral infusion

Animal protocols were approved by the University of Michigan Committee on the Use and Care of Animals. Male and female Sprague-Dawley rats (Charles River Laboratories, Portage, MI), each weighing 300 to 400 g, were used in the experiments. Rats were allowed free access to food and water. The animals were anesthetized with pentobarbital (40 mg/kg, i.p.) and the right femoral artery was catheterized to monitor arterial blood pressure and to sample blood for intracerebral infusion. Blood pH, PaO_2, $PaCO_2$, hematocrit, and glucose levels were monitored. Rectal temperature was maintained at 37.5 °C using a feedback-controlled heating pad. The rats were positioned in a stereotaxic frame (David Kopf Instruments, Tujunga, CA) and a cranial burr hole (1 mm) was drilled near the right coronal suture 3.5 mm lateral to the midline. A 26-gauge needle was inserted stereotaxically into the right basal ganglia (coordinates: 0.2 mm anterior, 5.5 mm ventral, and 3.5 mm lateral to the bregma). Autologous whole blood (100 μL) was infused at a rate of 10 μL/min with the use of a microinfusion pump (Harvard Apparatus, South Natick, MA). The needle was removed, the burr hole was filled with bone wax, and the skin incision closed with suture.

Experimental groups

This study was performed in four parts. With the exception of 1 experiment in part 2, all rats received an intracaudate injection of 100 μL autologous whole blood and were sacrificed 24 hours later. Part 1 evaluated ICH-induced changes in brain water content in male and female rats (n = 5 each group).

Part 2 investigated the effect of ER inhibitor ICI182,780 (Tocris Cookson Inc., Ellisville, MO) on ICH-induced brain edema (n = 5 each group) in male and female rats. ICI182,780 (10 μg dissolved in 10 μL saline plus 2.5% ethanol and 1% gelatin) was mixed with 100 μL autologous whole blood. Twenty rats had an intracaudate injection of 100 μL autologous blood with either 10 μg ICI182,780 or vehicle (2.5% ethanol and 1% gelatin in saline). The animals were divided into the following 4 groups: (1) ICI182,780 or (2) vehicle treatment in ICH male rats. (3) ICI182,780 or (4) vehicle treatment in ICH female rats. Animals were anesthetized and sacrificed 24 hours after ICH for brain edema measurement (n = 5 each group).

To examine whether the differential effect of ICI182,780 might reflect differences in the level of ERα expression between sexes, normal male and female rats were sacrificed for Western blot analysis and immunohistochemistry (n = 3 per gender).

Part 3 examined the effect of exogenous 17β-estradiol (Sigma-Aldrich, St. Louis, MO) as a therapeutic agent for ICH-induced brain damage in male rats. Thirty rats had an intracaudate injection of 100 μL autologous whole blood and were treated with either 17β-estradiol (5.0 mg/kg dissolved in saline plus 1% gelatin) or vehicle (1% gelatin in saline) subcutaneously.

The animals were divided into the following 6 groups according to sex and time of treatment after ICH: (1) 17β-estradiol or (2) vehicle administered 2 hours after ICH in male rats. (3) 17β-estradiol or (4) vehicle administered 6 hours after ICH in male rats. (5) 17β-estradiol or (6) vehicle administered 2 hours after ICH in female rats. Animals underwent behavioral testing before ICH, and 24 hours after ICH (n = 5 each group). Animals were then anesthetized and sacrificed for brain edema measurement (n = 5 each group).

Involvement of ER in the protective effects of exogenous 17β-estradiol given after 2 hours in male rats was examined using ICI182,780. Ten rats had an intracaudate injection of 100 μL autologous whole blood mixed with ICI182,780 (10 μg). Animals were treated with either 17β-estradiol (5.0 mg/kg dissolved in saline plus 1% gelatin; n = 5) or vehicle (1% gelatin in saline; n = 5) 2 hours later. Animals were re-anesthetized and sacrificed 24 hours after ICH for brain edema examination (n = 5 each group).

Part 4 examined the effect of 17β-estradiol (given 2 hours after ICH) on heme oxygenase-1 (HO-1) levels 24 hours following ICH in male rats. HO-1 levels were investigated by Western blot analysis and immunohistochemistry (n = 3 each group).

Brain water content

Animals were re-anesthetized (pentobarbital 60 mg/kg, i.p.) and decapitated 24 hours after ICH for brain water measurements. The brains were removed, and a coronal brain slice (approximately 3 mm thick) 4 mm from the frontal pole was cut with a blade. The brain slice was divided into 2 hemispheres along the midline, and each hemisphere was dissected into cortex and basal ganglia. The cerebellum served as a control. Five samples from each brain were obtained: ipsilateral and contralateral cortices, ipsilateral and contralateral basal ganglia, and the cerebellum. Brain samples were immediately weighed on an electric analytical balance (Model AE 100, Mettler Instrument, Highstown, NJ) to obtain the wet weight. Brain samples were then dried at 100 °C for 24 hours to obtain the dry weight. The formula for calculation was as follows: (Wet Weight − Dry Weight)/Wet Weight.

Western blot analysis

Animals were anesthetized before undergoing transcardial perfusion with saline. The brains were then removed and a 3 mm thick coronal brain slice was cut approximately 4 mm from the frontal pole. The slice was separated into ipsilateral and contralateral basal ganglia. Western blot analysis was performed as previously described. Briefly, 50 μg proteins for each were separated by sodium dodecyl sulfate polyacrymide gel electrophoresis and transferred to a Hybond-C pure nitrocellulose membrane (Amersham, Piscataway, NJ). The membranes were blocked in Carnation nonfat milk. Membranes were probed with a 1:2000 dilution of the primary antibody (rabbit anti-ERα antibody or rabbit anti-hemeoxygenase-1 antibody) and a 1:3000 dilution of the second antibody (peroxidase-conjugated horse anti-rabbit antibody, BIO-RAD Laboratories, Hercules, CA). The antigen-antibody complexes were visualized with a chemiluminescence system (Amersham) and exposed to film (Kodak X-OMAT, Rochester, NY). The relative densities of bands were analyzed with NIH Image (Version 1.61, National Institutes of Health, Bethesda, MD).

Behavioral tests

ICH-induced neurological deficits were assessed using forelimb-placing and corner-turn tests [11]. In the vibrissae-elicited forelimb-placing test, animals were held by their bodies to allow their forelimbs to hang free. Independent testing of each forelimb was induced by brushing the respective vibrissae on the corner of a table top once per trial for 10 trials. A score of 1 was given each time the rat placed its forelimb onto the edge of the table in response to the vibrissae stimulation. Percent successful placing responses were determined for impaired forelimb and non-impaired forelimb.

For the corner-turn test, each rat was allowed to proceed into a 30° corner. To exit the corner, the animal could turn either to the left or right, and this was recorded. This was repeated 10 to 15 times, and the percentage of right turns was calculated. Both forelimb-placing and corner-turn tests were performed and scored by an investigator blinded to treatment conditions.

Statistical analysis

All data are presented as mean ± SD. Data from water and ion contents, Western blot analysis, and behavioral tests were analyzed using Student t test or analysis of variance (ANOVA), followed by Scheffe post hoc test for multiple comparisons. Significance levels were measured at $p < 0.05$.

Results

Female rats had less brain edema in the ipsilateral basal ganglia 24 hours after ICH compared to male rats (80.1 ± 0.3% vs. 81.7 ± 0.4%, $p < 0.05$). Co-injection of the ER inhibitor ICI 182,780 along with the blood had no significant effect on ICH-induced brain edema formation in male rats (ipsilateral basal ganglia water content $81.3 \pm 1.0\%$ vs. $81.7 \pm 0.4\%$ in vehicle-treated rats). In contrast, ICI 182,780 co-infusion in female rats resulted in an increase in brain water content in the ipsilateral basal ganglia ($81.4 \pm 0.5\%$ vs. $80.1 \pm 0.3\%$ in vehicle-treated rats, $p < 0.05$). Interestingly, in the presence of ICI 182,780, ICH-induced brain edema was similar in male and female rats.

Whether the sex-dependent effects of ICI 182,780 were associated with differences in basal ganglia ER-α levels was examined by immunohistochemistry and Western blot. By Western blot analysis, ER-α levels in the ipsilateral basal ganglia were higher in female compared to male rats (2198 ± 228 pixels and 1487 ± 327 pixels, $p < 0.05$).

In male rats, subcutaneous administration of 17β-estradiol 2 hours after ICH reduced water content in the ipsilateral basal ganglia at 24 hours compared to vehicle-treated controls ($79.1 \pm 0.6\%$ vs. $81.8 \pm 0.9\%$, $p < 0.05$). 17β-estradiol treatment 2 hours after ICH also reduced ICH-induced neurological deficits. The forelimb-placing score was improved 24 hours after ICH compared with the vehicle group ($32 \pm 13\%$ vs. $0 \pm 0\%$, $p < 0.05$). There was also an improvement in ICH-induced corner-turn test scores compared with vehicle-treatment ($82 \pm 13\%$ vs. $98 \pm 4\%$, $p < 0.05$). Unlike males, 17β-estradiol treatment in female rats did not reduce brain edema in the ipsilateral basal ganglia when given 2 hours after ICH (brain water content 79.5 ± 0.3 vs. $80.1 \pm 0.1\%$ in vehicle-treated controls). Similarly, delaying 17β-estradiol treatment in male rats until 6 hours after ICH did not reduce ICH-induced edema in the ipsilateral basal ganglia ($81.2 \pm 0.7\%$ vs. $81.7 \pm 0.6\%$).

As described above, 17β-estradiol treatment 2 hours after ICH markedly reduced brain edema formation in male rats. This was still the case when the ER inhibitor, ICI 182,780, was co-injected with blood. Thus, even in the presence of this inhibitor, 17β-estradiol treatment reduced brain water content in ipsilateral basal ganglia ($79.0 \pm 0.3\%$) compared with vehicle-treated animals ($81.4 \pm 0.5\%$, $p < 0.05$). Indeed the degree of 17β-estradiol-induced protection was similar in ICI 182,780 and vehicle-treated male rats.

HO-1 protein levels were increased in the ipsilateral basal ganglia compared with the contralateral basal ganglia 24 hours after ICH in male rats (6738 ± 1231 pixels vs. 1538 ± 345, $p < 0.05$). Two-hour delayed 17β-estradiol treatment reduced HO-1 protein levels (2068 ± 457 vs. 6738 ± 1231 in the vehicle group, $p < 0.05$).

Discussion

The greater neuroprotection afforded to females in cerebral ischemia and trauma is likely due to the effects of circulating estrogens and progestins [9]. In the present study we found that brain injury in female, but not male, rats was exacerbated in the presence of the ER inhibitor, ICI182,780. Indeed, in the presence of this inhibitor there was no difference in brain edema between females and males, suggesting that reduced injury in female rats after ICH is due to ER activation. Interestingly, a recent report found that ICI182,780 only exacerbates ischemic injury in female but not male mice [10]. Western blot analysis and immunohistochemistry both indicated that ERα expression was greater in the basal ganglia of female compared to male rats. However, in accord with other studies, the receptor was present in the brain in males [12]. Thus, the absence of the effect of ICI 182,780 on brain edema in male rats may reflect differences in estrogen rather than receptor levels.

The current study showed that 17β-estradiol treatment 2 hours after ICH in male rats attenuated perihematomal edema. There was, however, no protection if treatment was delayed for 6 hours. These results indicate that 17β-estradiol could be a therapeutic agent for ICH during the acute phase after hemorrhage. In clinical cases, many patients with intracerebral hematoma deteriorate progressively because of brain edema formation [6]. In addition, we also found that 17β-estradiol reduced neurological deficits when given 2 hours after ICH in male rats. We used several sensorimotor behavioral tests to examine ICH-induced neurological deficits [7]. 17β-estradiol improved both forelimb-placing and corner-turn scores in the present study.

Heme oxygenase is a key enzyme in hemoglobin degradation, converting heme to iron, carbon monoxide, and biliverdin. HO-1 is markedly up-regulated in the brain after ICH [14] and heme oxygenase inhibition reduces brain injury after ICH [13]. The current study demonstrates that 17β-estradiol reduces perihematomal HO-1 levels. Although the reduced HO-1 levels with 17β-estradiol could reflect reduced brain injury, these results suggest that 17β-estradiol may, in part, reduce brain injury by limiting the rate of iron release from the hematoma, thereby limiting iron-induced toxicity. Recently, an iron chelator, deferoxamine, has been shown to limit ICH-induced brain injury in the rat [7].

The effects of exogenous 17β-estradiol on ICH-induced edema in male rats were not inhibited by ICI 182,780; i.e., it does not appear to be mediated by an ER-dependent mechanism. Evidence suggests that estrogens can induce neuroprotection by non-ER as well as ER-mediated mechanisms [4]. Estrogens possess anti-oxidative properties [1], and neuroprotection against oxidative damage by estrogens is most likely caused by the antioxidant properties of these steroids rather than through ER activation [2]. In contrast to the protective effect of 17β-estradiol in male rats, administration of this agent had no protective effect in female rats after ICH. Because of the high level of estrogens circulating in female rats, it may be that exogenous 17β-estradiol can add no further protection against ICH. The degree of brain edema in male rats treated with 17β-estradiol was similar to that of normal female rats.

This study showed that female rats have less ICH-induced injury than male rats via an ER-dependent mechanism. ICH-induced injury in male rats can, however, be reduced by administration of exogenous 17β-estradiol through an ER-independent mechanism. Since delayed 17β-estradiol treatment was effective in male rats, it could be a potential therapeutic agent for ICH.

Acknowledgments

This study was supported by grants NS-17760 (JTH), NS-039866, NS-047245 (GX) from the National Institutes of Health and the Scientist Development Grant 0435354Z from the American Heart Association (YH).

References

1. Behl C, Skutella T, Lezoualch F, Post A, Widmann M, Newton CJ, Holsboer F (1997) Neuroprotection against oxidative stress by estrogens: structure-activity relationship. Mol Pharmacol 51: 535–541
2. Culmsee C, Vedder H, Ravati A, Junker V, Otto D, Ahlemeyer B, Krieg JC, Krieglstein J (1999) Neuroprotection by estrogens in a mouse model of focal cerebral ischemia and in cultured neurons: evidence for a receptor-independent antioxidative mechanism. J Cereb Blood Flow Metab 19: 1263–1269
3. Harder DR, Coulson PB (1975) Estrogen receptors and effects of estrogen on membrane properties of coronary vascular smooth muscle. J Cell Physiol 100: 375–382
4. Hurn PD, Brass LM (2003) Estrogen and stroke: a balanced analysis. Stroke 34: 338–341
5. Hurn PD, Littleton-Kearney MT, Kirsch JR, Dharmarajan AM, Traystman RJ (1995) Postischemic cerebral blood flow recovery in the female: effect of 17 beta-estradiol. J Cereb Blood Flow Metab 15: 666–672
6. Kase CS, Caplan LR (1994) Intracerebral Hemorrhage. Butterworth-Heinemann, Boston
7. Nakamura T, Keep RF, Hua Y, Schallert T, Hoff JT, Xi G (2004) Deferoxamine-induced attenuation of brain edema and neurological deficits in a rat model of intracerebral hemorrhage. J Neurosurg 100: 672–678
8. Nakamura T, Xi G, Hua Y, Schallert T, Hoff JT, Keep RF (2004) Intracerebral hemorrhage in mice: model characterization and application for genetically modified mice. J Cereb Blood Flow Metab 24: 487–494
9. Roof RL, Hall ED (2000) Gender differences in acute CNS trauma and stroke: Neuroprotective effects of estrogen and progesterone. J Neurotrauma 17: 367–388
10. Sawada M, Alkayed NJ, Goto S, Crain BJ, Traystman RJ, Shaivitz A, Nelson RJ, Hurn PD (2000) Estrogen receptor antagonist ICI182,780 exacerbates ischemic injury in female mouse. J Cereb Blood Flow Metab 20: 112–118
11. Schallert T, Fleming SM, Leasure JL, Tillerson JL, Bland ST (2000) CNS plasticity and assessment of forelimb sensorimotor outcome in unilateral rat models of stroke, cortical ablation, parkinsonism and spinal cord injury. Neuropharmacology 39: 777–787
12. Sheng Z, Kawano J, Yanai A, Fujinaga R, Tanaka M, Watanabe Y, Shinoda K (2004) Expression of estrogen receptors (alpha, beta) and androgen receptor in serotonin neurons of the rat and mouse dorsal raphe nuclei; sex and species differences. Neurosci Res 49: 185–196
13. Wagner KR, Hua Y, de Courten-Myers GM, Broderick JP, Nishimura RN, Lu SY, Dwyer BE (2000) Tin-mesoporphyrin, a potent heme oxygenase inhibitor, for treatment of intracerebral hemorrhage: in vivo and in vitro studies. Cell Mol Biol (Noisy-le-grand) 46: 597–608
14. Wu J, Hua Y, Keep RF, Nakamura T, Hoff JT, Xi G (2003) Iron and iron-handling proteins in the brain after intracerebral hemorrhage. Stroke 34: 2964–2969

Correspondence: Ya Hua, Department of Neurosurgery, University of Michigan Medical School, 5550 Kresge I Bldg., Ann Arbor, MI 48109-0532, USA. e-mail: yahua@umich.edu

Dopamine changes in a rat model of intracerebral hemorrhage

J. R. Cannon[1,2], T. Nakamura[2], R. F. Keep[2,3], R. J. Richardson[1], Y. Hua[2], and G. Xi[2]

[1] Department of Environmental Health Sciences, University of Michigan Medical School, Ann Arbor, MI, USA
[2] Department of Neurosurgery, University of Michigan Medical School, Ann Arbor, MI, USA
[3] Department of Physiology, University of Michigan Medical School, Ann Arbor, MI, USA

Summary

Recent case reports suggest that dopamine (DA) replacement may reduce behavioral deficits resulting from hemorrhages along the nigrostriatal tract. In the rat model of intracerebral hemorrhage (ICH), behavioral deficits are first evident on day 1, with return to near control levels by day 28. The current study was conducted to determine if striatal dopamine alterations are correlated with behavioral deficits. Gamma-aminobutyric acid (GABA) levels were measured to determine selectivity. Striatal DA, DA metabolites, and GABA were determined at days 1, 3, 7, and 28 after ICH by high-pressure liquid chromatography with electrochemical detection. ICH resulted in significant increases above control in DA contralateral to the lesion (177 to 361% above control, days 1 to 28). There were also significant, but much less marked changes in GABA. In the ipsilateral striatum, significant DA increases also occurred (~200% at day 3 and ~275% day 28), while GABA alterations were not significant. These results indicate that the striatal DA system is selectively altered after ICH. Further studies will be needed to determine if regional dopamine alterations occur relative to the location of the hematoma.

Keywords: Intracerebral hemorrhage; dopamine; γ-aminobutyric acid.

Introduction

Intracerebral hemorrhage (ICH) accounts for roughly 37,000 cases per year in the United States and 10% of all strokes [3, 11]. Those that survive often experience severe neurological deficits. Measurable behavioral deficits also occur in a prominent rat model of ICH [8]. While the pathophysiology following ICH is becoming better understood, effective novel therapeutics have yet to be developed to treat persistent neurological deficits [22].

Certain neurodegenerative diseases are characterized by damage to specific neurotransmitter systems. The most prominent is Parkinson's disease (PD), where degeneration of dopaminergic nigrostriatal neurons and subsequent striatal dopamine depletion results. The primary therapy for PD-associated behavioral deficits is dopamine replacement in the form of L-dopa [15]. Specific neurotransmitter alterations after ICH in animals or humans have yet to be characterized and the identification of such alterations could represent a novel therapeutic approach to ICH.

Iron overload and oxidative stress are pathogenic features present in both PD and ICH [16, 20, 21]. Multiple groups have shown that thrombin, a key mediator of ICH secondary injury, injected into the substantia nigra results in degeneration of dopaminergic neurons [4, 5]. Therefore, it is possible that a hematoma along the nigrostriatal tract could selectively damage dopaminergic neurons. Indeed, there are multiple case reports of parkinsonism associated with midbrain hemorrhage [7, 9]. Additionally, at least 2 case reports exist where behavioral deficits after hemorrhage improved with L-dopa. First, after a large right temporal lobe hemorrhage, a patient developed many of the classic behavioral signs of parkinsonism, which improved significantly with L-dopa [10]. Second, a brief clinical report described a traumatic subarachnoid hemorrhage in left dorsolateral midbrain and the right cerebral peduncle, which the clinician believed to be affecting the nigrostriatal dopamine system. The patient remained in a vegetative state during several months of conservative management, but showed marked neurological improvement with the administration of L-dopa [12]. These findings prompt further studies regarding the ability of ICH to selectively damage dopaminergic neurons. Because a significant portion of ICHs occur in the basal ganglia, determination of dopamine levels in that area is warranted. The aims of the current study were to determine alterations in striatal dopamine after ICH in the rat, and determine if these alterations were specific to the dopamine system by evaluation of γ-aminobutyric acid (GABA).

Materials and methods

Materials

Adult male Sprague-Dawley rats (300 to 375 g; Charles River Laboratories, Portage, MI) were used for all experiments. All experiments were approved by the University of Michigan Committee on Use and Care of Animals. All chemicals were obtained from Sigma-Aldrich (St. Louis, MO) unless otherwise noted.

Surgery

Animals were anesthetized with pentobarbital (50 mg/kg, i.p.; Abbott Laboratories, Abbott Park, IL) and received striatal injection of 100 μL autologous blood as previously described [8].

Dopamine determination

Animals were deeply anesthetized and decapitated on days 1 (n = 5), 3 (n = 5), 7 (n = 5), or 28 (n = 4). Normal animals (n = 5) were also sampled as absolute controls. Both the ipsilateral and contralateral striata were dissected out on ice (0.0 to 0.2 mm, A/P to bregma). The striata were weighed and stored at −80 °C until sonication in ice-cold 0.1 N perchloric acid. Samples were then centrifuged at 16,000 g for 30 minutes followed by the collection of the supernatant and storage at −80 °C. The protein pellet underwent a standard Bradford protein assay (Bio-Rad, Hercules, CA). The supernatant was then injected into a high-pressure liquid chromatography (HPLC) pump (Waters, Milford, MA, Model 1525). The mobile phase consisted of the following: 0.06 mol sodium phosphate, 0.03 mol citric acid, 1.3 mmol 1-octanesulfonic acid, 0.1 mmol EDTA, 8% methanol in HPLC water, pH 3.5 with a flow rate of 1.0 mL/min. DA, 3,4-dihydroxyphenylacetic acid (DOPAC), and homovanillic acid (HVA) were separated on a Symmetry 4.6×150 mm C_{18} column with a 3.5 μmol particle size (Waters). Catecholamines were measured using an electrochemical detector with a glassy carbon working electrode set at 750 mV referenced to an AG/AgCl salt-bridge electrode (Waters, Model 2465) and quantified by comparison of the area under peak to a standard curve.

GABA determination

Following DA analysis, the supernatant was stored at −80 °C until GABA analysis. A portion of the samples were neutralized with NaOH (final concentration ∼0.1 mol) and underwent derivatization with o-phthaldialdehyde as previously described [18]. The derivatized supernatant was injected into the HPLC unit. The flow rate was 1.0 mL/min and the mobile phase contained the following: 0.1 mol sodium phosphate, 0.5 mmol EDTA, 25% methanol, pH 4.5 [17]. Amino acids were separated on a SunFire C_{18} 4.6×100 mm column with a 3.5 μmol particle size (Waters). GABA was measured using electrochemical detection with a glassy carbon working electrode set at 850 mV.

Statistics

ANOVA was used to determine the presence of significant differences between groups and the Tukey/Kramer post hoc test was used to determine specific significant differences. $P < 0.05$ was deemed significant for all tests.

Results

Control DA and GABA

Striatal DA and GABA levels in control animals were 53.9 ± 4.2 (ng/μg protein ± SEM) and 727.0 ± 58.6, respectively. DOPAC and HVA levels were 13.5 ± 1.1 and 5.5 ± 0.6, respectively.

Contralateral DA and GABA

Striatal ICH resulted in a significant ($p < 0.05$) up-regulation of contralateral DA (Fig. 1A) on days 1 ($177 \pm 24\%$ of control ± SEM), 3 ($313 \pm 15\%$), 7 ($220 \pm 9\%$), and 28 ($361 \pm 18\%$). Significant but less-marked increases were also observed in contralateral GABA (Fig. 1C) on days 1 ($154 \pm 5\%$) and 28 ($166 \pm 10\%$).

Ipsilateral DA and GABA

Significant DA increases also occurred ipsilateral to ICH (Fig. 1B) on days 3 ($212 \pm 25\%$ of control) and 28 ($276 \pm 12\%$). Differences in ipsilateral GABA were not significant (Fig. 1D).

DA metabolites

Significant differences in DOPAC were observed in the contralateral striatum (Fig. 1E) on day 1 ($57 \pm 4\%$ of control) and in the ipsilateral striatum (Fig. 1F) on day 7 ($29 \pm 10\%$). Additionally, significant differences in contralateral HVA (Fig. 1G) were noted on day 7 ($268 \pm 31\%$) and in ipsilateral HVA (Fig. 1H) on day 1 ($180 \pm 12\%$) and 7 ($175 \pm 17\%$).

Discussion

The case report by Matsuda et al. [12] raised the possibility that a hematoma along the nigrostriatal tract could induce behavioral deficits that improve with dopamine replacement therapy. Further, the report by Ling et al. [10] suggested that behavioral deficits indicative of parkinsonism associated with intracerebral hemorrhage improve with dopamine replacement. To the best of our knowledge, this is the first report examining alterations in striatal dopamine levels after ICH.

It may have been expected that dopamine levels would decrease after ICH because of iron release, oxidative stress, and subsequent damage to termi-

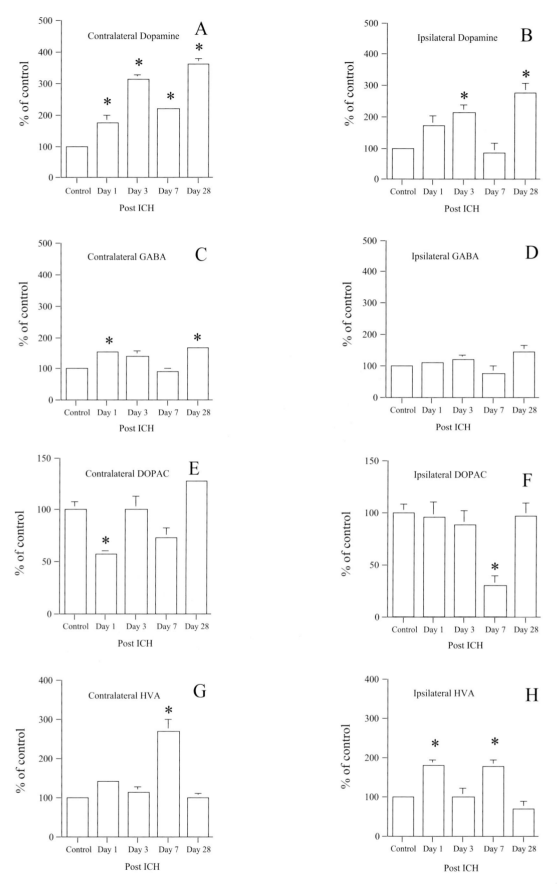

Fig. 1. Contralateral and ipsilateral alterations in striatal DA and GABA 1 to 28 days after ICH. (A) Contralateral dopamine; (B) Ipsilateral dopamine; (C) Contralateral GABA; (D) Ipsilateral GABA; (E) Contralateral DOPAC; (F) Ipsilateral DOPAC; (G) Contralateral HVA; (H) Ipsilateral HVA. Results expressed as % of control (normal animals). Error bars indicate SEM. *$p < 0.05$ vs. control

nals. However, dopamine levels significantly increased greater than 2-fold above control ipsilateral to ICH until at least 28 days. Ipsilateral GABA alterations were not significant, indicating that the DA system is preferentially affected after ICH. The changes in ipsilateral DA may reflect an attempt to compensate for damage to cell bodies, axons, or synaptic connections affecting dopaminergic transmission. Additionally, the presence of contralateral DA increases may represent compensation for damage to the ipsilateral striatum. Interpreting these alterations in DA will clearly require additional research, as the methods utilized here measured gross striatal neurotransmitter levels. It may be that there are regional differences in DA after ICH. Particularly, differences may exist between the necrotic core, penumbra, and the non-affected region.

In order to identify both regional and temporal changes to the DA system after ICH, positron emission tomography could be employed. The use of a scanner designed for rodents would provide sufficient resolution (~ 1–2 mm) to discern the hematoma. Overlay with MRI images would provide additional anatomical resolution. Specific tracers could be utilized to quantify damage to DA terminals and changes in DA receptor expression [6, 14]. Results from these studies would help determine if the DA system is a possible target for therapeutic modulation after ICH.

In animal models of PD, levels of both of the DA metabolites, DOPAC and HVA, are altered. Within the first few hours after 6-hydroxydopamine lesioning, both DOPAC and HVA increase in a similar fashion to DA [2]. However, in both the rat 6-hydroxydopamine and 1-methyl-4-phenyl-1,2,3,6-tetrahydropyridine models of PD, long-term decreases in striatal DOPAC and HVA occur, but are less marked than DA depletion [1, 13]. Additionally, these metabolites are also reduced in the human parkinsonian brain [19]. The ratio of DOPAC/DA increases after lesioning, where it is thought that surviving DA neurons increase DA production and resulting turnover as a compensatory mechanism [23]. Here we report significant decreases in contralateral DOPAC on day 1 and ipsilateral DOPAC on day 7. Additionally, significant increases in HVA occurred on day 1 (ipsilateral) and day 7 (ipsilateral and contralateral). While the neurochemical profile differs significantly from that in a 6-hydroxydopamine model, it is clear that long-term alterations in DA metabolism are evident. Since HVA is generally considered the final metabolite in DA catabolism, an increase on day 7 associated with DA and DOPAC decrease may indicate cell damage or altered metabolism. By day 28, DA metabolites return to control levels, with only DA remaining elevated. Here, the nigrostriatal DA system may be able to maintain long-term viability through increased production of DA, which is not the case in PD.

Because those who survive ICH often suffer severe neurological deficits, the identification of novel therapeutics remains a high priority. The current research has shown that the dopamine system is affected to a greater degree than the GABA system after striatal ICH. Further research is needed to determine if these changes in the DA system are specific and responsible for observed behavioral deficits.

Acknowledgments

This work was supported by grants NS-39866 and NS-47245 (GX) from the National Institutes of Health. JRC received support from the institutional training grant NIEHS T32ES07062 (RJR).

References

1. Altar CA, Marien MR, Marshall JF (1987) Time course of adaptations in dopamine biosynthesis, metabolism, and release following nigrostriatal lesions: implications for behavioral recovery from brain injury. J Neurochem 48: 390–399
2. Altar CA, O'Neil S, Marshall JF (1984) Sensorimotor impairment and elevated levels of dopamine metabolites in the neostriatum occur rapidly after intranigral injection of 6-hydroxydopamine or gamma-hydroxybutyrate in awake rats. Neuropharmacology 23: 309–318
3. Broderick JP, Adams HP Jr, Barsan W, Feinberg W, Feldmann E, Grotta J, Kase C, Krieger D, Mayberg M, Tilley B, Zabramski JM, Zuccarello M (1999) Guidelines for the management of spontaneous intracerebral hemorrhage: A statement for healthcare professionals from a special writing group of the Stroke Council, American Heart Association. Stroke 30: 905–915
4. Carreno-Muller E, Herrera AJ, de Pablos RM, Tomas-Camardiel M, Venero JL, Cano J, Machado A (2003) Thrombin induces in vivo degeneration of nigral dopaminergic neurones along with the activation of microglia. J Neurochem 84: 1201–1214
5. Choi SH, Joe EH, Kim SU, Jin BK (2003) Thrombin-induced microglial activation produces degeneration of nigral dopaminergic neurons in vivo. J Neurosci 23: 5877–5886
6. Cicchetti F, Brownell AL, Williams K, Chen YI, Livni E, Isacson O (2002) Neuroinflammation of the nigrostriatal pathway during progressive 6-OHDA dopamine degeneration in rats monitored by immunohistochemistry and PET imaging. Eur J Neurosci 15: 991–998
7. Defer GL, Remy P, Malapert D, Ricolfi F, Samson Y, Degos JD (1994) Rest tremor and extrapyramidal symptoms after midbrain haemorrhage: clinical and 18F-dopa PET evaluation. J Neurol Neurosurg Psychiatry 57: 987–989
8. Hua Y, Schallert T, Keep RF, Wu J, Hoff JT, Xi G (2002) Behavioral tests after intracerebral hemorrhage in the rat. Stroke 33: 2478–2484

9. Inoue H, Udaka F, Takahashi M, Nishinaka K, Kameyama M (1997) Secondary parkinsonism following midbrain hemorrhage. Rinsho Shinkeigaku 37: 266–269 [in Japanese]
10. Ling MJ, Aggarwal A, Morris JG (2002) Dopa-responsive parkinsonism secondary to right temporal lobe haemorrahage. Mov Disord 17: 402–404
11. Manno EM, Atkinson JL, Fulgham JR, Wijdicks EF (2005) Emerging medical and surgical management strategies in the evaluation and treatment of intracerebral hemorrhage. Mayo Clin Proc 80: 420–433
12. Matsuda W, Sugimoto K, Sato N, Watanabe T, Yanaka K, Matsumura A, Nose T (1999) A case of primary brain-stem injury recovered from persistent vegetative state after L-dopa administration. No To Shinkei 51: 1071–1074 [in Japanese]
13. Muramatsu Y, Kurosaki R, Watanabe H, Michimata M, Matsubara M, Imai Y, Araki T (2003) Cerebral alterations in a MPTP-mouse model of Parkinson's disease – an immunocytochemical study. J Neural Transm 110: 1129–1144
14. Nikolaus S, Larisch R, Beu M, Forutan F, Vosberg H, Muller-Gartner HW (2003) Bilateral increase in striatal dopamine D2 receptor density in the 6-hydroxydopamine-lesioned rat: a serial in vivo investigation with small animal PET. Eur J Nucl Med Mol Imaging 30: 390–395
15. Olanow CW (2004) The scientific basis for the current treatment of Parkinson's disease. Annu Rev Med 55: 41–60
16. Olanow CW, Tatton WG (1999) Etiology and pathogenesis of Parkinson's disease. Annu Rev Neurosci 22: 123–144
17. Rowley HL, Martin KF, Marsden CA (1995) Determination of in vivo amino acid neurotransmitters by high-performance liquid chromatography with o-phthalaldehyde-sulphite derivatisation. J Neurosci Methods 57: 93–99
18. Smith S, Sharp T (1994) Measurement of GABA in rat brain microdialysates using o-phthaldialdehyde-sulphite derivatization and high-performance liquid chromatography with electrochemical detection. J Chromatogr 652: 228–233
19. Sparks DL, Slevin JT (1985) Determination of tyrosine, tryptophan and their metabolic derivatives by liquid chromatography-electrochemical detection: application to post mortem samples from patients with Parkinson's and Alzheimer's disease. Life Sci 36: 449–457
20. Wu J, Hua Y, Keep RF, Nakamura T, Hoff JT, Xi G (2003) Iron and iron-handling proteins in the brain after intracerebral hemorrhage. Stroke 34: 2964–2969
21. Wu J, Hua Y, Keep RF, Schallert T, Hoff JT, Xi G (2002) Oxidative brain injury from extravasated erythrocytes after intracerebral hemorrhage. Brain Res 953: 45–52
22. Xi G, Fewel ME, Hua Y, Thompson J, BG, Hoff JT, Keep RF (2004) Intracerebral hemorrhage: Pathophysiology and therapy. Neurocritical Care 1: 5–18
23. Yuan H, Sarre S, Ebinger G, Michotte Y (2004) Neuroprotective and neurotrophic effect of apomorphine in the striatal 6-OHDA-lesion rat model of Parkinson's disease. Brain Res 1026: 95–107

Correspondence: Guohua Xi, Crosby Neurosurgical Laboratories, University of Michigan Medical School, R5550 Kresge I Bldg., 200 Zina Pitcher Place, Ann Arbor, MI 48109-0532, USA. e-mail: guohuaxi@umich.edu

Intracerebral hemorrhage in complement C3-deficient mice

S. Yang[1,4], T. Nakamura[1], Y. Hua[1], R. F. Keep[1,2], J. G. Younger[3], J. T. Hoff[1], and G. Xi[1]

[1] Department of Neurosurgery, University of Michigan Medical School, Ann Arbor, Michigan, USA
[2] Department of Physiology, University of Michigan Medical School, Ann Arbor, Michigan, USA
[3] Department of Emergency Medicine, University of Michigan Medical School, Ann Arbor, Michigan, USA
[4] Zhejiang University School of Medicine, Hangzhou, China

Summary

The complement cascade is activated and contributes to brain damage after intracerebral hemorrhage (ICH). The present study investigated ICH-induced brain damage in complement C3-deficient mice.

This study was divided into 2 parts. Male C3-deficient and C3-sufficient mice received an infusion of 30-μl autologous whole blood into the right basal ganglia. In the first part of our study, mice were killed 3 days later for brain water content measurement. Behavioral assessments including forelimb use asymmetry and corner turn tests were also preformed before and after ICH. In the second part of the study, brain heme oxygenase-1 (HO-1) was measured by Western blot analysis and immunohistochemistry 3 days after the infusion. We found that brain water content in the ipsilateral basal ganglia 3 days after ICH was less in C3-deficient mice compared to C3-sufficient mice ($p < 0.05$). The C3-deficient mice had reduced ICH-induced forelimb use asymmetry deficits compared with C3-sufficient mice ($p < 0.05$), although there was no significant difference in the corner turn test score. Western blot analysis showed that HO-1 contents were significantly lower in C3-deficient mice (day 3: 2024 ± 560 vs. 5140 ± 1151 pixels in the C3-sufficient mice, $p < 0.05$). We conclude that ICH causes less brain edema and behavioral deficits in complement C3-deficient mice. These results suggest that complement C3 is a key factor contributing to brain injury following ICH.

Keywords: Cerebral hemorrhage; complement C3; brain edema; hemeoxygenase-1; mice.

Introduction

Complement cascade activation occurs in the brain following intracerebral hemorrhage (ICH) and contributes to perihematomal edema formation. Complement activation and complement-mediated brain injury are also found in a variety of central nervous system diseases, including brain trauma, cerebral ischemia, and subarachnoid hemorrhage [6, 15]. Previous studies have demonstrated that complement depletion with cobra venom factor or complement inhibition with N-acetylheparin reduces brain edema and inflammatory responses after ICH [4, 23]. Recently, we developed behavioral tests for ICH in rats and mice to quantify the effects of ICH on neurological function [5, 16]. It remains unclear whether or not functional outcomes after ICH can be improved by manipulation of the complement system.

Heme oxygenase-1 (HO-1), also known as heat shock protein 32, can be induced in the brain by oxidative stress, heme, and hemoglobin, by both transcriptional and translational mechanisms [13, 20]. HO-1 expression is up-regulated after intracerebral hemorrhage and is primarily located in glial cells [7, 14, 21]. The biological significance of HO-1 up-regulation is still uncertain, but ICH-induced brain damage is attenuated by heme oxygenase inhibitors [7, 9, 19].

The aim of the present study was to examine ICH-induced brain damage in complement C3-deficient mice. Brain edema and functional outcomes were measured. Brain HO-1 levels around the clot were also examined.

Materials and methods

Animal preparation and ICH model

The University of Michigan Committee on the Use and Care of Animals approved the animal protocols. The experiments used 24 male C57BL/6 C3-deficient mice (from Dr. John Younger's laboratory, University of Michigan) and 24 male C57BL/6J C3-sufficient mice (from Jackson Laboratory, Bar Harbor, ME, USA), all approximately 2 to 3 months of age. Mice were allowed free access to food and water. The animals were anesthetized with ketamine (90 mg/kg, i.p.) and xylazine (5 mg/kg, i.p.), and the right femoral artery was catheterized using a PE10 tube to monitor arterial blood pressure and to sample blood for intracerebral infusion. Blood pH, PaO_2, $PaCO_2$, hematocrit, and glucose levels were monitored. Rectal temperature was maintained at 37.5 °C using a feedback-controlled heating pad. The mice were positioned in a stereotaxic frame (Model 500, Kopf Instruments, Tujunga, CA, USA) and a cranial burr hole (1 mm) was drilled near the right coronal suture 2.5 mm lateral to midline. A 26-gauge needle was inserted stereotaxically into the right basal ganglia (coordinates: 0.2 mm anterior, 3.5 mm ventral, and 2.5 mm lateral to the bregma). Autologous whole blood (30 μL) was infused at a rate of 2 μL/min using a micro-infusion pump (Harvard Apparatus Inc, South Natick, MA, USA). The needle was removed, the burr hole was filled with bone wax, and the skin incision was closed with suture.

Experimental group

There were 2 parts to this study. In the first part, mice were killed after 3 days for brain water and ion content measurement (n = 6). Behavioral assessment including forelimb use asymmetry and corner turn tests were performed before ICH and 1 and 3 days after ICH (n = 6). In the second part, animals were killed 1 and 3 days after ICH. The brains were prepared for Western blot analysis (n = 3 per time point) and immunohistochemistry (n = 3 per time point).

Brain water and ion content

Animals were anesthetized (ketamine 120 mg/kg and xylazine 5 mg/kg, i.p.) and decapitated 3 days after intracerebral blood injection to determine brain water and ion contents. The brains were removed and a blade was used to cut a coronal brain slice (about 2 mm thick) that was located 2 mm from the frontal pole. The brain slice was divided into 2 hemispheres along the midline. Each hemisphere was dissected into the cortex and the basal ganglia. The cerebellum served as a control. Five samples from each brain were obtained: ipsilateral and contralateral cortex, ipsilateral and contralateral basal ganglia, and cerebellum. Brain samples were weighed immediately on an electronic analytical balance (Model AE 100, Mettler Instrument Co, Highstown, NJ, USA) to obtain the wet weight. The brain samples were then dried at 100 °C for 24 hours to obtain the dry weight. The formula for our calculation was the following: (wet weight − dry weight)/wet weight. The dehydrated samples were digested in 1 mL of 1 mol/L nitric acid for 1 week, after which the Na^+ and K^+ contents of this solution were measured using a flame photometer (Model IL 943, Instrumentation Laboratory, Inc., Lexington, MA, USA). The ion content was expressed in micro-equivalents per gram of dehydrated brain tissue (μEq/g dry weight).

Behavioral tests

Two behavioral tests were used: a forelimb use asymmetry test and a corner turn test [5, 16, 25]. Forelimb use during exploratory activity was analyzed in a standing transparent cylinder. Behavior was scored according to the following criteria: independent use of the left or right forelimb for contacting the wall during a full rear to initiate a weight-shifting movement and simultaneous use of both the left and right forelimbs to contact the wall. Behavior was quantified by determining the number of times the normal ipsilateral (I) forelimb, the impaired contralateral (C) forelimb, and both (B) forelimbs were used as a percentage of total number of limb usage. A single, overall limb-use asymmetry score was calculated as follows: forelimb-use asymmetry score = $[I/(I + C + B)] - [C/(I + C + B)]$.

The second behavioral analysis involved a corner turn test. The mouse was allowed to proceed into a corner with an angle of 30°. To exit the corner, the animal could turn either left or right. When the mouse turned, its choice of direction was recorded. This was repeated 10 to 15 times, and the percentage of right turns was calculated.

Western blot analysis

Animals were anesthetized before undergoing intracardiac perfusion with 0.1 mol phosphate-buffered saline. The brains were then removed and a 2-mm thick coronal brain slice was cut approximately 2 mm from the frontal pole. The slice was separated into ipsilateral and contralateral basal ganglia. Western blot analysis was performed as previously described [22]. Briefly, 50 μg proteins for each were separated by sodium dodecyl sulfate polyacrylamide gel electrophoresis and transferred to a Hybond-C pure nitrocellulose membrane (Amersham Biosciences, Piscataway, NJ, USA). The membranes were blocked in Carnation nonfat milk. Membranes were probed with a 1:2000 dilution of the primary antibody, rabbit anti-heme-oxygenase-1 polyclonal antibody (Stressgen Biotechnologies, San Diego, CA) and a 1:2500 dilution of the second antibody, peroxidase-conjugated goat anti-rabbit antibody (Vector Laboratories, Inc., Burlingame, CA, USA). The antigen-antibody complexes were visualized with a chemiluminescence system (Amersham) and exposed to Kodak X-OMAT film (Rochester, NY, USA). The relative densities of bands were analyzed with the NIH Image (version 1.61, National Institutes of Health, Bethesda, MD).

Immunohistochemical staining

Mice were anesthetized and underwent intracardiac perfusion with 4% paraformaldehyde in 0.1 mol/L (pH 7.4) phosphate-buffered saline. The brains were removed and kept in 4% paraformaldehyde for 24 hours, then immersed in 30% sucrose for 3 to 4 days at 4 °C. Brains were then placed in embedding OCT compound (Sakura Finetek USA, Inc., Torrance, CA) and sectioned on a cryostat (18 μm thick). Using the avidin-biotin complex technique [22], sections were incubated overnight with the primary antibody. The primary antibodies were rabbit anti-heme-oxygenase-1 polyclonal antibody (Stressgen). Normal rabbit IgG was used as negative control.

Statistical analysis

All data in this study are presented as mean ± SD. Data were analyzed with Student *t* test and analysis of variance (ANOVA), followed by Scheffe post hoc test. Significance levels were measured at $p < 0.05$.

Results

Brain water contents in the ipsilateral basal ganglia and in the ipsilateral cortex were less in complement C3-deficient mice compared to complement C3-sufficient mice 3 days after ICH (basal ganglia: 79.5 ± 0.4 vs. 80.6 ± 1.0%, p < 0.05; cortex: 78.6 ± 0.6 vs. 79.6 ± 0.6%, p < 0.05; Fig. 1). Reduced brain edema in complement C3-deficient mice was associated with less accumulation of sodium ion (basal ganglia: 178 ± 17 vs. 238 ± 11 μEq/g dry weight,

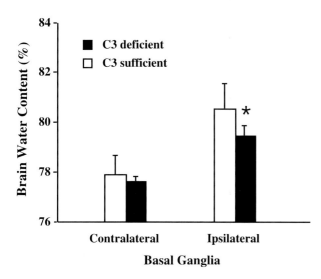

Fig. 1. Brain water contents in complement C3-deficient and C3-sufficient mice 3 days after intracerebral hemorrhage. Values are mean ± SD, n = 6. *$p < 0.05$ vs. C3-sufficient group

$p < 0.05$; cortex: 160 ± 15 vs. 202 ± 15 µEq/g dry weight, $p < 0.05$) and loss of potassium ion (basal ganglia: 370 ± 21 vs. 338 ± 28 µEq/g dry weight, $p < 0.05$; cortex: 408 ± 15 vs. 376 ± 18 µEq/g dry weight, $p < 0.05$) than in complement C3-sufficient mice. There was no difference in water or ion content in the contralateral basal ganglia, the contralateral cortex, and the cerebellum between complement C3-deficient and complement C3-sufficient mice.

Behavioral tests were performed before, as well as 1 day and 3 days after ICH. Complement C3-deficient mice had reduced ICH-induced forelimb use asymmetry deficits compared with complement C3-sufficient mice (day 1: 35.0 ± 5.0 vs. 58.1 ± 12.3% in the C3-sufficient mice, $p < 0.05$; day 3: 29.0 ± 8.2 vs. 47.9 ± 14.3% in the C3-sufficient mice, $p < 0.05$), although there was no difference in the corner turn test score (day 1: 90.0 ± 10.0 vs. 97.5 ± 14.0% in the C3-sufficient mice, $p > 0.05$; day 3: 82.0 ± 11.0 vs. 80.0 ± 10.7% in the C3-sufficient mice, $p > 0.05$).

HO-1 is an inducible enzyme for heme degradation. Brain HO-1 levels were examined by Western blot analysis and immunohistochemistry. Western blots showed that HO-1 contents in the ipsilateral basal ganglia 3 days after ICH were lower in the complement C3-deficient mice (2024 ± 560 vs. 5140 ± 1151 pixels in the complement C3-sufficient mice, $p < 0.05$). HO-1-positive cells in the perihematomal zone were also fewer in complement C3-deficient mice than those in complement-sufficient mice.

Discussion

We found less ICH-induced brain edema and fewer forelimb use asymmetry deficits in complement C3-deficient mice compared with C3-sufficient mice. Brain HO-1 levels after ICH were also lower in C3-deficient mice. These results indicate that complement C3 is an important factor causing brain damage following ICH.

Complement is normally excluded from the brain parenchyma by the blood-brain barrier (BBB), but entry can occur after ICH as part of the extravasated blood or later because of BBB disruption. Earlier we found that the complement cascade is activated in brain parenchyma after ICH [4]. Although the precise mechanisms are not clear, complement-related brain injury may be due to the classic inflammatory response and membrane attack complex (MAC) formation.

Inflammation in brain parenchyma adjacent to the hematoma has been found following experimental ICH and is probably complement-mediated [3]. After activation of the complement cascade, anaphylatoxins C3a and C5a are generated and may can cause BBB leakage by degranulating mast cells and leukocytes. We have demonstrated that systemic complement depletion by cobra venom factor, a non-toxic protein found in cobra venom, reduces tumor necrosis factor-alpha (TNF-α) levels and myeloperoxidase positive cells [23].

MAC is assembled following complement activation [2]. MAC formation can cause the formation of a pore in the cell membrane leading to early lysis of red blood cells (RBC). Our current study found that brain HO-1 levels around the clot are lower in complement C3-deficient mice. Hemoglobin and heme are important HO-1 inducers. A delay in RBC lysis in C3-deficient mice could result in lower brain HO-1 levels. However, whether RBC lysis is delayed in C3-deficient mice should be studied further.

We found that less perihematomal brain edema and reduced forelimb use asymmetry deficits were associated with lower brain HO-1 levels after ICH in the C3-deficient mice. Heme from hemoglobin, which is released after RBC lysis, is degraded by heme oxygenase in the brain into iron, carbon monoxide, and biliverdin. Biliverdin is then converted to bilirubin by biliverdin reductase [10]. Investigations have shown that hemoglobin causes brain injury through its degradation products and heme oxygenase is a key enzyme in ICH-induced injury [7, 9, 19].

The coagulation cascade, particularly thrombin activation, is another important factor causing early edema formation after ICH [11, 24]. Intraparenchymal infusion of thrombin causes BBB disruption [12] and inflammation [17]. However, it is not clear whether thrombin-induced brain injury is partly complement-mediated. Studies suggest that there is a close relationship between thrombin and the complement cascade. For example, thrombin can cleave and activate complement C3, and thrombin-cleaved C3a-like fragments are chemotactic for leukocytes and induce enzyme release from neutrophils [1, 8].

We found less ICH-mediated brain damage in complement C3 mice, but an earlier study found that brain edema induced by ICH was greater and neurological deficits were worse in complement C5-deficient mice [16]. These results indicate different complement factors have either beneficial or harmful effects after ICH. For instance, Pasinetti *et al.* [18] reported increased kainic acid-induced hippocampal damage in C5-deficient mice. It should be noted that C5-deficient mice prevent the formation of C5a fragments and MAC complex formation, but do not prevent formation of C3a and C3b. In contrast, the C3-deficient mice prevent the formation of C3a, C3b, C5a, and MAC.

In conclusion, ICH results in less HO-1 up-regulation and less brain damage in complement C3-deficient mice, suggesting that C3 is a key detrimental factor during complement activation.

Acknowledgments

This study was supported by grants NS-017760, NS-039866, and NS-047245 from the National Institutes of Health, and the Scientist Development Grant 0435354Z from American Heart Association.

References

1. Bokisch VA, Muller-Eberhard HJ, Cochrane CG (1969) Isolation of a fragment (C3a) of the third component of human complement containing anaphylatoxin and chemotactic activity and description of an anaphylatoxin inactivator of human serum. J Exp Med 129: 1109–1130
2. Esser AF (1991) Big MAC attack: complement proteins cause leaky patches. Immunol Today 12: 316–318
3. Gong C, Hoff JT, Keep RF (2000) Acute inflammatory reaction following experimental intracerebral hemorrhage. Brain Res 871: 57–65
4. Hua Y, Xi G, Keep RF, Hoff JT (2000) Complement activation in the brain after experimental intracerebral hemorrhage. J Neurosurg 92: 1016–1022
5. Hua Y, Schallert T, Keep RF, Wu J, Hoff JT, Xi G (2002) Behavioral tests after intracerebral hemorrhage in the rat. Stroke 33: 2478–2484
6. Huang J, Kim LJ, Mealey R, Marsh HC Jr, Zhang Y, Tenner AJ, Connolly ES Jr, Pinsky DJ (1999) Neuronal protection in stroke by an sLex-glycosylated complement inhibitory protein. Science 285: 595–599
7. Huang F, Xi G, Keep RF, Hua Y, Nemoianu A, Hoff JT (2002) Brain edema after experimental intracerebral hemorrhage: role of hemoglobin degradation products. J Neurosurg 96: 287–293
8. Hugli TE (1977) Complement factors and inflammation: Effects of thrombin on components of C3 and C5. In: Lundblad RL, Fenton JW, Mann KG (eds) Chemistry and biology of thrombin. Ann Arbor Science Publishers, Ann Arbor, MI, p 345–360
9. Koeppen AH, Dickson AC, Smith J (2004) Heme oxygenase in experimental intracerebral hemorrhage: the benefit of tin-mesoporphyrin. J Neuropathol Exp Neurol 63: 587–597
10. Kutty RK, Maines MD (1981) Purification and characterization of biliverdin reductase from rat liver. J Biol Chem 256: 3956–3962
11. Lee KR, Colon GP, Betz AL, Keep RF, Kim S, Hoff JT (1996) Edema from intracerebral hemorrhage: the role of thrombin. J Neurosurg 84: 91–96
12. Lee KR, Kawai N, Kim S, Sagher O, Hoff JT (1997) Mechanisms of edema formation after intracerebral hemorrhage: effects of thrombin on cerebral blood flow, blood-brain barrier permeability, and cell survival in a rat model. J Neurosurg 86: 272–278
13. Maines MD (1988) Heme oxygenase: function, multiplicity, regulatory mechanisms, and clinical applications. FASEB J 2: 2557–2568
14. Matz PG, Weinstein PR, Sharp FR (1997) Heme oxygenase-1 and heat shock protein 70 induction in glia and neurons throughout rat brain after experimental intracerebral hemorrhage. Neurosurgery 40: 152–160
15. Morgan BP, Gasque P, Singhrao S, Piddlesden SJ (1997) The role of complement in disorders of the nervous system. Immunopharmacology 38: 43–50
16. Nakamura T, Xi G, Hua Y, Schallert T, Hoff JT, Keep RF (2004) Intracerebral hemorrhage in mice: Model characterization and application for genetically modified mice. J Cereb Blood Flow Metab 24: 487–495
17. Nishino A, Suzuki M, Ohtani H, Motohashi O, Umezawa K, Nagura H, Yoshimoto T (1993) Thrombin may contribute to the pathophysiology of central nervous system injury. J Neurotrauma 10: 167–179
18. Pasinetti GM, Tocco G, Sakhi S, Musleh WD, DeSimoni MG, Mascarucci P, Schreiber S, Baudry M, Finch CE (1996) Hereditary deficiencies in complement C5 are associated with intensified neurodegenerative responses that implicate new roles for the C-system in neuronal and astrocytic functions. Neurobio Dis 3: 197–204
19. Wagner KR, Hua Y, de Courten-Myers GM, Broderick JP, Nishimura RN, Lu SY, Dwyer BE (2000) Tin-mesoporphyrin, a potent heme oxygenase inhibitor, for treatment of intracerebral hemorrhage: in vivo and in vitro studies. Cell Mol Biol (Noisy-le-grand) 46: 597–608
20. Wagner KR, Sharp FR, Ardizzone TD, Lu A, Clark JF (2003) Heme and iron metabolism: Role in cerebral hemorrhage. J Cereb Blood Flow Metab 23: 629–652
21. Wu J, Hua Y, Keep RF, Nakamura T, Hoff JT, Xi G (2003) Iron and iron-handling proteins in the brain after intracerebral hemorrhage. Stroke 34: 2964–2969
22. Xi G, Keep RF, Hua Y, Xiang JM, Hoff JT (1999) Attenuation of thrombin-induced brain edema by cerebral thrombin preconditioning. Stroke 30: 1247–1255

23. Xi G, Hua Y, Keep RF, Younger JG, Hoff JT (2001) Systemic complement depletion diminishes perihematomal brain edema. Stroke 32: 162–167
24. Xi G, Keep RF, Hoff JT (2002) Pathophysiology of brain edema formation. Neurosurg Clin N Am 13: 371–383
25. Zhang L, Schallert T, Zhang ZG, Jiang Q, Arniego P, Li Q, Lu M, Chopp M (2002) A test for detecting long-term sensorimotor dysfunction in the mouse after focal cerebral ischemia. J Neurosci Meth 117: 207–214

Correspondence: Guohua Xi, Department of Neurosurgery, University of Michigan Medical School, R5550 Kresge I, Ann Arbor, MI 48109-0532, USA. e-mail: guohuaxi@umich.edu

Systemic zinc protoporphyrin administration reduces intracerebral hemorrhage-induced brain injury

Y. Gong[1,2], H. Tian[1], G. Xi[1], R. F. Keep[1], J. T. Hoff[1], and Y. Hua[1]

[1] Department of Neurosurgery, University of Michigan Medical School, Ann Arbor, Michigan, USA
[2] Department of Neurosurgery, Huashan Hospital, Shanghai, China

Summary

Hemoglobin degradation products result in brain injury after intracerebral hemorrhage (ICH). Recent studies found that intracerebral infusion of heme oxygenase inhibitors reduces hemoglobin- and ICH-induced brain edema in rats and pigs. The present study examined whether systemic use of zinc protoporphyrin (ZnPP), a heme oxygenase inhibitor, can attenuate brain edema, behavioral deficits, and brain atrophy following ICH.

All rats had intracerebral infusion of 100-μL autologous blood. ZnPP (1 nmol/hour/rat) or vehicle was given immediately or 6 hours following ICH. ZnPP was delivered intraperitoneally up to 14 days through an osmotic mini-pump. Rats were killed at day 3 and day 28 after ICH for brain edema and brain atrophy measurements, respectively. Behavioral tests were performed. We found that ZnPP attenuated brain edema in animals sacrificed 3 days after ICH ($p < 0.05$). ZnPP also reduced ICH-induced caudate atrophy ($p < 0.05$) and ventricular enlargement ($p < 0.05$). In addition, ZnPP given immediately or 6 hours after ICH improved neurological deficits ($p < 0.05$). In conclusion, systemic zinc protoporphyrin treatment started at 0 or 6 hours after ICH reduced brain edema, neurological deficits, and brain atrophy after ICH. These results indicate that heme oxygenase may be a new target for ICH therapeutics.

Keywords: Intracerebral hemorrhage; heme oxygenase inhibitor; zinc protoporphyrin.

Introduction

Hemoglobin and its degradation product, iron, contribute to brain damage after intracerebral hemorrhage (ICH). Iron overload in the brain has been found in a rat ICH model, with a 3-fold increase in non-heme iron [13]. We also demonstrated that intracerebral infusion of iron causes brain edema and an iron chelator reduces hematoma- and hemoglobin-induced edema, suggesting that iron plays an important role in edema formation after ICH [4, 7]. Heme oxygenase is the key enzyme of hemoglobin degradation, which metabolizes heme to iron, carbon monoxide, and biliverdin in brain. Biliverdin is then converted to bilirubin by biliverdin reductase [6]. Recent studies found that heme oxygenase-1 is up-regulated in the brain after ICH [12] and intracerebral infusion of heme oxygenase inhibitors reduces hemoglobin- and ICH-induced brain edema in rats and pigs. The present study examined whether systemic use of zinc protoporphyrin (ZnPP), a heme oxygenase inhibitor, can attenuate brain edema, behavioral deficits, and brain atrophy following ICH.

Materials and methods

Animal preparation and intracerebral infusion

The University of Michigan Committee on the Use and Care of Animals approved the protocols for these animal studies. A total of 56 male Sprague-Dawley rats (weighing 300 to 400 g, Charles River Laboratories, Wilmington, MA) were used in the present study. Aseptic precautions were utilized in all surgical procedures. Animals were anesthetized with pentobarbital (50 mg/kg, i.p.). The right femoral artery was catheterized for continuous blood pressure monitoring and blood sampling. Blood was obtained from the catheter for analysis of pH, PaO_2, $PaCO_2$, hematocrit and glucose and as the source for the intracerebral blood infusion. Body temperature was maintained at 37.5 °C using a feedback-controlled heating pad. The animals were positioned in a stereotactic frame (David Kopf Instruments, Tujunga, CA) and a cranial burr hole (1 mm) was drilled on the right coronal suture 4.0 mm lateral to the midline. Autologous blood was withdrawn from the right femoral artery and infused (100 μL) immediately into the right caudate nucleus through a 26-gauge needle at a rate of 10 μL per minute using a microinfusion pump (Harvard Apparatus Inc., Holliston, MA). The coordinates were 0.2 mm anterior, 5.5 mm ventral, and 4.0 mm lateral to the bregma. After intracerebral infusion, the needle was removed and the skin incision closed with suture.

Experimental groups

All rats had intracerebral infusion of 100 μL autologous blood. An intraperitoneal osmotic mini-pump was implanted immediately or 6 hours after ICH to deliver vehicle or the heme oxygenase inhibitor, ZnPP (1 nmol/h), for up to 14 days. Rats were killed at days 3 (n = 6 for each group) for brain water content measurement and day 28 (n = 8 for each group) for brain atrophy examination. Behavioral tests were performed at days 1, 3, 5, 7, 14, 21, and 28.

Brain water content measurement

Rats were killed by decapitation under deep pentobarbital anesthesia (60 mg/kg, i.p.). The brains were removed immediately and a coronal brain slice (approximately 4 mm thick) 3 mm from the frontal pole was cut with a blade. The brain samples were then divided into cortex or basal ganglia (ipsilateral or contralateral). A total of 5 samples from each brain were obtained: the ipsilateral cortex and basal ganglia, the contralateral cortex and basal ganglia, and cerebellum. Brain water content was measured by the wet/dry weight method and sodium ion content was measured by flame photometry [14].

Histology

Rats were anesthetized with pentobarbital (60 mg/kg, i.p.) and perfused with 4% paraformaldehyde in 0.1 mol, pH 7.4 phosphate-buffered saline. The brains were removed and kept in 4% paraformaldehyde for 4 to 6 hours, then immersed in 25% sucrose for 3 to 4 days at 4 °C. The brains were embedded in O.C.T compound (Sakura Finetek U.S.A. Inc., Torrance, CA) and sectioned on a cryostat (18 μm thick).

We estimated brain atrophy morphometrically [1]. Coronal sections from 1 mm posterior to the blood injection site were stained with hematoxylin and eosin. The brain sections were scanned. The caudate, cortex, and lateral ventricle were outlined on a computer and the outlined areas were measured using an NIH Image program (version 1.62, National Institutes of Health, Bethesda, MD). All measurements were repeated 3 times and the average was used. To minimize the influence of tissue shrinkage, brain atrophy was expressed as a percentage of the contralateral area.

Behavioral tests

All animals were tested before and after surgery and scored by investigators who were blind to both neurological and treatment conditions. Three behavioral tests were used: the forelimb-placing test, the forelimb-use asymmetry (cylinder) test, and the corner-turn test [3].

Forelimb-placing test

Forelimb placing was scored using a vibrissae-elicited forelimb-placing test [9]. Independent testing of each forelimb was induced by brushing the vibrissae ipsilateral to that forelimb on the edge of a tabletop once per trial for 10 trials. Intact animals placed the forelimb quickly onto the countertop. Percent successful placing responses were determined. There was a reduction in successful responses in the forelimb contralateral to the site of injection after ICH [3].

Forelimb-use asymmetry test

Forelimb use during explorative activity was analyzed by videotaping rats in a transparent cylinder for 3 to 10 minutes, depending on the degree of activity during the trial [9]. Behavior was quantified first by determining the occasions when the non-impaired ipsilateral (I) forelimb was used as a percentage of total number of limb-use observations on the wall; second, the occasions when the impaired forelimb contralateral (C) to the blood-injection site was used as a percentage of total number of limb-use observations on the wall; and third, the occasions when both (B) forelimbs were used simultaneously as a percentage of total number of limb-use observations on the wall. A single overall limb-use asymmetry score was calculated as: Limb-use asymmetry score = $[I/(I + C + B)] - [C/(I + C + B)]$.

Corner-turn test

The rat was allowed to proceed into a corner, which was a 30° angle. To exit the corner, the rat could turn either to the left or the right and the direction was recorded [8]. The test was repeated 10 to 15 times, with at least 30 seconds between each trial, and the percentage of right turns was calculated. Only turns involving full rearing along either wall were included. The rats were not picked up immediately following each turn so that they did not develop an aversion for turning around.

Statistical analysis

Student t test and Mann-Whitney U test were used to compare brain edema, behavioral, and brain atrophy data. Values are mean ± SD. Statistical significance was set at $p < 0.05$.

Results

Physiological variables including mean arterial blood pressure, blood pH, PaO_2, $PaCO_2$, hematocrit, and blood glucose were measured and controlled within normal ranges (PaO_2, 70 to 120 mmHg; $PaCO_2$, 35 to 45 mmHg; pH, 7.40 to 7.50; mean arterial blood pressure, 70 to 110 mmHg; hematocrit, 35% to 45%; blood glucose 6 to 9 mmol/L).

Perihematomal brain edema peaked at 3 days after ICH. ZnPP attenuated brain edema in the ipsilateral basal ganglia 3 days after ICH (0-hour delay: 81.2 ± 0.6% vs. 82.1 ± 0.5% in the vehicle, $p < 0.05$ (Fig. 1); 6-hour delay: 81.1 ± 0.6% vs. 82.1 ± 0.8% in the vehicle, $p < 0.05$). ZnPP also reduced sodium accumulation in the ipsilateral basal ganglia (0-hour delay: 236.3 ± 64.2 vs. 427.0 ± 72.7 mEq/kg dry weight in the vehicle, $p < 0.05$; 6-hour delay: 345.6 ± 36.9 vs. 439.1 ± 107.6 mEq/kg dry weight in the vehicle, $p < 0.05$).

Brain atrophy developed in the ipsilateral basal ganglia several weeks after ICH. Caudate size and ventricle enlargement were measured 28 days after ICH. ZnPP reduced ICH-induced caudate atrophy (percentage of the contralateral caudate; 0-hour delay: 97.9 ± 4.0% vs. 88.0 ± 7.1% in the vehicle, $p < 0.05$; 6-hour delay: 95.1 ± 5.0% vs. 82.0 ± 7.9% in the vehicle, $p < 0.05$) and ventricular enlargement (percentage of the contralateral lateral ventricle; 0-hour delay: 198 ± 144% vs. 468 ± 281% in the vehicle, $p < 0.05$;

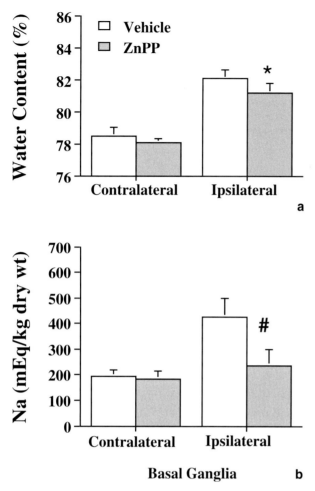

Fig. 1. Brain water (a) and sodium (b) contents in the basal ganglia 3 days after ICH. Values are mean ± SD. *p < 0.05; #p < 0.01 vs. vehicle

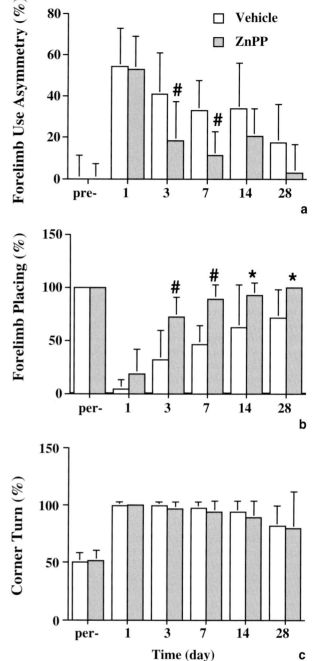

Fig. 2. (a–c) Forelimb-use asymmetry, forelimb-placing, and corner-turn score after ICH with or without ZnPP treatment. *p < .05; #p < 0.01 vs. vehicle

6-hour delay: 199 ± 121% vs. 651 ± 488% in the vehicle, p < 0.05).

ICH caused marked neurological deficits. ZnPP given immediately after ICH improved the forelimb-use asymmetry score (day 7: 9.9 ± 5.5% vs. 33.8 ± 9.2% in the vehicle group, p < 0.05) and forelimb-placing score (day 7: 95.0 ± 7.6% vs. 61.3 ± 36.8% in the vehicle group, p < 0.05). ZnPP did not improve ICH-induced corner-turn deficits. ZnPP given 6 hours after ICH also improved neurological deficits (Fig. 2).

Discussion

The present study demonstrated that systemic use of ZnPP, a heme oxygenase inhibitor, attenuates peri-hematomal edema, reduces ICH-induced brain atrophy, and improves functional outcomes following ICH. The results suggest that hemoglobin degradation products contribute to brain damage after ICH and heme oxygenase is a potential therapeutic target.

Hemoglobin and its degradation products are important factors causing perihematomal brain edema and behavioral deficits. Infusion of packed erythrocytes causes edema and neurological deficits several days later, suggesting that erythrocyte lysis and hemoglobin toxicity are associated with delayed brain injury [3, 14]. A clinical study of edema and ICH indicates that delayed brain edema is related to significant midline shift after ICH in man [16]. This delayed brain edema (in the second or third weeks after ictus in humans) is probably due to hemoglobin and its degradation products.

Intracerebral injection of heme oxygenase inhibitors reduces ICH-induced brain edema in rats and pigs. Our previous studies also demonstrate that an intracerebral infusion of hemoglobin and its degradation products, hemin, iron and bilirubin, cause brain edema formation within 24 hours. Hemoglobin itself induces heme oxygenase-1 (HO-1) up-regulation in the brain and heme oxygenase inhibition by the intracerebral infusion of tin-protoporphyrin (SnPP) reduces hemoglobin-induced brain edema [4]. Wagner et al. [11] found that an intracerebral injection of tin-mesoporphyrin (SnMP) reduced brain edema 24 hours after ICH in a pig model. The present study showed that ZnPP has similar effects when administered systemically. Our finding is supported by a very recent report in which systemic use of SnMP is neuroprotective in a rabbit ICH model [5]. It should be noted that metalloporphyrins may inhibit enzymes other than heme oxygenase [2]. However, a range of inhibitors including SnPP, SnMP, and ZnPP all can reduce ICH-induced brain damage, indicating that heme oxygenase inhibition is neuroprotective.

It has been difficult to obtain quantifiable markers of brain injury after ICH except perihematomal brain edema because neuronal injury appears to be diffuse (i.e., there is no defined infarct) and only a small cavity is found after the clot is absorbed. There have, however, been several recent approaches to improve quantification of therapeutic results after ICH. Behavioral tests can be used to assess brain injury after intracaudate injection of whole blood or blood components [3]. It has also been have demonstrated that brain atrophy occurs after ICH in rats and in humans [1, 10, 15]. Felberg et al. [1] reported that the volume of the ipsilateral striatum is reduced by 20% with an increase in ipsilateral ventricular size 100 days after experimental ICH. The present study showed that systemic use of ZnPP reduces brain atrophy and neurological deficits.

In conclusion, systemic zinc protoporphyrin treatment started at 0 or 6 hours after ICH reduced brain edema, neurological deficits, and brain atrophy after ICH. These results indicate that heme oxygenase may be a new target for ICH therapeutics.

Acknowledgments

This study was supported by grants NS-17760, NS-39866, and NS-47245 from the National Institutes of Health, and Scientist Development Grant 0435354Z from the American Heart Association.

References

1. Felberg RA, Grotta JC, Shirzadi AL, Strong R, Narayana P, Hill-Felberg SJ, Aronowski J (2002) Cell death in experimental intracerebral hemorrhage: the "black hole" model of hemorrhagic damage. Ann Neurol 51: 517–524
2. Grundemar L, Ny L (1997) Pitfalls using metalloporphyrins in carbon monoxide research. Trends Pharmacol Sci 18: 193–195
3. Hua Y, Schallert T, Keep RF, Wu J, Hoff JT, Xi G (2002) Behavioral tests after intracerebral hemorrhage in the rat. Stroke 33: 2478–2484
4. Huang F, Xi G, Keep RF, Hua Y, Nemoianu A, Hoff JT (2002) Brain edema after experimental intracerebral hemorrhage: role of hemoglobin degradation products. J Neurosurg 96: 287–293
5. Koeppen AH, Dickson AC, Smith J (2004) Heme oxygenase in experimental intracerebral hemorrhage: the benefit of tin-mesoporphyrin. J Neuropathol Exp Neurol 63: 587–597
6. Kutty RK, Maines MD (1981) Purification and characterization of biliverdin reductase from rat liver. J Biol Chem 256: 3956–3962
7. Nakamura T, Keep R, Hua Y, Schallert T, Hoff J, Xi G (2004) Deferoxamine-induced attenuation of brain edema and neurological deficits in a rat model of intracerebral hemorrhage. J Neurosurg 100: 672–678
8. Schallert T, Upchurch M, Wilcox RE, Vaughn DM (1983) Posture-independent sensorimotor analysis of inter-hemispheric receptor asymmetries in neostriatum. Pharmacol Biochem Behav 18: 753–759
9. Schallert T, Fleming SM, Leasure JL, Tillerson JL, Bland ST (2000) CNS plasticity and assessment of forelimb sensorimotor outcome in unilateral rat models of stroke, cortical ablation, parkinsonism and spinal cord injury. Neuropharmacology 39: 777–787
10. Skriver EB, Olsen TS (1986) Tissue damage at computed tomography following resolution of intracerebral hematomas. Acta Radiol Diag (Stockh) 27: 495–500
11. Wagner KR, Hua Y, de Courten-Myers GM, Broderick JP, Nishimura RN, Lu SY, Dwyer BE (2000) Tin-mesoporphyrin, a potent heme oxygenase inhibitor, for treatment of intracerebral hemorrhage: in vivo and in vitro studies. Cell Mol Biol (Noisy-le-grand) 46: 597–608
12. Wu J, Hua Y, Keep RF, Schallert T, Hoff JT, Xi G (2002) Oxidative brain injury from extravasated erythrocytes after intracerebral hemorrhage. Brain Res 953: 45–52
13. Wu J, Hua Y, Keep RF, Nakamura T, Hoff JT, Xi G (2003) Iron and iron-handling proteins in the brain after intracerebral hemorrhage. Stroke 34: 2964–29696
14. Xi G, Keep RF, Hoff JT (1998) Erythrocytes and delayed brain edema formation following intracerebral hemorrhage in rats. J Neurosurg 89: 991–996
15. Xi G, Fewel ME, Hua Y, Thompson BG, Hoff J, Keep R (2004) Intracerebral hemorrhage: pathophysiology and therapy. Neurocrit Care 1: 5–18

16. Zazulia AR, Diringer MN, Derdeyn CP, Powers WJ (1999) Progression of mass effect after intracerebral hemorrhage. Stroke 30: 1167–1173

Correspondence: Ya Hua, Department of Neurosurgery, University of Michigan Medical School, R5550 Kresge I Bldg., Ann Arbor, MI 48109-0532, USA. e-mail: yahua@umich.edu

Experimental Cerebral Ischemia

Restitution of ischemic injuries in penumbra of cerebral cortex after temporary ischemia

U. Ito[1,3], E. Kawakami[1], J. Nagasao[1], T. Kuroiwa[2], I. Nakano[3], and K. Oyanagi[1]

[1] Department of Neuropathology, Tokyo Metropolitan Institute for Neuroscience, Tokyo, Japan
[2] Department of Neuropathology, Medical Research Institute, Tokyo Medical and Dental University, Tokyo, Japan
[3] Department of Neurology, Jichi Medical School, Tochigi, Japan

Summary

We investigated, at both light and ultrastructural levels, the fate of swollen astrocytes and remodeling of neurites connected to disseminated, dying neurons in the ischemic neocortical penumbra. Specimens from left cerebral cortex were cut coronally at the infundibulum and observed by light and electron microscopy. We measured synapses and spines, and the thickness of neuritic trunks in the neuropil on electron microscopy photos. We also determined percent volume of axon terminals and spines by Weibel's point-counting method. Astrocytic swelling gradually subsided from day 4 after the ischemic insult, with increases in cytoplasmic glial fibrils and GFAP-positive astrocytes. Disseminated dying electron-dense neurons were fragmented by invading astrocytic cell processes and accumulated as granular pieces. The number of synapses and spines and total percent volume of axon terminals and spines decreased with an increasing sparsity of synaptic vesicles until day 4. One to 12 weeks after the ischemic insult, these values increased to or exceeded control values, and sprouting and increased synaptic vesicles were seen. Axons that had been attached to the dying neurons appeared to have shifted their connections to the spines and the neurites of the surviving neurons, increasing their thickness. Astrocytic restitution and neuronal remodeling processes started at 4 days continuing until 12 weeks after ischemic insult.

Keywords: Maturation phenomenon; cerebral ischemia; neuronal remodeling.

Introduction

Cerebral infarction develops rapidly after a major ischemic insult. Earlier, we developed a model to induce an ischemic penumbra around a small focal infarction in the cerebral cortex of Mongolian gerbils [5, 8] by giving a threshold amount of ischemic insult to induce cerebral infarction. The histopathology of this model revealed disseminated eosinophilic ischemic neurons by light microscopic observation, and disseminated electron-dense neurons seen ultrastructurally (disseminated selective neuronal necrosis) increased in number in the penumbra of the cerebral cortex after restoration of blood flow. A focal infarction developed later in a part of this area of disseminated selective neuronal necrosis within 12 to 24 hours after ischemic insult, due to massive astrocytic death. This area expanded gradually, involving dead and still-living eosinophilic neurons, and normal-looking neurons progressing to death 4 days after the ischemic insult [8, 9, 11]. No additional new infarction (pan necrosis) was found later than 4 days after the ischemic insult in our coronal as well as para-sagittal sections of the forebrain [3, 19].

In previous studies of the cortical penumbra [7, 9, 11], we found that the cytoplasm and cell processes of living astrocytes in the penumbra were actively swollen and that brain edema, determined by tissue gravimetry, was maximum around 3 days after ischemic insult, subsiding gradually by 7 days [4, 11]. Isolated dark neurons with different grades of high-electron density increased in number among the normal-looking neurons from 5 to 24 hours. These dark neurons were surrounded by severely swollen astrocytic cell processes. As a general pathological sign of irreversible cellular damage, granular chromatin condensation was apparent in the nuclear matrix and along the nuclear membrane of some of these dark neurons [12]. The dark neurons increased in number rapidly until day 4, and new ones continued to appear 12 weeks after the ischemic insult. These observations correspond to the maturation phenomenon of ischemic injuries [3, 6, 19], which is the same as the delayed neuronal death described for CA1 neurons [3, 15, 19].

In the present study, we investigated the fate of swollen edematous astrocytes and dead neurons at the ultrastructural level, as well as remodeling of axons connected to the dead neurons in the ischemic penumbra.

Materials and methods

Stroke-positive Mongolian gerbils were selected according to their stroke index score [18] during left carotid clipping for 10 minutes, followed by another 10 minutes of clipping with a 5-hour interval between the 2 occlusions. The gerbils were sacrificed at 5, 12, and 24 hours, at 4 days, and at 3, 5, 8, 12, and 24 weeks following the last ischemic insult by intracardiac perfusion with cacodylate-buffered glutaraldehyde fixative (3 animals in each group) for electron microscopy and with 10% phosphate-buffered formaldehyde fixative for light microscopy (5 animals in each group).

Ultrathin sections including the second through fifth cortical layers were obtained from the neocortex at the mid-point between the interhemispheric and rhinal fissures on the left coronal face sectioned at the infundibular level, in which only the penumbra appeared. The sections were double-stained with uranyl acetate and lead solution, and observed with a Hitachi electron microscope (H9000). Separate paraffin sections were stained with hematoxylin-eosin, periodic acid fuchsin Schiff, or by Bodian silver impregnation or immuno-histochemistry for glial fibrillary acidic protein (GFAP).

Placing 1 cm × 1 cm lattices on the 5,000 × 2.67 enlarged EM photographs, we measured the number of synapses and spines in the neuropil in a 100-square cm (56 sq.μ, by real size), and determined the percent volume of the axon terminals and spines using the point-counting method [22] by counting intersections of the lattice dropped on the axon terminals and/or spines. We also measured neuritic thickness as the maximal diameter perpendicular to their neurofilaments and/or microtubules on the same EM pictures.

Results

Astrocytic swelling gradually subsided starting on day 4 after ischemic insult, then an increase was observed in the number of cytoplasmic glial fibrils in astrocytes seen ultrastructurally and in GFAP-positive cells seen by light microscopy. Astrocytes in mitosis or with 2 nuclei were occasionally seen.

The disseminated dying electron-dense neurons had been fragmented into granular pieces by invading astrocytic cell processes (Fig. 1A). These accumulations of fragmented dark neurons were observed as eosinophilic ghost cells by light microscopy. The electron-dense granular pieces were dispersed around the extracellular spaces and phagocytized by microglia, astrocytes, and neurons. There was no evidence of macrophages in the penumbra.

The number of synapses and spines, and the percent volume of the axon terminals and spines (Table 1) decreased with an increase in a sparsity of synaptic vesicles until day 4 (Fig. 2A). From 1 to 12 weeks after ischemic insult, however, they recovered to or exceeded the control values and were found surrounding the thickened neurites of the surviving neurons (Fig. 2B).

From 4 days to 8 weeks after the ischemic insult, most axon terminals that had been attached to dying neurons were found around the fragmented dead dark neurons. Some of them were separated from the dead neurons, being attached by a crust of granular electron-dense fragments (Fig. 1A). From 24 hours to 8 weeks after ischemic insult, some axons attached to dying neurons showed globular or spindle-shaped distension of their terminals, as seen by Bodian silver impregnation (Fig. 1B). Electron microscopic observation of these distensions showed amplified axon terminals containing degenerated mitochondria, lamellated dense bodies, and irregularly located neurofilaments and microtubules. They were frequently observed around accumulations of fragmented electron-dense granular pieces of dead neurons.

From 1 to 12 weeks, some axon terminals associated with crusts of electron-dense granular pieces became newly connected to the spines and neurites of the surviving neurons.

Neuronal death continued in the penumbra during these periods (maturation phenomenon). From 8 to 24 weeks after the ischemic insult, these structures and the accumulation of eosinophilic ghost cells remained confined to the third cortical layer, especially in some portions of the lateral part of the left coronal face sectioned at the infundibular level. Cortical thickness and cortical neuronal density were reduced evenly in the face during these periods.

Discussion

Astrocytes swell in the acute phase after an ischemic insult, showing increases in the number of glycogen granules and mitochondrial size and number, indicating an active reaction of astrocytes to prevent ischemic neuronal injury [7, 11]. Four days after ischemic insult, astrocytic swelling subsides and glial fibrils, stained by GFAP antibodies, increase in number. These GFAP-positive reactive astrocytes are increased in number by mitotic division, especially those surrounding the focal infarction (pan necrosis), which evolves and develops from 12 hours to 4 days after the insult. Necrotic tissue is then scavenged by macrophages and

Restitution of ischemic injuries in penumbra of cerebral cortex after temporary ischemia

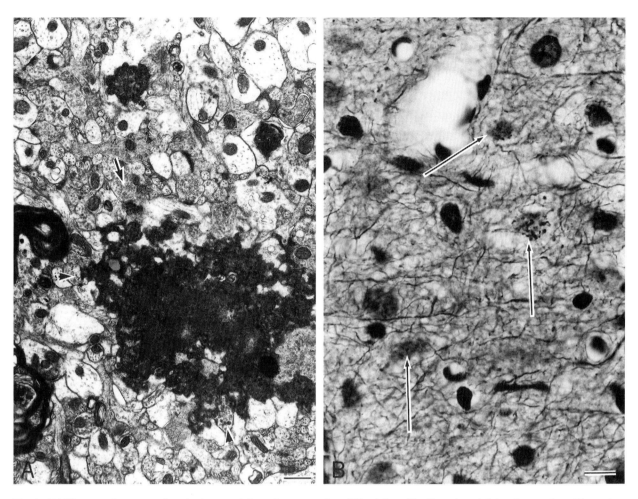

Fig. 1. (A) Electron microscopy of cerebral cortex 4-days after restoration of blood flow. The disseminated dying electron-dense neuron has been fragmented into granular pieces by invading astrocytic cell processes. Some axon terminals that were attached to the dying neurons are found around the fragmented dead dark neurons (arrowheads). Bottom bar 0.1 micron. (B) Light microscopy of cerebral cortex 24 hours after restoration of blood flow. Some thick axons attached to dying neurons showed globular or spindle-shaped distension of their terminals surrounding the dead neurons (arrows). Bodian silver impregnation. Bottom bar 2 micron

Table 1. *Data showing astrocytic restitution and neuronal remodeling processes up to 12 weeks after ischemic injury.*

Average value	Time after last ischemic insult						
	Control	5 hours	4 days	1 week	5 weeks	8 weeks	12 weeks
% vol. of axon terminal	17.75 ± 1.32	14.69 ± 5.79†	5.48 ± 1.71*	14.89 ± 1.69†	19.55 ± 2.62†	21.62 ± 2.24†	28.90 ± 3.55*†
% vol. of spine	4.80 ± 1.12	3.09 ± 0.89*†	1.19 ± 0.20*	1.70 ± 0.25*	2.25 ± 0.45*	2.91 ± 0.42*†	4.16 ± 1.04†
No. of synapses/56 sq μ	17.21 ± 1.09	23.24 ± 4.52*†	12.65 ± 1.67*	13.80 ± 1.92†	14.71 ± 3.26	16.68 ± 1.55†	19.48 ± 3.10†
No. of spines/56 sq μ	11.89 ± 2.50	11.20 ± 2.33†	4.26 ± 0.40*	7.14 ± 0.78*	6.69 ± 0.63*	7.10 ± 0.68*	10.93 ± 3.04†
Thickness of neuritis (μ)	0.61 ± 0.02	–	0.59 ± 0.02	0.60 ± 0.02	–	0.67 ± 0.02*†	0.93 ± 0.03*†

Average ± standard error, $p < 0.05$: * compared with control; † compared with 4 days.

becomes liquefied [3, 8, 19]. The infarcted focus is surrounded by gliosis induced by reactive astrocytes. GFAP-positive reactive astrocytes increase moderately in number, but do not induce gliosis in the penumbra. These are the restitutional processes of astrocytes in the ischemic tissue [8].

It has been thought that dead neurons and ischemically injured tissue are scavenged by macrophage inva-

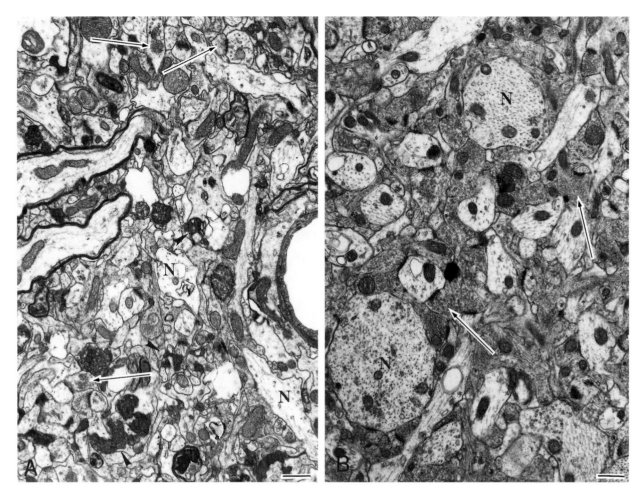

Fig. 2. (A) Electron microscopy of cerebral cortex 4 days after restoration of blood flow. The number of synapses and spines, and the volume of axon terminals and spines decreased with increase in sparsity of synaptic vesicles (arrows). Neurites (*N*) are degenerative. Electron-dense granular pieces are dispersed in the extracellular spaces of the neuropil (arrowheads). Bottom bar 0.1 micron. (B) Electron microscopy of cerebral cortex 12 weeks after restoration of blood flow. The number of synapses and spines, and volume of axon terminals and spines recovered, with increase in number of synaptic vesicles (arrows). Neurites (*N*) are thickened, surrounded and/or synapsed by axon terminals. Bottom bar 0.1 micron

sion into the injured tissue from the blood stream. However, in the present study, dead neurons were found disseminated among surviving neurons in the cortical penumbra. The axons and dendritic processes of the dying neurons were still connected to axon terminals and neurites of surviving neurons. Solitary dying neurons, which were connected by neuritic networks, were not phagocytized by a single macrophage. In contrast to infarction, i.e., massive necrosis, macrophages did not enter the neuropil of the penumbra where the network of the neuropil was still tight. In this situation, it is reasonable to assume that shrunken dead neurons become fragmented into granular debris (eosinophilic ghost cells seen by the light microscopy) and are removed by astrocytes, neurons, and perivascularly located microglia [10, 14, 16]. However, the tattered central cytosol of shrunken neurons remained for more than 5 weeks. No inflammatory cells or macrophages appeared in the ischemic penumbra wandering in the neuropils [10, 16].

We found a marked decrease in the number of synapses and volume of the axon terminals in the entire neuropil of the ischemic penumbra from 5 hours to 4 days after start of recirculation, along with marked shrinkage of axon terminals, which contained a decreased number of synaptic vesicles. These changes seemed to be due to calcium-dependent neuronal hyperexcitation [21] and were reduced by N-methyl-D-aspartate receptor antagonists in a morphological study recording excitatory postsynaptic potential from hippocampal slice cultures subjected to brief anoxia-hypoglycemia [13].

The number of synapses increased gradually from 1 to 12 weeks after the ischemic insult, associated with an increase in the volume of axon terminals showing sprouting [20] and paralleling a marked increase in the number of synaptic vesicles. The number and volume of spines also increased in parallel. Axons that had been attached to the dying neurons were considered to have shifted their connections to the spines and the neurites of the surviving neurons, increasing their thickness associated with synaptogenesis in the neuropil [1, 2, 17]. The neuronal remodeling process progressed in the ischemic penumbra from its early stage to 12 weeks after the start of recirculation.

References

1. Crepel V, Epsztein J, Ben-Ari Y (2003) Ischemia induces short- and long-term remodeling of synaptic activity in the hippocampus. J Cell Mol Med 7: 401–407
2. Frost SB, Barbay S, Friel KM, Plautz EJ, Nudo RJ (2003) Reorganization of remote cortical regions after ischemic brain injury: a potential substrate for stroke recovery. J Neurophysiol 89: 3205–3214
3. Graham DI, Lantos PL (2002) Greenfield's neuropathology illustrated. Oxford University Press, London, pp 230–280
4. Hakamata Y, Hanyu S, Kuroiwa T, Ito U (1997) Brain edema associated with progressive selective neuronal death or impending infarction in the cerebral cortex. Acta Neurochir [Suppl] 70: 20–22
5. Hanyu S, Ito U, Hakamata Y, Nakano I (1997) Topographical analysis of cortical neuronal loss associated with disseminated selective neuronal necrosis and infarction after repeated ischemia. Brain Res 767: 154–157
6. Ito U, Spatz M, Walker JT Jr, Klatzo I (1975) Experimental cerebral ischemia in mongolian gerbils. I. Light microscopic observations. Acta Neuropathol Berl 32: 209–223
7. Ito U, Hanyu S, Hakamata Y, Nakamura M, Arima K (1997) Ultrastructure of astrocytes associated with selective neuronal death of cerebral cortex after repeated ischemia. Acta Neurochir [Suppl] 70: 46–49
8. Ito U, Hanyu S, Hakamata Y, Arima K, Oyanagi K, Kuroiwa T, Nakano I (1999) Temporal profile of cortical injury following ischemic insult just-below and at the threshold level for induction of infarction – light and electron microscopic study. In: Ito U, Fieschi C, Orzi F, Kuroiwa T, Klatzo I (eds) Maturation phenomenon in cerebral ischemia III. Springer, Berlin New York, pp 227–235
9. Ito U, Kuroiwa T, Hanyu S, Hakamata Y, Nakano I, Oyanagi K (2000) Ultrastructural behavior of astrocytes to singly dying cortical neurons. In: Krieglstein J, Klumpp S (eds) Pharmacology of cerebral ischemia. Medpharm Science Publications, Stuttgart, pp 285–291
10. Ito U, Kuroiwa T, Hakamata Y, Kawakami E, Nakano I, Oyana K (2002) How are ischemically dying eosinophilic neurons scavenged in the penumbra? An ultrastructural study. In: Krieglstein J, Klumpp S (eds) Pharmacology of cerebral ischemia. Medpharm Science Publication, Stuttgart, pp 261–265
11. Ito U, Kuroiwa T, Hanyu S, Hakamata Y, Kawakami E, Nakano I, Oyanagi K (2003) Temporal profile of experimental ischemic edema after threshold amount of insult to induce infarction – ultrastructure, gravimetry and Evans' blue extravasation. Acta Neurochir [Suppl] 86: 131–135
12. Ito U, Kuroiwa T, Hanyu S, Hakamata Y, Kawakami E, Nakano I, Oyanagi K (2003) Ultrastructural temporal profile of the dying neuron and surrounding astrocytes in the ischemic penumbra: apoptosis or necrosis? In: Buchan AM, Ito U, Colbourne F, Kuroiwa T, Klatzo I (eds) Maturation phenomenon in cerebral ischemia V. Springer, Berlin Heidelberg, pp 189–196
13. Jourdain P, Nikonenko I, Alberi S, Muller D (2002) Remodeling of hippocampal synaptic networks by a brief anoxia-hypoglycemia. J Neurosci 22: 3108–3116
14. Kalmar B, Kittel A, Lemmens R, Kornyei Z, Madarasz E (2001) Cultured astrocytes react to LPS with increased cyclooxygenase activity and phagocytosis. Neurochem Int 38: 453–461
15. Kirino T, Tamura A, Sano K (1984) Delayed neuronal death in the rat hippocampus following transient forebrain ischemia. Acta Neuropathol Berl 64: 139–147
16. Lemkey-Johnston N, Butler V, Reynolds WA (1976) Glial changes in the progress of a chemical lesion. An electron microscopic study. J Comp Neurol 167: 481–501
17. Nudo RJ, Larson D, Plautz EJ, Friel KM, Barbay S, Frost SB (2003) A squirrel monkey model of poststroke motor recovery. ILAR J 44: 161–174
18. Ohno K, Ito U, Inaba Y (1984) Regional cerebral blood flow and stroke index after left carotid artery ligation in the conscious gerbil. Brain Res 297: 151–157
19. Rosenblum WI (1997) Histopathologic clues to the pathways of neuronal death following ischemia/hypoxia. J Neurotrauma 14: 313–326
20. Stroemer RP, Kent TA, Hulsebosch CE (1995) Neocortical neural sprouting, synaptogenesis, and behavioral recovery after neocortical infarction in rats. Stroke 26: 2135–2144
21. von Lubitz DK, Diemer NH (1983) Cerebral ischemia in the rat: ultrastructural and morphometric analysis of synapses in stratum radiatum of the hippocampal CA-1 region. Acta Neuropathol (Berl) 61: 52–60
22. Weibel ER (1963) Morphometry of the human lung. Springer, Berlin, pp 19–20

Correspondence: Umeo Ito, 4-22-24, Zenpukuji, Suginami-ku, Tokyo 167-0041 Japan. e-mail: umeo-ito@nn.iij4u.or.jp

Inhibition of Na^+/H^+ exchanger isoform 1 attenuates mitochondrial cytochrome C release in cortical neurons following in vitro ischemia

J. Luo[1,2], H. Chen[2,3], D. B. Kintner[2], G. E. Shull[4], and D. Sun[1,2]

[1] Department of Physiology, University of Wisconsin Medical School, Madison, Wisconsin, USA
[2] Department of Neurosurgery, University of Wisconsin Medical School, Madison, Wisconsin, USA
[3] Department of Neuroscience Training Program, University of Wisconsin Medical School, Madison, Wisconsin, USA
[4] Department of Molecular Genetics, Biochemistry, and Microbiology, University of Cincinnati, Cincinnati, Ohio, USA

Summary

Na^+/H^+ exchanger isoform 1 (NHE1) is a major acid extrusion mechanism following intracellular acidosis. We hypothesized that stimulation of NHE1 after cerebral ischemia contributes to disruption of Na^+ homeostasis and neuronal death. In the present study, expression of NHE1 was detected in cultured mouse cortical neurons. Oxygen and glucose deprivation (OGD) for 3 hours followed by 21 hours of reoxygenation (REOX) led to $68 \pm 10\%$ cell death. Inhibition of NHE1 with the potent inhibitor HOE 642 or genetic ablation of NHE1 reduced OGD-induced cell death by $\sim 40\%$ to 50% ($p < 0.05$). In NHE1$^{+/+}$ neurons, OGD/REOX triggered significant increases in Na_i^+ and Ca_i^{2+}. Genetic ablation of NHE1 and HOE 642 treatment reduced the rise of Na_i^+ by $\sim 40\%$ to 50% and abolished the OGD/REOX-mediated Ca^{2+} accumulation. Moreover, mitochondrial cytochrome C release was significantly attenuated by inhibition of NHE1 activity. These results imply that NHE1 activity disrupts Na^+ and Ca^{2+} homeostasis and contributes to ischemic neuronal damage.

Keywords: Oxygen and glucose deprivation; HOE 642; sodium; calcium; Na^+/Ca^{2+} exchange; edema.

Introduction

Na^+/H^+ exchangers (NHEs) catalyze the electroneutral exchange of protons and sodium ions across cellular membranes and down their concentration gradients [19], thereby regulating the pH of the cytoplasm or organelle lumen [5]. To date, 9 NHE family members have been identified in mammals. NHE1-5 are expressed on the plasma membrane in various cell types [9]. NHE6-9 reside on intracellular organelle membranes of the endosomal/trans-Golgi network [18], although there is evidence that NHE8 is also located on brush border membranes of the renal proximal tubule [11].

NHE1 is a ubiquitously expressed plasma membrane protein and the most abundant NHE isoform in the rat central nervous system [16]. NHE1 serves the crucial function of protecting cells from internal acidification. In addition to an established role in intracellular pH and cell volume homeostasis, NHE1 can serve as a structural anchor for actin filaments [10] and as a plasma membrane scaffold in the assembly of signal complexes that are independent of its function as an ion exchanger [2].

In cardiac myocytes, NHE1 activity has been shown to contribute to the ionic imbalances occurring during ischemia-reperfusion injury. However, the functions of NHE1 in ion homeostasis in the central nervous system are not well understood. NHE1 is essential for the maintenance and regulation of pH_i in cortical astrocytes [13, 17, 21]. Our recent study of mouse cortical astrocytes demonstrated that NHE1 activity is significantly elevated following in vitro ischemia and leads to overloading of intracellular Na^+ and Ca^{2+} [13, 14]. Thus, we hypothesized that activation of NHE1 following ischemia may subsequently cause ischemic cerebral damage via Na^+- and Ca^{2+}-mediated toxic effects [6, 8].

In the current study, we used a pharmacological approach, utilizing the potent NHE1 inhibitor, HOE 642, and genetic ablation of NHE1 to elucidate the function of NHE1 in ischemic neuronal cell damage. We found that inhibition of NHE1 activity reduced neuronal cell death in an in vitro ischemic model, thereby showing that NHE1 activity contributes to mitochondrial dysfunction in cortical neurons.

Materials and methods

Cortical neuron cultures

NHE1 null mutant (NHE1$^{-/-}$) mice were established previously [4]. Timed-pregnant NHE1$^{+/-}$ mice were created by pairing male and female heterozygous mutant (NHE1$^{+/-}$) mice for 48 hours. The mice were then separated. As described in our recent study [7], E14-16 pregnant mice were anesthetized with 5% halothane and euthanized. Fetuses were removed and rinsed in cold Hanks balanced salt solution. Each mouse fetus was genotyped. The tails were removed from the fetus and polymerase chain reaction was performed as described previously [13].

Cortices were removed and minced as described previously [3]. Tissues were treated with 0.2 mg/mL trypsin at 37 °C for 25 minutes. Cells were centrifuged at 350 g for 4 minutes. The cell suspension was diluted in Eagle's minimal essential media (EMEM) containing 5% fetal bovine and 5% horse sera. The cells from individual fetal cortices were seeded separately in 24-well plates or on glass coverslips coated with poly-D-lysine and incubated at 37 °C in an incubator with 5% CO_2 and atmospheric air. After 96 hours in culture, 1 mL of fresh media containing 8 μmol cytosine 1-b-D arabinofuranoside was added. The media were replaced as described before [7]. Cultures from 10 to 15 days in vitro were used in the study.

Oxygen and glucose deprivation (OGD) treatment

Neuronal cultures (10 to 15 days in vitro) grown in 24-well plates were rinsed with an isotonic OGD solution (pH 7.4) containing (in mmol): 0 glucose, 20 $NaHCO_3$, 120 NaCl, 5.36 KCl, 0.33 Na_2HPO_4, 0.44 KH_2PO_4, 1.27 $CaCl_2$, 0.81 $MgSO_4$. Cells were incubated in 0.5 mL of the OGD solution in a hypoxic incubator (model 3130, Thermo Forma, Marietta, OH) containing 94% N_2, 1% O_2, and 5% CO_2. The oxygen level in the medium of cultured cells in 24-well plates was monitored with an oxygen probe (Model M1-730, Microelectrodes, Bedford, NH) and decreased to ~2% to 3% after 60 minutes in the hypoxic incubator. The OGD incubation was 3 hours. For reoxygenation (REOX), the cells were incubated for 21 hours in 0.5 mL of EMEM containing 5.5 mmol glucose at 37 °C in the incubator with 5% CO_2 and atmospheric air. Normoxic control cells were incubated in 5% CO_2 and atmospheric air in a buffer identical to the OGD solution except for the addition of 5.5 mmol glucose. In the drug treatment studies, cells were pretreated with 1 μmol HOE 642 and remained present in all subsequent washes and incubations. Ionic changes were measured in neurons following 2 hours OGD and 1 hour REOX.

Measurement of cell death

Cell viability was assessed by propidium iodide (PI) uptake and retention of calcein using a Nikon TE 300 inverted epifluorescence microscope. Cultured neurons were rinsed with the isotonic control buffer and incubated with 1 μg/mL calcein-AM and 10 μg/mL PI in the same buffer at 37 °C for 30 minutes. For cell counting, cells were rinsed with the isotonic control buffer and visualized using a Nikon 20X objective lens. Calcein and PI fluorescences were visualized using FITC filters and Texas Red filters as described previously [3]. Images were collected using a Princeton Instruments MicroMax CCD camera. In a blind manner, a total of 1000 cells/condition were counted using MetaMorph image-processing software (Universal Imaging Corp., Downingtown, PA). Cell mortality was expressed as the ratio of PI-positive cells to the sum of calcein-positive and PI-positive cells.

Cytochrome C immunofluorescence

Colocalization of cytochrome C (CytC) release and neuronal death were determined by double-staining with PI in conjunction with a specific antibody against CytC [20]. Briefly, cells on coverslips were first stained with 5 μmol PI for 5 minutes and washed with phosphate-buffered saline (PBS) 3 times. They were fixed in 4% paraformaldehyde in PBS for 10 minutes. After rinsing, cells were incubated with a blocking solution for 20 minutes followed by application of anti-CytC antibody (1:100 diluted in blocking buffer) for 1 hour at room temperature. After rinsing in PBS, slices were incubated with goat anti-mouse fluorescein isothiocyanate-conjugated IgG (1:100) for 1 hour at 37 °C. Fluorescence images were obtained using a Leica DMIRE2 inverted confocal laser-scanning microscope (63X) and Leica confocal software (Leica Microsystems Inc., Mannheim, Germany).

Neurons were excited sequentially at 488 nm (argon/krypton) and 543 nm (Gre/Ne) and the emission fluorescence was recorded at 500 to 535 nm for CytC staining, and at 550 to 620 nm for PI staining.

Materials

EMEM and Hanks balanced salt solution were from Mediatech Cellgro (Herndon, VA). Fetal bovine serum was obtained from Hyclone Laboratories (Logan, UT). HOE 642 was a kind gift from Aventis Pharma (Frankfurt, Germany). Anti-cytochrome C antibody (clone 6H2.B4) was from Pharmingen (San Jose, CA).

Results

Inhibition of NHE1 activity significantly reduces OGD-induced cell death

A low level of cell death occurred in control NHE1$^{+/+}$ neurons (<15%). After 3 hours of OGD and 21 hours REOX, NHE1$^{+/+}$ neurons exhibited a significant increase in cell death (68 ± 10%, $p < 0.05$), compared to basal levels of cell death under normoxic conditions. In contrast, inhibition of NHE1 activity with 1 μmol HOE 642 significantly attenuated OGD-mediated cell death in NHE1$^{+/+}$ neurons (33 ± 14%, $p < 0.05$). This finding suggests that NHE1 plays a role in ischemic neuronal damage.

Inhibition of NHE1 activity reduces OGD/REOX-mediated CytC release

We investigated whether mitochondrial CytC release, a hallmark of mitochondrial damage, is altered following OGD/REOX. In normoxic control NHE1$^{+/+}$ neurons, CytC staining showed abundant and punctated perinuclear expression patterns (Fig. 1A). Some punctated CytC staining suggesting mitochondrial CytC expression was also observed in neurites. No PI-positive staining was found in these corre-

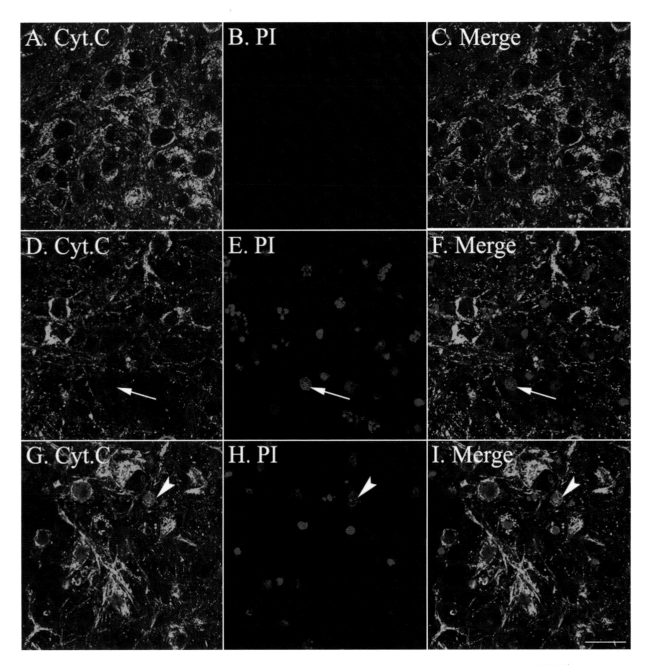

Fig. 1. Inhibition of NHE1 activity reduces OGD/REOX-mediated CytC release. The localization pattern of CytC in NHE1$^{+/+}$ neurons was visualized by labeling with an anti-CytC antibody (1:100 dilution). Cell death was determined by PI staining. For HOE 642 treatment, NHE1$^{+/+}$ neurons were incubated in EMEM in the presence of 1 μmol HOE 642 at 37 °C during 2 hours of OGD and 22 hours REOX. Sister NHE1$^{+/+}$ cultures were incubated for 24 hours in normoxic control buffers (CON). (A–C): CON; (D–F): 2 hours OGD/22 hours REOX; (G–I): 2 hours OGD/22 hours REOX plus 1 μmol HOE 642. (A, D, G) CytC staining; (B, E, H) PI staining. (C, F, and I) merged images. *Arrow*: PI-positive cells without CytC staining. *Arrowhead*: Diffused CytC staining in dead cells. This is a representative of 4 experiments. Scale bar = 20 μm

sponding normoxic NHE1$^{+/+}$ neurons (Fig. 1B). In contrast, after 2 hours OGD/22 hours REOX, many cells stained positively with PI (Fig. 1E, arrow). Immunoreactivity for CytC was lost in these dying neurons (Fig. 1D, 1F, arrow). The loss of CytC immunoreactive signals suggests that CytC was released from mitochondria and subsequently degraded following OGD/REOX. Neurons containing CytC did not take up PI (Fig. 1F). This demonstrates a positive correlation between CytC release and cell death. However, when NHE1 activity was inhibited with 1 μmol HOE 642, cell mortality was substantially reduced (Fig. 1H) and the intensity of CytC staining remained high (Fig. 1G). The merged image shows that the NHE1$^{+/+}$ neurons containing abundant CytC did not take up PI (Fig. 1I). However, some dying cells stained positive for PI, containing a diffused, reduced staining for CytC (Fig. 1G, arrowhead). Taken together, these findings further suggest that NHE1 activity plays a role in ischemic mitochondrial damage in neurons.

Discussion

Inhibition of NHE1 activity is neuroprotective following both in vitro and in vivo ischemia

Stimulation of NHE1 activity is detrimental to the ischemic myocardium [1], and we recently reported that both pharmacological inhibition and genetic ablation of NHE1 reduced the impairment of Na$^+$ and Ca^{2+} homeostasis in cortical astrocytes following OGD [13, 14]. However, little is known about the role of NHE1 in neuronal ischemic damage. In our study, we found that the selective NHE1 inhibitor HOE 642 significantly reduced OGD-mediated neuronal cell death by ~50% in NHE1$^{+/+}$ mice. Genetic ablation of NHE1 attenuated neuronal cell death by ~37%. Moreover, a significant reduction in infarction in cerebral focal ischemia occurred when NHE1 activity was inhibited with HOE 642 in NHE1$^{+/+}$ mice or in NHE1 knockdown (NHE1$^{+/-}$) mice [15].

The role of NHE1 activity in ischemic mitochondrial damage

Mitochondrial Ca^{2+} overload is a major trigger of the mitochondrial death pathway, which features the loss of mitochondrial membrane potential, opening of the mitochondrial permeability transition pore, release of CytC, and enhanced generation of reactive oxygen species [6, 12]. Inhibition of NHE1 activity remarkably suppressed cytosolic Na$^+$ and Ca^{2+} accumulation and mitochondrial Ca^{2+} overload in cardiomyocytes following H$_2$O$_2$-induced oxidative stress [22]. HOE 642 also reduced caspase-3 activity and annexin V fluorescence in cardiomyocytes stimulated by H$_2$O$_2$ [22]. This suggests that NHE1 function not only affects cytosolic Na$^+$ and Ca^{2+} homeostasis, but also alters mitochondrial Ca^{2+} homeostasis and functions. In our recent study, we found that OGD/REOX caused a significant increase in mitochondrial Ca^{2+} accumulation [15]. However, this increase in mitochondrial Ca^{2+} overload was attenuated in NHE1$^{-/-}$ neurons. We found that inhibition of NHE1 activity with HOE 642 prevented the OGD/REOX-mediated CytC release and death of NHE1$^{+/+}$ neurons. Taken together, our findings support the view that NHE1 activity plays a role in mitochondrial dysfunction in oxidative cell damage. Therefore, preserving mitochondrial integrity by NHE1 inhibition may, in part, contribute to the neuroprotection we found in both in vitro and in vivo ischemia models.

In summary, we found that NHE1 activity contributed to ischemic neuronal death following in vitro ischemia. Genetic ablation or pharmacological inhibition of NHE1 activity was neuroprotective following OGD. The underlying mechanisms included activation of NHE1 activity in conjunction with the Na$^+$/Ca^{2+} exchanger leading to disruption of intracellular Na$^+$ and Ca^{2+} homeostasis. In addition to cytosolic Ca^{2+} overload, NHE1 also played a detrimental role in mitochondrial Ca^{2+} overload and mitochondrial dysfunction.

Acknowledgments

This work was supported in part by National Institutes of Health grants R01NS048216 (D. Sun) and R01HL61974 (G. E. Shull), and AHA Established-Investigator award 0540154N (D. Sun).

References

1. Avkiran M (2001) Protection of the ischaemic myocardium by Na$^+$/H$^+$ exchange inhibitors: potential mechanisms of action. Basic Res Cardiol 96: 306–311
2. Baumgartner M, Patel H, Barber DL (2004) Na$^+$/H$^+$ exchanger NHE1 as plasma membrane scaffold in the assembly of signaling complexes. Am J Physiol Cell Physiol 287: C844–C850
3. Beck J, Lenart B, Kintner DB, Sun D (2003) Na-K-Cl cotransporter contributes to glutamate-mediated excitotoxicity. J Neurosci 23: 5061–5068
4. Bell SM, Schreiner CM, Schultheis PJ, Miller ML, Evans RL, Vorhees CV, Shull GE, Scott WJ (1999) Targeted disruption of the murine Nhe1 locus induces ataxia, growth retardation, and seizures. Am J Physiol 276: C788–C795
5. Brett CL, Donowitz M, Rao R (2005) Evolutionary origins of eukaryotic sodium/proton exchangers. Am J Physiol Cell Physiol 288: C223–C239
6. Brookes PS, Yoon Y, Robotham JL, Anders MW, Sheu SS (2004) Calcium, ATP, and ROS: a mitochondrial love-hate triangle. Am J Physiol Cell Physiol 287: C817–C833
7. Chen H, Luo J, Kintner DB, Shull GE, Sun D (2005) Na$^+$-dependent chloride transporter (NKCC1)-null mice exhibit less gray and white matter damage after focal cerebral ischemia. J Cereb Blood Flow Metab 25: 54–66

8. Choi DW (1995) Calcium: still center-stage in hypoxic-ischemic neuronal death. Trends Neurosci 18: 58–60
9. Counillon L, Pouyssegur J (2000) The expanding family of eucaryotic Na^+/H^+ exchangers. J Biol Chem 275: 1–4
10. Denker SP, Huang DC, Orlowski J, Furthmayr H, Barber DL (2000) Direct binding of the Na–H exchanger NHE1 to ERM proteins regulates the cortical cytoskeleton and cell shape independently of H^+ translocation. Mol Cell 6: 1425–1436
11. Goyal S, Mentone S, Aronson PS (2005) Immunolocalization of NHE8 in rat kidney. Am J Physiol Renal Physiol 288: F530–F538
12. Hoyt KR, Stout AK, Cardman JM, Reynolds IJ (1998b) The role of intracellular Na^+ and mitochondria in buffering of kainate-induced intracellular free Ca^{2+} changes in rat forebrain neurones. J Physiol 509: 103–116
13. Kintner DB, Su G, Lenart B, Ballard AJ, Meyer JW, Ng LL, Shull GE, Sun D (2004) Increased tolerance to oxygen and glucose deprivation in astrocytes from Na^+/H^+ exchanger isoform 1 null mice. Am J Physiol Cell Physiol 287: C12–C21
14. Kintner DB, Look A, Shull GE, Sun D (2005) Activation of extracellular signal-regulatory kinase (ERK1/2) stimulates Na^+/H^+ exchange activity in astrocytes in response to in vitro ischemia. Am J Physiol Cell Physiol [submitted]
15. Luo J, Chen H, Kintner DB, Shull GE, Sun D (2005) Decreased neuronal death in Na^+/H^+ exchanger isoform 1-null mice following in vitro and in vivo ischemia [submitted]
16. Ma E, Haddad GG (1997) Expression and localization of Na^+/H^+ exchangers in rat central nervous system. Neuroscience 79: 591–603
17. Mellergard P, Ouyang YB, Siesjo BK (1993) Intracellular pH regulation in cultured rat astrocytes in $CO_2/HCO3$-containing media. Exp Brain Res 95: 371–380
18. Nakamura N, Tanaka S, Teko Y, Mitsui K, Kanazawa H (2005) Four Na^+/H^+ Exchanger isoforms are distributed to Golgi and post-Golgi compartments and are involved in organelle pH regulation. J Biol Chem 280: 1561–1572
19. Orlowski J, Grinstein S (1997) Na^+/H^+ exchangers of mammalian cells. J Biol Chem 272: 22373–22376
20. Pei W, Liou AK, Chen J (2003) Two caspase-mediated apoptotic pathways induced by rotenone toxicity in cortical neuronal cells. FASEB J 17: 520–522
21. Shrode LD, Putnam RW (1994) Intracellular pH regulation in primary rat astrocytes and C6 glioma cells. Glia 12: 196–210
22. Teshima Y, Akao M, Jones SP, Marban E (2003) Cariporide (HOE642), a selective Na^+-H^+ exchange inhibitor, inhibits the mitochondrial death pathway. Circulation 108: 2275–2281

Correspondence: Dandan Sun, Department of Neurosurgery, University of Wisconsin Medical School, H4/332 Clinical Sciences Center, 600 Highland Ave., Madison, WI 53792, USA. e-mail: sun@neurosurg.wisc.edu

Controlled normothermia during ischemia is important for the induction of neuronal cell death after global ischemia in mouse

H. Ohtaki[1], T. Nakamachi[1], K. Dohi[1,2], S. Yofu[1,3], K. Hodoyama[1], M. Matsunaga[3], T. Aruga[2], and S. Shioda[1]

[1] Department of Anatomy, Showa University School of Medicine, Tokyo, Japan
[2] Department of Emergency and Clinical Care Medicine, Showa University School of Medicine, Tokyo, Japan
[3] Gene Trophology Research Institute, Tokyo, Japan

Summary

A stable model of neuronal damage after ischemia is needed in mice to enable progression of transgenic strategies. We performed transient global ischemia induced by common carotid artery occlusions with and without maintaining normal rectal temperature (Trec) in order to determine the importance of body temperature control during ischemia. We measured brain temperature (Tb) during ischemia/reperfusion. Mice with normothermia (Trec within $\pm 1\,°C$) had increased mortality and neuronal cell death in the CA1 region of hippocampus, which did not occur in hypothermic animals. If the Trec was kept within $\pm 1\,°C$, the Tb decreased during ischemia. After reperfusion, Tb in the normothermia group developed hyperthermia, which reached $>40\,°C$ and was $>2\,°C$ higher than Trec. We suggest that tightly controlled normothermia and prevention of hypothermia (Trec) during ischemia are important factors in the development of a stable neuronal damage model in mice.

Keywords: Global ischemia; mouse; hypothermia; neuronal cell death.

Introduction

Controlled hypothermia is known to reduce neuronal cell death after ischemic [4, 12, 19, 32, 34, 38], hypoxic [35], traumatic [14], and hemorrhagic [18] brain damage. Several reports show that depression of body temperature decreases activation of caspase-3 [2, 27, 34, 38], phosphorylation of stress-activated protein kinase/c-jun N-terminal kinase [15, 34], release of glutamate [5, 13], generation of toxic reactive oxygen species [8, 9, 33], and prevents suppression of bcl-2 [12] and anti-oxidants [6]. Hypothermia also suppresses the decrease of regional cerebral blood flow, which follows vascular occlusion [17]. The effect of hypothermia is considered to decrease apoptotic and necrotic neuronal cell death.

Management of body temperature is a means to stabilize neuronal damage after ischemia and hypoxia in animal models [31, 36]. Transgenic/knockout mice variants allow the roles of specific proteins to be studied in cerebral ischemia, using stable ischemia models in mice developed over the last decade. The development of mouse global ischemia models still needs improvement. We performed transient global ischemia with and without maintaining rectal temperature (Trec) to reveal the importance of the management of body temperature during ischemia in order to develop the stable ischemia model further. We measured the changes of brain temperature (Tb) and brain pO$_2$ during early ischemia/reperfusion periods.

Materials and methods

Animals

All experimental procedures involving animals were approved by the Institutional Animal Care and Use Committee of Showa University (#03018). Male C57BL/6 mice aged 8 weeks were purchased from Charles River Japan (Kanagawa, Japan). All mice were maintained on a 12-hour light/12-hour dark cycle at $24 \pm 2\,°C$ with constant humidity ($50 \pm 15\%$) for 1 week, then were used in the studies described below.

Production of transient forebrain ischemia

The following procedures, including transient forebrain ischemia, were performed at room temperature (27.0 to 28.0 °C). Trec was monitored in all animals. Normothermia (38.0 °C) was maintained with a heating blanket in 1 group of animals. A second group initially at Trec 38.0 °C constituted the hypothermia group (Fig. 1).

Using 2.5% sevoflurane in 70% N$_2$O/30% O$_2$ with facemask inhalation, anesthetized mice were subjected to transient occlusion of the common carotid artery (tCCAO) to create forebrain ischemia [22, 24]. In brief, bilateral common carotid arteries were carefully exposed and isolated, then occluded with clips (Zen temporary clip,

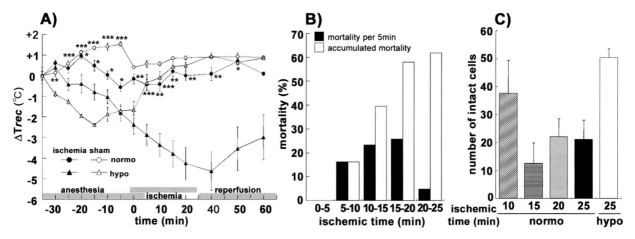

Fig. 1. Change of rectal temperature (Trec; A), mortality (B), induction of neuronal cell death (C) after global ischemia in the mouse with normothermia (normo) or hypothermia (hypo) maintaining Trec with a heating mat. (A) Trec was monitored with or without a heating mat to control Trec during anesthesia, ischemia and reperfusion. Hypothermia decreased Trec by inhalation anesthesia and recovered by stopping anesthesia. After ischemia, hypothermia was not returned to baseline. In contrast, the Trec in normothermia was kept ±1 °C during the experimental periods. (B) Mortality at 5-minute intervals during the ischemic periods (black bar) and the accumulated mortality (white bar) for 25 minutes of ischemia in the normothermia group. (C) Number of intact cells after ischemia in normothermia (normo) and hypothermia (hypo) groups. Data are expressed as mean ± SE. *p < 0.05, **p < 0.01, ***p < 0.001 compared with hypothermia group using Student t test

Oowa-tusho, Tokyo, Japan). Then the animals were left off inhalation anesthesia and kept on a heat mat for 10, 15, 20, or 25 minutes. After ischemia, the animals were re-anesthetized and reperfused by removal of the clips. After recovering from anesthesia on the heat mat, animals were kept for a night in a chamber at 35 °C in order to maintain body temperature. Animals were excluded from analysis on the basis of technical error if an occlusion clip became dislodged from an artery during ischemia or if an animal had excessive blood loss.

Histological examination

Four days after tCCAO, animals were anesthetized with sodium pentobarbital (50 mg/kg, i.p.), and fixed by perfusion with 10% buffered formalin. Brains were then paraffin embedded. The hippocampal regions were cut into 4-μm-thick coronal sections for histological examination. The paraffin-embedded sections were evaluated morphologically after toluidine blue staining.

The number of dead cells in the CA1 region (2 areas each per hemisphere) were compared with the number of morphologically-intact cells in accordance with previous studies [22]. The numbers of cells were averaged.

Measurement of brain pO_2, Tb, and Trec

Animals were anesthetized with inhaled sevoflurane (2.5%) in N_2O/O_2 (70%/30%). The mice were fixed in a stereotaxic frame and a burr hole of 0.5 mm in diameter was drilled in the skull over the right hemisphere 2 mm posterior to the bregma and 3 mm lateral to the midline with the aid of a surgical microscope. After removal from the frame, animals were placed prone. A pO_2 and Tb sensor (digital pO_2 monitor model POG-203, Unique Medical, Tokyo, Japan) was placed 3 mm into brain tissue through the burr hole. Trec was monitored as well. Ten minutes after equilibration, pre-ischemia baseline values were measured 4 times for 4 minutes. Animals were then subjected to CCAO and monitored for Tb, Trec, and brain pO_2 for 60 minutes (25 minutes of ischemia and 35 minutes of reperfusion).

Statistical analysis

Data are expressed as mean ± SEM. Student t test was used to analyze the control and hypothermia groups. Values of p < 0.05 were considered significant.

Results

Change of Trec and mortality during ischemia

There was no difference in Trec between the normothermia (37.7 ± 0.2 °C) and hypothermia (37.4 ± 0.3 °C) groups immediately after anesthesia. As shown in Fig. 1A, Trec in the hypothermia group decreased approximately 1.5 °C (36.0 ± 0.3 °C) 30 minutes after anesthesia compared with Trec immediately post-anesthesia. Animals were then subjected to global ischemia and inhalation anesthesia was discontinued. Sham-operated control animals in the hypothermia group recovered Trec to post-anesthesia values within 10 minutes of anesthesia cessation. Trec in hypothermia animals subjected to ischemia decreased approximately 4 °C (33.1 ± 0.8 °C) during the 25-minute ischemia period in which no anesthesia was administered. Hypothermia gradually resolved to −2 °C (34.3 ± 1.1 °C) by 30 minutes after reperfusion. Animals that were maintained at near-normal Trec by heating mat (normothermia) were kept within ±1 °C during the experimental period. Animals in the normo-

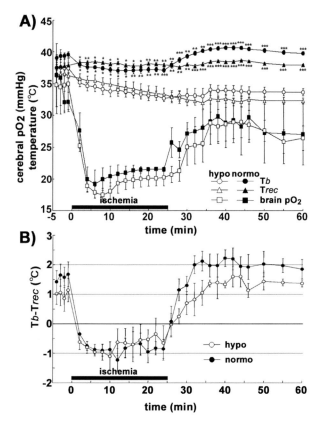

Fig. 2. Change in T*rec*, T*b*, and brain pO$_2$ after global ischemia (A) and the differences of T*rec* and T*b* in acute periods of ischemic reperfusion (B) during normothermia (normo) or hypothermia (hypo). Data are expressed as mean ± SE. *p < 0.05, **p < 0.01, ***p < 0.001 normothermia compared with hypothermia using Student *t* test

thermia group that were subjected to ischemia deteriorated 5 minutes after ischemic induction and died after 25 minutes of ischemia (Fig. 1B). The accumulated mortality reached about 60% 20 minutes after ischemia.

Neuronal cell death in CA1 region of the hippocampus 4 days after ischemia

To evaluate the relationship of T*rec* during ischemia and the induction of neuronal cell death, morphologically intact cells in the CA1 region of the hippocampus were counted after toluidine blue staining 4 days after ischemia [22]. In order to determine the relationship of the ischemic interval and neuronal damage in the normothermia group, the same evaluation was performed 4 days after tCCAO at intervals of 10, 15, 20, and 25 minutes (Fig. 1C). The cell number in the hippocampal CA1 region post-ischemia was 63.6 ± 4.5 (n = 10, data not shown). Most animals in the hypothermia group had few dead neurons. Some animals did show severe neuronal cell death; in them the number of cells was 50.4 ± 3.1 (n = 29). In contrast, the number of intact cells in the normothermia group was 21.1 ± 6.8 (n = 7), 22.1 ± 6.5 (n = 4), 12.6 ± 7.3 (n = 4), and 37.6 ± 11.8 (n = 3) after 25, 20, 15, and 10 minutes after tCCAO, respectively.

*Changes in T*rec*, T*b*, and brain pO$_2$ after tCCAO*

The results described above suggest that maintaining T*rec* at normothermia and preventing hypothermia during ischemia is important for the induction of neuronal cell death after ischemia. However, T*rec* does not coincide with T*b* during ischemia [21]. Hence, we studied T*rec*, T*b*, and brain pO$_2$ during ischemia and after reperfusion (Fig. 2). During the experimental periods, T*rec* in the hypothermia group gradually decreased and T*rec* in the normothermia group was kept at ±1 °C as in Fig. 1. Baseline brain pO$_2$ was about 35 mmHg. Brain pO$_2$ decreased dramatically (50% to 60% of baseline) by 6 minutes after ischemia, then recovered to 70% to 80% of pre-ischemia values by 5 minutes after reperfusion. There were no differences in brain pO$_2$ during the experimental periods in the 2 groups. On the other hand, the baseline T*b* was 39.1 ± 0.5 °C and 37.5 ± 0.8 °C 10 minutes after induction of anesthesia in the normothermia and hypothermia animals, respectively, and was 1 to 1.5 °C higher than T*rec* (Fig. 2B). T*b* in normothermia and hypothermia animals was lower (−1.2 [37.1 ± 0.5] °C and −1.1 [33.8 ± 0.9] °C, respectively) than T*rec* 10 to 12 minutes after ischemic induction. After reperfusion, T*b* in the hypothermia group returned to the baseline, which was 1 to 1.5 °C higher than T*rec*. However, T*b* in the normothermia group reached >40 °C and was 2 °C higher than T*rec*.

Discussion

We determined that maintaining T*rec* during acute periods of ischemia is important in order to induce neuronal cell death. We also determined that T*rec*, T*b*, and brain pO$_2$ change during and after global ischemia induced by CCAO in mice.

Several studies report that spontaneous hyperthermia during ischemia in brain exaggerates brain damage after focal and forebrain ischemia in the rat model [1, 9, 28]. In the present study, we did not observe spontaneous hyperthermia in the brain in our normothermia group during ischemia.

Tb is controlled by a balance of heat production and heat consumption or loss. The mechanisms of heat loss involve heat transfer from the brain core through blood flow, evaporation by breathing and sweating, non-evaporating heat loss through breathing, and is determined by the blood flow in brain, species differences, area of body surface including body mass, respiration rate, venous plexuses in the nasal mucosa, etc. [3, 10, 20]. Some mammals, including humans [10, 11, 21] and rats [1, 28], have increased Tb during ischemia due to heat transfer from the core of brain through blood flow. Thus, prevention of hyperthermia in brain by regional cooling and the suppression of pyrogenic cytokines and cyclo-oxygenases (COX) are measures to reduce neuronal damage after ischemia [11]. The mechanism of heat loss differs according to animal species and is not understood in mice. Because the Trec in our hypothermia group kept decreasing, the Tb in mice may be lost by evaporation of breathing and sweating due to their small body mass which is easily influenced by air temperature and humidity.

Tb after reperfusion in controlled normothermia increased, reaching >40 °C, which was 2 °C or more than Trec. Inducible factors for hyperthermia after ischemia are pyrogenic cytokines such as interleukin-1 (IL-1) β [1, 7, 29], and tumor necrosis factor (TNF) α [29], COX-2 [11, 16], and prostaglandin E2 [30]. We did not measure these in our study. However, we have reported that the IL-1β gene and protein are induced after transient focal and global ischemia, and brain damage in IL-1α/β gene-deficient mice is lower than in wild-type mice [23, 25]. Moreover, the gene and protein of TNFα and its receptor is induced 30 to 60 minutes after focal ischemia in the mouse [26, 37]. We also determined clinically that indomethacin, a COX-2 inhibitor, decreases Tb and the expression of IL-1β in patients with brain injury [11].

Our study indicates that controlling Trec during ischemia is necessary to develop a stable neuronal damage model in the mouse because the mouse easily loses Tb during ischemia, perhaps due to its small body mass.

References

1. Abraham H, Somogyvari-Vigh A, Maderdrut JL, Vigh S, Arimura A (2003) Rapidly activated microglial cells in the preoptic area may play a role in the generation of hyperthermia following occlusion of the middle cerebral artery in the rat. Exp Brain Res 153: 84–91
2. Adachi M, Sohma O, Tsuneishi S, Takada S, Nakamura H (2001) Combination effect of systemic hypothermia and caspase inhibitor administration against hypoxic-ischemic brain damage in neonatal rats. Pediatr Res 50: 590–595
3. Andrews PJ, Harris B, Murray GD (2005) Randomized controlled trial of effects of the airflow through the upper respiratory tract of intubated brain-injured patients on brain temperature and selective brain cooling. Br J Anaesth 94: 330–335
4. Aoki M, Tamatani M, Taniguchi M, Yamaguchi A, Bando Y, Kasai K, Miyoshi Y, Nakamura Y, Vitek MP, Tohyama M, Tanaka H, Sugimoto H (2001) Hypothermic treatment restores glucose regulated protein 78 (GRP78) expression in ischemic brain. Brain Res Mol Brain Res 95: 117–128
5. Arai H, Uto A, Ogawa Y, Sato K (1993) Effect of low temperature on glutamate-induced intracellular calcium accumulation and cell death in cultured hippocampal neurons. Neurosci Lett 163: 132–134
6. Binienda Z, Virmani A, Przybyla-Zawislak B, Schmued L (2004) Neuroprotective effect of L-carnitine in the 3-nitropropionic acid (3-NPA)-evoked neurotoxicity in rats. Neurosci Lett 367: 264–267
7. Cartmell T, Luheshi GN, Rothwell NJ (1999) Brain sites of action of endogenous interleukin-1 in the febrile response to localized inflammation in the rat. J Physiol 518: 585–594
8. Chiueh CC (2001) Iron overload, oxidative stress, and axonal dystrophy in brain disorders. Pediatr Neurol 25: 138–147
9. Dietrich WD, Busto R, Valdes I, Loor Y (1990) Effects of normothermic versus mild hyperthermic forebrain ischemia in rats. Stroke 21: 1318–1325
10. Dohi K, Jimbo H, Abe T, Aruga T (2005) A novel and simple selective brain cooling method by nasopharyngeal cooling (positive selective brain cooling method – a technical note). Acta Neurochir Suppl [submitted]
11. Dohi K, Jimbo H, Ikeda Y, Fujita S, Ohtaki H, Shioda S, Abe T, Aruga T (2005) Pharmacological brain cooling using indomethacin in acute hemorrhagic stroke – anti-inflammatory cytokines and anti-oxidative effects. Acta Neurochir [Suppl] [submitted]
12. Eberspacher E, Werner C, Engelhard K, Pape M, Laacke L, Winner D, Hollweck R, Hutzler P, Kochs E (2005) Long-term effects of hypothermia on neuronal cell death and the concentration of apoptotic proteins after incomplete cerebral ischemia and reperfusion in rats. Acta Anaesthesiol Scand 49: 477–487
13. Eilers H, Bickler PE (1996) Hypothermia and isoflurane similarly inhibit glutamate release evoked by chemical anoxia in rat cortical brain slices. Anesthesiology 85: 600–607
14. Fritz HG, Bauer R (2004) Traumatic injury in the developing brain – effects of hypothermia. Exp Toxicol Pathol 56: 91–102
15. Hicks SD, Parmele KT, DeFranco DB, Klann E, Callaway CW (2000) Hypothermia differentially increases extracellular signal-regulated kinase and stress-activated protein kinase/c-Jun terminal kinase activation in the hippocampus during reperfusion after asphyxial cardiac arrest. Neuroscience 98: 677–685
16. Hoffmann C (2000) COX-2 in brain and spinal cord implications for therapeutic use. Curr Med Chem 7: 1113–1120
17. Jenkins LW, DeWitt DS, Johnston WE, Davis KL, Prough DS (2001) Intraischemic mild hypothermia increases hippocampal CA1 blood flow during forebrain ischemia. Brain Res 890: 1–10
18. Kawanishi M (2003) Effect of hypothermia on brain edema formation following intracerebral hemorrhage in rats. Acta Neurochir Suppl 86: 453–456
19. Kil HY, Zhang J, Piantadosi CA (1996) Brain temperature

alters hydroxyl radical production during cerebral ischemia/reperfusion in rats. J Cereb Blood Flow Metab 16: 100–106

20. Mariak Z, White MD, Lewko J, Lyson T, Piekarski P (1999) Direct cooling of the human brain by heat loss from the upper respiratory tract. J Appl Physiol 87: 1609–1613

21. Marion DW (2004) Controlled normothermia in neurologic intensive care. Crit Care Med 32: S43–S45

22. Matsunaga M, Ohtaki H, Takaki A, Iwai Y, Yin L, Mizuguchi H, Miyake T, Usumi K, Shioda S (2003) Nucleoprotamine diet derived from salmon soft roe protects mouse hippocampal neurons from delayed cell death after transient forebrain ischemia. Neurosci Res 47: 269–276

23. Mizushima H, Zhou CJ, Dohi K, Horai R, Asano M, Iwakura Y, Hirabayashi T, Arata S, Nakajo S, Takaki A, Ohtaki H, Shioda S (2002) Reduced postischemic apoptosis in the hippocampus of mice deficient in interleukin-1. J Comp Neurol 448: 203–216

24. Ohtaki H, Funahashi H, Dohi K, Oguro T, Horai R, Asano M, Iwakura Y, Yin L, Matsunaga M, Goto N, Shioda S (2003) Suppression of oxidative neuronal damage after transient middle cerebral artery occlusion in mice lacking interleukin-1. Neurosci Res 45: 313–324

25. Ohtaki H, Yin Li, Nakamachi T, Dohi K, Kudo Y, Makino R, Shioda S (2004) Expression of tumor necrosis factor α in nerve fibers and oligodendrocytes after transient focal ischemia in mice. Neurosci Lett 368: 162–166

26. Ohtaki H, Dohi K, Nakamachi T, Yofu S, Endo S, Kudo Y, Shioda S (2005) Evaluation of brain ischemia in mice. Acta Histochem Cytochem 38: 99–106

27. Phanithi PB, Yoshida Y, Santana A, Su M, Kawamura S, Yasui N (2000) Mild hypothermia mitigates post-ischemic neuronal death following focal cerebral ischemia in rat brain: immunohistochemical study of Fas, caspase-3 and TUNEL. Neuropathology 20: 273–282

28. Reglodi D, Somogyvari-Vigh A, Maderdrut JL, Vigh S, Arimura A (2000) Postischemic spontaneous hyperthermia and its effects in middle cerebral artery occlusion in the rat. Exp Neurol 163: 399–407

29. Roth J, De Souza GE (2001) Fever induction pathways: evidence from responses to systemic or local cytokine formation. Braz J Med Biol Res 34: 301–314

30. Satinoff E, Peloso E, Plata-Salamn CR (1999) Prostaglandin E2-induced fever in young and old Long-Evans rats. Physiol Behav 67: 149–152

31. Sheng H, Laskowitz DT, Pearlstein RD, Warner DS (1999) Characterization of a recovery global cerebral ischemia model in the mouse. J Neurosci Methods 88: 103–109

32. Tsuchiya D, Hong S, Suh SW, Kayama T, Panter SS, Weinstein PR (2002) Mild hypothermia reduces zinc translocation, neuronal cell death, and mortality after transient global ischemia in mice. J Cereb Blood Flow Metab 22: 1231–1238

33. Uemura K, Hoshino S, Uchida K, Tsuruta R, Maekawa T, Yoshida K (2003) Hypothermia attenuates delayed cortical cell death and ROS generation following CO inhalation. Toxicol Lett 145: 101–106

34. Xu L, Yenari MA, Steinberg GK, Giffard RG (2002) Mild hypothermia reduces apoptosis of mouse neurons in vitro early in the cascade. J Cereb Blood Flow Metab 22: 21–28

35. Yager JY, Armstrong EA, Jaharus C, Saucier DM, Wirrell EC (2004) Preventing hyperthermia decreases brain damage following neonatal hypoxic-ischemic seizures. Brain Res 1011: 48–57

36. Yang G, Kitagawa K, Matsushita K, Mabuchi T, Yagita Y, Yanagihara T, Matsumoto M (1997) C57BL/6 strain is most susceptible to cerebral ischemia following bilateral common carotid occlusion among seven mouse strains: selective neuronal death in the murine transient forebrain ischemia. Brain Res 752: 209–218

37. Yin L, Ohtaki H, Nakamachi T, Kudo Y, Makino R, Shioda S (2004) Delayed expressed TNFR1 co-localize with ICAM-1 in astrocyte in mice brain after transient focal ischemia. Neurosci Lett 370: 30–35

38. Zhu C, Wang X, Cheng X, Qiu L, Xu F, Simbruner G, Blomgren K (2004) Post-ischemic hypothermia-induced tissue protection and diminished apoptosis after neonatal cerebral hypoxia-ischemia. Brain Res 996: 67–75

Correspondence: Hirokazu Ohtaki, Department of Anatomy, Showa University School of Medicine, 1-5-8 Hatanodai, Shinagawa-ku, Tokyo 142-8555, Japan. e-mail: taki@med.showa-u.ac.jp

Ex vivo measurement of brain tissue viscoelasticity in postischemic brain edema

T. Kuroiwa[1], I. Yamada[2], N. Katsumata[1], S. Endo[3], and K. Ohno[4]

[1] Department of Neuropathology, Tokyo Medical and Dental University, Tokyo, Japan
[2] Department of Radiology, Tokyo Medical and Dental University, Tokyo, Japan
[3] Animal Research Center, Tokyo Medical and Dental University, Tokyo, Japan
[4] Department of Neurosurgery, Tokyo Medical and Dental University, Tokyo, Japan

Summary

Knowledge of the biomechanical properties of postischemic brain tissue is important for understanding the mechanisms of postischemic secondary brain tissue injury. We describe the method and results of biomechanical property measurement in ex vivo postischemic brain tissue by applying an indentation method.

Mongolian gerbils were subjected to a transient unilateral hemispheric ischemia. At day 1 after ischemia, multi-parametric MRI was performed, the brain was removed under anesthesia, sliced, and kept in a container with silicone oil for the measurement. A compression probe attached to a pressure transducer was inserted to a predetermined depth at the regions of interest and maintained at a constant speed. A pressure relaxation curve was recorded for the calculation of elasticity modulus (E) and viscosity modulus (η) according to Maxwell-Voigt's 3-element model. One day after ischemia, E and η decreased to 78.7% and 73.1% of the control level, respectively. This decrease corresponded to a mild decrease in apparent diffusion coefficient (ADC) and magnetization transfer ratio, and an increase in T2 value. Tissue water content increased to 105.1% of control. Microvacuolation with demyelination and axonal disruption was evident in the postischemic brain tissue.

Keywords: Cerebral infarction; elasticity; viscosity; Mongolian gerbils; indentation method; multi-parametric MRI.

Introduction

Brain tissue injury secondary to ischemic brain edema and swelling is an important factor in determining the prognosis of stroke patients. It is well known that postischemic brain edema and swelling potentially induces fatal brain herniation at the acute phase. Brain herniation signifies brain mass expansion and protrusion into the free residual spaces through a notch or opening in rigid structures such as the cerebellar tentorium and the foramen magnum. The biomechanical properties of the brain tissue are obviously important in this process. Biomechanical data on brain tissue, both in the normal and pathological state, are also indispensable for the computer simulation of intracranial disease processes.

In previous studies we have quantified the biomechanical properties of brain with vasogenic edema [3–5]. Here, we describe a method for the measurement of biomechanical properties of ex vivo brain tissue by using an indentation method. This technique was used to examine brains subjected to unilateral hemispheric ischemia and the results compared to morphological and magnetic resonance imaging (MRI) tissue changes. Maxwell-Voigt's 3-element model [1, 2] (Fig. 1b) was applied for the calculation of elasticity modulus (E) and the viscosity modulus (η) of the postischemic tissue.

Materials and methods

Adult Mongolian gerbils of both sexes were used in the experiment. For induction of unilateral hemispheric ischemia, the left common carotid artery was occluded for 10 minutes by using a miniature vascular clip under 2% isoflurane anesthesia. After vascular occlusion, the animal was allowed to recover from anesthesia; the stroke index [7] was measured during the initial occlusion. Animals (n = 4) with stroke index equal to or more than 10 were selected as symptom-positive animals, and subjected to the second ischemia for 10 minutes after a 5-hour interval. At day 1 after ischemia, multi-parametric MRI (mapping of apparent diffusion coefficient [ADC], T2 value, and magnetization transfer ratio [MTR]) was performed using an experimental MRI (4.7 T, Unity-INOVA, Varian, Inc., Palo Alto, CA), and the animal was then sacrificed with an overdose of anesthesia. The brain was removed, cut coronally with 3 mm thickness at the chiasma level, and kept in a container with silicone oil heated to 37 °C for the viscoelasticity and tissue water content measurement.

Ex vivo viscoelastic response of the brain tissue was measured by applying an indentation method (Fig. 1). For the measurement, a compression probe attached to a pressure transducer was inserted

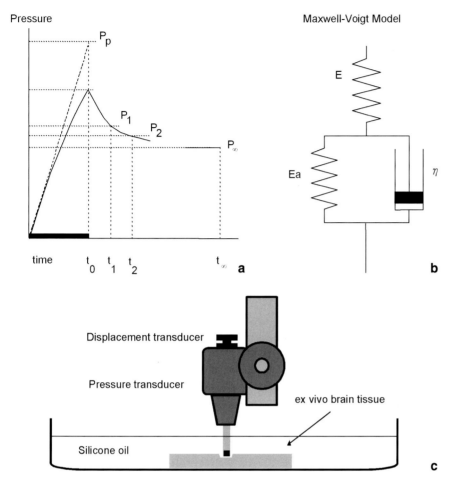

Fig. 1. (a) Schematic drawing of the indentation method using pressure-time relationship; (b) Maxwell-Voigt 3-element model; and (c) a pressure-displacement transducer system

to a predetermined depth (400 µm) at the regions of interest at a constant speed (25 µm/s). The time course of pressure response was schematically drawn as shown in Fig. 1a.

Assuming that biomechanical property of the brain tissue is well simulated by the Maxwell-Voigt 3-element model (Fig. 1b), tissue elasticity modulus (E) and viscosity modulus (η) were calculated according the following equations [1].

$$E = \frac{P}{u} \times \frac{\pi(1-\mu^2)b^2}{2a\{1 - \sqrt{1-(b/a)^2}\}} \quad (1)$$

$$\eta = E \times \frac{t_1 - t_2}{\log_e \frac{P_1 - P_\infty}{P_2 - P_\infty}} \times \left(1 + \frac{P_\infty/P_p}{1 - P_\infty/P_p}\right) \quad (2)$$

E: elasticity modulus
µ: Poisson's ratio
P_p: Estimated peak pressure at t_0
P_1: pressure at t_1
P_∞: pressure after tissue relaxation
a: radius of compression probe

η: viscosity modulus
u: insertion depth

P_2: pressure at t_2

b: radius of pressure transducer

After the indentation method, tissue from the region of interest was cut out for specific gravity measurement [6]. The coronal section with a mirror surface of the above measurement was immersion-fixed in 4% buffered formalin and prepared for light microscopy examination. The results were compared to those obtained from sham operated animals (n = 4).

Results

Histological examination revealed postischemic brain edema with neuropil microvacuolation in the postischemic cerebral hemisphere at day 1 postischemia. Demyelination and axonal disruption were also evident. Specific gravity measurement revealed tissue water content increased to 105.1% of the control level. E and η decreased to 78.7% and 73.1% of the control level, respectively. These decreases corresponded to a decrease in ADC and magnetization transfer ratio (70.6% and 85.5% of the control level, respectively), and to an increase in T2 value to 151.9% of the control (Fig. 2).

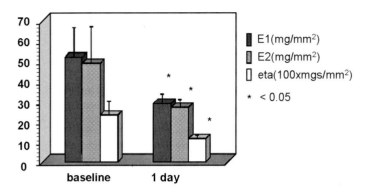

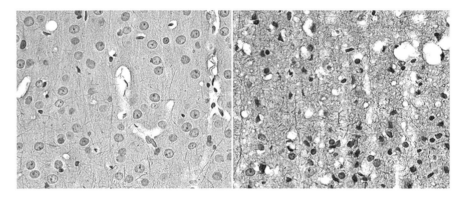

Fig. 2. (a) Change of tissue water content (% water), elasticity modulus (E: mg/mm^2) and viscosity modulus (η: 10^2 mgs/mm^2) of the cortex day 1 after a transient cerebral ischemia (* = p < 0.05). (b) Light microscopy of the cortex in sham operated animal (upper), and at day 1 after transient ischemia (lower) (Bodian staining)

Discussion

Biomechanical properties of brain tissue have been examined both in physiological and pathological conditions. Walsh et al. [8] measured in vivo brain tissue elastic response under physiological conditions. The measurement was done using a pressure-displacement transducer system attached to the skull of dogs. They observed that the elastic response of the brain was sensitive to systemic factors such as ventilation.

The elastic property of the brain under pathological conditions has also been measured. Aoyagi et al. [1, 2] measured brain tissue compliance in dogs subjected to traumatic brain edema, hydration, and dehydration. They found a significant decrease in tissue elasticity in the brain with traumatic brain edema. Thus, in vivo measurement is useful for understanding the overall pressure response of the brain tissue located inside the skull; however, it is not possible to discriminate the influence of systemic factors such as blood pressure and cerebrospinal fluid pressure using this procedure. Meningeal membranes also influence the compressibility of the brain tissue in the in vivo measurement. However, for understanding the mechanism of secondary brain tissue injury, the biomechanical property of the brain tissue *per se* as a material composing brain structure is important. The influences of cerebrospinal fluid pressure or blood pressure should be excluded in the measurement. This is also true in the biomechanical data for the computer simulation of intracranial disease processes, where biomechanical data of tissue elasticity and viscosity independent from the systemic factors in the pathological as well as normal conditions are essential for simulation procedures such as the finite element method.

We applied an indentation method to ex vivo brain tissue slices. It is well known that a shift of tissue water from the extracellular compartment to the intracellular compartment takes place immediately after energy failure. To minimize this influence and also to prevent tissue from drying, we kept the ex vivo brain in silicone oil during the measurement. By using the indentation method, the size for measurement was approximately 1 mm^2, which was small enough for regional comparison to tissue water content and histological changes.

Maxwell-Voigt's 3-element model has been used to simulate the biomechanical properties of various materials to a combination of 3 elements, *i.e.*, 1 dash-pot (viscosity) connected in parallel to 1 spring (elasticity), and in series with another spring (elasticity). We calculated the elasticity modulus and the viscosity modulus of the brain tissue under the assumption that the 3-element model simulates biomechanical aspects of brain tissue. We observed that both elasticity and viscosity had significantly decreased already by 1 day postischemia.

Histological examination revealed significant neuropil microvacuolation with ischemic neuronal changes. Bodian staining revealed disruption of neuronal processes in the ischemic tissue. Tissue-specific gravity measurement revealed significant edema. Several factors including sparsely-fibered neuropil, expansion of the extracellular space, loss of cellular integrity, and neural fiber connections in the postischemic brain tissue appear to be responsible for the decrease in tissue viscoelasticity.

Multi-parametric MRI revealed a mild decrease in ADC and MTR and an increase in T2 value. The increase of net tissue water content reflected by T2 increase, and the decrease of tissue water integrated in the macromolecular structure reflected by MTR decrease, appear to coordinate with the decrease in tissue viscoelasticity [9].

References

1. Aoyagi N, Masuzawa H, Sano K (1980) Compliance of brain [in Japanese]. No To Shinkei 32: 47–56
2. Aoyagi N, Masuzawa H, Sano K, Kihara M, Kobayashi S (1982) Compliance of brain – Part 2, Approach from the local elastic and viscous moduli [in Japanese]. No To Shinkei 34: 509–516
3. Kuroiwa T, Taniguchi I, Okeda R (1993) Regional tissue compliance of edematous brain after cryogenic injury in cats. In: Avezaat CJJ, Van Eijndhoven JHM, Maas AIR (eds) Intracranial pressure VIII. Springer Berlin Heidelberg New York Tokyo, pp 127–129
4. Kuroiwa T, Ueki M, Suemasu H, Taniguchi I, Okeda R (1994) Biomechanical characteristics of brain edema: the difference between vasogenic-type and cytotoxic-type edema. Acta Neurochir [Suppl] 60: 158–161
5. Kuroiwa T, Ueki M, Ichiki H, Kobayashi M, Suemasu H, Taniguchi I, Okeda R (1997) Time course of tissue elasticity and fluidity in vasogenic brain edema. Acta Neurochir [Suppl] 70: 87–90
6. Marmarou A, Tanaka K, Schulman K (1982) An improved gravimetric measure of cerebral edema. J Neurosurg 56: 246–253
7. Ohno K, Ito U, Inaba Y (1984) Regional cerebral blood flow and stroke index after left carotid artery ligation in the conscious gerbil. Brain Res 297: 151–157
8. Walsh EK, Schettini A (1976) Elastic behavior of brain tissue in vivo. Am J Physiol 230: 1058–1062
9. Wolff SD, Balaban RS (1989) Magnetization transfer contrast (MTC) and tissue water proton relaxation in vivo. Magn Reson Med 10: 134–144

Correspondence: Toshihiko Kuroiwa, Department of Neuropathology, Medical Research Institute, Tokyo Medical and Dental University, Yushima 1-5-45, Bunkyo-ku, Tokyo 113-8510, Japan. e-mail: t.kuroiwa.npat@mri.tmd.ac.jp

Effect of dimethyl sulfoxide on blood-brain barrier integrity following middle cerebral artery occlusion in the rat

A. Kleindienst, J. G. Dunbar, R. Glisson, K. Okuno, and A. Marmarou

Department of Neurosurgery, Medical College of Virginia, Virginia Commonwealth University, Richmond, VA, USA

Summary

Dimethyl sulfoxide (DMSO) is widely used as a solvent for other drugs, i.e., for the protein kinase C activator phorbol 12-myristate 13-acetate (PMA) and the V1a receptor-antagonist SR49059, to reduce brain edema. We studied the effect of DMSO on blood-brain barrier (BBB) integrity following middle cerebral artery occlusion (MCAO) and the consequences on brain edema development.

Male Sprague-Dawley rats were randomly assigned to sham procedure or infusion of 1% DMSO, PMA (230 µg/kg in 1% DMSO), or SR49059 (1 mg/kg in 1% DMSO) followed by MCAO (each group n = 10). After a 2-hour period of ischemia and 2 hours reperfusion, the animals were sacrificed for assessment of brain water content, sodium, and potassium concentration. BBB integrity was assessed by Evans blue extravasation. Statistical analysis was performed by ANOVA followed by a Tukey post hoc test.

Low-dose DMSO treatment following MCAO significantly opened the BBB on the ischemic side ($p < 0.037$). PMA and SR49059 did not have any additional effect on BBB compromise compared to DMSO ($p = 1.000$, $p < 0.957$, respectively).

We conclude that DMSO as a vehicle for drug administration may increase the drug concentration into the extracellular space, but since BBB permeability is increased, it may also provide an avenue for development of vasogenic edema.

Keywords: Dimethyl sulfoxide; blood-brain barrier; Evans blue extravasation; middle cerebral artery occlusion; brain edema.

Introduction

Dimethyl sulfoxide (DMSO, [(CH$_3$)$_2$SO]), a by-product of the wood industry, has been used as a commercial solvent since 1953. DMSO is an amphipathic molecule with a highly polar domain and 2 apolar groups, making it soluble in both aqueous and organic media. Penetration of most tissue membranes occurs within minutes due to a reversible change in the protein configuration when DMSO substitutes for water [12, 17]. Accordingly, DMSO passes the blood-brain barrier (BBB) [4], and is supposed to increase BBB permeability for other drugs, although controversial results have been obtained [3, 10]. DMSO has been found to lower intracranial pressure following injury and to be an effective treatment in different types of brain insults [5, 6]. Different mechanisms of protection against ischemic injury by DMSO have been proposed. DMSO has been suggested to be involved in the prostaglandin/thromboxane system restoring cerebral blood flow (CBF) [11], to protect cell membranes [8], to serve as a hydroxyl radical scavenger in mitochondria [16], or to stabilize the intracellular Ca^{2+} concentration [18].

Because of its chemical properties, DMSO is widely used as a solvent for other drugs in concentrations of 0.5% to 90%, i.e., for the protein kinase C (PKC)-activator phorbol 12-myristate 13-acetate (PMA) and the V1a receptor antagonist SR49059. The purpose of this study was to examine the effect of DMSO used as a solvent in a concentration of 1% on BBB integrity following middle cerebral artery occlusion (MCAO) and the consequences on brain edema development. Since we found that both PMA and SR49059 reduce brain edema and subsequent electrolyte imbalances following cortical contusion injury and MCAO, the present study investigated whether DMSO as the vehicle solution altered the drug effects. Specifically, we assessed the brain water, sodium, and potassium content as well as BBB integrity by Evans blue extravasation following an intravenous saline, vehicle (DMSO), PMA, or SR49059 infusion after MCAO.

Materials and methods

Animals and surgical procedure

The studies were conducted with the approval of the Institutional Animal Care and Use Committee using National Institutes of Health guidelines. Experiments were carried out on 350 to 400 g adult, male Sprague-Dawley rats (Harlan, Indianapolis, IN). Rats were housed at $22 \pm 1\,°C$ at 60% humidity, with a 12-hour light/12-hour dark cycle, and pellet food and water ad libitum. Surgery was performed after intubation under halothane anesthesia and controlled ventilation (1.3% halothane in 70% nitrous oxide and 30% oxygen). Rectal temperature was maintained at $36.5 \pm 0.5\,°C$ using a heating lamp. The left femoral artery and vein were cannulated with polyethylene tubing (P.E. 50, Becton Dickinson, Sparks, MD) for continuous monitoring of mean arterial blood pressure (MABP), blood sampling, or for drug infusion. Adequate ventilation was verified by an arterial blood gas measurement after 1 hour of anesthesia.

CBF over the supply territory of the right middle cerebral artery was continuously monitored by laser Doppler flowmetry (LaserFlo, Vasamedics, St Paul, MN) through a burr hole located 1 mm posterior and 5 mm lateral to bregma, leaving the dura mater intact. Animals were placed in a supine position over the laser Doppler probe, and CBF as well as MABP were recorded continuously using a data acquisition system (ADInstruments, Colorado Springs, CO).

MCAO was induced using the intraluminal suture method described elsewhere [2], slightly modified. Through a midline neck incision, bifurcation of the right common carotid artery was exposed and branches of the external carotid artery (ECA) and internal carotid artery (ICA) including the occipital, lingual, and maxillary arteries were microsurgically separated and coagulated. The ECA was ligated with a 4-0 silk suture, and after temporary occlusion of the ICA and common carotid artery with vascular mini-clips, a 4-0 monofilament nylon suture (4-0 SN-644 MONOSOF nylon Polyamide) with a silicon tip of 0.3 mm diameter was inserted through the ECA stump and secured by a suture. The clips were removed and the filament was advanced through the ICA into the circle of Willis while occluding the pterygopalatine artery with a forceps. A reduction in CBF between 70 and 80% of baseline was observed when the suture was advanced a distance of 22 to 24 mm from the carotid bifurcation, thereby verifying proper MCAO. Two hours after occlusion, a 2-hour period of reperfusion in the middle cerebral artery territory was performed by withdrawing the suture into the ECA stump. Reperfusion was confirmed by an increase in CBF.

Study protocol and drug preparation

The objective of these experiments was to assess the effect of DMSO as a solvent for drugs known to affect brain swelling and BBB integrity following MCAO. Animals were randomly assigned to sham procedure or an infusion of 1% DMSO (Sigma-Aldrich, St Louis, MO), PMA (Sigma-Aldrich, St Louis, MO; 230 µg/kg in 1% DMSO), or SR49059 (Sanofi Recherche, Montpellier, France; 1 mg/kg in 1% DMSO) followed by MCAO. The drugs were intravenously administered using a continuous infusion pump (sp210w syringe pump, KD Scientific, New Hope, PA). After the experiments were completed, animals were either sacrificed by an overdose of halothane, decapitated and the brains removed for assessment of brain water, sodium, and potassium contents, or rats were transcardially perfused for assessment of BBB integrity.

Tissue processing

Cerebral tissue was immediately cut into 4 consecutive 4 mm coronal sections excluding the most rostral and caudal sections from further analysis. After division into the right and left hemispheres along the anatomic midline, the 4 regional samples obtained were processed for water content, measured by the wet/dry weight method. The wet weight of each sample was measured using an electronic analytical balance before drying the sample at $95\,°C$ for 5 days and reweighing to obtain the dry weight. The water content of each sample is given as percentage of total tissue weight. For measurement of brain sodium and potassium concentrations, the dried samples were placed in a furnace for 24 hours at $400\,°C$ and reduced to ashes. The ash was then extracted with distilled water, and the concentrations of sodium and potassium were determined using a flame photometer (943 nm; Instrument Laboratory, Anaheim, CA) with caesium as an internal standard.

Evaluation of BBB integrity

Integrity of the BBB was evaluated by assessing the extravasation of Evans blue dye as described previously [1]. Before the end of the respective experiments, a 2% solution of Evans blue dye (E515; Fisher Scientific, Fairlawn, NJ) in 0.9% NaCl was administered intravenously for 1 minute at a dose of 7 mL/kg body weight, and then allowed to circulate for 10 minutes prior to sacrifice. A thoracotomy was performed after an overdose of halothane. Brains were perfused with approximately 700 mL of saline via catheter inserted into the left ventricle of the heart and rapidly removed. The cerebrum of the brain was cut with a blade into right and left hemispheres along the anatomic midline. Each hemispheric section was placed separately in a precisely-measured volume (2 mL) of formamide and allowed to soak for 48 hours at room temperature. The supernatant solution was transferred to a microcuvette, and the absorbance of each solution was measured against a pure formamide standard at 625 nm using a Shimadzu UV 1600 Spectrophotometer (Shimadzu Instruments, Columbia, MD). The tissue was then dried in the oven at $95\,°C$ for 5 days. Data were expressed as the relative absorbance (unit/g dry weight).

Statistical analysis

SPSS software (SigmaStat, Chicago, IL) was used for statistical analysis. Data were analyzed by a randomized one-way analysis of variance (ANOVA) for group variations followed by a Tukey post hoc analysis. Statistical significance was accepted at $p < 0.05$.

Results

The injury-induced mortality was 14% following MCAO. MABP and arterial blood gases were kept within physiological limits throughout the experimental procedure, requiring few adjustments in the halothane concentration and ventilation parameters.

The ANOVA of brain water content produced a significant group effect ($F_{7,88} = 15.12$, $p < 0.001$) and the Tukey post hoc analysis indicated that a DMSO infusion increased the brain water content on the ischemic side significantly compared to sham procedure or either PMA or SR49059 treatment following MCAO ($p < 0.001$, $p < 0.001$, and $p \leq 0.001$, respectively). Data are presented in Table 1. The ANOVA of brain sodium content produced a significant group effect

Table 1. *Effect of drug infusion on brain water, sodium, and potassium content following middle cerebral artery occlusion*

Differences in brain water content between ischemic and contralateral area

Groups n = 6 per group	Tissue water	Tissue sodium	Tissue potassium
Sham	−0.02 ± 0.06	−5.81 ± 6.0	−3.17 ± 8.0
DMSO	2.16 ± 0.06	109.01 ± 12.0	−42.87 ± 7.5
PMA	0.78 ± 0.22	42.61 ± 11.5	−12.27 ± 7.5
SR49059	0.87 ± 0.20	34.25 ± 11.0	1.89* ± 10.0

Values shown are average ± SEM; *$p < 0.05$ as compared with Sham.

Table 2. *Effect of drug infusion on blood-brain barrier integrity assessed by Evans blue extravasation following middle cerebral artery occlusion*

Absorbance/dry weight in ischemic area

n = 4 each group	Average	SEM	Significance (p value)
Sham	0.41	0.05	–
DMSO	1.64	0.37	0.037 compared with contralateral
PMA	1.51	0.31	1.000 compared to DMSO
SR49059	1.30	0.29	0.957 compared to DMSO

DMSO dimethyl sulfoxide, *PMA* phorbol 12-myristate 13-acetate.

($F_{7,88} = 14.11$, $p < 0.001$) and the Tukey post hoc analysis indicated that a DMSO infusion increased the brain sodium concentration on the ischemic side significantly compared to sham procedure or either PMA or SR49059 treatment following MCAO ($p < 0.001$, $p < 0.001$, and $p < 0.001$, respectively). The ANOVA of brain potassium content produced a significant group effect ($F_{7,88} = 5.90$, $p < 0.001$) and the Tukey post hoc analysis indicated that a DMSO infusion decreased the brain potassium concentration on the ischemic side compared to sham procedure or either PMA or SR49059 treatment following MCAO ($p < 0.001$, $p < 0.001$, and $p < 0.001$, respectively).

ANOVA assessment of BBB integrity in the ischemic and non-ischemic sides produced a significant group effect ($F_{7,32} = 4.70$, $p \leq 0.002$). Data are presented in Table 2. The Tukey post hoc analysis indicated that low-dose DMSO treatment following MCAO significantly opened the BBB in the ischemic hemisphere compared to the non-ischemic hemisphere ($p \leq 0.037$). PMA and SR49059 did not have any additional effects on BBB compromise compared to DMSO ($p = 1.000$ and $p \leq 0.957$, respectively).

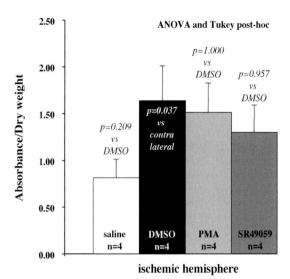

Fig. 1. Drug effect on BBB integrity following MCAO. Following MCAO, a significant opening of the BBB occurs in animals infused with 1% DMSO but not in saline-infused animals. Treatment with either the PKC-activator PMA or the V1a-antagonist SR49059, both dissolved in 1% DMSO, did not have any additional effect compared with 1% DMSO alone. Thus, DMSO may contribute to a vasogenic component of ischemia-induced brain edema

Discussion

In an experimental model of ischemic injury following MCAO, we found DMSO used as a solvent to increase BBB permeability thereby augmented drug access to the extracellular space of the brain, but also possibly enhanced vasogenic brain edema development (Fig. 1). We demonstrated that the predominately cellular cerebral edema resulting from MCAO is reduced by treatment with the PKC-activator, PMA, or the selective V1a antagonist, SR49059, when compared with DMSO (Fig. 2). The drugs were administered intravenously for 5 hours starting 1 hour before occlusion. Parallel results were obtained regarding brain sodium and potassium shift after ischemic injury. PMA and SR49059 treatment were able to reduce sodium uptake and increase potassium levels in the ischemic area, findings which are consistent with the generally-accepted opinion that water and sodium tend to coexist and transfer together through the plasma membrane under physiological and pathological conditions [9, 15, 19].

DMSO has been found to pass the BBB [4] and is supposed to increase BBB permeability for other drugs, although controversial results have been obtained [3, 10]. Both PMA and SR49059 were dissolved in a 1% DMSO solution. Following MCAO in the rat,

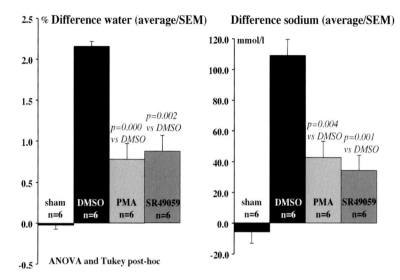

Fig. 2. Drug effect on brain edema development following MCAO. Following vehicle infusion (1% DMSO), a pronounced increase of brain water and sodium occurs in the infarct area compared with sham procedure. This effect is reduced by treatment with either the PKC-activator PMA or the V1a-antagonist SR49059, both dissolved in 1% DMSO. Thus, both drugs can be assumed to attenuate ischemia-induced cellular brain edema

others demonstrated that DMSO does not affect neurological outcome or infarct volume when compared with saline [7]. Similarly, in a baboon model of temporary MCAO or a cat model of right MCAO, DMSO had no effect when compared with untreated controls [13, 14]. Thus, DMSO was considered to be an appropriate vehicle for our experiments of ischemic injury.

We demonstrated that low-dose DMSO treatment following MCAO opened the BBB in the infarct area, while PMA and SR49059 did not cause any additional effects on BBB compromise. Thus, DMSO may have increased penetration of PMA and SR49059 into the extracellular space following MCAO. However, we cannot exclude the possibility that by increasing BBB permeability, DMSO may have also provided an avenue for development of a vasogenic edema component in the infarct area.

Acknowledgments

This research was supported by grants NS12587 and NS19235 from the National Institutes of Health.

References

1. Beaumont A, Marmarou A, Hayasaki K, Barzo P, Fatouros P, Corwin F, Marmarou C, Dunbar J (2000) The permissive nature of blood brain barrier (BBB) opening in edema formation following traumatic brain injury. Acta Neurochir [Suppl] 76: 125–129
2. Belayev L, Alonso OF, Busto R, Zhao W, Ginsberg MD (1996) Middle cerebral artery occlusion in the rat by intraluminal suture. Neurological and pathological evaluation of an improved model. Stroke 27: 1616–1623
3. Brink JJ, Stein DG (1967) Pemoline levels in brain: enhancement by dimethyl sulfoxide. Science 158: 1479–1480
4. Broadwell RD, Salcman M, Kaplan RS (1982) Morphologic effect of dimethyl sulfoxide on the blood-brain barrier. Science 217: 164–166
5. de la Torre JC (1970) Relative penetration of L-dopa and 5-HTP through the brain barrier using dimethyl sulfoxide. Experientia 26: 1117–1118
6. de la Torre JC, Kawanaga HM, Johnson CM, Goode DJ, Kajihara K, Mullan S (1975) Dimethyl sulfoxide in central nervous system trauma. Ann N Y Acad Sci 243: 362–389
7. Ginsberg MD, Becker DA, Busto R, Belayev A, Zhang Y, Khoutorova L, Ley JJ, Zhao W, Belayev L (2003) Stilbazulenyl nitrone, a novel antioxidant, is highly neuroprotective in focal ischemia. Ann Neurol 54: 330–342
8. Gollan F (1967) Effect of DMSO and THAM on ionizing radiation in mice. Ann N Y Acad Sci 141: 63–64
9. Gotoh O, Asano T, Koide T, Takakura K (1985) Ischemic brain edema following occlusion of the middle cerebral artery in the rat. I: The time courses of the brain water, sodium and potassium contents and blood-brain barrier permeability to 125I-albumin. Stroke 16: 101–109
10. Greig NH, Sweeney DJ, Rapoport SI (1985) Inability of dimethyl sulfoxide to increase brain uptake of water-soluble compounds: implications to chemotherapy for brain tumors. Cancer Treat Rep 69: 305–312
11. Johnson M, Jessup R, Ramwell PW (1974) The significance of protein disulfide and sulfhydryl groups in prostaglandin action. Prostaglandins 5: 125–136
12. Kolb KH, Jaenicke G, Kramer M, Schulz PE (1967) Absorption, distribution and elimination of labeled dimethyl sulfoxide in man and animals. Ann N Y Acad Sci 141: 85–95
13. Little JR, Cook A, Lesser RP (1981) Treatment of acute focal cerebral ischemia with dimethyl sulfoxide. Neurosurgery 9: 34–39

14. Little JR, Spetzler RF, Roski RA, Selman WR, Zabramski J, Lesser RP (1983) Ineffectiveness of DMSO in treating experimental brain ischemia. Ann N Y Acad Sci 411: 269–277
15. Loo DD, Zeuthen T, Chandy G, Wright EM (1996) Cotransport of water by the Na+/glucose cotransporter. Proc Natl Acad Sci U S A 93: 13367–13370
16. Panganamala RV, Sharma HM, Heikkila E, Geer JC, Cornwell DG (1976) Role of hydroxyl radical scavengers dimethyl sulfoxide, alcohols and methional in the inhibition of prostaglandin biosynthesis. Prostaglandins 11: 599–607
17. Rammler DH (1967) The effect of DMSO on several enzyme systems. Ann N Y Acad Sci 141: 291–299
18. Ruigrok TJ, de Moes D, Slade AM, Nayler WG (1981) The effect of dimethylsulfoxide on the calcium paradox. Am J Pathol 103: 390–403
19. Wright EM, Loo DD (2000) Coupling between Na+, sugar, and water transport across the intestine. Ann N Y Acad Sci 915: 54–66

Correspondence: Anthony Marmarou, Department of Neurosurgery, Virginia Commonwealth University Medical Center, 1101 East Marshall Street, Box 980508, Richmond, VA 23298-0508, USA. e-mail: marmarou@abic.vcu.edu

Increased substance P immunoreactivity and edema formation following reversible ischemic stroke

R. J. Turner[1], P. C. Blumbergs[1,2], N. R. Sims[3], S. C. Helps[3], K. M. Rodgers[1], and R. Vink[1]

[1] Department of Pathology, University of Adelaide, Adelaide, SA, Australia
[2] Centre for Neurological Diseases, Hanson Institute, Adelaide, SA, Australia
[3] Department of Medical Biochemistry and Centre for Neuroscience, School of Medicine, Flinders University, Adelaide, SA, Australia

Summary

Previous results from our laboratory have shown that neurogenic inflammation is associated with edema formation after traumatic brain injury (TBI). This neurogenic inflammation was characterized by increased substance P (SP) immunoreactivity and could be attenuated with administration of SP antagonists with a resultant decrease in edema formation. Few studies have examined whether neurogenic inflammation, as identified by increased SP immunoreactivity, occurs after stroke and its potential role in edema formation. The present study examines SP immunoreactivity and edema formation following stroke.

Experimental stroke was induced in halothane anaesthetized male Sprague-Dawley rats using a reversible thread model of middle cerebral artery occlusion. Increased SP immunoreactivity at 24 hours relative to the non-infarcted hemisphere was observed in perivascular, neuronal, and glial tissue, and within the penumbra of the infarcted hemisphere. It was not as apparent in the infarct core. This increased SP immunoreactivity was associated with edema formation. We conclude that neurogenic inflammation, as reflected by increased SP immunoreactivity, occurs following experimental stroke, and that this may be associated with edema formation. As such, inhibition of neurogenic inflammation may represent a novel therapeutic target for the treatment of edema following reversible, ischemic stroke.

Keywords: Ischemia; edema; neuropeptides; inflammation.

Introduction

Stroke is a major cause of morbidity and the third leading cause of death worldwide [19, 20]. Currently there are no effective treatments, with the exception of thrombolytic recanalization with tissue plasminogen activator; however, use of tissue plasminogen activator is limited to administration within 3 hours of stroke onset and is therefore only available to a subset of stroke patients [1].

Many deleterious injury cascades are initiated following ischemia, one of which is edema formation [8]. Cerebral edema may lead to increased intracranial pressure and decreased cerebral blood flow that may further aggravate edema formation [4]. Edema is a leading cause of death within the first week of stroke and is therefore a life-threatening issue that needs to be addressed with timely therapy [8]. In particular, vasogenic edema predominates within penumbral tissue [15], the potentially viable tissue that surrounds the infarct core and the target of neuroprotective strategies. This vasogenic edema occurs in the setting of blood-brain barrier disruption, where fluid escapes from the vasculature into the intercellular space [13]. The exact mechanism by which ischemia disrupts the blood-brain barrier and leads to edema formation is unclear. However, vasogenic edema appears to be the most important contributor to ischemic brain swelling [4]. Hence, identification of the factors associated with vasogenic edema formation may provide insight into the mechanisms of neuronal cell death within the penumbra.

Neuropeptides, and in particular, substance P (SP), have long been known to contribute to the genesis of edema in the periphery [6, 16]. When released from C-fibers, SP causes neurogenic inflammation, a local inflammatory response to certain types of injury or infection that is characterized by vasodilation and increased vascular permeability [16]. These changes in blood vessel size and permeability lead to edema formation [11, 12]. Hence, SP is implicated in the control of plasma extravasation and edema formation in the periphery [2]. Recent studies in our laboratory have demonstrated a central role for SP in the formation of edema following traumatic brain injury (TBI) [18].

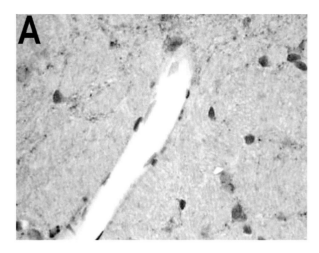 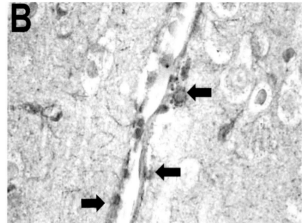

Fig. 1. (A) SP immunoreactivity in sham animals. (B) Infarcted animals at 24 hours post-stroke. Note the increased perivascular SP immunoreactivity (arrows) in the penumbral tissue surrounding the infarct

Given the incomplete investigation of SP in cerebral ischemia, and in particular, edema, our present study examines the relationship between neurogenic inflammation, as indicated by increased SP, and edema formation following reversible ischemic stroke.

Materials and methods

Reversible middle cerebral artery occlusion

Adult male Sprague-Dawley rats (n = 29; 265–295 g) were fed and watered *ad libitum* then fasted overnight before surgery. Anesthesia was induced with halothane (3%, 1.5 L/min O_2) and animals were intubated and mechanically ventilated with 1.5–2% halothane (1 L/min O_2). Middle cerebral artery occlusion was performed as described in detail elsewhere [3, 5]. Briefly, a 4-0 monofilament nylon suture with a tip rounded by heating near a flame and coated with 0.1% Poly-L-lysine (Sigma-Aldrich Co., St. Louis, MO) was introduced into the lumen of the external carotid artery through a puncture and subsequently advanced into the internal carotid artery. The suture was then advanced 17 mm past the external carotid/internal carotid artery bifurcation to occlude the origin of the middle cerebral artery. Lignocaine was applied to the surgical area and the wound closed with wound clips (9 mm Autoclip wound clips, Becton Dickinson, Franklin Lakes, NJ). Anesthesia was discontinued. When animals were able to breathe spontaneously they were extubated and allowed to recover. Reperfusion of the ischemic territory was achieved via withdrawal of the suture into the external carotid artery under halothane anesthesia. Body temperature was measured with a rectal probe and temperature maintained at 37 °C with a thermostatically controlled heating pad throughout all procedures. Any animals that did not show circling toward the side of the middle cerebral artery occlusion at 2 hours after thread insertion, indicative of successful stroke, were excluded from further study.

Brain water content

The amount of brain water was calculated using the wet weight/dry weight method as previously described [18]. Briefly, at 24 hours post-reperfusion, animals (n = 14) were killed by decapitation under halothane anesthesia and their brains rapidly removed. The left and right hemispheres were dissected and weighed to obtain wet weight. Hemispheres were then dried at 100 °C for 72 hours and reweighed to obtain dry weight. The percentage of brain water was subsequently calculated using the wet/dry method formula: % water = (wet weight − dry weight/wet weight) × 100.

Immunohistochemistry for SP

At 24 hours following reperfusion or sham surgery, animals (n = 15) were transcardially perfused with 10% formalin. Brains were then removed, processed and embedded in paraffin wax. Sections 5 μm thick were cut and processed for SP immunoreactivity (polyclonal N-18, Santa Cruz Biotechnology, Inc., Santa Cruz, CA).

Statistical analysis

All data are expressed as mean ± SEM. Statistical differences were determined using a two-tailed *t* test. A p value of 0.05 was considered significant.

Results

SP immunoreactivity

Increased SP immunoreactivity was observed within the penumbral tissue of the infarcted hemisphere at 24 hours post-stroke (Fig. 1). This increase was relative to the contralateral (non-infarcted) hemisphere and to sham animals and was evident within neuronal, glial, and perivascular tissue.

Brain water content

Hemispheric water content in sham animals was 80.27 ± 0.70% (Fig. 2). Following stroke, there was a

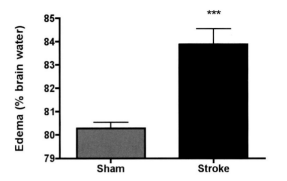

Fig. 2. Edema in sham and infarcted animals measured at 24 hours post-stroke (n = 7/group) (*** = p < 0.001)

significant (p < 0.001) increase in water content of the ipsilateral (infarcted) hemisphere of vehicle animals to 83.87 ± 1.79%. A non-significant increase in brain water content of the contralateral hemisphere was observed in vehicle animals.

Discussion

The present study has demonstrated that neurogenic inflammation, as indicated by increased SP immunoreactivity, occurs following reversible ischemic stroke in rats. Such increased SP immunoreactivity was associated with significant edema formation within the infarcted hemisphere. While recent investigations have focused on the role of neurogenic inflammation in the genesis of edema following TBI, this is the first study to examine the relationship between neurogenic inflammation and edema formation following stroke.

In the periphery, neurogenic inflammation is a known contributor to edema formation with SP identified as the most potent initiator [10, 14]. However, it has recently been shown that neurogenic inflammation, as indicated by increased SP immunoreactivity, is also involved in the formation of cerebral edema following TBI, in particular, vasogenic edema [18]. Our findings of increased SP immunoreactivity and edema formation following stroke are consistent with these results and reveal a key role for neuropeptides such as SP in the pathophysiology of edema formation following stroke. Specifically, they indicate that post-ischemic increased SP immunoreactivity influences the formation of edema.

Activation of neuropeptide receptors within the brain has been shown to contribute to cerebral edema formation; however, work focused on the role of neuropeptides in stroke has been rare [17]. SP can cause the formation of focal reversible endothelial gaps between endothelial cells of vasculature that is responsible for an increase in vascular permeability, one process in the formation of edema [9]. This may be one mechanism by which increased SP immunoreactivity and neurogenic inflammation lead to the formation of edema. It has been proposed that edema formation and plasma extravasation observed in response to SP is due to direct/indirect mechanisms such as the release of calcitonin-gene related peptide, histamine, serotonin, prostanoids, and nitric oxide [7]. It is of particular note that the penumbra has been identified as showing an edema profile consistent with vasogenic edema [15]. Our findings of increased SP immunoreactivity within the penumbra of the infarcted hemisphere suggest the edema observed may be of the vasogenic type.

In conclusion, we have shown that increased SP immunoreactivity after induction of ischemia is associated with edema formation. Further studies will examine whether inhibition of neurogenic inflammation with SP antagonists will provide a novel therapeutic intervention for the treatment of edema following ischemic stroke.

References

1. Alberts MJ (2003) Update on the treatment and prevention of ischaemic stroke. Curr Med Res Opin 19: 438–441
2. Alves RV, Campos MM, Santos AR, Calixto JB (1999) Receptor subtypes involved in tachykinin-mediated edema formation. Peptides 20: 921–927
3. Anderson MF, Sims NR (1999) Mitochondrial respiratory function and cell death in focal cerebral ischemia. J Neurochem 73: 1189–1199
4. Ayata C, Ropper AH (2002) Ischaemic brain oedema. J Clin Neurosci 9: 113–124
5. Belayev L, Alonso OF, Busto R, Zhao W, Ginsberg MD (1996) Middle cerebral artery occlusion in the rat by intraluminal suture. Neurological and pathological evaluation of an improved model. Stroke 27: 1616–1622; discussion 1623
6. Black PH (2002) Stress and the inflammatory response: a review of neurogenic inflammation. Brain Behav Immun 16: 622–653
7. Campos MM, Calixto JB (2000) Neurokinin mediation of edema and inflammation. Neuropeptides 34: 314–322
8. Gartshore G, Patterson J, Macrae IM (1997) Influence of ischemia and reperfusion on the course of brain tissue swelling and blood-brain barrier permeability in a rodent model of transient focal cerebral ischemia. Exp Neurol 147: 353–360
9. Green PG, Luo J, Heller PH, Levine JD (1993) Neurogenic and non-neurogenic mechanisms of plasma extravasation in the rat. Neuroscience 52: 735–743
10. Holzer P (1998) Neurogenic vasodilatation and plasma leakage in the skin. Gen Pharmacol 30: 5–11
11. Kuroiwa T, Cahn R, Juhler M, Goping G, Campbell G, Klatzo I (1985) Role of extracellular proteins in the dynamics of vasogenic brain edema. Acta Neuropathol (Berl) 66: 3–11

12. Kuroiwa T, Ting P, Martinez H, Klatzo I (1985) The biphasic opening of the blood-brain barrier to proteins following temporary middle cerebral artery occlusion. Acta Neuropathol (Berl) 68: 122–129
13. Lo EH, Singhal AB, Torchilin VP, Abbott NJ (2001) Drug delivery to damaged brain. Brain Res Brain Res Rev 38: 140–148
14. Otsuka M, Yoshioka K (1993) Neurotransmitter functions of mammalian tachykinins. Physiol Rev 73: 229–308
15. Quast MJ, Huang NC, Hillman GR, Kent TA (1993) The evolution of acute stroke recorded by multimodal magnetic resonance imaging. Magn Reson Imaging 11: 465–471
16. Severini C, Improta G, Falconieri-Erspamer G, Salvadori S, Erspamer V (2002) The tachykinin peptide family. Pharmacol Rev 54: 285–322
17. Stumm R, Culmsee C, Schafer MK, Krieglstein J, Weihe E (2001) Adaptive plasticity in tachykinin and tachykinin receptor expression after focal cerebral ischemia is differentially linked to GABAergic and glutamatergic cerebrocortical circuits and cerebrovenular endothelium. J Neurosci 21: 798–811
18. Vink R, Young A, Bennett CJ, Hu X, Connor CO, Cernak I, Nimmo AJ (2003) Neuropeptide release influences brain edema formation after diffuse traumatic brain injury. Acta Neurochir [Suppl] 86: 257–260
19. Warlow CP (1998) Epidemiology of stroke. Lancet 352 [Suppl] 3: SIII1–4
20. Williams LS, Weinberger M, Harris LE, Biller J (1999) Measuring quality of life in a way that is meaningful to stroke patients. Neurology 53: 1839–1843

Correspondence: Robert Vink, Department of Pathology, University of Adelaide, Adelaide SA, Australia, 5005. e-mail: Robert.Vink@adelaide.edu.au

Micro-blood-brain barrier openings and cytotoxic fragments of amyloid precursor protein accumulation in white matter after ischemic brain injury in long-lived rats

R. Pluta, M. Ułamek, and S. Januszewski

Department of Neurodegenerative Disorders, Medical Research Centre, Polish Academy of Sciences, Warsaw, Poland

Summary

Our study demonstrates that ischemia-reperfusion brain injury induces an increase in blood-brain barrier (BBB) permeability in the periventricular white matter. This chronic insufficiency of BBB may allow entry of neurotoxic fragments of amyloid precursor protein (APP) and other blood components such as platelets into the perineurovascular white matter tissue. These components may have secondary and chronic harmful effects on the ischemic myelin and axons and can intensify the phagocytic activity of microglial cells. Pathological accumulation of toxic fragments of APP in myelinated axons and oligodendrocytes appears after ischemic BBB injury and seem to be concomitant with, but independent of neuronal injury. It seems that ischemia-reperfusion disturbances may play important roles, both directly and indirectly, in the pathogenesis of white matter lesions. This pathology appears to have distribution similar to that of sporadic Alzheimer's disease. We noted micro-BBB openings in ischemic white matter lesions that probably would act as seeds of future Alzheimer's-type pathology.

Keywords: Blood-brain barrier; brain ischemia; horseradish peroxidase; white matter lesions; β-amyloid peptide; leukoaraiosis; Alzheimer's disease.

Introduction

Brain ischemia-reperfusion injury is a neurodegenerative disease that affects cognition, behavior, and function. In these types of functional changes, white matter lesions have been implicated in the neuropathogenesis seen in the subcortical and periventricular areas [4, 5, 12, 15]. These white matter lesions are referred to as leukoaraiosis [15]. Leukoaraioses have been found in the brains of patients with ischemic stroke and Alzheimer's disease [14]. The study of brain white matter changes in experimental ischemia has long been neglected because white matter was considered less vulnerable [3]. It is now known that the incidence of ischemic stroke involving white matter is relatively high [3] and an effect of this incidence is irreversible dysfunction [15].

Recent findings propose an early, silent, cumulative, and significant role for ischemia-reperfusion factors contributing to the development of sporadic Alzheimer's disease [6]. The profile of white matter neuropathology that is observed in brain ischemia probably shares a commonality with the same changes in Alzheimer's disease brain. In this study, we explored the role and impact of ischemia-reperfusion on brain white matter neurodegeneration and its connection with possible development of Alzheimer's disease injury. Because it is not clear whether the blood-brain barrier (BBB) in ischemic white matter lesions is altered in long-lived animals, we investigated BBB insufficiency and staining of different fragments of amyloid precursor protein (APP) in the perineurovascular space. In view of the potential late effects of various extravasated proteins on the development of white matter neurodegenerative changes [13], we discuss the possible influence of chronic ischemic BBB dysfunction on Alzheimer's-type cognitive impairment. We also focus on the question of whether or not the neuropathological mechanism(s) observed in ischemic white matter lesions are the same as those observed in Alzheimer's disease.

Materials and methods

Using female Wistar rats (n = 5), BBB dysfunction [10], distributions of APP around BBB vessels [5, 8], and platelet pathology [5, 9] were examined in white matter after 10 minutes of brain ischemia [7] then the rats were allowed to survive for 1 year. As controls (n = 3), sham operated rats were sacrificed at the appropriate time. Rat brains were perfusion-fixed for light and electron microscopic analy-

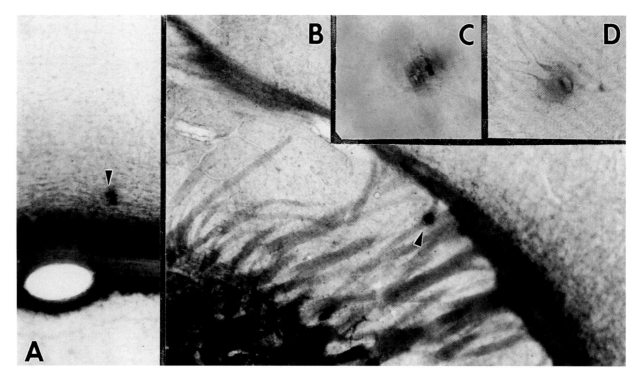

Fig. 1. Vibratome sections reacted for HRP histochemistry. (A) Extravasated HRP (arrowhead) in subcortical white matter (×40). (B) Extravasated HRP (arrowhead) in white matter close to lateral ventricle (×40). (C) High magnification of leaking area noted by arrowhead in Fig. 1B (×200). (D) Evidence of leakage of HRP in vessel branches/bifurcations in area of lateral ventricle (×200)

sis [5, 10]. Horseradish peroxidase (HRP) (Type VI, Sigma-Aldrich Co., St. Louis, MO) served as an indicator of BBB changes [5, 10]. HRP was injected into the femoral vein and allowed to circulate for 30 minutes [10]. One hemisphere was cut at 50–80 μm in the coronal plane with a Vibratome (Vibratome, St. Louis, MO) [5]. For demonstration of HRP, all brain tissue slices were incubated in a solution of 3,3′-diaminobenzidine tetra-HCl [10]. After mounting slices on microscope slides, they were examined using light microscopy. Some sections from the second hemisphere were selected for ultrastructural studies [5, 10]. Electron microscopy was performed using a Hitachi H-7000 transmission electron microscope (Hitachi High-Technologies, Krefeld, Germany) [5, 10]. For immunocytochemistry, we used monoclonal antibody (mAb) 22C11 against the N-terminal of APP, mAb 6E10 and polyclonal antibodies (pAb) SP28 against different parts of β-amyloid peptide, and pAb RAS 57 against the C-terminal of APP [4, 5, 8, 12].

Results

Postischemic white matter demonstrated chronic insufficiency of BBB. These micro-BBB lesions predominated in periventricular white matter and were random and spotty (Fig. 1A–D). HRP extravasations involved veins, venules, capillaries, and arterioles, and were restricted to branches and bifurcations of these vessels (Fig. 1D). Frequently we observed damaged endothelial cells and pericytes filled with HRP. The outer walls of the micro-vessels were also labeled with HRP. At the same time, diffuse deposits of neurotoxic fragments of C-terminals of APP surrounded the BBB vessels, forming perivascular cuffs with rarefaction of neighboring tissue and parallel staining of oligodendrocytes (Fig. 2A–C). In this investigation, we noted immunoreactivity for C-terminals of APP and β-amyloid peptide inside vessels predominantly in capillaries. C-terminals of APP and β-amyloid peptide deposits dominated in the paramedian part of the corpus callosum (Fig. 2A). Perivascular deposits of cytotoxic proteins from APP took the same form as extravasated HRP. Additionally, our study revealed numerous platelet aggregates of varying sizes inside and outside brain veins, venules, microcirculation, and arterioles. Aggregating platelets were in various stages of disintegration. The platelets inside and outside vessels were irregularly shaped and had large numbers of pseudopodia. The endothelial cell surfaces of the leaking microvessels were consistently abnormal, having attached platelets and occasional remnants of platelets with or without contact with endotheium. Examination of vessels demonstrated pseudopodial projections extending from platelets that were inserted into the luminal endothelial

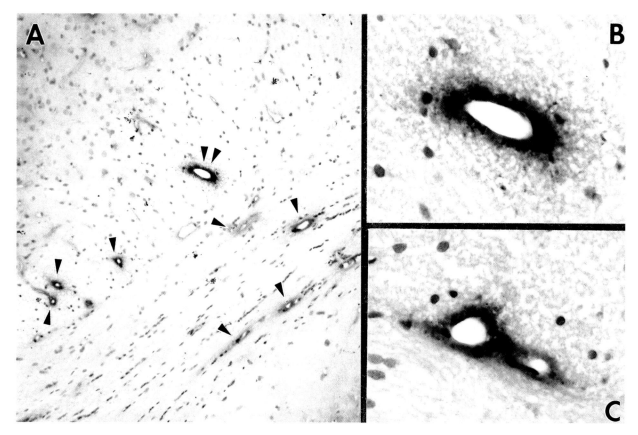

Fig. 2. (A) Perivascular C-terminal of APP-positive material in subcortical white matter and in corpus callosum (arrowheads) (×100). (B) High magnification of vessel noted with 2 arrowheads in Fig. 2A. Rarefaction of white matter in vicinity of vessel (×400). (C) β-amyloid peptide-positive material surrounding 2 vessels in border between subcortical white matter and corpus callosum. Rarefaction of white matter in vicinity of vessels (×400)

cell surface. Endothelial microvillus projections were also seen extending toward the luminal platelets. Some vessels were partially or completely plugged by aggregating platelets and their membranous remnants. Endothelial cells demonstrated both functional activation and pathological changes with clear evidence of perivascular edema. Platelet aggregates, like BBB changes and C-terminals of APP deposits, were focal, random, and dispersed. C-terminals of APP deposits and platelet pathology correlated well with BBB insufficiency. Control rat brains showed no HRP leakage, and APP staining and platelet deposition inside and outside vessels.

Discussion

This study implicated chronic insufficiency of BBB in white matter after a period of ischemia followed by long reperfusion as a secondary event in the pathological changes during ischemic neurodegeneration. The diffuse distribution of neurotoxic fragments of APP [16] around BBB vessels suggests an uncontrolled passage of these proteins through the endothelial cell body. This was confirmed by intravenous injection of human β-amyloid peptide into ischemic rats [11]. On other hand, diffuse leakage of HRP and C-terminals of APP through the endothelial cell cytoplasm may indicate chronic endothelial cell damage. Extravasations of cytotoxic parts of APP [16], which was most marked in the paramedian part of the corpus callosum facing the lateral ventricle, may suggest a vulnerability of the BBB vessels in this region, or it may indicate the evacuation of C-terminals of APP fragments through the ependymal cells of the lateral ventricle to cerebrospinal fluid.

Our finding of pseudopodial projections in both platelets and endothelial cells directed toward each other provides physical evidence for endothelial cell/platelet adhesion and attachment [9]. This may have an important functional role in transendothelial platelet passage and destruction [5, 9]. Chronic micro-BBB lesions [5] and platelets in the perivascular space [5, 9] combined with extravasated neurotoxic fragments of APP [16] may be involved in the gradual maturation of an injurious process in ischemic white matter, which may lead to severe and progressive dementia over a lifetime. Progressive damage of the white matter after ischemia may be caused not only by degeneration of axons and neurons destroyed during ischemic injury, but also by pathological changes in BBB vessels with deposition of neurotoxic fragments of APP. Accumulation of the cytotoxic fragments of APP in ischemic axons as granule-like deposits probably represents inhibition of axoplasmic flow with eventual axotomy and irreversible injury. On the other hand, the presence of blood components outside BBB vessels [2, 5, 9, 11] may have harmful effects on myelin directly [13] and may enhance the phagocytic activity of microglial cells [1].

The exact mechanism(s) by which the white matter injury occurs in ischemic stroke and Alzheimer's disease is unknown; the trigger can be brain ischemia with secondary chronic BBB dysfunction. This idea is supported by random and diffuse white matter changes that are related to a BBB injury [14, 15]. We further examined the role of cerebral ischemia injury with an alternative hypothesis proposing that repetitive and silent periods of micro-ischemic-reperfusion may form the basis for development of chronic neurodegenerative disorders such as sporadic Alzheimer's disease [6]. This process may occur by increasing the sensitivity of ischemic neurons and ischemic white matter to β-amyloid peptide formation and aberrant APP processing and/or extravasations of neurotoxic fragments of APP [16] through ischemic BBB from blood [2, 5, 11]. The BBB acts as a doorkeeper of the neurons' internal milieu. It is recognized that the BBB protects neuronal cell bodies and axons from injury, not only by stopping toxic substances but also by regulating the optimal extracellular environment required for proper functioning of the neuronal network in the brain. The neuropathology in our study appears to have a similar distribution to that of sporadic Alzheimer's disease [2, 6, 14]. We found chronic micro-BBB openings in ischemic white matter that probably would act as seeds of future Alzheimer's-type pathology.

Conclusion

Our data indicates that chronic insufficiency of BBB in white matter following ischemia-reperfusion injury is a fundamental event in the maturation of neurodegenerative processes in the neuronal network.

Acknowledgments

This work was supported in part by funds from European Community, Polish Community of Scientific Research and Medical Research Centre.

References

1. D'Andrea MR, Cole GM, Ard MD (2004) The microglial phagocytic role with specific plaque types in the Alzheimer disease brain. Neurobiol Aging 25: 675–683
2. Mehta PD, Pirttila T (2002) Biological markers of Alzheimer's disease. Drug Dev Res 56: 74–84
3. Pantoni L, Garcia JH, Gutierrez JA (1996) Cerebral white matter is highly vulnerable to ischemia. Stroke 27: 1641–1647
4. Pluta R (2000) The role of apolipoprotein E in the deposition of β-amyloid peptide during ischemia-reperfusion brain injury. A model of early Alzheimer's disease. Ann NY Acad Sci 903: 324–334
5. Pluta R (2003) Blood-brain barrier dysfunction and amyloid precursor protein accumulation in microvascular compartment following ischemia-reperfusion brain injury with 1-year survival. Acta Neurochir [Suppl] 86: 117–122
6. Pluta R (2004) From brain ischemia-reperfusion injury to possible sporadic Alzheimer's disease. Curr Neurovasc Res 1: 441–453
7. Pluta R, Lossinsky AS, Mossakowski MJ, Faso L, Wisniewski HM (1991) Reassessment of a new model of complete cerebral ischemia in rats. Method of induction of clinical death, pathophysiology and cerebrovascular pathology. Acta Neuropathol (Berl) 83: 1–11
8. Pluta R, Kida E, Lossinsky AS, Golabek AA, Mossakowski MJ, Wisniewski HM (1994) Complete cerebral ischemia with short-term survival in rats induced by cardiac arrest. I. Extracellular accumulation of Alzheimer's β-amyloid protein precursor in the brain. Brain Res 649: 323–328
9. Pluta R, Lossinsky AS, Walski M, Wisniewski HM, Mossakowski MJ (1994) Platelet occlusion phenomenon after short- and long-term survival following complete cerebral ischemia in rats produced by cardiac arrest. J Himforsch 35: 463–471
10. Pluta R, Lossinsky AS, Wisniewski HM, Mossakowski MJ (1994) Early blood-brain barrier changes in the rat following transient complete cerebral ischemia induced by cardiac arrest. Brain Res 633: 41–52
11. Pluta R, Barcikowska M, Januszewski S, Misicka A, Lipkowski AW (1996) Evidence of blood-brain barrier permeability/leakage for circulating human Alzheimer's β-amyloid-(1–42)-peptide. Neuroreport 7: 1261–1265
12. Pluta R, Barcikowska M, Kida E, Zelman I, Mossakowski MJ (1997) Late extracellular deposits of β-amyloid precursor protein in ischaemic rat brain show different immunoreactivity to the N- and C-terminal. Alzheimer Res 3: 51–57
13. Silberberg DH, Manning MC, Schreiber AD (1984) Tissue culture demyelination by normal human serum. Ann Neurol 15: 575–580
14. Tomimoto H, Akiguchi I, Suenaga T, Nishimura M, Wakita H, Nakamura S, Kimura J (1996) Alterations of the blood-brain barrier and glial cells in white-matter lesions in cerebrovascular and Alzheimer's disease patients. Stroke 27: 2069–2074

15. Ueno M, Tomimoto H, Akiguchi I, Wakita H, Sakamoto H (2002) Blood-brain barrier disruption in white matter lesions in a rat model of chronic cerebral hypoperfusion. J Cereb Blood Flow Metab 22: 97–104
16. Yokota M, Saido TC, Tani E, Yamaura I, Minami N (1996) Cytotoxic fragment of amyloid precursor protein accumulates in hippocampus after global forebrain ischemia. J Cereb Blood Flow Metab 16: 1219–1223

Correspondence: Ryszard Pluta, Department of Neurodegenerative Disorders, Medical Research Centre, Polish Academy of Sciences, Pawinskiego 5 Str., 02-106 Warsaw, Poland. e-mail: pluta@medres.cmdik.pan.pl

Time profile of eosinophilic neurons in the cortical layers and cortical atrophy

L. Sun[1], T. Kuroiwa[2], S. Ishibashi[1], N. Katsumata[2], S. Endo[3], and H. Mizusawa[1]

[1] Department of Neurology and Neurological Science, Tokyo Medical and Dental University, Tokyo, Japan
[2] Department of Neuropathology, Tokyo Medical and Dental University, Tokyo, Japan
[3] Department of Animal Research Center, Tokyo Medical and Dental University, Tokyo, Japan

Summary

Eosinophilic neurons (ENs) appear in the post-ischemic cortex; however, whether there are differences in the time profile for different cortical layers and the fate of the cortex with ENs is largely unknown. We examined the time profile of ENs in different cortical layers and evolution of cortical atrophy after transient cerebral ischemia in Mongolian gerbils. Unilateral forebrain ischemia was induced twice by 10-minute unilateral common carotid artery occlusions. Brains at 24 hours, 4 days, and 2, 4, and 16 weeks post-ischemia were prepared for morphometric analysis.

Quantitative analysis of ENs in regions of interest in the rostral and caudal cortex showed the highest number of ENs at 4 days post-ischemia in layers 3 and 6. Reduction in ENs after this peak was slower in layer 6 than in layer 3 in both rostral and caudal cortex, and this difference was significant in layer 6 of the caudal cortex. Infarcts with significant atrophy appeared in the rostral cortex. In the caudal cortex, only selective neuronal death with mild but distinct atrophy was observed.

We observed a significant difference between cortical layers in the time profile of ENs in the post-ischemic cortex. Selective neuronal death without infarction was sufficient to induce cortical atrophy after transient cerebral ischemia.

Keywords: Eosinophilic neuron; cortical layer 3 and 6; cortical atrophy; Mongolian gerbil; global ischemia.

Introduction

Focal brain ischemia induces 2 types of morphological injury: selective neuronal death and infarction. The area of selective neuronal death has been shown to be located around the ischemic center (infarction) in both clinical and laboratory investigations [3, 11]. Eosinophilic neurons (ENs) signifying neurons with homogeneous eosinophilic cytoplasm and pyknotic nuclei are called alternatively ischemic neurons and thought to be a form of neurons injured by ischemia. ENs are often found at the middle and deep layer of the post-ischemic cortex.

It has been shown that the evolution of both selective neuronal death and infarction is slow when the ischemic insult is mild. Both changes appear several days after transient cerebral ischemia [1, 4–8]. We examined whether there are differences in the time profile of ENs in different layers of the post-ischemic cortex, and whether post-ischemic cortex with only ENs and no infarcts evolves atrophy or not.

Materials and methods

We used adult male Mongolian gerbils (n = 36), 16 to 20 weeks of age. Our study was conducted in accordance with National Institutes of Health guidelines for the care and use of animals in research [10].

Surgical procedures

For induction of ischemia, the animals were anesthetized with 2% isoflurane, the left common carotid artery was occluded with a miniature vascular clip, the clip was removed after 10 minutes of occlusion, and the animals were allowed to recover from anesthesia. During the initial carotid artery occlusion, stroke symptoms were evaluated according to a stroke index (SI) of 0 to 25 points [12], by which the animals with SIs greater than 10 points have been shown to evolve brain infarction. In this study, animals manifesting SI scores of between 14 and 17 points were selected as "post-ischemic animals" to minimize the variation of the extent of ischemic tissue injury. In the post-ischemic animals, a second 10-minute period of ischemia was induced similarly 5 hours later. Sham-operated animals (n = 4) were operated upon in the same manner except for the left carotid artery occlusion.

Histological preparation

At 24 hours, 4 days, 2 weeks, 4 weeks, and 16 weeks after ischemia, animals (n = 4, each group) were anesthetized deeply with diethylether and then fixed by transcardiac perfusion with 10% phosphate-buffered formalin fixative. The brain was cut into 2 serial coronal sections at 0.5 mm anterior to the bregma and 3 mm posterior to bregma, which corresponds to chiasmal level (rostral face) and superior colliculus level (caudal face). These sections were

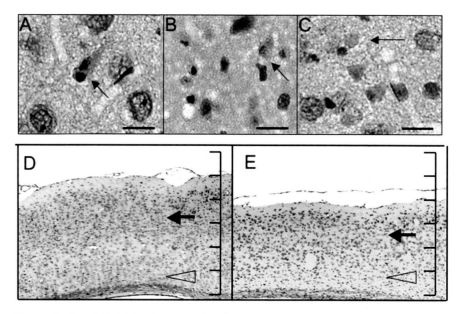

Fig. 1. (A, B, and C): Light microscopy of various types of ENs (arrows) in the post-ischemic cortex. (A) ENs showing pyknosis; (B) ENs showing karyorrhexis or karyolysis; (C) ENs showing loss of nuclear hematoxylinophilia (HE staining; bars = 20 μm). (D and E): Typical cortical atrophy without infarction, stained with Klüver-Barrera. On caudal face, the dorsolateral cortex without infarction (E) showed mild atrophy compared to the contralateral cortex (D). (Arrows: area of layer 3; arrowheads: area of layer 6. Scale: 200 μm)

embedded in paraffin and sliced at a thickness of 4 μm, and stained with hematoxylin and eosin for light microscopic examination. EN was identified as a neuron exhibiting pyknosis, karyorrhexis, karyolysis, and loss of affinity for hematoxylin, with cytoplasmic eosinophilia (Fig. 1A, 1B, 1C).

Morphometric analysis

The vertical distribution of ENs in the cortex was examined by counting the number in each cortical layer at regions of interest (ROIs) set in the coronal plane, using a drawing tube (Nikon, Tokyo, Japan) attached to a microscope at a magnification of ×40. The ROIs, cortical columns 0.5 mm in width placed perpendicular to the cortical surface, were the areas that showed consistently evolving infarction and only EN change, which correspond to 5.3 mm left of midline in the rostral face and 3.5 mm left of midline in the caudal face, respectively (Fig. 2A). Cortical atrophy was calculated as the ratio of the ipsilateral to contralateral cortical thickness at the sites that showed consistently evolving infarction.

Statistical analysis

Data were analyzed by repeated-measures analysis of variance with independent variables of treatment group and day of testing, followed by the Bonferroni post hoc test for multiple comparisons between groups. The level of statistical significance was set at $p < 0.05$. All values are presented as mean ± SD.

Results

On the rostral face, EN areas appeared in dorsolateral cortical layers 3, 5, and 6 at 24 hours post-ischemia in all animals. The EN areas enlarged to cover the whole layer at 4 days and localized laminar infarction was found in the centers of the high EN density areas. By 2 and 4 weeks, infarction had extended to the whole cortical layer. Atrophy significantly progressed from 77.4% of the control at 2 weeks to 57.1% at 16 weeks post-ischemia.

On the caudal face, areas of EN appeared in layers 3, 5, and 6 of the dorsolateral cortex at 24 hours, increased in size by 4 days, and then gradually decreased in size from 2 weeks to 16 weeks. No infarction was observed in the cortex of any animals throughout the post-ischemic period. Mild but distinct atrophy to 80.5% of the control was observed from 2 to 16 weeks post-ischemia (Fig. 1D, 1E).

Both in cortical layers 3 and 6 in the ROIs of the rostral and caudal face, ENs appeared by 24 hours, peaked at 4 days post-ischemia, and decreased over 2 to 16 weeks. However, the decrease was slower in layer 6 than in layer 3 in both rostral and caudal faces, and this difference was significant in layer 6 of the caudal face (Fig. 2B, 2C).

Discussion

We used a gerbil model of repeated unilateral carotid artery occlusion. This model induces widespread selective neuronal death mainly in the cerebral cortex in addition to infarction. The distribution of infarction and selective neuronal death in the present study are

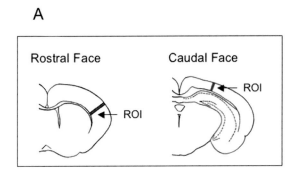

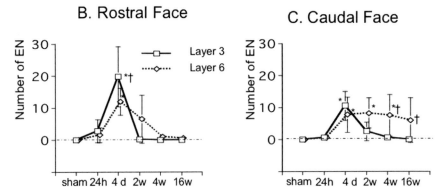

Fig. 2. Quantitative analysis of ENs in layers 3 and 6 of the cortex on rostral and caudal faces. (A) ROI were set in cortical columns 0.5 mm wide: 5.3 mm left of midline in the rostral face and 3.5 mm left of midline in the caudal face. (B) On the rostral face, the number of ENs peaked 4 days after ischemia in both layers 3 and 6. The rate of disappearance of ENs showed no distinct difference between layers 3 and layer 6 from 2 to 16 weeks after ischemia. (C) On the caudal face, the number of ENs peaked at 4 days post-ischemia in both layers 3 and 6; however, the disappearance of ENs was significantly slower in layer 6. *$p < 0.05$, compared with sham; †$p < 0.05$, compared between layers 3 and 6

similar to those in our previous study [8]. This distribution in the cerebral cortex is suitable for examination of the vulnerability of different cortical layers and whether cortical atrophy may result from selective neuronal death.

The pyramidal cells in layers 3 and 6 of the dorsolateral cortex are known to be vulnerable to ischemia, whereas the neurons in layers 2, 4, and 5 are more resistant. More pyramidal cells are killed in layer 3 than in layer 6 [9]. Our results are in accordance with these findings. ENs were detected more frequently in layers 3, 5, and 6 than in layers 2 and 4 at 24 hours and at 4 days post-ischemia (data not shown). Cortical laminar infarction preferentially evolved in the middle layer. A recent study of distribution of glutamic acid decarboxylase (GAD)-mRNA positive cells showed that GAD-positive cells, which are known to be resistant to ischemia, are present in all cortical layers with a slightly higher prevalence in layers 2 and 4 after photothrombosis [2]. This is a possible mechanism of neuronal selective vulnerability in layers 3, 5, and 6.

The reason why ENs were present longest in layer 6 is unknown. Differences in the axonal projections of the cortex may influence the vulnerability of neurons to ischemia by the retrograde or trans-synaptic degeneration mechanisms. For example, pyramidal neurons in layer 3 have either cortico-cortical projections or callosal projections, but those in layer 6 have thalamic projections.

In conclusion, we observed a significant difference in cortical layers in the time profile of ENs in the post-ischemic cortex. Selective neuronal death without infarction was sufficient to induce cortical atrophy after transient cerebral ischemia.

References

1. Du C, Hu R, Csemansky CA, Hsu CY, Choi DW (1996) Very delayed infarction after mild focal cerebral ischemia: a role for apoptosis? J Cereb Blood Flow Metab 16: 195–201
2. Frahm C, Haupt C, Witte OW (2004) GABA neurons survive focal ischemic injury. Neuroscience 127: 341–346

3. Garcia JH, Lossinsky AS, Kauffman FC, Conger KA (1978) Neuronal ischemic injury: light microscopy, ultrastructure and biochemistry. Acta Neuropathol 43: 85–95
4. Garcia JH, Liu KF, Ye ZR, Gutierrez JA (1997) Incomplete infarct and delayed neuronal death after transient middle cerebral artery occlusion in rats. Stroke 28: 2303–2309; discussion 2310
5. Hanyu S, Ito U, Hakamata Y, Nakano I (1997) Topographical analysis of cortical neuronal loss associated with disseminated selective neuronal necrosis and infarction after repeated ischemia. Brain Res 767: 154–157
6. Ito U, Spatz M, Walker JT Jr, Klatzo I (1975) Experimental cerebral ischemia in Mongolian gerbils. I. Light microscopic observations. Acta Neuropathol 32: 209–223
7. Kirino T (1982) Delayed neuronal death in the gerbil hippocampus following ischemia. Brain Res 239: 57–69
8. Kuroiwa T, Mies G, Hermann D, Hakamata Y, Hanyu S, Ito U (2000) Regional differences in the rate of energy impairment after threshold level ischemia for induction of cerebral infarction in gerbils. Acta Neuropathol 100: 587–594
9. Lin CS, Polsky K, Nadler JV, Crain BJ (1990) Selective neocortical and thalamic cell death in the gerbil after transient ischemia. Neuroscience 35: 289–299
10. National Research Council (1996) Guide for the Care and Use of Laboratory Animals. Washington DC, National Academy Press, pp 1–125
11. Nedergaard M (1988) Mechanisms of brain damage in focal cerebral ischemia. Acta Neurol Scand 77: 81–101
12. Ohno K, Ito U, Inaba Y (1984) Regional cerebral blood flow and stroke index after left carotid artery ligation in the conscious gerbil. Brain Res 297: 151–157

Correspondence: T. Kuroiwa, Department of Neuropathology, Medical Research Institute, Tokyo Medical and Dental University, 1-5-45, Yushima, Bunkyo-ku, Tokyo 113-8510, Japan. e-mail: t.kuroiwa.npat@mri.tmd.ac.jp

Forebrain ischemia and the blood-cerebrospinal fluid barrier

S. R. Ennis[1] and R. F. Keep[1,2]

[1] Department of Neurosurgery, University of Michigan, Ann Arbor, Michigan, USA
[2] Department of Physiology, University of Michigan, Ann Arbor, Michigan, USA

Summary

Although the effects of cerebral ischemia on the blood-brain barrier have been extensively studied, the effects on the blood-cerebrospinal fluid barrier (BCSFB) at the choroid plexuses have received much less attention. This paper reviews evidence on the effects of cerebral ischemia on the choroid plexus, particularly focusing on the degree of blood flow reduction required to damage the lateral ventricle choroid plexuses during transient forebrain ischemia, and whether disruption of the BCSFB might affect nearby tissues.

Studies have shown that 2 common models of forebrain ischemia (4-vessel and 2-vessel with hypotension) cause damage to the lateral ventricle choroid plexus via necrosis and apoptosis. We have found that bilateral common carotid artery occlusion with hypotension causes an 87% reduction in lateral ventricle choroid plexus blood flow during ischemia and an approximate tripling of the permeability of the BCSFB to inulin after 6 hours of reperfusion. Interestingly, evidence suggests that this disruption of the BCSFB rather than disruption to the blood-brain barrier is the major cause of enhanced inulin entry into the hippocampus. The hippocampus undergoes selective delayed neuronal loss in that model of forebrain ischemia and the BCSFB disruption may participate in or modulate that delayed injury.

Keywords: Carotid artery occlusion; hypotension; stroke; choroid plexus.

Introduction

Two barrier systems serve as an interface between blood and brain. The blood-brain barrier (BBB), formed by cerebral endothelial cells and their linking tight junctions, and the blood-cerebrospinal fluid barrier (BCSFB), formed by the choroid plexus (CP) epithelial cells with their linking tight junctions and the arachnoid membrane. The effects of cerebral ischemia (focal or global) on the BBB have been extensively studied. Thus, ischemia causes BBB disruption with the extravasation of plasma proteins and vasogenic edema formation. In addition, ischemia triggers changes in the endothelium that result in the migration of leukocytes into the injured brain, contributing to inflammation and brain injury. The effects of cerebral ischemia on CP and BCSFB function has been much less studied [6], even though the early studies of Pulsinelli *et al.* [9] found that transient forebrain ischemia caused CP necrosis after 6 hours of reperfusion. This pattern of injury contrasts to nearby brain parenchyma where there is delayed cell death. We review the evidence of the effects of forebrain ischemia on the CP with a particular focus on what degree of CP ischemia is necessary to cause CP injury, and the effect of CP injury on tissues bordering the cerebrospinal fluid (CSF) system.

Forebrain ischemia-induced CP injury in rat

Pulsinelli *et al.* [9] first examined the effects of transient forebrain ischemia (bilateral carotid and vertebral artery occlusion, 4VO, with reperfusion of the carotid arteries after 10, 20, or 30 minutes) on lateral ventricle CP morphology in the rat. They found evidence of necrosis after 6 hours. Johanson *et al.* [5], using bilateral carotid occlusion (2VO) with hypotension in the rat, also found early damage to the epithelial brush border, organelles, and nucleus following reperfusion (and restoration of blood pressure) with rapid necrosis. Although there is evidence of necrosis in these models, Ferrand-Drake and Wieloch [3] have also reported evidence supporting CP apoptosis after 18–24 hours of reperfusion in the rat 2VO with hypotension model, and Kitagawa *et al.* [7] found evidence of apoptosis as a result of reperfusion after 5 minutes of 2VO in the gerbil.

We have examined the effects of ischemia on CP epithelial function by measuring CP glutamine transport, a sodium-dependent process which indirectly utilizes adenosine triphosphate. CP glutamine transport was reduced by 45% and 72% after 10 and 30 minutes of

Table 1. *Changes in blood flow during ischemia and the influx of [^{3}H] inulin during reperfusion in different regions of the brain [2]*

	Blood flow (ml/g/min)		Inulin K_i (µl/g/min)	
	Control	Ischemia	Control	Ischemia + reperfusion
CSF	–	–	0.15 ± 0.01	0.43 ± 0.10* (287%)
Choroid plexus	2.46 ± 0.22	0.33 ± 0.06*** (13%)	–	–
Anterior cortex	0.94 ± 0.08	0.07 ± 0.03*** (7%)	0.041 ± 0.002	0.077 ± 0.014* (188%)
Hippocampus	0.92 ± 0.05	0.19 ± 0.05*** (21%)	0.048 ± 0.002	0.16 ± 0.035** (333%)

Measurements of blood flow were made in control rats and animals subjected to 10 minutes of ischemia. Measurements of the influx rate constant (K_i) for inulin were made in control rats and animals subjected to 30 minutes of ischemia with 6 hours of reperfusion. Values are mean ± SE, n = 7–8 for the influx rate constants, and n = 4 for blood flows.
*, ** and *** indicate a difference from control at $p < 0.05$, $p < 0.01$, and $p < 0.001$ levels, respectively. Numbers in parentheses are the ischemia or ischemia with reperfusion values expressed as a % of control. *CSF* cerebrospinal fluid.

permanent 2VO with hypotension in the rat [2]. Although CP glutamine transport returned to control values if the brain was reperfused after 10 minutes of ischemia, there was a residual deficit following reperfusion after 30 minutes [2]. Dienel [1] also found a long-term derangement in CP calcium homeostasis after 30 minutes of 4VO with reperfusion in the rat.

CP blood flow during ischemia

The levels of CP blood flow required to induce injury have received little study. Ten minutes of 2VO with hypotension caused a marked reduction in blood flow to the lateral ventricle CPs in the rat (~87%; [2]). By contrast, in the brain parenchyma, there was a 93% reduction in the anterior cortex and a 79% reduction in the hippocampus (Table 1; [2]).

BCSFB disruption during reperfusion

The effects of CP injury on BCSFB function merit further investigation. Ikeda *et al.* [4] found increased blood to CSF calcium flux after 5 minutes of 2VO with reperfusion in the gerbil. In our studies, rats receiving 6 hours of reperfusion after 30 minutes of 2VO with hypotension showed a marked increase in the influx rate constant for [^{3}H] inulin entry into CSF (to ~300% of control; [2]). Those animals also showed a marked increase of entry into hippocampus and a more modest increase into anterior cortex (Table 1).

Discussion

A number of studies have now shown that the lateral ventricle CPs are damaged by forebrain ischemia (reviewed in greater detail in [6]), and a number of these papers have also shown a fairly rapid recovery in CP function following the ischemic event [6]. Although not the subject of this review, an understanding of those mechanisms may be important for understanding BBB as well as BCSFB function after a stroke.

Studies on the CP blood flows necessary to induce such injury have been few. We have found that 2VO occlusion with hypotension induces a very marked 87% reduction in lateral CP blood flow [2]. This model of ischemia has been shown to induce apoptosis and necrosis in the CP [6]. The percentage reduction in blood flow that induced CP injury is similar to that which is known to induce parenchymal damage. It should be noted, however, that because control CP blood flows are much higher than in other brain regions like cerebral cortex, the absolute blood flows (~33 ml/100 g/min) that induced CP injury would likely not induce injury in the brain parenchyma. Thus, in terms of absolute blood flows, the CP may actually be selectively vulnerable to ischemia. This is also suggested by the work of Pulsinelli *et al.* [9], who found that forebrain ischemia caused early CP injury but much later damage in the hippocampus.

The lateral ventricle CPs receive blood from both the anterior and posterior choroidal arteries. This raises the question of whether blood flows from both sources would have to be compromised (such as in a heart attack) to reach the flows necessary to cause CP injury. Recently, however, Liebeskind and Hurst [8] have reported infarction of a lateral CP following a posterior choroidal artery stroke in man.

Ischemic damage to the CP might affect parenchymal injury by a number of different mechanisms [6].

For example, the CP may produce growth factors that would normally protect periventricular tissues from ischemic damage [5]. Another mechanism by which CP damage might affect parenchymal damage is by allowing the entry of blood components into the CSF. For this to be relevant pathophysiologically, however, the entry of such compounds across the damaged BCSFB would have to be greater than across the BBB. Recent evidence from our laboratory suggests that this is the case for the hippocampus following 2VO with hypotension [2]. In that model, despite the hippocampus blood flows being less affected than those of the anterior cortex (a 79% reduction in flow vs. a 93% reduction), there was a greater uptake of inulin into hippocampus from blood than into the anterior cortex (Table 1). The likely cause of this disparity is not differences in the degree of BBB disruption within the 2 tissues, but rather differences in proximity to the CSF system. For the hippocampus, which has close proximity to the CSF system, tissue inulin entry closely followed that into CSF [2]. In contrast, for the anterior cortex, which is seldom influenced by CSF, there was no correlation between tissue inulin entry with that into CSF [2].

In conclusion, the CP may be selectively vulnerable to ischemia in terms of absolute blood flows required to induce ischemic damage. In addition, ischemia-induced disruption of the CP may impact tissues adjacent to the ventricular system and may be a major cause of increased entry of blood-borne solutes into those tissues.

Acknowledgments

This work was supported by a grant from the National Institutes of Health (NS 34709 to RFK).

References

1. Dienel GA (1984) Regional accumulation of calcium in postischemic rat brain. J Neurochem 43: 913–925
2. Ennis SR, Keep RF (2005) The effects of cerebral ischemia on the rat choroid plexus. J Cereb Blood Flow Metab [in press]
3. Ferrand-Drake M, Wieloch T (1999) The time-course of DNA fragmentation in the choroid plexus and the CA1 region following transient global ischemia in the rat brain. The effect of intra-ischemic hypothermia. Neuroscience 93: 537–549
4. Ikeda J, Mies G, Nowak TS, Joo F, Klatzo I (1992) Evidence for increased calcium influx across the choroid plexus following brief ischemia of gerbil brain. Neurosci Lett 142: 257–259
5. Johanson CE, Palm DE, Primiano MJ, McMillan PN, Chan P, Knuckey NW, Stopa EG (2000) Choroid plexus recovery after transient forebrain ischemia: role of growth factors and other repair mechanisms. Cell Mol Neurobiol 20: 197–216
6. Keep RF, Xiang J, Ennis SR (2005) The blood-CSF barrier and cerebral ischemia. In: Zheng W, Chodobski A (eds) The blood-cerebrospinal fluid barrier. Taylor & Francis, Boca Raton, pp 245–260
7. Kitagawa H, Setoguchi Y, Fukuchi Y, Mitsumoto Y, Koga N, Mori T, Abe K (1998) DNA fragmentation and HSP72 gene expression by adenovirus-mediated gene transfer in postischemic gerbil hippocampus and ventricle. Metab Brain Dis 13: 211–223
8. Liebeskind DS, Hurst RW (2004) Infarction of the choroid plexus. AJNR Am J Neuroradiol 25: 289–290
9. Pulsinelli WA, Brierley JB, Plum F (1982) Temporal profile of neuronal damage in a model of transient forebrain ischemia. Ann Neurol 11: 491–498

Correspondence: Richard F. Keep, Department of Neurosurgery, University of Michigan, R5550 Kresge I, Ann Arbor, Michigan 48109-0532, USA. e-mail: rkeep@umich.edu

Neurological dysfunctions versus apparent diffusion coefficient and T2 abnormality after transient focal cerebral ischemia in Mongolian gerbils

N. Katsumata[1], T. Kuroiwa[1], I. Yamada[2], Y. Tanaka[3], S. Ishibashi[4], S. Endo[5], and K. Ohno[3]

[1] Department of Neuropathology, Tokyo Medical and Dental University, Tokyo, Japan
[2] Department of Radiology, Tokyo Medical and Dental University, Tokyo, Japan
[3] Department of Neurosurgery, Tokyo Medical and Dental University, Tokyo, Japan
[4] Department of Neurology & Neurological Science, Tokyo Medical and Dental University, Tokyo, Japan
[5] Animal Research Center, Graduate School of Medicine, Tokyo Medical and Dental University, Tokyo, Japan

Summary

We examined temporal profiles of neurological dysfunctions and compared them with apparent diffusion coefficient (ADC) and T2 changes in ischemic cortical regions after transient focal cerebral ischemia in Mongolian gerbils.

Mongolian gerbils (n = 7) underwent right common carotid artery occlusion for 20 minutes. Asymmetric motor behavior and unilateral somatosensory dysfunction were quantified by the elevated body swing test and the bilateral asymmetry test at 0, 2, 3, and 8 days after ischemia. The results were compared to the ADC and T2 changes in the primary motor cortex and the somatosensory cortex.

Transient motor dysfunction was observed at day 2 after ischemia. MRI revealed transient and mild ADC decrease without T2 increase at day 2 after ischemia in the primary motor cortex. Persistent somatosensory dysfunction was observed at 2, 3, and 8 days after ischemia, which corresponded to a moderate ADC decrease, and a mild T2 increase in the primary somatosensory cortex at days 2 and 3 after ischemia.

Time profiles of neurological deficits concurred with ADC changes of the post-ischemic cortex responsible for the deficits. The post-ischemic lesions responsible for the neurological deficits were detectable by using ADC mapping in the acute phase after transient focal cerebral ischemia.

Keywords: Cerebral ischemia; ADC; T2; behavior.

Introduction

Detection of post-ischemia lesions responsible for neurological deficits is one of the major concerns in the acute phase of stroke. Persistent neurological dysfunctions generally appear in association with infarction, which corresponds to significant apparent diffusion coefficient (ADC) decrease with high-intensity T2. However, recent studies indicate that the areas with transient or mild decrease in ADC mapping do not always signify infarction [2, 8]. Magnetic resonance imaging (MRI) detection of mildly injured tissue responsible for neurological deficit is important for the treatment of injury, because any substantial amount of tissue that suffers from mild ischemia is known to be reversible with treatment [5]. However, studies on the coordination between transient neurological dysfunctions and milder tissue changes detected by MRI ADC/T2 mappings are scarce.

We used an animal model of cerebral ischemia, a condition in which widespread areas of selective neuronal death evolves in the cortex. We then examined temporal changes of the neurological deficits, and correlated the changes with ADC and T2 changes.

Materials and methods

Animals

Studies were approved by the animal experiment committee of Tokyo Medical and Dental University and carried out in accordance with the National Institutes of Health Guide for the Care and Use of Laboratory Animals. Adult male Mongolian gerbils (n = 27, weight 55–76 g) were housed 3 or 4 per cage on a 14:10-hour light/dark cycle.

Surgical procedures

The gerbils were divided into 2 groups: an ischemic group and a control group. The ischemic group underwent right common carotid artery occlusion with a miniature vascular clip under 2% isoflurane anesthesia, and were allowed to recover from anesthesia immediately after the occlusion. During occlusion, stroke signs were evaluated according to the Stroke Index (SI: 0 to 25 points) of Ohno et al. [7]. Animals manifesting 10–16 SI points (n = 20) were selected as mild post-ischemic animals. Twenty minutes after initiation of the occlusion, the clip was removed to restore blood flow (n = 7).

The control group animals (n = 7) underwent the same operative procedures except for the right common carotid artery occlusion. After surgery, all animals were kept in their home cages until the following behavioral analysis.

Behavioral tests

The mild ischemia animals and the control animals were subjected to a series of behavioral tests during the 8 days after ischemia. The elevated body swing test (EBST) and bilateral asymmetry test (BAT) were conducted at 0, 2, 3, and 8 days post-ischemia.

EBST

The EBST was used to evaluate asymmetrical motor behavior [4]. Animals were held by the base of the tail and elevated approximately 10 cm above a tabletop. The direction of body swing, defined as an upper body turn of 10 degrees to either side, was recorded for 1 minute during each of 3 trials per day. The numbers of left and right turns were counted, and the percentage of turns made contralateral to the ischemic hemisphere side was determined.

BAT

The BAT is a test of unilateral somatosensory dysfunction [4]. Two small adhesive-backed paper dots (each 60 mm^2) were used as tactile stimuli on the distal-radial region of the wrist of each forelimb. The time, to a maximum of 3 minutes, that it took each gerbil to remove each stimulus from the forelimb (removal time) was recorded in 3 trials per day. Individual trials were separated by at least 3 minutes.

MRI measurement

Studies were performed using a 4.7 tesla superconducting MRI with a 33 cm horizontal bore magnet and a 67 mT/m maximum gradient capability (Unity INOVA, Varian, Palo Alto, CA). Diffusion weighted imaging was performed using a multi-section, spin-echo sequence with parameters of 1500/80/1 (TR/TE/excitations), a matrix of 128 × 64, a field of view of 30 × 30 mm, and a section thickness of 2 mm without an intersection gap. The diffusion gradients were applied along the 3 orthogonal directions (x-, y-, and z-axes). The resulting values for the gradient factor were 0, or 1200 s/mm^2. For T2 calculation, multi-section spin-echo imaging without diffusion gradient was performed with parameters of 1500/20, 50, and 80 (TR/TE). Other imaging parameters included a matrix of 128 × 64, a field of view of 30 × 30 mm, a section thickness of 2 mm without intersection gap, and 1 signal acquired.

MRI data analysis

All analysis was performed using a Sun Sparc 10 workstation (Sun Microsystems Inc., Santa Clara, CA) and image analysis software (XDS software; Davis Bioengineering, St. Louis, MO). ADC maps were calculated on a pixel by pixel basis using the following equation: $ADC = \ln(S_o/S)/(b-b_o)$, and T2 maps were calculated on a pixel by pixel basis using the following equation: $SI = K \exp(-TE/T2)$.

Regions of interest

The regions of interest were set at the primary motor cortex and the somatosensory cortex. The ADC % decrease and the T2 % increase of each region of interest were calculated. ADC and T2 % decrease/increase was defined as the ADC and T2 decreased/increased ratio compared to the homologous region of the contralateral hemisphere.

Results

Time course of behavioral dysfunctions

EBST

The percentage of right-biased swings was significantly increased in the ischemia group on the day of ischemia compared to the control group. This changed to a significant left-biased swing percentage at 2 days post-ischemia in the ischemia group, and gradually normalized through 8 days (Fig. 1A).

BAT

The contralateral removal time was significantly increased in the ischemia group on the day of ischemia compared to the control group. This significant increase persisted through 8 days (Fig. 1B).

Time course of MRI changes

ADC findings

The ADC of the primary motor cortex and the somatosensory cortex of the ipsilateral cerebral cortex started to decrease on day 2 post-ischemia (Fig. 2B, 2F). The lowest ADC of the primary motor cortex was observed at day 2 post-ischemia (Fig. 2B, 2I), and gradually normalized through 8 days (Fig. 2C, 2D, 2I). The lowest ADC of the somatosensory cortex was observed at day 3 post-ischemia (Fig. 2G, 2I), and gradually normalized through 8 days (Fig. 2I). ADC decrease of the primary motor cortex was small compared to ADC decreases of the somatosensory cortex (Fig. 2I). No ADC decreases were observed in the contralateral cerebral hemisphere.

T2 findings

T2 of the somatosensory cortex of the ipsilateral cerebral cortex started to increase on day 2 post-ischemia, and high-intensity T2 was observed on day 3 post-ischemia (Fig. 2H, J). No T2 increase was ob-

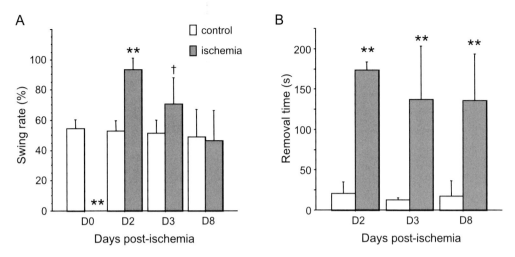

Fig. 1. Time course of behavioral dysfunctions. Contralateral turn was significantly increased at day 2 post-ischemia and gradually decreased to day 8 (A). Contralateral removal time was significantly increased at day 2 post-ischemia and persisted to day 8 (B). (**p < .01; †p < .10)

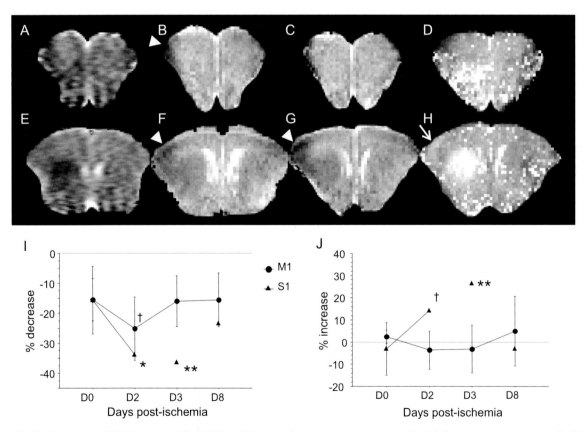

Fig. 2. Time course of MRI changes. The ADC and T2 maps of the primary motor cortex (A–D), the somatosensory cortex (E–H), and the time profile of ADC (I) and T2 (J) signal intensification. The primary motor cortex ADC decrease did not appear on the day of ischemia (A), started to appear at day 2 (B), gradually normalized from day 3 (C). No T2 increase was observed in the primary motor cortex (D). The somatosensory cortex ADC decrease did not appear on the day of ischemia (E), started to appear at day 2 (F), peaked at day 3 (G). T2 increase was observed in the somatosensory cortex at day 3 (H). Arrowheads indicate the ADC decrease, and an arrow indicates the T2 increase in the somatosensory cortex. ADC % decrease was significant in somatosensory cortex at 2 and 3 days after ischemia (I). T2 increase was significant at 2 days after ischemia in somatosensory cortex (J). (**p < .01; *p < .05; †p < .10)

served in the ipsilateral primary motor cortex (Fig. 2D, J) or the contralateral cerebral hemisphere.

Discussion

It has been shown that areas of infarction are responsible for post-ischemia neurological deficits in both clinical and laboratory investigations; however, recent evidence indicates that post-ischemia neural tissue not evolving to infarction is also responsible for post-ischemia neurological deficits [1]. The duration of neurological symptoms appears to be an indicator of the severity of tissue injury in the cerebral ischemia. Our focus in this study was to elucidate the relationships between the time profile of mild neurological deficits and changes in the ADC and T2 mapping of the cortical regions involved after transient cerebral ischemia.

Mild ADC decrease and transient motor dysfunction

A left-biased swing in the EBST, signifying transient motor dysfunction, was observed on day 2 after ischemia. A mild and transient decrease of ADC was also observed in the ipsilateral primary motor cortex on day 2 after ischemia, but returned to the baseline level 3 days after ischemia. The time course of motor dysfunction paralleled the ADC decrease as shown in Figs. 1A and 2I. A T2 change was not observed at the same region of interest. These findings complement recent reports that an area of transient and mild ADC decrease does not necessarily evolve to infarction [2, 6, 8]. In a previous report, Desmond *et al.* [2] stated regions of an ADC less than 75% indicates an infarct core, but regions of an ADC greater than 90% are unlikely to be infarcted. Transient neurological deficit concurring with transient and mild ADC decrease appears to be post-ischemia tissue without infarction, indicating that post-ischemia mild tissue injury causing transient neurological deficits is detectable by ADC mapping.

Moderate ADC decrease and persistent somatosensory dysfunction

In this study, persistent somatosensory dysfunction signified by the prolongation of tape removal time in the BAT was observed at 2, 3, and 8 days after ischemia. ADC mapping revealed moderate and prolonged decrease in the ipsilateral somatosensory cortex at 2 and 3 days after ischemia and normalized at 8 days post-ischemia. Concurrent changes in the neurological deficits and ADC abnormality were also found in the somatosensory cortex suffering a more severe ischemic insult.

We do not know if the normalization at day 8 after ischemia is "pseudonormalization" indicating evolution of infarction [6, 8] or not. The T2 increase in the somatosensory cortex was very mild and transient, and it returned to normal level 8 days after ischemia. These findings indicate that the lesion in the somatosensory cortex was not an infarction. Histological findings in previous reports using a similar ischemia model indicated that the areas with ADC recovery correspond to the areas of disseminated selective neuronal necrosis [3].

References

1. Croquelois A, Wintermark M, Reichhart M, Meuli R, Bogousslavsky J (2003) Aphasia in hyperacute stroke: language follows brain penumbra dynamics. Ann Neurol 54: 321–329
2. Desmond PM, Lovell AC, Rawlinson AA, Parsons MW, Barber PA, Yang Q, Li T, Darby DG, Gerraty RP, Davis SM, Tress BM (2001) The value of apparent diffusion coefficient maps in early cerebral ischemia. AJNR Am J Neuroradiol 22: 1260–1267
3. Hanyu S, Ito U, Hakamata Y, Yoshida M (1993) Repeated unilateral carotid occlusion in Mongolian gerbils: quantitative analysis of cortical neuronal loss. Acta Neuropathol (Berl) 86: 16–20
4. Ishibashi S, Kuroiwa T, Endo S, Okeda R, Mizusawa H (2003) Neurological dysfunctions versus regional infarction volume after focal ischemia in Mongolian gerbils. Stroke 34: 1501–1506
5. Kundrotiene J, Cebers G, Wagner A, Liljequist S (2004) The NMDA NR2B subunit-selective receptor antagonist, CP-101,606, enhances the functional recovery the NMDA NR2B subunit-selective receptor and reduces brain damage after cortical compression-induced brain ischemia. J Neurotrauma 21: 83–93
6. Miyasaka N, Nagaoka T, Kuroiwa T, Akimoto H, Haku T, Kubota T, Aso T (2000) Histopathologic correlates of temporal diffusion changes in a rat model of cerebral hypoxia/ischemia. ANJR Am J Neuroradiol 21: 60–66
7. Ohno K, Ito U, Inaba Y (1984) Regional cerebral blood flow and stroke index after left carotid artery ligation in the conscious gerbil. Brain Res 297: 151–157
8. Tanaka Y, Kuroiwa T, Miyasaka N, Tanabe F, Nagaoka T, Ohno K (2003) Recovery of apparent diffusion coefficient after embolic stroke does not signify complete salvage of post-ischemic neuronal tissue. Acta Neurochir [Suppl] 86: 141–145

Correspondence: T. Kuroiwa, Department of Neuropathology, Graduate School of Medicine, Tokyo Medical and Dental University, 1-5-45, Yushima, Bunkyo-ku, Tokyo 113-8519, Japan. e-mail: t.kuroiwa.npat@mri.tmd.ac.jp

Progressive expression of vascular endothelial growth factor (VEGF) and angiogenesis after chronic ischemic hypoperfusion in rat

H. Ohtaki[1,2], T. Fujimoto[1], T. Sato[1], K. Kishimoto[1], M. Fujimoto[1], M. Moriya[1], and S. Shioda[2]

[1] Department of Neurosurgery, Showa University, Fujigaoka Hospital, Kanagawa, Japan
[2] Department of Anatomy, Showa University School of Medicine, Tokyo, Japan

Summary

Cerebrovascular stenosis caused by arteriosclerosis induces failure of the cerebral circulation. Even if chronic cerebral hypoperfusion does not induce acute neuronal cell death, cerebral hypoperfusion may be a risk factor for neurodegenerative diseases. The purpose of this study was to determine if vasodilation, expression of VEGF, and neovascularization are homeostatic signs of cerebral circulation failure after permanent common carotid artery occlusion (CCAO) in the rat.

Neuronal cell death in neocortex was observed 2 weeks after CCAO and gradually increased in a time-dependent manner. The diameter of capillaries and expression of VEGF also increased progressively after CCAO. Moreover, we observed unusual irregular angiogenic vasculature at 4 weeks.

In conclusion, chronic hypoperfusion results in mechanisms to compensate for insufficiency in blood flow including vasodilation, VEGF expression, and neovascularization in the ischemic region. These results suggest that angiogenesis might be induced in adult brain through the support of growth factors and transplantation of vascular progenitor cells, and that neovascularization might be a therapeutic strategy for children and adults with diseases such as vascular dementia.

Keywords: Global ischemia; hypoperfusion; vascular endothelial growth factor; angiogenesis.

Introduction

Cerebrovascular stenosis caused by arteriosclerosis induces cerebral circulation failure. Even if chronic cerebral hypoperfusion might not induce acute neuronal cell death, magnetic resonance imaging and positron emission tomography suggest it participates in the development of vascular dementia [4, 10, 12, 20, 24, 31]. Moreover, it has been reported that cerebral hypoperfusion occurs in Alzheimer's disease and Binswanger's disease (subcortical arteriosclerotic encephalopathy) [2, 8, 15, 26, 30]. Therefore, cerebral hypoperfusion is suggested as a risk factor for neurodegenerative diseases.

Therapeutic angiogenesis is a strategy where blood vessel formation is induced for the purposes of treating and/or preventing ischemic disease such as myocardial, hind-limb, and cerebral ischemia [7, 9, 22, 23, 32]. Moyamoya disease, a cerebrovascular disease that occurs mostly in children, features angiogenesis in the brain and an increase of growth factors in the cerebrospinal fluid [19, 21, 28, 29, 34]. Angiogenesis (neovascularization) is thought to be induced by chronic cerebral hypoperfusion and, thus, is a homeostatic sign of failure of the cerebral circulation that is involved in compensation for the impaired circulation. Neovascularization through surgery is one therapeutic strategy to compensate for impaired circulation in Moyamoya disease [6, 7, 17].

It is still uncertain whether therapeutic neovascularization is appropriate for adult cerebral hypoperfusion. The purpose of this study was, therefore, to measure vasodilation by vessel diameter and the expression of vascular endothelial growth factor (VEGF) as homeostatic signs of cerebral circulation failure in young adult rats during chronic hypoperfusion.

Materials and methods

Production of permanent forebrain ischemia

Male Slc/Wistar rats aged 13 to 15 weeks (SLC, Shizuoka, Japan) were anesthetized with sodium pentobarbital (50 mg/kg, i.p.). Anesthetized rats were laid on their back and a midline neck incision made. The common carotid arteries (CCAs) were carefully exposed and isolated. Then, the CCAs were doubly ligated with 3-0 silk suture and the arteries cut between the sutures (CCA occlusion; CCAO). After the surgical operation, the rats were maintained under an infrared heat lamp until awake to avoid a decline in body temperature. All experimental procedures involving animals were approved by the Institutional Animal Care and Use Committee of Showa University.

Histology and measurement of capillary diameter

At 0, 1, 2, and 4 weeks after ischemia, the animals were re-anesthetized with sodium pentobarbital (50 mg/kg, i.p.), and the brains were removed. The brains were immediately immersed in 10% buffered formalin for fixation for 1 week and decalcified with 1% formic acid. After embedding in paraffin, 10 μm sections were cut. These were deparaffinized and used for either hematoxylin and eosin staining for morphological evaluation, or for VEGF and factor VIIIa (FVIIIa, a vascular marker) immunostaining. After boiling in 10 mmol sodium citrate buffer (pH 6.0) for 15 minutes, the sections were preincubated in 0.3% H_2O_2 to inhibit endogenous peroxidase activity. After washing, the sections were incubated in normal goat serum to block the non-specific reaction and were then incubated with a polyclonal rabbit anti-VEGF (Ab-2) antibody (1:100; Oncogene Research Products, Cambridge, MA) or rabbit anti-human FVIIIa polyclonal antibody (1:200; DakoCytomation, Glostrup, Denmark). They were developed with Histofine SAB-PO (R) kit (Nichirei, Tokyo, Japan) with diaminobenzidine (DAB) as the chromogen. Sections were counterstained with hematoxylin after the DAB reaction for cell identification. After staining, the sections were examined and images taken with the aid of light microscopy (Olympus AX-70; Olympus, Tokyo, Japan).

Capillary diameters in hippocampus and neocortex were measured in a coronal section from the bregma (-1.0 to -5.0 mm) immunostained for FVIIIa. Images (440×320 μm^2) were examined and the diameters of 5 to 7 capillaries within that area were measured. A total of 64 to 81 (n = 3–4 animals) capillaries were measured in each region of the hippocampus and neocortex at 0, 1, 2, and 4 weeks after CCAO. The quantitative determination of capillary diameter was performed using NIH Image, version 1.62 (National Institutes of Health, Bethesda, MD).

All data are expressed the mean ± standard error. Statistical comparisons were made using Dunnet's post hoc test followed by one-way ANOVA as compared to sham-operated control (0 week). A p-value of < 0.05 was considered statistically significant.

Results

Progressively increased neuronal cell death in cerebral cortex (Fig. 1)

In the absence of hypotension, occlusion of CCAs generally does not cause neuronal cell death. However, it is reported that CCAO in Slc/Wistar rats induces neuronal cell death due to the patency of the posterior communicating arteries [16]. One week after CCAO, approximately 30% of animals died and 36.4% (4 of 12) of the surviving animals showed neuronal cell death in the hippocampal CA1 region. The percentage of animals with neuronal cell death in the hippocampus increased to 57.1% (4 of 7) at 2 weeks and did not increase further at 4 weeks. In contrast, there was no marked neuronal cell death observed in the neocortex

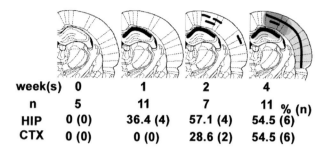

Fig. 1. Morphological changes in hippocampus (*HIP*) and neocortex (*CTX*) after chronic cerebral hypoperfusion induced by 0, 1, 2, or 6 weeks of permanent CCAO. There was a time-dependent gradual increase in neuronal cell damage in HIP and CTX, as indicated by *gray* and/or *black*. The number of animals examined (*n*) and percentage of animals with HIP and CTX neuronal damage (absolute number in parentheses) is given. By 4 weeks after CCAO, 50% or more of surviving animals had neuronal cell death in HIP and CTX

1 week after CCAO. Neuronal cell death in the neocortex gradually increased to 28.6% (2 of 7) and 54.5% (6 of 11) at 2 and 4 weeks after CCAO, respectively.

Homeostatic signs of failure of cerebral circulation

Vasodilation is thought to compensate for insufficient blood flow in brain [13]. Thus, we determined the diameter of capillaries in the hippocampus and neocortex following CCAO (Fig. 2A). The average capillary diameters were 1.7 ± 0.07 and 1.6 ± 0.05 μm in sham-operated controls (0 week) in hippocampal and neocortical regions, respectively. After ischemia, capillary diameter clearly increased and the diameter was significantly greater than pre-ischemia values 1 week after CCAO. Four weeks after CCAO, the diameters of capillaries in the hippocampus and neocortex were 2.4 (4.1 ± 0.29 μm) and 2.5 (4.0 ± 0.19 μm) fold greater than sham-operated controls, respectively.

VEGF expression was examined using immunohistochemistry. Few immunopositive reactions for VEGF were observed preischemia (Fig. 2B). After ischemia, the positive reaction for VEGF was increased in a time-dependent manner and was obvious adjacent to the vasculature (Fig. 2C–E). Immunopositive cells for VEGF were mainly observed around the infarction, but not within the infarction. VEGF was expressed in astrocytes, as demonstrated by double-immunostaining (data not shown).

VEGF is well known to participate in angiogenesis [5, 22]. Therefore, the angiogenic vasculature was determined by staining for FVIIIa. As shown in Fig. 2F and G, an unusual irregularly-constructed vasculature

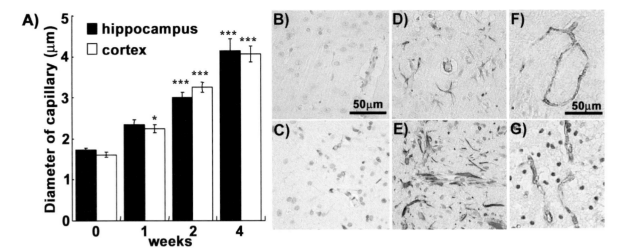

Fig. 2. Homeostatic sign of failure of cerebral circulation after chronic cerebral hypoperfusion. (A) Semi-quantification of mean capillary diameter in hippocampal (*black bar*) and neocortical (*white bar*) regions. There was a time-dependent increase in capillary diameter in both regions. Data are presented as mean ± SE. * = $p < 0.05$, *** = $p < 0.001$, as compared with sham operated control (0 week) using by Dunnett's post hoc test followed by one-way ANOVA. (B–E) VEGF-like immunoreactivity (ir) 0 (B), 1 (C), 2 (D), and 4 (E) weeks after CCAO. VEGF-ir progressively increased in a time-dependent manner. VEGF-ir was observed adjacent to the capillary as shown in (E). (F) and (G) show the brain microvasculature 4 weeks after CCAO. Unusual irregular constructed vasculature such as loops (F) and meandering (G) were often noted at 4 weeks after CCAO. This suggests that neovascularization was occurring in the brain parenchyma during chronic hypoperfusion

(e.g., loops and meandering vessels) was often noted in the peri-infarct region of cerebral cortex 4 weeks after CCAO. These were angiogenic vasculature. This implies that neovascularization was increased in the brain parenchyma during forebrain ischemia.

Discussion

The present study indicates that there is sustained vasodilation and expression of VEGF after chronic forebrain ischemia in our rat model. Moreover, we also found evidence of neovascularization by FVIIIa immunostaining at 4 weeks after CCAO.

Numerous reports on rodents, such as gerbil, rat, and mouse, have shown that global ischemia induces neuronal cell death in the hippocampus 2 or 3 days after induction [11, 14, 18, 33]. Such cell death is known as delayed neuronal cell death and is considered a target for therapy of cerebrovascular diseases. However, in most strains of rats, CCAO alone does not induce neuronal cell death due to the patency of the posterior communicating artery and few papers have reported cortical neuronal cell death in rat global ischemia in the chronic phase [1, 3, 16, 33]. Slc/Wistar rat is one strain of rats where cortical neuronal cell death is induced by CCAO alone [16]. In the present study, we reconfirmed that CCAO induces neuronal cell death in the hippocampus and neocortex in that strain. In particular, neocortical neuronal cell death was observed 2 weeks after CCAO and the infarction gradually extended in a time-dependent manner. At 4 weeks after CCAO, the neocortex had evident parenchymal atrophy and proliferation of fibroblast-like cells in the core of the infarction (data not shown). These results indicate that the CCAO model in Slc/Wistar rats might be a good model for studies of chronic hypoperfusion, such as vascular dementia.

We found that both the diameter of capillaries and the expression of VEGF increased in the ischemic region during the 4 weeks after CCAO. We hypothesize that this is a compensatory response to insufficiency of blood flow in brain. In addition, at 4 weeks after CCAO, there was evidence of revascularization in the ischemic region.

Injection of VEGF induces angiogenesis after cardiac, hind-limb, or cerebral ischemia [9, 23, 27, 32]. On the other hand, VEGF has also been reported to increase vascular permeability and inflammatory responses [5, 22, 25]. Although we need to clarify the relationship between VEGF and angiogenesis further, we hypothesize that neovascularization occurs in the adult brain to increase blood flow to ischemic regions during chronic hypoperfusion.

In conclusion, this study provides evidence that chronic brain hypoperfusion elicits mechanisms to compensate for the insufficiency of blood flow, i.e., it induces vasodilation, VEGF expression, and neovascularization in the ischemic region. These results suggest that angiogenesis might be induced in the adult brain by the application of growth factors and transplantation of vascular progenitor cells. Thus, neovascularization might be a therapeutic strategy for children and adults with diseases such as vascular dementia.

Acknowledgments

This study was supported in part by grants from the Ministry of Education, Science, Sports and Culture (TF).

References

1. Cheung WM, Chen SF, Nian GM, Lin TN (2000) Induction of angiogenesis related genes in the contralateral cortex with a rat three-vessel occlusion model. Chin J Physiol 43: 119–124
2. de la Torre JC (2004) Alzheimer's disease is a vasocognopathy: a new term to describe its nature. Neurol Res 26: 517–524
3. de la Torre JC, Fortin T, Park GA, Butler KS, Kozlowski P, Pappas BA, de Socarraz H, Saunders JK, Richard MT (1992) Chronic cerebrovascular insufficiency induces dementia-like deficits in aged rats. Brain Res 582: 186–195
4. De Reuck J, Decoo D, Marchau M, Santens P, Lemahieu I, Strijckmans K (1998) Positron emission tomography in vascular dementia. J Neurol Sci 154: 55–61
5. Ferrara N (2000) VEGF: an update on biological and therapeutic aspects. Curr Opin Biotechnol 11: 617–624
6. Fujimoto T, Mukoyama M, Asai J, Fukushima Y (1993) Reversed durapexia (RDP): A new surgical treatment of childhood Moyamoya disease. Video J Jpn Neurosurg 1
7. Fujimoto T, Asai J, Miyo T, Takahashi M, Suzuki R, Nagashima G, Hokaku H, Sato T (1997) Effects of a new treatment, reversed durapexia (RDP), for ischemic lesions. 11th International Congress of Neurological Surgery. Monduzzi Editore, Bologne, pp 1031–1035
8. Hanyu H, Shimuzu S, Tanaka Y, Takasaki M, Koizumi K, Abe K (2004) Cerebral blood flow patterns in Binswanger's disease: a SPECT study using three-dimensional stereotactic surface projections. J Neurol Sci 220: 79–84
9. Isner JM (1996) Therapeutic angiogenesis: a new frontier for vascular therapy. Vasc Med 1: 79–87
10. Kawamura J, Meyer JS, Terayama Y, Weathers S (1991) Cerebral hypoperfusion correlates with mild and parenchymal loss with severe multi-infarct dementia. J Neurol Sci 102: 32–38
11. Kirino T, Tamura A, Sano K (1986) A reversible type of neuronal injury following ischemia in the gerbil hippocampus. Stroke 17: 455–459
12. Kurz AF (2001) What is vascular dementia? Int J Clin Pract Suppl 120: 5–8
13. Lin TN, Sun SW, Cheung WM, Li F, Chang C (2002) Dynamic changes in cerebral blood flow and angiogenesis after transient focal cerebral ischemia in rats. Evaluation with serial magnetic resonance imaging. Stroke 33: 2985–2991
14. Matsunaga M, Ohtaki H, Takaki A, Iwai Y, Yin L, Mizuguchi H, Miyake T, Usumi K, Shioda S (2003) Nucleoprotamine diet derived from salmon soft roe protects mouse hippocampal neurons from delayed cell death after transient forebrain ischemia. Neurosci Res 47: 269–276
15. Meyer JS, Muramatsu K, Mortel KF, Obara K, Shirai T (1995) Prospective CT confirms differences between vascular and Alzheimer's dementia. Stroke 26: 735–742
16. Nanri M, Watanabe H (1999) Availability of 2VO rats as a model for chronic cerebrovascular disease. Nippon Yakurigaku Zasshi 13: 85–95
17. Nariai T, Suzuki R, Matsushima Y, Ichimura K, Hirakawa K, Ishii K, Senda M (1994) Surgically induced angiogenesis to compensate for hemodynamic cerebral ischemia. Stroke 25: 1014–1021
18. Ohtaki H, Dohi K, Nakamachi T, Yofu S, Endo S, Kudo Y, Shioda S (2005) Evaluation of brain ischemia in mice. Acta Histochem Cytochem 38: 99–106
19. Ono K, Fujimoto T, Komatsu K, Inaba Y, Ida T (1977) Case of occlusion of the circle of Willis associated with the growth of abnormal vascular networks at the base of the brain; its relation to the pathogenesis of so-called "Moyamoya" disease. No To Shinkei 29: 37–43 [in Japanese]
20. Parnetti L, Mari D, Mecocci P, Senin U (1994) Pathogenetic mechanisms in vascular dementia. Int J Clin Lab Res 24: 15–22
21. Perren F, Meairs S, Schmiedek P, Hennerici M, Horn P (2005) Power Doppler evaluation of revascularization in childhood moyamoya. Neurology 64: 558–560
22. Plate KH (1999) Mechanisms of angiogenesis in the brain. J Neuropathol Exp Neurol 58: 313–320
23. Pratt PF, Medhora M, Harder DR (2004) Mechanisms regulating cerebral blood flow as therapeutic targets. Curr Opin Investig Drugs 5: 952–956
24. Scheel P, Puls I, Becker G, Schoning M (1999) Volume reduction in cerebral blood flow in patients with vascular dementia. Lancet 354: 2137
25. Shibuya M (2001) Structure and function of VEGF/VEGF-receptor system involved in angiogenesis. Cell Struct Funct 26: 25–35
26. Shyu WC, Lin JC, Shen CC, Hsu YD, Lee CC, Shiah IS, Tsao WL (1996) Vascular dementia of Binswanger's type: clinical, neuroradiological and 99mTc-HMPAO SPET study. Eur J Nucl Med 23: 1338–1344
27. Sun Y, Jin K, Xie L, Childs J, Mao XO, Logvinova A, Greenberg DA (2003) VEGF-induced neuroprotection, neurogenesis, and angiogenesis after focal cerebral ischemia. J Clin Invest 111: 1843–1851
28. Takahashi A, Sawamura Y, Houkin K, Kamiyama H, Abe H (1993) The cerebrospinal fluid in patients with moyamoya disease (spontaneous occlusion of the circle of Willis) contains high level of basic fibroblast growth factor. Neurosci Lett 160: 214–216
29. Takahashi M, Fujimoto T, Suzuki R, Asai J, Miyo T, Hokaku H (1997) A case of spontaneous middle cerebral artery occlusion associated with a cerebral aneurysm angiographically disappearing after STA-MCA anastomosis. No Shinkei Geka 25: 727–732 [in Japanese]
30. Tanoi Y, Okeda R, Budka H (2000) Binswanger's encephalopathy: serial sections and morphometry of the cerebral arteries. Acta Neuropathol (Berl) 100: 347–355
31. Varma AR, Adams W, Lloyd JJ, Carson KJ, Snowden JS, Testa HJ, Jackson A, Neary D (2002) Diagnostic patterns of regional atrophy on MRI and regional cerebral blood flow change on SPECT in young onset patients with Alzheimer's disease, frontotemporal dementia and vascular dementia. Acta Neurol Scand 105: 261–269

32. Ware JA, Simons M (1997) Angiogenesis in ischemic heart disease. Nat Med 3: 158–164
33. Weinstock M, Shoham S (2004) Rat models of dementia based on reductions in regional glucose metabolism, cerebral blood flow and cytochrome oxidase activity. J Neural Transm 111: 347–366
34. Yoshimoto T, Houkin K, Takahashi A, Abe H (1996) Angiogenic factors in moyamoya disease. Stroke 27: 2160–2165

Correspondence: Hirokazu Ohtaki, Department of Anatomy, Showa University School of Medicine, 1-5-8 Hatanodai Shinagawa-ku, Tokyo, 142-8555, Japan. e-mail: taki@med.showa-u.ac.jp

Intracerebral administration of neuronal nitric oxide synthase antiserum attenuates traumatic brain injury-induced blood-brain barrier permeability, brain edema formation, and sensory motor disturbances in the rat

Hari S. Sharma[1,2], L. Wiklund[1], R. D. Badgaiyan[2], S. Mohanty[3], and P. Alm[4]

[1] Laboratory of Cerebrovascular Research, Department of Surgical Sciences, Anesthesiology and Intensive Care Medicine, University Hospital, Uppsala University, Uppsala, Sweden
[2] Laboratory of Neuroanatomy, Department of Medical Cell Biology, Biomedical Center, Uppsala University, Uppsala, Sweden
[3] Department of Neurosurgery, Institute of Medical Sciences, Banaras Hindu University, Varanasi, India
[4] Department of Pathology, University Hospital, Lund University, Lund, Sweden

Summary

The role of nitric oxide (NO) in traumatic brain injury (TBI)-induced sensory motor function and brain pathology was examined using intracerebral administration of neuronal nitric oxide synthase (nNOS) antiserum in a rat model. TBI was produced by a making a longitudinal incision into the right parietal cerebral cortex limited to the dorsal surface of the hippocampus. Focal TBI induces profound edematous swelling, extravasation of Evans blue dye, and up-regulation of nNOS in the injured cerebral cortex and the underlying subcortical areas at 5 hours. The traumatized animals exhibited pronounced sensory motor deficit, as seen using Rota-Rod and grid-walking tests. Intracerebral administration of nNOS antiserum (1:20) 5 minutes and 1 hour after TBI significantly attenuated brain edema formation, Evans blue leakage, and nNOS expression in the injured cortex and the underlying subcortical regions. The nNOS antiserum-treated rats showed improved sensory motor functions. However, administration of nNOS antiserum 2 hours after TBI did not influence these parameters significantly. These novel observations suggest that NO participates in blood-brain barrier disruption, edema formation, and sensory motor disturbances in the early phase of TBI, and that nNOS antiserum has some potential therapeutic value requiring additional investigation.

Keywords: Traumatic brain injury; nitric oxide; neuronal nitric oxide synthase; edema; blood-brain barrier; sensory motor functions.

Introduction

The role of nitric oxide (NO) in traumatic brain injury (TBI) is still not well understood. However, recent studies suggest that NO is involved in various forms of ischemic, excitotoxic, and hypoxic brain injuries, as well as in several other neurodegenerative diseases [1–4, 13]. NO influences neuronal communication by diffusing from one cell to another within a few seconds [11, 14, 19, 25]. However, NO is a potent free radical gas, which can induce direct damage to cell membranes if produced in abnormally high quantities within the central nervous system [3, 8, 33–35].

NO is synthesized by the enzyme nitric oxide synthase (NOS), which occurs in constitutive neuronal (nNOS) and inducible (iNOS) isoforms [13, 19, 25, 27]. Up-regulation of nNOS or iNOS results in increased NO production [2, 11, 14, 19]. While nNOS up-regulation results in mild production of NO for a short period of time, iNOS activation results in massive NO production for a long duration [11, 14].

Previous studies from our laboratory showed increased nNOS expression following hyperthermic brain injury in several brain regions associated with breakdown of blood-brain barrier (BBB) permeability, edema formation, and cell damage [17, 19, 24–28, 30]. Furthermore, increased expression of nNOS is found in lesioned cortical neurons and in Purkinje cells 3 to 42 days after injury [2–4]. These observations suggest that production of NO from nNOS is harmful to the neurons. However, the effects of pharmacological blockade of NOS in ischemia or other models of neuronal injuries are still controversial [13, 19]. This is mainly because selective blockers of different NOS isoforms are still not available. Thus, the role played by nNOS in the pathophysiology of brain injury is still speculative and requires further investigation.

An up-regulation of nNOS is seen 5 hours after spinal cord injury, and this correlates well with cell and tissue damage [22]. Interestingly, topical or systemic administration of a potent non-specific NOS inhibitor, L-NAME, did not reduce spinal cord pathology [22, 32]. On the other hand, topical application of nNOS antiserum over the traumatized spinal cord 5 minutes after injury markedly attenuated microvascular permeability disturbances, edema formation, and cell injury [15, 18, 19, 22, 32]. In antiserum-treated traumatized rats, the expression of nNOS is also blocked [19, 22]. These observations suggest that up-regulation of nNOS is neurotoxic and contributes to cell and tissue damage.

Since basic mechanisms of cell and tissue injury in brain or spinal cord appear to be similar in nature [20], it appears that NO is somehow involved in TBI-induced BBB disruption, brain edema formation, and sensory motor dysfunction. In this investigation, the role of nNOS in brain pathology and sensory motor dysfunction following brain injury was examined using intracerebral administration of nNOS antiserum at various time intervals in a TBI rat model.

Materials and methods

Animals

Experiments were carried out on male Sprague-Dawley rats (350–450 g) housed at a controlled ambient temperature ($21 \pm 1\,°C$) with a 12-hour light, 12-hour dark schedule. Food pellets and tap water were provided ad libitum before the experiments.

Traumatic brain injury

Under Equithesin anesthesia (0.3 mL/100 g, i.p.), a 4-mm^2 burr hole was made in the right and left parietal bones and the dura carefully removed to expose the underlying cerebral cortex [5, 6, 21, 29]. A longitudinal incision on the right parietal cerebral cortex (about 3 mm deep and 3 mm long) was made under stereotaxic guidance using a sharp sterile scalpel blade [5, 29]. The lesion was limited to the cerebral cortex and/or the superficial parts of the subcortical white matter [21]. The exposed brain areas in both hemispheres were covered with cotton soaked in saline to prevent drying of the exposed brain tissues [5, 21, 29]. The animals were allowed to survive 5 hours after TBI. Normal animals served as controls. These experiments were approved by the Ethics Committee of Uppsala University and Banaras Hindu University.

Sensory motor deficits

Sensory motor functions were analyzed using Rota-Rod and grid-walking tests in a blinded fashion. Rats were trained on the Rota-Rod at a 16 RPM setting for 10 minutes for 4 days prior to the experiment [32]. Rats that did not fall off the Rota-Rod for 2 minutes were considered normal during a 3-minute session and counted manually [16]. To determine changes in locomotor behavior, gait, and overall walking skill, an elevated (30°) stainless steel grid was used with a mesh size of 30 mm [30]. Rats were trained for 1 minute twice every day for 4 days prior to the experiments. The animals were placed on the grid for 1 minute and the total number of paired steps (i.e., placement of both forelimbs) was counted. During this period, the number of misplaced limb errors (i.e., the forelimbs fell through the grid) was recorded. The total number of errors for each forelimb was also counted manually [16, 30, 32].

Intracortical administration of nNOS antiserum

The nNOS antiserum (1:20) was administered into the injured cerebral cortex 5, 10, and 120 minutes after the lesion using a microliter syringe [17, 19, 22, 26, 27].

BBB permeability

BBB permeability in the cerebral cortex of both hemispheres was measured using Evans blue albumin (3 mL/kg of a 2% solution, i.v.) and [131]Iodine as described previously [17, 25, 27, 29, 31]. In brief, both tracers were administered 5 hours after trauma via the right external branch of the jugular vein through a needle puncture [23, 25, 27]. These tracers were allowed to circulate for 5 minutes. The animals then underwent transcardiac perfusion with 0.9% saline to wash out remaining intravascular tracer. Immediately before perfusion, about 1 mL of arterial blood was withdrawn via heart puncture for later determination of whole blood radioactivity. The BBB permeability of radiotracer was determined by percentage increase in the radioactivity in the brain over the whole blood radioactivity [15].

Brain water content

The water contents of the right and left cerebral cortices were determined using the difference in sample wet and dry weights [5, 6, 29]. In brief, the cortices were dissected out after perfusion and weighed immediately. The samples were then placed in an oven maintained at 90 °C for 72 hours or until sample dry weight became constant in at least 2 determinations [29]. Percentage of brain swelling was calculated from the changes in the brain water content as described previously [5, 6, 18, 21, 29].

Perfusion and fixation

At the end of the experiment, animals were deeply anesthetized with Equithesin and the chest rapidly opened. The right auricle was cut and a 21-gauge butterfly needle was inserted into the left ventricle, which was connected to the perfusion apparatus [15, 16, 19]. About 50 mL of phosphate-buffered saline (0.1 mol, pH 7.0) was perfused (at 90 torr) to wash out the remaining blood followed by 4% paraformaldehyde in 0.1 mol phosphate-buffered saline [17]. The animals were wrapped in aluminum foil and kept overnight in a refrigerator at 4 °C. On the next day, brain samples were dissected out and kept into the same fixative at 4 °C for 1 week [29].

Measurement of NOS immunoreactivity

NOS immunoreactivity was examined using polyclonal antibodies directed against nNOS [17, 22, 28] in the cortical brain regions on 3 μm thick paraffin sections using standard procedures [29]. The antibodies of nNOS were diluted 1:5000 and applied for 48 hours at room temperature with continuous shaking [17]. The immune reaction was developed using the peroxidase-antiperoxidase technique. In a few sections, the primary antibody step was omitted and the reaction product was developed as usual. The number of nNOS-positive cells in each group was counted in a blinded fashion in all animals [17, 22, 24, 28].

Statistical analysis

Quantitative data obtained were analyzed using ANOVA followed by Dunnet's test for multiple group comparison with 1 control group. The Chi-square test was used to find statistical significance between control and experimental groups, where the control value is zero. A p-value less than 0.05 was considered significant.

Results

Sensory motor dysfunction in TBI

There was a significant decline in Rota-Rod performance 3 hours after TBI compared to the control group. This decline in performance was progressive in nature. The traumatized rats were also unable to walk normally during a grid-walking session 4 hours after TBI. The number of steps taken during a 60-second grid-walking session was significantly reduced compared to the control group. There were a greater number of forelimb placement errors 4 hours after TBI, and this deficit was progressive in nature.

Intracerebral administration of nNOS antiserum 5 minutes and 60 minutes after TBI significantly improved animal performance on the Rota-Rod and grid-walking tests. However, antiserum administered 2 hours after TBI did not improve these sensory motor performances (results not shown).

nNOS expression in TBI

Focal TBI resulted in a marked up-regulation of nNOS expression in the distorted and damaged neurons located in the edematous cortical regions around the lesion site and in the underlying subcortical region (Fig. 1). The intensity of nNOS immunostaining was greater in the ipsilateral injured cortex than in the contralateral hemisphere (Fig. 1). nNOS antiserum treatment given 5 or 60 minutes after TBI markedly attenuated the number of nNOS-positive cells in the cortex and subcortical areas. This effect was less pronounced in the contralateral uninjured cerebral cortex compared to the ipsilateral injured side (Fig. 1; Table 1). Administration of nNOS antiserum 2 hours after TBI did not influence nNOS expression significantly.

BBB permeability in TBI

Focal TBI induced profound extravasation of Evans blue dye in the injured cerebral cortex at 5 hours (Fig. 1). Measurement of Evans blue and [131]Iodine showed increased tracer permeability in both traumatized and untraumatized cerebral cortices compared to intact control animals (Table 1). The magnitude of these changes was most pronounced in the ipsilateral cortex (Table 1). Intracerebral administration of nNOS antiserum 5 minutes or 1 hour after TBI significantly attenuated Evans blue leakage, as seen in the injured cortex (Fig. 1, Table 1). A significant reduction in tracer extravasation was also seen in nNOS antiserum treated injured rats (Table 1). Antiserum administered 2 hours after TBI did not influence BBB permeability to these tracers (results not shown).

Brain edema formation in TBI

Brain water content increased significantly in both the traumatized and untraumatized cerebral cortices at 5 hours. This increase in brain water content was most pronounced in the ipsilateral hemisphere (Table 1). Intracerebral administration of nNOS antiserum 5 and 60 minutes after TBI significantly reduced the increase in brain water content at 5 hours (Table 1). Antiserum treatment given 2 hours after TBI was ineffective in attenuating brain edema formation (results not shown).

Discussion

The new finding of this study is the marked up-regulation of nNOS following TBI that was largely seen in areas showing BBB disruption and edema formation. Furthermore, nNOS up-regulation, BBB breakdown, and edema formation were significantly attenuated by intracortical administration of nNOS antiserum. These observations suggest that nNOS up-regulation is injurious to cells and tissues, and that local administration of nNOS antiserum has remarkable neuroprotective capabilities in TBI, a novel finding. Our observations further show that focal TBI influences sensory motor function of the rat, which is also markedly reduced by nNOS antiserum. This indicates that NO is somehow involved in behavioral dysfunction in TBI.

The most marked neuroprotective effects on behavioral changes and brain pathology were seen when the antiserum was administered either 5 minutes or 1 hour after TBI. On the other hand, administration of nNOS antiserum 2 hours after TBI was ineffective. This effect of nNOS antiserum was closely related to its capacity to attenuate nNOS expression after TBI,

Intracerebral administration brain injury-induced blood-brain barrier permeability 291

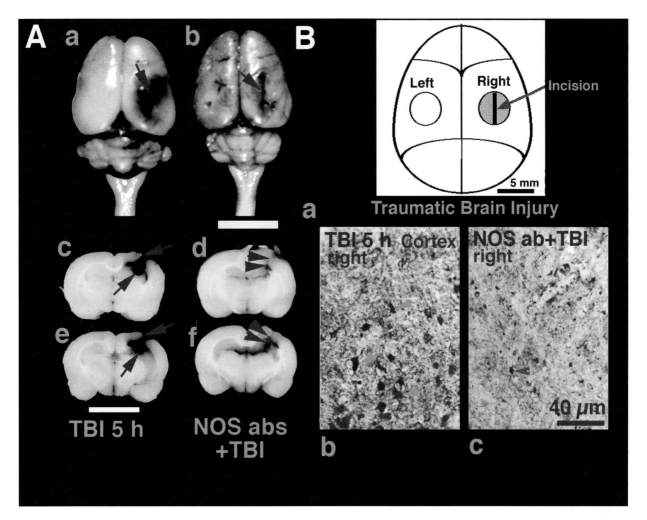

Fig. 1. Extravasation of Evans blue dye, visual swelling, and nNOS expression in the TBI rat model and its modification with intracortical administration of nNOS antiserum. Diagrammatic representation of the incision into the right parietal cerebral cortex is shown by arrow (B:a). Five hours after TBI, extravasation of Evans blue on the dorsal superficial cortical surface (A:a) and the underlying subcortical regions (A:c,e) following incision is clearly visible (arrows). Low power light micrograph showing profound up-regulation of nNOS expression (B:b) in the ipsilateral cortex below the lesioned area (arrows). Treatment with nNOS antiserum either 5 minutes (A:b,d) or 1 hour (A:f) after TBI markedly attenuated extravasation of Evans blue and visual swelling 5 hours after TBI. In rats treated with nNOS 5 minutes after TBI (B:c), expression of nNOS is largely absent. Bar: A:a,b = 5 mm, c–f = 3 mm; B:a = 5 mm, b = 40 μm. Data (A:a–f, B:b,c) modified after Sharma and Alm 2004 [19]

indicating that blockade of nNOS expression and consequently the production of NO during the early phase of TBI is neuroprotective. It appears that administration of neuroprotective compounds (e.g., nNOS antiserum) locally within 1 hour of TBI has some potential therapeutic value. However, this is a new subject that requires further investigation.

We observed significant breakdown of the BBB, development of edema, and nNOS expression in the contralateral cortex compared to control rats. This observation suggests that secondary injury-inducing factors released in the ipsilateral cortex may have migrated to the contralateral hemisphere through the cerebral and/or systemic circulation. Previous studies from our laboratory that showed a significant increase in serotonin concentration in both the injured and uninjured hemispheres 5 hours after TBI are in line with this concept [5, 6, 21, 29].

Trauma to the cerebral cortex physically damages microvessels resulting in micro-hemorrhages and leakage of serum components into the cerebral compartment [5, 6, 30]. Obviously, leakage of serum proteins caused by direct damage of microvessels will to some extent contribute to edema formation [15, 19, 20].

Table 1. *Sensory motor dysfunction, BBB permeability, edema formation, and nNOS expression in rats before and after treatment for TBI with nNOS antiserum*

Parameters measured	n	Control	5 h TBI	nNOS antiserum treatment in 5 h TBI§	
				+5 min	+60 min
Sensory motor dysfunction					
Rota-Rod performance, sec	5	120 ± 0	$70 \pm 6^{**}$	94 ± 8^{aa}	85 ± 8^{aa}
Grid walking, no. steps	5	40 ± 4	$18 \pm 5^{**}$	33 ± 7^{a}	24 ± 6^{a}
Placement error forelimbs, %	5	100	$38 \pm 5^{\#}$	16 ± 4^{bb}	26 ± 3^{b}
BBB Permeability					
Evans blue, R, mg%	6	0.24 ± 0.04	$1.78 \pm 0.12^{**}$	0.54 ± 0.08^{aa}	0.78 ± 0.14^{aa}
Evans blue, L, mg%		0.26 ± 0.06	$0.67 \pm 0.08^{**c}$	$0.38 \pm 0.06^{aa,c}$	$0.44 \pm 0.06^{aa,c}$
[131]Iodine, R, %	5	0.35 ± 0.06	$1.98 \pm 0.11^{**}$	0.62 ± 0.04^{aa}	0.80 ± 0.06^{aa}
[131]Iodine, L, %	5	0.33 ± 0.08	$0.86 \pm 0.09^{**}$	$0.42 \pm 0.04^{aa,c}$	$0.61 \pm 0.07^{aa,c}$
Edema Formation					
Water content, R %	5	76.12 ± 0.18	$80.34 \pm 0.21^{**}$	76.84 ± 0.14^{aa}	77.48 ± 0.22^{aa}
Water content, L %	5	76.16 ± 0.09	$78.64 \pm 0.13^{**}$	$76.60 \pm 0.08^{aa,c}$	$77.08 \pm 0.08^{aa,c}$
nNOS expression					
No. nNOS-positive cells, R	5	12 ± 6	$66 \pm 12^{**}$	20 ± 4^{aa}	38 ± 12^{aa}
No. nNOS-positive cells, L	5	14 ± 4	$36 \pm 8^{**c}$	$16 \pm 4^{aa,c}$	$26 \pm 8^{aa,c}$

§ nNOS antiserum (1:20 in PBS) was administered either 5 min or 60 min after TBI. The cell changes were graded from 1 (minimum) to 4 (maximum) and analyzed in blinded fashion [31].
R Right; *L* left; *BBB* blood-brain barrier; *nNOS* neuronal nitric oxide synthase; *TBI* traumatic brain injury.
Values are Mean ± SD of 5–6 rats in each group.
* = $p < 0.05$; ** $p < 0.01$ (compared from control); a = $p < 0.05$; aa = $p < 0.01$ (compared from 5 h SCI), ANOVA followed by Dunnett's test from one control. # = $p < 0.05$, Chi-square test from control group; b = $p < 0.05$; bb = $p < 0.01$; from 5 h TBI group; c = $p < 0.05$, compared from right injured half (from 5 h TBI group).

However, spread of such edema fluid to the contralateral side within 5 hours appears unlikely. Thus, the profound swelling in the contralateral hemisphere is likely due to the spread of several chemical mediators that are released within the circulation as well as in brain following trauma. Since BBB permeability is disrupted, neurochemicals gain access into the microcirculation from the brain. Since passage of molecules through the cerebral endothelium is restricted from both luminal and abluminal directions, it is quire likely that TBI may also disrupt the brain-blood barrier, a subject that is currently being investigated in our laboratory.

The mechanisms by which severe TBI induces NOS up-regulation are still unclear [1–4]. Traumatic insults to the brain cause profound cellular and oxidative stresses associated with the release of several neurochemicals, lipid peroxidation, and generation of free radicals [13, 15, 24, 30, 32, 34]. Several neurochemicals that are released following TBI, such as serotonin, prostaglandins, histamine, and opioids, are known mediators of BBB disruption and vasogenic edema formation [5, 6, 10, 13, 15, 20, 29, 34]. Generation of free radicals triggers opening of cation-permeable channels resulting in an increased accumulation of intracellular Ca^{2+} [1, 4, 7, 12, 13, 33, 34]. Intracellular Ca^{2+} then binds to calmodulin, a cofactor of NOS, and stimulates nNOS activity [3, 8, 13, 19, 32]. An up-regulation of nNOS will further stimulate NO production [11].

Overproduction of NO contributes to neurotoxicity through generation of oxidant compounds [7–9, 13, 33]. NO generated by activated NOS reacts with superoxide anion to produce the potent oxidant, peroxynitrite [7, 10, 12], which inhibits DNA synthesis, liberates iron from the iron storage protein ferritin, and influences iron metabolism at the post-transcription level [33, 34]. Alternatively, NO and peroxynitrite-induced activation of the nuclear enzyme poly (ADP ribose) synthetase (PARS) may case DNA damage [32–35]. Activation of PARS causes cell death by rapid depletion of cell energy, and PARS inhibition protects cortical cell cultures from glutamate and NO neurotoxicity [19, 25, 32].

There are reasons to believe that NO contributes to BBB disruption [4, 7–10, 12, 17, 19, 22, 32], probably through its binding to the heme-iron protein in guanylyl cyclase to elicit cGMP formation. Increased cGMP contributes to microvascular permeability disruption [1, 4, 7, 9, 35].

Breakdown of the BBB will allow serum components or other vasoactive compounds to enter the brain [15, 19, 20]. Profound alterations in extracellular fluid bathing the cellular elements of the brain by ionic, chemical, immunological, and metabolic reactions may contribute to vasogenic edema formation and neuronal, glial, or axonal injury [5, 6, 15–17, 19–22, 25, 28–31]. Reduction in nNOS expression by intracerebral administration of nNOS antiserum both attenuates BBB dysfunction and reduces edema formation, supporting this hypothesis.

Conclusion

In conclusion, our results suggest that TBI has the capacity to induce nNOS up-regulation and that this is instrumental in BBB disruption, edema formation, and sensory motor disturbances. Intracerebral application of nNOS antiserum within 1 hour after TBI is able to thwart NOS expression and thus reduce BBB leakage and edema formation. Taken together, these observations indicate that nNOS up-regulation is injurious to the brain and that nNOS antiserum may have some potential therapeutic value in the clinical setting. This requires additional investigation.

Acknowledgments

This investigation was supported by grants from the Swedish Medical Research Council (2710 HSS); "Laerdal Foundation for Acute Medicine"; Alexander von Humboldt Foundation (HSS), Germany; The University Grants Commission (HSS), New Delhi, India; and The Indian Council of Medical Research (HSS), New Delhi, India. The expert technical assistance of Kärstin Flink, Kerstin Rystedt, Franziska Drum, and Katherin Kern, and the secretarial assistance of Aruna Sharma are greatly appreciated.

References

1. Ahn MJ, Sherwood ER, Prough DS, Lin CY, DeWitt DS (2004) The effects of traumatic brain injury on cerebral blood flow and brain tissue nitric oxide levels and cytokine expression. J Neurotrauma 21: 1431–1442
2. Bayir H, Kagan VE, Borisenko GG, Tyurina YY, Janesko KL, Vagni VA, Billiar TR, Williams DL, Kochanek PM (2005) Enhanced oxidative stress in iNOS-deficient mice after traumatic brain injury: support for a neuroprotective role of iNOS. J Cereb Blood Flow Metab 25: 673–684
3. Calingasan NY, Park LC, Calo LL, Trifiletti RR, Gandy SE, Gibson GE (1998) Induction of nitric oxide synthase and microglial responses precede selective cell death induced by chronic impairment of oxidative metabolism. Am J Pathol 153: 599–610
4. Cipolla MJ, Crete R, Vitullo L, Rix RD (2004) Transcellular transport as a mechanism of blood-brain barrier disruption during stroke. Front Biosci 9: 777–785
5. Dey PK, Sharma HS (1983) Ambient temperature and development of traumatic brain oedema in anaesthetized animals. Indian J Med Res 77: 554–563
6. Dey PK, Sharma HS (1984) Influence of ambient temperature and drug treatments on brain oedema induced by impact injury on skull in rats. Indian J Physiol Pharmacol 28: 177–186
7. Hooper DC, Scott GS, Zborek A, Mikheeva T, Kean RB, Koprowski H, Spitsin SV (2000) Uric acid, a peroxynitrite scavenger, inhibits CNS inflammation, blood-CNS barrier permeability changes, and tissue damage in a mouse model of multiple sclerosis. FASEB J 14: 691–698
8. Hurst RD, Clark JB (1997) Nitric oxide-induced blood-brain barrier dysfunction is not mediated by inhibition of mitochondrial respiratory chain activity and/or energy depletion. Nitric Oxide 1: 121–129
9. Hurst RD, Azam S, Hurst A, Clark JB (2001) Nitric-oxide-induced inhibition of glyceraldehyde-3-phosphate dehydrogenase may mediate reduced endothelial cell monolayer integrity in an in vitro model blood-brain barrier. Brain Res 894: 181–188
10. Jaworowicz DJ Jr, Korytko PJ, Singh Lakhman S, Boje KM (1998) Nitric oxide and prostaglandin E2 formation parallels blood-brain barrier disruption in an experimental rat model of bacterial meningitis. Brain Res Bull 46: 541–546
11. Martinez-Lara E, Canuelo AR, Siles E, Hernandez R, Del Moral ML, Blanco S, Pedrosa JA, Rodrigo J, Peinado MA (2005) Constitutive nitric oxide synthases are responsible for the nitric oxide production in the ischemic aged cerebral cortex. Brain Res 1054: 88–94
12. Mayhan WG (1995) Role of nitric oxide in disruption of the blood-brain barrier during acute hypertension. Brain Res 686: 99–103
13. Munoz-Fernandez MA, Fresno M (1998) The role of tumour necrosis factor, interleukin 6, interferon-gamma and inducible nitric oxide synthase in the development and pathology of the nervous system. Prog Neurobiol 56: 307–340
14. Park EM, Cho S, Frys KA, Glickstein SB, Zhou P, Anrather J, Ross ME, Iadecola C (2005) Inducible nitric oxide synthase contributes to gender differences in ischemic brain injury. J Cereb Blood Flow Metab [Epub ahead of print]
15. Sharma HS (2005) Pathophysiology of blood-spinal cord barrier in traumatic injury and repair. Curr Pharm Des 11: 1353–1389
16. Sharma HS (2005) Alterations of amino acid neurotransmitters in hyperthermic brain injury. J Neural Transm [in press]
17. Sharma HS, Alm P (2002) Nitric oxide synthase inhibitors influence dynorphin A (1–17) immunoreactivity in the rat brain following hyperthermia. Amino Acids 23: 247–259
18. Sharma HS, Winkler T (2002) Assessment of spinal cord pathology following trauma using spinal cord evoked potentials: a pharmacological and morphological study in the rat. Muscle Nerve [Suppl] 11: S83–S91
19. Sharma HS, Alm P (2004) Role of nitric oxide on the blood-brain and the spinal cord barriers. In: Sharma HS, Westman J (eds) Blood-spinal cord and brain barriers in health and disease. Elsevier Academic Press, San Diego, pp 191–230
20. Sharma HS, Westman J (2004) Blood-spinal cord and brain barriers in health and disease. Academic Press, San Diego, pp 1–617
21. Sharma HS, Cervós-Navarro J, Gosztonyi G, Dey PK (1992) Role of serotonin in traumatic brain injury. An experimental study in the rat. In: Globus M, Dietrich WD (eds) The role of neurotransmitters in brain injury. Plenum Press, New York, pp 147–152
22. Sharma HS, Westman J, Olsson Y, Alm P (1996) Involvement of

nitric oxide in acute spinal cord injury: an immunocytochemical study using light and electron microscopy in the rat. Neurosci Res 24: 373–384
23. Sharma HS, Nyberg F, Gordh T, Alm P, Westman J (1997) Topical application of insulin like growth factor-1 reduces edema and upregulation of neuronal nitric oxide synthase following trauma to the rat spinal cord. Acta Neurochir [Suppl] 70: 130–133
24. Sharma HS, Westman J, Alm P, Sjoquist PO, Cervos-Navarro J, Nyberg F (1997) Involvement of nitric oxide in the pathophysiology of acute heat stress in the rat. Influence of a new antioxidant compound H-290/51. Ann NY Acad Sci 813: 581–590
25. Sharma HS, Alm P, Westman J (1998) Nitric oxide and carbon monoxide in the brain pathology of heat stress. Prog Brain Res 115: 297–333
26. Sharma HS, Nyberg F, Westman J, Alm P, Gordh T, Lindholm D (1998) Brain derived neurotrophic factor and insulin like growth factor-1 attenuate upregulation of nitric oxide synthase and cell injury following trauma to the spinal cord. An immunohistochemical study in the rat. Amino Acids 14: 121–129
27. Sharma HS, Nyberg F, Gordh T, Alm P, Westman J (1998) Neurotrophic factors attenuate neuronal nitric oxide synthase upregulation, microvascular permeability disturbances, edema formation and cell injury in the spinal cord following trauma. In: Stålberg E, Sharma HS, Olsson Y (eds) Spinal cord monitoring: basic principles, regeneration, pathophysiology, and clinical aspects. Springer, New York, pp 118–148
28. Sharma HS, Drieu K, Alm P, Westman J (2000) Role of nitric oxide in blood-brain barrier permeability, brain edema and cell damage following hyperthermic brain injury. An experimental study using EGB-761 and Gingkolide B pretreatment in the rat. Acta Neurochir [Suppl] 76: 81–86
29. Sharma HS, Winkler T, Stalberg E, Mohanty S, Westman J (2000) p-Chlorophenylalanine, an inhibitor of serotonin synthesis reduces blood-brain barrier permeability, cerebral blood flow, edema formation and cell injury following trauma to the rat brain. Acta Neurochir [Suppl] 76: 91–95
30. Sharma HS, Sjoquist PO, Alm P (2003) A new antioxidant compound H-290151 attenuates spinal cord injury induced expression of constitutive and inducible isoforms of nitric oxide synthase and edema formation in the rat. Acta Neurochir [Suppl] 86: 415–420
31. Sharma HS, Winkler T, Stalberg E, Gordh T, Alm P, Westman J (2003) Topical application of TNF-alpha antiserum attenuates spinal cord trauma induced edema formation, microvascular permeability disturbances and cell injury in the rat. Acta Neurochir [Suppl] 86: 407–413
32. Sharma HS, Badgaiyan RD, Mohanty S, Alm P, Wiklund L (2005) Neuroprotective effects of nitric oxide synthase inhibitors in spinal cord injury induced pathophysiology and motor functions. An experimental study in the rat. Ann NY Acad Sci 1053: 422–434
33. Tan KH, Harrington S, Purcell WM, Hurst RD (2004) Peroxynitrite mediates nitric oxide-induced blood-brain barrier damage. Neurochem Res 29: 579–587
34. Warner DS, Sheng H, Batinic-Haberle I (2004) Oxidants, antioxidants and the ischemic brain. J Exp Biol 207: 3221–3231
35. Zhou P, Qian L, Iadecola C (2005) Nitric oxide inhibits caspase activation and apoptotic morphology but does not rescue neuronal death. J Cereb Blood Flow Metab 25: 348–357

Correspondence: Hari Shanker Sharma, Department of Surgical Sciences, Anesthesiology & Intensive Care Medicine, University Hospital, SE-75185 Uppsala, Sweden. e-mail: Sharma@surgsci.uu.se

Effects of 2,4-dinitrophenol on ischemia-induced blood-brain barrier disruption

S. R. Ennis[1] and R. F. Keep[1,2]

[1] Department of Neurosurgery, University of Michigan Medical School, Ann Arbor, Michigan, USA
[2] Department of Molecular and Integrative Physiology, University of Michigan Medical School, Ann Arbor, Michigan, USA

Summary

This study examines the effect of 2,4-dinitrophenol (DNP), a mitochondrial uncoupling agent, during focal brain ischemia induced by middle cerebral artery (MCA) occlusion. Blood-brain barrier (BBB) disruption was assessed after 2 hours of occlusion with 2 hours of reperfusion or 4 hours of permanent occlusion by measurement of the influx rate constant (K_i) for ^{3}H-inulin in the MCA territory ipsi- and contralateral to the occlusion. Three experimental groups were examined: vehicle and 1 and 5 mg/kg DNP treated animals (given 30 minutes prior to occlusion). Four hours of permanent MCA occlusion only induced a modest increase in the K_i for inulin in vehicle-treated animals (0.09 ± 0.01 vs. 0.07 ± 0.01 µL/g/min in contralateral tissue). Although 5 mg/kg DNP significantly increased this disruption ($p < 0.01$), this effect was relatively minor (0.14 ± 0.02 µL/g/min). In contrast, DNP treatment in transient ischemia markedly increased barrier disruption. The ipsilateral K_i for ^{3}H-inulin were 0.15 ± 0.04, 0.37 ± 0.06, and 0.79 ± 0.17 µL/g/min in vehicle, 1 mg/kg DNP and 5 mg/kg DNP groups, respectively. DNP did not induce barrier disruption in the contralateral hemisphere. Thus, while there is evidence that DNP can be neuroprotective, it has adverse effects on the BBB during ischemia, particularly with reperfusion. Considering the importance of naturally- or therapeutically-induced reperfusion in limiting brain damage, this may limit the utility of DNP and mitochondrial uncouplers as therapeutic agents.

Keywords: Cerebral ischemia; metabolic uncoupler; DNP; permeability; inulin; blood-brain barrier.

Introduction

A number of recent studies have suggested that mitochondrial uncoupling agents, such as 2,4-dinitrophenol (DNP), are neuroprotective during cerebral ischemia and other forms of brain injury (reviewed in [4]). Maragos and Korde [4] postulated that DNP-induced mitochondrial uncoupling would lead to a decrease in the mitochondrial membrane potential, reduce Ca^{2+} entry into mitochondria, and decrease free radical generation. These studies have not examined whether such agents might have beneficial or deleterious effects on blood-brain barrier (BBB) disruption during cerebral ischemia.

We have previously shown that ischemic preconditioning, which may also act by preserving mitochondrial function [7], results in both reduced brain edema and BBB opening [5]. In the current study, we investigated the effect of DNP, at the previously recommended dose of 5 mg/kg [4], on BBB permeability, water and ion content, and infarct volume during middle cerebral artery (MCA) occlusion.

Materials and methods

The University of Michigan Committee on the Use and Care of Animals approved the protocol for these animal studies. Adult male Sprague-Dawley rats (Charles River Laboratories, Wilmington, MA) weighing 275–350 g were used for all experiments. Rats were allowed free access to food and water before the experiment.

Rats underwent left MCA occlusion, with and without reperfusion, using the intra-luminal thread method [3, 5] under pentobarbital anesthesia (50 mg/kg). The experiments consisted of 5 parts. In the first part, rats underwent 2 hours of MCA occlusion with 2 hours of reperfusion or 4 hours of permanent occlusion. For the last 15 minutes of the experiment, BBB disruption was assessed, as previously described [5], by measurement of the influx rate constant (K_i) for [^{3}H]inulin in the MCA territory, ipsi- and contralateral to the occlusion. The plasma volume was measured during the last 0.5 minute, using [^{14}C]inulin to correct for intravascular content. Three experimental groups were examined, vehicle (50 mM Tris in 0.9% saline, 5 mL/kg) and 1 and 5 mg/kg DNP treated animals (given 30 minutes prior to occlusion).

In the second part, brain water, sodium, and potassium contents were determined in vehicle and DNP-treated animals (5 mg/kg; 30 minutes prior to occlusion) after 2 hours of ischemia with 2 hours of reperfusion or after 4 hours of permanent occlusion. Water content was measured by the wet weight/dry weight method. Brain ions were measured by flame photometry.

In the third and fourth parts, BBB disruption and brain water and ion contents were assessed in vehicle and DNP groups (5 mg/kg; 30 minutes prior to occlusion) after 2 hours of occlusion followed by 22 hours of reperfusion.

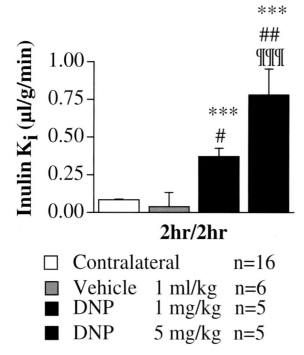

Fig. 1. The inulin influx rate constant (K_i) in the core of the ischemic hemisphere after 2 hours of ischemia with 2 hours of reperfusion. Animals were treated 30 minutes before occlusion of the MCA with vehicle, 1 or 5 mg/kg DNP. Values are mean ± SE. One, two, or three symbols indicate a significant difference at the $p < 0.05$, $p < 0.01$, or $p \leq 0.001$ levels, respectively. * Contralateral vs. all other groups; # vehicle 1 mL/kg vs. 1 mg/kg DNP or 5 mg/kg DNP; ¶ 1 mg/kg DNP vs. 5 mg/kg DNP

In the fifth part, infarct volume was measured in vehicle and DNP groups (5 mg/kg; 30 minutes prior to occlusion) after 2 hours of occlusion followed by 22 hours of reperfusion. Infarct volume was measured by 2,3,5-triphenyltetrazolium staining.

Statistical analysis

All data in this study are presented as mean ± SE. Data were analyzed using Student *t*-test except when multiple comparisons were made to a control using ANOVA, which was followed by a Newman-Keuls test. Statistical significance was accepted at $p < 0.05$.

Results

Figure 1 presents the inulin K_i after 2 hours of ischemia with 2 hours of reperfusion. Animals were treated 30 minutes before occlusion of the MCA with vehicle, 1 or 5 mg/kg DNP. Unexpectedly, the treatment of animals with both 1 and 5 mg/kg DNP in transient ischemia markedly increased barrier disruption. The ipsilateral K_i for [^{3}H]inulin was 0.15 ± 0.04, 0.37 ± 0.06, and 0.79 ± 0.17 μL/g/min in vehicle, 1 mg/kg DNP, and 5 mg/kg DNP groups, respectively. DNP did not induce barrier disruption in the contralateral hemisphere, which is presented as the combined average of the non-stroked hemisphere from the 3 groups.

Four hours of permanent MCA occlusion resulted in a modest but significant increase ($p \leq 0.01$) in the K_i for inulin in vehicle-treated animals (0.09 ± 0.01 vs. 0.07 ± 0.01 μL/g/min in contralateral tissue). DNP (5 mg/kg) significantly increased this disruption ($p \leq 0.05$), but the effect was relatively minor (0.14 ± 0.02 μL/g/min).

These unexpected adverse effects of DNP on the BBB, especially during reperfusion, caused us to examine the effects of DNP on brain edema. In vehicle-treated rats, the water content after 2 hours of ischemia with 2 hours of reperfusion was increased in the ischemic core to 3.72 ± 0.09 mL/g dry weight compared to 3.42 ± 0.03 mL/g dry weight in the contralateral tissue ($p \leq 0.01$). The animals treated with DNP also had increased ($p \leq 0.001$) water content in the ischemic core (3.93 ± 0.10 vs. 3.38 ± 0.03 mL/g dry weight in contralateral tissue). Although the increase in water content in the ischemic core of the DNP-treated animals was slightly larger, the difference failed to reach statistical significance ($p \geq 0.13$). However, there was a significantly enhanced sodium content in the ischemic core of DNP-treated animals (256 ± 13 vs. 206 ± 15 mEq/kg/dry weight in controls, n = 8 per group; $p \leq 0.05$). There were no differences in the contralateral hemisphere between the 2 groups (148 ± 4 and 150 ± 6 mEq/kg/dry weight in the DNP- and vehicle-treated animals, respectively). Thus, the net sodium gain in the ischemic core was approximately double in the DNP-treated rats (107 ± 13 vs. 56 ± 15 mEq/kg dry wt in the vehicle-treated animals, $p \leq 0.05$; Fig. 2). This increased sodium gain may be due to the enhanced BBB disruption and vasogenic edema.

The potassium content in vehicle- and DNP-treated rats were not significantly different in either ipsilateral (vehicle: 368 ± 16 DNP: 349 ± 24 mEq/kg dry weight) or contralateral (vehicle, 392 ± 15; DNP, 391 ± 13 mEq/kg dry weight) tissues. The net potassium losses for vehicle- and DNP-treated animals were not significantly different (24 ± 14 and 43 ± 14 mEq/kg dry weight, respectively; $p \geq 0.38$; Fig. 2).

After 4 hours, permanent ischemia produced small increases in water content in both vehicle and DNP treated animals ($p \leq 0.01$). However, the water contents of the vehicle and DNP animals in the ischemic

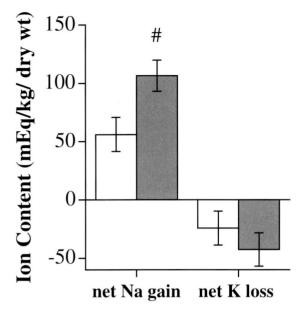

Fig. 2. The net change in either sodium or potassium in the core of the ischemic hemisphere after 2 hours of ischemia with 2 hours of reperfusion. Animals were treated 30 minutes before occlusion of the MCA with either vehicle or DNP. Values are mean ± SE; # indicate significant differences between 1 mL/kg vehicle and 5 mg/kg DNP at the $p < 0.05$ level

core (3.79 ± 0.10 and 3.87 ± 0.08 mL/g dry weight, respectively) were not different from each other ($p \geq 0.53$).

The duration of the adverse effects of DNP on BBB disruption was examined by measuring the inulin K_i in animals after 2 hours of ischemia with 22 hours of reperfusion. Rats were treated 30 minutes before MCA occlusion with either vehicle or 5 mg/kg DNP. Ischemia produced approximately 3-fold increases in inulin permeability in the ischemic core, without any significant difference between the vehicle and DNP groups. Thus, K_i increased from 0.06 ± 0.01 μL/g/min in the contralateral core to 0.2 ± 0.02 μL/g/min in the ischemic core in vehicle controls ($p \leq 0.01$), and from 0.07 ± 0.01 μL/g/min in the contralateral core to 0.21 ± 0.02 μL/g/min in the ischemic core ($p \leq 0.01$) in DNP-treated rats.

Similarly, pretreatment with DNP (5 mg/kg) before 2 hours of ischemia with 22 hours of reperfusion had no effect on infarct volume (247 ± 26 vs. 240 ± 61 mm^3 in vehicle controls, n = 6 per group). DNP also had no significant effect on brain water content in the contralateral hemisphere (3.54 ± 0.04 vs. 3.56 ± 0.04 mL/g dry weight in vehicle controls) or the ischemic core (4.48 ± 0.16 vs. 4.93 ± 0.29 mL/g dry weight in vehicle controls). Even though the increase in the water content of the ischemic core in the DNP treated animals tended to be smaller, the difference failed to reach statistical significance ($p \geq 0.21$).

Discussion

Recent reports have suggested that mitochondrial uncouplers such as DNP [2, 4, 6] may be neuroprotective in stroke and traumatic CNS injury. It has been postulated that DNP will reduce brain damage by uncoupling mitochondrial electron transport from ATP production and decreasing free radical production. Previous studies have focused on the neuronal effects of DNP. The current study was undertaken to examine the effects of DNP on the BBB during ischemia.

Unexpectedly, we found that pretreatment with DNP produced a small but significant increase in BBB inulin permeability after 4 hours of permanent ischemia, and a greater than 5-fold increase after 2 hours of ischemia with 2 hours of reperfusion. This acute effect of DNP was associated with increased brain sodium content in the ischemic hemisphere, suggestive of greater vasogenic edema secondary to BBB disruption. Although there was a tendency for the total brain edema to be greater in the DNP group, this did not quite reach significance.

Other recent evidence has suggested that mitochondrial uncoupling may have adverse effects during ischemia. de Bilbao et al. [1] found that knockout mice without uncoupling protein 2 (UCP2), an endogenous uncoupler, had smaller infarct sizes than wild-type mice. This was attributed to the effects of UCP2 on mitochondrial glutathione levels in microglia.

Although DNP caused an opening in the BBB after either 4 hours of permanent ischemia or 2 hours of ischemia and 2 hours of reperfusion, we found that DNP pretreatment did not increase either permeability, water content, or stroke volume after 2 hours of ischemia and 22 hours of reperfusion. These data leave open the possibility that dosing with DNP after ischemia onset or during reperfusion might prove beneficial.

In conclusion, while there is evidence that DNP can be neuroprotective [4], it has adverse acute effects on the BBB during ischemia, particularly in the setting of reperfusion. Considering the importance of naturally- or therapeutically-induced reperfusion in limiting brain damage, this may limit the utility of DNP and mitochondrial uncouplers as therapeutic agents. In this occlusive model of stroke, the enhanced BBB disruption with DNP was associated with greater sodium accumulation, which may reflect increased vasogenic edema. Finally, DNP pretreatment provided no benefit (infarct volume, water content) to the brain, nor did it harm the BBB (inulin permeability) during longer periods (22 hours) of reperfusion.

Acknowledgments

This work was supported in part by Grants R01 NS034709 and P01 HL018575 from the National Institutes of Health.

References

1. de Bilbao F, Arsenijevic D, Vallet P, Hjelle OP, Ottersen OP, Bouras C, Raffin Y, Abou K, Langhans W, Collins S, Plamondon J, Alves-Guerra MC, Haguenauer A, Garcia I, Richard D, Ricquier D, Giannakopoulos P (2004) Resistance to cerebral ischemic injury in UCP2 knockout mice: evidence for a role of UCP2 as a regulator of mitochondrial glutathione levels. J Neurochem 89: 1283–1292
2. Jin Y, McEwen ML, Nottingham SA, Maragos WF, Dragicevic NB, Sullivan PG, Springer JE (2004) The mitochondrial uncoupling agent 2,4-dinitrophenol improves mitochondrial function, attenuates oxidative damage, and increases white matter sparing in the contused spinal cord. J Neurotrauma 21: 1396–1404
3. Longa EZ, Weinstein PR, Carlson S, Cummins R (1989) Reversible middle cerebral artery occlusion without craniectomy in rats. Stroke 20: 84–91
4. Maragos WF, Korde AS (2004) Mitochondrial uncoupling as a potential therapeutic target in acute central nervous system injury. J Neurochem 91: 257–262
5. Masada T, Hua Y, Xi G, Ennis SR, Keep RF (2001) Attenuation of ischemic brain edema and cerebrovascular injury after ischemic preconditioning in the rat. J Cereb Blood Flow Metab 21: 22–33
6. Mattiasson G, Shamloo M, Gido G, Mathi K, Tomasevic G, Yi S, Warden CH, Castiho RF, Melcher T, Gozalez-Zulueta M, Nikolich K, Wieloch T (2003) Uncoupling protein-2 prevents neuronal death and diminishes brain dysfunction after stroke and brain trauma. Nature Med 9: 1062–1068
7. Minners J, McLeod CJ, Sack MN (2003) Mitochondrial plasticity in classical ischemic preconditioning – moving beyond the mitochondrial KATP channel. Cardiovasc Res 59: 1–6

Correspondence: Richard F. Keep, Department of Neurosurgery, University of Michigan Medical School, R5550 Kresge I, Ann Arbor, MI 48109-0532, USA. e-mail: rkeep@umich.edu

Long-term cognitive and neuropsychological symptoms after global cerebral ischemia in Mongolian gerbils

S. Ishibashi[1], T. Kuroiwa[2], S. LiYuan[1], N. Katsumata[2], S. Li[3], S. Endo[4], and H. Mizusawa[1]

[1] Department of Neurology & Neurological Science, Tokyo Medical and Dental University, Tokyo, Japan
[2] Department of Neuropathology, Tokyo Medical and Dental University, Tokyo, Japan
[3] Department of Neurosurgery, Tokyo Medical and Dental University, Tokyo, Japan
[4] Animal Research Center, Tokyo Medical and Dental University, Tokyo, Japan

Summary

The objective of this study was to establish a rodent model of vascular dementia that showed long-term cognitive and neuropsychological deficits, and to correlate those behavioral deficits with the patterns of ischemic lesions, thus providing a platform for future testing of potential therapeutic agents. In Mongolian gerbils, either 5-minute single bilateral common carotid artery occlusion (SBCCAO) or repetitive bilateral common carotid artery occlusion (two 7-minute occlusions, RBCCAO) was induced, and the behavioral deficits were evaluated using 2 tests: a modified open-field test with an escape zone to evaluate changes in anxiety and locomotor activity, and a T-maze test to assess cognitive dysfunction.

SBCCAO did not induce anxiety changes but caused transient locomotor hyperactivity and mild cognitive deficits. Only pyramidal neuronal death was found in the bilateral CA1 sector of the hippocampus following SBCCAO. In contrast, RBCCAO induced persistent locomotor hyperactivity, reduced anxiety, and caused severe cognitive deficits at 4 weeks post-ischemia. RBCCAO caused significant atrophy associated with diffuse selective neuronal death in the bilateral cerebral cortex and caudate nucleus, as well as the CA1 region. The repetitive ischemia model appears to be a potentially useful platform for the long-term analysis of cognitive and neuropsychological symptoms associated with vascular dementia.

Keywords: Global ischemia; gerbil; behavior.

Introduction

Vascular dementia is one of the major causes of cognitive decline in the elderly [7]. In addition to cognitive dysfunction, neuropsychological symptoms such as anxiety, aggression, delusion, and depression are often observed in vascular dementia patients and cause additional disability [7]. These symptoms result from various ischemic lesions such as cortical ischemic atrophy, subcortical lesion, and multi-infarction [7]. However, this lesion pattern is not well-replicated in animal models. Current rodent models of global ischemia often involve a single occlusion of the bilateral common carotid artery [3], leading to selective hippocampal-related memory deficits and pyramidal cell death in the CA1 region of the hippocampus. Establishment of a model that shows diffuse lesions and additional behavioral deficits would provide a better platform for evaluation of therapeutic strategies in vascular dementia.

In Mongolian gerbils, depending on the total length of the ischemic period, the extent of ischemic lesion after temporary carotid artery occlusion is not limited to selective neuronal death in the CA1 sector with neuronal death extending to the cortex [4]. By applying single and repetitive occlusion of the bilateral common carotid artery in gerbils, we aimed to elucidate the characteristics of the cognitive and neuropsychological dysfunctions and the relationship to the cerebral distribution of ischemic tissue injury.

Materials and methods

We used 32 male Mongolian gerbils (age, 14 to 18 weeks; weight, 60 to 72 g). The animals first were divided into 2 groups: bilateral common carotid artery-occluded (BCCAO) and sham-operated animals. For the induction of ischemia, the animals were anesthetized with 2% isoflurane, and the bilateral common carotid artery was occluded with a miniature vascular clip. BCCAO animals were randomly divided into 2 groups, single BCCAO (SBCCAO, n = 10) and repeated BCCAO (RBCCAO, n = 12), by removal of the clip after either 5 or 7 minutes of occlusion. The RBCCAO animals underwent a second, similarly induced 7-minute period of ischemia 8 hours after the first occlusion. Sham-operated animals (n = 10) underwent the same procedures except for carotid artery occlusion.

All animals underwent a series of behavioral tests during the 4-week period after ischemia. Spontaneous locomotion was analyzed using a video-tracking system (PanLab, Barcelona, Spain).

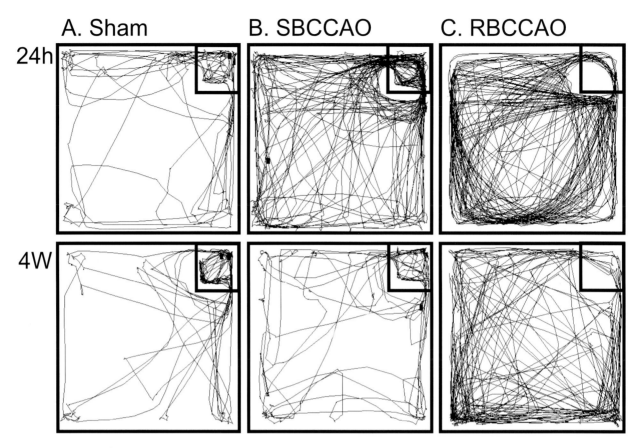

Fig. 1. Trace lines of representative animal movement during a 10-minute modified open-field test, either 24 hours (upper row) or 4 weeks (lower row) after ischemic surgery. (A) The sham-operated animal was predisposed to enter the escape zone (upper right corner). (B) The SBCCAO animal showed transiently increased locomotion 24 hours after ischemia, which recovered by 4 weeks. (C) The RBCCAO animal had no tendency to rest in the escape zone and showed increased locomotion at both 24 hours and 4 weeks after ischemia

A modified open-field test was used to evaluate locomotor activity and anxiety behavior [8]. The apparatus consisted of a white acrylic floor, 85 × 85 cm in size, with white acrylic walls 20 cm in height. An escape zone enclosed by black acrylic walls but freely accessible through 2 entrances was placed at a corner of the open field (Fig. 1). The animals were placed individually in the field, and spontaneous locomotion was recorded for 10 minutes. Total distance (cm) and resting time (sec) in the escape zone were calculated. Resting time was defined as the time in the escape zone with a velocity less than 0.5 mm/s.

The T-maze spontaneous alternation task is a method to test exploratory behavior and working memory [3]. Animals were allowed to alternate between the left and right goal arms of a T-shaped maze throughout a 15-trial continuous alternation session. Upon entering a particular goal arm, a door is lowered to block entry to the opposite arm. The door is re-opened only after animals returned to the start arm, thus allowing a new alternation trial to be started. The spontaneous alternation rate was calculated as the ratio between the alternating choices and the total number of choices.

At the end of the observation period, animals were deeply anesthetized with diethylether, sacrificed, and brains fixed by perfusion with 4% paraformaldehyde. The brains were cut into 6 serial 2.0 mm thick coronal blocks extending from the level of anterior pole of caudate nucleus posteriorly. Each of the 6 coronal sections were embedded in paraffin, cut into 6 μm slices, and stained with Klüver-Barrera for evaluation of the ischemic lesions.

The results were analyzed using repeated-measures analysis of variance, with treatment group and days of testing as independent variables, followed by Bonferroni post hoc test for multiple comparisons between groups. The level of statistical significance was set at $p < 0.05$. All values are presented as the mean ± SD.

Results

All of the sham-operated and SBCCAO animals survived, but 17% (2/12) of the RBCCAO animals died during the observation period. The SBCCAO (n = 10), RBCCAO (n = 10) and sham-operated animals (n = 10) underwent behavioral and histological analysis.

Histology

No histological changes were found in sham-operated animals. In the SBCCAO animals, pyramidal neuronal death was confined to the bilateral CA1 sector of the hippocampus. In the RBCCAO animals,

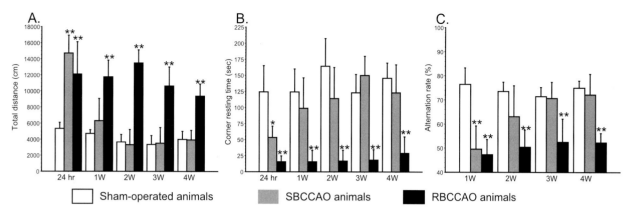

Fig. 2. Time course of behavioral tests from sham-operated, SBCCAO, and RBCCAO animals. (A) The total distance traveled by both the SBCCAO and RBCCAO animals were significantly increased 24 hours after ischemia compared with that of the sham-operated animals. The total distance traversed remained persistently elevated in RBCCAO but not SBCCAO animals. (B) The corner resting time of RBCCAO animals was significantly decreased during observation period. (C) The spontaneous alternation rate in SBCCAO animals was transiently but significantly decreased to values representing random alternation of choice (50%). RBCCAO animals showed a persistent decrease in spontaneous alternation rate. $*p < 0.05$; $**p < 0.01$, versus sham-operated animals

significant atrophy associated with diffuse selective neuronal death was evident in the bilateral cerebral cortex and caudate nucleus in addition to CA1 neuronal death. In 2 of the 10 RBCCAO animals, small laminar necrosis was confined in the rostral middle cortical layer.

Behavioral tests

Modified open-field test

The total distances traversed by both SBCCAO and RBCCAO animals were significantly increased 24 hours after ischemia compared to sham-operated animals (Fig. 1). The hyperactivity in SBCCAO animals was maximal 24 hours after occlusion, with activity returning to levels similar to the sham-operated group within 1 week. In contrast, the hyperactivity of the RBCCAO animals persisted for 4 weeks (Fig. 2A).

Sham-operated and SBCCAO animals had a tendency to rest in the escape zone (Fig. 1). Resting time in the RBCCAO animals was significantly shorter than sham-operated animals during the 4 weeks (Fig. 2B).

T-maze test

Sham-operated animals tended to chose a goal arm alternatively in the T-maze. One week after ischemia, however, the spontaneous alternation rate in SBCCAO and RBCCAO animals was significantly lower than in the sham-operated animals and approached approximately 50% (random choice). In SBCCAO animals, this rate was recovered within 1 week after ischemia. In RBCCAO animals, this rate remained persistently reduced to the level of random choice (Fig. 2C).

Discussion

Our experiment revealed 2 important differences in the behavioral changes that follow SBCCAO and RBCCAO in gerbils: 1) Transient locomotor hyperactivity and mild cognitive dysfunction is transiently induced by bilateral CA1 injury, and 2) persistent hyper-locomotion, reduced anxiety, and severe cognitive dysfunction is induced by bilateral cortical atrophy combined with basal ganglia lesion.

Hyper-locomotion and anxiety-related behavior in modified open-field test

We observed 2 different-types of locomotor hyperactivity after global ischemia: transient hyper-locomotion without anxiety changes and persistent hyper-locomotion with reduced anxiety. The time profile of hyper-locomotion observed in SBCCAO animals, which peaked 24 hours after ischemia and decreased thereafter, is similar to previous findings in gerbil models of global ischemia [6]. Results of a previous study on the histological changes after bilateral common carotid artery occlusion in gerbils indicated that CA1 neuronal death is sufficient to induce locomotor hyperactivity [6]. The mechanism might be related to the excitotoxicity of post-ischemic neural tissue, because the CA1 sector after transient global ischemia has a high extracellular glutamate level [2].

In contrast, the fact that the additional lesions in the cortex and striatum seen in the RBCCAO animals led to persistent behavioral changes suggests that ischemic lesions in these areas may be responsible for anxiety-related behavior. Changes in anxiety-related behavior are often observed in both clinical and experimental strokes, and is evaluated in rodents by the elevated plus maze [1] and measuring corner resting time in open-field tests [5, 8]. Anxiety-related behavior is reported to be regulated mainly by the amygdala, cerebral cortex, and nucleus accumbens [1, 5, 8]. Thus, the cortical lesions observed in our RBCCAO model may be responsible for reduced anxiety.

Spatial cognitive dysfunction in the T-maze test

The spontaneous alternation rate in the T-maze measures exploratory behavior and spatial memory [3]. The hippocampus is widely recognized to have an important role in learning and memory [3]. It is, therefore, reasonable to assume that animals with intact CA1 function have high spontaneous alternation rates reflecting good recent memory, and that SBCCAO animals with CA1 injury have a transient decrease in alternation rate of approximately 50% because of recent memory disturbances. In contrast, RBCCAO animals showed persistent cognitive dysfunction, indicating that cortical and striatal atrophy may also disrupt spatial cognitive function. Further experiments will be needed to elucidate the mechanism and underlying lesion(s) responsible for this long-term spatial cognitive dysfunction.

In conclusion, bilateral cortical atrophy was induced after repetitive common carotid artery occlusion in gerbils. Long-term cognitive and neuropsychological dysfunction was found in the animals after RBCCAO. Thus, RBCCAO appears to be a suitable model of vascular dementia, and may provide a significant contribution to developing therapies for this condition.

References

1. Andersen SL, Teicher MH (1999) Serotonin laterality in amygdala predicts performance in the elevated plus maze in rats. Neuroreport 10: 3497–3500
2. Drejer J, Benveniste H, Diemer NH, Schousboe A (1985) Cellular origin of ischemia-induced glutamate release from brain tissue in vivo and in vitro. J Neurochem 45: 145–151
3. Gerlai R (1998) A new continuous alternation task in T-maze detects hippocampal dysfunction in mice: A strain comparison and lesion study. Behav Brain Res 95: 91–101
4. Ito U, Spatz M, Walker JT Jr, Klatzo I (1975) Experimental cerebral ischemia in Mongolian gerbils. I. Light microscopic observations. Acta Neuropathol (Berl) 32: 209–223
5. Katsumata N, Kuroiwa T, Ishibashi S, LiYuan S, Li S, Endo S, Okazawa H (2004) Two types of locomotor hyperactivity induced by focal ischemia in Mongolian gerbils. Neuropathology 24: A37 [Abstract]
6. Kuroiwa T, Bonnekoh P, Hossmann KA (1991) Locomotor hyperactivity and hippocampal CA1 injury after transient forebrain ischemia of gerbils. Neurosci Lett 122: 141–144
7. Rockwood K, Howard K, MacKnight C, Darvesh S (1999) Spectrum of disease in vascular cognitive impairment. Neuroepidemiology 18: 248–254
8. Rosen JB (2004) The neurobiology of conditioned and unconditioned fear: a neurobehavioral system analysis of the amygdala. Behav Cogn Neurosci Rev 3: 23–41

Correspondence: T. Kuroiwa, Department of Neuropathology, Medical Research Institute, Tokyo Medical and Dental University, 1-5-45, Yushima, Bunkyo-ku, Tokyo 113-8510, Japan. e-mail: t.kuroiwa.npat@mri.tmd.ac.jp

Protective effect of the V1a receptor antagonist SR49059 on brain edema formation following middle cerebral artery occlusion in the rat

A. Kleindienst, G. Fazzina, J. G. Dunbar, R. Glisson, and **A. Marmarou**

Department of Neurosurgery, Medical College of Virginia, Virginia Commonwealth University, Richmond, VA, USA

Summary

There exists no pharmacological treatment for fulminating brain edema. Since evidence indicates that brain aquaporin-4 (AQP4) water channels are modulated by vasopressin V1a receptors, we examined the edema-reducing properties of the selective V1a receptor antagonist, SR49059, following middle cerebral artery occlusion (MCAO).

Male Sprague-Dawley rats were randomly assigned to sham procedure, vehicle, or SR49059 infusion at different dosages (each n = 6, 480 µL/hr, 640 µL/hr, 720 µL/hr) and starting 60 minutes before or after MCAO. After a 2-hour period of ischemia and 2 hours of reperfusion, the animals were sacrificed for assessment of brain water content, sodium, and potassium concentration. Statistics were performed using an ANOVA followed by a Tukey post hoc analysis.

SR049059 treatment reduced brain water content in the infarcted area given at 640 µL/hr ($p = 0.036$), 720 µL/hr 60 minutes before ($p = 0.002$) or 60 minutes after ($p = 0.005$) MCAO. The consecutive sodium shift into the brain was prevented ($p = 0.001$), while the potassium loss was inhibited only by pre-treatment ($p = 0.003$).

These findings imply that in ischemia-induced brain edema, the selective V1a receptor-antagonist SR49059 inhibits brain edema and the subsequent sodium shift into brain. This substance offers a new avenue in brain edema treatment and prompts further study into AQP4 modulation.

Keywords: AQP water channel; vasopressin; middle cerebral artery occlusion; brain edema.

Introduction

Brain edema following many types of brain insult causes an increase in intracranial pressure, thereby contributing to the high rate of secondary complications and consecutive mortality found in these patients. Aquaporins (AQP) are a family of water-selective transporting proteins and the subtype, aquaporin-4 (AQP4), is abundant in astrocytes and ependymal cells, having a highly polarized distribution in glial membranes facing capillaries and the pia mater [1, 15]. Many authors have described the important role of AQP4 in water homeostasis during traumatic and ischemic brain edema development [13, 14], although controversial results have been obtained.

Recent evidence suggests that AQP4 may be regulated by arginine vasopressin. In vitro, the AQP4-mediated water flux is facilitated by vasopressin V1a receptor agonists [12], and the non-peptide V1 receptor antagonist OPC-21268 significantly reduces brain edema after cold injury [3]. Therefore, we assessed the efficacy of different dosages of the selective vasopressin V1a antagonist SR49059 on brain edema development when given intravenously before middle cerebral artery occlusion (MCAO). Specifically, we measured water content, sodium, and potassium concentrations in the ischemic and non-ischemic hemispheres. We determined whether treatment started after MCAO was as effective as treatment started before induction of ischemia.

Materials and methods

Animals and surgical procedure

The studies were conducted under approval of the Institutional Animal Care and Use Committee using National Institutes of Health guidelines. Experiments were carried out on 330 to 400 g adult male Sprague-Dawley rats (Harlan, Indianapolis, IN). Rats were housed at $22 \pm 1\,°C$ with 60% humidity, 12-hour light/12-hour dark cycles, and pellet food and water ad libitum. Surgery was performed after intubation under halothane anesthesia and controlled ventilation (1.3% halothane in 70% nitrous oxide and 30% oxygen). Rectal temperature was maintained at $36.5 \pm 0.5\,°C$ using a heat lamp. The left femoral artery and vein were cannulated with polyethylene tubing (P.E. 50, Becton Dickinson and Company, Sparks, MD) for continuous monitoring of mean arterial blood pressure (MABP), blood sampling, and drug infusion. Adequate ventilation was verified by arterial blood gas measurement after 1 hour of anesthesia.

Cerebral blood flow (CBF) to the territory of the right middle cerebral artery was continuously monitored by Laser Doppler Flowmetry (LaserFlo Vasamedics Inc., St Paul, MN) through a burr hole located 1 mm posterior and 5 mm lateral to bregma leaving the dura intact. Animals were placed in a supine position over the laser Doppler probe, and CBF as well as MABP were recorded continuously using a data acquisition system (ADInstruments, Colorado Springs, CO).

MCAO was induced using a slightly modified version of the intraluminal suture method described elsewhere [2]. Through a midline neck incision, the bifurcation of the right common carotid artery was exposed, and branches of the external carotid artery (ECA) and internal carotid artery (ICA) including the occipital, lingual, and maxillary arteries were microsurgically separated and coagulated. The ECA was ligated with a 4-0 silk suture, and after temporary occlusion of the ICA and common carotid artery with vascular mini-clips, a 4-0 monofilament nylon suture (4-0 SN-644 MONOSOF nylon Polyamide) with a silicon tip of 0.30 to 0.35 mm diameter was inserted through the ECA stump and secured with a suture. The clips were removed and the filament was advanced through the ICA into the circle of Willis while occluding the pterygopalatine artery with a forceps. A CBF reduction between 70 and 80% to the baseline was observed when the suture was advanced at a distance of 22 to 24 mm from the carotid bifurcation, thereby verifying proper MCAO. Two hours after occlusion, the middle cerebral artery territory was perfused by withdrawing the suture into the ECA stump, confirmed by an increase in CBF.

Study protocol and drug preparation

The objective of these experiments was to assess the effect of intravenous infusion of SR49059 on brain edema at different concentrations (83 mmol at 720 µL/hr, n = 6; 73 mmol at 640 µL/hr, n = 6; 56 mmol at 480 µL/hr, n = 6) and started at different time points (60 minutes pre-ischemia, n = 6; 60 minutes post-ischemia, n = 6) following MCAO. The animals were randomly assigned to sham procedure (n = 4), vehicle infusion (n = 18), or intravenous SR49059 at different doses and time points after MCAO. SR49059 (Sanofi Recherche, Montpellier, France) was dissolved in 1% dimethyl sulfoxide as vehicle solution (Sigma-Aldrich, St Louis, MO). The drug was intravenously administered using a continuous infusion pump (sp210w syringe pump, KD Scientific, Holliston, MA). At the end of the experiments, the animals were sacrificed by an overdose of halothane, decapitated, and the brains removed.

Tissue processing

Cerebral tissue was immediately cut into 4 consecutive 4 mm coronal sections excluding the most rostral and caudal sections from further analysis. After division into the right and left hemispheres along the anatomic midline, the 4 regional samples obtained were processed for water content measured by the wet/dry weight method. The wet weight of each sample was measured using an electronic analytical balance before drying the sample at 95 °C for 5 days and reweighing to obtain the dry weight. The water content of each sample is given as percentage of total tissue weight. For measurement of brain sodium and potassium concentrations, the dried samples were ashed in a furnace for 24 hours at 400 °C. The ash was then extracted with distilled water, and the concentrations of sodium and potassium were determined using a flame photometer (943 nm; Instrument Laboratory, San Jose, CA) with caesium as an internal standard.

Statistical analysis

SPSS software (SigmaStat, Chicago, IL) was used for statistical analysis. The data were analyzed by a randomized one-way ANOVA for group variations followed by a Tukey post hoc analysis. Statistical significance was accepted at $p < 0.05$.

Table 1. *Effects of different doses of SR49059 on brain water, sodium, and potassium after middle cerebral artery occlusion**

Groups	% Tissue water	Tissue sodium mEq/kg dry wt	Tissue potassium mEq/kg dry wt
Vehicle	2.3 ± 0.0	116.1 ± 8.8	−43.2 ± 6.0
SR49059 480 µL/hr	1.8 ± 0.4	69.2 ± 19.0	−26.0 ± 12.6**
SR49059 640 µL/hr	1.2 ± 0.3**	39.9 ± 13.5**	−27.5 ± 6.3
SR49059 720 µL/hr	0.8 ± 0.2**	36.7 ± 6.2**	2.1 ± 11.3**

* Data shown are average ± SEM, ** $p < 0.05$ as compared with vehicle.

Results

The injury-induced mortality was 14% following MCAO. MABP and arterial blood gases were kept within physiological limits throughout the experimental procedure, requiring few adjustments in halothane concentration and ventilation parameters.

The comparison of different doses of SR49059 by ANOVA produced a significant group effect for the reduction of brain water content in the ischemic area ($F_{3,66} = 9.34$, $p < 0.001$), the reduction of sodium accumulation ($F_{3,66} = 10.42$, $p < 0.001$) and the reduction of potassium loss ($F_{3,66} = 4.40$, $p \leq 0.001$) (Table 1). Tukey post hoc analysis indicated that the effect of SR49059 treatment on brain water content, as compared to vehicle infusion, was more effective at an infusion rate of 720 µL/hr ($p \leq 0.002$), than at 640 µL/hr ($p \leq 0.036$) or 480 µL/hr ($p \leq 0.870$) when started 60 minutes before MCAO. Accordingly, the Tukey post hoc analysis for the effect of different doses of SR49059 on sodium accumulation and potassium loss confirmed the protective effect of the 720 µL/hr infusion rate.

In order to test a clinically relevant time point for SR49059 application, we compared the infusion rate of 720 µL/hr started either 60 minutes before or 60 minutes after MCAO to a vehicle infusion (Table 2). Comparison by ANOVA of SR49059 infusion started before or after MCAO produced a significant group effect for the reduction of brain water content in the ischemic area ($F_{3,66} = 8.30$, $p < 0.001$), the reduction of sodium accumulation ($F_{3,66} = 11.53$, $p < 0.001$) and the reduction of potassium loss ($F_{3,66} = 5.09$, $p \leq 0.001$).

Table 2. *Effect of SR49059 on brain water, sodium, and potassium at different times after middle cerebral artery occlusion**

Groups	% Tissue water	Tissue sodium mEq/kg dry wt	Tissue potassium mEq/kg dry wt
Vehicle (n = 18)	2.2 ± 0.1	111.3 ± 6.5	−43.2 ± 6.0
SR49059 60 min before (n = 6)	0.9 ± 0.2**	38.2 ± 8.0**	3.6 ± 21.2**
SR49059 60 min after (n = 6)	1.1 ± 0.3**	39.9 ± 8.3**	−14.5 ± 12.6

* Data shown are average ± SEM, ** $p < 0.05$ as compared with vehicle.

The effects of SR49059 treatment according to the Tukey post hoc analysis follow. 1) Brain water content compared to vehicle infusion: treatment was equally effective when started 60 minutes before ($p < 0.001$) or 60 minutes after ($p < 0.003$) MCAO. 2) Brain sodium accumulation compared to vehicle infusion: treatment was equally effective when started 60 minutes before ($p < 0.001$) or 60 minutes after ($p < 0.001$) MCAO. 3) Brain potassium loss compared to vehicle infusion: treatment was only effective when started 60 minutes before ($p < 0.003$) but not 60 minutes after ($p < 0.385$) MCAO.

Discussion

To the best of our knowledge, this study is the first to demonstrate that brain edema resulting from MCAO is reduced by treatment with SR49059, the selective vasopressin V1a receptor antagonist. SR49059 was administered intravenously for 5 hours starting 1 hour before occlusion, testing with different concentrations of the drug. Afterwards, the most effective concentration was used starting before or after MCAO. We demonstrated that SR49059 caused a significant dose-dependent reduction in brain water content and subsequent electrolyte shift, and was still effective when given 1 hour after onset of ischemia. The most significant effect in the ischemic area was observed using the highest dose of 83 mmol SR49059, starting either 1 hour before or after MCAO. Similar results were obtained regarding brain sodium and potassium shift after ischemic injury. SR49059 treatment was able to reduce sodium uptake and increase potassium levels in a dose-dependent manner. These findings are consistent with the generally accepted opinion that water and sodium tend to coexist and transfer together through the plasma membrane under physiological and pathological conditions [6, 10, 17].

Water can cross cell membranes through different pathways: specific water channels (aquaporins), the lipid bilayer [5], or through ion-water cotransport proteins [19, 20]. Because of its specific anatomical and cellular localization in the central nervous system, AQP4 has been suggested to play a role in cerebral water balance. According to this hypothesis, AQP4-deficient mice developed less brain edema after acute water intoxication and MCAO [11]. Regarding the function and regulation of AQP4 following injury, there exist conflicting results. In different models inducing neuronal degeneration, AQP4 mRNA was up-regulated following blood-brain barrier disruption [16]. In a combined head injury model, AQP4 immunostaining was negative and the AQP4 mRNA down-regulated in areas with impaired blood-brain barrier [7]. Following cortical impact injury, hemispheric ipsilateral AQP4 was progressively down-regulated within the first 48 hours [8]. Similar results were found following ischemia [13] and hypoxia [18], all known to produce a predominately cytotoxic edema.

Evidence indicates that AQP4 is regulated by vasopressin, a neuropeptide endogenous to the brain [9] that includes a vasopressin-containing fiber system [4]. In vitro experiments on rat neocortical slices suggest that there exists a tonic, vasopressin V1a receptor-mediated facilitation of water permeability [12]. V1a receptors are coupled via G-proteins to phospholipase C, resulting in an IP3-dependent Ca^{2+} release from internal stores. V1a receptor stimulation also causes activation of protein kinase C, and evidence suggests that vasopressin exerts part of its facilitatory effect through this signaling pathway [12]. One possible explanation for the protective effects on brain edema development following ischemic injury found in this study may be the vasopressin-dependent inhibition of AQP4 activity or expression through V1a receptors. However, additional studies are necessary to clarify the precise biochemical pathway by which the selective V1a antagonist SR49059 prevents brain edema development after MCAO.

Acknowledgments

This research was supported by grants NS 12587 and NS 19235 from the National Institutes of Health, Bethesda, MD.

References

1. Badaut J, Verbavatz JM, Freund-Mercier MJ, Lasbennes F (2000) Presence of aquaporin-4 and muscarinic receptors in astrocytes and ependymal cells in rat brain: a clue to a common function? Neurosci Lett 292: 75–78
2. Belayev L, Alonso OF, Busto R, Zhao W, Ginsberg MD (1996) Middle cerebral artery occlusion in the rat by intraluminal suture. Neurological and pathological evaluation of an improved model. Stroke 27: 1616–1623
3. Bemana I, Nagao S (1999) Treatment of brain edema with a nonpeptide arginine vasopressin V1 receptor antagonist OPC-21268 in rats. Neurosurgery 44: 148–155
4. de Vries GJ, Miller MA (1998) Anatomy and function of extrahypothalamic vasopressin systems in the brain. Prog Brain Res 119: 3–20
5. Finkelstein A (1987) Water movements through lipid bilayers, pores and plasma membranes: theory and reality. John Wiley & Sons, New York
6. Gotoh O, Asano T, Koide T, Takakura K (1985) Ischemic brain edema following occlusion of the middle cerebral artery in the rat. I: The time courses of the brain water, sodium and potassium contents and blood-brain barrier permeability to 125I-albumin. Stroke 16: 101–109
7. Ke C, Poon WS, Ng HK, Pang JC, Chan Y (2001) Heterogeneous responses of aquaporin-4 in oedema formation in a replicated severe traumatic brain injury model in rats. Neurosci Lett 301: 21–24
8. Kiening KL, van Landeghem FK, Schreiber S, Thomale UW, von Deimling A, Unterberg AW, Stover JF (2002) Decreased hemispheric Aquaporin-4 is linked to evolving brain edema following controlled cortical impact injury in rats. Neurosci Lett 324: 105–108
9. Landgraf R (1992) Central release of vasopressin: stimuli, dynamics, consequences. Prog Brain Res 91: 29–39
10. Loo DD, Zeuthen T, Chandy G, Wright EM (1996) Cotransport of water by the Na+/glucose cotransporter. Proc Natl Acad Sci USA 93: 13367–13370
11. Manley GT, Fujimura M, Ma T, Noshita N, Filiz F, Bollen AW, Chan P, Verkman AS (2000) Aquaporin-4 deletion in mice reduces brain edema after acute water intoxication and ischemic stroke. Nat Med 6: 159–163
12. Niermann H, Amiry-Moghaddam M, Holthoff K, Witte OW, Ottersen OP (2001) A novel role of vasopressin in the brain: modulation of activity-dependent water flux in the neocortex. J Neurosci 21: 3045–3051
13. Sato S, Umenishi F, Inamasu G, Sato M, Ishikawa M, Nishizawa M, Oizumi T (2000) Expression of water channel mRNA following cerebral ischemia. Acta Neurochir [Suppl] 76: 239–241
14. Taniguchi M, Yamashita T, Kumura E, Tamatani M, Kobayashi A, Yokawa T, Maruno M, Kato A, Ohnishi T, Kohmura E, Tohyama M, Yoshimine T (2000) Induction of aquaporin-4 water channel mRNA after focal cerebral ischemia in rat. Brain Res Mol Brain Res 78: 131–137
15. Venero JL, Vizuete ML, Machado A, Cano J (2001) Aquaporins in the central nervous system. Prog Neurobiol 63: 321–336
16. Vizuete ML, Venero JL, Vargas C, Ilundain AA, Echevarria M, Machado A, Cano J (1999) Differential upregulation of aquaporin-4 mRNA expression in reactive astrocytes after brain injury: potential role in brain edema. Neurobiol Dis 6: 245–258
17. Wright EM, Loo DD (2000) Coupling between Na+, sugar, and water transport across the intestine. Ann NY Acad Sci 915: 54–66
18. Yamamoto N, Yoneda K, Asai K, Sobue K, Tada T, Fujita Y, Katsuya H, Fujita M, Aihara N, Mase M, Yamada K, Miura Y, Kato T (2001) Alterations in the expression of the AQP family in cultured rat astrocytes during hypoxia and reoxygenation. Brain Res Mol Brain Res 90: 26–38
19. Zeuthen T (1994) Cotransport of K+, Cl– and H_2O by membrane proteins from choroid plexus epithelium of Necturus maculosus. J Physiol 478: 203–219
20. Zeuthen T, Meinild AK, Klaerke DA, Loo DD, Wright EM, Belhage B, Litman T (1997) Water transport by the Na+/glucose cotransporter under isotonic conditions. Biol Cell 89: 307–312

Correspondence: Anthony Marmarou, Department of Neurosurgery, Virginia Commonwealth University Medical Center, 1101 East Marshall Street, Box 980508, Richmond, VA 23298-0508, USA. e-mail: marmarou@abic.vcu.edu

Experimental Spinal Cord Injury

Topical application of dynorphin A (1-17) antibodies attenuates neuronal nitric oxide synthase up-regulation, edema formation, and cell injury following focal trauma to the rat spinal cord

H. S. Sharma[1], F. Nyberg[2], T. Gordh[1], and P. Alm[3]

[1] Laboratory of Cerebrovascular Research, Department of Anesthesiology and Intensive Care, Institute of Surgical Sciences, University Hospital, Uppsala University, Uppsala, Sweden
[2] Department of Pharmaceutical Biosciences, Division of Drug Dependence, Biomedical Center, Uppsala University, Uppsala, Sweden
[3] Department of Pathology, University Hospital, Lund, Sweden

Summary

Previous investigations from our laboratory show that up-regulation of neuronal nitric oxide synthase (NOS) following spinal cord injury (SCI) is injurious to the cord. Antiserum to dynorphin A (1-17) induces marked neuroprotection in our model of SCI, indicating an interaction between dynorphin and NOS regulation. The present investigation was undertaken to find out whether topical application of dynorphin A (1-17) antiserum has some influence on neuronal NOS up-regulation in the traumatized spinal cord.

SCI was produced in anesthetized animals by making a unilateral incision into the right dorsal horn of the T10–11 segments. The antiserum to dynorphin A (1-17) was applied (1 : 20, 20 μL in 10 seconds) 5 minutes after trauma over the injured spinal cord and the rats were allowed to survive 5 hours after SCI. Topical application of dynorphin A (1-17) antiserum significantly attenuated neuronal NOS up-regulation in the adjacent T9 and T12 segments. In the antiserum-treated group, spinal cord edema and cell injury were also less marked. These observations provide new evidence that the opioid active peptide dynorphin A may be involved in the mechanisms underlying NOS regulation in the spinal cord after injury, and confirms our hypothesis that up-regulation of neuronal NOS is injurious to the cord.

Keywords: Dynorphin A (1-17); spinal cord injury; nitric oxide synthase; blood-spinal cord barrier.

Introduction

Spinal cord injury (SCI)-induced cell and tissue damage is a complex process that involves several neurotransmitters, free radicals, vasoactive compounds, neurochemicals, growth factors, cytokines, and other proteins/factors in a cascade of events leading to cell and tissue injury [5, 19–21, 36]. These endogenous compounds/factors released after trauma may interact with each other to induce their neurodegenerative and/or neuroprotective effects in vivo [19, 20]. Thus, further research is needed to clarify the potential interaction among different compounds/factors that could synergistically potentiate or neutralize the neurodestructive and/or neuroprotective capabilities in SCI.

Previous reports from our laboratory suggest that dynorphin is involved in the pathophysiology of SCI [25, 28, 39]. The peptide is present in the dorsal horn of the normal spinal cord, particularly in Rexed's lamina II and I, and occasionally in lamina V and VI [2, 4, 8, 16]. Intrathecal administration of dynorphin A (1-17 or 2-17) induces motor dysfunction in rats [6, 7, 9, 13–15]. On the other hand, intrathecal administration of dynorphin antibodies (dynorphin A 1-17 antiserum, 10 μL given 15 minutes before and 4 hours after injury) improved walking ability without affecting the angle difference on the inclined plane test [5]. However, the molecular mechanisms involved in the dynorphin-induced pathophysiology of SCI are largely unknown.

Trauma to the spinal cord induces breakdown of the blood-spinal cord barrier (BSCB) and is largely responsible for edema formation and cell injury [17–21]. There are reasons to believe that dynorphin, through mechanisms that probably involve other neurochemicals such as nitric oxide, influence spinal cord pathophysiology [10, 11, 22]. Previous studies from our laboratory demonstrate that the dynorphin A (1-17) content of the spinal cord is increased in the T9 segment following an incision into the right dorsal horn of the T10–11 segments [25]. In these animals, edema formation and cell injury are quite prominent in the T9 segment [19, 25], indicating that increased dynorphin levels somehow contribute to cell injury in the spinal cord [25]. This idea is further supported by the observation that topical application of dynorphin A (1-17) antiserum (1:20 dilution in the phosphate buffer) 2 minutes after SCI significantly reduces the edema formation and cell injury in the cord [27, 28, 39], suggesting that dynorphin A antiserum is neuroprotective in SCI.

There is evidence that 2 major products of dynorphin A (dynorphin A 1-13 and dynorphin A 1-17) cause neuronal injury that can be prevented by N-methyl-D-aspartate (NMDA) glutaminergic receptor antagonist MK-801, but not by the opioid antagonists [3, 5, 9–12, 37], indicating that dynorphin A-induced neurotoxicity is mediated by non-opioid mechanisms [3, 37]. In spinal nerve ligation injury models, antiserum to dynorphin A has the same profile of actions as the NMDA receptor antagonists MK-801 in blocking thermal hyperalgesia and restoring the efficacy of morphine against allodynia [15]. Thus, antibodies to dynorphin A appear to have potential therapeutic value in SCI that require additional investigation.

Recent investigations from our laboratory show that up-regulation of neuronal nitric oxide synthase (nNOS) following SCI is injurious to the cord [23, 29, 33]. This was demonstrated by topical application of neuronal NOS antiserum over the traumatized cord, which clearly attenuated edema formation and cell injury [29]. Since an interaction between dynorphin and NOS regulation was recently documented [10, 11, 22, 23], it is quite likely that dynorphin may interact with nitric oxide to induce cell and tissue injury following trauma. The present investigation was undertaken to determine whether topical application of dynorphin A (1-17) antiserum has some influence on neuronal NOS up-regulation following trauma to the spinal cord in the rat.

Materials and methods

Animals

Experiments were carried out on 64 male Sprague-Dawley rats (250 to 300 g) housed at controlled room temperature ($21 \pm 1\,°C$) with a 12-hour dark/12-hour light schedule. Food and tap water were supplied ad libitum before the experiments.

Spinal cord injury

Under Equithesin anesthesia (3 mL/kg, i.p.), a 1-segment laminectomy was done at the T10–11 level. SCI was inflicted by making a longitudinal incision into the right dorsal horn using a sterile scalpel blade under aseptic conditions [24–26]. Animals were allowed to survive for 5 hours after injury. This experiment was approved by the Ethics Committee of Uppsala University, Uppsala, Sweden.

Control group

Normal animals under Equithesin anesthesia were used as controls. Since no differences between normal and sham-operated animals were observed with regard to edema formation or cell injury [19–21], all comparisons between untreated injured animals and antiserum-treated rats were made in reference to intact control rats.

Treatment with dynorphin A (1-17) antiserum

A commercial dynorphin A (1-17) polyclonal antiserum (Calbiochem, San Diego, CA) was used in this investigation (dilution 1:20 in phosphate-buffered saline (PBS), 0.1 mol, pH 7.0 [22, 28]. The dynorphin A antiserum was applied (20 μL in 10 seconds) over the exposed spinal cord 2 minutes after injury. Animals were allowed to survive 5 hours after SCI [27, 28, 39].

Perfusion and fixation

At the end of the experiments, intravascular blood was washed out with about 50 mL of cold 0.1 mol PBS followed by 4% paraformaldehyde solution (150 mL) in PBS containing 2.5% picric acid [30–32]. The animals were wrapped in aluminum foil and kept overnight at $4\,°C$. On the next day, the spinal cord was dissected out and kept in the same fixative at $4\,°C$ for 1 week.

NOS immunohistochemistry

NOS immunostaining was examined in the T9 and T12 segments using a monoclonal antiserum to the constitutive isoform of NOS [29, 30, 33] as described earlier [29]. In brief, free-floating 40 μm thick Vibratome sections were cut from the T9 segment and processed for NOS immunostaining. The immunoreactivity was developed using a peroxidase-antiperoxidase technique [29].

Quality control

To examine the specificity of immunohistochemical methods, the primary antiserum was omitted and the rest of the processing was done as described above. These negative controls did not show immunostaining of nerve cells or nerve fibers [22].

Image processing

Microphotographs were taken on Kodak Supra 100 ASA negative film and digital images (size 52 cm × 32 cm, 80 pixels/inch) were processed by Kodak Colour Laboratories (Stockholm, Sweden). The digital images (size 8 × 13 cm, 300 pixels/inch) were modified using Adobe Photoshop 3.5 software program on a G-4 Macintosh computer. Identical color filter or color balance was used on the images obtained from the control, injured, or drug-treated controls, or spinal cord-injured animals [22].

Spinal cord edema

The spinal cord was examined visually for gross swelling, then edema formation was determined in selected segments (T9 or T12) by measuring water content according to differences in wet and dry weights of the samples [18, 34].

Morphology of the spinal cord

Structural changes in the spinal cord tissue pieces obtained from the T9 and T12 segments were processed for standard light and transmission electron microscopy. The ultra-thin sections were contrasted with uranyl acetate and lead citrate and viewed under a Phillips or Hitachi electron microscope [18–20, 23, 29, 31].

BSCB permeability

BSCB permeability was examined in different spinal cord segments using Evans blue and ^{131}Iodine as protein tracers [18, 29, 32].

Statistical evaluation

The differences between control, injured, and antiserum-treated groups were analyzed using ANOVA followed by Dunnett's test for multiple group comparison. A p-value less than 0.05 was considered significant.

Results

Effect of dynorphin A antiserum on gross morphology

Topical application of dynorphin A antiserum on the traumatized spinal cord resulted in a considerable reduction in visual swelling, micro-hemorrhages, and edema compared to the untreated injured group seen 5 hours after SCI (Fig. 1). Treatment with dynorphin A antiserum alone did not influence spinal cord morphology compared to the control group (results not shown).

Effect of dynorphin A antiserum on NOS immunohistochemistry

Focal trauma to the spinal cord resulted in marked up-regulation of neuronal NOS in the T9 and T12 segments (Fig. 1). The up-regulation of NOS was mainly confined within the edematous region of the spinal cord gray matter (Fig. 1). Topical application of dynorphin A antiserum significantly attenuated NOS up-regulation in these spinal cord segments after injury (Fig. 1, Table 1). Application of dynorphin A antiserum alone did not influence NOS expression in the normal spinal cord (Table 1).

Expression of dynorphin A antiserum-induced reduction in neuronal NOS was most marked in the contralateral side of the T9 (rostral) and T12 (caudal) segments of the spinal cord compared to the ipsilateral cord (Table 1, Fig. 1).

Effect of dynorphin A antiserum on BSCB permeability

Topical application of dynorphin A antiserum markedly reduced extravasation of Evans blue and radiotracer within the spinal cord segments compared to untreated injured rats (Table 1). Antiserum alone did not influence BSCB permeability to these tracers compared to the normal control group (Table 1). However, pretreatment with dynorphin A antiserum significantly attenuated the trauma-induced extravasation of Evans blue and ^{131}Iodine tracers in the spinal cord at 5 hours. The magnitude of reduction in tracer extravasation was most marked for Evans blue dye compared to the radioactive iodine.

Effect of dynorphin A antiserum on edema formation

Treatment with dynorphin A antiserum significantly attenuated trauma-induced edema formation in the T9 and T12 segments of the spinal cord at 5 hours (Table 1). However, dynorphin A antiserum treatment alone did not alter the spinal cord water content compared to the control group (Table 1).

Effect of dynorphin A antiserum on ultrastructure of the cord

Pretreatment with dynorphin A antiserum markedly reduced trauma-induced ultrastructural changes in the spinal cord. Myelin vesiculation, edema, vacuolation, and membrane disruption were much less frequent in the antiserum-treated injured group (Fig. 1). The effect of dynorphin A antiserum was most marked in the contralateral T9 and T12 segments of the cord. Treatment of spinal cord with dynorphin A antiserum alone did not show any morphological alterations in the uninjured cord (Table 1).

Discussion

Our investigation shows that topical application of dynorphin A antiserum over the traumatized spinal cord attenuates SCI-induced expression of neuronal NOS. This observation suggests that endogenous dynorphin is involved in trauma-induced neuronal NOS expression in the spinal cord.

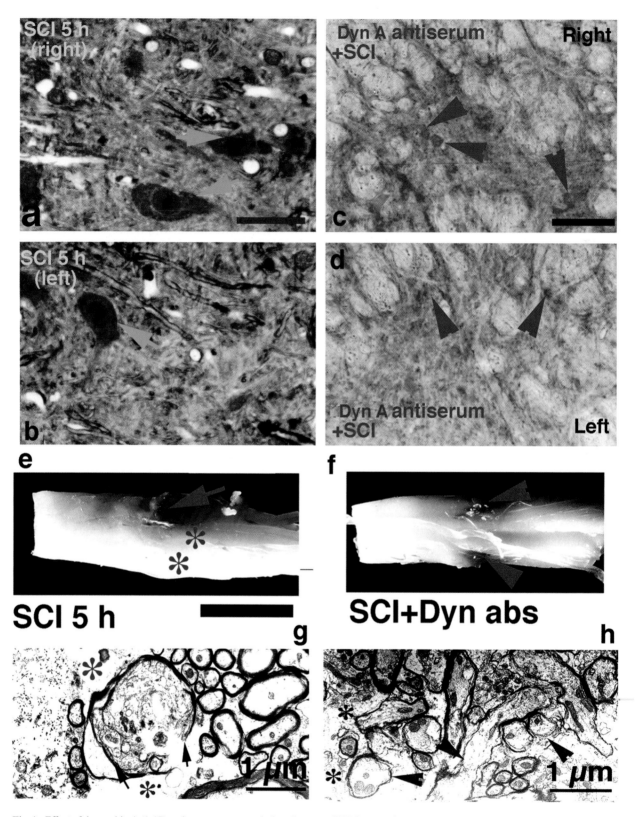

Fig. 1. Effect of dynorphin A (1-17) antiserum on trauma-induced neuronal NOS expression (arrowheads) (a–d), gross edematous swelling (*) of the spinal cord (e, f), and ultrastructural changes in the neuropil (g, h). Post-trauma treatment with dynorphin A (1-17) antiserum markedly attenuated edema formation, neuronal NOS expression, and membrane damage [e–h, modified after 28, 39]

Table 1. *Effects of treatment with dynorphin A (1-17) antiserum on BSCB permeability, edema formation, and nNOS expression after SCI in rats*

Parameters measured	n	Control		5 h SCI		Dyn A (1-17) antiserum + SCI§	
		T9	T12	T9	T12	T9	T12
BSCB permeability							
Evans blue mg, %	6	0.24 ± 0.04	0.28 ± 0.02	1.54 ± 0.28**	1.78 ± 0.24**	0.46 ± 0.08^{aa}	0.54 ± 0.10^{aa}
[131]Iodine, %	5	0.35 ± 0.06	0.38 ± 0.10	1.89 ± 0.14**	1.96 ± 0.16**	0.56 ± 0.14^{aa}	0.68 ± 0.16^{aa}
Edema formation							
Water content, Right, %	5	66.32 ± 0.18	66.34 ± 0.21	69.14 ± 0.24**	69.48 ± 0.34**	67.18 ± 0.22^{aa}	67.43 ± 0.13^{aa}
nNOS expression							
No. of nNOS-positive cells, Right	5	4 ± 2	4 ± 2	28 ± 4##	26 ± 8##	8 ± 4^{aa}	7 ± 3^{aa}
No. of nNOS-positive cells, Left	5	5 ± 3	3 ± 2	14 ± 6##,b	14 ± 6##,b	6 ± 3^{aa}	4 ± 2^{aa,b}

Values are mean ± SD of 5–6 rats in each group. * = p < 0.05; ** p < 0.01 (compared from control); a = p < 0.05; aa = p < 0.01 (compared from 5 h SCI), ANOVA followed by Dunnett's test from 1 control. ## = p < 0.01, Chi-square test from control group; b = p < 0.05, from 5 h SCI right side.
BSCB Blood-spinal cord barrier; *nNOS* neuronal nitric oxide synthase; *SCI* spinal cord injury, § Dynorphin A (1-17) antiserum (1:20 in phosphate-buffered saline) was administered 5 minutes after SCI. Cell changes were analyzed in blinded fashion [28].

Topical application of dynorphin A antiserum on the traumatized spinal cord binds endogenous dynorphin antigens within the spinal cord in vivo, resulting in neutralization of the neurotoxic effects of the peptide [31]. This is evident from our finding that application of dynorphin A antiserum induces marked neuroprotection in the spinal cord. Increased dynorphin A expression in the spinal cord occurs 4 hours after injury and continues until 24 hours after trauma [5–7, 25, 40], indicating a potential role of the peptide in cell and tissue injury [19, 25]. Previous observations from our laboratory, which demonstrated increased dynorphin A content and cell injury in the T9 segment of the spinal cord 5 hours after injury [25], are consistent with this hypothesis. This increase in dynorphin A content in the cord was prevented by pretreatment with a potent serotonin synthesis inhibitor drug, p-chlorophenylalanine (p-CPA) [25], which induces marked neuroprotection in SCI [19, 20]. These observations suggest that elevation of dynorphin A in the cord following SCI is injurious to the cell, and topical application of dynorphin A antiserum neutralizes the harmful effects of the peptide in vivo. Further studies using immunohistochemical expression of dynorphin A in the cord in untreated injured and the antiserum-treated traumatized group are needed to confirm this point.

There is evidence that dynorphin-induced neurotoxicity is mediated through both opioid and non-opioid mechanisms [5–7]. The peptide is known to stimulate glutamate release in the CNS by acting through NMDA receptors [38]. This hypothesis is supported by the fact that antiserum to dynorphin antagonizes NMDA receptors in vivo [9]. Furthermore, effects of the dynorphin A antiserum are far more superior in attenuating allodynia-induced pain than that of the classical NMDA receptor antagonist, MK-801 [15].

The mechanisms by which dynorphin stimulates NOS remains speculative. It is reasonable to believe that activation of glutamate receptors through dynorphin contributes to increased nitric oxide production. Stimulation of glutamate receptors are known to induce generation of free radicals and formation of nitric oxide [1]. It appears that dynorphin up-regulation in SCI can influence nitric oxide formation. Inhibition of SCI-induced NOS expression and cell injury following pretreatment with the antioxidant compound H-290/51 further supports this hypothesis [20, 35]. That interaction between dynorphin and nitric oxide exists in vivo is supported by the fact that pretreatment with NOS inhibitors significantly attenuates dynorphin immunoreactivity in hyperthermia-induced brain injury [22]. Taken together, it appears that dynorphin and nitric oxide synergistically play important roles in cell injury and cell death.

It appears that dynorphin stimulates NOS via accumulation of intracellular Ca^{2+} [23, 36]. This up-regulation of NOS requires activation of intracellular Ca^{2+} [9]. Increased intracellular accumulation of Ca^{2+} is responsible for cell death [6, 7, 20]. In cell culture studies, dynorphin A-induced motor neuron death is accompanied by an increase in intracellular Ca^{2+} [9, 38]. Thus, it seems that dynorphin and/or nitric oxide in SCI synergistically induces intracellular accumulation of Ca^{2+} leading to cell death. It would be interesting to examine whether pretreatment with a potent calcium channel blocker, such as nimodipine that induces neuroprotection in SCI [19–21, 36], influences expression of neuronal NOS and/or dynorphin A in this model.

Our results show that dynorphin A antiserum attenuated BSCB disturbances, edema formation, and cell injury, supporting a role of dynorphin A in spinal cord pathology. However, dynorphin A (1-17), but not enkephalins and β-endorphins, accumulate at the lesion site following spinal trauma [5–7]. Intrathecal administration of dynorphin induces paralysis of hindlimbs and tail, which correlates well with the severe reductions in local blood flow, widespread ischemic cell damage, neuronal loss, necrosis, gliosis, cavitation, and vascular injury within 72 hours [12–14]. Furthermore, infusion of antiserum to dynorphin A (2-17) improves neurological function in animals following SCI [5], mediated through both opioid and non-opioid mechanisms [5, 6]. Our study has shown that dynorphin A induces neurotoxicity by mechanisms involving up-regulation of nitric oxide. Obviously, up-regulation of nitric oxide is injurious to cell and tissue [22, 23]. It remains to be seen whether blockade of κ-opioid receptors, to which dynorphin A binds in vivo, also attenuate NOS up-regulation in SCI, a feature currently being investigated in our laboratory.

Conclusion

Dynorphin antiserum inhibits neuronal NOS activity in SCI that could be instrumental in the attenuation of cell and tissue injury. These findings suggest that dynorphin A influences NOS regulation in the spinal cord following trauma and supports our hypothesis that up-regulation of neuronal NOS is injurious to the cord. Taken together, our observations indicate that dynorphin-induced neurotoxicity in SCI is mediated via mechanisms involving nitric oxide.

Acknowledgments

This investigation was supported by grants from Swedish Medical Research Council (2710); Astra-Zeneca, Mölndal, Sweden; Alexander von Humboldt Foundation, Germany; The University Grants Commission, New Delhi, India; and The Indian Council of Medical Research, New Delhi, India. The expert technical assistance of Kärstin Flink, Kerstin Rystedt, Franziska Drum, and Katherin Kern, and the secretarial assistance of Aruna Sharma are greatly appreciated.

References

1. Azbill RD, Mu X, Bruce-Keller AJ, Mattson MP, Springer JE (1997) Impaired mitochondrial function, oxidative stress and altered antioxidant enzyme activities following traumatic spinal cord injury. Brain Res 765: 283–290
2. Chavkin C, Goldstein A (1981) Specific receptors for the opioid peptide dynorphin: structure-activity relationships. Proc Natl Acad Sci USA 78: 6543–6547
3. Chen L, Gu Y, Huang LY (1995) The opioid peptide dynorphin directly blocks NMDA receptor channels in the rat. J Physiol 482: 575–581
4. Cho HJ, Basbaum AI (1989) Ultrastructural analysis of dynorphin B-immunoreactive cells and terminals in the superficial dorsal horn of the deafferented spinal cord of the rat. J Comp Neurol 281: 193–201
5. Faden AI (1990) Opioid and non-opioid mechanisms may contribute to dynorphin's pathophysiological actions in the spinal cord injury. Ann Neurol 27: 64–74
6. Faden AI (1993) Role of endogenous opioids and opioid receptors in central nervous system injury. Handbook Exp Pharmacol 104, Part I: 325–341
7. Faden AI, Takemori AE, Portoghese PS (1987) Kappa-selective opiate antagonist nor-binaltorphimine improves outcome after traumatic spinal cord injury in rats. Cent Nerv Syst Trauma 4: 227–237
8. Fallon JH, Ciofi P (1990) Dynorphin-containing neurons. Handbook Chem Neuroanat 9: 1–286
9. Hauser KF, Knapp PE, Turbek CS (2001) Structure-activity analysis of dynorphin A toxicity in spinal cord neurons: intrinsic neurotoxicity of dynorphin A and its carboxyl-terminal, non-opioid metabolites. Exp Neurol 168: 78–87
10. Hu WH, Li F, Qiang WA, Liu N, Wang GQ, Xiao J, Liu JS, Liao WH, Jen MF (1999) Dual role for nitric oxide in dynorphin spinal neurotoxicity. J Neurotrauma 16: 85–98
11. Hu WH, Qiang WA, Li F, Liu N, Wang GQ, Wang HY, Wan XS, Liao WH, Liu JS, Jen MF (2000) Constitutive and inducible nitric oxide synthases after dynorphin-induced spinal cord injury. J Chem Neuroanat 17: 183–197
12. Isaac L, VanZandt O'Malley T, Ristic H, Stewart P (1990) MK-801 blocks dynorphin A (1-13)-induced loss of the tail-flick reflex in the rat. Brain Res 531: 83–87
13. Long JB, Martinez-Arizala A, Echevarria EE, Tidwell RE, Holaday JW (1988) Hindlimb paralytic effects of prodynorphin-derived peptides follwoing spinal subarachnoid injection in rats. Eur J Pharmacol 153: 45–54
14. Long JB, Rigamonti DR, deCosta B, Rice KC, Martinez-Arizala A (1989) Dynorphin A-induced rat hindlimb paralysis and spinal cord injury are not altered by the κ-opioid antagonist nor-binaltorphimine. Brain Res 497: 155–162
15. Nichols ML, Lopez Y, Ossipov MH, Bian D, Porreca F (1997) Enhancement of the antiallodynic and antinociceptive efficacy of spinal morphin by antisera to dynorphin A (1-13) or MK-801 in nerve-ligation model of peripheral neuropathy. Pain 69: 317–322
16. Nyberg F, Sharma HS, Wiesenfeld-Hallin Z (1995) Neuropeptides in the spinal cord. Prog Brain Res 104: 1–416
17. Sharma HS (2000) A bradykinin BK2 receptor antagonist HOE-140 attenuates blood-spinal cord barrier permeability following a focal trauma to the rat spinal cord. An experimental study using Evans blue, ^{131}I-sodium and lanthanum tracers. Acta Neurochir [Suppl] 76: 159–163
18. Sharma HS (2003) Neurotrophic factors attenuate microvascular permeability disturbances and axonal injury following trauma to the rat spinal cord. Acta Neurochir [Suppl] 86: 383–388

19. Sharma HS (2004) Pathophysiology of the blood-spinal cord barrier in traumatic injury. In: Sharma HS, Westman J (eds) Blood-spinal cord and brain barriers in health and disease. Elsevier Academic Press, San Diego, pp 437–518
20. Sharma HS (2005) Pathophysiology of blood-spinal cord barrier in traumatic injury and repair. Curr Pharm Des 11: 1353–1389
21. Sharma HS (2005) Neuroprotective effects of neurotrophins and melanocortins in spinal cord injury. An experimental study in the rat using pharmacological and morphological approaches. Ann NY Acad Sci [in press]
22. Sharma HS, Alm P (2002) Nitric oxide synthase inhibitors influence dynorphin A (1-17) immunoreactivity in the rat brain following hyperthermia. Amino Acids 23: 247–259
23. Sharma HS, Alm P (2004) Role of nitric oxide on the blood-brain and the spinal cord barriers. In: Sharma HS, Westman J (eds) Blood-spinal cord and brain barriers in health and disease. Elsevier Academic Press, San Diego, pp 191–230
24. Sharma HS, Nyberg F, Olsson Y, Dey PK (1990) Alteration in substance P in brain and spinal cord following spinal cord injury. An experimental study in the rat. Neuroscience 38: 205–212
25. Sharma HS, Nyberg F, Olsson Y (1992) Dynorphin A content in the rat brain and spinal cord after a localized trauma to the spinal cord and its modification with p-chlorophenylalanine. An experimental study using radioimmunoassay technique. Neurosci Res 14: 195–203
26. Sharma HS, Nyberg F, Thörnwall M, Olsson Y (1993) Met-enkephalin-Arg[6]-Phe[7] in spinal cord and brain following traumatic injury of the spinal cord: influence of p-chlorophenylalanine. An experimental study in the rat using radioimmunoassay technique. Neuropharmacology 32: 711–717
27. Sharma HS, Nyberg F, Olsson Y (1994) Topical application of dynorphin antibodies reduces edema and cell changes in traumatized rat spinal cord. Regul Pept 1: S91–S92
28. Sharma HS, Olsson Y, Nyberg F (1995) Influence of dynorphin-A antibodies on the formation of edema and cell changes in spinal cord trauma. In: Nyberg F, Sharma HS, Wissenfeld-Hallin Z (eds) Neuropeptides in the spinal cord. Elsevier, Amsterdam, pp 401–416
29. Sharma HS, Westman J, Olsson Y, Alm P (1996) Involvement of nitric oxide in acute spinal cord injury: an immunohistochemical study using light and electron microscopy in the rat. Neurosci Res 24: 373–384
30. Sharma HS, Westman J, Alm P, Sjöquist PO, Cervós-Navarro J, Nyberg F (1997) Involvement of nitric oxide in the pathophysiology of acute heat stress in the rat. Influence of a new antioxidant compound H-290/51. Ann NY Acad Sci 813: 581–590
31. Sharma HS, Westman J, Nyberg F (1997) Topical application of 5-HT antibodies reduces edema and cell changes following trauma to the rat spinal cord. Acta Neurochir [Suppl] 70: 155–158
32. Sharma HS, Westman J, Cervós-Navarro J, Dey PK, Nyberg F (1997) Opioid receptor antagonists attenuate heat stress-induced reduction in cerebral blood flow, increased blood-brain barrier permeability, vasogenic brain edema and cell changes in the rat. Ann NY Acad Sci 813: 559–571
33. Sharma HS, Nyberg F, Gordh T, Alm P, Westman J (1997) Topical application of insulin like growth factor-1 reduces edema and upregulation of neuronal nitric oxide synthase following trauma to the rat spinal cord. Acta Neurochir [Suppl] 70: 130–133
34. Sharma HS, Lundstedt T, Flärdh M, Westman J, Post C, Skottner A (2003) Low molecular weight compounds with affinity to melanocortin receptors exert neuroprotection in spinal cord injury – an experimental study in the rat. Acta Neurochir [Suppl] 86: 399–405
35. Sharma HS, Sjöquist PO, Alm P (2003) A new antioxidant compound H-290/51 attenuates spinal cord injury induced expression of constitutive and inducible isoforms of nitric oxide synthase and edema formation in the rat. Acta Neurochir [Suppl] 86: 415–420
36. Sharma HS, Badgaiyan RD, Mohanty S, Alm P, Wiklund L (2005) Neuroprotective effects of nitric oxide synthase inhibitors in spinal cord injury induced pathophysiology and motor functions. An experimental study in the rat. Ann NY Acad Sci [in press]
37. Smith AP, Lee MN (1988) Pharmacology of dynorphin. Ann Rev Pharmacol Toxicol 28: 123–140
38. Tang Q, Lynch RM, Porreca F, Lai J (2000) Dynorphin A elicits an increase in intracellular calcium in cultured neurons via a non-opioid, non-NMDA mechanims. J Neurophysiol 83: 2610–2615
39. Winkler T, Sharma HS, Gordh T, Badgaiyan RD, Stålberg E, Westman J (2002) Topical application of dynorphin A (1-17) antiserum attenuates trauma induced alterations in spinal cord evoked potentials, microvascular permeability disturbances, edema formation and cell injury. An experimental study in the rat using electrophysiological and morphological approaches. Amino Acids 23: 273–281
40. Yakovlev AG, Faden AI (1994) Sequential expression of c-fos protooncogene, TNF-alpha, and dynorphin genes in spinal cord following experimental traumatic injury. Mol Chem Neuropathol 23: 179–190

Correspondence: Hari Shanker Sharma, Sweden. e-mail: Sharma@surgsci.uu.se

Histamine receptors influence blood-spinal cord barrier permeability, edema formation, and spinal cord blood flow following trauma to the rat spinal cord

H. S. Sharma[1,2], P. Vannemreddy[2], R. Patnaik[2,3], S. Patnaik[4], and S. Mohanty[5]

[1] Department of Surgical Sciences, Anesthesiology and Intensive Care, University Hospital, Uppsala University, Uppsala, Sweden
[2] Laboratory of Neuroanatomy, Department of Medical Cell Biology, Biomedical Center, Uppsala University, Uppsala, Sweden
[3] Department of Biomedical Engineering, Institute of Technology, Banaras Hindu University, Varanasi, India
[4] Department of Pharmacy, Institute of Technology, Banaras Hindu University, Varanasi, India
[5] Department of Neurosurgery, Institute of Medical Sciences, Banaras Hindu University, Varanasi, India

Summary

The role of histamine in edema formation, blood-spinal cord barrier (BSCB) permeability, and spinal cord blood flow (SCBF) following spinal cord injury (SCI) was examined using modulation of histamine H_1, H_2, and H_3 receptors in the rat. Focal trauma to the spinal cord at the T10–11 level significantly increased spinal cord edema formation, BSCB permeability to protein tracers and SCBF reduction in the T9 and T12 segments. Pretreatment with histamine H_1 receptor antagonist mepyramine (1 mg, 5 mg, and 10 mg/kg, i.p.) did not attenuate spinal pathophysiology following SCI. Blockade of histamine H_2 receptors with cimetidine or ranitidine (1 mg, 5 mg, or 10 mg/kg 30 minutes before injury) significantly reduced early pathophysiological events in a dose dependent manner. The effects of ranitidine were far superior to cimetidine in identical doses. Pretreatment with a histamine H_3 receptor agonist α-methylhistamine (1 mg and 2 mg/kg/i.p.), that inhibits histamine synthesis and release in the CNS, thwarted edema formation, BSCB breakdown, and SCBF disturbances after SCI. The lowest dose of histamine H_3 agonist was most effective. Blockade of histamine H_3 receptors with thioperamide (1 mg, 5 mg/kg, i.p.) exacerbated spinal cord pathology. These observations suggest that stimulation of histamine H_3 receptors and blockade of histamine H_2 receptors is neuroprotective in SCI.

Keywords: Histamine; spinal cord injury; edema.

Introduction

The role of histamine in the pathophysiology of spinal cord injury (SCI) is still speculative [15–17, 28]. Trauma to the spinal cord induces widespread alterations in blood-spinal cord barrier (BSCB) permeability, vasogenic edema formation, and cell injury [26–29]. Since histamine is one of the chemical mediators of blood-brain barrier (BBB) function and edema formation in brain injuries [2, 20, 27, 37, 38], a possibility exists that the amine participates in the pathophysiology of spinal cord injuries as well.

The spinal cord dorsal horn [39, 40] is moderately to densely innervated by histaminergic fibers mainly in the cervical region [1, 22–24, 39]. The amine is largely localized in neurons and in vascular walls of the spinal cord [22–24]. However, the histamine concentration is much lower in the spinal cord (20 to 30 ng/g wet tissue) compared to the whole brain (50 to 60 ng/g) [34]. Release of histamine in the brain causes arterial dilation, increased vascular permeability, and alterations in behavior and temperature regulation [6, 21, 23, 24]. These effects are blocked with specific histamine receptor antagonists [20, 21, 38, 40]. Three kinds of histamine receptors have been identified in the central nervous system (CNS) [3, 4, 13, 23]. The histamine H1 receptor interacts with the classical antihistamines such as mepyramine and chlorpheniramine [24]. The histamine H2 receptor is blocked by agents like burimamide, metiamide, and cimetidine [5, 8, 23, 24, 27]. The histamine H3 receptor, recently identified, largely functions as an autoreceptor that regulates the release of amine in the CNS [3, 13, 27]. The details of histamine receptor involvement in CNS injury are still not well known.

Naftchi *et al.* [16] was first to report an elevation of the histamine level in the injured spinal cord segments following impact trauma in cats [16]. It is believed that release of histamine from the injured neurons in the central gray matter and/or an increase in blood histamine content due to secondary injury mechanisms are responsible for the hemorrhagic lesion in the cord [12]. These authors found that SCI-induced hyperemia is blocked by chlorpheniramine and metiamide, respectively [12, 27].

Later, a potent histamine H2 receptor antagonist, burimamide, was found to significantly reduce irradiation-induced brain edema and extravasation of Evans blue [5, 9]. Histamine infusion revealed that the amine-induced increased BBB permeability and brain edema formation were significantly attenuated by pretreatment with the H2 receptor blocker, metiamide, but not with the H1 receptor antagonist, mepyramine [8]. Dux et al. [7] demonstrated that cerebral edema, induced by experimental pneumothorax in newborn piglets, can be significantly reduced by pretreatment with a combination of histamine H1 and H2 receptor antagonists [7]. Histamine increases the permeability of pial microvessels in cell culture, probably via increased intracellular calcium ions [4, 14, 19] that are also mediated through histamine H2 receptors [18, 21, 27, 32, 38]. These observations indicate that histamine increases the BBB permeability leading to edema development mediated by specific histamine receptors. The role of histamine or histamine receptors in BSCB function and SCI is still unknown.

In the present investigation, we wanted to know whether drugs influencing histamine H1, H2 and H3 receptors could modify SCI induced alterations in local blood flow, microvascular permeability, and edema formation in the rat.

Materials and methods

Animals

Experiments were carried out on male Wistar rats (250 to 350 g) under urethane anesthesia (1.5 g/kg, i.p.). The animals were housed at controlled ambient temperature (21 ± 1 °C) with a 12-hour light/12-hour dark schedule. Food and tap water were supplied ad libitum before the experiments.

These experiments were approved by the Ethics Committee of Uppsala University and Banaras Hindu University.

Spinal cord injury

One segment laminectomy was done over the T10–11 segments and a 2 mm deep and 5 mm long incision was made in the right dorsal horn using a scalpel blade [26, 29, 31]. The deepest part of the lesion was limited to Rexed's laminae VII–VIII [31] (Fig. 1). The wound was covered with cotton soaked in 0.9% saline to avoid direct exposure to air. Intact control rats served as controls [26, 28, 33].

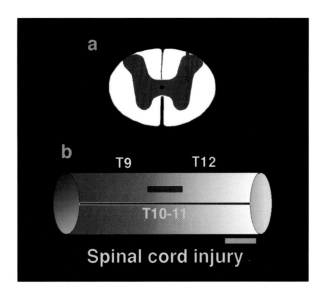

Fig. 1. Model of SCI and measurement of parameters in tissue samples. The spinal cord T9 (rostral), T12 (caudal), and the injured segment (T10–11) were used to determine BSCB permeability, edema formation, and SCBF changes

Treatment with histamine receptor modulating compounds

Mepyramine (1 mg, 5 mg, and 10 mg/kg) was used as an histamine H_1 receptor antagonist, whereas cimetidine or ranitidine (1 mg, 5 mg, and 10 mg/kg) were employed to block histamine H_2 receptors [17, 18, 27, 32, 35]. To influence histamine H_3 receptors, a potent H_3 receptor agonist, α-methylhistamine (1 and 2 mg/kg) and a powerful histamine H_3 receptor antagonist, thioperamide (1 and 5 mg/kg) was used [3, 13, 27, 36]. All histamine receptor agonists or antagonists were administered intraperitoneally 30 minutes before SCI. Animals were allowed to survive for 5 hours after injury.

BSCB permeability

The BSCB was determined in T9, T10–11, and T12 spinal cord segments (Fig. 1) using Evans blue and [131]Iodine tracers [25, 30].

Spinal cord edema

Spinal cord edema formation was examined in different spinal cord segments (Fig. 1, Table 1) using water content measured from the differences in the wet and dry weight of the samples [26, 31, 32].

Spinal cord blood flow

Spinal cord blood flow (SCBF) in various cord segments (Table 1) was assessed using tracer microspheres radiolabelled with [125]Iodine [25]. In brief, about 10^6 microspheres were injected into the left cardiac ventricle using a catheter implanted into the common carotid artery in a retrograde direction toward the heart [25]. Peripheral blood samples (40 μL every 15 minutes) were withdrawn from the right femoral artery (at the rate of 0.8 mL/min) starting 30 minutes before injection of microspheres and continued until 90 minutes after termination of the injection [25, 32]. After 90 minutes, the rats were decapitated. The spinal cord was dissected out and placed on a cold saline-wetted filter paper. Large superficial blood-vessels were removed. Spinal cord segments were dissected, weighed, and the

Table 1. *Effect of histamine receptor modulating agents on changes in BSCB permeability, edema formation, and SCBF induced by SCI at the T9 segment in rats*

Parameters measured	n	Control	5 h SCI	Histamine receptor modulating agents				
				Mepyramine 1 mg/kg	Cimetidine 10 mg/kg	Ranitidine 10 mg/kg	α-Methylhistamine 1 mg/kg	Thioperamide 1 mg/kg
BSCB permeability								
Evans blue mg %	6	0.18 ± 0.08	1.56 ± 0.12**	1.72 ± 0.08^{aa}	0.64 ± 0.20^{aa}	0.58 ± 0.21^{aa}	0.66 ± 0.14^{aa}	1.58 ± 0.14^{ns}
[131]Iodine %	7	0.38 ± 0.08	2.06 ± 0.14**	2.34 ± 0.10^{ns}	0.84 ± 0.12^{aa}	0.72 ± 0.26^{aa}	0.71 ± 0.12^{aa}	1.97 ± 0.16^{ns}
Edema formation								
Water content %	6	65.34 ± 0.21	68.76 ± 0.12**	68.94 ± 0.23^{ns}	66.13 ± 0.08^{aa}	65.87 ± 0.32^{aa}	66.34 ± 0.23^{aa}	68.58 ± 0.14^{ns}
Volume Swelling %		–	+9%	+9%	+2%	+1%	+2%	+1%
Spinal cord blood flow								
SCBF mL/g/min	5	0.72 ± 0.04	0.64 ± 0.10**	0.62 ± 0.11^{ns}	0.68 ± 0.08^{aa}	0.69 ± 0.04^{aa}	0.68 ± 0.04^{a}	0.62 ± 0.05^{ns}

Values are mean ± SD of 5–7 rats in each group. * = p < 0.05; ** p < 0.01 (compared from control); a = p < 0.05; aa = p < 0.01 (compared from 5 h SCI), ANOVA followed by Dunnett's test from one control. ns = not significant (from 5 h SCI group).
BSCB Blood-spinal cord barrier; *SCBF* spinal cord blood flow; *SCI* spinal cord injury.

radioactivity determined in a 3-inch well type Packard gamma counter at the energy window 25–50 keV [25]. Whole blood radioactivity obtained from the reference samples was counted at the end of the experiment. SCBF was calculated as: SCBF (mL/mg/min) = CPM$^{-1\,mg}$ spinal cord tissue × RBF/CR, where CPM = counts per minute, CR = total counts in reference blood samples, RBF = 0.8 mL/min [25, 27, 32].

Statistical analyses of the data obtained

ANOVA followed by Dunnett's test for multiple group comparison from 1 control group was used to evaluate the data. P-value < 0.05 was considered significant.

Results

Pathophysiology of SCI

After 5 hours, rats subjected to SCI exhibited marked increase in BSCB permeability to Evans blue and [131]Iodine tracers in the T9, T10–11, and T12 segments compared to controls (Table 1). At this time, a significant increase in water content was observed in these segments. A considerable reduction in SCBF was seen that was most marked in the traumatized cord compared to the adjacent rostral or caudal segments (Table 1).

Effect of histamine H_1 receptor antagonist on SCI

Pretreatment with histamine H1 receptor antagonist mepyramine (1 mg, 5 mg, and 10 mg/kg, i.p.) did not attenuate spinal cord edema formation, BSCB, or SCBF disturbances following SCI (Table 1).

Effect of histamine H_2 receptor antagonists on SCI

Blockade of histamine H2 receptors with cimetidine or ranitidine (1 mg, 5 mg, or 10 mg/kg) significantly reduced BSCB permeability, edema formation, and SCBF disturbances (Table 1). The most powerful neuroprotective effects were found at 10 mg/kg doses of ranitidine or cimetidine. The effects of ranitidine appeared to be far superior to cimetidine in attenuating edema formation and microvascular permeability disturbances (Table 1).

Effect of histamine H_3 receptor agonist/antagonist on SCI

Pretreatment with histamine H3 receptor agonist α-methylhistamine (1 mg and 2 mg/kg, i.p.), was able to thwart edema formation, BSCB permeability, and SCBF disturbances significantly in the non-injured T9 and T12 segments. The lowest dose of histamine H3 agonist appeared to be the most effective in reducing edema formation. On the other hand, blockade of histamine H3 receptors with thioperamide (1 mg, 5 mg/kg, i.p.) exacerbated spinal cord pathology. This effect was dose-dependent.

Discussion

Histamine plays an important role in the pathophysiology of SCI. This is apparent from the findings that pretreatment with cimetidine and ranitidine (histamine H_2 receptor antagonists) and α-methylhistamine

(histamine H3 receptor agonist) significantly attenuated increased BSCB permeability, edema formation, and reduction in the SCBF induced by SCI. These observations suggest that *stimulation* of histamine H3 receptors and *blockade* of histamine H2 receptors are neuroprotective in SCI.

We examined the role of histamine in SCI using pharmacological manipulation of histamine H1, H2 and H3 receptors. Low doses of histamine H2 receptor antagonists, cimetidine or ranitidine (1 mg or 5 mg/kg, 30 minutes before SCI) did not reduce trauma-induced edema formation. On the other hand, high doses (10 mg/kg, i.p.) of histamine H2 receptor antagonists attenuated spinal cord edema following trauma, indicating that histamine participates in edema formation via histamine H2 receptors.

The anti-edematous effect of histamine H2 receptor antagonists appears to be related to their ability to reduce BSCB permeability. Vasogenic edema formation following SCI is mediated by leakage of serum proteins across the microvascular endothelium [11]. A close parallelism between reduction in spinal cord water content and BSCB leakage to Evans blue and radioiodine supports this idea. Far superior effects of ranitidine on reduction of spinal cord edema formation and BSCB disruption in SCI, compared to cimetidine, suggest that specificity and selectivity of compounds is important in neuroprotective effects of histaminergic H2 receptor antagonists. However, we did not use equimolar concentrations of cimetidine and ranitidine in this study. Thus, bioavailability, dose responses, or other actions of these compounds in relation to their neuroprotective effects on SCI require further investigation.

Histamine H1 receptor antagonists do not have anti-edematous roles in SCI. An increase in spinal cord water content was observed following 10 mg/kg dose of mepyramine, indicating that blockade of histamine H1 receptors exacerbates edema formation. However, the functional significance of this finding is still obscure. It would be interesting to see whether histamine H1 agonist may have a role in reducing trauma-induced edema formation in the spinal cord [10], a question currently being investigated in our laboratory.

Our investigation suggests that drugs able to influence histamine H3 receptors can influence the pathophysiological outcome of SCI. Thus, pretreatment with low doses of histamine H3 receptor agonist α-methylhistamine (1 mg/kg, i.p.) significantly attenuated SCI-induced edema formation. A high dose of the compound was ineffective, however. Histamine H3 receptor agonists inhibit the release of histamine in the CNS by activating histamine autoreceptors [4, 13, 27]. Thus, blockade of histamine release following trauma by histamine H3 agonist might be primarily responsible for the neuroprotective effects of the compound [13]. On the other hand, overdose of the compound may desensitize or inhibit autoreceptors, resulting in enhanced release of the amine causing an adverse reaction.

The antiedematous effect of H3 agonist was most pronounced in the T12 segment below the lesion site, indicating that release of histamine from the cord is attenuated in the caudal region below the lesion site (results not shown). To confirm this finding, a study using histamine immunoreactivity in the cord after SCI is needed.

Pretreatment with the histamine H3 receptor antagonist thioperamide (1 mg and 5 mg/kg) exacerbated trauma-induced spinal cord edema formation in a dose-related manner. This adverse effect was most marked in the rostral segment (T9) of the cord (results not shown). Increased production of histamine following blockade of histamine autoreceptor by a histamine H3 receptor antagonist [13, 24, 33] is likely to enhance SCI-induced edema formation. Our observations indicate that histamine H3 receptor agonists may have a potential therapeutic role in SCI.

It appears that histamine influences microvascular reactions in SCI from our SCBF studies. Pretreatment with histamine H2 receptor antagonists (cimetidine and ranitidine) and a histamine H3 receptor agonist (α-methylhistamine) partially attenuated the trauma-induced decline in SCBF. These observations support the idea that histamine H2 and H3 receptors influence trauma-induced changes in SCBF [32]. SCI-induced vasodilatation is attenuated by histamine H2 receptor antagonists [6, 10], further supporting our hypothesis. However, little is known about stimulation of histamine H3 receptors on blood flow changes in the spinal cord. Inhibition of histamine release by activation of autoreceptors somehow influences SCI-induced vasodilatation. Further studies are needed to clarify these points.

Conclusion

Histamine participates actively in the pathophysiology of SCI. Blockade of histamine H2 receptors and stimulation of histamine H3 receptors prior to injury are neuroprotective. Whether these compounds are effective when administered as post-injury treatment remains to be seen.

Acknowledgments

This investigation was supported by grants from Swedish Medical Research Council (2710); Astra-Zeneca, Mölndal, Sweden; Alexander von Humboldt Foundation, Germany; The University Grants Commission, New Delhi, India; and The Indian Council of Medical Research, New Delhi, India. The expert technical assistance of Kärstin Flink, Kerstin Rystedt, Franziska Drum, and Katherin Kern, and the secretarial assistance of Aruna Sharma are greatly appreciated.

References

1. Airaksinen MS, Panula P (1988) The histaminergic system in the guinea pig central nervous system: an immunocytochemical mapping study using an antiserum against histamine. J Com Neurol 273: 163–186
2. Black KL (1995) Biochemical opening of the blood-brain barrier. Adv Drug Del Rev 15: 37–52
3. Cannon KE, Nalwalk JW, Stadel R, Ge P, Lawson D, Silos-Santiago I, Hough LB (2003) Activation of spinal histamine H3 receptors inhibits mechanical nociception. Eur J Pharmacol 470: 139–147
4. Canonaco M, Madeo M, Alo R, Giusi G, Granata T, Carelli A, Canonaco A, Facciolo RM (2005) The histaminergic signaling system exerts a neuroprotective role against neurodegenerative-induced processes in the hamster. J Pharmacol Exp Ther [Epub ahead of print]
5. Casanda E (1980) Radiation brain edema. In: Cervos-Navarro J, Ferszt R (eds) Brain edema: pathology, diagnosis and therapy. Raven Press, New York, pp 125–146
6. Dacey RG, Bassett JE (1987) Histaminergic vasodilatation of intracerebral arterioles in the rat. J Cereb Blood Flow Metab 7: 327–331
7. Dux E, Temesvári P, Szerdahelyi P, Nagy Á, Kovács P, Joó F (1987) The protective effect of antihistamines on cerebral oedema induced by experimental pneumothorax in newborn piglets. Neuroscience 22: 317–321
8. Gross PM, Teasdale GM, Angerson WJ, Harper AM (1981) H2-receptor mediate increase in permeability of the blood-brain barrier during arterial histamine infusion. Brain Res 210: 436–440
9. Joó F, Szücz A, Casanda E (1976) Metiamide-treatment of brain oedema in animals exposed to 90yttrium irradiation. J Pharm Pharmacol 28: 162–163
10. Karlstedt K, Sallmen T, Eriksson KS, Lintunen M, Couraud PO, Joo F, Panula P (1999) Lack of histamine synthesis and down-regulation of H1 and H2 receptor mRNA levels by dexamethasone in cerebral endothelial cells. J Cereb Blood Flow Metab 19: 321–330
11. Klatzo I (1987) Pathophysiological aspects of brain edema. Acta Neuropathol (Berl) 72: 236–239
12. Kobrine AI, Doyle TF, Rizzoli HV (1976) The effect of antihistamines on experimentally posttraumatic edema of the spinal cord. Surg Neurol 5: 307–309
13. Leurs R, Bakker RA, Timmerman H, de Esch IJ (2005) The histamine H3 receptor: from gene cloning to H3 receptor drugs. Nat Rev Drug Discov 4: 107–120
14. Mayhan WG, Joyner WL (1984) The effect of altering the external calcium concentration and a calcium channel blocker, verapamil, on microvascular leaky sites and dextran clearance in the hamster cheek pouch. Microvasc Res 28: 159–179
15. Mohanty S, Dey PK, Sharma HS, Singh S, Chansouria JP, Olsson Y (1989) Role of histamine in traumatic brain edema. An experimental study in the rat. J Neurol Sci 90: 87–97
16. Naftchi NE, Demeny M, DeCrescito V, Tomasula JJ, Flamm ES, Campbell JB (1974) Biogenic amine concentrations in traumatized spinal cords of cats. Effect of drug therapy. J Neurosurg 40: 52–57
17. Patnaik R, Sharma HS, Westman J (1997) Histamine H2 receptors influence hyperthermia, brain edema, blood-brain barrier permeability, cerebral blood flow and cell changes in the rat brain following heat stress. In: Nielsen-Johanssen B, Nielsen R (eds) Thermal physiology 1997. The August Krogh Institute, Copenhagen, pp 131–134
18. Patnaik R, Mohanty S, Sharma HS (2000) Blockade of histamine H2 receptors attenuate blood-brain barrier permeability, cerebral blood flow disturbances, edema formation and cell reactions following hyperthermic brain injury in the rat. Acta Neurochir [Suppl] 76: 535–539
19. Sarker MH, Fraser PA (2002) The role of guanylyl cyclases in the permeability response to inflammatory mediators in pial venular capillaries in the rat. J Physiol 540: 209–218
20. Schilling L, Wahl M (1999) Mediators of cerebral edema. Adv Exp Med Biol 474: 123–141
21. Schilling L, Ksoll E, Wahl M (1987) Vasomotor and permeability effects of histamine in cerebral vessels. Int J Microcirc Clin Exp 6: 70
22. Schwartz JC, Barbin G, Baudry M, Garbarg M, Martres MP, Pollard H, Verdiere M (1979) Metabolism and functions of histamine in the brain. In: Essman WB (ed) Current developments in psychopharmacology. Spectrum Publishers, New York, pp 173–262
23. Schwartz JC, Pollard H, Quach TT (1980) Histamine as a neurotransmitter in mammalian brain: neurochemical evidence. J Neurochem 35: 26–33
24. Schwartz JC, Arrang JM, Garbarg M, Pollard H, Ruat M (1991) Histaminergic transmission in the mammalian brain. Physiol Rev 71: 1–51
25. Sharma HS (1987) Effect of captopril (a converting enzyme inhibitor) on blood-brain barrier permeability and cerebral blood flow in normotensive rats. Neuropharmacology 26: 85–92
26. Sharma HS (2000) A bradykinin BK2 receptor antagonist HOE-140 attenuates blood-spinal cord barrier permeability following a focal trauma to the rat spinal cord. An experimental study using Evans blue, ^{131}I-sodium and lanthanum tracers. Acta Neurochir [Suppl] 76: 159–163
27. Sharma HS (2004) Histamine influences the blood-spinal cord and brain barriers following injuries to the central nervous system. In: Sharma HS, Westman J (eds) Blood-spinal cord and brain barriers in health and disease. Elsevier Academic Press, San Diego, pp 159–190
28. Sharma HS (2005) Pathophysiology of blood-spinal cord barrier in traumatic injury and repair. Curr Pharm Des 11: 1353–1389
29. Sharma HS (2005) Neuroprotective effects of neurotrophins and melanocortins in spinal cord injury. An experimental study in the rat using pharmacological and morphological approaches. Ann NY Acad Sci [in press]
30. Sharma HS, Dey PK (1986) Probable involvement of 5-hydroxytryptamine in increased permeability of blood-brain barrier under heat stress. Neuropharmacology 25: 161–167
31. Sharma HS, Olsson Y (1990) Edema formation and cellular al-

terations following spinal cord injury in the rat and their modification with p-chlorophenylalanine. Acta Neuropathologica (Berl) 79: 604–610
32. Sharma HS, Nyberg F, Cervós-Navarro J, Dey PK (1992) Histamine modulates heat stress induced changes in blood-brain barrier permeability, cerebral blood flow, brain oedema and serotonin levels: an experimental study in conscious young rats. Neuroscience 50: 445–454
33. Sharma HS, Gordh T, Wiklund L, Mohanty S, Sjöquist PO (2005) Spinal cord injury induced heat shock protein expression is reduced by an antioxidant compound H-290/51. An experimental study using light and electron microscopy in the rat. J Neural Transm [in press]
34. Steinbush HW, Mulder AH (1984) Immunohistochemical localization of histamine neurons and mast cells in the rat brain. Handbook Chem Neuroanat 3: 126–140
35. Tosaki A, Szerdahelyi P, Joo F (1994) Treatment with ranitidine of ischemic brain edema. Eur J Pharmacol 264: 455–458
36. Tosco P, Bertinaria M, Di Stilo A, Marini E, Rolando B, Sorba G, Fruttero R, Gasco A (2004) A new class of NO-donor H3-antagonists. Farmaco 59: 359–371
37. Unterberg AW, Stover J, Kress B, Kiening KL (2004) Edema and brain trauma. Neuroscience 129: 1021–1029
38. Wahl M, Unterberg A, Baethmann A, Schilling L (1988) Mediators of blood-brain barrier dysfunction and formation of vasogenic brain edema. J Cereb Blood Flow Metab 8: 621–634
39. Wahlestedt C, Skagerberg G, Håkanson R, Sundler F, Wada H, Watanabe T (1985) Spinal projections of hypothalamic histidine decarboxylase-immunoreactive neurones. Agents Actions 16: 231–233
40. Watanabe T, Taguchi Y, Shiosaka S, Tanaka J, Kubota H, Terano Y, Tohyama M, Wada H (1984) Distribution of the histaminergic neuron system in the central nervous system of rats: a fluorescent immunohistochemical analysis with histidine decarboxylase as a marker. Brain Res 295: 13–25

Correspondence: Hari Shanker Sharma, Department of Surgical Sciences, Anesthesiology & Intensive Care Medicine, University Hospital, SE-75185 Uppsala, Sweden. e-mail: Sharma@surgsci.uu.se

Post-injury treatment with a new antioxidant compound H-290/51 attenuates spinal cord trauma-induced c-*fos* expression, motor dysfunction, edema formation, and cell injury in the rat

H. S. Sharma[1], P. O. Sjöquist[2], S. Mohanty[3], and L. Wiklund[1]

[1] Laboratory of Cerebrovascular Research, Department of Surgical Sciences, Anesthesiology and Intensive Care Medicine, University Hospital, Uppsala University, Uppsala, Sweden
[2] Department of Integrative Pharmacology, Astra-Zeneca, Mölndal, Sweden
[3] Department of Neurosurgery, Institute of Medical Sciences, Banaras Hindu University, Varanasi, India

Summary

The neuroprotective efficacy of post-injury treatment with the antioxidant compound H-290/51 (10, 30, and 60 minutes after trauma) on immediate early gene expression (c-*fos*), blood-spinal cord barrier (BSCB) permeability, edema formation, and motor dysfunction was examined in a rat model of spinal cord injury (SCI). SCI was produced by a longitudinal incision into the right dorsal horn of the T10–11 segment under Equithesin anesthesia. Focal SCI in control rats resulted in profound up-regulation of c-*fos* expression, BSCB dysfunction, edema formation, and cell damage in the adjacent T9 and T12 segments at 5 hours. Pronounced motor dysfunction was present at this time as assessed using the Tarlov scale and the inclined plane test. Treatment with H-290/51 (50 mg/kg, p.o.) 10 and 30 minutes after SCI (but not after 60 minutes) markedly attenuated c-*fos* expression and motor dysfunction. In these groups, BSCB permeability, edema formation, and cell injuries were mildly but significantly reduced. These observations suggest that (i) antioxidants are capable of attenuating cellular and molecular events following trauma, and (ii) have the capacity to induce neuroprotection and improve motor function if administered during the early phase of SCI, a novel finding.

Keywords: Spinal cord injury; gene expression; c-fos; blood-spinal cord barrier; edema; oxidative stress; antioxidants; H-290/51.

Introduction

Oxidative stress appears to play an important role in inflammatory damage to myelin sheaths and axons following spinal cord injury (SCI), multiple sclerosis, and in pathogenesis of several other neurodegenerative diseases, e.g., hypertension, stroke, Alzheimer's, and Parkinson's disease [3, 4, 6–9]. Micro-hemorrhages and extravasation of blood components into the central nervous system compartment is associated with oxidative stress and generation of free radicals causing cell injury [11]. In addition, alterations in the balance between cellular oxidants and antioxidants in chronic diseases are also responsible for neurodegenerative changes [6]. Altered expression of several antioxidant enzymes, such as superoxide dismutase, glutathione peroxidase, gamma-glutamylcysteine synthase, catalases, glutathione S-transferase, and quinone reductase in central nervous system injuries and in neurodegenerative diseases are in line with this hypothesis [6–10, 13–15].

Oxidative stress induces breakdown of the blood-spinal cord barrier (BSCB) and up-regulates heat shock protein (HSP 72) response [16, 25, 26]. Release of cytokines and immunoglobulins following oxidative and/or cellular stress contributes to cell injury or cell death through mechanisms involving apoptosis and/or necrosis [27, 31, 36]. Furthermore, oxidants are capable of enhancing expression and/or DNA binding of several immediate early genes (IEG) and transcription factors that are involved in inflammation and DNA damage including *fos*, *jun*, *myc*, *erg*-1, heat shock factor, and nuclear factor kappa-B [5–7, 10]. One of the IEG, *cellular*-fos (c-*fos*), is a primary response gene that can be detected within 20 to 90 minutes after neuronal excitation [3, 4, 6, 16, 23]. Prolonged expression of c-*fos* precedes programmed cell death *in vitro* [23, 27, 34, 36]. Thus, it is likely that c-*fos* up-regulation following trauma may represent a neuronal marker for cell injury [23]. However, the detailed involvement of oxidative stress-induced IEG expression and cell death in SCI is unclear.

There are reasons to believe that antioxidants and lipid peroxidation inhibitors play important roles in attenuating spinal cord cell and tissue injury following SCI [12, 16, 18, 20–22, 24–26]. A significant reduction in BSCB permeability, edema formation, and cell damage in SCI in animals pretreated with a potent chain-breaking new antioxidant compound, H-290/51 (Astra-Zeneca, Mölndal, Sweden), further supports this hypothesis [24, 26]. However, it is unclear whether the compound is still effective in reducing trauma-induced IEG expression, motor dysfunction, and spinal cord pathology, if given at various time intervals after SCI.

The first few hours following SCI are crucial for outcome as the early events following the primary insult set the stage for later development of spinal cord cell and tissue injury leading to long-term deficit and disability [16–19, 26]. Suitable therapeutic intervention initiated within the first 3 hours in spinal cord injury victims improves functional recovery, whereas delayed pharmacological treatment beyond 3 hours is largely ineffective [16, 18]. Thus, further studies on the cellular and molecular mechanisms of early events following SCI are necessary to explore new therapeutic strategies to minimize later development of spinal cord cell and tissue injury.

The present study was undertaken to investigate the effects of a potent antioxidant compound H-290/51 [28, 30, 33] on *c-fos* expression, motor dysfunction, and cord pathology when given 10, 30, and 60 minutes after SCI in a rat model.

Materials and methods

Animals

Experiments were carried out on 60 male Sprague Dawley rats (200–250 g) housed at controlled room temperature of $21 \pm 1\,°C$ with a 12-hour light, 12-hour dark schedule. Food and tap water were supplied *ad libitum* before the experiment.

Spinal cord injury

SCI was induced by making a longitudinal incision (about 5 mm) over the right dorsal horn of the T10–11 segments under Equithesin anesthesia (0.3 mL/100 g, i.p.). The deepest part of the lesion was mainly located around the Rexed's laminae VIII to X [18, 19, 26]. Experiments were approved by the Ethical Committee of Uppsala University, Uppsala, Sweden, and Banaras Hindu University, Varanasi, India.

H-290/51 treatment

H-290/51 (Astra-Zeneca, Mölndal, Sweden) was dissolved in water and administered 50 mg/kg, p.o. by gastric tube [20, 24–26] in separate groups of rats (n = 5) at 10, 30, or 60 minutes after SCI.

Functional paralysis

Functional paralysis of the hind limb was determined using a semiquantitative analysis during open field walking using the modified Tarlov scale: 0 = total paraplegia; 1 = no spontaneous movement but responds to pinch; 2 = spontaneous movement; 3 = able to support weight but unable to walk; 4 = walking with gross deficits; 5 = walking with mild deficits; 6 = normal walking [19, 26, 29].

Inclined plane test

Motor disturbances in the rat after SCI were determined using the inclined plane test. Each rat was trained on a plane using an angle in such a way that the rats could stay on it for 5 seconds without falling [19, 26].

Perfusion and fixation

Five hours after SCI, rats were perfused through the heart with 0.1 mol phosphate buffer (pH 7.0) followed by 4% buffered paraformaldehyde in 0.1 mol phosphate buffer. Perfusion pressure was maintained at 90 torr throughout the process [17–19].

c-fos immunohistochemistry

Immunohistochemistry for c-*fos* was performed on free-floating vibratome sections obtained from the T9 segment of the cord using monoclonal c-*fos* antiserum (Calbiochem, Boston, MA) according to the manufacturer's protocol [23].

BSCB permeability and edema formation

BSCB permeability was measured using Evans blue (0.3 ml/100 g) and $^{[131]}$Iodine (10 μCi/100 g) as described previously [26]. Spinal cord edema formation was examined by measurement of the spinal cord water content [16, 20, 24, 25].

Spinal cord pathology

Spinal cord pathology was examined by light and electron microscopy. Spinal cord tissue pieces were embedded in paraffin or Epon and examined by light microscopy. Epon-embedded tissue pieces were processed for transmission electron microscopy [16, 20]. The cell changes were graded from 1 (minimum) to 4 (maximum) and analyzed [17].

Statistical evaluation

The quantitative or semiquantitative data obtained were analyzed using ANOVA followed by Dunnet's test for multiple group comparison from one control group. A p-value less than 0.05 was considered significant.

Table 1. *Post-trauma treatment with H-290/51 on the SCI-induced motor dysfunction, c-fos expression, BSCB permeability, spinal cord edema formation, and cell injury in T9 segment in rats*

Parameters measured	n	Control	5 h SCI	H-290/51 treatment in 5 h SCI§		
				+10 min	+30 min	+60 min
Motor dysfunction						
Tarlov scale	5	6 ± 0	$2 \pm 1^{**}$	5 ± 1^a	4 ± 1^{aa}	2 ± 1^{ns}
Capacity angle	5	60 ± 0	$30 \pm 2^{**}$	43 ± 4^a	40 ± 6^{aa}	32 ± 6^{aa}
BSCB permeability						
Evans blue mg%	6	0.24 ± 0.04	$1.65 \pm 0.12^{**}$	0.68 ± 0.12^{aa}	0.79 ± 0.12^{aa}	1.58 ± 0.42^{ns}
$^{[131]}$Iodine %	5	0.35 ± 0.06	$1.96 \pm 0.14^{**}$	0.72 ± 0.08^{aa}	0.91 ± 0.09^{aa}	1.88 ± 0.48^{ns}
Edema formation						
Cord width mm	6	3 ± 0.5	$5 \pm 0.5^{**}$	4 ± 0.5^a	4 ± 1^{aa}	5 ± 1^{ns}
Water content %	5	66.12 ± 0.18	$69.34 \pm 0.23^{**}$	66.64 ± 0.18^a	67.38 ± 0.12^a	69.73 ± 0.28^{ns}
Structural changes						
c-*fos* positive cells	6	nil	34 ± 8	$8 \pm 4\#$	$12 \pm 6\#$	28 ± 8^{ns}
Neuronal damage	5	nil	4	$2 \pm 1\#$	$3 \pm 1\#$	4 ± 1^{ns}
Glial cell injury	5	nil	4	$1 \pm 1\#$	$3 \pm 1\#$	3 ± 1^{ns}
Myelin damage	5	nil	4	$2 \pm 1\#$	$2 \pm 1\#$	4 ± 1^{ns}
Endothelial injury	6	nil	4	$2 \pm 1\#$	$2 \pm 2\#$	4 ± 1^{ns}

Cord width was measured in formalin fixed spinal cord specimens before embedding in paraffin [16]. § H-290/51 (50 mg/kg, p.o.) was administered 10 min, 30 min or 60 min after SCI. The cell changes were graded from 1 (minimum) to 4 (maximum) and analyzed in blinded fashion [see 16].
Values are Mean ± SD of 5–6 rats in each group.
BSCB Blood-spinal cord barrier; *SCI* spinal cord injury; * = $p < 0.05$; ** $p < 0.01$ (compared from control); a = $p < 0.05$; aa = $p < 0.01$ (compared with 5-hour SCI), ANOVA followed by Dunnett's test from one control. # = $p < 0.05$, Chi-square test from 5-hour SCI group; ns = not significant (from 5-hour SCI group).

Results

Effect of H-290/51 on motor function

Animals subjected to SCI showed profound motor dysfunction at 5 hours. Administration of H-290/51 to rats either 10 or 30 minutes after SCI significantly improved hind-limb function using the Tarlov scale or the capacity angle derived from the inclined plane test (Table 1). However, when the compound was administered 60 minutes after SCI, no significant improvement in motor function was seen.

Effect of H-290/51 on c-fos immunohistochemistry

Untreated SCI rats exhibited marked up-regulation of c-*fos* expression in neurons of the injured as well as adjacent segments in the edematous region of the spinal cord. Administration of H-290/51 10 or 30 minutes after SCI significantly attenuated c-*fos* expression in the cord. This reduction in c-*fos* expression was most marked in the ventral gray matter on the contralateral side. In contrast, no apparent reduction in c-*fos* expression was seen in rats treated with H-290/51 60 minutes after injury (Fig. 1).

Effect of H-290/51 on BSCB permeability

SCI rats that received H-290/51 either 10 or 30 minutes after injury showed a significant reduction in Evans blue, radioiodine, or lanthanum extravasation across the BSCB. However, treatment received 60 minutes after SCI did not reduce BSCB breakdown to these tracers.

Effect of H-290/51 on spinal cord edema formation

Rats that received H-290/51 either 10 or 30 minutes after injury did not exhibit much swelling and/or increase in spinal cord water content. However, drug administration 60 minutes after SCI failed to reduce spinal cord water content or swelling.

Effect of H-290/51 on cell injury

Treatment of rats with H-290/51 either 10 or 30 min following SCI reduced the gross expansion of the cord

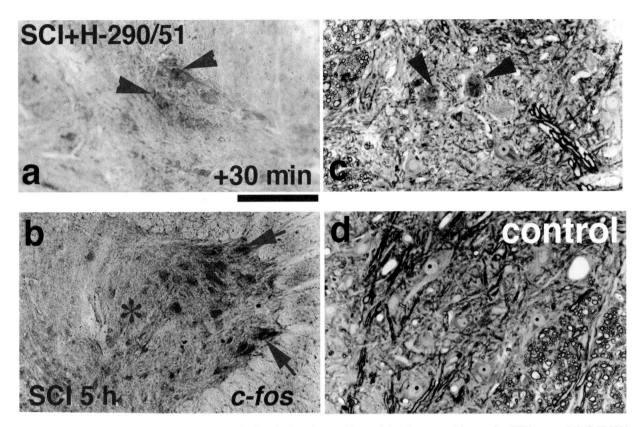

Fig. 1. c-*fos* expression (a,b) and cell changes (c,d) in the spinal cord ventral horn of the T9 segment 5 hours after SCI in control (b,d) H-290/51 treated rats (a,c). H-290/51 was given 30 minutes after SCI. Administration of H-290/51 was neuroprotective. Marked up-regulation of c-*fos* expression (arrows) was seen in the edematous area (*) after 5 hours SCI in control rats (b), but this was attenuated by H-290/51 (arrowheads) (a). H-290/51 also induced neuroprotection (c). Only a few damaged nerve cells can be seen (arrowheads) after drug treatment compared to no treatment (d)

edema, and micro-hemorrhages and damage to neuropil were considerably attenuated. A clear distinction between the gray and white matter was visible in these rats. Several nerve cells with distinct were present and swelling of neurons, astrocytes, and damage to myelin were much less evident in the treatment groups than in controls (Fig. 1, Table 1). At the ultrastructural level, signs of vacuolation, perivascular edema and myelin vesiculation were much less evident. On the other hand, H-290/51 failed to exert any neuroprotective effects on the spinal cord cell injury seen either at the light or electron microscopic level 60 minutes after SCI (results not shown).

Discussion

The salient new findings of the present study are that the chain-breaking antioxidant compound H-290/51, if administered within 30 minutes after SCI, attenuates motor dysfunction and spinal cord pathology at 5 hours. These observations suggest that oxidative stress during the early hours of SCI plays an important role in the secondary injury cascade that results in sensory motor dysfunction and spinal cord cell and tissue injury. However, when the antioxidant compound is administered 60 minutes after SCI, no significant neuroprotection or improvement in motor function was noted. This suggests that blockade of lipid peroxidation and/or generation of free radicals within the first hour after trauma is neuroprotective in SCI.

Previous reports from our laboratory indicate that treatment with H-290/51 requires 30 minutes to inhibit lipid peroxidation and, thus, block generation of free radicals [12, 20, 24]. The compound H290/51 in the present study has a short time to onset, reaching protective concentrations in the central nervous system within 30 minutes, with maximal effect lasting more than 6 hours [12]. Thus, administration of H-290/51 10 and 30 minutes after SCI is likely to inhibit further production of free radicals around 40 minutes to 1 hour after trauma. On the other hand, administration of the compound 1 hour after SCI exerts its influence on free radical generation around 90 minutes after injury, suggesting that blockade of lipid peroxidation and/or generation of free radicals within 1 hour after SCI is beneficial in nature, whereas later blockade of free radical formation is ineffective. The detailed cellular and molecular mechanisms of such time-related neuroprotection with the antioxidant in SCI are not known and require additional investigation.

The antioxidant compound H-290/51 was able to attenuate IEG expression in the cord as seen using c-*fos* immunostaining. Since, the up-regulation of c-*fos* was mainly located within the edematous expansion of the spinal cord [23], a reduction in cell and tissue injury with H-290/51 is probably responsible for the diminished IEG expression in drug-treated spinal cord injured rats. These findings indicate that generation of free radicals and lipid peroxidation are important factors in the c-*fos* up-regulation, a novel finding. The fact that no reduction in c-*fos* expression was found in spinal cord injured animals that received H-290/51 treatment 60 minutes after injury is in line with this hypothesis.

Trauma to the spinal cord results in the release of numerous molecules, free radicals, vasoactive compounds, neurochemicals, growth factors, cytokines and other proteins/factors in a cascade of events leading to cell and tissue injury [16–19, 25, 26]. It is believed that several endogenous compounds/factors released after trauma may have the ability to induce neuroprotection, whereas numerous other endogenous factors/elements increased after injury are likely to have neurodestructive capabilities [16, 18]. Thus, a balance between endogenous neuroprotective and neurodestructive elements is crucial for cell injury and/or survival. It remains to be seen whether an interaction among different compounds/factors will synergistically potentiate or neutralize the neurodestructive and/or neuroprotective capabilities of certain elements *in vivo*. Thus, pharmacological blockade of release and/or synthesis of endogenous neurodestructive elements *before* the injury are likely to achieve neuroprotection [1–3, 7]. However, when the same compound is administered *after* the insult, the neuroprotective effect is either diminished or neutralized, as several other endogenous compounds/factors are likely to influence the outcome [32, 35]. Time-related neuroprotection induced by H-290/51 treatment *after* SCI in our present investigation is consistent with this hypothesis.

Our results suggest that generation of free radicals and oxidative stress plays an important role in SCI-induced cell and tissue injury. Recent evidence has shown that oxidative stress, generation of free radicals, and nitric oxide can up-regulate vascular endothelial growth factor (VEGF) expression in the microvascular endothelium [1, 2, 13, 14, 27, 32, 34–36]. An up-regulation of VEGF is associated with breakdown of microvascular permeability [32]. VEGF is a 45-kD glycoprotein secreted in the vascular wall by endothelial and smooth muscle cells [14, 32]. VEGF is a major regulator of angiogenesis and increased microvascular permeability [5, 7, 8]. Thus, VEGF up-regulation by oxidative stress and generation of free radicals could be one of the important factors in BSCB disruption in SCI. The BSCB breakdown following SCI in this investigation and its amelioration with H-290/51 is in line with this idea. However, further studies on expression of VEGF in SCI and its modification with H-290/51 are needed to confirm this hypothesis.

A reduction in BSCB permeability to macromolecules in the spinal cord microenvironment results in either quick resolution of edema or prevention of water accumulation in the spinal cord [21, 24]. Alternatively, in the absence of direct cell membrane damage, accumulation of water in the spinal cord extra- or intracellular compartments is less likely [24]. Obviously, a reduction in the BSCB permeability and edema formation by H-290/51 will induce neuroprotection [1, 2, 20].

Our observations show a close parallelism between improvement in motor function and spinal cord cell and tissue injury, indicating improvements in motor function are related to spinal cord pathology [19, 26]. Improvement in motor function by antioxidants suggests that cell and membrane damage by free radicals is likely to contribute to functional paralysis. The mechanisms by which antioxidants improve sensory and motor functions are not known. However, it appears that antioxidant-induced stimulation of neurotrophins and/or growth factor receptors might play some role.

Oxidative stress is known to stimulate VEGF and other neurotrophins that are involved in necrosis/apoptosis and to down-regulate neuroprotective neurotrophins, such as brain-derived neurotrophic factor, glial-derived neurotrophic factor, and nerve growth factor [3, 5–8, 10]. Exogenous supplement of neurotrophins, e.g., brain-derived neurotrophic factor and glial-derived neurotrophic factor in SCI, improves motor function and cell injury and is consistent with this idea [17]. However, further studies on expression of neurotrophins and/or their receptors in H-290/51-treated SCI animals are needed to clarify these points.

Conclusion

The results presented in this investigation show for the first time marked neuroprotection and improvement of motor functions when the antioxidant compound H-290/51 is administered 10 to 30 minutes after SCI. These observations suggest that blockade of lipid peroxidation and/or generation of free radicals within the first hours after trauma is crucial for spinal cord function. However, the antioxidant was ineffective when administered 60 minutes after SCI, suggesting that late blockade of lipid peroxidation and/or generation of free radicals are incapable of attenuating spinal cord pathology. Understanding the cellular and molecular mechanisms of time-related neuroprotection with the antioxidant requires additional investigation.

Acknowledgments

This investigation was supported by grants from Swedish Medical Research Council (2710); Astra-Zeneca, Mölndal, Sweden; Laerdal Foundation for Acute Medicine; Alexander von Humboldt Foundation, Germany; The University Grants Commission, New Delhi, India; and The Indian Council of Medical Research, New Delhi, India. The expert technical assistance of Kärstin Flink, Kerstin Reystedt, Franzisca Drum, and Katherin Kern, and secretarial assistance of Aruna Sharma are greatly appreciated.

References

1. Alm P, Sharma HS, Hedlund S, Sjoquist PO, Westman J (1998) Nitric oxide in the pathophysiology of hyperthermic brain injury. Influence of a new anti-oxidant compound H-290/51. A pharmacological study using immunohistochemistry in the rat. Amino Acids 14: 95–103
2. Alm P, Sharma HS, Sjoquist PO, Westman J (2000) A new antioxidant compound H-290/51 attenuates nitric oxide synthase and heme oxygenase expression following hyperthermic brain injury. An experimental study using immunohistochemistry in the rat. Amino Acids 19: 383–394
3. Bazan NG (2005) Lipid signaling in neural plasticity, brain repair, and neuroprotection. Mol Neurobiol 32: 89–104
4. Calabrese V, Scapagnini G, Colombrita C, Ravagna A, Pennisi G, Giuffrida Stella AM, Galli F, Butterfield DA (2003) Redox regulation of heat shock protein expression in aging and neurodegenerative disorders associated with oxidative stress: a nutritional approach. Amino Acids 25: 437–444
5. Caldwell RB, Bartoli M, Behzadian MA, El-Remessy AE, Al-Shabrawey M, Platt DH, Liou GI, Caldwell RW (2005) Vascular endothelial growth factor and diabetic retinopathy: role of oxidative stress. Curr Drug Targets 6: 511–524
6. Chan PH (2005) Mitochondrial dysfunction and oxidative stress as determinants of cell death/survival in stroke. Ann N Y Acad Sci 1042: 203–209
7. Chong ZZ, Li F, Maiese K (2005) Oxidative stress in the brain: novel cellular targets that govern survival during neurodegenerative disease. Prog Neurobiol 75: 207–246
8. Chong ZZ, Li F, Maiese K (2005) Stress in the brain: novel cellular mechanisms of injury linked to Alzheimer's disease. Brain Res Brain Res Rev 49: 1–21
9. Kirby J, Halligan E, Baptista MJ, Allen S, Heath PR, Holden H, Barber SC, Loynes CA, Wood-Allum CA, Lunec J, Shaw PJ (2005) Mutant SOD1 alters the motor neuronal transcriptome: implications for familial ALS. Brain 128: 1686–1706
10. Kowaltowski AJ, Fiskum G (2005) Redox mechanisms of cytoprotection by Bcl-2. Antioxid Redox Signal 7: 508–514
11. Laplace C, Huet O, Vicaut E, Ract C, Martin L, Benhamou D, Duranteau J (2005) Endothelial oxidative stress induced by serum from patients with severe trauma hemorrhage. Intensive Care Med [Epub ahead of print]
12. Mustafa A, Sharma HS, Olsson Y, Gordh T, Thoren P, Sjoquist PO, Roos P, Adem A, Nyberg F (1995) Vascular permeability to growth hormone in the rat central nervous system after focal spinal cord injury. Influence of a new anti-oxidant H 290/51 and age. Neurosci Res 23: 185–194
13. Poulet R, Gentile MT, Vecchione C, Distaso M, Aretini A, Fratta L, Russo G, Echart C, Maffei A, De Simoni MG, Lembo G (2005) Acute hypertension induces oxidative stress in brain tissues. J Cereb Blood Flow Metab [Epub ahead of print]
14. Rodriguez JA, Nespereira B, Perez-Ilzarbe M, Eguinoa E, Paramo JA (2005) Vitamins C and E prevent endothelial VEGF and VEGFR-2 overexpression induced by porcine hypercholesterolemic LDL. Cardiovasc Res 65: 665–673
15. Rogerio F, Teixeira SA, de Rezende AC, de Sa RC, de Souza Queiroz L, De Nucci G, Muscara MN, Langone F (2005) Superoxide dismutase isoforms 1 and 2 in lumbar spinal cord of neonatal rats after sciatic nerve transection and melatonin treatment. Brain Res Dev Brain Res 154: 217–225
16. Sharma HS (2004) Pathophysiology of the blood-spinal cord barrier in traumatic injury. In: Sharma HS, Westman J (eds) Blood-spinal cord and brain barriers in health and disease. Elsevier Academic Press, San Diego, pp 437–518
17. Sharma HS (2005) Post-traumatic application of brain derived neurotrophic factor and glia derived neurotrophic factor in combination over the traumatized rat spinal cord enhances neuroprotection and improves motor functions. Acta Neurochir [Suppl] 96: 359–364
18. Sharma HS (2005) Pathophysiology of blood-spinal cord barrier in traumatic injury and repair. Curr Pharm Des 11: 1353–1389
19. Sharma HS (2005) Neuroprotective effects of neurotrophins and melanocortins in spinal cord injury. An experimental study in the rat using pharmacological and morphological approaches. Ann NY Acad Sci 1053: 407–421
20. Sharma HS, Sjöquist PO (2002) A new antioxidant compound H-290/51 modulates glutamate and GABA immunoreactivity in the rat spinal cord following trauma. Amino Acids 23: 261–272
21. Sharma HS, Winkler T (2002) Assessment of spinal cord pathology following trauma using early changes in the spinal cord evoked potentials: a pharmacological and morphological study in the rat. Muscle Nerve [Suppl] 11: S83–S91
22. Sharma HS, Westman J, Alm P, Sjoquist PO, Cervos-Navarro J, Nyberg F (1997) Involvement of nitric oxide in the pathophysiology of acute heat stress in the rat. Influence of a new antioxidant compound H-290/51. Ann N Y Acad Sci 813: 581–590
23. Sharma HS, Alm P, Sjoquist PO, Westman J (2000) A new antioxidant compound H-290/51 attenuates upregulation of constitutive isoform of heme oxygenase (HO-2) following trauma to the rat spinal cord. Acta Neurochir [Suppl] 76: 153–157

24. Sharma HS, Sjöquist PO, Westman J (2001) Pathophysiology of the blood-spinal cord barrier in spinal cord injury. Influence of a new antioxidant compound H-290/51. In: Kobiler D, Lustig S, Shapira S (eds) Blood-brain barrier: drug delivery and brain pathology. Kluwer Academic/Plenum Publishers, New York, pp 401–416
25. Sharma HS, Sjoquist PO, Alm P (2003) A new antioxidant compound H-290/51 attenuates spinal cord injury induced expression of constitutive and inducible isoforms of nitric oxide synthase and edema formation in the rat. Acta Neurochir [Suppl] 86: 415–420
26. Sharma HS, Gordh T, Wiklund L, Mohanty S, Sjoquist PO (2005) Spinal cord injury induced heat shock protein expression is reduced by an antioxidant compound H-290/51. An experimental study using light and electron microscopy in the rat. J Neural Transm [in press]
27. Strosznajder RP, Jesko H, Zambrzycka A (2005) Poly(ADP-ribose) polymerase: the nuclear target in signal transduction and its role in brain ischemia-reperfusion injury. Mol Neurobiol 31: 149–167
28. Svensson L, Borjesson I, Kull B, Sjoquist PO (1993) Automated procedure for measuring TBARS for in vitro comparison of the effect of antioxidants on tissues. Scand J Clin Lab Invest 53: 83–85
29. Tariq M, Morais C, Kishore PN, Biary N, Al Deeb S, Al Moutaery K (1998) Neurological recovery in diabetic rats following spinal cord injury. J Neurotrauma 15: 239–251
30. Thornwall M, Sharma HS, Gordh T, Sjoquist PO, Nyberg F (1997) Substance P endopeptidase activity in the rat spinal cord following injury: influence of the new anti-oxidant compound H 290/51. Acta Neurochir [Suppl] 70: 212–215
31. Tormos C, Javier Chaves F, Garcia MJ, Garrido F, Jover R, O'Connor JE, Iradi A, Oltra A, Oliva MR, Saez GT (2004) Role of glutathione in the induction of apoptosis and c-fos and c-jun mRNAs by oxidative stress in tumor cells. Cancer Lett 208: 103–113
32. Valable S, Montaner J, Bellail A, Berezowski V, Brillault J, Cecchelli R, Divoux D, Mackenzie ET, Bernaudin M, Roussel S, Petit E (2005) VEGF-induced BBB permeability is associated with an MMP-9 activity increase in cerebral ischemia: both effects decreased by Ang-1. J Cereb Blood Flow Metab [Epub ahead of print]
33. Westerlund C, Ostlund-Lindqvist AM, Sainsbury M, Shertzer HG, Sjoquist PO (1996) Characterization of novel indenoindoles. Part I. Structure-activity relationships in different model systems of lipid peroxidation. Biochem Pharmacol 51: 1397–1402
34. Xu W, Chi L, Xu R, Ke Y, Luo C, Cai J, Qiu M, Gozal D, Liu R (2005) Increased production of reactive oxygen species contributes to motor neuron death in a compression mouse model of spinal cord injury. Spinal Cord 43: 204–213
35. Yokoi M, Yamagishi SI, Takeuchi M, Ohgami K, Okamoto T, Saito W, Muramatsu M, Imaizumi T, Ohno S (2005) Elevations of AGE and vascular endothelial growth factor with decreased total antioxidant status in the vitreous fluid of diabetic patients with retinopathy. Br J Ophthalmol 89: 673–675
36. Yu F, Sugawara T, Maier CM, Hsieh LB, Chan PH (2005) Akt/Bad signaling and motor neuron survival after spinal cord injury. Neurobiol Dis (Epub ahead of print)

Correspondence: Hari Shanker Sharma, Department of Surgical Sciences, Anesthesiology & Intensive Care Medicine, University Hospital, SE-75185 Uppsala, Sweden. e-mail: Sharma@surgsci.uu.se

Post-traumatic application of brain-derived neurotrophic factor and glia-derived neurotrophic factor on the rat spinal cord enhances neuroprotection and improves motor function

H. S. Sharma

Laboratory of Cerebrovascular Biology, Department of Surgical Sciences, Anesthesiology and Intensive Care Medicine, University Hospital, Uppsala University, Uppsala, Sweden

Summary

We examined the potential efficacy of brain-derived neurotrophic factor (BDNF) and glial-derived neurotrophic factor (GDNF) applied over traumatized spinal cord, alone or in combination, for attenuating motor dysfunction, blood-spinal cord barrier (BSCB) breakdown, edema formation, and cell injury in a rat model. Under Equithesin anesthesia, spinal cord injury (SCI) was performed by making a unilateral incision into the right dorsal horn of the T10–11 segment. The rats were allowed to survive 5 hours after trauma. The BDNF or GDNF was applied (0.1 to 1 µg/10 µl in phosphate buffer saline) 30, 60, or 90 minutes after SCI. Topical application of BDNF or GDNF 30 minutes after SCI in high concentration (0.5 µg and 1 µg) significantly improved motor function and reduced BSCB breakdown, edema formation, and cell injury at 5 hours. These beneficial effects of neurotrophins were markedly absent when administered separately either 60 or 90 minutes after injury. However, combined application of BDNF and GDNF at 60 or 90 minutes after SCI resulted in a significant reduction in motor dysfunction and spinal cord pathology. These novel observations suggest that neurotrophins in combination have potential therapeutic value for the treatment of SCI in clinical situations.

Keywords: Spinal cord injury; neurotrophins, brain-derived neurotrophic factor; glial-derived neurotrophic factor; blood-spinal cord barrier; spinal cord edema; motor dysfunction; spinal cord pathology.

Introduction

Victims of spinal cord injury (SCI) include young men aged 20 to 30 years that develop quadriplegia followed by paraplegia [20, 22, 23, 26]. These patients are often without any sign of voluntary motor or sensory perception below the level of the lesion [20, 22, 23]. Thus, efforts should be made to improve the quality of life for these young victims and to reduce the clinical burden on society. One way to enhance spinal cord neuroprotection and/or regeneration is to use neurotrophins. Neurotrophins are a family of growth factors consisting of nerve growth factor (NGF), brain-derived neurotrophic factor (BDNF), insulin-like growth factor-1, neurotrophin-3, neurotrophin-4/5, glial-derived neurotrophic factor (GDNF), ciliary neurotrophic factor, and transforming growth factor-β [1, 2, 5, 11, 13, 34]. Neurotrophins promote survival and rescue nerve cells from death in trauma, ischemia, or hypoxia [13, 19, 21, 22, 24], and promote neurite extension, neuronal survival, and differentiation [2, 3, 20, 34]. However, their roles in the pathophysiology of SCI and blood-spinal cord barrier (BSCB) dysfunction are not well described.

Recently, GDNF was found to induce neuroprotection in an animal model of ischemic injury when administered several hours after the insult [2, 5]. Application of other neurotrophins, such as NGF and neurotrophin-3, in animal models of CNS injury were not effective [7, 16, 20, 29]. The reasons behind such diverse effects of neurotrophins in influencing cell injury and cell survival in traumatic injury models are not known and require additional investigation.

Previous works from our laboratory showed that topical application of BDNF and insulin-like growth factor-1, when given separately, are able to attenuate BSCB breakdown, edema formation, and cell injury in a rat model of SCI [21, 23, 24, 27–30]. However, these beneficial effects of neurotrophins are limited to short periods ranging from 5 to 30 minutes after SCI [21]. Thus, in order to discover the possible clinical significance of neurotrophins in the treatment of SCI victims, further studies are needed. One point of interest is whether a combination of neurotrophins will enhance or neutralize the neuroprotective efficacy of the growth factors in trauma models.

The present investigation was, therefore, carried-out to examine effects of BDNF or GDNF, either alone or in combination, at various time intervals after SCI-induced breakdown of BSCB permeability, edema formation, cell injury, and motor function in a rat model.

Materials and methods

Animals

Experiments were carried out on 56 adult male Wistar rats (200–250 g body weight, aged 20 to 28 weeks) housed at a controlled ambient temperature ($21 \pm 1\,°C$) with a 12-hour light, 12-hour dark schedule. Food and tap water were provided ad libitum before the experiments.

Spinal cord injury

SCI was produced under Equithesin anesthesia (0.3 ml/100 g body weight, i.p.) by making an incision into the right dorsal horn of the T10–11 segments using a scalpel blade [21]. The deepest part of the lesion was close to the lamina VII–VIII [24]. The animals were allowed to survive 5 hours after SCI. This method allowed study of morphological changes in the ipsi- and contralateral sides of the cord in several segments located rostrally and caudally from the lesion site. This experiment was approved by the Ethics Committee of Uppsala University, Uppsala, Sweden.

Treatments with neurotrophins

The BDNF or GDNF (Sigma-Aldrich, St. Louis, MO) was applied (0.1 to 1 µg/10 µl in phosphate buffer saline) over the traumatized cord in separate groups of rats (n = 5) 30, 60, or 90 minutes after injury [21, 24]. In another group of animals (n = 5), BDNF and GDNF was applied in combination 60 or 90 minutes after SCI.

Functional paralysis

Functional paralysis of the hind-limb was determined using a semiquantitative analysis using modified Tarlov scale [24, 25, 31, 32]. These experiments were conducted according to the National Institutes of Health (USA) guidelines on the care and use of animals, and approved by Animal Care and Experimental Committee of Banaras Hindu University, Varanasi, India. The following score for hind-limb function was used: 0 = total paraplegia; 1 = no spontaneous movement but responds to pinch; 2 = spontaneous movement; 3 = able to support weight but unable to walk; 4 = walking with gross deficits; 5 = walking with mild deficits; 6 = normal walk [31, 32].

BSCB permeability and edema formation

BSCB permeability was examined using Evans blue and ^{131}I-sodium in the perifocal T9 and T12 segments as described previously [24, 25]. Spinal cord edema formation in the same segments was examined using water content. The water content of the cord was calculated from the differences in the wet and dry weight. The dry weight was obtained by placing the samples in an oven maintained at 90 °C for 72 hours [22, 23].

Spinal cord pathology

In a separate group of rats, the animals 5 hours after SCI were perfused with Somogyi fixative and the T9 and T12 spinal cord segments were taken out, photographed for visual swelling, and processed for paraffin embedding. About 3 µm thick sections were stained with hematoxylin and eosin, or Nissl for light microscopy, and examined for sponginess, edema, cell injury, cell loss, and/or cell death in a blinded fashion. A rough score of 1 (least) to 4 (maximum) was assigned for each parameter in individual animals for semiquantitative analysis [22–24].

Statistical analysis

ANOVA followed by Dunnett's test for multiple group comparison with one control group was used to evaluate statistical significance of quantitative data obtained. The Chi-square test was applied to evaluate semiquantitative data. A p-value < 0.05 was considered significant.

Results

Effect of neurotrophins on functional paralysis

Untreated traumatized rats showed functional paralysis of the ipsilateral hind-limb at 5 hours after SCI [24, 32]. Application of BDNF or GDNF alone (in high concentration, 0.5 and 1 µg) 30 minutes after SCI markedly improved motor function in a dose-dependent manner. This effect was not seen when neurotrophins were applied individually 60 or 90 minutes after trauma. On the other hand, when BDNF and GDNF were applied in combination (0.5 µg each) 60 or 90 minutes after SCI, the functional outcome was significantly improved compared to neurotrophins given alone.

Effect of neurotrophins on BSCB permeability and edema formation

Measurement of BSCB permeability showed profound increase in Evans blue and radioiodine extravasation in the T9 and T12 segments [21]. Topical application of BDNF or GDNF significantly attenuated the leakage of tracers across the BSCB when given 30 minutes after SCI, a feature not seen in animals receiving neurotrophins alone either 60 or 90 min after trauma. However, a marked reduction in BSCB to these protein tracers was observed when BDNF and GDNF were co-administered over the injured spinal cord either 60 or 90 min after the lesion.

Edema measurement showed a close parallelism between increased water content and leakage of tracers. Thus, a significant reduction in water content was ob-

Table 1. *Effects of post-trauma treatment with BNDF or GDNF either alone or in combination on the SCI-induced motor dysfunction, BSCB permeability, spinal cord edema formation, and cell injury in T9 segment in rats*

Parameters measured	n	Control	5 h SCI	Neurotrophins treatment alone (1 µg)§			Neurotrophins combination (1 µg)§	
				BDNF +30 min	GDNF +30 min	GDNF +60 min	BDNF+ GDNF +60 min	BDNF+ GDNF +90 min
Motor dysfunction								
Tarlov scale	5	6 ± 0	2 ± 1**	4 ± 1^a	5 ± 2^{aa}	2 ± 2^{ns}	5 ± 1^{aa}	4 ± 1^{aa}
Capacity angle	5	60 ± 0	30 ± 2**	40 ± 3^a	42 ± 4^{aa}	32 ± 3^{ns}	48 ± 4^{aa}	42 ± 6^{aa}
BSCB permeability								
Evans blue mg %	6	0.24 ± 0.04	1.65 ± 0.12**	0.87 ± 0.32^a	0.79 ± 0.22^{aa}	1.47 ± 0.43^{ns}	0.72 ± 0.18^{aa}	0.89 ± 0.23^a
[131]Iodine %	5	0.35 ± 0.06	1.96 ± 0.14**	0.94 ± 0.12^{aa}	0.89 ± 0.12^{aa}	1.78 ± 0.56^{ns}	0.81 ± 0.16^{aa}	1.07 ± 0.24^a
Edema formation								
Cord width mm	5	3 ± 0.5	5 ± 0.5**	4 ± 1^a	4 ± 0.5^{aa}	5 ± 1^{ns}	4 ± 0.5^{aa}	4 ± 1^a
Water content %	5	66.12 ± 0.18	69.34 ± 0.23**	67.67 ± 0.21^a	67.23 ± 0.18^a	68.76 ± 0.44^{ns}	67.16 ± 0.12^{aa}	68.21 ± 0.12^a
Cell injury								
Neuronal damage	5	nil	4	2 ± 1#	2 ± 1#	4 ± 1^{ns}	2 ± 2#	2 ± 2#
Glial cell injury	5	nil	4	2 ± 2#	2 ± 1#	3 ± 1^{ns}	2 ± 2#	3 ± 1
Myelin damage	5	nil	4	2 ± 2#	2 ± 2#	4 ± 1^{ns}	2 ± 1#	4 ± 1

The SCI was performed by making a longitudinal incision into the right dorsal horn of the T10–11 segments and the animals were allowed to survive 5 h after trauma. Cord width was measured in formalin-fixed spinal cord specimens before embedding in paraffin [24].
§ BDNF or GDNF (Total amount, 1 µg in 10 µl) was applied topically in separate group of animals after SCI. In combination, BDNF and GDBF (0.5 µg each, total dose 1 µg) was used in identical manner.
BDNF Brain-derived neurotrophic factor; *BSCB* blood-spinal cord barrier; *GDNF* glial-derived neurotrophic factor; *SCI* spinal cord injury.
Values are Mean ± SD of 5–6 rats in each group. * = p < 0.05; ** p < 0.01 (compared from control); a = p < 0.05; aa = p < 0.01 (compared from 5 h SCI), ANOVA followed by Dunnett's test from one control); # = p < 0.05, Chi-square test from 5 h SCI group; ns = not significant (from 5 h SCI group).

served after individual application of BDNF or GDNF 30 minutes after SCI, and co-application of BDNF and GDNF in high concentrations 60 or 90 min after trauma (Table 1).

Effect of neurotrophins on spinal cord pathology

Marked neuroprotection is seen 5 hours after SCI in rats that received either BDNF or GDNF in high concentration 30 minutes after injury. However, application of neurotrophins separately 60 or 90 minutes after SCI did not attenuate cell injury in the spinal cord. When BDNF and GDNF were applied in combination 60 or 90 min after SCI, profound neuroprotection was observed in the perifocal segments of the traumatized cord (Fig. 1). The neurons, astrocytes, and the myelin damage were minimal in this group of rats compared with the untreated injured group (results not shown).

Discussion

The salient new findings of the present investigation showed that a combination of growth factors derived from nerve cells (BDNF) [11, 36] and glial cells (GDNF) [5, 7], when applied over the traumatized spinal cord after 60 or 90 minutes, was able to attenuate motor dysfunction and the pathophysiology of spinal cord cell and tissue injury. However, application of BDNF or GDNF alone at these time periods was ineffective. To our knowledge, these observations are the first to show that BDNF and GDNF in combination potentiated the beneficial effects of neurotrophins in SCI, indicating that a suitable combination of neurotrophins works in synergy to enhance neuroprotection in CNS trauma, a finding that has not been reported previously. Whether the additive effects of neurotrophins on neuroprotection are limited to a select combination of growth factors is still not known. Additional studies using various combinations of neurotrophins are needed to clarify this point.

A decrease in endogenous neurotrophins in the spinal cord following trauma deprives neurons and or glial cells of their trophic support, resulting in atrophy or cell death [35, 36]. Thus, exogenous supplementation of BDNF and GDNF in combination will likely to rescue the nerve cells and glial cells from damage. These observations suggest that both nerve cell and

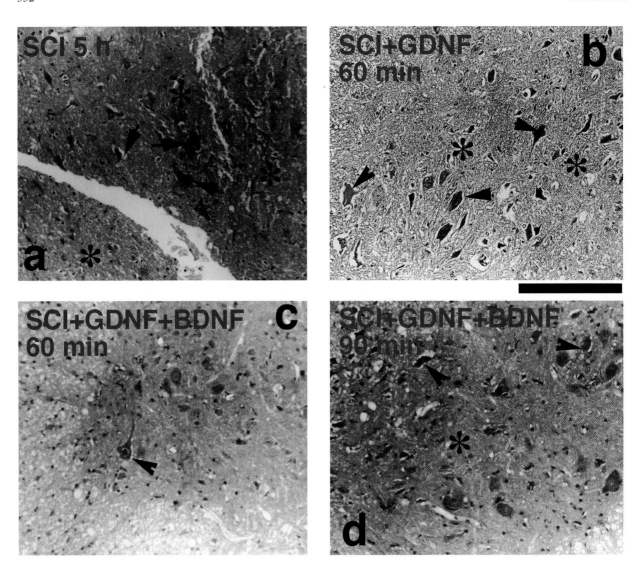

Fig. 1. Structural changes in the contralateral ventral horn of the T9 segment following SCI (a) and its modification with GDNF alone (+60 minutes, b) or in combination with BDNF 60 minutes (c) or 90 minutes (d) after trauma. SCI was performed on the right dorsal horn of the T10–11 segments (see text for details). Trauma to the spinal cord resulted in profound nerve cell damage (arrows), edema, and sponginess (*) in the untreated cord (a). Treatment with GDNF alone 60 minutes after SCI did not reduce trauma-induced cell and tissue injury (b). However, when GDNF and BDNF was given in combination, the neurotrophins were able to induce the most marked neuroprotective effect in animals treated 60 minutes after SCI (c). This protective effect of neurotrophins was considerably reduced when GDNF and BDNF were administered 90 minutes after trauma (d). Bar: 80 μm

glial cells actively participate in cellular and molecular mechanisms of spinal cord cell and tissue injury during the first hours of trauma.

Neurotrophic factors and their receptors are present in the developing and adult spinal cord. The spinal cord content of BDNF, GDNF, ciliary neurotrophic factor, and NGF is very low under normal conditions [3, 4, 11–13]. Alterations in neurotrophins or their receptors occur after SCI [20, 23, 29]. Increased expression of basic fibroblast growth factor mRNA in moto-neurons and astrocytes is seen 6 hours after contusion injury in rat that continued up to 2 days [6, 9]. Contusion injury to the cord in cats and rats up-regulates low-affinity NGF receptors in motoneurons, microvessels, and in ventral funiculus [17, 18]. These observations suggest that neurotrophins participate in trauma-induced alterations in spinal cord function.

Neurotrophin receptors influence neuronal survival by modulation of neurotransmitters, neuropeptide synthesis, and/or their release in the spinal cord [12,

14, 15, 20, 33]. There are indications that communication occurs between neurotrophins, dynorphin, and cytokines with their receptors located on neurons, glial cells, inflammatory cells, meninges, blood vessels in scar tissue, or perifocal edematous tissue [10, 20, 33]. Obviously, some signals are beneficial following an up-regulation of certain kinds of neurotrophins, whereas, increased expression of other types of neurotrophins may have adverse effects [7, 19, 20]. Our study suggests that BDNF and GDNF work together to enhance neuroprotection, probably by influencing beneficial cell signaling pathways. The exact nature of these pathways requires further investigation.

In animals treated with neurotrophins, trauma-induced edema formation and cell injuries are considerably attenuated and the distortion of nerve cells, glial cells, and myelin vesiculation is much less apparent. In addition, neurotrophins attenuated BSCB disturbances in SCI, indicating that neurotrophins influence cell and membrane functions after injury. Treatment of spinal cord cell and tissue injury with neurotrophins followed by improvement in motor function suggests that survival of nerve cells and myelin play an important role in motor function [31, 32]. Neurotrophin-induced reduction in disturbances of the spinal cord cell and tissue micro-fluid environment and/or modification of intracellular signal transduction cascades such as Ca^{2+} injury signals could be responsible for neuroprotective effects [14, 16, 22]. Modulation of sensory information processes by neurotrophins further supports this idea [3, 8, 11, 12].

An inhibitory influence of neurotrophins on nitrous oxide synthase up-regulation and/or cellular or oxidative stress may also contribute to neuroprotection in SCI [31, 32]. Improvement of motor function and reduction in cell and tissue injury following early blockade of oxidative stress and generation of free radicals with the antioxidant H-290/51 are in line with this assumption [32]. It remains to be seen whether combinations of other growth factors will further potentiate neuroprotective effects of neurotrophins in attenuating cellular stress and disturbances in the spinal cord fluid microenvironment, a feature currently being investigated in our laboratory.

Conclusion

In conclusion, our study demonstrates that a combination of BDNF and GDNF, when administered 60 or 90 minutes after SCI, attenuates motor dysfunction, BSCB breakdown, edema formation, and cell injury in a rat model, a feature not seen when these neurotrophins are applied individually at these time periods. These observations suggest that BDNF and GDNF in combination act in synergy to potentiate their neuroprotective effects in SCI during the early hours after trauma, indicating a potential therapeutic value for growth factors in a clinical setting in the future.

Acknowledgments

This investigation is supported by grants from Swedish Medical Research Council (2710); Astra-Zeneca, Mölndal, Sweden; Alexander von Humboldt Foundation, Germany; The University Grants Commission, New Delhi, India; and The Indian Council of Medical Research, New Delhi, India. Expert technical assistance of Kärstin Flink, Kerstin Rystedt, Franziska Drum, and Katherin Kern, and secretarial assistance of Aruna Sharma are greatly appreciated.

References

1. Barde YA (1994) Neurotrophins: a family of proteins supporting the survival of neurons. Prog Clin Biol Res 390: 45–56
2. Cheng H, Huang SS, Lin SM, Lin MJ, Chu YC, Chih CL, Tsai MJ, Lin HC, Huang WC, Tsai SK (2005) The neuroprotective effect of glial cell line-derived neurotrophic factor in fibrin glue against chronic focal cerebral ischemia in conscious rats. Brain Res 1033: 28–33
3. Davies AM (1994) The role of neurotrophins in the developing nervous system. J Neurobiol 25: 1334–1348
4. Ernfors P, Persson H (1991) Developmentally regulated expression of HDNF/NT-3 mRNA in rat spinal cord motoneurons and expression of BDNF mRNA in embryonic dorsal root ganglion. Eur J Neurosci 3: 953–961
5. Eves EM, Tucker MS, Roback JD, Downen M, Rosner MR, Wainer BH (1992) Immortal rat hippocampal cell lines exhibit neuronal and glial lineages and neurotrophin gene expression. Proc Natl Acad Sci U S A 89: 4373–4377
6. Follesa P, Wrathal JR, Mocchetti I (1994) Increased basic fibroblast growth factor mRNA following contusive spinal cord injury. Mol Brain Res 22: 1–8
7. Harrington AW, Leiner B, Blechschmitt C, Arevalo JC, Lee R, Morl K, Meyer M, Hempstead BL, Yoon SO, Giehl KM (2004) Secreted proNGF is a pathophysiological death-inducing ligand after adult CNS injury. Proc Natl Acad Sci U S A 101: 6226–6230
8. Koh JY, Gwag BJ, Lobner D, Choi DW (1995) Potentiated necrosis of cultured cortical neurons by neurotrophins. Science 268: 573–575
9. Koshinaga M, Sanon HR, Whitemore SR (1993) Altered acidic and basic fibroblast growth factor expression following spinal cord injury. Exp Neurol 120: 32–48
10. Lambiase A, Micera A, Sgrulletta R, Bonini S (2004) Nerve growth factor and the immune system: old and new concepts in the cross-talk between immune and resident cells during pathophysiological conditions. Curr Opin Allergy Clin Immunol 4: 425–430
11. Lindsay RM (1996) Role of neurotrophins and trk receptors in

the development and maintenance of sensory neurons: an overview. Philos Trans R Soc Lond B Biol Sci 351: 365–373
12. Lindsay RM, Harmar AJ (1989) Nerve growth factor regulates expression of neuropeptide genes in adult sensory neurons. Nature 337: 362–364
13. Lindvall O, Kokaia Z, Bengzon J, Elmer E, Kokaia M (1994) Neurotrophins and brain insults. Trends Neurosci 17: 490–496
14. Mattson MP, Lovell MA, Furukawa K, Markesbery WR (1995) Neurotrophic factors attenuate glutamate-induced accumulation of peroxides, elevation of intracellular Ca2+ concentration, and neurotoxicity and increase antioxidants enzyme activities in hippocampal neurons. J Neurochem 65: 1740–1751
15. Nawa H, Bassho Y, Carnahan J, Nakanishi S, Mizuno K (1993) Regulation of neuropeptide expression in cultured cerebral cortical neurons by brain-derived neurotrophic factor. J Neurochem 60: 772–775
16. Nicole O, Ali C, Docagne F, Plawinski L, MacKenzie ET, Vivien D, Buisson A (2001) Neuroprotection mediated by glial cell line-derived neurotrophic factor: involvement of a reduction of NMDA-induced calcium influx by the mitogen-activated protein kinase pathway. J Neurosci 21: 3024–3033
17. Reynolds ME, Brunello N, Mocchetti I, Wrathall JR (1991) Localization of nerve growth factor receptor mRNA in contused rat spinal cord by in situ hybridisation. Brain Res 559: 149–153
18. Risling M, Fried K, Linda H, Carlstedt T, Cullheim S (1993) Regrowth of motor axons following spinal cord lesions: distribution of laminin and collagen in CNS scar tissue. Brain Res Bull 30: 405–414
19. Samdani AF, Newcamp C, Resink A, Facchinetti F, Hoffman BE, Dawson VL, Dawson TM (1997) Differential susceptibility to neurotoxicity mediated by neurotrophins and neuronal nitric oxide synthase. J Neurosci 17: 4633–4641
20. Schwab ME, Bartholdi D (1996) Degeneration and regeneration of axons in the lesioned spinal cord. Physiol Rev 76: 319–370
21. Sharma HS (2003) Neurotrophic factors attenuate microvascular permeability disturbances and axonal injury following trauma to the rat spinal cord. Acta Neurochir [Suppl] 86: 383–388
22. Sharma HS (2004) Pathophysiology of the blood-spinal cord barrier in traumatic injury. In: Sharma HS, Westman J (eds) Blood-spinal cord and brain barriers in health and disease. Elsevier Academic Press, San Diego, pp 437–518
23. Sharma HS (2005) Pathophysiology of blood-spinal cord barrier in traumatic injury and repair. Curr Pharm Des 11: 1353–1389
24. Sharma HS (2005) Neuroprotective effects of neurotrophins and melanocortins in spinal cord injury. An experimental study in the rat using pharmacological and morphological approaches. Ann NY Acad Sci 1053 [in press]
25. Sharma HS (2005) Alterations of amino acid neurotransmitters in hyperthermic brain injury. J Neural Transm [in press]
26. Sharma HS, Westman J (2004) Blood-spinal cord and brain barriers in health and disease. Elsevier Academic Press, San Diego, pp 1–617
27. Sharma HS, Westman J, Nyberg F (1997) Topical application of 5-HT antibodies reduces edema and cell changes following trauma to the rat spinal cord. Acta Neurochir [Suppl] 70: 155–158
28. Sharma HS, Nyberg F, Westman J, Alm P, Gordh T, Lindholm D (1998) Brain derived neurotrophic factor and insulin like growth factor-1 attenuate upregulation of nitric oxide synthase and cell injury following trauma to the spinal cord. Amino Acids 14: 121–130
29. Sharma HS, Nyberg F, Gordh T, Alm P, Westman, J (1998) Neurotrophic factors attenuate neuronal nitric oxide synthase upregulation, microvascular permeability disturbances, edema formation and cell injury in the spinal cord following trauma. In: Stålberg E, Sharma HS, Olsson Y (eds) Spinal cord monitoring: basic principles, regeneration, pathophysiology and clinical aspects. Springer, New York, pp 118–148
30. Sharma HS, Nyberg F, Gordh T, Alm P, Westman J (2000) Neurotrophic factors influence upregulation of constitutive isoform of heme oxygenase and cellular stress response in the spinal cord following trauma. An experimental study using immunohistochemistry in the rat. Amino Acids 19: 351–361
31. Sharma HS, Gordh T, Wiklund L, Mohanty S, Sjoquist PO (2005) Spinal cord injury induced heat shock protein expression is reduced by an antioxidant compound H-290/51. An experimental study using light and electron microscopy in the rat. J Neural Transm [in press]
32. Sharma HS, Sjöquist PO, Mohanty S, Wiklund L (2006) Post-injury treatment with a new antioxidant compound H-290/51 attenuates spinal cord trauma induced c-*fos* expression, motor dysfunction, edema formation and cell injury in the rat. Acta Neurochir [Suppl] 96: 351–357
33. Sharma HS, Nyberg F, Gordh T, Alm P (2006) Topical application of dynorphin A (1-17) antibodies attenuate neuronal nitric oxide synthase up-regulation, edema formation and cell iInjury following a focal trauma to the rat spinal cord. Acta Neurochir [Suppl] 96: 337–343
34. Thoenen H (1995) Neurotrophins and neuronal plasticity. Science 270: 593–598
35. Yao DL, Liu X, Hudson LD, Webster HD (1995) Insulin-like growth factor I treatment reduces demyeaination and up-regulates gene expression of myelin-related proteins in experimental autoimmune encephalomyelitis. Proc Natl Acad Sci USA 92: 6190–6194
36. Yin QW, Johnson J, Prevette D, Oppenheim RW (1994) Cell death of spinal motoneurons in the chick embryo following deafferentation: rescue effects of tissue extracts, soluble proteins, and neurotrophic agents. J Neurosci 14: 7629–7640
37. Zaheer A, Zhong W, Lim R (1995) Expression of mRNAs of multiple growth factors and receptors by neuronal cell lines: detection with RT-PCR. Neurochem Res 20: 1457–1463

Correspondence: Hari Shanker Sharma, Dept. of Surgical Sciences, Anesthesiology & Intensive Care Medicine, University Hospital, SE-75185 Uppsala, Sweden. e-mail: Sharma@surgsci.uu.se

Chronic spinal nerve ligation induces microvascular permeability disturbances, astrocytic reaction, and structural changes in the rat spinal cord

T. Gordh and H. S. Sharma

Laboratory of Pain Research, Department of Surgical Sciences, Anesthesiology and Intensive Care Medicine, University Hospital, Uppsala University, Uppsala, Sweden

Summary

The possibility that a chronic nerve ligation impairs the spinal cord cellular microenvironment was examined using leakage of endogenous albumin, reaction of astrocytes, and structural changes in a rat model. Rats subjected to 8 weeks of unilateral L4/L5 nerve ligation (a model of neuropathic pain) showed leakage of albumin, up-regulation of glial fibrillary acidic protein (GFAP) immunoreaction, and abnormal cell reaction. Distortion and loss of nerve cells as well as general sponginess of the gray matter was clearly evident. Cell changes were present in both dorsal and ventral horns and were most marked on the ipsilateral side compared to the contralateral cord. Nerve cell and glial cell changes are normally present in the regions showing intense albumin immunoreactivity, indicating disruption of the blood-spinal cord barrier (BSCB). Our observations indicate that a chronic nerve lesion has the capacity to induce selective breakdown of the BSCB that could be responsible for activation of astrocytes and abnormal cell reaction. These findings enhance our understanding of the pathophysiology of neuropathic pain and/or other spinal cord disorders.

Keywords: Spinal nerve lesion; blood-spinal cord barrier; immunoreactivity; pain.

Introduction

Peripheral neuropathy, nerve lesion, or spinal nerve ligation induces profound changes in the spinal cord microenvironment and selective neuronal damage [9, 18, 34]. Alterations in the fluid microenvironment of the spinal cord participate in the slow degenerative changes in the cord [22, 24, 36]. The spinal cord is equipped with a blood-spinal cord barrier (BSCB) that restricts the passage of proteins and other harmful molecules within the spinal cord under normal conditions [21, 22, 24]. However, in several spinal cord disorders including multiple sclerosis, experimental allergic encephalomyelitis, and traumatic or ischemic injuries to the cord, a breakdown of the BSCB to protein is commonly seen [22, 24]. Passage of proteins and other unwanted molecules that gain access to the brain or spinal cord microenvironment are likely to result in a series of events leading to abnormal cell reactions [21, 23]. Breakdown of the BSCB is associated with activation of astrocytes, vesiculation of myelin, and nerve cell reaction in several acute and chronic spinal cord disorders, e.g., multiple sclerosis, experimental allergic encephalomyelitis, or spinal cord injury [5, 6, 19, 23]. Alteration in several neurochemical mediators of microvascular permeability within the cord, including opioids, serotonin, prostaglandins, histamine, nitric oxide, carbon monoxide, and cytokines, appear to play important roles in BSCB disruption and cell injury [6, 19, 22, 23, 25, 32]. Since neuropathic pain is also known to alter the metabolism of these neurochemicals [9, 11, 13], a possibility exists that breakdown of the SBCB could contribute to neurodegenerative changes in the cord caused by the nerve lesion.

Previous studies suggest that models of chronic neuropathic pain can be achieved in experimental models, e.g., lesion or ligation of peripheral spinal nerves of L5 and L6 segments in the rat [7, 8, 15, 16]. Animals exhibit symptoms of chronic neuropathic pain, such as hyperalgesia, 4 to 8 weeks after receiving the nerve lesion [7, 8, 11, 13, 15, 16]. Using animal models, we found profound up-regulation of the enzymes nitric oxide synthase (NOS) and heme oxygenase (HO-2), which are responsible for production of nitric oxide and carbon monoxide, respectively, in the spinal cord in areas showing marked cellular changes [11, 13]. Since up-regulation of both NOS and HO-2 in spinal cord injury is often associated with breakdown of microvascular permeability and cell changes [20, 22, 23, 27–32], the present investigation was undertaken to examine the BSCB in the rat with neuropathic pain. Structural changes and activation of astrocytes in the spinal cord were also studied.

Materials and methods

Animals

Experiments were carried out on adult male Sprague-Dawley rats weighing 270–310 g housed in controlled ambient temperature ($21 \pm 1\,^\circ\text{C}$) with a 12-hour light and dark schedule. Food and tap water were supplied ad libitum.

Neuropathic pain models

Under inhalation anesthesia (a mixture of 2% enflurane and a 1:1 flow ratio of O2 and N2O), the L4 spinal nerve was exposed [11–13]. In a separate group of rats, the spinal nerve was ligated as described earlier [7, 9, 11, 13, 15, 16]. A sham group received the same surgical procedures, except for the nerve injury. Sham and nerve-lesioned rats were allowed to survive for 8 weeks after surgery [13]. Care was taken according to the National Institutes of Health Guidelines (USA) so that animals did not suffer pain during this period. This experimental protocol was approved by the Medical Ethics Committee for Animal Studies at Uppsala University.

Immunohistochemistry

Using standard immunohistochemical techniques, endogenous serum proteins (albumin and fibronectin) and glial fibrillary acidic protein (GFAP) were examined in spinal cord L4/L5 segments according to a standard protocol [25, 26]. The integrity of the BSCB was studied using polyclonal albumin and fibronectin antibodies (Sigma-Aldrich, St. Louis, MO). The state of astrocyte activation was examined using polyclonal GFAP antibodies [26].

The reaction was visualized using 3-amino-9-ethycarbazole (Vector Laboratories, Burlingame, CA) and counterstained with hematoxylin. Reagent controls (omitting the primary antibody or substituting nonimmune serum for the primary antibody in the staining protocol) on tissue sections revealed no staining, thus confirming the specificity of the primary antibodies used.

The changes in immunohistochemical staining were assessed using semiquantitative analysis in the dorsal and ventral horns in both the ipsilateral (right) and contralateral (left) sides (Table 1).

Morphological study

Tissue pieces from the L5 segment of the cord were embedded in Epon resin (Resolution Europe BV, The Netherlands) for routine light and electron microscopy for structural investigation [13, 27, 28, 33]. In brief, for high resolution light microscopy, about 1 µm thick sections were cut and stained with toluidine blue and examined with a light microscope for gross pathology. Ultrathin sections from the dorsal and ventral horns were cut, counterstained with lead acetate and uranyl citrate, and examined under a Phillips Transmission Electron Microscope [11, 13]. For semiquantitative analysis, the number of distorted nerve cells was counted in dorsal and ventral horn in both the ipsilateral and contralateral sides of the L5 spinal cord segment [11, 13, 28].

Statistical analysis

Quantitative or semiquantitative data were analyzed using ANOVA followed by Dunnett's test for multiple group comparison. P-value less that 0.05 was considered significant.

Results

Spinal nerve ligation and spinal cord morphology

Marked neurodegenerative changes in the spinal cord were most pronounced in the ipsilateral dorsal and ventral horns (Fig. 1). These changes include vacuolation of neuronal cytoplasm, degeneration of myelin, and distorted neurons. Epon sections (1 µm thick) of the spinal cord L5 segment stained with toluidine blue exhibited pronounced structural changes. Thus, dark and distorted nerve cells, vacuolation in neuronal cytoplasm, and degeneration of myelin were frequent in the ipsilateral L5 segment of the cord (Fig. 1e). These nerve cell changes were most marked in the ipsilateral side compared to the contralateral cord (Fig. 1). Mild degenerative changes around the perivascular regions were prominent by electron microscopy (Fig. 1f). Vesication of myelinated nerves and damaged synapses were also present (Fig. 1). The sham-operated group did not show any abnormal nerve cell reaction in the cord compared to control group.

Spinal nerve ligation and extravasation of endogenous serum proteins

Profound extravasation of albumin and fibronectin was observed in the spinal cord gray matter of the nerve-lesioned animals in the regions associated with nerve cell damage (Table 1). This increase in albumin and fibronectin immunoreactivity was most pronounced in the ipsilateral cord compared to the contralateral side (Table 1; Fig. 1a,b). The sham-operated group did not exhibit any increase in endogenous albumin or fibronectin immunostaining.

Spinal nerve ligation and activation of astrocytes

The spinal nerve lesion showed a marked increase in GFAP 8 weeks after the nerve lesion (Fig. 1c,d, Table

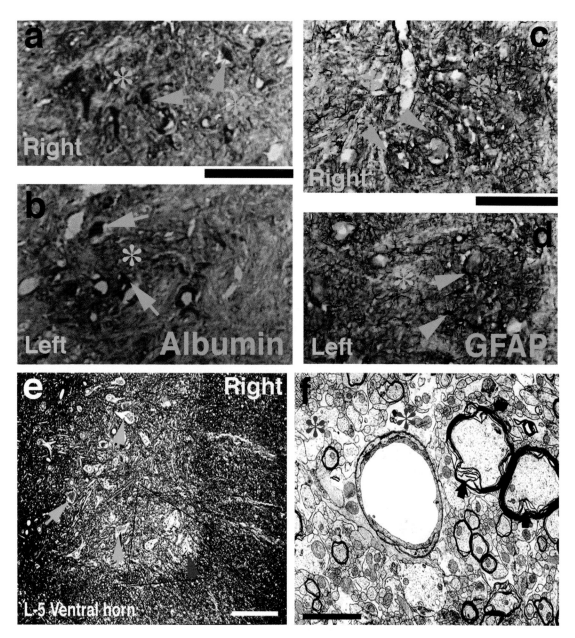

Fig. 1. Leakage of albumin (a,b), activation of astrocytes (c,d), and morphological alterations within the spinal cord neuropil at light (e) and electron (f) microscopy 8 weeks after nerve lesion (for details, see text). Extravasation of albumin can be seen within the neuropil and a few damaged and distorted nerve cells are infiltrated with albumin (arrows). The intensity of albumin extravasation and damaged nerve cells is most prominent in the ipsilateral (right) side compared to the contralateral cord (a,b). Over-expression of GFAP is seen in the ventral horns of ipsilateral and contralateral cord (d). The magnitude and intensity of GFAP immunostaining can be seen around the damaged nerve cells (arrows) located in the edematous regions (*). Several distorted nerve cells (arrows) are present in the spinal cord in the ventral horn (e). A few normal nerve cells (arrowhead) are also seen. At the ultrastructural level (f), perivascular edema (*) and damage to astrocytic end-feet is visible. Vesiculation of myelin (arrows) and degeneration of neuropil is also seen. Bar: a,b = 100 μm; c,d = 60 μm; e = 25 μm; f = 1 μm. Data (e,f) modified after [13]

1). This increase in GFAP-immunoreactive cells was significantly higher in the ipsilateral cord compared to the contralateral side (Table 1; Figs. 1c,d). Sham-operated animals did not show increased GFAP immunoreactivity compared to normal rats [25].

Discussion

The salient new finding in the present investigation is a marked change in the spinal cord environment 8 weeks after a nerve ligation, which correlates well with

Table 1. *Leakage of albumin, activation of astrocytes, and morphological alterations 8 weeks after nerve ligation in the rat spinal cord*

Parameters measured	n	Control		Sham		Nerve lesion 8 weeks§	
		Right	Left	Right	Left	Right	Left
BSBC permeability							
Number of albumin-positive cells	6	nil	nil	nil	nil	68 ± 8	46 ± 12a
Astrocytic activation							
Number of GFAP-positive cells	5	6 ± 2	8 ± 3	10 ± 4	8 ± 5	108 ± 23**	68 ± 16**a
Structural changes	5						
Nerve cell reaction		nil	nil	±?	nil	++	+
Glial cell reaction		nil	nil	±?	nil	+++	+
Endothelial cell reaction		nil	nil	nil	nil	++	+

Albumin extravasation was used to detect blood-spinal cord barrier breakdown. Activation of astrocytes was assessed by immunostaining of glial fibrillary acidic protein (GFAP) immunoreactivity. Morphological changes were examined using standard light and electron microscopy [for details see text].
§ Spinal nerve lesion was produced at L5 level and the animals were allowed to survive 8 weeks [11, 13].
BSBC Blood-spinal cord barrier; *GFAP* glial fibrillary acidic protein; *nil* absent; ±? uncertain; + mild; ++ moderate; +++ considerable [see text].
Values are Mean ± SD of 5–6 rats in each group.
** = $p < 0.01$ (compared from control); a = $p < 0.05$; (compared from right side), ANOVA followed by Dunnett's test from one control.

leakage of endogenous albumin, activation of astrocytes, and structural changes in the cord. Our results show that an experimental model of chronic neuropathic pain produced by a peripheral nerve lesion is associated with neurodegenerative changes in the spinal cord.

The structural changes can be seen 2 weeks after nerve lesion and are progressive up to 8 weeks. Since our observations are limited to the survival period of 8 weeks, it is unclear whether neurodegenerative changes in the spinal cord are maximal at this time or are reversible. To confirm this point, studies beyond 8 weeks (10 to 20 weeks) are needed.

The other important finding of this investigation is that the contralateral side also shows neurodegenerative changes. However, the magnitude and intensity of changes in the contralateral cord are less compared to the ipsilateral side. This indicates that a peripheral nerve lesion or chronic neuropathic pain-induced widespread alterations in the spinal cord neurochemical environment could contribute to some of these structural changes. Previous reports showing pronounced up-regulation of NOS and HO-2 expression in both ipsi- and contralateral cord at a similar time period following nerve lesion is in line with this idea [11–13].

Increased production of nitric oxide and carbon monoxide is evident with over-expression of the enzymes NOS and HO-2, respectively, in the brain and spinal cord following hyperthermia or trauma and is associated with breakdown of BSCB permeability, activation of astrocytes, and nerve cell injury [20, 21, 27, 28, 31–33]. These observations suggest that a chronic nerve lesion induces alteration in the neurochemical environment and/or release of secondary injury signals responsible for neurodegenerative changes in the cord. Reduction in nerve lesion-induced spinal cord cell damage by pharmacological inhibition of NOS in this model further supports this hypothesis.

Our observations showing extravasation of endogenous albumin and fibronectin in the regions associated with cell damage confirm that breakdown of the BSCB contributes to secondary cell and tissue injury following a spinal nerve lesion. Leakage of albumin and fibronectin was most pronounced in the ipsilateral cord compared to the contralateral side. These observations indicate that breakdown of the BSCB is one of the important factors in cell and tissue injury in the chronic neuropathic pain model. Breakdown of the SBCB exposes the spinal cord microenvironment to restricted molecules and serum proteins that are normally excluded by the intact barrier leading to abnormal cell reactions [21–24].

Disruption of the BSCB will also expose astrocytes to serum components. Astrocytes are important constructional elements of the spinal cord and have many additional functions under normal and pathological conditions [1, 2, 6]. These cells play important roles for normal BSCB functions and for homeostasis of the extracellular environment of the parenchyma [19]. Astrocytes are potential targets for chemical signals released from neurons and possess binding sites for neuro-active peptides, amino acids, amines, and eicosanoids [17]. Pathological processes affecting the spinal cord are associated with swelling of astrocytes and the formation of gliosis that involves proliferation and hypertrophy of astrocytes [1–4, 10, 14]. Gliosis involves activation of metabolic processes with production of cytoskeletal components including GFAP [3, 4]. It seems likely that alterations in the microenvironment of the spinal cord by metabolic, traumatic, or ischemic insults could play an important role in activating astrocytes either directly or by altered neurochemical metabolism [2, 5, 6, 10, 26, 35]. Our results show that the chronic neuropathic pain model up-regulates GFAP, indicating activation of astrocytes and formation of gliosis. The detailed mechanism of gliosis in chronic nerve lesion remains unclear, however.

When cells and tissues are exposed to serum components, several chemical, ionic, or immunologic reactions take place [6]. Since astrocytes participate in BSCB function and homeostasis of the spinal cord, leakage of serum proteins and other factors activate astrocytes resulting in activation of GFAP [5]. The close correlation of albumin and GFAP immunoreactivity supports this hypothesis.

That leakage of serum proteins is related to nerve cell damage is evident from our findings, showing profound uptake of albumin by several neurons in the neuropil [21, 22, 24]. Uptake of serum proteins by neurons indicates that nerve cells are going to die from ischemia [19, 21, 23]. It may be that similar mechanisms are operating in ischemia and chronic nerve lesion-induced cell damage. To confirm these findings further, studies using specific markers of nerve cell death or DNA damage are needed in the neuropathic pain model.

Conclusion

Our results demonstrate that a chronic nerve ligation is associated with breakdown of BSCB permeability and activation of astrocytes. Glial cell activation is related to alterations in the brain fluid microenvironment. These observations are in line with our hypothesis that alteration in the brain fluid microenvironment following a chronic nerve lesion plays a key role in spinal cord neurodegeneration.

Acknowledgments

This investigation is supported by grants from the Swedish Medical Research Council (2710); Astra-Zeneca, Mölndal, Sweden; Alexander von Humboldt Foundation, Germany; The University Grants Commission, New Delhi, India; and The Indian Council of Medical Research, New Delhi, India. Expert technical assistance of Inga Hörte, Kärstin Flink, Kerstin Rystedt, and the secretarial assistance of Aruna Sharma are greatly appreciated.

References

1. Aquino DA, Chiu FC, Brosnan CF, Norton WT (1988) Glial fibrillary acidic protein increases in the spinal cord of Lewis rats with acute experimental autoimmuno encephalomyelitis. J Neurochem 51: 1085–1096
2. Beck DW, Roberts RL, Olson JJ (1986) Glial cells influence membrane-associated enzyme activity at the blood-brain barrier. Brain Res 381: 131–137
3. Bernstein JJ, Goldberg WJ (1987) Injury-related spinal cord astrocytes are immunoglobulin-positive (IgM and/or IgG) at different time periods in the regenerative process. Brain Res 426: 112–118
4. Bignami A, Dahl D, Rueger DC (1980) Glial fibrillary acidic protein (GFA) in normal neural cells and in pathological conditions. Adv Cell Neurobiol 1: 285–319
5. Bologa L, Cole R, Chiappelli F, Saneto RP, De Villis J (1988) Expression of glial fibrillary acidic protein by differentiated astrocytes is regulated by serum antagonistic factors. Brain Res 457: 295–302
6. Cervos-Navarro J, Sharma HS, Westman J, Bongcam-Rudloff E (1998) Glial reactions in the central nervous system following heat stress. Prog Brain Res 115: 241–274
7. Chung JM, Choi Y, Yoon YW, Na HS (1995) Effects of age on behavioral signs of neuropathic pain in an experimental rat model. Neurosci Lett 183: 54–57
8. Chung JM, Kim HK, Chung K (2004) Segmental spinal nerve ligation model of neuropathic pain. Methods Mol Med 99: 35–45
9. Dubner R, Ruda MA (1992) Activity-dependent neuronal plasticity following tissue injury and inflammation. Trends Neurosci 15: 96–103
10. Goldman JE, Abramson B (1990) Cyclic AMP-induced shape changes of astrocytes are accompanied by rapid depolymerization of actin. Brain Res 528: 189–196
11. Gordh T, Sharma HS, Alm P, Westman J (1998) Spinal nerve lesion induces upregulation of neuronal nitric oxide synthase in the spinal cord. An immunohistochemical investigation in the rat. Amino Acids 14: 105–112
12. Gordh T, Sharma HS, Azizi M, Alm P, Westman J (2000) Spinal nerve lesion induces upregulation of constitutive isoform of heme oxygenase in the spinal cord. An immunohistochemical investigation in the rat. Amino Acids 19: 373–381
13. Gordh T, Chu H, Sharma HS (2005) Spinal nerve lesion alters blood-spinal cord barrier function and activates astrocytes. An immunohistochemical study using albumin and glial fibrillary acidic protein in the rat. Pain (in press)
14. Goshgarian HG, Yu XJ, Rafols JA (1989) Neuronal and glial changes in the rat phrenic nucleus occurring within hours after spinal cord injury. J Comp Neurol 284: 519–533
15. Kim SH, Chung JM (1992) An experimental model for peripheral neuropathy produced by segmental spinal nerve ligation in the rat. Pain 50: 355–363

16. LaBuda CJ, Little PJ (2005) Pharmacological evaluation of the selective spinal nerve ligation model of neuropathic pain in the rat. J Neurosci Methods 144: 175–181
17. Murphy S, Pearce B (1987) Functional receptors for neurotransmitters on astroglial cells. Neuroscience 22: 381–394
18. Polgar E, Hughes DI, Arham AZ, Todd AJ (2005) Loss of neurons from laminas I–III of the spinal dorsal horn is not required for development of tactile allodynia in the spared nerve injury model of neuropathic pain. J Neurosci 25: 6658–6666
19. Schmidt-Kastner R, Freund TF (1991) Selective vulnerability of the hippocampus in brain ischemia. Neuroscience 40: 599–636
20. Sharma HS (1998) Neurobiology of the nitric oxide in the nervous system. Basic and clinical perspectives. Amino Acids 14: 83–85
21. Sharma HS (2004) Blood-brain and spinal cord barriers in stress. In: Sharma HS, Westman J (eds) Blood-spinal cord and brain barriers in health and disease. Elsevier Academic Press, San Diego, pp 231–298
22. Sharma HS (2004) Pathophysiology of the blood-spinal cord barrier in traumatic injury. In: Sharma HS, Westman J (eds) Blood-spinal cord and brain barriers in health and disease. Elsevier Academic Press, San Diego, pp 437–518
23. Sharma HS (2005) Pathophysiology of blood-spinal cord barrier in traumatic injury and repair. Curr Pharm Des 11: 1353–1389
24. Sharma HS, Westman J (2004) Blood-spinal cord and brain barriers in health and disease. Elsevier Academic Press, San Diego, pp 1–617
25. Sharma HS, Zimmer C, Westman J, Cervos-Navarro J (1992) Acute systemic heat stress increases glial fibrillary acidic protein immunoreactivity in brain: experimental observations in conscious normotensive young rats. Neuroscience 48: 889–901
26. Sharma HS, Olsson Y, Cervós-Navarro J (1993) p-Chlorophenylalanine, a serotonin synthesis inhibitor, reduces the response of glial fibrillary acidic protein induced by trauma to the spinal cord. An immunohistochemical investigation in the rat. Acta Neuropathol (Berl) 86: 422–427
27. Sharma HS, Alm P, Westman J (1998) Nitric oxide and carbon monoxide in the brain pathology of heat stress. Prog Brain Res 115: 297–333
28. Sharma HS, Nyberg F, Westman J, Alm P, Gordh T, Lindholm D (1998) Brain derived neurotrophic factor and insulin like growth factor-1 attenuate upregulation of nitric oxide synthase and cell injury following trauma to the spinal cord. An immunohistochemical study in the rat. Amino Acids 14: 121–129
29. Sharma HS, Nyberg F, Gordh T, Alm P, Westman J (2000) Neurotrophic factors influence upregulation of constitutive isoform of heme oxygenase and cellular stress response in the spinal cord following trauma. An experimental study using immunohistochemistry in the rat. Amino Acids 19: 351–361
30. Sharma HS, Westman J, Gordh T, Alm P (2000) Topical application of brain derived neurotrophic factor influences upregulation of constitutive isoform of heme oxygenase in the spinal cord following trauma an experimental study using immunohistochemistry in the rat. Acta Neurochir [Suppl] 76: 365–369
31. Sharma HS, Badgaiyan RD, Mohanty S, Alm P, Wiklund L (2005) Neuroprotective effects of nitric oxide synthase inhibitors in spinal cord injury induced pathophysiology and motor functions. An experimental study in the rat. Ann NY Acad Sci 1053 (in press)
32. Sharma HS, Wiklund L, Badgaiyan RD, Mohanty S, Alm P (2006) Intracerebral administration of neuronal nitric oxide synthase antiserum attenuates traumatic brain injury induced blood-brain barrier permeability, brain edema and sensory motor disturbances in the rat. Acta Neurochir [Suppl] 96: 288–294
33. Sharma HS, Nyberg F, Gordh T, Alm P (2006) Topical application of dynorphin A (1-17) antibodies attenuate neuronal nitric oxide synthase up-regulation, edema formation, and cell injury following a focal trauma to the rat spinal cord. Acta Neurochir [Suppl] 96: 309–315
34. Solodkin A, Traub RJ, Gebhart GF (1992) Unilateral hindpaw inflammation produces a bilateral increase in NADPH-diaphorase histochemical staining in the rat lumbar spinal cord. Neuroscience 51: 495–499
35. Tao-Cheng JH, Brightman MW (1988) Development of membrane interactions between brain endothelial cells and astrocytes in vitro. Int J Dev Neurosci 6: 25–37
36. Yamamoto T, Shimoyama N (1995) Role of nitric oxide in the development of thermal hyperesthesia induced by sciatic nerve constriction injury in the rat. Anesthesiology 82: 1266–1273

Correspondence: Hari Shanker Sharma, Dept. of Surgical Sciences, Anesthesiology & Intensive Care Medicine, University Hospital, SE-75185 Uppsala, Sweden. e-mail: Sharma@surgsci.uu.se

Hydrocephalus

Gravitational valves: relevant differences with different technical solutions to counteract hydrostatic pressure

M. Kiefer[1], **U. Meier**[2], and **R. Eymann**[1]

[1] Saarland University, Medical School, Department of Neurosurgery, Homburg-Saar, Germany
[2] Unfallkrankenhaus Berlin, Department of Neurosurgery, Berlin, Germany

Summary

Two different technical principles of gravitational valves (G-valves) have been presented: counterbalancer and switcher G-valves. The objective of our prospective study was to look for clinically relevant differences between both.

A total of 54 patients with normal-pressure hydrocephalus (NPH) were treated; 30 patients received an Aesculap-Miethke GA-Valve (GAV; counterbalancer), and in 24 patients an Aesculap-Miethke Dualswitch-Valve (DSV; switcher) was implanted. The opening pressure of the posture-independent valve was 5 cm H_2O in both devices. The outcome was clearly better with the usage of the GAV than with the DSV. The frequency and severity of complications was pronounced in the DSV group.

We recommend the Aesculap-Miethke-GAV valve with a low opening pressure in a posture-independent valve for patients with NPH.

Keywords: Hydrocephalus; normal-pressure hydrocephalus; overdrainage; shunt.

Introduction

It has recently been shown that low-pressure valves may result in a better clinical outcome in patients suffering from normal-pressure hydrocephalus (NPH) than medium-pressure valves [4]. The disadvantage with low-pressure valves is their more than 70% overdrainage compared to about 30% with medium-pressure valves [4, 15]. These findings represent a clinical dilemma when choosing a shunt [16]. With a higher opening pressure, clinical results are probably worse, but the overdrainage risk is less. In contrast, better clinical results may be expected using a low-pressure valve, but at the cost of higher overdrainage rates.

Accordingly, shunts such as gravitational valves (G-valves), which allow sufficient drainage in the prone position yet also prevent overdrainage in the upright position, could be a good solution to this problem.

Up until now, 2 different technical solutions for G-valves have been introduced: the switcher type and the counterbalancer type. To date, there have been no reports in the literature comparing the 2 constructions.

Materials and methods

Patients

The study was performed at 2 neurosurgical centers. Center A used a counterbalance G-valve (n = 30), and Center B a switcher G-valve (n = 24). A total of 54 patients (38 men, 16 women) with NPH (89% idiopathic NPH) were evaluated. The average age was 66 ± 12 years (range: 29 to 83 years). Each patient was required to have at least 2 symptoms of Hakim's Triad and an Evans Index > 0.3. The average follow-up was 12 ± 4 months.

Indications

We performed a constant-volume infusion test in each patient [9, 10, 13, 14]. If the patients had an outflow resistance > 13 mmHg/ml × min, they became candidates for shunting.

Treatment

Each patient received a non-adjustable, low-pressure G-valve with an opening pressure of 5 cm H_2O in the prone position. The opening pressure selected for the upright position depended on the level of hydrostatic pressure to be compensated.

G-valves are posture-dependent because they change their total resistance based on the patient's posture. In contrast, conventional differential-pressure valves are posture-independent because their opening pressure remains the same regardless of the posture and hydrostatic pressure.

Outcomes

Preoperatively and at 1 and 12 months postoperatively, each patient was examined clinically and radiologically by magnetic resonance imaging (MRI). To document the clinical state, we used the Kiefer Index and Recovery Index and its adaptation to Black's Grading [10, 11]. The ventricular size was estimated using the Evans Index.

Statistics

We used the Mann-Whitney U test and the Wilcoxon matched pair test at a significance level of $\alpha = 0.05$.

Technical note (Fig. 1)

Aesculap-Miethke Dualswitch-Valve (DSV) (Christoph Miethke GMBH & Co. KG, Potsdam, Germany)

The DSV represents the switcher-type of G-valve [10]. It has 2 valves in 1 housing: A low-pressure valve for the lying position and a high-pressure valve for the upright position. A tantalum ball switches between both in accordance to posture. Typically at an angle of 60 to 70° the DSV inactivates the low-pressure valve and only the high-pressure valve may be passed by cerebrospinal fluid (CSF). However, a person's posture is not restricted by 2 extremes: lying flat and sitting or standing upright. If the patient is in an inclined position between 0 to 60°, the hydrostatic pressure rises while the DSV still allows passage of CSF via the low-pressure chamber. At first view, this may result in overdrainage; however, the tantalum ball can close the low-pressure chamber at an inclined position < 60° if the differential pressure is high, the CSF flow rate is high, and the position approaches 60°. In other words, at an incline angle > 0° and < 60°, the DSV acts as a flow-regulated valve. Its flow characteristics are determined by the temporal proportion while the low-pressure and the high-pressure valve are activated.

At an incline angle > 70° the low-pressure valve is permanently closed. However, the opening pressure of the high-pressure valve has to be selected according to the hydrostatic pressure in the completely upright position, so at an angle > 70° and < 90° the DSV may cause slight underdrainage.

To summarize, the DSV may have a tendency for overdrainage at an incline angle < 60° and underdrainage at an incline angle > 70° if the patient is not in the extreme positions of lying flat or being totally upright. In daily practice, this is not relevant except in immobilized persons, because time intervals of slight over- or underdrainage may compensate for each other resulting in an overall physiological CSF drainage.

The advantage of this valve is that it is not as sensitive to underdrainage for wrong implantation (not in exact alignment with the vertical body axis, which may be difficult in some instances) compared to the counterbalancer type.

Aesculap-Miethke Gravity-Assisted-Valve (GAV) (Christoph Miethke GMBH & Co. KG, Potsdam, Germany)

The GAV is a counterbalancer G-valve [15]. A normal differential pressure valve acts independently from posture and a second valve in the same housing changes its characteristics depending on posture. The opening pressure of the posture sensitive valve depends on the incline angle and the weight of its tantalum ball. The weight of the tantalum ball corresponds exactly to the weight of the hanging CSF column, which has to be compensated for in the upright position. If the patient lies flat, it is inactive. But even at a minimal incline position, the tantalum ball begins to close the posture-sensitive valve. The closing pressure of the posture-sensitive part of the device can be calculated as: sin (incline angle) × weight of tantalum ball.

So, the counterbalancer G-valve adjusts its resistance analogous to the changing hydrostatic pressure. Therefore, it may be assumed that the GAV compensates hydrostatic pressure more exactly in each position beyond extremes than the DSV.

One disadvantage is that a correct implantation is critical for correct function of the valve. If it is not implanted exactly in alignment with the vertical body axis, the GAV has a tendency toward underdrainage.

Results

Preoperative state

The average Kiefer Index of the GAV-group was with 8.0 ± 4.0 points (range: 1 to 18), not significantly different from the DSV group with 9.7 ± 4.3 points

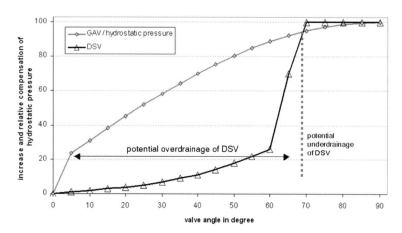

Fig. 1. Position-dependent relative increase in hydrostatic pressure and alteration of the valve's total resistance. The increase in hydrostatic pressure as a patient stands up is equal to the increase and relative compensation of hydrostatic pressure by GAV; therefore, both curves are represented by one line

(range: 1 to 19) (p = 0.140). The same is true for the preoperative ventricular size, which was 0.42 ± 0.08 (range: 0.3 to 0.66) on average for the GAV group and 0.39 ± 0.06 (range: 0.33 t 0.55) for the DSV group (p = 0.114).

Other clinical parameters such as Body Mass Index (p = 0.188), resistance to outflow (p = 0.124), and distribution of idiopathic or secondary NPH did not differ in either group.

However, the DSV group was 9 years older on average (71 ± 9 years) than the GAV group (62 ± 13 years) (p = 0.009). Although the age difference did not create a difference in patient outcome (p = 0.064), a trend towards a worse outcome with higher age should be kept in mind when interpreting the totality of our results.

Postoperative findings

At the time of the last examination, the average Kiefer Index decreased in both groups (DSV, 5.5 ± 3.8 points; GAV, 2.3 ± 2.8 points) compared to preoperative condition (DSV, p < 0.000; GAV, p < 0.000). A striking difference became obvious when comparing the number of shunt responders (Recovery Index > 2), which was 94% in the GAV group compared to 62% in the DSV group (p < 0.000).

After shunt implantation, ventricular size decreased in both groups (DSV, p = 0.035; GAV, p = 0.012). However, ventricular shrinking was more pronounced in the GAV group ($11 \pm 12\%$ Evans Index decrease on average) than in the DSV group ($5 \pm 6\%$ Evans Index decrease on average) (p = 0.016); however, correlation between the extent of ventricular decrease and clinical benefit was not obvious (DSV, p = 0.339; GAV, p = 0.328).

Complications

The overall complication rate was 20% (11% requiring operative revision). Nine patients (16%) experienced some sign (clinical and/or radiological) of overdrainage. However, only 4 (7%) required further intervention: 1 due to intractable orthostatic headache; 2 due to chronic subdural hematoma > 1 cm; and 1 due to an acute subdural hematoma, which required shunt occlusion and burr-hole trephination. In all operative revisions due to overdrainage the hydrostatic compensation was increased (using G-valves with higher hydrostatic compensation or by adding an Aesculap-Miethke Shunt-Assistant into the existing shunt).

In 3 patients, an asymptomatic chronic subdural hematoma < 5 mm could be detected at the first postoperative MRI, which absorbed spontaneously in all 3 patients within 12 weeks without specific treatment. Two patients presented with asymptomatic slit ventricles, but no proximal shunt occlusions have been necessary.

Some clear differences between the 2 groups can be recognized. The relative frequency of overdrainage was 21% in the DSV group, whereas it was 13% in the GAV group. Three of 5 patients in the DSV group who experienced overdrainage required re-operation, while only 1 of 4 patients in the GAV group required an operation to treat overdrainage.

Shunt-infection and underdrainage did not occur. The latter was excluded in all non-responders by infusion tests.

In each group, 1 patient required surgical revision as a result of shunt failure due to distal tube kinking.

Discussion

The optimal opening pressure for NPH is still under discussion [15]. Recently, a study showed better clinical results with low-pressure valves compared to medium-pressure valves [4], but with a higher overdrainage rate (70% versus 30%). In Boon's study [4], conventional differential-pressure valves without any feature to counteract overdrainage were used. The question resulting from this study was whether there are valves available that provide sufficient drainage in the prone position, and also counteract the influence of siphoning in the upright position.

The best technical solution to counteract siphoning is still a matter of debate. CSF flow restriction by very thin tubes [1, 2], flow-regulated conventional high-pressure shunts [7], or on-off valves [2] may reduce overdrainage, but at the cost of sufficient drainage in the lying position [2]. The antisiphon device was the first technical solution to prevent overdrainage [2], but its sensitivity to external pressure (*e.g.* occlusion due to increased subcutaneous pressure temporarily while lying on it or permanently from scars) is critical [2]. Another solution, which does not carry the inherent risk of occlusion by external forces yet also allows sufficient drainage in the prone position and prevents siphoning, is the G-valve [10, 12, 14, 15]. Previous experiences were confined to medium-pressure G-valves only.

Based on Boon's findings [4], it was our hope that G-valves with low prone opening pressures would still get good clinical results without high overdrainage rates.

Overall we had a better clinical outcome in 80% of shunt responders in contrast to earlier studies [3–5, 8, 17, 18]. Accordingly, our study shows that a better outcome can result from improved shunt technology, a fact which has not been found with several other shunt designs [6]. Only the Orbis-Sigma study [7] and studies with medium-pressure G-valves [10, 12, 14, 15] provide similarly favorable data.

The overall complication rate of 20% was at the lower end of the published outcomes [18]. While long-term results were not presented, and with the knowledge that further complications may yet occur, the first postoperative year appears most susceptible to complications [6]. Hopefully, these values will not rise dramatically.

Overdrainage occurred in 16% (9% asymptomatic and only radiologically detected) of our patients with low-pressure G-valves overall. This result is favorable compared with studies of other low-pressure shunts (70%) [4]. It is only slightly better than the 20% value provided for conventional medium- to high-pressure shunts [18], and it is worse than the results with medium-pressure G-valves (4%) [10, 12, 14].

Differences between both constructions

The low-pressure DSV results correspond well with those of medium-pressure G-valves [10, 12, 14], with both groups being among the best mentioned in the NPH literature, while the low-pressure GAV provides an even better outcome.

Frequency and seriousness of overdrainage were different also. Though still better than the values of conventional low-pressure (non-gravity-assisted) valves [4], the low-pressure DSV had an overdrainage rate similar to non-gravity-assisted medium- to high-pressure valves [5, 8, 18]. While not as outstanding as the overdrainage rates of medium-pressure G-valves [10, 12, 14], overdrainage frequency of the low-pressure GAV was significantly lower than those values typically mentioned for non-gravity-assisted medium-pressure valves [2, 4, 5, 18]. The superiority of the low-pressure GAV compared to the low-pressure DSV also holds true regarding the seriousness of overdrainage sequels. Despite the overall better avoidance of overdrainage, ventricular size reduction was more pronounced with the GAV.

The interpretation of these values is difficult. None of these facts alone can explain the different outcomes of both groups. Ventricular shrinking has no correlation between complication rates and outcome, and increasing age provides only a trend toward worse outcome.

The fact that the DSV leads to a lower diminution of the ventricular size, but on the other hand has a higher overdrainage rate, seems contradictory at first view. If we believe excessive ventricular shrinkage is based on excessive CSF drainage and underdrainage results in shunt non-responders, then the GAV tends to drain more physiologically.

Less physiological drainage can result in 2 more frequently occurring situations. First, overdrainage occurs with a greater reduction in ventricular size. Second, shunt non-responders occur at a higher rate (due to underdrainage) with little to no ventricular reduction.

Physiological drainage results in fewer patients with underdrainage and overdrainage complications. With physiological drainage, the change in ventricular size is based on the underlying physiology of the specific patient, and thus it occurs at varying levels. Previous works with endoscopic third ventriculostomy and gravitational shunts show that clinical improvement does not correlate with changes in ventricular size [10].

From a statistical standpoint, because the level of shunt responders was 32% better with the GAV group and the difference in overdrainage was significantly less between the groups, it makes sense to assume that the GAV group would have a greater size reduction. This occurs because the difference in the number of underdraining patients is greater than the number of overdraining patients. By definition, the shunt responders' ventricle reduction is random and does not impact the equation.

Since underdrainage and physiological drainage look the same radiologically, the percentage of non-responders is the best way to tell them apart. If these hypotheses are correct, then differences between the GAV and DSV are to be expected, and are not contradictory. Only functional differences between the valve types can explain this superficial contradiction.

If a patient remains long-term in an intermediate posture between lying flat and <60° upright, the positioning is steep enough to evoke a hydrostatic pressure but not steep enough to close the low-pressure valve of the DSV permanently. The angle of 60° is so critical for the function of the DSV because the low-pressure valve closes at 60–70° and the high-pressure valve remains open for CSF. Lying nocturnally with an elevated head, which may be more common in elderly patients, may be a critical posture. The low-pressure valve of the DSV was selected in accordance with the requirements of the drainage for lying flat, when no hydrostatic pressure occurs. For an intermediate posture, which must evoke hydrostatic pressure, the low opening pressure of the DSV may be so low that a relevant negative intracranial pressure may result.

This is completely different than with the GAV. The counterbalancer technology compensates its opening pressure analogously to the hydrostatic pressure. Accordingly, in every conceivable posture with its distinct hydrostatic pressure, the siphoning is counteracted completely.

Conclusion

Low-pressure counterbalancer G-valves provide better outcomes in NPH than conventional low-pressure valves without gravitational compensation, which, in turn, have been shown to have far better outcomes than higher opening pressure conventional valves. The cost, however, is a slightly higher (but mostly asymptomatic) overdrainage rate than that of the outstanding overdrainage prevention of medium-pressure G-valves. Low-pressure switcher G-valves provide no better clinical results, while providing increased overdrainage rates, than its medium-pressure counterparts. As a result, we recommend the low-pressure GAV as the standard NPH valve.

References

1. Arriada N, Sotelo J (2002) Review: treatment of hydrocephalus in adults. Surg Neurol 58: 377–384
2. Aschoff A, Kremer P, Benesch C, Fruh K, Klank A, Kunze S (1995) Overdrainage and shunt technology. A critical comparison of programmable, hydrostatic and variable-resistance valve and flow-reducing devices. Childs Nerv Syst 11: 193–202
3. Bech RA, Waldemar G, Gjerris F, Klinken L, Juhler M (1999) Shunting effects in patients with idiopathic normal pressure hydrocephalus; correlation with cerebral and leptomeningeal biopsy findings. Acta Neurochir (Wien) 141: 633–639
4. Boon AJ, Tans JT, Delwel EJ, Egeler-Peerdeman SM, Hanlo PW, Wurzer HA, Avezaat CJ, de Jong DA, Gooskens RH, Hermans J (1998) Dutch Normal-Pressure Hydrocephalus Study: randomized comparison of low- and medium-pressure shunts. J Neurosurg 88: 490–495
5. Dauch WA, Zimmermann R (1990) Normal pressure hydrocephalus. An evaluation 25 years following the initial description [in German]. Fortschr Neurol Psychiatr 58: 178–190
6. Drake JM, Kestle JR, Milner R, Cinalli G, Boop F, Piatt J Jr, Haines S, Schiff SJ, Cochrane DD, Steinbok P, MacNeil N (1998) Randomized trial of cerebrospinal fluid shunt valve design in pediatric hydrocephalus. Neurosurgery 43: 294–303; discussion 303–305
7. Hanlo PW, Cinalli G, Vandertop WP, Faber JA, Bogeskov L, Borgensen SE, Boschert J, Chumas P, Eder H, Pople IK, Serlo W, Vitzthum E (2003) Treatment of hydrocephalus determined by the European Orbis Sigma Valve II survey: a multicenter prospective 5-year shunt survival study in children and adults in whom a flow-regulating shunt was used. J Neurosurg 99: 52–57
8. Hebb AO, Cusimano MD (2001) Idiopathic normal pressure hydrocephalus: a systematic review of diagnosis and outcome. Neurosurgery 49: 1166–1184; discussion 1184–1186
9. Kiefer M, Eymann R, Steudel WI (2000) The dynamic infusion test in rats. Childs Nerv Syst 16: 451–456
10. Kiefer M, Eymann R, Meier U (2002) Five years experience with gravitational shunts in chronic hydrocephalus of adults. Acta Neurochir (Wien) 144: 755–767; discussion 767
11. Kiefer M, Eymann R, Komenda Y, Steudel WI (2003) A grading system for chronic hydrocephalus [in German]. Zentralbl Neurochir 64: 109–115
12. Kiefer M, Eymann R, Steudel WI, Strowitzki M (2005) Gravitational shunt management of long-standing overt ventriculomegaly in adult (LOVA) hydrocephalus. J Clin Neurosci 12: 21–26
13. Meier U, Zeilinger FS, Kintzel D (1999) Diagnostic in normal pressure hydrocephalus: A mathematical model for determination of the ICP-dependent resistance and compliance. Acta Neurochir (Wien) 141: 941–947; discussion 947–948
14. Meier U, Kiefer M, Sprung C (2004) Evaluation of the Miethke dual-switch valve in patients with normal pressure hydrocephalus. Surg Neurol 61: 119–127; discussion 127–128
15. Meier U, Kiefer M, Lemcke J (2005) On the optimal opening pressure of hydrostatic valves in cases of idiopathic normal pressure hydrocephalus. Neurosurg Q 15: (in press)
16. Portnoy HD, Amirjamshidi A, Hoffman HJ, Levy LP, Haase J, Scott RM, Zhao YD, Peter J, Krivoy A, Sotelo J (1998) Shunts: which one, and why? Surg Neurol 49: 8–13
17. Richards HK, Seeley HM, Pickard JD (2000) Shunt revisions: Data from the UK shunt registry. Eur J Pediatr Surg [Suppl] 10: 59
18. Vanneste JA (2000) Diagnosis and management of normal-pressure hydrocephalus. J Neurol 247: 5–14

Correspondence: Michael Kiefer, Saarland University, Medical School, Department of Neurosurgery, 66421 Homburg-Saar, Germany. e-mail: ncmkie@med-rz.uni-saarland.de

Brain tissue water content in patients with idiopathic normal pressure hydrocephalus

G. Aygok[1], A. Marmarou[1], P. Fatouros[2], and H. Young[1]

[1] Department of Neurosurgery, Medical College of Virginia Commonwealth University, Richmond, VA
[2] Department of Neuroradiology, Medical College of Virginia Commonwealth University, Richmond, VA

Summary

Relatively little is known regarding the water content of brain tissue in idiopathic normal-pressure hydrocephalus (NPH) patients. The objective of our study was to determine absolute water content non-invasively in hydrocephalic patients, particularly in the anterior and posterior ventricular horns and in the periventricular white matter.

Ten patients who were diagnosed and treated for idiopathic NPH in our clinic were selected for study. Magnetic resonance imaging (MRI) techniques were used to obtain anatomical image slices for quantitative brain water measurements. Apparent diffusion coefficient measures were also extracted from regions of interest.

To our knowledge, this is the first study to confirm that periventricular lucency seen on MRI represents increased water content in the extracellular space that is markedly elevated prior to shunting.

Keywords: Periventricular lucency; NPH; hydrocephalus; vasogenic edema.

Introduction

Morphological changes associated with normal-pressure hydrocephalus (NPH) with attendant ventriculomegaly are poorly understood. In idiopathic NPH, periventricular lucency (PVL) of several degrees is seen in frontal and posterior horns, and although many authors have presumed that this hypodensity on computed tomography (CT) occurs as a result of increased edema, no studies have confirmed that PVL is caused by increased tissue water. Mori *et al.* [5] examined CT images of patients with dilated ventricles in childhood and adult hydrocephalus of various causes and concluded that since the periventricular low density zone disappears shortly after a shunting operation, the lucent zone must represent an increase in tissue water content rather than loss of lipids and protein. Experimental and clinical studies have shown that PVL is reduced following shunting and represents acute edema or chronic cerebrospinal fluid (CSF) retention in the periventricular white matter [6]. Other investigations have shown that there is usually no detectable disturbance of the blood-brain barrier associated with PVL, suggesting that the CSF flows through the intact or disrupted ependyma into the periventricular white matter and gliosis ensues. Moreover, the degree of active tissue disruption is reduced in chronic stages of hydrocephalus when CSF is often reduced [9]. In contrast, other investigators used spin-echo magnetic resonance imaging (MRI) methods in patients with multiple sclerosis and hydrocephalus. They found that some degree of PVL is present in most patients with no other evidence of intracranial pathology and concluded that mild PVL is a normal finding and should not be considered indicative of either demyelinating disease or hydrocephalus [10].

In light of this controversy, the objective of this study was to utilize a specialized MRI technique that is capable of measuring brain tissue water in absolute terms, and to determine the brain tissue water level in areas of PVL in patients with idiopathic NPH. A second objective was to characterize PVL using apparent diffusion coefficient (ADC) to identify whether hyperdense regions are associated with either a vasogenic or cellular component of edema.

Methods

Obtaining the water map

As part of the management protocol, a standard head MRI was obtained for diagnosis and possible shunt treatment for all patients admitted to the NPH program. Ventriculomegaly was evident in all patients with associated elements of the NPH triad of gait disturbance, incontinence, and dementia. During the procedure, a water

CT, T1 and T2 MRI

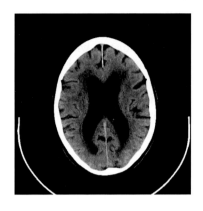 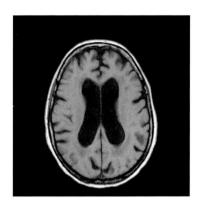 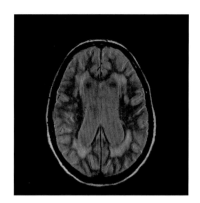

MRI Water, DWI-MRI and ADC-MAP

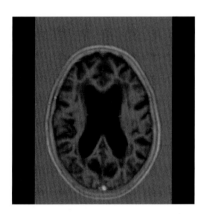 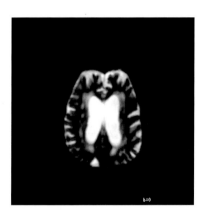 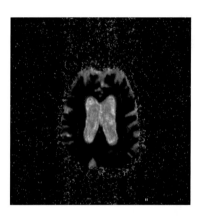

Fig. 1. CT, TI, and T2 MRI, MRI water, diffusion-weighted imaging (*DWI*) MRI, and ADC map of idiopathic normal-pressure hydrocephalus patient

map [2] was obtained from the region of brain exhibiting maximum PVL. Briefly, this method is based on the principle that nuclear magnetic resonance relaxation times T_1 and T_2 in brain tissue are influenced by the dynamic structure and amount of water present. Water occurring in the so-called "free" phase is associated with a long relaxation time. Motion-restricted or "bound" water constituting approximately 20% of the total tissue water has a much shorter relaxation time. The measured longitudinal relaxation time T_1 is a weighted average of the "free" and "bound" water. The patient imaging protocol involves acquisition of 5 phase-sensitive inversion recovery MRI images of a selected anatomical slice. The standard imaging head coil of a 1 T MRI unit (Siemens Magnetom, Erlanger, Germany) was used with TR/TE = 2.5 s/20 ms, TI = 150, 400, 800, 1300, 1900 ms, 5 mm slice thickness, and 128 × 256 matrix. From the sequential inversion recovery images, a pure T_1 image was calculated by a fitting procedure as described elsewhere [2]. The accuracy of this imaging protocol for determining brain tissue T_1 values was checked with calibration standards of known and comparable relaxation times. From the T_1 map, a water image or "water map" is calculated by means of the following equation: $1/f_w = A + B/T_1$ which is derived from the fast exchange 2-state model where T_1 is the local tissue longitudinal relaxation time and f_w is the corresponding total tissue water fraction defined as f_w = (water weight/total brain weight) [1]. This model assigns the tissue water in either a "free" or a "bound" state. The constants A and B are dependent on the hydration fraction k – the ratio of bound water fraction to solid tissue component – as well as the relaxation properties of the free and bound water fractions. In the resulting water map, the intensity of each pixel provides the water content at that location expressed in g water/g tissue.

Measurement of ADC

Diffusion-weighted imaging

Diffusion-weighted imaging was performed on a 1.0 T whole body clinical imager (Siemens Vision MR system, Siemens Medical Solutions, Malvern, PA) equipped with 15 mT/m gradients using spin-echo sequences with and without diffusion sensitizing gradients. These pulse sequences generated an ADC trace image using a single shot technique with 3 b values: 0, 500 and 1000 s/mm^2. Typically, 20 slices were generated, each with 5 mm slice thickness, 1.5 mm gap, 230 mm field of view, 96 × 128 matrix, and TE 100 ms. The ADC trace was generated to obviate issues arising from tissue anisotropy and head orientation. Regional measurements of ADC were extracted from the ADC maps.

Brain measurements

After positioning in the magnet, standard T1- and T2-weighted pulse sequences were used to produce images in the axial, coronal, and sagittal planes in order to identify the slice with maximum PVL. Regions of interest (ROIs) were positioned in the regions of the anterior and posterior horns of the lateral ventricles (Fig. 1). In addition, ROIs were positioned in the white matter distant from regions of PVL.

Water content in %gmH$_2$O/g of tissue was measured in each ROI for all patients. Similarly, ADC values for the same ROIs were extracted.

Results

Patient demographics

The study cohort consisted of 10 patients, 6 female and 4 male, with a clinical diagnosis of idiopathic NPH and an average age of 75 ± 4 years. Seven patients had all 3 elements of the triad, and 3 patients had gait and incontinence disturbance only. All patients were eventually shunted with a programmable ventriculo-atrial or ventriculo-peritoneal shunt.

The severity of ventricular enlargement was assessed using the frontal horn index, a radiographic calculation. The index of our patient group ranged from 0.34 to 0.49 and averaged 0.41 ± 0.05.

Periventricular water content, anterior and posterior horns

Regions of PVL were evident on all scans but most prominent in the anterior and posterior horns of the lateral ventricles (Fig. 1). Water content of the left anterior and posterior horns averaged 79.00 ± .06 and 77.40 ± 0.39% g H$_2$O/g tissue respectively. This difference was not significant. Similarly, on the right anterior and posterior horns, the water content averaged 80.90 ± 0.048 and 76.3 ± 0.04 respectively. The difference between anterior and posterior was also not significant.

White matter tissue distal to the horns was used as control, and averages were similar in left and right hemispheres. Distal white matter water content measured 72.00 ± 0.3 in the left hemisphere and 72.4 ± 0.027 in the right hemisphere. The water content of left anterior and posterior horns (79.0 ± 0.06, 77.4 ± .039) was greater than the distal white matter (72.00 ± 0.02) in the same hemisphere, and these differences were highly significant ($p < 0.005$). Water content of the right anterior and posterior horns (80.90 ± 0.048, 76.3 ± 0.04) was also higher than the same hemisphere distal white matter (72.4 ± 0.027) ($p < 0.005$).

Periventricular water content, central region of lateral ventricle

PVL was also evident in the central region of the lateral ventricles but was not as marked as the anterior and posterior horns. Water content in left and right central regions of the lateral ventricles averaged 74.50 ± 0.02 and 75.5 ± 0.02% g H$_2$O/g tissue, respectively. Although water content was increased in these central regions, the difference between left and right central regions versus the left and right distal white matter tissue (72.0 ± 0.03, 72.4 ± 0.02) was not different. In summary, the regions of PVL in the anterior and posterior horns of the lateral ventricles were associated with high water content.

ADC in the regions of high PVL

Control values for ADC in the distal white matter equaled 0.72 ± 0.07, which was within the normal tissue range. In contrast, ADC in regions of PVL were high, averaging 1.45 ± 0.15 in the left anterior horn and 1.34 ± 0.21 in the left posterior horn. The left central regions were also high and averaged 1.14 ± 0.24. Similarly, right anterior and posterior horns (1.53 ± 0.18, 1.25 ± 0.2) and the central periventricular regions (1.16 ± 0.22) had higher ADC values compared with distal white matter ADC values (0.79 ± 0.13). The difference between regions of PVL and distal white matter ADC's was highly significant ($p < 0.0005$), indicating a predominantly extracellular edema.

Discussion

Our study shows that tissue water content is increased in regions of PVL by as much as 7% compared to distal white matter and the increase is associated with increased ADC signifying predominantly extracellular edema. Milhorat et al. [4] described edemas associated with hydrocephalus as follows: osmotic edema, which results from an unfavorable osmotic gradient between the plasma and interstitial fluid across an intact blood-brain barrier; compressive edema, which results from obstruction of the interstitial fluid bulk-flow pathways; and hydrocephalic edema, which results from obstruction of CSF bulk-flow pathways. In the hydrocephalic type of edema, distension of the collecting channels proximal to the block leads to retrograde flooding of the extracellular compartment with formation of periventricular edema. It is not clear which components of edema account for the PVL because it may consist of both compressive edema and hydrocephalic edema, as characterized by Milhorat. It is clear that edema resides in the extracellular space based on our ADC values. In the kaolin model of obstructive hydrocephalus in newborn rabbits, there was no difference between hydrocephalic animals and normal controls in experiments involving penetration of Evans blue into the tissue [7]. Thus, retrograde flow of CSF in the case of obstructive hydrocephalus could not be confirmed in that study. On the other hand, James et al. [3] showed loss of cilia and ependymal cells over the ventricular surface and abnormal small supra-ependymal cells in experimental studies of Silastic-induced hydrocephalus in primates. Transmission and scanning electron microscopy markedly increased communicating pathways between the ventricular lumen and the brain parenchyma. The study suggested that PVL in patients is caused by increased extracellular space; however, the lack of recognition of PVL by CT in chronic NPH may be due to changes in distribution and limits of resolution. There is no detectable disturbance of the blood-brain barrier in regions of PVL, according to Weller et al. [9]. Experimental histologic and ventriculo-perfusion studies suggest that CSF flows through the disrupted ependyma into the periventricular white matter when the normal pathways of drainage are occluded. Weller proposed that absorption of fluid into the blood probably occurs through the periventricular blood vessels as an alternative pathway of CSF drainage. Tissue damage in the edematous periventricular white matter occurs and gliosis ensues.

There is reasonable consensus that shunting patients with PVL gradually reduces the PVL, suggesting that blockage of CSF absorption is responsible for accumulation of periventricular edema. However, Tamaki et al. [8] studied NPH patients using MRI and found that there was no change in T1 and T2 in PVL regions after shunting. He concluded that MRI findings could not predict outcome of shunt surgery in patients with NPH. However, clinical improvement after surgery was associated with reduction in the irregular type of PVL located around the frontal horns.

In summary, the significance, development, and resolution of periventricular edema is still debatable. Further studies are necessary to define the mechanisms of PVL more clearly and why PVL changes occur after shunting. Nevertheless, our studies confirm that PVL represents an increase in water and ADC measures are consistent with extracellular edema.

References

1. Bell BA, Smith MA, Kean DM, McGhee CN, MacDonald HL, Miller JD, Barnett GH, Tocher JL, Douglas RH, Best JJ (1987) Brain water measured by magnetic resonance imaging. Correlation with direct estimation and changes after mannitol and dexamethasone. Lancet 1: 66–69
2. Fatouros PP, Marmarou A, Kraft KA, Inao S, Schwarz FP (1991) In vivo brain water determination by T1 measurements: effect of total water content, hydration fraction, and field strength. Magn Reson Med 17: 402–413
3. James AE Jr, Flor WJ, Novak GR, Ribas JL, Parker JL, Sickel WL (1980) The ultrastructural basis of periventricular edema: preliminary studies. Radiology 135: 747–750
4. Milhorat TH (1992) Classification of the cerebral edemas with reference to hydrocephalus and pseudotumor cerebri. Childs Nerv Syst 8: 301–306
5. Mori KT, Murata T, Nakano Y, Handa H (1977) Periventricular lucency in hydrocephalus on computerized tomography. Surg Neurol 8: 337–340
6. Murata T, Handa H, Mori T, Nakano Y (1981) The significance of periventricular lucency on computed tomography: experimental study with canine hydrocephalus. Neuroradiology 20: 221–227
7. Nyberg-Hansen R, Torvik A, Bhatia R (1975) On the pathology of experimental hydrocephalus. Brain Res 95: 343–350
8. Tamaki N, Nagashima T, Ehara K, Shirakuni T, Matsumoto S (1990) Hydrocephalic oedema in normal-pressure hydrocephalus. Acta Neurochir [Suppl] 51: 348–350
9. Weller RO, Mitchell J (1980) Cerebrospinal fluid edema and its sequelae in hydrocephalus. Adv Neurol 28: 111–123
10. Zimmerman RD, Fleming CA, Lee BC, Saint-Louis LA, Deck MD (1986) Periventricular hyperintensity as seen by magnetic resonance: prevalence and significance. AJR Am J Roentgenol 146: 443–450

Correspondence: Anthony Marmarou, Department of Neurosurgery, Virginia Commonwealth University Medical Center, 1001 East Broad Street, Suite 235, Richmond, VA 23298-0508, USA. e-mail: amarmaro@vcu.edu

Predictors of outcome in patients with normal-pressure hydrocephalus

U. Meier[1], J. Lemcke[1], and U. Neumann[2]

[1] Department of Neurosurgery, Unfallkrankenhaus Berlin, Berlin, Germany
[2] Institute of Radiology, Unfallkrankenhaus Berlin, Berlin, Germany

Summary

From 1982 until 2000 we examined 200 patients diagnosed with normal-pressure hydrocephalus (NPH) in a prospective study. From the patients who were surgically treated by a shunt implantation we could re-examine 155 (78%) at a mean time interval of 7 months after the operation. NPH differed in severity according to the results of the intrathecal infusion test in an early state NPH (without brain atrophy) and late state NPH (with brain atrophy).

In our study, we focused on the possible predictors: patient age; length of disease; clinical signs including gait ataxia, dementia, and bladder incontinence; idiopathic vs. secondary origin; implanted valve type and the resistance of the valve to cerebrospinal fluid outflow. In 80 patients without cerebral atrophy and a short course of disease (<1 year), a slight amount of dementia and an implanted Miethke Dualswitch-Valve were significant predictors for a positive postoperative outcome. The outflow resistance measured in the intrathecal infusion test showed only minimal relevance for outcome. Seventy-five patients with cerebral atrophy had a better outcome when dementia was not present, outflow resistance was above 20 mmHg/mL/min, the CSF tap-test was positive, and a Miethke Dualswitch-Valve was implanted.

Keywords: Hydrocephalus; outflow resistance; Miethke Dualswitch-Valve; shunt.

Introduction

This reappraisal of the clinical features of normal-pressure hydrocephalus (NPH), which was first described by Hakim and Adams [1, 17] in 1965, is based on physiological and pathophysiological findings pertaining to cerebrospinal fluid dynamics, various diagnostic methods for the detection and characterization of NPH, as well as clinical findings. The classic therapeutic approach, which is to implant a shunt that siphons excess cerebrospinal fluid, is associated with a host of complications. Consequently, indications for surgery need to be reevaluated and reliable predictors of postoperative improvement need to be defined. A review of the literature indicates that, despite improved shunt valve technology, treatment outcomes have not improved to any significant degree, and the prognosis of NPH is still associated with unsatisfactory treatment outcomes. Nonetheless, an increased rate of improvement in the clinical course of NPH can still be achieved through careful preoperative diagnosis, individualized and appropriate therapies, and preoperative modeling of valve characteristics and properties.

Materials and methods

Two hundred patients with proven NPH were treated with shunt surgery and evaluated prospectively over 12 years (1982–2000) in the Department of Neurosurgery at the Berlin-Friedrichshain Hospital and the Neurosurgery Clinic of the Unfallkrankenhaus, Berlin. The workup consisted of clinical examinations, computed tomography and/or magnetic resonance imaging studies, intrathecal infusion tests (measurement of resistance to cerebrospinal fluid outflow [Rout] and intracranial compliance), and cerebrospinal fluid tap tests. In the intrathecal infusion test, intracranial pressure (ICP) is measured continuously, while defined changes are made in cerebrospinal fluid volume [13, 41]. Our patients were classified as early stage or late stage cases of hydrocephalus in accordance with the results of the intrathecal infusion test [21–28]. While no cerebral atrophy was detected in the patient group with early stage NPH (n = 80), cerebral atrophy was noted in patients with late stage disease (n = 75), a sign of advanced disease for this group. At the time of the investigations, the mean age of the 122 men and 78 women was 52 years.

Predictors and statistics

In 155 cases (78%), a follow-up examination as well as an analysis of predictors of prognosis was performed 7 months postoperatively. The following potential predictors of prognosis were evaluated during the study: age, length of case history, idiopathic vs. secondary origin, gait difficulty, dementia, urinary incontinence, level of Rout, level of compliance, results of the cerebrospinal tap test, type and pressure level of implanted valve, possible valve infections, and postoperative changes in ventricular size. The predictors of prognosis were compared statistically to the course of the disease using Pearson's χ^2 test.

Clinical grading

The results of the follow-up examinations of 155 patients were evaluated 7 months postoperatively using the Black Grading Scale for Shunt Assessment [2] and Kiefer and Steudel's Clinical Grading Scale for NPH [19]. The graduated study results were divided into 3 groups as follows: The group with a positive and very good clinical course (NPH recovery rate of ≥5 points); the group with a satisfactory clinical course (NPH recovery rate of ≥2 points); the group with a poor clinical course (NPH recovery rate of <2 points).

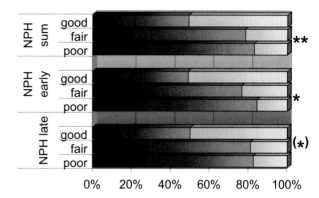

Fig. 2. Outcome versus resistance to CSF outflow (Rout in mmHg/mL/min) (χ^2 Pearson; **p = 0.01; *p = 0.05; (*)p = 0.1). *Rout* Resistance to cerebrospinal fluid outflow. ■ R(out) 13–15; ■ R(out) 15–20; □ R(out) higher than 20

Results

Figure 1 summarizes the clinical course, as indicated by the NPH recovery rate, for the 155 NPH patients who underwent a follow-up examination. Overall improvement rate was 81%. In patients with early stage NPH (i.e., no cerebral hypertrophy), a case history of less than 12 months duration was associated with a more positive clinical course (p = 0.01) than a history extending over more than 12 months.

The level of Rout was a key predictor in classifying our patients as early stage or late stage cases of hydrocephalus (Fig. 2). In the group as a whole (n = 155), patients with Rout > 15 mmHg/mL/min showed a significantly better clinical course (p = 0.01) than did patients with lower Rout. Although patients in the early stage NPH group (n = 80) manifested Rout exceeding 15 mmHg/mL/min and had a better prognosis following shunt surgery, this difference was not statistically significant (p = 0.1). Patients with late stage NPH (n = 75) (i.e., patients with cerebral atrophy) in whom Rout > 20 mmHg/mL/min was recorded on an intrathecal infusion test, showed a significantly better prognosis (p = 0.05) following shunt surgery than did patients with lower Rout (Fig. 2). In neither patient group were the results of the cerebrospinal fluid tap significant predictors of postoperative therapeutic outcome. This was also true of the working pressure of the implanted valve (above vs. below 100 mm H$_2$O), postoperative decrease in ventricular dilatation (as measured with the Evans index), and adequate treatment of infections. On the other hand, the presence (p = 0.01) and severity (p = 0.01) of dementia constituted significant predictors of prognosis for both patient groups. The prognosis for patients with no memory deficit was more favorable than for patients with short-term memory deficit, whose prognosis, in turn, was more positive than for patients with acute dementia.

Hydrostatic valves constitute an advance in valve technology. Patients with a Miethke Dualswitch-Valve (Christoph Miethke GMBH & Co. KG, Berlin, Germany) (M-DSV; n = 61) showed a more favorable clinical course than did patients with Cordis standard valves (n = 76) (Cordis Corporation, Miami, FL) or Cordis Orbis Sigma valves (n = 18) (Cordis Corporation, Miami, FL). This trend was evident in the group as a whole (n = 155) and in the patient group with late stage NPH (n = 75). In both of these groups, the prognosis following placement of an M-DSV was more favorable (p = 0.01) than was the case with the other implants (Fig. 3). This same phenomenon was also noted (p = 0.05) in patients with early stage NPH (n = 80). Thus, the type of valve implanted (M-DSV) has predictive value for the postoperative prognosis of patients with NPH (Fig. 3).

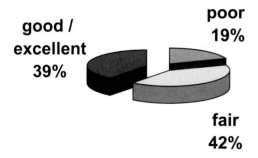

Fig. 1. Patient outcome with NPH

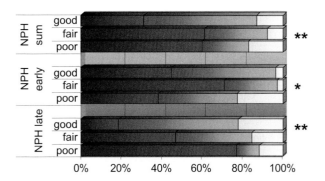

Fig. 3. Outcome versus valve mechanism outflow (χ^2 Pearson; **p = 0.01; *p = 0.05). *M-DSV* Miethke Dualswitch-valve; *C-OSV* Cordis Orbis Sigma valve. ■ standard valve; ▨ M-DSV */**; ☐ C-OSV

Discussion

Review of the literature revealed that no data on the incidence and prevalence of NPH has been published to date. Clarfield [7] evaluated 32 studies with a total of 2889 patients in whom the clinical features of dementia were observed, but only 1.6% of these patients had NPH. A number of studies have found NPH to be more prevalent in males. There is little data available for age distribution in NPH. An analysis by Dauch and colleagues [8] revealed that the oldest NPH patients are in the seventh decade, but that 25% are under 50 years of age. In other words, NPH can occur in adults at any age. Several authors have published reports on NPH in children and teenagers, although its existence in these age groups is disputed [14, 20, 31–33].

Outcomes

The overall improvement rate in patients with NPH ranges from 31% to 96% (mean: 53%) [15, 34, 39]. One multicenter study of 166 patients with NPH [38] found that in the long-term, 36% of patients showed permanent clinical improvement while 21% had only moderate remission of symptoms following shunt placement. The overall improvement in our own patients following shunt placement was 81% (Fig. 1). A recent meta-analysis [18] demonstrated an improvement rate of 24% to 100% (mean = 59%) in idiopathic NPH and improved outcome in a significant mean of 29% of cases (range 10% to 100%). These results are consistent with the positive clinical courses observed in our patient group as a whole (mean 39%).

Length of case history

The mean age of our patient cohort was 52 years, making this a relatively young group for NPH. This may be attributable to the referral criteria used by our colleagues and to a relatively large proportion of patients (48%) with secondary hydrocephalus. In contrast to the findings of other authors [32], the clinical picture of the NPH we saw was not primarily that of a morphological disorder occurring mostly in older patients, but was instead a functional disorder that can easily develop as a secondary disorder at any age. One review of the literature [28] found that NPH is twice as prevalent in males as in females. No published data is available pertaining to duration of symptoms that could be used for comparison with our own patient groups. Thus, there is no basis for comparing the more positive postoperative prognosis in early stage NPH patients whose case histories are less than 1 year in duration.

Resistance to outflow (Rout)

In our patients, all Rout of <10 mmHg/mL/min was categorized as physiological, whereas Rout measuring between 10 and 13 mmHg/mL/min was classified as a limit value, and all Rout measuring > 13 mmHg/mL/min was classified as pathological. The disproportionately low normal values for Rout obtained by Ekstedt [11], Fuhrmeister [13] and Shapiro et al. [34] should be reevaluated in the light of recent studies. Our own assessment criteria were confirmed by the results obtained in 1993 by Morgan et al. [30], which were consistent with the findings of Børgesen and Gjerris [6] as well as Tans and Poortvliet [36]. Boon et al. [5] demonstrated that the positive predictive value of Rout increases as Rout rises, which is consistent with our data (Fig. 2). These authors propose that Rout of 18 mmHg/mL/min be adopted in order to avoid reducing measurement sensitivity of the intrathecal infusion test. We found that classifying patients into an early or late stage NPH group allowed more individualized values for Rout as predictors of prognosis.

Dementia

The presence and acuteness of dementia constitutes a significant predictor of clinical outcome for all groups of NPH patients with shunts. Dementia has been reported in 80% of cases of NPH [3, 8, 12, 15, 35]. Although acute dementia is more prevalent in late stage NPH and in older published reports is characterized as the cardinal symptom of this disorder, dementia is nonetheless of little relevance in reaching a differential diagnosis of NPH. In the literature, gait difficulty occurs in 86% of cases of NPH [3, 8, 12, 15, 35]. In our own experience, this symptom has proven to be the most reliable sign for the diagnosis of NPH [26]. It is not, however, a valid predictor of therapeutic outcome following shunt placement.

Valve mechanisms

No overdrainage complications were observed in our early stage NPH patients in whom Cordis standard valves were implanted. In addition, prolonged bench testing [37] has shown the Cordis standard valve to be stable, although some have reported overdrainage problems with this valve. The results of a Dutch multicenter study [4] revealed a better clinical course for NPH patients with Cordis standard low-pressure valves than for patients with Cordis standard medium-pressure valves; however, this advantage came at the cost of a high overdrainage rate (71%) for the low-pressure valves as compared with 34% for their medium-pressure counterparts. These postoperative complications had no clinical repercussions for either of the groups studied, although the rate of complications was disproportionately high. The Cordis standard valves are relatively inexpensive, but even a single complication resulting from overdrainage can result in burdensome expenditures, thereby negating any potential savings.

In terms of ensuring a positive prognosis for patients with early stage NPH, the M-DSV is the treatment of choice (Fig. 3). In our own patients with late stage NPH, predominantly Orbis Sigma valves, Cordis standard valves, and the Medos Hakim valve (Codman & Shurtleff, Inc., Raynham, MA) were implanted, and more recently, the M-DSV. Unlike the Cordis standard valves, the Orbis Sigma valves provide flow-based drainage utilizing physiological cerebrospinal hydrodynamics within a range of pressure gradients. In patients with hydrocephalic atrophy, this is particularly important in preventing overdrainage. In the Orbis Sigma valves, overdrainage occurred in 5 patients (25%), 3 of whom (15%) also presented with subdural hematoma. These results are consistent with another study [40] in which subdural hematoma was found to be a sign of overdrainage following placement of Orbis Sigma valves in patients with NPH. On the other hand, Decq *et al.* [9] observed a lower rate of overdrainage with Orbis Sigma valves relative to conventional differential pressure valves; however, as this study was not confined to patients with NPH, it is of limited relevance. It remains to be seen whether the Orbis Sigma II valve will deliver better postoperative results than its predecessor.

The Cordis Orbis Sigma valve is not suitable for implantation in patients with NPH owing to its association with high rates of overdrainage and subdural hematoma. The disadvantage of the Cordis standard valves for patients with late stage NPH is that if the patient stands or sits up abruptly, the valve responds accordingly, thereby inducing a siphoning effect in the already atrophic brain. An anti-siphon device would theoretically alleviate this problem, but the flow resistance of the entire valve system would then increase. In a randomized pediatric study, Drake *et al.* [10] noted no significant change in therapeutic outcome following implantation of either Delta, Orbis Standard, or Sigma valves. This study generated considerable controversy, but it is not relevant to the specific clinical features of NPH. Overdrainage was seen in 3 of our patients (6%) in whom the Cordis standard valve was implanted. Two of these patients (4%) manifested the clinical signs associated with this complication.

With its 2 different closure mechanisms that operate in parallel, the M-DSV [20, 29] constitutes an important step forward in solving the problems associated with the position of the patient's craniospinal axis. In our own series, implantation of the M-DSV has thus far resulted in only 2 cases (5%) of chronic subdural hematoma. In 1 of these patients, the hematoma was resorbed within 1 month, and a revision operation had to be performed on the other patient. The post-implantation results obtained with the M-DSV demonstrate its high degree of suitability for the management of NPH in general, and late stage NPH in particular. For the latter disorder, the valve is highly reliable and safe due to the fact that problems of overdrainage are rarely seen (Fig. 3).

Conclusions

- NPH should be the suspected differential diagnosis in patients who present with dementia (cardinal symptom: gait ataxia) [16].
- A case history of less than 1-year duration, as well as

relatively mild dementia or the absence of memory deficit, should be regarded as predictors of a positive prognosis following shunt therapy in patients with NPH.
- The prognosis is positive for early stage NPH patients in whom the intrathecal infusion test shows Rout in excess of 15 mmHg/mL/min, or in late stage NPH patients (with cerebral atrophy) with Rout greater than 20 mmHg/mL/min on the intrathecal infusion test.
- In view of the superior therapeutic outcomes, hydrostatic valves should be implanted in patients with NPH [20, 28].

References

1. Adams RD (1975) Recent observations on normal pressure hydrocephalus. Schweiz Arch Neurol Neurochir Psychiatr 116: 7–15
2. Benzel EC, Pelletier AL, Levy PG (1990) Communicating hydrocephalus in adults: Prediction of outcome after ventricular shunting procedures. Neurosurgery 26: 655–660
3. Black PM, Ojemann RG, Tzouras A (1985) CSF shunts for dementia, incontinence, and gait disturbance. Clin Neurosurg 32: 632–651
4. Boon AJW, Tans JThJ (1998) Dutch normal pressure hydrocephalus (NPH) study. Randomised comparison of low and medium pressure shunts. Acta Neurochir [Suppl] 71: 401 (Abstract PO-3-103)
5. Boon AJ, Tans JT, Delwel EJ, Egeler-Peerdeman SM, Hanlo PW, Wurzer HA, Avezaat CJ, de Jong DA, Gooskens RH, Hermans J (1997) Dutch normal-pressure hydrocephalus study: prediction of outcome after shunting by resistance to outflow of cerebrospinal fluid. J Neurosurg 87: 687–693
6. Børgesen SE, Gjerris F (1982) The predictive value of conductance to outflow of CSF in normal pressure hydrocephalus. Brain 105: 65–86
7. Clarfield AM (1988) The reversibile dementias: do they reverse? Ann Intern Med 109: 476–486
8. Dauch WA, Zimmermann R (1990) Normal pressure hydrocephalus. An evaluation 25 years following the initial description [in German]. Fortschr Neurol Psychiatr 58: 178–190
9. Decq P, Barat JL, Duplessis E, Leguerinel C, Gendrault P, Keravel Y (1995) Shunt failure in adult hydrocephalus: flow-controlled shunt versus differential pressure shunts – a cooperative study in 289 patients. Surg Neurol 43: 333–339
10. Drake JM, Kestle JR, Milner R, Cinalli G, Boop F, Piatt J Jr, Haines S, Schiff SJ, Cochrane DD, Steinbok P, MacNeil N (1998) Randomized trial of cerebrospinal fluid shunt valve design in pediatric hydrocephalus. Neurosurgery 43: 294–303; discussion 303–305
11. Ekstedt J (1978) CSF hydrodynamic studies in man. 2 Normal hydrodynamic variables related to CSF pressure and flow. J Neurol Neurosurg Psychiatry 41: 345–353
12. Fisher CM (1982) Hydrocephalus as a cause of disturbances of gait in the elderly. Neurology 32: 1358–1363
13. Fuhrmeister U (1985) Liquorabflußwiderstand und intrakranielle Elastizität bei akuten und chronischen Erkrankungen des Subarachnoidalraums. Habilitationsschrift. Würzburg, Germany
14. Gaab MR, Koos WT (1984) Hydrocephalus in infancy and childhood: diagnosis and indication for operation. Neuropediatrics 15: 173–179
15. Gustafson L, Hagberg B (1978) Recovery in hydrocephalic dementia after shunt operation. J Neurol Neurosurg Psychiatry 41: 940–947
16. Gutzmann H (1988) Senile Demez vom Alzheimer-Typ. Enke Verlag, Stuttgart
17. Hakim S, Adams RD (1965) The special clinic problem of symptomatic hydrocephalus with normal cerebrospinal fluid pressure. Observations on cerebrospinal fluid hydrodynamics. J Neurol Sci 2: 307–327
18. Hebb AO, Cusimano MD (2001) Idiopathic normal pressure hydrocephalus: a systematic review of diagnosis and outcome. Neurosurgery 49: 1166–1184; discussion 1184–1186
19. Kiefer M, Steudel WI (1994) Moderne diagnostik und therapie des hydrocephalus beim älteren menschen. Saarländ Ärzteblatt 47: 498–505
20. Kiefer M, Eymann R, Mascaros V, Walter M, Steudel WI (2000) Significance of hydrostatic valves in therapy of chronic hydrocephalus [in German]. Nervenarzt 71: 975–986
21. Meier U (2002) The grading of normal pressure hydrocephalus. Biomed Tech (Berl) 47: 54–58
22. Meier U, Bartels P (2002) The importance of the intrathecal infusion test in the diagnosis of normal pressure hydrocephalus. J Clin Neurosci 9: 260–267
23. Meier U, Michalik M, Reichmuth B (1987) Computerized tomography and infusion test as a simultaneous study method in post-traumatic hydrocephalus [in German]. Psychiatr Neurol Med Psychol (Leipz) 39: 754–758
24. Meier U, Reichmuth B, Zeilinger FS, Lehmann R (1996) The importance of xenon-computed tomography in the diagnosis of normal pressure hydrocephalus. Int J Neuroradiol 2: 153–160
25. Meier U, Zeilinger FS, Kintzel D (1999) Diagnostic in normal pressure hydrocephalus: A mathematical model for determination of the ICP-dependent resistance and compliance. Acta Neurochir (Wien) 141: 941–947; discussion 947–948
26. Meier U, Zeilinger FS, Kintzel D (1999) Signs, symptoms and course of normal pressure hydrocephalus in comparison with cerebral atrophy. Acta Neurochir (Wien) 141: 1039–1048
27. Meier U, Kiefer M, Bartels P (2002) The ICP-dependency of resistance to cerebrospinal fluid outflow: a new mathematical method for CSF-parameter calculation in a model with H-TX rats. J Clin Neurosci 9: 58–63
28. Meier U, Kiefer M, Sprung C (2002) The Miethke Dual-Switch valve in patients with normal pressure hydrocephalus. Neurosurg Q 12: 114–121
29. Miethke C, Affeld K (1994) A new valve for the treatment of hydrocephalus. Biomed Tech (Berl) 39: 181–187
30. Morgan MK, Johnston IH, Spittaler PJ (1993) A ventricular infusion technique for the evaluation of treated and untreated hydrocephalus. In: Avezaat CJJ, van Eijndhoven JHM, Maas AIR, Tans JTJ (eds) Intracranial pressure VIII. Springer, Berlin Heidelberg New York Tokyo, pp 821–823
31. Mori K (1990) Hydrocephalus – revision of its definition and classification with special reference to "intractable infantile hydrocephalus." Childs Nerv Syst 6: 198–204
32. Nichtweiss M, Heetderks G, Rosenthal D (1988) Diagnosis of idiopathic so-called normal pressure hydrocephalus. Data from clinical findings, computerized tomography and epidural pressure measurement [in German]. Nervenarzt 59: 267–273
33. Raftopoulos C, Massager N, Balériaux D, Deleval J, Clarysse S, Brotchi J (1996) Prospective analysis by computed tomography and long-term outcome of 23 adult patients with chronic idiopathic hydrocephalus. Neurosurgery 38: 51–59

34. Shapiro K, Marmarou A, Shulman K (1980) A method for predicting PVI in normal patients. In: Shulman K, Marmarou A, Miller JD, Becker DP, Hochwald GM, Brock M (eds) Intracranial pressure IV. Springer, Berlin Heidelberg New York Tokyo pp 85–90
35. Spanu G, Sangiovanni G, Locatelli D (1986) Normal-pressure hydrocephalus: twelve years experience. Neurochirurgia 29: 15–19
36. Tans JTJ, Poortvliet DCJ (1989) Significance of compliance in adult hydrocephalus. In: Gjerris F, Børgesen SE, Sørensen PS (eds) Outflow of Cerebrospinal Fluid, Alfred Benzon Symposium 27. Munksgaard, Copenhagen, pp 272–279
37. Trost HA, Claussen G, Heissler HE, Gaab MR (1993) Long term in vitro test results of various new and explanted hydrocephalus shunt valves. In: Avezaat CJJ, van Eijndhoven JHM, Maas AIR, Tans JTJ (eds) Intracranial pressure VIII. Springer, Berlin Heidelberg New York Tokyo, pp 901–904
38. Vanneste J, Augustijn P, Dirven C, Tan WF, Goedhart ZD (1992) Shunting normal-pressure hydrocephalus: do the benefits outweigh the risks? A multicenter study and literature review. Neurology 42: 54–59
39. Vila F, Ferrer E, Isamaat F (1980) ICP monitoring in benign intracranial hypertension. In: Shulman K, Marmarou A, Miller JD, Becker DP, Hochwald GM, Brock M (eds) Intracranial pressure IV. Springer, Berlin Heidelberg New York Tokyo, pp 549–552
40. Weiner HL, Constantini S, Cohen H, Wisoff JH (1995) Current treatment of normal-pressure hydrocephalus: comparison of flow-regulated and differential-pressure shunt valves. Neurosurgery 37: 877–884
41. Zeilinger FS (1996) Der Normaldruck-Hydrocephalus – Eine Literaturstudie. Medizinische Dissertation. Humboldt-Universität, Berlin, Germany

Correspondence: Ullrich Meier, Department of Neurosurgery, Unfallkrankenhaus Berlin, Warener Strasse 7, 12683 Berlin, Germany. e-mail: ullrich.meier@ukb.de

On the optimal opening pressure of hydrostatic valves in cases of idiopathic normal-pressure hydrocephalus: a prospective randomized study with 123 patients

U. Meier[1], M. Kiefer[2], U. Neumann[3], and J. Lemcke[1]

[1] Department of Neurosurgery, Unfallkrankenhaus Berlin, Berlin, Germany
[2] Department of Neurosurgery, University Homburg/Saar, Homburg, Germany
[3] Institute of Radiology, Unfallkrankenhaus Berlin, Berlin, Germany

Summary

Does the opening pressure of hydrostatic shunts influence the clinical outcome for patients suffering from idiopathic normal-pressure hydrocephalus (NPH)? Between September 1997 and January 2003, 123 patients with idiopathic NPH were surgically treated by implanting a hydrostatic shunt at the Departments of Neurosurgery of the Unfallkrankenhaus Berlin and the University Homburg/Saar. As part of a prospective randomized study, all patients were examined preoperatively, postoperatively, and 1 year after the intervention.

Forty-three percent of the patients showed a very good outcome, 25% good outcome, 20% fair outcome, and 12% poor outcome 1 year after the shunt implantation. Patients treated with an opening pressure rating of 50 mmH$_2$O in the low-pressure stage of the gravitational valve showed a better outcome than those with an opening pressure of 100 or 130 mmH$_2$O. According to present knowledge, hydrostatic shunts with an opening pressure of 50 mmH$_2$O for the low-pressure stage are the best option for patients with idiopathic NPH. Due to the prompt switching function when the patient changes posture (lying down, standing, sitting, slanting etc.), the Miethke gravity-assisted valve (GAV) is more suitable in such cases than the Miethke Dual-Switch valve (DSV).

Keywords: Hydrocephalus; outcome; shunt operation; valves.

Introduction

Even after an exact diagnosis, surgical therapy remains complication-prone, which has a strong effect on the outcome and improvement rate for normal-pressure hydrocephalus (NPH) patients. The "hydraulic mismanagement" [16] of hydrocephalus shunts involves non-physiological functioning either in the form of overdrainage in a patient standing upright or relative underdrainage when the patient is lying down. These effects are technically unavoidable with conventional shunts, since they only allow a compromise setting between the requirements for the different postures [2]. While most younger patients show a considerable tolerance to non-physiological intracranial pressure (ICP), this presents a major problem for more elderly patients, many of whom already suffer from vasculopathy [13, 30, 31].

Clinical materials and methods

Between September 1997 and January 2003, 123 patients with idiopathic NPH were treated surgically at the neurosurgical clinics of the Unfallkrankenhaus Berlin-Marzahn and the University Homburg/Saar by implanting hydrostatic shunts (96 × Miethke Dual-Switch valve [DSV] Aesculap, and 26 × Miethke gravity-assisted valve [GAV] Aesculap, Ch. Miethke GmbH & Co. KG, Berlin, Germany). Over the period 1997 to 1999, mainly Miethke DSVs with a 130 mmH$_2$O opening pressure stage were implanted at the Unfallkrankenhaus Berlin-Marzahn. In 2000 and 2001, the same valves were implanted, but with an opening pressure rating of 100 mmH$_2$O. In 2002 and 2003, DSVs with a 50 mmH$_2$O low-pressure stage were implanted at Berlin-Marzahn, while Miethke GAVs with a 50 mmH$_2$O low-pressure stage were implanted at the University Clinics Homburg/Saar. Thus, we did not introduce any selection of the 3 shunt groups according to clinical and/or radiological criteria. The mean age of the 67 male (55%) and 55 female patients (45%) was 67 years (29 to 83 years).

Diagnostics

Following clinical examination with a diagnosis of gait ataxia and additional symptoms [27] and the detection of extended ventricles by means of neuroradiological imaging procedures, the intrathecal infusion test was carried out. The computer-aided dynamic infusion test, which was performed using the constant-flow technique at an infusion rate of 2 mL/min via lumbar puncture with the patient lying down, was used for calculating the ICP-dependent resistance to outflow (R$_{out}$) of cerebrospinal fluid (CSF) [28, 29]. A R$_{out}$ of more than

13 mmHg/min/mL was defined as pathological [24]. Immediately after the infusion test, a diagnostic drainage of at least 60 mL CSF was carried out. If clinical symptoms improved within the following 2 to 3 days, shunt implantation was indicated. If symptoms, especially gait ataxia, did not improve, external lumbar drainage was carried out for 2 to 3 days. If this led to a symptomatic improvement, a gravitational valve was implanted as a ventriculo-peritoneal shunt [26, 32].

Clinical grading

Results of the clinical examinations were graded according to the Black grading scale for shunt assessment and the NPH recovery rate based on the clinical grading for NPH introduced by Kiefer [19]. All graded clinical results were classed into 4 groups. The group with very good outcomes was characterized by an NPH recovery rate ≥ 7 points, good outcomes ≥ 5 points, fair outcomes ≥ 2 points, and poor outcomes < 2 points.

Statistical analysis

Statistical evaluations were carried out with a Fisher exact test at an error probability of p = < 0.05.

Results

Preoperatively, the average figure on the Kiefer scale (Fig. 1) did not show any statistically significant difference between the 3 valve groups, meaning that the severity of the symptoms caused by idiopathic NPH can be considered as roughly uniform.

Outcome

Results of the shunt operation for the 123 patients showed very good outcome in 43%, good outcome in 25%, fair outcome in 20%, and poor outcome in 12% one year after shunt implantation. Thus, 83 patients had good to very good success from therapy, 25 patients had fairly successful treatment, 15 patients did not benefit from surgery.

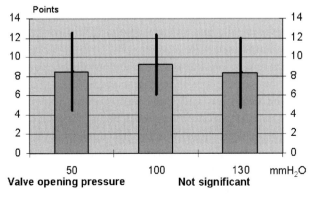

Fig. 1. Clinical symptoms, preoperative

Complications

We found an infection rate of 2%. All patients that experienced a shunt infection had the implant removed. After remediation of infection, all 3 patients had a new shunt implanted. Dislocation of a ventricular catheter necessitated revision in 3 patients, and in 5 patients the drainage tube needed revision because of dislocation into the abdominal wall (overall dislocation rate: 7%). The abdominal catheter was torn off just under or out of the DSV in 3 patients (2%), which necessitated a shunt revision. Thus, the valve-independent complication rate was 11% (14 patients).

Four patients (3%) receiving a DSV showed an apparent reduction of ventricular size by computed tomography (CT), as well as narrow subdural hygromas as a sign of overdrainage. In all these patients, the subdural hygromas were resorbed within 6 to 8 weeks without clinical symptoms. Another 4 patients (3%) developed chronic subdural hematomas with clinical symptoms such as nausea, vomiting, and headache. The chronic subdural hematomas were surgically treated through burr hole trepanation before new shunt systems were implanted. One patient received a shunt revision involving the implantation of a DSV with a higher low-pressure rating. Two patients underwent revision with a Codman Medos-Hakim programmable valve plus a shunt-assistant as a gravitational unit, and through implantation of a Codman Medos-Hakim programmable valve without a gravitational unit in another hospital, which led to a rapid improvement of the above symptoms. Four patients (3%) showed deterioration despite shunt implantation, and a slight increase in ventricle size on CT imaging. The intrathecal infusion test confirmed underdrainage from the perspective of CSF dynamics. For these 4 patients, a shunt revision was performed using a DSV of a lower low-pressure rating. After that, 2 patients showed significant improvement of symptoms. The other 2 patients did not improve, even after the second intervention. Thus, the rate of valve-related complications was 10% (12 patients).

Overall shunt-related morbidity was 21%. Within the study period, 4 patients died for reasons unrelated to shunt implantation (coronary thrombosis or tumor illness unknown at the time of the intervention, 5 to 7 months postoperatively). Another patient died from pulmonary embolism 6 days after the shunt operation, despite application of the correct thrombosis prophylaxis. Operation-related mortality amounted to 1%,

Table 1. *Valve-related complications versus valve configuration*

	Valve opening pressure rating mmH$_2$O		
	50	100	130
Underdrainage, n	0	2	2
Overdrainage, asymptomatic, n	2	2	0
Overdrainage, symptomatic, n	3	1	0

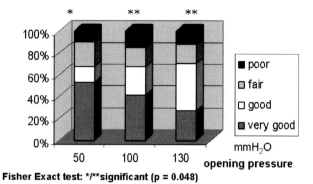

Fig. 2. Outcome versus valve configuration

while shunt-related mortality was 0. Despite the frequency of symptomatic overdrainage with a DSV with a valve pressure rating of 50 mmH$_2$O (6% DSV+GAV 50 mmH$_2$O or 12% DSV 50 mmH$_2$O) and of cases of underdrainage with a DSV with valve pressure ratings of 100 and 130 mmH$_2$O (5% DSV 100 mmH$_2$O and 6% DSV 130 mmH$_2$O), these clusters are not significant (Table 1). However, it must be noted that overdrainage only occurred with the DSV but never with the GAV. The valve-independent complications, too, did not correlate with the valve opening pressure ratings.

Valve opening pressure rating

Outcome after 1 year was evaluated in relationship to the opening pressure rating of the implanted valves (Fig. 2). Twenty-seven patients (53%) with a gravitational valve rating of 50 mmH$_2$O for the low-pressure stage showed a very good outcome, and 7 patients (14%) showed a good outcome. Twelve patients (23%) and 5 patients (10%), respectively, showed a fair or poor outcome. In the patient group with DSVs with a 100 or 130 mmH$_2$O low-pressure stage, 25 patients (35%) experienced very good outcome, 24 patients (33%) good outcome, 13 patients (18%) fair outcome, and 10 patients (14%) showed no improvement 1 year after operation. Thus, outcome for idiopathic NPH patients with a 50 mmH$_2$O GAV was significantly better than for patients that had a GAV with a opening pressure rating of 100 or 130 mmH$_2$O (Fig. 2).

Discussion

GAVs are currently the gold standard in the surgical treatment of chronic hydrocephalus patients [3, 9, 40]. However, which valve opening pressure rating is best for patients with idiopathic NPH remains debatable.

Outcome

The general improvement rates for patients with NPH after a shunt operation, as quoted in the literature [1, 4, 5, 7, 8, 14, 18, 20–37, 42, 46], varies between 31% and 96%, with an average of 53%.

Complications

Grumme *et al.* [15] reported a mortality of 0% to 6% and an incidence of overdrainage effects in 6% to 20% of NPH patients that received a shunt. Hebb *et al.* [18] gave a complication rate of 38%, a revision rate of 22%, and a combined figure of 6% for the incidence of lasting neurological deficits and death. The data from our study (1% mortality and 7% overdrainage) are below the international findings [10, 36]. Our results concerning the peri- and postsurgical complication rate (21%), too, are at the lower end of the range of 20% to 40% given by Vanneste [43]. The low incidence of overdrainage (7% according to radiological criteria, 3% symptomatic) compared to the results achieved with conventional shunts should be emphasized. Similar good results are reported in a European multicenter study of patients with hydrocephalus internus of varying causes, who had an Orbis Sigma II flow-controlled valve implanted [17].

Review of the literature on underdrainage following shunts produced inconsistencies and discrepancies in the definition and classification of postoperative complications. Some authors calculated the total number of complications; others took into account only the mechanical complications, neglecting infections. Drake *et al.* [12] clarified underdrainage as complications involving occlusions and overdrainage as subdural hygromas and slit-ventricle syndrome. In our opinion, one should only refer to a functional underdrainage if the selected valve pressure rating was too high, the actual opening pressure is higher than the nominal pressure rating due to manufacturing errors or changes of the valve functionality in vivo, or if CSF drainage is reduced due to an increase in intraabdominal pressure caused by adiposity or CSF resorption insufficiency, e.g., in a pseudocyst. In the British Shunt Registry [38], underdrainage is the dominant (52%) complication. In the same report, overdrainage as a cause for postoperative complications is rare (3%). In contrast, Scandinavian groups assert [7] that 80% of all shunt complications are caused by overdrainage. Concerning underdrainage, comparison with other valves is difficult because only a few publications offer a clear statement about this complication, which is not sufficiently well-defined or recorded in most studies of NPH. Clinically, cases of torn-off abdominal catheters just under or out of the DSV, as found in 3 patients (2%), can only be discussed as a material and/or technological problem. Explanations for this problem should be a matter for the developer or manufacturer.

Valve opening pressure rating

For patients undergoing a second operation because of overdrainage or suspected underdrainage, the implantation of a Codman Medos-Hakim programmable valve with an additional anti-siphon device as a shunt-assistant for the vertical posture of the patient could be the alternative. This treatment was chosen for the majority of our revision operations because of symptomatic overdrainage. For NPH patients, too, Codman Medos-Hakim valves have received positive assessments [35, 39, 41]. A Dutch multicenter study [6, 11] reported significant improvement in patients with NPH if low-pressure standard valves had been implanted compared to implantation of medium-pressure standard valves. However, the price for this advantage was a higher overdrainage rate (73% compared to 34%). No statement was made concerning the clinical relevance of this postoperative complication, which is described as subdural hygromas in that study. Such rates of complications are exceptionally high. In line with the findings of the Dutch multicenter study [6, 11], the outcome for patients with idiopathic NPH who had implantation of a GAV with a 50 mmH$_2$O low-pressure stage turned out better than for patients provided with a 100 or 130 mmH$_2$O low-pressure stage GAV. At 10% versus 4%, the advantage of the GAVs implanted by us compared to the Dutch study is apparent.

Complications versus valve type

We were surprised that all overdrainage-type complications in patients of the 50 mmH$_2$O valve group occurred only with the DSV, but never with the GAV. We believe the difference is related to functional properties of the gravitational components of the 2 valves. The gravitational unit of the GAV is already activated at a posture of 30° to 40° against horizontal, whereas the DSV only activates at about 60° to 70°. Despite having informed them otherwise, patients with a DSV implanted probably used 2 pillows for sleep or took afternoon naps in an inclined posture. Therefore, the gravitational component of the DSV was not activated, resulting in a higher incidence of overdrainage with this type of valve.

Conclusion

Hydrostatic shunt systems with an opening pressure of 50 mmH$_2$O in the low-pressure stage are the optimal therapy for patients suffering from idiopathic NPH. Due to the prompt switching function as soon as the patient changes posture (lying down, standing, sitting, slanting etc.), the Miethke GAV is more suitable in such cases than the Miethke DSV. Can programmable gravitational valves contribute to even better outcomes while minimizing valve-related complications?

References

1. Aguas-Valiente J, Martinez-Manas R, Ferrer-Rodriguez E (1997) Diagnostic and therapeutic criteria of adult normal-pressure hydrocephalus. Descriptive multicenter national study. Rev Neurol 25: 27–36 [in Spanish]
2. Aschoff A (1994) In-vitro-Testung von Hydrocephalus-Ventilen. [Dissertation] (Habilitationsschrift) Heidelberg, Germany
3. Aschoff A, Kremer P, Benesch C, Fruh K, Klank A, Kunze S (1995) Overdrainage and shunt-technology. A critical comparison of programmable, hydrostatic and variable-resistance valves and flow-reducing devices. Childs Nerv Syst 11: 193–202
4. Bech RA, Waldemar G, Gjerris F, Klinken L, Juhler M (1999) Shunting effects in patients with idiopathic normal pressure hydrocephalus; correlation with cerebral and leptomeningeal biopsy findings. Acta Neurochir (Wien) 141: 633–639
5. Benzel EC, Pelletier AL, Levy PG (1990) Communicating hydrocephalus in adults: prediction of outcome after ventricular shunting procedures. Neurosurgery 26: 655–660
6. Boon AJ, Tans JT, Delwel EJ, Egeler-Peerdeman SM, Hanlo PW, Wurzer HA, Hermans J (1999) Dutch Normal-Pressure

Hydrocephalus study: the role of cerebrovascular disease. J Neurosurg 90: 221–226
7. Borgesen SE (1984) Conductance to outflow of CSF in normal pressure hydrocephalus. Acta Neurochir (Wien) 71: 1–45
8. Caruso R, Cervoni L, Vitale AM, Salvati M (1997) Idiopathic normal-pressure hydrocephalus in adults: result of shunting correlated with clinical findings in 18 patients and review of the literature. Neurosurg Rev 20: 104–107
9. Czosnyka Z, Czosnyka M, Richards HK, Pickard JD (2002) Laboratory testing of hydrocephalus shunts – conclusion of the UK shunt evaluation programme. Acta Neurochir (Wien) 144: 525–538
10. Decq P, Barat JL, Duplessis E, Leguerinel C, Gendrault P, Keravel Y (1995) Shunt failure in adult hydrocephalus: flow-controlled shunt versus differential pressure shunts – a cooperative study in 289 patients. Surg Neurol 43: 333–339
11. de Jong DA, Delwel EJ, Avezaat CJ (2000) Hydrostatic and hydrodynamic considerations in shunted normal pressure hydrocephalus. Acta Neurochir (Wien) 142: 241–247
12. Drake JM, Kestle JR, Milner R, Cinalli G, Boop F, Piatt J Jr, Haines S, Schiff SJ, Cochrane DD, Steinbok P, MacNeil N (1998) Randomized trial of cerebrospinal fluid shunt valve design in pediatric hydrocephalus. Neurosurgery 43: 294–305
13. Frim DM, Goumnerova LC (2000) In vivo intracranial pressure dynamics in patients with hydrocephalus treated by shunt placement. J Neurosurg 92: 927–932
14. Fukuhara T, Luciano MG, Brant CL, Klauscie J (2001) Effects of ventriculoperitoneal shunt removal on cerebral oxygenation and brain compliance in chronic obstructive hydrocephalus. J Neurosurg 94: 573–581
15. Grumme T, Kolodziejczyk D (1995) Komplikationen in der Neurochirurgie. Band 2 Kraniale, zerebrale und neuropädiatrische Chirurgie. Blackwell, Berlin, pp 534–540
16. Hakim CA (1985) The physics and physicopathology of the hydraulic complex of the central nervous system. The mechanics of hydrocephalus and normal pressure hydrocephalus. [Dissertation] Massachusetts Institute of Technology
17. Hanlo PW, Cinalli G, Vandertop WP, Faber JA, Bogeskov L, Borgesen SE, Boschert J, Chumas P, Eder H, Pople IK, Serlo W, Vitzthum E (2003) Treatment of hydrocephalus determined by the European Orbis Sigma Valve II survey: a multicenter prospective 5-year shunt survival study in children and adults in whom a flow-regulating shunt was used. J Neurosurg 99: 52–57
18. Hebb AO, Cusimano MD (2001) Idiopathic normal pressure hydrocephalus: a systematic review of diagnosis and outcome. Neurosurgery 49: 1166–1186
19. Kiefer M, Eymann R, Meier U (2002) Five years experience with gravitational shunts in chronic hydrocephalus of adults. Acta Neurochir (Wien) 144: 755–767
20. Kosteljanetz M, Nehen AM, Kaalund J (1990) Cerebrospinal fluid outflow resistance measurements in the selection of patients for shunt surgery in the normal pressure hydrocephalus syndrome. A controlled trial. Acta Neurochir (Wien) 104: 48–53
21. Krauss JK, Droste DW, Vach W, Regel JP, Orszagh M, Borremans JJ, Tietz A, Seeger W (1996) Cerebrospinal fluid shunting in idiopathic normal-pressure hydrocephalus of the elderly: effect of periventricular and deep white matter lesions. Neurosurgery 39: 292–300
22. Lee EJ, Hung YC, Chang CH, Pai MC, Chen HH (1998) Cerebral blood flow velocity and vasomotor reactivity before and after shunting surgery in patients with normal pressure hydrocephalus. Acta Neurochir (Wien) 140: 599–605
23. Meier U (1998) Der intrathekale Infusionstest als Entscheidungshilfe zur Shunt-Operation beim Normaldruckhydrozephalus. [Dissertation] (Habilitation) Medizinische Fakultät (Charite) der Humboldt-Universität zu Berlin, Germany
24. Meier U, Bartels P (2001) The importance of the intrathecal infusion test in the diagnostic of normal pressure hydrocephalus. Eur Neurol 46: 178–186
25. Meier U, Kintzel D (2002) Clinical experiences with different valve systems in patients with normal-pressure hydrocephalus: evaluation of the Miethke dual-switch valve. Childs Nerv Syst 18: 288–294
26. Meier U, Mutze S (2004) Correlation between decreased ventricular size and positive clinical outcome following shunt placement in patients with normal-pressure hydrocephalus. J Neurosurg 100: 1036–1040
27. Meier U, Zeilinger FS, Kintzel D (1999) Signs, symptoms and course of disease in normal pressure hydrocephalus in relation to cerebral atrophy. Acta Neurochir (Wien) 141: 1039–1048
28. Meier U, Zeilinger FS, Kintzel D (1999) Diagnostic in normal pressure hydrocephalus: A mathematical model for determination of the ICP-dependent resistance and compliance. Acta Neurochir (Wien) 141: 941–948
29. Meier U, Kiefer M, Bartels P (2002) The ICP-dependency of resistance to cerebrospinal fluid outflow: A new mathematical method for CSF-parameter calculation in a model with H-Tx rats. J Clin Neurosci 9: 58–63
30. Meier U, Kiefer M, Sprung C (2002) The Miethke Dual-Switch valve in patients with normal pressure hydrocephalus. Neurosurg Q 12: 114–121
31. Meier U, Kiefer M, Sprung C (2003) Normal-Pressure Hydrocephalus: Pathology, Pathophysiology, Diagnostics, Therapeutics, and Clinical Course. PVV Science Publications, Ratingen, Germany
32. Meier U, Kiefer M, Sprung C (2004) Evaluation of the Miethke Dual-Switch valve in patients with normal pressure hydrocephalus. Surg Neurol 61: 119–128
33. Meier U, König A, Miethke C (2004) Predictors of outcome in patients with normal pressure hydrocephalus. Eur Neurol 51: 59–67
34. Miethke C, Affeld K (1994) A new valve for the treatment of hydrocephalus. Biomed Tech (Berl) 39: 181–187
35. Miyake H, Ohta T, Kajimoto Y, Nagao K (2000) New concept for the pressure setting of a programmable pressure valve and measurement of in vivo shunt flow performed using a microflowmeter. Technical note. J Neurosurg 92: 181–187
36. Pollack IF, Albright AL, Adelson PD (1999) A randomized, controlled study of a programmable shunt valve versus a conventional valve for patients with hydrocephalus. Hakim-Medos Investigator Group. Neurosurgery 45: 1399–1411
37. Raftopoulos C, Massager N, Baleriaux D, Deleval J, Clarysse S, Brotchi J (1996) Prospective analysis by computed tomography and long-term outcome of 23 adult patients with chronic idiopathic hydrocephalus. Neurosurgery 38: 51–59
38. Richards HK, Seeley HM, Pickard JD (2000) Shunt revisions: Data from the UK shunt registry. Eur J Pediatr Surg 10 [Suppl] I: 59
39. Rohde V, Mayfrank L, Ramakers VTh, Gilsbach JM (1998) Four-year experience with the routine use of the programmable Hakim valve in management of children with hydrocephalus. Acta Neurochir (Wien) 140: 1127–1134
40. Schwerdtfeger K, Mautes A, Kiefer M, et al (2003) Fortschritte in der Neurochirurgie durch kontrollierte klinische Studien [Conference paper] Dtsch Ärzteblatt 100: B2823–2828
41. Sindou M, Guyotat-Pelissou I, Chidiac A, Goutelle A (1993) Transcutaneous pressure adjustable valve for the treatment of hydrocephalus and arachnoid cysts in adults. Experiences with 75 cases. Acta Neurochir (Wien) 121: 135–139

42. Sotelo J, Rubalcava MA, Gomez-Liata S (1995) A new shunt for hydrocephalus that relies on CSF production rather than on ventricular pressure: initial clinical experiences. Surg Neurol 43: 324–331
43. Vanneste JA (2000) Diagnosis and managment of normal-pressure hydrocephalus. J Neurol 247: 5–14

Correspondence: Ullrich Meier, Department of Neurosurgery, Unfallkrankenhaus Berlin, Warener Strasse 7, 12683 Berlin, Germany. e-mail: ullrich.meier@ukb.de

Outcome predictors for normal-pressure hydrocephalus

M. Kiefer, R. Eymann, and **W. I. Steudel**

Saarland University, Medical School, Department of Neurosurgery, Homburg-Saar, Germany

Summary

The objective of this prospective study was to find outcome predictors for better selection for treatment of normal-pressure hydrocephalus (NPH) patients. A total of 125 patients were evaluated and provided with a gravitational shunt.

Cerebrospinal fluid hydrodynamics provided better predictive values if an algorithm to shunt all patients with a pressure/volume index of <30 mL or resistance to outflow > 13 mmHg/mL × min was used. In general, outcome became worse with increasing anamnesis duration, worse preoperative clinical state, and increasing comorbidity. If one of these parameters was lower than a critical value, the shunt-responder rate was about 90% and the normally negative influence of older age was not seen. The well-known paradigm of a worse prognosis with NPH is not the result of the hydrocephalus etiology itself, but the consequence of a typical accumulation of negative outcome predictors as a consequence of the misinterpretation of normal aging and delayed adequate treatment.

Keywords: Normal-pressure hydrocephalus; outcome; resistance to outflow; pressure/volume index; comorbidity; shunts.

Introduction

Responder rates after treatment of normal-pressure hydrocephalus (NPH) are suboptimal [9, 22]. Up to 70% of patients with NPH suffer further neurodegenerative diseases, which may cause symptoms similar to those of NPH [1, 20]. On one hand these patients may benefit from shunting, but on the other, the decision whether to shunt or not is more demanding.

One objective of this prospective study was to find outcome predictors which improve with the indication for shunting. Another objective was to re-evaluate the paradigm that NPH, and especially idiopathic normal-pressure hydrocephalus (iNPH), has a worse outcome than other forms of chronic hydrocephalus [22].

Materials and methods

Patients, clinical management

A total of 125 patients (68 female, 57 male) were included in the study: 64 iNPH, 19 secondary NPH (sNPH), 42 non-communicating hydrocephalus (aqueduct stenosis: n = 40; Chiari I malformation: n = 2). Patients with at least 2 symptoms of Hakim's triad and an Evans Index > 0.3 received gravitational valves: 82 Aesculap-Miethke Dual-Switch valve (Christoph Miethke GMBH & Co. KG, Potsdam, Germany), 18 Aesculap-Miethke gravity-assisted valve (Christoph Miethke GMBH & Co. KG, Potsdam, Germany), and 25 received a combination of an adjustable Codman-Hakim valve (Codman and Shurtleff, Inc., Raynham, MA) and Aesculap-Miethke Shunt-Assistant (Christoph Miethke GMBH & Co. KG, Potsdam, Germany) [12, 17]. Complications were treated as described earlier [12]. Average follow-up was 4.3 ± 2.4 years.

Indication policy

Shunt indication was based on intracranial pressure monitoring. If mean intracranial pressure was >20 mmHg or B-wave frequency > 50%, shunt implantation was indicated. Additionally, a constant-volume infusion test was performed, but without influence on indication for surgery [12, 16].

Documentation

Each patient was examined clinically and by magnetic resonance imaging or cranial computed tomography preoperatively, at 1 and 12 months postoperatively, and yearly thereafter. To document the clinical state and the ventricular size, we used the Kiefer Index (KI) [12, 16], the Recovery Index [12, 16] and the Evans Index. Comorbidity was documented according to a new grading scale (Table 1).

Statistics

Mann-Whitney U, Spearman, analysis of variance, and Kruskal-Wallis tests at a significance level of α = 0.05.

Table 1. *Comorbidity Index. Each mentioned symptom or disease has to be assigned according to the indicated parameter-values (1–3 points). The sum represents the individual comorbidity index*

	1 Point	2 Points	3 Points
Vascular risk factors	– Hypertension	– Diabetes mellitus	
Peripheral vascular occlusions	– Aortofemoral bypass – stent – ICA stenosis	– Peripheral vascular occlusion	
Cerebrovascular disease	– Posterior circulation insufficiency	– Vascular encephalopathy – TIA – PRIND	– Cerebral infarct
Heart	– Arrhythmia – Valvular disease – Heart failure (coronal) – Stent – Aortocoronary bypass – Infarction		
Others		– Parkinson's disease	

ICA Internal carotid artery; *PRIND* prolonged reversible ischemic neurologic deficit; *TIA* transient ischemic attack.

Results

Preoperative clinical state

The worse the clinical state was at admission, the worse the clinical outcome ($p = 0.003$). A mild clinical obstruction at admission (0–5 KI points) indicated an excellent prognosis (89% shunt-responder rate). In contrast, if severe preoperative obstructions were present (>12 KI points), responder rate dropped to 64%.

Comorbidity

From a statistical viewpoint, hypertension ($p = 0.015$), cerebrovascular diseases ($p < 0.001$), peripheral/coronary vascular occlusion ($p < 0.001$), diabetes mellitus ($p < 0.001$), and Parkinson's disease ($p < 0.003$) had a negative influence on outcome, while non-coronary heart failure ($p = 0.226$) and a history of alcohol abuse ($p = 0.738$) had none. However, each disease alone could not be taken as an independent variable of outcome, but only the combination of several diseases. To value the influence of all comorbidities, the usage of the Comorbidity Index (CMI) has been valuable. Three CMI points seemed to represent a critical value. Patients with 0–3 CMI points had a shunt-responder rate > 90% (age-independent), while beyond 3 CMI points the chance for a clinical benefit from shunting decreased to 65% ($p = 0.002$) overall with worse values for older patients.

Age

In general, outcome was worse with increasing age ($p < 0.001$); however, age was not an independent outcome predictor. A clear correlation between comorbidity and age was found ($p = 0.002$), because comorbidity normally increased with age. The influence of age on outcome has been mediated mainly by comorbidities.

Anamnesis duration

Shunt-responder rate decreased the longer the period between first hydrocephalus symptoms and treatment initiation ($p = 0.002$). Shunt-responder rates were higher (86%) for a shorter anamnesis (critical value: 1 year) than with a longer one ($p < 0.001$). Comorbidity plays an important role: at 0–3 CMI points the typical influence of the anamnesis duration seemed meaningless (responder rate > 90% independent from the anamnesis duration). In contrast, patients with >3 CMI points and a short anamnesis had responder rates of 80%, while those with an anamnesis of >1 year had responder rates < 60%.

Cerebrospinal fluid (CSF) hydrodynamics

Using CSF resistance to outflow (ROF) > 13 mmHg/mL × min as an independent outcome predictor alone, the positive predictive value (PPV) was 75%, the negative predictive value (NPV) 40%, with a sensitivity of 96% and a specificity of 7%. At a critical value > 18 mmHg/mL × min, the specificity increased to 33%, PPV did not increase, while NPV and sensitivity decreased. Similar predictive values could be found regarding pressure/volume index (PVI) alone at varying critical values as an independent outcome predictor. However, combining both, according to an algorithm whereupon all patients with a PVI < 30 mL or a ROF > 13 mmHg/mL × min are shunted, provided a specificity and sensitivity > 90%, PPV ~ 80%. Only NPV remained at a clinically unsatisfying 60%.

Impact of hydrocephalus etiology

NPH patients responded to shunt surgery in 71% of the cases (iNPH, 66% responders; sNPH, 82% responders), while 87% of "non-NPH" responded; however, this narrowed perspective may lead to a wrong assumption about the meaning of hydrocephalus etiology. Shunt-responder rates of NPH and non-NPH with similar favorable preconditions such as mild (KI < 6 points) preoperative obstruction (p = 0.643), short (<1 year) anamnesis (p = 0.114), mild (CMI < 3 points) comorbidity state (p = 0.082), were not significantly different. NPH patients with favorable preconditions had a similar or better prognosis than non-NPH patients with worse preconditions (responder rates, CMI value: iNPH 83%, sNPH 85%, non-NPH 66%; anamnesis duration: iNPH/sNPH 82%, non-NPH 80%; KI value: iNPH 84%, sNPH 86%, non-NPH 81%). Some influence of the hydrocephalus etiology could be seen with worse preconditions only. With a longer anamnesis (p = 0.0109) and a worse preoperative clinical condition (p = 0.021), NPH had a 20% lower responder rate than non-NPH sufferers with similarly worse preconditions. However, for a worse CMI (>3 points), NPH and non-NPH shared the same worse prognosis of 66% and 65% of responders (p = 0.856), respectively. Considering independent outcome predictors, the paradigm of a worse NPH prognosis no longer held true.

We found a worse NPH prognosis compared to non-NPH simply from an accumulation of negative influences in the NPH group such as older age (p < 0.001), longer anamnesis (p < 0.001), worse clinical state at admission (p < 0.001), more comorbidity (p < 0.001).

Discussion

An important finding of this study is that iNPH does not mean a worse prognosis, as is often assumed [9, 22]. When taking into account similar favorable preconditions, the post-interventional outcome of NPH may be as good as with non-communicating hydrocephalus or better compared to non-communicating hydrocephalus with worse preconditions. From a pathophysiological viewpoint, the worse prognosis of NPH remained an enigma. Apparently it is not the hydrocephalus type which results in worse clinical outcome in NPH patients, but the generally worse precondition they have when first seen. Because the first symptoms of NPH are often neglected or misinterpreted as a natural consequence of older age, a drop in rehabilitation chances occurs.

Whether old age automatically results in a worse prognosis is controversial [7, 10, 18, 19]. According to our data, age must not be the determining factor for outcome; rather, it is a pseudo-correlation between age and outcome mediated by comorbidity, which normally becomes worse with older age. Accordingly, elderly persons (>80 years) with a low CMI (0–3 points) could have an excellent prognosis (responder rates > 90%).

The clinical state at admission is typically not a discussion point. Mostly, specific symptoms are mentioned as good or bad outcome predictors instead of a global approach to the clinical state. We showed that grading allows an outcome prediction. A milder preoperative obstruction (5 KI points seems to represent a critical value) due to hydrocephalus allows better rehabilitation chances. Our grading system contrasts with earlier assumptions focusing on complete Hakim's triad and outcomes related to purity of the triad [2, 14, 20, 22]. However, other findings point in the same direction as ours in that a worse prognosis occurs with advanced mental deficits or the presence of urinary incontinence [2, 14, 20].

Additional diseases, especially cerebrovascular diseases, vascular occlusions, and Parkinson's disease, are important comorbidities [3, 13, 21], which can be seen in our data as well. There was previously no method established to value comorbidities as a prognostic instrument such as our Comorbidity Index, which allows gathering and valuing the influence on outcome of all additional diseases. Beyond a critical value of 3 CMI points, prognosis becomes worse even if other outcome predictors point to a favorable prognosis.

Our data indicates that both disputed viewpoints, an existing [3, 19] and a non-existing [2] influence of anamnesis duration on outcome, can be correct. Generally, prognosis becomes worse with longer anamnesis; however, this no longer holds true under favorable preoperative preconditions (CMI value < 4 points, KI < 6 points).

While the value of the ROF or PVI has been studied extensively with inconsistent results [4–6, 9, 11, 15, 16, 22], the recent trend to elevate the critical ROF value [3] may not be supported by our data. The infrequent use of PVI is astonishing, because recent data suggests that compliance is the initially disturbed parameter at the beginning of hydrocephalus, while ROF is only an epiphenomena [8]. Against this background, the mentioned algorithm may be found to be a better outcome predictor than those typically mentioned [9].

References

1. Bech RA, Waldemar G, Gjerris F, Klinken L, Juhler M (1999) Shunting effects in patients with idiopathic normal pressure hydrocephalus; correlation with cerebral and leptomeningeal biopsy findings. Acta Neurochir (Wien) 141: 633–639
2. Black PM (1980) Idiopathic normal-pressure hydrocephalus. Results of shunting in 62 patients. J Neurosurg 52: 371–377
3. Boon AJ, Tans JT, Delwel EJ, Egeler-Peerdeman SM, Hanlo PW, Wurzer HA, Avezaat CJ, de Jong DA, Gooskens RH, Hermans J (1997) Dutch normal-pressure hydrocephalus study: prediction of outcome after shunting by resistance to outflow of cerebrospinal fluid. J Neurosurg 87: 687–693
4. Czosnyka M, Whitehouse H, Smielewski P, Simac S, Pickard JD (1996) Testing of cerebrospinal compensatory reserve in shunted and non-shunted patients: a guide to interpretation based on an observational study. J Neurol Neurosurg Psychiatry 60: 549–558
5. Delwel EJ, De Jong DA, Avezaat CJJ (1993) The relative prognostic value of CSF outflow resistance measurement in shunting for normal pressure hydrocephalus. In: Avezaat CJJ, van Eijndhoven JHM, Maas AIR, Tans JTJ (eds) Intracranial pressure VIII. Springer, Berlin Heidelberg New York, pp 816–820
6. Gjerris F, Borgesen SE (1992) Pathophysiology of the CSF circulation. In: Crockard A, Hayward R, Hoff JT (eds) Neurosurgery: the scientific basis of clinical practice, 2nd edn. Blackwell Scientific Publications, Boston, pp 146–175
7. Greenberg JO, Shenkin HA, Adam R (1977) Idiopathic normal pressure hydrocephalus – a report of 73 patients. J Neurol Neurosurg Psychiatry 40: 336–341
8. Greitz D (2004) Radiological assessment of hydrocephalus: new theories and implications for therapy. Neurosurg Rev 27: 145–165; discussion 166–167
9. Hebb AO, Cusimano MD (2001) Idiopathic normal pressure hydrocephalus: a systematic review of diagnosis and outcome. Neurosurgery 49: 1166–1184; discussion 1184–1186
10. Hughes CP, Siegel BA, Coxe WS, Gado MH, Grubb RL, Coleman RE, Berg L (1978) Adult idiopathic communicating hydrocephalus with and without shunting. J Neurol Neurosurg Psychiatry 41: 961–971
11. Kahlon B, Sundbarg G, Rehncrona S (2002) Comparison between the lumbar infusion and CSF tap tests to predict outcome after shunt surgery in suspected normal pressure hydrocephalus. J Neurol Neurosurg Psychiatry 73: 721–726
12. Kiefer M, Eymann R, Meier U (2002) Five years experience with gravitational shunts in chronic hydrocephalus of adults. Acta Neurochir (Wien) 144: 755–767; discussion 767
13. Krauss JK, Regel JP, Vach W, Orszagh M, Jungling FD, Bohus M, Droste DW (1997) White matter lesion in patients with idiopathic normal pressure hydrocephalus and in an age-matched control group: a comparative study. Neurosurgery 40: 491–495; discussion 495–496
14. Larsson A, Wikkelsö C, Bilting M, Stephensen H (1991) Clinical parameters in 74 consecutive patients shunt operated for normal pressure hydrocephalus. Acta Neurol Scand 84: 475–482
15. Malm J, Kristensen B, Karlsson T, Fagerlund M, Elfverson J, Ekstedt J (1995) The predictive value of cerebrospinal fluid dynamic tests in patients with idiopathic adult hydrocephalus syndrome. Arch Neurol 52: 783–789
16. Meier U, Kiefer M, Sprung C (2003) Normal-pressure hydrocephalus: pathology, pathophysiology, diagnostics, therapeutics and clinical course. PVV Science Publications, Ratingen
17. Meier U, Kiefer M, Lemcke J (2005) On the optimal opening pressure of hydrostatic valves in cases of idiopathic normal-pressure-hydrocephalus. Neurosurg Q [in press]
18. Nacmias B, Tedde A, Guarnieri BM, Petruzzi C, Ortenzi L, Serio A, Amaducci L, Sorbi S (1997) Analysis of apolipoprotein E, alpha 1-antichymotrypsin and presenilin-1 genes polymorphisms in dementia caused by normal pressure hydrocephalus in man. Neurosci Lett 229: 177–180
19. Rem JA (1986) Dementia – normal pressure hydrocephalus [in German]. Schweiz Rundsch Med Prax 75: 569–570
20. Savolainen S, Hurskainen H, Paljarvi L, Alafuzoff I, Vapalahti M (2002) Five-year outcome of normal pressure hydrocephalus with or without a shunt: predictive value of the clinical signs, neuropsychological evaluation and infusion test. Acta Neurochir (Wien) 144: 515–523; discussion 523
21. Tanaka A, Kimura M, Nakayama Y, Yoshinaga S, Tomonaga M (1997) Cerebral blood flow and autoregulation in normal pressure hydrocephalus. Neurosurgery 40: 1161–1165; discussion 1165–1167
22. Vanneste J, Augustijn P, Dirven C, Tan WF, Goedhart ZD (1992) Shunting normal-pressure hydrocephalus: do the benefits outweigh the risks? Multicenter study and literature review. Neurology 42: 54–59

Correspondence: Michael Kiefer, Saarland University, Medical School, Department of Neurosurgery, 66421 Homburg-Saar, Germany. e-mail: ncmkie@uniklinik-saarland.de

First clinical experiences in patients with idiopathic normal-pressure hydrocephalus with the adjustable gravity valve manufactured by Aesculap (proGAV$^{Aesculap®}$)

U. Meier and J. Lemcke

Department of Neurosurgery, Unfallkrankenhaus Berlin, Berlin, Germany

Summary

Objective. Improved clinical outcomes after implantation of a low pressure valve in patients with idiopathic normal-pressure hydrocephalus is usually achieved at the expense of a higher overdrainage rate. Can an adjustable valve with a gravitational unit provide optimal results?

Method. In a prospective clinical outcome study conducted in the Unfallkrankenhaus Berlin, 30 patients with idiopathic normal-pressure hydrocephalus were treated surgically between June 2004 and May 2005 with the valve combination described above, and re-examination 3 months or 6 months postoperatively.

Results. Clinical outcome correlates with opening pressure level of the valve. Controlled adjustment of the valve from 100 mmH$_2$O to 70 mmH$_2$O, and then to 50 mmH$_2$O after 3 months, permits optimum adaptation of the brain to the implanted valve without overdrainage complications.

Conclusions. Advantages of this programmable gravity valve include: 1) the absence of unintentional readjustment through external magnets, and 2) the possibility of controlling the valve setting using an accessory instrument without the need for x-ray monitoring. A significant disadvantage is adjusting the valve after implantation. From the clinical point of view, this new "proGAV$^{Aesculap®}$" valve is a necessary development in the right direction, but at the moment it is still beset with technical problems.

Keywords: Normal-pressure hydrocephalus; idiopathic; programmable valve; gravitational valve; shunt.

Introduction

Despite modern diagnostic methods, the accurate diagnosis and treatment of idiopathic normal-pressure hydrocephalus (iNPH) continues to present a challenge to the clinician [10]. Even after an exact diagnosis has been made, surgical treatment is full of potential complications, which substantially influences the clinical outcome and the improvement rate of patients with iNPH. Non-physiological "hydraulic management" [10] of hydrocephalus shunts consists of either overdrainage when the patient is standing, or relative underdrainage when the patient is lying down. This problem is technically unavoidable with conventional valves, since they merely allow a compromise between the requirements of the different body positions [1]. While the majority of younger patients show considerable tolerance for non-physiological intracranial pressure, it becomes a major problem for older iNPH patients, who often also have pre-existing vascular damage [20, 21].

Materials and methods

From June 2004 to May 2005, 30 patients with iNPH were treated surgically in the Department of Neurosurgery at the Unfallkrankenhaus Berlin-Marzahn by implantation of a "proGAV$^{Aesculap®}$" adjustable valve with gravitational unit (Chrisohph Miethke GmbH & Co. KG, Berlin, Germany). The 18 men and 12 women had an average age of 68 (27 to 83) years.

Diagnostics

After identification of gait disturbance and additional symptoms [18] during clinical examination, as well as evidence of ventricular enlargement from neuroradiological imaging techniques, the intrathecal infusion test was performed. A computer-assisted dynamic infusion test using the constant flow technique with an infusion rate of 2 mL/min via lumbar puncture served to calculate the individual parameters of the cerebrospinal fluid (CSF) dynamics [17, 19]. CSF resistance to outflow greater than 13 mmHg/mL/min was defined as pathological [15]. The infusion test was directly followed by diagnostic CSF drainage of at least 60 mL. If the clinical symptoms improved in the following 2 to 3 days, shunt operation was indicated. If there was no improvement in symptoms, particularly in gait disturbance, external lumbar drainage for 2 to 3 days was carried out. If the symptoms improved after this, the valve combination described above was implanted as a ventriculoperitoneal shunt [22].

Investigation protocol of the prospective study

Controlled adjustment of the valve from 100 mmH$_2$O to 70 mmH$_2$O in the first week postoperatively and to 50 mmH$_2$O after 3 months was carried out in order to compensate for the relative underdrainage produced iatrogenically by CSF loss during the operation. For the selection of the gravity valve, however, the height of the patient is decisive (shorter than 160 cm = 250 mmH$_2$O; 160 to 180 cm = 300 mmH$_2$O; taller than 180 cm = 350 mmH$_2$O).

Clinical grading

Clinical findings were graded according to the Black grading scale for shunt assessment (Table 1) and the normal-pressure hydrocephalus (NPH) recovery rate [14] based on the Kiefer clinical score for NPH [13].

NPH recovery rate

$$= \frac{\text{NPH grading}_{\text{preoperative}} - \text{NPH grading}_{\text{postoperative}}}{\text{Kiefer NPH score [13]}_{\text{preoperative}}} \times 10$$

Table 1. *Black grading scale for shunt assessment*

Grading	Description	NPH recovery rate
Excellent	same level of activity as before the illness	≥7 points
Good	slight impairment	≥5 points
Fair	gradual improvement	≥3 points
Transient	temporary improvement	≥2 points
Poor	unchanged or deterioration	<2 points

NPH Normal-pressure hydrocephalus.

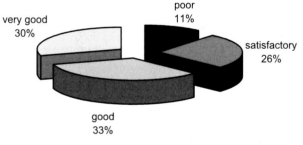

Fig. 1. Clinical outcome 3 or 6 months after shunt operation

Results

Outcome

Therapeutic results 3 or 6 months after shunt implantation are illustrated in Fig. 1. To summarize, we were able to record good to excellent results for 19 patients, fair for 8, and a poor therapeutic outcome in 3 patients. Average Kiefer scores (Fig. 2) corresponding to severity of the symptoms caused by iNPH were reduced by half, from 7.8 points preoperatively, to 4.2 points postoperatively, to 4.0 points at 3-month follow-up. Since the Kiefer clinical score for NPH [13] is more suited to observing individual outcome than to group comparison [14], we compared the NPH recovery rate values postoperatively with those at follow-up 3 and 6 months after shunt implantation (Fig. 3). The average score postoperatively was 5.1, borderline between a fair and a good clinical outcome. Good therapeutic results were also recorded at 3- and 6-month follow-ups, with scores of 5.5 and 7.0, respectively.

Complications

Complications after shunt implantation that were not related to the valve showed an infection rate of 0%. No revision operations were necessary owing to incorrect placement of ventricular catheters or dislocation of peritoneal catheters in the abdominal wall (dislocation: 0%). Thus, in this relatively small patient series, the rate of complications not related to the valve was 0%.

In 6 patients, wound swelling and a suture made it impossible to adjust the valve immediately after surgery or up to day 10 postoperatively. Following suture removal and with normal wound conditions it was

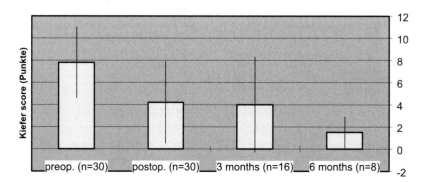

Fig. 2. Clinical symptoms according to the Kiefer score

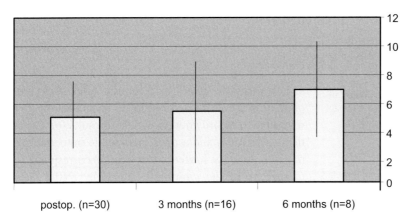

Fig. 3. Outcome according to NPH recovery rates

possible to adjust the valves in these patients without any difficulty.

On neuroradiological imaging, one patient showed chronic subdural hematomas on both sides as direct evidence of overdrainage (overdrainage rate: 3%). In this same patient, the valve had locked at an opening pressure of 20 mmH$_2$O during one month, when we tried to change the opening pressure due to the hematomas (intermittent valve adjustment problems: 7 patients, 23%). First, we ligated the valve and evacuated the chronic subdural hematomas via burr holes.

The shunt-dependent morbidity was 3%. No patient died in the study period, giving a surgery-related fatality rate of 0%. No problems arose in handling the magnet instrument to read the valve setting. If the reading pen was correctly positioned following palpatory location of the valve and the inflow spout that serves as a directional guide, the measured pressure level always matched the radiologically-defined value. The representation on x-ray under image intensifier was problem-free; in the cCT-Scout a definite reading is only possible since the manufacturer improved the side marking on the valve.

Discussion

Various strategies have been attempted in the past to solve the basic problem that the pressure gradient as the body axis becomes vertical is additionally increased by the action of the hydrostatic pressure components. It became apparent at an early stage that overdrainage, which occurs particularly in the upright position, could be prevented by the use of standard valves with a high opening pressure. However, this only succeeded at the price of permanent underdrainage in the lying position, when by virtue of the high valve opening pressure and the fact that the hydrostatic pressure components were no longer operating, CSF drainage no longer took place [3], making it impossible to compensate for the occurrence of nightly pressure peaks in CSF. The use of standard valves with low opening pressures led to a better outcome, but also to a substantially higher overdrainage rate [4]. Later designs, such as the antisiphon device, suffered malfunctions after becoming completely encased in scar tissue [12]. The development of adjustable valves opened up the possibility of programming the opening pressure so high that no more overdrainage took place, meaning that it was also no longer possible to guarantee sufficient CSF outflow. Bergsneider [3] pointed out that when using programmable valves without gravitational components, a partial compensation of the hydrostatic pressure difference that occurs in the vertical position could be observed through increasing pressure when the body axis was brought into the upright position; however, this effect is scarcely calculable and even less able to be regulated. It was only with the development of gravity-controlled valves capable of switching between different pressure levels for the horizontal and upright positions that a decisive step toward solving this problem was taken.

Outcome

The general improvement rates stated in the literature for patients with NPH after a shunt operation vary between 31% and 96%, with an average value of 53% [2, 6]. Following a meta-analysis, Vanneste [24] cites improvement rates of 30% to 50% for iNPH. In another meta-analysis, Hebb *et al.* [11] cite improvement rates of 59% after shunt implantation in iNPH and 29% long-term. The results of our study reflect an overall improvement rate of 72%, falling within the range cited in the literature. In children, Drake *et al.* [8] found no differences in clinical outcome after shunt operation between different valve groups.

Complications

Grumme *et al.* [9] cited a fatality rate of 0% to 6% and a frequency of overdrainage phenomena of 6% to 20% following shunt operations in patients with NPH. Hebb *et al.* [11] report a complication rate of 38%, revision rate of 22%, and the combination of remaining neurological deficits and surgery-related fatality at 6%. In the English Shunt Registry [23], with its large patient collective of more than 9000 cases, underdrainage is the major complication at 52%. As a postoperative cause of complications, overdrainage is cited in this paper at 3%. By contrast, Scandinavian working parties [7] claim that 80% of all shunt complications arise from overdrainage. A Dutch multicenter study [5, 7] reported a significantly better clinical outcome for patients with NPH after implantation of low pressure rather than medium pressure standard valves, but this advantage was gained at the expense of a higher overdrainage rate of 73% as opposed to 34%. Our investigation protocol confirms that setting the valve opening pressure at 50 mmH$_2$O makes for a better clinical outcome than opening pressures set at 100 mmH$_2$O or more [16].

Conclusions

The advantages of this programmable valve with gravitational unit are the absence of unintentional readjustment through external magnets, and the ability to control the valve setting using an accessory instrument without the need for x-ray. The most significant disadvantage is valve adjustment problems after implantation. Despite the implanted gravitational unit, overdrainage complications arose in 3% of patients.

References

1. Aschoff A (1994) In-vitro-Testung von Hydrocephalus-Ventilen. Habilitationsschrift, Heidelberg, Germany
2. Bech RA, Waldemar G, Gjerris F, Klinken L, Juhler M (1999) Shunting effects in patients with idiopathic normal pressure hydrocephalus; correlation with cerebral and leptomeningeal biopsy findings. Acta Neurochir (Wien) 141: 633–639
3. Bergsneider M, Yang I, Hu X, McArthur DL, Cook SW, Boscardin WJ (2004) Relationship between valve opening pressure, body position, and intracranial pressure in normal pressure hydrocephalus: paradigm for selection of programmable valve pressure setting. Neurosurgery 55: 851–858; discussion 858–859
4. Boon AJ, Tans JT, Delwel EJ, Egeler-Peerdeman SM, Hanlo PW, Wurzer HA, Avezaat CJ, de Jong DA, Gooskens RH, Hermans J (1998) Dutch normal-pressure hydrocephalus study: randomized comparison of low- and medium-pressure shunts. J Neurosurg 88: 490–495
5. Boon AJ, Tans JT, Delwel EJ, Egeler-Peerdeman SM, Hanlo PW, Wurzer HA, Hermans J (1999) Dutch normal-pressure hydrocephalus study: The role of cerebrovascular disease. J Neurosurg 90: 221–226
6. Borgesen SE (1984) Conductance to outflow of CSF in normal pressure hydrocephalus. Acta Neurochir (Wien) 71: 1–45
7. de Jong DA, Delwel EJ, Avezaat CJ (2000) Hydrostatic and hydrodynamic considerations in shunted normal pressure hydrocephalus. Acta Neurochir (Wien) 142: 241–247
8. Drake JM, Kestle JR, Milner R, Cinalli G, Boop F, Piatt J Jr, Haines S, Schiff SJ, Cochrane DD, Steinbok P, MacNeil N (1998) Randomized trial of cerebrospinal fluid shunt valve design in pediatric hydrocephalus. Neurosurgery 43: 294–303; discussion 303–305
9. Grumme T, Kolodziejczyk D (1995) Komplikationen in der Neurochirurgie. Blackwell Wissenschafts-Verlag, Berlin, pp 534–540
10. Hakim CA (1985) The physics and physicopathology of the hydraulic complex of the central nervous system. The mechanics of hydrocephalus and normal-pressure hydrocephalus. Massachusetts Institute of Technology [dissertation]
11. Hebb AO, Cusimano MD (2001) Idiopathic normal pressure hydrocephalus: a systematic review of diagnosis and outcome. Neurosurgery 49: 1166–1184; discussion 1184–1186
12. Kiefer M, Eymann R, Mascaros V, Walter M, Steudel WI (2000) Significance of hydrostatic valves in therapy of chronic hydrocephalus. Nervenarzt 71: 975–986 [in German]
13. Kiefer M, Eymann R, Meier U (2002) Five years experience with gravitational shunts in chronic hydrocephalus of adults. Acta Neurochir (Wien) 144: 755–767; discusion 767
14. Meier U (2002) The grading of normal pressure hydrocephalus. Biomed Tech (Berl) 47: 54–58
15. Meier U, Bartels P (2001) The importance of the intrathecal infusion test in the diagnostic of normal-pressure hydrocephalus. Eur Neurol 46: 178–186
16. Meier U, Kiefer M (2004) Zum optimalen Öffnungsdruck hydrostatischer Ventile beim idiopathischen Normaldruckhydrozephalus: Eine prospektive Studie mit122 Patienten. Akt Neurol 31: 216–222
17. Meier U, Zeilinger FS, Kintzel D (1999) Diagnostic in normal pressure hydrocephalus: A mathematical model for determination of the ICP-dependent resistance and compliance. Acta Neurochir (Wien) 141: 941–947; discussion 947–948
18. Meier U, Zeilinger FS, Kintzel D (1999) Signs, symptoms and course of disease in normal pressure hydrocephalus in comparison with cerebral atrophy. Acta Neurochir (Wien) 141: 1039–1048
19. Meier U, Kiefer M, Bartels P (2002) The ICP-dependency of resistance to cerebrospinal fluid outflow: a new mathematical method for CSF-parameter calculation in a model with H-TX rats. J Clin Neurosci 9: 58–63
20. Meier U, Kiefer M, Sprung C (2002) The Miethke Dual-Switch valve in patients with normal pressure hydrocephalus. Neurosurg Q 12: 114–121
21. Meier U, Kiefer M, Sprung C (2003) Normal-pressure hydroce-

phalus: Pathology, pathophysiology, diagnostics, therapeutics and clinical course. PVV Science Publications, Ratingen
22. Meier U, Kiefer M, Sprung C (2004) Evaluation of the Miethke dual-switch valve in patients with normal pressure hydrocephalus. Surg Neurol 61: 119–127; discussion 127–128
23. Richards HK, Seeley HM, Pickard JD (2000) Shunt revisions: Data from the UK shunt registry. Eur J Pediatr Surg [Suppl] I: 10–59
24. Vanneste JA (2000) Diagnosis and management of normal-pressure hydrocephalus. J Neurol 247: 5–14

Correspondence: Ullrich Meier, Department of Neurosurgery, Unfallkrankenhaus Berlin, Warener Strasse 7, 12683 Berlin, Germany. e-mail: ullrich.meier@ukb.de

Decompressive craniectomy for severe head injury in patients with major extracranial injuries

U. Meier, J. Lemcke, T. Reyer, and A. Gräwe

Department of Neurosurgery, Unfallkrankenhaus Berlin, Berlin, Germany

Summary

Neurosurgical therapy aims to minimize secondary brain damage after a severe head injury. This includes the evacuation of intracranial space-occupying hematomas, the reduction of intracranial volumes, external ventricular drainage, and aggressive therapy in order to influence increased intracranial pressure (ICP) and decreased $P(ti)O_2$. When conservative treatment fails, a decompressive craniectomy might be successful in lowering ICP.

From September 1997 until December 2004, we operated on 836 patients with severe head injuries, of whom 117 patients (14%) were treated by means of a decompressive craniectomy. The prognosis after decompression depends on the clinical signs and symptoms at admission, patient age, and the existence of major extracranial injuries. Our guidelines for decompressive craniectomy after failure of conservative interventions and evacuation of space-occupying hematomas include: patient age below 50 years without multiple trauma, patient age below 30 years in the presence of major extracranial injuries, severe brain swelling on CT scan, exclusion of a primary brainstem lesion or injury, and intervention before irreversible brainstem damage.

Keywords: Head injury; edema; craniectomy; intracranial pressure.

Introduction

Bergmann described decompressive craniectomy in 1880 and Cushing published a case report about a subtemporal decompressive craniectomy for relief of intracranial pressure in 1908. There is still a controversy going on about the value of operative decompression after severe head injuries with traumatic brain edema [1, 2, 5, 6].

The aim of neurosurgical therapy after severe head injuries is the minimization of secondary brain damage. General principles of neurosurgical therapy are the evacuation of space-occupying hematomas, the reduction of intracranial volume, the drainage of hematocephalus, and conservative therapy focused on intracranial pressure (ICP), cerebral perfusion pressure, and brain tissue PO_2. In intractable intracranial hypertension that is refractory to conservative interventions, a decompressive craniectomy is indicated in a few patients. Indications for decompressive craniectomy, course of disease, and prognostic criteria are analyzed and compared with the literature [16–18].

Patients and methods

All patients with a severe head injury at the Unfallkrankenhaus are treated by the neurosurgical service in the interdisciplinary intensive care unit. Standard management includes an initial computed tomography scan in the emergency room. Patients were included in our prospective study when we saw an indication for a neurosurgical operation.

Patients with severe head injuries (n = 836) were operatively treated in the Department of Neurosurgery of the Unfallkrankenhaus Berlin between September 1997 to December 2004. The average age of the 674 male and 162 female patients was 41 years. Decompressive craniectomy was performed in 117 patients (14%). In 74 patients, craniectomy was performed in addition to removal of a space-occupying hematoma (subdural or epidural hematoma, contusion hemorrhage) because generalized brain edema was found intraoperatively. The second group included 43 patients in whom conservative treatment in the intensive care unit was therapy-resistant when neuro-monitoring with ICP, cerebral perfusion pressure, mean arterial pressure, and $P(ti)O_2$ measurement revealed intractable brain edema. The average age of the decompression group of patients was 35 years and the male/female ratio was 3:1 (87 male, 30 female). In 105 patients we performed a unilateral craniectomy (right side 48, left side 57), and in 12 patients a bilateral craniectomy was necessary.

Results

In accordance with the results of other groups [16], the majority of head injuries with indication for craniectomy was due to motor vehicle accidents (58%), followed by falls (28%), attempted suicide (6%), and violence (8%). Fifty-five percent of patients had a craniectomy after diffuse brain injury, 27% after an acute subdural hematoma, 9% after an acute epidural hematoma, and 9% after an open head injury. In patients with a diffuse brain injury, the primary operative intervention was performed for evacuation of a space-occupying contusion hemorrhage.

Forty-seven patients (40%) died despite decompression, 16 patients (14%) remained in a vegetative state, 24 patients (20%) had persistent severe neurological deficits, and 30 patients (26%) reached a Glasgow Outcome Score (GOS) of 4 to 5. The course of disease in patients who were decompressed is illustrated in Fig. 1.

A comparison of patient age and postoperative results is not possible because of the small number of patients (Fig. 2). Altogether, patients younger than 40 years have a better prognosis than the older ones. In the fourth and fifth decades, satisfactory results are obtainable in 80%. According to our results, a decompression is less favorable in patients over 60 years of age, and our patients in this group remained in a vegetative state or died.

There is evidence for an important influence on the outcome when major extracranial injuries are present in contrast to an isolated head injury (Fig. 3). We found increased lethality of 53% when there was multiple trauma, in contrast to 34% with an isolated head injury. Furthermore, the outcome is influenced by the existence of an extracranial injury, with positive results in only 10% compared to 34% in isolated head injuries. In our small series, we could not find a difference in the time span between accident and surgical decompression. This is not contradictory for early decompression when indicated.

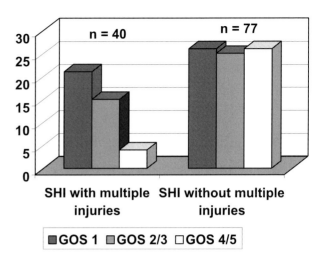

Fig. 3. Comparison between severe head injury (*SHI*) patients with and without major extracranial injuries

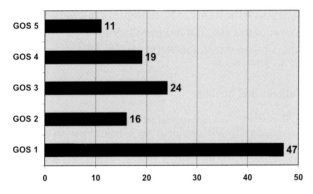

Fig. 1. Prognosis in severe head injuries with decompressive craniectomy

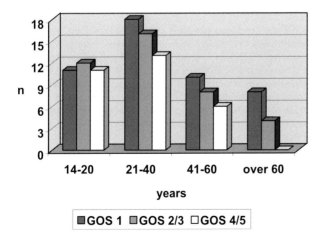

Fig. 2. Outcome versus patient age

Discussion

Prognosis after severe head injury depends on the clinical status at admission, intracranial lesions, patient age, and the existence of accompanying injuries. Patients with epidural hematomas and open head injuries have a better prognosis than those with acute subdural hematomas. A poor prognosis is given in patients with brain contusions, diffuse brain injuries, traumatic injuries of the venous sinuses, less than 8 points on the Glasgow Coma Scale, and advanced age [16–18].

The pathophysiology of posttraumatic, primarily

vasogenic and cytotoxic, cerebral edema is mainly a disturbance of the functional entity consisting of capillary, astroglia, and neuron. The driving force for the formation of edema is the pressure gradient across the injured capillary with loss of blood-brain barrier function and an accumulation of extracellular fluid. Fluid accumulation is not caused by diffusion, but by a hydrostatic pressure gradient in the interstitium of the white matter. This global, mostly hemispherical blood-brain barrier disturbance induces generalized brain edema, which can only be marginally influenced by conservative treatment. To a degree, an intracranial volume increase can be compensated for by shifts in cerebrospinal fluid or blood volume. According to the intracranial pressure-volume relationship, when the intracranial reserve areas are exhausted, any further increase in volume is responsible for a considerable change in pressure. When the ICP increase decompensates, a decompressive craniectomy must be considered. Decompression changes the pressure-volume relationship and causes an increase in the compliance and, therefore, a shift to the right on the pressure-volume curve [1, 2, 6, 8, 10, 15].

In animal experiments, it was found by other groups [1, 2, 4, 11] that after decompression the injured brain exhibits an ICP decrease, an increase in the pressure-volume index, and a decrease in interstitial fluid accumulation. There are contradictory statements in the literature concerning an improvement in the regional cerebral blood flow and cerebrovascular resistance [11, 24]. In clinical trials with PET, Xenon-CT, and MR-spectroscopy, Yamakami et al. [24] and Yoshida et al. [25] found a hemispheric CBF increase as well as an increase in the cerebral metabolism after the craniectomy in comparison to the preoperative results. This requires early decompression and the craniotomy must be wide enough and include an opening and prolongation of the dura. Craniectomy should be performed when conservative treatment fails to influence an ICP between 25 to 40 mmHg and before the cerebral perfusion is irreversibly disturbed [2].

There are different methods described in the literature. Clark et al. [3] prefer circumferential craniotomy, and other groups [13, 20, 23] advise bifrontal craniotomy. Gerl et al. [6] recommend bilateral craniotomy, and Gaab et al. [5, 12] recommend wide unilateral or bilateral frontotemporal-parietal craniectomy with dura opening and duraplasty. According to the recommendations of Gaab et al. [5, 9] we have performed the wide frontotemporal-parietal craniectomy with duraplasty, for which we used the temporal muscle and its fascia or a Neuropatch (Braun, Melsungen, Germany) as a graft. The bone flap is preserved by subcutaneous implantation by storage under sterile conditions [19]. In accordance with Yoshida et al. [25], who found that regional cerebral blood flow and cerebral metabolism is reduced in the area of craniectomy, we performed the reimplantation after the regression of cerebral edema and when physiological ICP was present. Other than decompressions in patients with diffuse brain injuries, we performed a craniectomy more often in patients with acute subdural hematomas when the development of cerebral edema appeared during the operation and was resistant to conservative interventions.

The course of disease after craniectomy depends on the clinical status at admission, which was also found by other authors [5, 9, 21, 22]. In the literature, the postoperative results differ. In an analysis of the literature, there are between 11% and 71% dead despite decompression, 8% to 27% were in a vegetative state, 8% to 27% survived with severe neurological deficits, 0% to 21% survived with mild neurological deficits, and 0% to 41% had good recovery [5–7, 9, 12, 14, 20]. In another large analysis, Guerra et al. [9] found 49% of patients died, 13% survived with severe neurological deficits corresponding to a GOS of 2 and 3, and 32% had a favorable outcome with a GOS of 4 or 5. Our results are in accordance with those reported above, but we had a slightly larger number of patients in a vegetative state (Fig. 1). Polin et al. [20] found significantly better results after decompression, a finding that may be due to fewer patients with diffuse brain injuries who did not have surgery. In our comparable group, all were operatively treated patients after severe head injuries.

The influence of patient age on outcome after severe head injury and craniectomy (Fig. 2) was also found by other authors [5, 9, 14, 20, 21]. We need to critically re-evaluate the indication for decompressive craniectomy in patients over 60 years of age. Postoperative results in patients with a GOS of 3 is one reason for our higher rate of patients in a vegetative state, and we postulate that there should be an age restriction. Like Karlen et al. [12], the prognosis was worse when there were major extracranial injuries present (Fig. 3).

Under the influence of the results of Gaab et al. [5, 9, 21], we conclude with the following guidelines as an indication for decompressive craniotomy after severe head injuries:

1. Patient age below 50 years without multiple trauma.
2. Patient age below 30 years in the presence of major extracranial injuries.
3. Severe brain swelling on CT scan.
4. Exclusion of a primary brainstem lesion/injury.
5. Intervention before irreversible brainstem damage.
6. ICP increase up to 40 mmHg and unsuccessful conservative therapy.
7. Primarily for space-occupying hematomas with hemispheric brain edema.
8. Rising ICP and falling tissue oxygenation $P(ti)O_2$ before irreversible brain damage has occurred.

References

1. Burkert W (1985) Zeitlicher Verlauf des posttraumatischen Hirnödems nach Ultraschalleinwirkung – Experimentelle Untersuchungen und Modellvorstellungen zum Zeitpunkt der operativen Dekompression und zur Größe der Trepanationsfläche. [Dissertation B] Martin-Luther-Universität Halle
2. Burkert W, Plaumann H (1989) The value of large pressure-relieving trepanation in treatment of refractory brain edema. Animal experiment studies, initial clinical results. Zentralbl Neurochir 50: 106–108 [in German]
3. Clark K, Nash TM, Hutchinson GC (1968) The failure of circumferential craniotomy in acute traumatic cerebral swelling. J Neurosurg 29: 367–371
4. Cooper PR, Hagler H, Clark WK, Barnett P (1979) Enhancement of experimental cerebral edema after decompressive craniectomy: implications for the management of severe head injuries. Neurosurgery 4: 296–300
5. Gaab MR, Rittierodt M, Lorenz M, Heissler HE (1990) Traumatic brain swelling and operative decompression: a prospective investigation. Acta Neurochir [Suppl] 51: 326–328
6. Gerl A, Tavan S (1980) Bilateral craniectomy in the treatment of severe traumatic brain edema. Zentralbl Neurochir 41: 125–138 [in German]
7. Gower DJ, Lee KS, McWhorter JM (1988) Role of subtemporal decompression in severe closed head injury. Neurosurgery 23: 417–422
8. Grände PO, Asgeirsson B, Nordström CH (1997) Physiologic principles for volume regulation of a tissue enclosed in a rigid shell with application to the injured brain. J Trauma 42: S23–S31
9. Guerra KW, Gaab MR, Dietz H, Mueller JU, Piek J, Fritsch MJ (1999) Surgical decompression for traumatic brain swelling: indications and results. J Neurosurg 90: 187–196
10. Hase U, Reulen HJ, Meinig G, Schürmann K (1978) The influence of the decompressive operation on the intracranial pressure and the pressure-volume relation in patients with severe head injuries. Acta Neurochir (Wien) 45: 1–13
11. Hatashita S, Hoff JT (1987) The effect of craniectomy on the biomechanics of normal brain. J Neurosurg 67: 573–578
12. Karlen J, Stula D (1987) Decompressive craniotomy in severe craniocerebral trauma following unsuccessful treatment with barbiturates. Neurochirurgia (Stuttg) 30: 35–39 [in German]
13. Kjellberg RN, Prieto A Jr (1971) Bifrontal decompressive craniotomy for massive cerebral edema. J Neurosurg 34: 488–493
14. Kunze E, Meixensberger J, Janka M, Sörensen N, Roosen K (1998) Decompressive craniectomy in patients with uncontrollable intracranial hypertension. Acta Neurochir [Suppl] 71: 16–18
15. Meier U (1985) Computertomographische Untersuchungen zum generalisierten Hirnödem nach Schädel-Hirn-Trauma und Subarachnoidalblutung. [Dissertation A] Humboldt-Universität, Berlin
16. Meier U, Gärtner F, Knopf W, Klötzer R, Wolf O (1992) The traumatic dural sinus injury – a clinical study. Acta Neurochir (Wien) 119: 91–93
17. Meier U, Heinitz A, Kintzel D (1994) Surgical outcome after severe craniocerebral trauma in childhood and adulthood. A comparative study. Unfallchirurg 97: 406–409 [in German]
18. Meier U, Zeilinger F, Kintzel D (1996) Prognose und Ergebnisse der operativen Versorgung von schweren Schädel-Hirn-Traumen. Akt Traumatol 26: 1–5
19. Pasaoglu A, Kurtsoy A, Koe RK, Kontas O, Akdemir H, Öktem IS, Selcuklu A, Kavuncu IA (1996) Cranioplasty with bone flaps preserved under the scalp. Neurosurg Rev 19: 153–156
20. Polin RS, Shaffrey ME, Bogaev CA, Tisdale N, Germanson T, Bocchicchio B, Jane JA (1997) Decompressive bifrontal craniectomy in the treatment of severe refractory posttraumatic cerebral edema. Neurosurgery 41: 84–94
21. Rittierodt M, Gaab MR, Lorenz M (1991) Decompressive craniectomy after severe head injury. Useful therapy in pathophysiologically guided indications. In: Bock WJ (ed) Advances in neurosurgery, vol 19. Springer, Berlin Heidelberg New York Tokyo, pp 265–273
22. Shigemori M, Syojima K, Nakayama K, Kojima T, Watanabe M, Kuramoto S (1979) Outcome of acute subdural haematoma following decompressive hemicraniectomy. Acta Neurochir [Suppl] 28: 195–198
23. Venes JL, Collins WF (1975) Bifrontal decompressive craniectomy in the management of head trauma. J Neurosurg 42: 429–433
24. Yamakami I, Yamaura A (1993) Effects of decompressive craniectomy on regional cerebral blood flow in severe head trauma patients. Neurol Med Chir (Tokyo) 33: 616–620
25. Yoshida K, Furuse M, Izawa A, Iizima N, Kuchiwaki H, Inao S (1996) Dynamics of cerebral blood flow and metabolism in patients with cranioplasty as evaluated by ^{133}Xe CT and ^{31}P magnetic resonance spectroscopy. J Neurol Neurosurg Psychiatry 61: 166–171

Correspondence: Ullrich Meier, Department of Neurosurgery, Unfallkrankenhaus Berlin, Warener Strasse 7, 12683 Berlin, Germany. e-mail: ullrich.meier@ukb.de

Clinical outcome of patients with idiopathic normal pressure hydrocephalus three years after shunt implantation

U. Meier and J. Lemcke

Department of Neurosurgery, Unfallkrankenhaus Berlin, Berlin, Germany

Summary

Objective. To investigate the outcomes of surgical treatment of idiopathic normal pressure hydrocephalus (iNPH).

Patients and methods. We prospectively investigated 51 patients treated for iNPH by insertion of a ventriculoperitoneal shunt with gravitational valve.

Results. The proportion of excellent, good, and satisfactory outcomes immediately following surgery was 80%; the same clinical outcome was later verified in 67% of patients on average of 34 months postoperatively. These results are similar to those reported in the literature. Ventricle volume decreased minimally during the course of treatment using Evans' index, which was concordant with recent literature and with current understanding of hydrocephalus treatment.

Conclusions. A gravitational valve for treatment of iNPH is the logical implementation of current knowledge of the pathophysiology of this illness enabling us to solve the problem of cerebrospinal fluid drainage with the lowest possible opening pressure while simultaneously protecting from overdrainage. The use of programmable valves, in combination with a gravitational component, is the next evolutionary stage, potentially making revision operations unnecessary.

Keywords: Normal pressure hydrocephalus; idiopathic; outcome; gravitational valve.

Introduction

The treatment of normal pressure hydrocephalus (NPH) continues to present a challenge for the clinically active physician. The path to good and excellent long-term outcomes begins with the correct indication, leading through to the selection of the individually appropriate method for controlling internal cerebrospinal fluid (CSF) drainage, and ending in careful postoperative treatment of patients over the course of many years. Our objective was to prospectively study the outcomes of the surgical treatment for idiopathic normal pressure hydrocephalus (iNPH) using hydrostatic valves on average 3 years after implantation.

Materials and methods

This prospective study of clinical outcome included all patients who, within a 6-year period (1997 to 2003), were diagnosed in the Neurosurgery Clinic of the Unfallkrankenhaus Berlin, Germany, and treated surgically for iNPH by insertion of a ventriculoperitoneal shunt using a hydrostatic gravitational valve (Miethke dual-switch valve, Christoph Miethke GMBH & Co. KG, Berlin, Germany). The progression of the disease was recorded before and after surgery, and at follow-up on average 34 months (14 to 82 months) after surgery. The follow-up investigations took place between July 2003 and January 2005 and included detailed questioning of patients regarding symptoms as well as the evaluation of neurological status and assessment using the Homburg scale devised by Kiefer. A computed tomography (CT) examination of the head was performed on all patients before and after surgery and in the follow-up period to establish the Evans index. In some cases, recent cerebral CT scans produced elsewhere were available for the follow-up examination.

Diagnostics

In our clinic, magnetic resonance imaging CSF flow examination, lumbar infusion test, and spinal tap test were carried out for routine diagnosis. In our opinion, a high level of diagnostic certainty is provided through this testing and through calculating the compliance of the CSF spaces and CSF outflow resistance [8–10]. A ventricular infusion test for continuous CSF drainage over 3 days is only performed in cases where it has not been possible to establish the indication for shunt therapy with sufficient certainty using the preceding diagnostic techniques.

Examination criteria

The Homburg score [6] records the following 5 criteria: amnestic disorders, gait disturbances, urinary incontinence, giddiness symptoms, and intensity of headaches. The NPH recovery rate (NPH-R-R) was used in an attempt to formulate postoperative development of clinical symptoms in patients with different intensities of initial symptoms so as to allow comparison.

$$NPH - RR = \frac{KieferScore_{preoperative} - KieferScore_{postoperative}}{KieferScore_{preoperative}} \cdot 10$$

Outcome was then evaluated as follows: scores between 7 and 10 points were rated as excellent, scores ≥ 5 points were rated as good, scores ≥ 2 points were rated as satisfactory, and scores under 2 points were defined as poor outcome.

Statistical analysis

Descriptive statistics were performed using SPSS for Windows (SPSS Inc., Chicago, IL) and Microsoft Excel for Windows (Microsoft Corp., Redmond, WA). Explorative statistics were carried out using SPSS for Windows with the Wilcoxon test for paired samples. The error probability was $p < 0.05$.

Results

Preoperative investigations

In a 6-year period (1997 to 2003), 107 patients were diagnosed with suspected iNPH in the Neurosurgery Clinic of the Unfallkrankenhaus Berlin. In 66 patients, the indication was made for surgical treatment with ventriculoperitoneal shunt implantation of a Miethke dual-switch valve. These patients satisfied the inclusion criteria for the study described in this paper. Three patients did not consent to the planned operation after the preoperative diagnosis had been performed. These patients were not treated surgically and were excluded from the study.

Age and sex distribution

At the time of operation, there was a clear preponderance of males among the 63 patients (35 men, 28 women). The average age of the patients at the time of surgery was 68 years (32 to 84 years). It is noticeable within the age distribution that almost three-quarters of the client group (46 patients, 73%) were recruited from the 60 to 79 year age group.

Follow-up investigations

From the original 63 patients admitted into the study, it was only possible to conduct follow-up investigations on 51, despite intensive efforts. A total of 9 patients had already died at the time of follow-up. One patient died 6 days after the shunt operation from a pulmonary embolism despite correct thrombosis prophylaxis. The cause of death for another 5 patients was established through questioning their relatives or general practitioners. From this we learned that 2 patients died of pneumonia, 1 patient of heart failure, 1 patient from kidney failure, and 1 patient from a malignant disease not identified at the time of shunt implantation. For the remaining 3 deaths no data could be collected, making it impossible to establish whether the causes of death were independent of the valve implant. The current place of residence of the 3 patients was not known and could not be established even with the help of the Residents' Registration Office in Berlin. The authors were also unsuccessful in obtaining further information about these patients using the addresses of general practitioners and relatives contained in the treatment notes. The average age of the patients at the time of follow-up was 70 years (34 to 86 years, n = 51).

Course of disease

Figure 1 shows a lower Kiefer score immediately following surgery compared with preoperative assessment. The measured improvement is statistically significant ($p < 0.0005$). These values were essentially reproduced on follow-up an average of 3 years later. Here, too, there is a statistically significant difference between the preoperative values and the values an average of 3 years after implantation ($p < 0.0005$). The minimal increase in arithmetical mean value between the postoperative values and the values after an average of 3 years is not significant ($p = 0.691$).

The clinical outcomes presented by the NPH-R-R show that in comparing the preoperative and postoperative symptoms it was possible to achieve an excellent outcome in 16% of patients, a good outcome in 27%, a satisfactory outcome in 37%, and a poor clinical outcome in 20% of patients. The comparison of preoperative investigation findings with those after 3 years shows the outcome to be excellent in 33% of the patients, good in 18%, satisfactory in 16%, and unsatisfactory in 33% of patients (Fig. 2). To summarize, a positive clinical outcome (excellent, good, or satisfactory) could be measured in 67% of patients 3 years after insertion of the ventriculoperitoneal shunt.

Discussion

Epidemiology

The literature on iNPH provides no information about the frequency of the disease based on the analysis of patient groups of epidemiologically relevant sizes. The incidences established using smaller collectives differ considerably from each other. Vanneste [13] gives values of 0.13 to 0.22 per 100,000 people

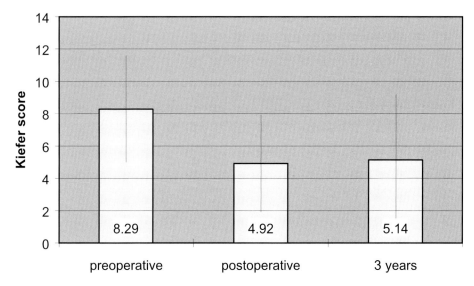

Fig. 1. Kiefer score (arithmetical mean and standard deviation)

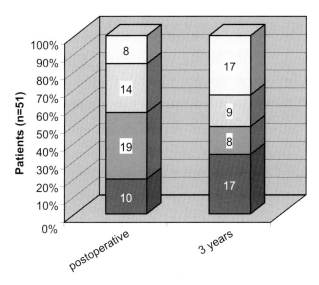

Fig. 2. Clinical outcome according to NPH-R-R. ☐ excellent; ☐ good; ▨ satisfactory; ■ poor

per year. In a more recent work, Krauss and Halve [7] assume an annual incidence of 1.8 per 100,000 people. In a survey of 982 inhabitants of Starnberg over 65 years old, Trenkwalder et al. [12] established a prevalence of 0.41% for NPH.

Outcome

The improvement rates cited in the available reviews of original publications on NPH show a relatively broad distribution. In 1984, Børgesen [2] established average improvement rates of 52% (42% to 67%) in an analysis of 6 original publications written between 1972 and 1980. Hebb and Cusimano [3] analyzed 44 original publications in 2001 and cited an average improvement rate of 59% and a long-term improvement rate of 29%. The following values can be taken from large single studies: for 147 operated patients with iNPH, Romner and Zemack [14] cite an improvement rate of 79% after an average of 27 months (3 months to 9 years); Boon et al. [1] established for 96 patients in both parts of the study (low pressure vs. medium pressure standard valves) an average improvement rate of 53% and 47%, respectively; Mori et al. [11] established an average improvement rate after 3 years of 73% for 120 patients with iNPH; Kiefer et al. [5] observed an improvement rate of 70% after an average of 26 months in 122 patients with NPH.

In our present study, figures similar to those cited in the literature could generally be repeated in our own patient cohort. Whereas in the postoperative investigation of patients, excellent, good, and satisfactory clinical outcomes were measured in 80% of the cases, 3 years after the operation it was only possible to verify an excellent, good, or satisfactory outcome in 67% of the patients. On closer observation, the results disclose a trend towards the two extremes. The patient groups with satisfactory or good outcomes decreased markedly (from 37% to 16% and from 27% to 18%, respectively), whereas the groups with excellent or poor outcomes both increased considerably. However, in comparing these figures with those of other authors it should be noted that, in our working group, an improvement of less than 20% (NPH-R-R < 2) is already classified as a poor clinical outcome.

Overall it can be seen that a proportion of the patients investigated experienced further improvement in their symptoms during the 3 years following surgery. This speaks for the reversibility of certain damage processes within a section of the patient collective. A further explanation may be offered by the clinical improvement in the postoperative period as part of the adaptation to unphysiological (over-)drainage. Kiefer [4] sees this possibility for example in the framework of reactive congestion, but he is referring to the pre-hydrostatic valve era. On the other hand, some of the patients we investigated experienced further deterioration of their symptoms after shunt implantation. This also corresponds to expectations, for various reasons. First, even with accurate diagnosis a proportion of non-responders did not show any clinical improvement at postoperative examination. The obvious attempts at explanation here would be that these patients were either wrongly diagnosed or were in the advanced stage of disease and whose atrophic cranial components were no longer reversible by shunt therapy. After unsatisfactory postoperative results, it is probable that no improvement can be expected for these non-responders in the further progression of the illness. Second, when observing patients with iNPH, a high comorbidity of cardiovascular and musculoskeletal systems must be assumed by reason of patient age alone. Even if the Kiefer score claims a specific orientation toward the symptoms of NPH, it cannot be completely ruled out that some of the symptoms, for instance gait disturbances, are the result of other coexisting causes. For this reason it must be assumed that patients with increasing age will tend to have poorer ratings. Third, iNPH is by its nature a progressive disease, so that a stagnating or at least slowed down deterioration of the clinical symptoms represents an advantage against the hypothetical progression of the untreated disease.

References

1. Boon AJ, Tans JT, Delwel EJ, Egeler-Peerdeman SM, Hanlo PW, Wurzer HA, Avezaat CJ, de Jong DA, Gooskens RH, Hermans J (1998) Dutch Normal-Pressure Hydrocephalus Study: randomized comparison of low- and medium-pressure shunts. J Neurosurg 88: 490–495
2. Borgesen SE (1984) Conductance to outflow of CSF in normal pressure hydrocephalus. Acta Neurochir (Wien) 71: 1–45
3. Hebb AO, Cusimano MD (2001) Idiopathic normal pressure hydrocephalus: a systematic review of diagnosis and outcome. Neurosurgery 49: 1166–1186
4. Kiefer M, Eymann R, Mascaros V, Walter M, Steudel WI (2000) Significance of hydrostatic valves in therapy of chronic hydrocephalus. Nervenarzt 71: 975–986 [in German]
5. Kiefer M, Eymann R, Meier U (2002) Five years experience with gravitational shunts in chronic hydrocephalus of adults. Acta Neurochir (Wien) 144: 755–767
6. Kiefer M, Eymann R, Komenda Y, Steudel WI (2003) A grading system for chronic hydrcephalus. Zentralbl Neurochir 64: 109–115 [in German]
7. Krauss JK, Halve B (2004) Normal pressure hydrocephalus: survey on contemporary diagnostic algorithms and therapeutic decision-making in clinical practice. Acta Neurochir (Wien) 146: 379–388
8. Meier U (2004) Gravity valves for idiopathic normal pressure hydrocephalus. A prospective study of 60 patients. Nervenarzt 75: 577–583 [in German]
9. Meier U, Bartels P (2001) The importance of the intrathecal infusion test in the diagnostic of normal-pressure hydrocephalus. Eur Neurol 46: 178–186
10. Meier U, Kunzel B, Zeilinger FS, Riederer A (2000) Pressure-dependent flow resistance in craniospinal cerebrospinal fluid dynamics: a calculation model for diagnosis of normal pressure hydrocephalus. Biomed Tech (Berl) 45: 26–33 [in German]
11. Mori K (2001) Management of idiopathic normal-pressure hydrocephalus: a multiinstitutional study conducted in Japan. J Neurosurg 95: 970–973
12. Trenkwalder C, Schwarz J, Gebhard J, Ruland D, Trenkwalder P, Hense HW, Oertel WH (1995) Starnberg trial on epidemiology of Parkinsonism and hypertension in the elderly. Prevalence of Parkinson's disease and related disorders assessed by a door-to-door survey of inhabitants older than 65 years. Arch Neurol 52: 1017–1022
13. Vanneste JA (2000) Diagnosis and management of normal-pressure hydrocephalus. J Neurol 247: 5–14
14. Zemack G, Romner B (2002) Adjustable valves in normal-pressure hydrocephalus: a retrospective study of 218 patients. Neurosurgery 51: 1392–1402

Correspondence: Ullrich Meier, Department of Neurosurgery, Unfallkrankenhaus Berlin, Warener Str 7, 12683 Berlin, Germany. e-mail: ullrich.meier@ukb.de

Is it possible to optimize treatment of patients with idiopathic normal pressure hydrocephalus by implanting an adjustable Medos Hakim valve in combination with a Miethke shunt assistant?

U. Meier and **J. Lemcke**

Department of Neurosurgery, Unfallkrankenhaus Berlin, Berlin, Germany

Summary

A better course of the disease after implantation of a low-pressure valve in patients with idiopathic normal pressure hydrocephalus normally comes at the cost of a distinctly higher rate of overdrainage. Can combining an adjustable valve with a gravity unit produce optimization of treatment results?

In a prospective observation of the course of the disease, 18 patients with idiopathic normal pressure hydrocephalus were surgically treated with the aforementioned valve combination during the period January to June 2004 at the Unfallkrankenhaus Berlin and examined after 6 and 12 months.

The course of the disease correlates with the opening pressure level of the valve. The controlled setting of the valve from 100 mmH$_2$O to 70 mmH$_2$O, then to 50 mmH$_2$O after 3 months permits the brain to adapt optimally to the implanted valve without complications from overdrainage.

In our view, combining an adjustable differential pressure valve with a gravity unit currently represents the optimal treatment variant for patients with idiopathic normal pressure hydrocephalus. In the future, the gravity valve should also be adjustable.

Keywords: Normal pressure hydrocephalus; idiopathic; outcome; programmable valves.

Introduction

Despite modern diagnostic methods, the accurate diagnosis and treatment of idiopathic normal pressure hydrocephalus (iNPH) represents a challenge for the clinician [10]. Even after a precise diagnosis is made, surgical treatment is rife with complications, which has a considerable influence on the clinical progress and rate of improvement in patients with iNPH. The so-called "hydraulic mismanagement" [10] of hydrocephalus shunts includes a non-physiological function either from overdrainage while standing or from relative underdrainage while lying down. This situation is technically unavoidable in the case of conventional valves, since the latter only permits a compromise between the requirements of the various body positions [1]. While the majority of younger patients clinically show a considerable tolerance for non-physiological intracranial pressure, this is a great problem for older patients with iNPH and vascular compromise [20, 21].

Materials and methods

From January to June 2004, 18 patients with iNPH at the neurosurgical clinics of the Unfallkrankenhaus Berlin-Marzahn were treated surgically by the implantation of an adjustable Medos Hakim valve (Codman/Johnson & Johnson, Raynham, MA) combined with a Miethke shunt assistant, (Christoph Miethke GmbH & Co. KG, Berlin, Germany). The average age of the 10 men (56%) and 8 women (44%) was 65 years (33 to 82 years).

Diagnostics

After a clinical examination for gait ataxia and additional symptoms [18], and the verification of ventricular enlargement using neuroradiology imaging methods, an intrathecal infusion test was conducted. A computer-assisted dynamic infusion test using the constant flow technique with an infusion rate of 2 mL/min via lumbar puncture was used to calculate the individual parameters of cerebrospinal fluid dynamics [17, 19]. A resistance to outflow of greater than 13 mmHg * min/mL was defined as pathological [15]. The diagnostic fluid drainage of at least 60 mL fluid took place immediately following the infusion test. With improvement in clinical symptomatology in the following 2 to 3 days, the indication for a shunt operation was present. If symptomatology, particularly gait ataxia, did not improve, external lumbar drainage was performed for 2 to 3 days. If symptomatology improved after this, the described valve combination was implanted as a ventriculo-peritoneal shunt [22].

Investigation protocol of the prospective study

The controlled setting of the valve from 100 mmH$_2$O to 70 mmH$_2$O during the first postoperative week, then 50 mmH$_2$O

Table 1. *Black grading scale for shunt assessment*

Grading	Description	NPH recovery rate
Excellent	activity level same as before illness	≥7 points
Good	slight limitations	≥5 points
Fair	gradual improvement	≥3 points
Transient	temporary improvement	≥2 points
Poor	no change or worsening	<2 points

NPH Normal pressure hydrocephalus.

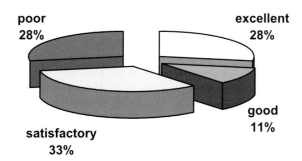

Fig. 1. Disease progression 1 year after the shunt operation

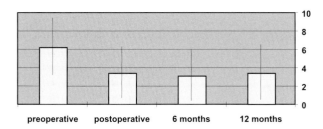

Fig. 2. Clinical signs and symptoms in accordance with the Kiefer scale

after 3 months was undertaken with the point of view that relative underdrainage is iatrogenically generated due to fluid loss during the operation, for which the aforementioned valve setting is intended to compensate. On the other hand, the height of the patient is crucial for selecting the gravity valve (heights less than 160 cm = 250 mmH$_2$O; 160 to 180 cm = 300 mmH$_2$O; greater than 180 cm = 350 mmH$_2$O for the shunt assistant). This combination of an adjustable Medos Hakim valve and a Miethke shunt assistant regulates intracranial pressure using the differential pressure valve when lying down and the gravitation valve when standing up.

Clinical grading

The results of the clinical study were evaluated in accordance with the Black grading scale for shunt assessment (Table 1) and the normal pressure hydrocephalus (NPH) recovery rate [14] based on Kiefer's [13] clinical grading for NPH.

All graduated study results were collected into 4 groups. In this case, groups were identified with NPH recovery rates as follows: very good outcome, ≥7 points; good outcome, ≥5 points; satisfactory outcome, ≥2 points; poor outcome, <2 points.

$$\text{NPH Recovery Rate} = \frac{\text{NPH Grading}_{\text{preoperative}} - \text{NPH Grading}_{\text{postoperative}}}{\text{NPH Grading as per Kiefer [13]}_{\text{preoperative}}} \times 10$$

Statistics

Statistical analyses were conducted using Fisher exact test with a probability of error of p = 0.05.

Results

Outcome

Treatment results of the 18 patients 1 year after shunt implantation are depicted in Fig. 1. In summary, we were able to report good to very good treatment success with 7 patients, satisfactory treatment success with 6 patients, and poor treatment success with 5 patients. The average value on the Kiefer scale (Fig. 2) corresponding to the severity of the symptomatology induced by the iNPH dropped from 6 points preoperatively to 3 points postoperatively and 3 points during follow-up examinations; a significant reduction by one-half. Since clinical grading for NPH per Kiefer [13] is better suited for observing individual progress than for comparing groups [14], the values of the NPH recovery rate were compared postoperatively with those of the follow-up examination 6 months and 1 year after shunt implantation (Fig. 3). The average proportional value postoperatively was 5.2, which is the transitional area between a satisfactory and good treatment result, followed by values of 5.0 after 6 months and 4.3 after a year, again, good or satisfactory treatment results. Average improvement in clinical symptomatology was 50% or 43%.

Complications

In regard to the valve-independent complications after shunt implantation, we found an infection rate of 0%. In addition, no ventricular catheters had to be revised due to incorrect placement or peritoneal catheters due to dislocation in the abdominal wall (dislocation: 0%). Thus, the valve-independent complications in this relatively small patient group were 0%. Using neuroradiology imaging methods, no patient showed a clear reduction in the ventricular size or subdural hygromas or chronic subdural hematomas as a direct sign of overdrainage (overdrainage rate: 0%).

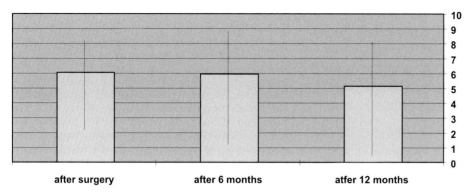

Fig. 3. Outcome in accordance with the NPH recovery rate

A 78-year-old female patient showed deterioration of symptomatology 3 months after the shunt operation. The cause was diagnosed as an inadvertent valve readjustment to 200 mmH$_2$O, a quasi-closure of the valve caused by a magnetic bracelet that was acquired during a sightseeing bus excursion. The fact that this patient places a hearing aid in her right ear multiple times each day, and in doing so, moves the magnetic bracelet on her wrist near the adjustable valve, conclusively explained the inadvertent valve readjustment. After the valve was readjusted to 50 mmH$_2$O, clinical symptomatology improved considerably, particularly the gait ataxia. In another patient, there were difficulties adjusting the valve in accordance with the investigation protocol. By using a stronger magnet it was possible to readjust this valve. All patients were x-rayed after valve adjustment or readjustment. There were no problems identifying the current valve pressure level in the process. The valve-dependent complication rate from unintentional valve adjustment is 6% (1 patient), and shunt-dependent morbidity was 6%. No patient died during the study period. As a result, the operation-related fatality rate was 0%.

Discussion

The fundamental problem of pressure gradient increases when the body's axis is vertical created by additional hydrostatic pressure has been countered in the past using various strategies. It was shown early on that overdrainage particularly occurring in an upright position could be prevented by using standard valves with high opening pressure. However, this came at the price of permanent underdrainage in a horizontal position, when fluid drainage no longer occurred because the hydrostatic pressure component had been eliminated and because of the high valve opening pressure [3]. The occurrence of nocturnal pressure peaks in cerebrospinal fluid could not be compensated for in this way. Although using standard valves with low opening pressures produced a better outcome, this was counteracted by a substantially higher overdrainage rate [4]. Malfunctions from surrounding scar tissue occurred with subsequent designs such as the anti-siphon device [12]. While the development of adjustable valves opened up the possibility of programming the opening pressure so high that overdrainage no longer occurred, sufficient fluid drainage was still not guaranteed. Bergsneider [3] called attention to the fact that when using programmable valves without a gravitational component, partial compensation of the drop in hydrostatic pressure that occurs in a vertical position could be observed because of pressure increases when the body's axis is brought upright. However, this effect is hardly calculable and has even less potential for being regulated. Not until the development of gravity-controlled valves, which can switch between one pressure level for a lying position and another for a standing position, was a decisive step made toward eradicating this problem.

Outcome

The literature shows that the general rate of improvement in patients with NPH after a shunt operation varies between 31% and 96% with an average value of 53% [2, 6]. According to a meta-analysis, Vanneste [24] cites improvement rates of 30 to 50% for iNPH. In a meta-analysis, Hebb *et al.* [11] indicate improvement rates of 59% after a shunt operation for iNPH and long-term improvement rates of 29%. The results of our study, with a general improvement rate of 72%, lie within the range of the citations in the literature. Drake *et al*. [8] found no difference in the course of the disease between different valve groups of children after shunt operation.

Complications

After shunt operations in patients with NPH, Grumme *et al*. [9] reported a fatality rate of 0% to 6% and overdrainage phenomena in 6% to 20%. Hebb *et al*. [11] placed the complication rate at 38%, the revision rate at 22%, and the combination of a lasting neurological deficit and fatality from the operation at 6%. In England's [23] shunt registry with a very large patient number of over 9,000 cases, the most frequent complication is underdrainage with an incidence of 52%. Overdrainage as a postoperative cause for complications is reported in this study as very low (3%). This is in contrast with the Scandinavian working groups [7], which make the statement that 80% of all shunt complications arise due to overdrainage. A Dutch multicenter study [5, 7] achieved a significantly better outcome in patients with NPH if standard valves were implanted in the low-pressure range as opposed to implantation of standard valves in the mid-pressure range, but this advantage came at the cost of a higher overdrainage rate of 73% versus 34%. Our investigation protocol takes into account that adjusting the valve opening pressure to 50 mmH$_2$O makes a better outcome possible than with valve opening pressures of 100 mmH$_2$O or greater [16].

Conclusions

In our view, combining an adjustable Medos Hakim valve with a Miethke shunt assistant currently represents the optimal treatment variant for patients with iNPH. Advantages of the programmable Medos Hakim valve are good intraoperative handling, extensive clinical experience [25] over many years, and simple valve adjustment (pressure setting). Considered disadvantageous are inadvertent readjustments of the valve and the necessity for radiological checks after every valve readjustment. In the case of precise vertical implantation, the Miethke shunt assistant functions smoothly as a gravity unit. There were no overdrainage complications.

References

1. Aschoff A (1994) In-vitro-Testung von Hydrocephalus-Ventilen. Habilitationsschrift, Heidelberg, Germany
2. Bech RA, Waldemar G, Gjerris F, Klinken L, Juhler M (1999) Shunting effects in patients with idiopathic normal pressure hydrocephalus; correlation with cerebral and leptomeningeal biopsy findings. Acta Neurochir (Wien) 141: 633–639
3. Bergsneider M, Yang I, Hu X, McArthur DL, Cook SW, Boscardin WJ (2004) Relationship between valve opening pressure, body position, and intracranial pressure in normal pressure hydrocephalus: paradigm for selection of programmable valve pressure setting. Neurosurgery 55: 851–859
4. Boon AJ, Tans JT, Delwel EJ, Egeler-Peerdeman SM, Hanlo PW, Wurzer HA, Avezaat CJ, de Jong DA, Gooskens RH, Hermans J (1998) Dutch Normal-Pressure Hydrocephalus Study: randomized comparison of low- and medium-pressure shunts. J Neurosurg 88: 490–495
5. Boon AJ, Tans JT, Delwel EJ, Egeler-Peerdeman SM, Hanlo PW, Wurzer HA, Hermans J (1999) Dutch Normal Pressure Hydrocephalus Study: the role of cerebrovascular disease. J Neurosurg 90: 221–226
6. Borgesen SE (1984) Conductance to outflow of CSF in normal pressure hydrocephalus. Acta Neurochir (Wien) 71: 1–45
7. de Jong DA, Delwel EJ, Avezaat CJ (2000) Hydrostatic and hydrodynamic considerations in shunted normal pressure hydrocephalus. Acta Neurochir (Wien) 142: 241–247
8. Drake JM, Kestle JR, Milner R, Cinalli G, Boop F, Piatt J Jr, Haines S, Schiff SJ, Cochrane DD, Steinbok P, MacNeil N (1998) Randomized trial of cerebrospinal fluid shunt valve design in pediatric hydrocephalus. Neurosurgery 43: 294–305
9. Grumme T, Kolodziejczyk D (1995) Komplikationen in der Neurochirurgie. Band 2 Kraniale, zerebrale und neuropädiatrische Chirurgie. Blackwell, Berlin, pp 534–540
10. Hakim CA (1985) The physics and physicopathology of the hydraulic complex of the central nervous system. The mechanics of hydrocephalus and normal pressure hydrocephalus. [Dissertation] Massachusetts Institute of Technology
11. Hebb AO, Cusimano MD (2001) Idiopathic normal pressure hydrocephalus: a systematic review of diagnosis and outcome. Neurosurgery 49: 1166–1186
12. Kiefer M, Eymann R, Mascaros V, Walter M, Steudel WI (2000) Significance of hydrostatic valves in therapy of chronic hydrocephalus. Nervenarzt 71: 975–986 [in German]
13. Kiefer M, Eymann R, Meier U (2002) Five years experience with gravitational shunts in chronic hydrocephalus of adults. Acta Neurochir (Wien) 144: 755–767
14. Meier U (2002) The grading of normal pressure hydrocephalus. Biomed Tech (Berl) 47: 54–58
15. Meier U, Bartels P (2001) The importance of the intrathecal infusion test in the diagnostic of normal-pressure hydrocephalus. Eur Neurol 46: 178–186
16. Meier U, Kiefer M (2004) Zum optimalen Öffnungsdruck hydrostatischer Ventile beim idiopathischen Normaldruckhydrozephalus: Eine prospektive Studie mit 122 Patienten. Akt Neurol 31: 216–222
17. Meier U, Zeilinger FS, Kintzel D (1999) Diagnostic in normal pressure hydrocephalus: A mathematical model for determination of the ICP-dependent resistance and compliance. Acta Neurochir (Wien) 141: 941–948
18. Meier U, Zeilinger FS, Kintzel D (1999) Signs, symptoms and course of normal pressure hydrocephalus in comparison with cerebral atrophy. Acta Neurochir (Wien) 141: 1039–1048
19. Meier U, Kiefer M, Bartels P (2002) The ICP-dependency of resistance to cerebrospinal fluid outflow: a new mathematical

method for CSF-parameter calculation in a model with H-TX rats. J Clin Neurosci 9: 58–63
20. Meier U, Kiefer M, Sprung C (2002) The Miethke Dual-Switch valve in patients with normal pressure hydrocephalus. Neurosurg Q 12: 114–121
21. Meier U, Kiefer M, Sprung C (2003) Normal-pressure hydrocephalus: pathology, pathophysiology, diagnostics, therapeutics and clinical course. PVV Science Publications, Ratingen
22. Meier U, Kiefer M, Sprung C (2004) Evaluation of the Miethke dual-switch valve in patients with normal pressure hydrocephalus. Surg Neurol 61: 119–128
23. Richards HK, Seeley HM, Pickard JD (2000) Shunt revisions: data from the UK shunt registry. Eur J Pediatr Surg (Suppl I) 10: 59
24. Vanneste JA (2000) Diagnosis and managment of normal-pressure hydrocephalus. J Neurol 247: 5–14
25. Zemack G, Romner B (2000) Seven years of clinical experience with the programmable Codman Hakim valve: a retrospective study of 583 patients. J Neurosurg 92: 941–948

Correspondence: Ullrich Meier, Department of Neurosurgery, Unfallkrankenhaus Berlin, Warener Str. 7, 12683 Berlin, Germany. e-mail: ullrich.meier@ukb.de

Aquaporins

Increased seizure duration in mice lacking aquaporin-4 water channels

D. K. Binder[1], X. Yao[1], A. S. Verkman[2], and G. T. Manley[1]

[1] Department of Neurological Surgery, University of California, San Francisco, CA, USA
[2] Department of Medicine and Physiology, University of California, San Francisco, CA, USA

Summary

Aquaporins are intrinsic membrane proteins involved in water transport in fluid-transporting tissues. In the brain, aquaporin-4 (AQP4) is expressed widely by glial cells, but its function is unclear. Extensive basic and clinical studies indicate that osmolarity affects seizure susceptibility, and in our previous studies we found that AQP4 −/− mice have an elevated seizure threshold in response to the chemoconvulsant pentylenetetrazol. In this study, we examined the seizure phenotype of AQP4 −/− mice in greater detail using in vivo electroencephalographic recording. AQP4 −/− mice were found to have dramatically longer stimulation-evoked seizures following hippocampal stimulation as well as a higher seizure threshold. These results implicate AQP4 in water and potassium regulation associated with neuronal activity and seizures.

Keywords: Epilepsy; brain edema; water transport; glial cells; aquaporin-4.

Introduction

Relative cellular and extracellular space (ECS) volume has been demonstrated to play an important role in propensity to seizures in vitro. In particular, decreasing ECS volume in hippocampal slice preparations by hypotonic exposure produces hyperexcitability and enhanced epileptiform activity [8, 16, 20], whereas hyperosmolar medium attenuates epileptiform activity [8, 21]. These experimental data parallel extensive clinical experience indicating that hypoosmolar states, such as hyponatremia, lower seizure threshold while hyperosmolar states elevate seizure threshold [4]. However, the role of osmolarity, water transport, and seizure susceptibility in vivo has not been well-studied.

The aquaporins are a family of membrane proteins that function as "water channels" in many cell types and tissues in which fluid transport is crucial [22]. Aquaporin-4 (AQP4) is abundantly expressed by glial cells lining the ependymal and pial surfaces that are in contact with cerebrospinal fluid in the ventricular system and subarachnoid space [19], and highly polarized AQP4 expression is also found in astrocytic foot processes near or in direct contact with blood vessels [19]. Recently, we found that mice deficient in AQP4 (AQP4 −/−) had decreased cerebral edema and improved neurological outcome following water intoxication and focal cerebral ischemia [14, 15].

In view of the potential role of AQP4 in mediating water fluxes in response to neuronal activity and perhaps in seizure-induced edema, we examined seizure susceptibility in AQP4 −/− mice and found elevated seizure threshold in response to the chemoconvulsant pentylenetetrazol (PTZ) [5]. In the current study, we examined the seizure phenotype of AQP4 −/− mice in greater detail using in vivo electroencephalographic (EEG) recording following hippocampal stimulation-evoked seizures.

Materials and methods

AQP4 −/− mice

AQP4 −/− mice, in a CD1 genetic background, were generated as previously described [13]. These mice lack detectable AQP4 protein by immunoblot and immunocytochemical analysis, and phenotypically have normal growth, development, survival, and neuromuscular function except for a mild defect in maximal urine-concentrating ability produced by decreased water permeability in the inner medullary collecting duct.

Electrode implantation

Male AQP4 −/− mice (n = 8) and wild-type (WT) littermates (n = 10) were anesthetized with 2,2,2-tribromoethanol (125 mg/kg,

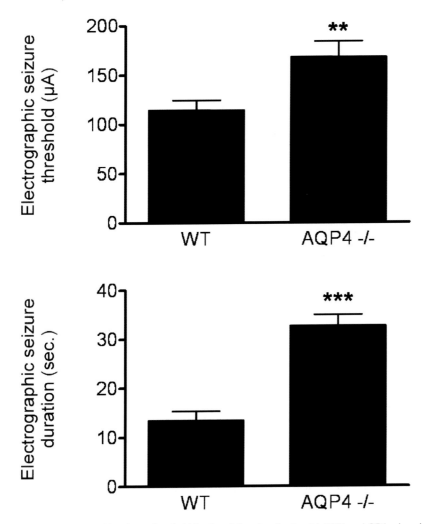

Fig. 1. Electrographic seizure threshold (top) and duration (bottom) in WT vs. AQP4 −/− mice. AQP4 −/− mice had a higher electrographic seizure threshold (167 ± 17 μA) than WT controls (114 ± 10 μA) (p < 0.01). Interestingly, AQP4 −/− mice were found to have dramatically longer stimulation-evoked seizures (32.6 ± 2.3 seconds) compared to WT controls (13.4 ± 1.9 seconds) (p < 0.001)

i.p.) and placed in a standard mouse stereotaxic frame. Bipolar electrodes made from Teflon-coated stainless steel wire were implanted in the right dorsal hippocampus (bregma as reference: 2.0 mm posterior; 1.5 mm lateral; 1.8 mm below dura) [9]. Electrodes were secured firmly to the skull with dental cement and anchor screws, and a ground wire was attached to 1 anchor screw. Mice were then allowed to recover for at least 3 days prior to stimulation.

Seizure stimulation procedure

Following postoperative recovery, electrical stimulations were given to assess electrographic seizure threshold and duration. Each stimulation consisted of a 60-Hz 1-sec train of 1 msec biphasic rectangular pulses delivered by a digital stimulator (BIOPAC Systems Inc., Goleta, CA). To determine electrographic seizure threshold (EST), stimulation intensity was increased by 10 μA increments starting at 20 μA. Presence or absence of electrographic seizure discharge was recorded for each stimulation by an observer blinded to genotype. All EEG data were recorded by a digital signal acquisition system (BIOPAC Systems Inc.) and analyzed by an observer blinded to genotype. EST was recorded as the threshold at which at least 3 seconds of hippocampal afterdischarge (seizure) was recorded. Electrographic seizure duration was recorded as the total duration of electrographic seizure at the EST.

Results

Baseline EEG in WT and AQP4 −/− mice was indistinguishable (not shown). AQP4 −/− mice had a higher mean electrographic seizure threshold (167 ± 17 μA) than WT controls (114 ± 10 μA) (Fig. 1, top). This is consistent with our previous results using the PTZ model [5]. Interestingly, AQP4 −/− mice were found to have dramatically longer stimulation-evoked seizures (32.6 ± 2.3 seconds) compared to WT controls (13.4 ± 1.9 seconds) (Fig. 1, bottom).

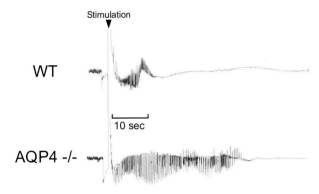

Fig. 2. Representative hippocampal stimulation-induced electrographic seizures are shown for a WT mouse (top) and an AQP4 −/− mouse (bottom). The WT mouse had an 11-second seizure (stimulation threshold: 100 μA), whereas the AQP4 −/− mouse had a much longer seizure (37 seconds) (stimulation threshold: 170 μA). Behavioral arrest was observed in both animals during the seizure

Representative examples of hippocampal stimulation-evoked seizures for WT vs. AQP4 −/− mice are shown in Fig. 2.

Discussion

These results demonstrate that absence of AQP4 increases EST and also dramatically increases seizure duration. Together with our previous results in the PTZ model [5], these results implicate the glial water channel AQP4 in the modulation of brain excitability in vivo. Importantly, developmental effects of the AQP4-null mutation cannot be excluded.

There is increasing evidence that water movement in the brain involves aquaporin channels [1, 15]. Since brain tissue excitability is very sensitive to ECS volume, AQP4 deletion may alter ECS volume or composition at baseline and/or following neuronal activity. A larger ECS volume fraction prior to seizure-inducing stimuli and/or a blunted reduction in ECS volume during neuronal activity via abrogation of water influx through glial AQP4 may limit neuronal excitability and synchrony. Indeed, using a novel in vivo cortical photobleaching paradigm, we have recently found evidence for a larger ECS volume fraction at baseline in AQP4 −/− mice [6].

Impaired sequestration and/or redistribution of K^+ may underlie the increased seizure duration in AQP4 −/− mice. Indeed, altered glial/neuronal water and K^+ recycling in response to neuronal activity has already been proposed to underlie impaired hearing in AQP4 −/− mice [12]. In support of this possibility is the colocalization of AQP4 with the inwardly-rectifying potassium channel, Kir4.1 [7, 17], which is thought to mediate spatial K^+ buffering by astrocytes [10, 11].

A recent model for the role of AQP4 in neural signal transduction comes from study of mice deficient in the gene α-syntrophin. Deletion of α-syntrophin, an adapter protein in the dystrophin-containing protein complex required for anchoring AQP4 to specialized membrane domains [1, 18], also leads to attenuated brain edema in response to transient cerebral ischemia [2]. In a recent study, Amiry-Moghaddam et al. [3] studied hippocampal slices from α-syntrophin-deficient mice and found a deficit in extracellular K^+ clearance following evoked neuronal activity. In addition, using a hyperthermia model of seizure induction, they found more of the α-syntrophin-deficient mice had more severe seizures than WT mice. These data are consistent with the idea that AQP4 and its molecular partners (e.g. Kir4.1, α-syntrophin, dystrophin) together comprise a multifunctional 'unit' responsible for clearance of K^+ and/or H_2O following neural activity. Thus, further understanding of the glial modulation of ECS ion and water homeostasis in epileptic tissue may lead to novel concepts and targets for anticonvulsant drug development.

References

1. Amiry-Moghaddam M, Ottersen OP (2003) The molecular basis of water transport in the brain. Nat Rev Neurosci 4: 991–1001
2. Amiry-Moghaddam M, Otsuka T, Hurn PD, Traystman RJ, Haug FM, Froehner SC, Adams ME, Neely JD, Agre P, Ottersen OP, Bhardwaj A (2003) An alpha-syntrophin-dependent pool of AQP4 in astroglial end-feet confers bidirectional water flow between blood and brain. Proc Natl Acad Sci USA 100: 2106–2111
3. Amiry-Moghaddam M, Williamson A, Palomba M, Eid T, de Lanerolle NC, Nagelhus EA, Adams ME, Froehner SC, Agre P, Ottersen OP (2003) Delayed K+ clearance associated with aquaporin-4 mislocalization: phenotypic defects in brains of alpha-syntrophin-null mice. Proc Natl Acad Sci USA 100: 13615–13620
4. Andrew RD, Fagan M, Ballyk BA, Rosen AS (1989) Seizure susceptibility and the osmotic state. Brain Res 498: 175–180
5. Binder DK, Oshio K, Ma T, Verkman AS, Manley GT (2004) Increased seizure threshold in mice lacking aquaporin-4 water channels. Neuroreport 15: 259–262
6. Binder DK, Papadopoulos MC, Haggie PM, Verkman AS (2004) In vivo measurement of brain extracellular space diffusion by cortical surface photobleaching. J Neurosci 24: 8049–8056
7. Dalloz C, Sarig R, Fort P, Yaffe D, Bordais A, Pannicke T, Grosche J, Mornet D, Reichenbach A, Sahel J, Nudel U, Re-

ndon A (2003) Targeted inactivation of dystrophin gene product Dp71: phenotypic impact in mouse retina. Hum Mol Genet 12: 1543–1554
8. Dudek FE, Obenhaus A, Tasker JG (1990) Osmolality-induced changes in extracellular volume alter epileptiform bursts independent of chemical synapses in the rat: importance of nonsynaptic mechanisms in hippocampal epileptogenesis. Neurosci Lett 120: 267–270
9. Franklin KBJ, Paxinos G (1997) The mouse brain in stereotaxic coordinates. Academic Press, San Diego
10. Horio Y (2001) Potassium channels of glial cells: distribution and function. Jpn J Pharmacol 87: 1–6
11. Kofuji P, Connors NC (2003) Molecular substrates of potassium spatial buffering in glial cells. Mol Neurobiol 28: 195–208
12. Li J, Verkman AS (2001) Impaired hearing in mice lacking aquaporin-4 water channels. J Biol Chem 276: 31233–31237
13. Ma T, Yang B, Gillespie A, Carlson EJ, Epstein CJ, Verkman AS (1997) Generation and phenotype of a transgenic knockout mouse lacking the mercurial-insensitive water channel aquaporin-4. J Clin Invest 100: 957–962
14. Manley GT, Fujimura M, Ma T, Noshita N, Filiz F, Bollen AW, Chan P, Verkman AS (2000) Aquaporin-4 deletion in mice reduces brain edema after acute water intoxication and ischemic stroke. Nat Med 6: 159–163
15. Manley GT, Binder DK, Papadopoulos MC, Verkman AS (2004) New insights into water transport and edema in the central nervous system from phenotype analysis of aquaporin-4 null mice. Neuroscience 129: 983–991
16. McBain CJ, Traynelis SF, Dingledine R (1990) Regional variation of extracellular space in the hippocampus. Science 249: 674–677
17. Nagelhus EA, Veruki ML, Torp R, Haug FM, Laake JH, Nielsen S, Agre P, Ottersen OP (1998) Aquaporin-4 water channel protein in the rat retina and optic nerve: polarized expression in Muller cells and fibrous astrocytes. J Neurosci 18: 2506–2519
18. Neely JD, Amiry-Moghaddam M, Ottersen OP, Froehner SC, Agre P, Adams ME (2001) Syntrophin-dependent expression and localization of Aquaporin-4 water channel protein. Proc Natl Acad Sci USA 98: 14108–14113
19. Nielsen S, Nagelhus EA, Amiry-Moghaddam M, Bourque C, Agre P, Ottersen OP (1997) Specialized membrane domains for water transport in glial cells: high-resolution immunogold cytochemistry of aquaporin-4 in rat brain. J Neurosci 17: 171–180
20. Roper SN, Obenhaus A, Dudek FE (1992) Osmolality and nonsynaptic epileptiform bursts in rat CA1 and dentate gyrus. Ann Neurol 31: 81–85
21. Traynelis SF, Dingledine R (1989) Role of extracellular space in hyperosmotic suppression of potassium-induced electrographic seizures. J Neurophysiol 61: 927–938
22. Verkman AS, Mitra AK (2000) Structure and function of aquaporin water channels. Am J Physiol Renal Physiol 278: F13–F28

Correspondence: Geoffrey T. Manley, Department of Neurological Surgery, University of California, San Francisco, 1001 Potrero Avenue, Room 101, San Francisco, CA 94110, USA. e-mail: manley@itsa.ucsf.edu

Modulation of AQP4 expression by the protein kinase C activator, phorbol myristate acetate, decreases ischemia-induced brain edema

A. Kleindienst, G. Fazzina, A. M. Amorini, J. G. Dunbar, R. Glisson, and A. Marmarou

Department of Neurosurgery, Medical College of Virginia, Virginia Commonwealth University, Richmond, VA, USA

Summary

The protein kinase C activator, phorbol 12-myristate 13-acetate (PMA), is known to interact with aquaporin-4 (AQP4), a water-selective transporting protein abundant in astrocytes and ependymal cells, that has been found to decrease osmotically-induced swelling. The purpose of this study was to examine whether PMA given at different time points following focal ischemia induced by middle cerebral artery occlusion (MCAO) reduces brain edema by AQP4 modulation.

Male Sprague-Dawley rats were randomly assigned to sham procedure, vehicle, or PMA infusion (230 μg/kg), starting either 60 minutes before, or 30 or 60 minutes after MCAO (each group n = 12). After a 2-hour period of ischemia and 2 hours of reperfusion, the animals were sacrificed for assessment of brain water content, sodium, and potassium concentrations. AQP4 expression was assessed by immunoblotting. Statistical analysis was performed by ANOVA followed by Tukey's post hoc test.

PMA treatment significantly reduced brain water content concentration in the infarcted area when started before or 30 minutes post-occlusion ($p < 0.001$, $p = 0.022$) and prevented the subsequent sodium shift ($p < 0.05$). Furthermore, PMA reduced ischemia-induced AQP4 up-regulation ($p \leq 0.05$).

Attenuation of the ischemia-induced AQP4 up-regulation by PMA suggests that the reduction in brain edema formation following PMA treatment was at least in part mediated by AQP4 modulation.

Keywords: Phorbol ester; PMA; middle cerebral artery occlusion; brain edema; AQP4.

Introduction

Aside from the initial extent of brain damage caused by any type of brain insult, the resulting brain edema and consecutive increase in intracranial pressure contributes substantially to mortality in these patients. The cellular localization of the aquaporin-4 (AQP4) water channels in the central nervous system suggests the importance of AQP4 in maintaining cerebral water balance [22]. Recently, an important role of AQP4 in water homeostasis during traumatic and ischemic brain edema development has been described [17, 20], although controversial results have been obtained.

The exact mechanisms of AQP4 regulation have not yet been identified. However, direct and indirect evidence suggests that protein kinase C (PKC) is involved in AQP4 regulation [14, 15, 25]. PKC serves as a second messenger for G-protein receptors that are coupled to the phosphoinositide pathway, causing either a transient rise in intracellular Ca^{2+} through inositol-triphosphate or activating PKC through diacylglycerol. Furthermore, the PKC activator, phorbol 12-myristate 13-acetate (PMA), is known to interact with AQP4, and has been shown to decrease osmotically-induced swelling in AQP4-transfected *Xenopus laevis* oocytes [5]. PKC activation has been found to decrease AQP4 mRNA in cultured astrocytes [25], but prolonged treatment eliminated the subsequent decrease in AQP4 mRNA [14].

Since we found PMA can reduce brain edema and subsequent electrolyte imbalance following cortical contusion injury and middle cerebral artery occlusion (MCAO) when PMA was given before induction of ischemia, the present study investigated whether PMA is as effective given after the onset of ischemia. Specifically, we assessed the brain water, sodium, and potassium content following an intravenous PMA infusion started at different time points after MCAO. The AQP4 expression was assessed by immunoblotting and quantified by densitometry.

Materials and methods

Animals and surgical procedure

The studies were conducted under approval of the Institutional Animal Care and Use Committee using National Institutes of Health guidelines. Experiments were carried out on 350 to 400 g adult male Sprague-Dawley rats (Harlan, Indianapolis, IN). Rats were housed at $22 \pm 1\,°C$ with 60% humidity, with a 12-hour light/12-hour dark cycle, and pellet food and water ad libitum. Surgery was performed after intubation under halothane anesthesia and controlled ventilation (1.3% halothane in 70% nitrous oxide and 30% oxygen). Rectal temperature was maintained at $36.5 \pm 0.5\,°C$ using a heat lamp. The left femoral artery and vein were cannulated with polyethylene tubing (P.E. 50, Becton Dickinson, Sparks, MD) for continuous monitoring of mean arterial blood pressure (MABP), arterial blood sampling, or intravenous drug infusion. Adequate ventilation was verified by an arterial blood gas measurement after 1 hour of anesthesia.

Cerebral blood flow (CBF) to tissue perfused by the right middle cerebral artery was continuously monitored by Laser Doppler Flowmetry (LaserFlo Vasamedics Inc., St Paul, MN) through a burr hole located 1 mm posterior and 5 mm lateral to bregma leaving the dura mater intact. Animals were placed in a supine position over the laser Doppler probe, and CBF as well as MABP were recorded continuously using a data acquisition system (ADInstruments, Colorado Springs, CO).

MCAO was induced using the intraluminal suture method described elsewhere [1], slightly modified. Through a midline neck incision, the bifurcation of the right common carotid artery was exposed, and branches of the external carotid artery (ECA) and internal carotid artery (ICA) including the occipital, lingual, and maxillary arteries were microsurgically separated and coagulated. The ECA was ligated with a 4-0 silk suture, and after temporary occlusion of the ICA and common carotid artery with vascular mini-clips, a 4-0 monofilament nylon suture (4-0 SN-644 MONOSOF nylon Polyamide) with a silicon tip of 0.3 mm diameter was inserted through the ECA stump and secured with a suture. The clips were removed and the filament was advanced through the ICA into the circle of Willis while occluding the pterygopalatine artery with a forceps. A CBF reduction between 70 and 80% to baseline was observed when the suture was advanced at a distance of 22 to 24 mm from the carotid bifurcation, thereby verifying proper MCAO. Two hours after occlusion, a 2-hour period of reperfusion was performed by withdrawing the suture into the ECA stump, confirming a consecutive increase in CBF.

Study protocol and drug preparation

Our objective was to assess the effect of intravenous PMA infusion on brain swelling started at different intervals after MCAO. The animals were randomly assigned to vehicle or PMA infusion starting 60 minutes before or either 30 minutes or 60 minutes after induction of ischemia, with each group consisting of 6 animals. According to the literature [12], 230 µg/kg PMA (Sigma-Aldrich, St Louis, MO) was dissolved in 1% dimethyl sulfoxide as vehicle solution (Sigma-Aldrich). The drug was intravenously administered using a continuous infusion pump (sp210w syringe pump, KD Scientific, Holliston, MA). To keep the total drug concentration administered constant in different treatment groups, the infusion rate was adapted to either 480 µL/hr, 640 µL/hr, or 720 µL/hr. After the experiments were completed, the animals were sacrificed by an overdose of halothane, decapitated, and the brains were removed.

Tissue processing

The cerebral tissue was immediately cut into 4 consecutive 4 mm coronal sections excluding the most rostral and caudal sections from further analysis. After division into the right and left hemispheres along the anatomic midline, the 4 regional samples obtained were processed for water content measured by the wet/dry weight method. The wet weight of each sample was measured using an electronic analytical balance before drying the sample at $95\,°C$ for 5 days and reweighing to obtain the dry weight. The water content of each sample was given as percentage of total tissue weight. For measurement of brain sodium and potassium concentrations, the dried samples were placed in a furnace for 24 hours at $400\,°C$ and reduced to ashes. The ash was then extracted with distilled water, and the concentrations of sodium and potassium were determined using a flame photometer (943 nm; Instrument Laboratory, San Jose, CA) with caesium as an internal standard.

Immunoblot

After sacrifice, brains were removed and immediately cut on dry ice into two 2 mm sections excluding 4 mm of the rostral tissue. Slices were cut into the right and left hemisphere, then the striatum and the parietal cortex were isolated, minced finely at $4\,°C$ with a Potter Elvehjem tissue grinder, and each sample was homogenized in 800 µL of radioimmunoprecipitation buffer (50 mmol Tris, 150 mmol NaCl, 1% NP40, 1% deoxycholic acid, 0.5% sodium n-dodecyl sulfate, pH 7.2). This homogenate was centrifuged in a Thermo MicroMax RF centrifuge (Thermo Electron Corporation, Needham Heights, MA) at $14000\times g$ for 30 minutes at $4\,°C$ to remove nuclei and mitochondria. From the resultant pellet, gel samples in 2% sodium n-dodecyl sulfate were made. Supernatant samples were run on 12% Bis-Tris gels (Invitrogen Nu-Page, Invitrogen, Carlsbad, CA). After transfer by electroelution to nitrocellulose membranes, blots were blocked with 5% milk powder in tris buffered saline plus 0.1% Tween 20 (pH 7.5) for 1 hour and incubated with primary antibodies (monoclonal mouse anti-AQP4, 1:1000; Ab-Cam, Cambridge, MA). The labeling was visualized with horseradish peroxidase-conjugated secondary antibody (goat anti-mouse, 1:5000; Rockland, Gilbertsville, PA) using an enhanced chemiluminescence system (Amersham, Buckinghamshire, UK). Controls were made by replacing primary antibody with non-immune IgG (cyclopylin). Quantification was performed by densitometry.

Statistical analysis

SPSS software (SigmaStat, Chicago, IL) was used for statistical analysis. The data were analyzed by a randomized one-way ANOVA for group variations followed by a Tukey post hoc analysis. Statistical significance was accepted at $p < 0.05$.

Results

Injury-induced mortality was 14% following MCAO. MABP and arterial blood gases were kept within physiological limits throughout the experimental procedure, requiring few adjustments in the halothane concentration and ventilation parameters.

Drug effect on brain edema

The ANOVA of brain water content comparing vehicle and PMA infusion started at different time points following MCAO produced a significant group effect ($F_{4,79} = 11.25$, $p < 0.001$) (Table 1). Tukey post hoc

Table 1. *Effect of PMA on brain water, sodium, and potassium content**

Groups	% Tissue water	Tissue sodium mEq/kg dry wt	Tissue potassium mEq/kg dry wt
Vehicle (n = 18)	2.2 ± 0.1	115.1 ± 11.2	−43.2 ± 6.0
PMA 60 min before (n = 6)	0.8 ± 0.3	48.2 ± 9.2**	−12.6 ± 5.2**
PMA 30 min after (n = 6)	1.2 ± 0.1	51.8 ± 8.1**	−20.7 ± 14.1**
PMA 60 min after (n = 6)	1.5 ± 0.4	59.9 ± 21.3	−27.5 ± 16.2**

* Data are presented as average ± SEM, ** $p < 0.05$ as compared with vehicle.
PMA Phorbol 12-myristate 13-acetate.

Table 2. *Effect of PMA on aquaporin-4 expression in the ischemic area assessed by immunoblot following middle cerebral artery occlusion*

Group	Average	SEM	n	Significance (p-value)
Sham	0.68	0.20	3	
DMSO	183.59	20.06	4	0.050 compared to sham
PMA	4.76	2.12	4	0.999 compared to sham

DMSO Dimethyl sulfoxide; *PMA* phorbol 12-myristate 13-acetate.

analysis indicated that PMA treatment started either 60 minutes before or 30 minutes after MCAO significantly reduced brain edema in the ischemic area compared to vehicle infusion ($p < 0.001$ and $p \leq 0.022$, respectively), while a PMA infusion started 60 minutes after MCAO failed to suppress edema formation.

The ANOVA of brain sodium content comparing vehicle and PMA infusion started at different time points following MCAO produced a significant group effect ($F_{4,79} = 10.01$, $p < 0.001$). Tukey post-hoc analysis indicated that PMA treatment started either 60 minutes before, or 30 or 60 minutes after MCAO significantly reduced brain sodium accumulation in the ischemic area compared to vehicle infusion ($p \leq 0.004$, $p \leq 0.008$, and $p \leq 0.045$, respectively).

The ANOVA of brain potassium content comparing vehicle and PMA infusion started at different time points following MCAO produced a significant group effect ($F_{4,79} = 3.42$, $p \leq 0.013$). Tukey post-hoc analysis indicated that PMA treatment started either 60 minutes before, or 30 or 60 minutes after MCAO did not reduce the brain potassium loss in the ischemic area compared to vehicle infusion ($p = 0.079$, $p = 0.284$, and $p = 0.542$, respectively).

Drug effect on AQP4 expression

The ANOVA of AQP4 expression in the ischemic hemisphere comparing sham treatment, vehicle, and PMA infusion started 60 minutes before MCAO produced a significant group effect ($F_{3,64} = 3.59$, $p \leq 0.05$) (Table 2). Tukey post hoc analysis indicated that AQP4 was significantly up-regulated after MCAO following vehicle infusion compared to sham animals ($p = 0.050$). PMA treatment started 60 minutes before MCAO reduced the AQP4 up-regulation accompanying the brain edema, as indicated by the lack of a difference in AQP4 expression as compared to sham treatment ($p = 0.999$).

Discussion

This study confirms that brain edema following cerebral ischemia is reduced by treatment with PMA. We demonstrated that PMA, started either 60 minutes before or 30 minutes after MCAO, significantly reduced the brain water content in the ischemic hemisphere and prevented the shift of electrolytes accompanying brain edema development. These findings are consistent with the generally accepted opinion that water and sodium tend to coexist and transfer together through the plasma membrane in physiological and pathological conditions [4, 10, 24].

Phorbol esters are tumor-promoting and inflammatory agents [7, 23]. Some toxic effects have been reported when PMA was administered in dogs, rabbits, and rats [9, 16, 21]. However, PMA has been used without any irreversible toxic effect [6, 18], and a species difference in susceptibility to PMA-induced toxicity has been suggested [12]. The physiological parameters monitored continuously throughout our study confirmed that blood pressure and pulmonary function were not affected by PMA treatment. However, more detailed pathological studies are required to verify that PMA administration does not induce any relevant toxicity.

In vitro studies demonstrated an inhibition of AQP4 activity [5] as well as a down-regulation of AQP4 expression [14, 25] by PMA. Changes in AQP4 mRNA levels in rat brain after permanent focal ischemia or cortical contusion injury have been described [17, 19, 20]. Furthermore, deletion of the AQP4 gene in knockout mice reduced brain edema and perivascular foot process swelling after acute water intoxication and ischemic stroke [13]. We utilized a model of focal ischemia, where the cytotoxic edema seems to be predominant [3, 11]. Thus, the reduction in brain edema after MCAO observed in our experiments may have been the result of a PMA-dependent regulation of AQP4.

We found that ischemia-induced AQP4 up-regulation was prevented by PMA. Thus, PMA-treated animals did not express more AQP4 following MCAO than animals subjected to sham occlusion. The down-regulation of AQP4 following ischemia and PMA treatment occurs simultaneously with the reduction in brain edema and suggests these effects are related. Although the exact biochemical pathways of AQP4 regulation are not yet known, experimental evidence suggests the involvement of PKC and, in view of our experiments, AQP4 expression seems to be altered by the PKC-activator, PMA. However, further studies are necessary to elucidate the precise mechanism of AQP4 regulation and subsequent brain edema development following ischemia.

Conclusion

Different mechanisms have been proposed to explain the origin of brain swelling during the early stages of cerebral ischemia, but this issue remains unresolved [2, 8]. We demonstrated that intravenous infusion of PMA decreases brain water content and subsequent shift of electrolytes in a rat model of MCAO. Since the attenuation of brain edema development by PMA is accompanied by an attenuation of ischemia-induced AQP4 up-regulation, this study suggests that the effect of PMA on brain edema is mediated by AQP4, possibly utilizing PKC-dependent receptor signaling.

Acknowledgments

This research was supported by grants NS 12587 and NS 19235 from the National Institutes of Health, Bethesda, MD.

References

1. Belayev L, Alonso OF, Busto R, Zhao W, Ginsberg MD (1996) Middle cerebral artery occlusion in the rat by intraluminal suture. Neurological and pathological evaluation of an improved model. Stroke 27: 1616–1623
2. Chen Y, Swanson RA (2003) Astrocytes and brain injury. J Cereb Blood Flow Metab 23: 137–149
3. Dijkhuizen RM, de Graaf RA, Tulleken KA, Nicolay K (1999) Changes in the diffusion of water and intracellular metabolites after excitotoxic injury and global ischemia in neonatal rat brain. J Cereb Blood Flow Metab 19: 341–349
4. Gotoh O, Asano T, Koide T, Takakura K (1985) Ischemic brain edema following occlusion of the middle cerebral artery in the rat. I: The time courses of the brain water, sodium and potassium contents and blood-brain barrier permeability to 125I-albumin. Stroke 16: 101–109
5. Han Z, Wax MB, Patil RV (1998) Regulation of aquaporin-4 water channels by phorbol ester-dependent protein phosphorylation. J Biol Chem 273: 6001–6004
6. Han ZT, Tong YK, He LM, Zhang Y, Sun JZ, Wang TY, Zhang H, Cui YL, Newmark HL, Conney AH, Chang RL (1998) 12-O-Tetradecanoylphorbol-13-acetate (TPA)-induced increase in depressed white blood cell counts in patients treated with cytotoxic cancer chemotherapeutic drugs. Proc Natl Acad Sci USA 95: 5362–5365
7. Hussaini IM, Karns LR, Vinton G, Carpenter JE, Redpath GT, Sando JJ, VandenBerg SR (2000) Phorbol 12-myristate 13-acetate induces protein kinase ceta-specific proliferative response in astrocytic tumor cells. J Biol Chem 275: 22348–22354
8. Kimelberg HK, Rutledge E, Goderie S, Charniga C (1995) Astrocytic swelling due to hypotonic or high K+ medium causes inhibition of glutamate and aspartate uptake and increases their release. J Cereb Blood Flow Metab 15: 409–416
9. Lafuze JE, Baker MD, Oakes AL, Baehner RL (1987) Comparison of in vivo effects of intravenous infusion of N-formyl-methionyl-leucyl-phenylalanine and phorbol myristate acetate in rabbits. Inflammation 11: 481–488
10. Loo DD, Zeuthen T, Chandy G, Wright EM (1996) Cotransport of water by the Na+/glucose cotransporter. Proc Natl Acad Sci USA 93: 13367–13370
11. Loubinoux I, Volk A, Borredon J, Guirimand S, Tiffon B, Seylaz J, Meric P (1997) Spreading of vasogenic edema and cytotoxic edema assessed by quantitative diffusion and T2 magnetic resonance imaging. Stroke 28: 419–427
12. Manabe S, Lin YC, Takaoka M, Yamoto T, Yanai T, Yamashita K, Tarumi C, Matsunuma N, Masuda H (1992) Species difference in susceptibility to phorbol myristate acetate-induced leukopenia and lung injury: rat vs. dog. J Toxicol Sci 17: 211–223
13. Manley GT, Fujimura M, Ma T, Noshita N, Filiz F, Bollen AW, Chan P, Verkman AS (2000) Aquaporin-4 deletion in mice reduces brain edema after acute water intoxication and ischemic stroke. Nat Med 6: 159–163
14. Nakahama K, Nagano M, Fujioka A, Shinoda K, Sasaki H (1999) Effect of TPA on aquaporin 4 mRNA expression in cultured rat astrocytes. Glia 25: 240–246
15. Neely JD, Christensen BM, Nielsen S, Agre P (1999) Heterotetrameric composition of aquaporin-4 water channels. Biochemistry 38: 11156–11163
16. O'Flaherty JT, Cousart S, Lineberger AS, Bond E, Bass DA, DeChatelet LR, Leake ES, McCall CE (1980) Phorbol myristate acetate: in vivo effects upon neutrophils, platelets, and lung. Am J Pathol 101: 79–92
17. Sato S, Umenishi F, Inamasu G, Sato M, Ishikawa M, Nishizawa M, Oizumi T (2000) Expression of water channel mRNA following cerebral ischemia. Acta Neurochir [Suppl] 76: 239–241
18. Strair RK, Schaar D, Goodell L, Aisner J, Chin KV, Eid J, Senzon R, Cui XX, Han ZT, Knox B, Rabson AB, Chang R, Conney A (2002) Administration of a phorbol ester to patients with hematological malignancies: preliminary results from a phase I clinical trial of 12-O-tetradecanoylphorbol-13-acetate. Clin Cancer Res 8: 2512–2518
19. Sun MC, Honey CR, Berk C, Wong NL, Tsui JK (2003) Regu-

lation of aquaporin-4 in a traumatic brain injury model in rats. J Neurosurg 98: 565–569
20. Taniguchi M, Yamashita T, Kumura E, Tamatani M, Kobayashi A, Yokawa T, Maruno, Kato A, Ohnishi T, Kohmura E, Tohyama M, Yoshimine T (2000) Induction of aquaporin-4 water channel mRNA after focal cerebral ischemia in rat. Brain Res Mol Brain Res 78: 131–137
21. Taylor RG, McCall CE, Thrall RS, Woodruff RD, O'Flaherty JT (1985) Histopathologic features of phorbol myristate acetate-induced lung injury. Lab Invest 52: 61–70
22. Venero JL, Vizuete ML, Ilundain AA, Machado A, Echevarria M, Cano J (1999) Detailed localization of aquaporin-4 messenger RNA in the CNS: preferential expression in periventricular organs. Neuroscience 94: 239–250
23. Won JS, Song DK, Kim YH, Huh SO, Suh WH (1998) The stimulation of rat astrocytes with phorbol-12-myristate-13-acetate increases the proenkephalin mRNA: involvement of proto-oncogenes. Brain Res Mol Brain Res 54: 288–297
24. Wright EM, Loo DD (2000) Coupling between Na+, sugar, and water transport across the intestine. Ann N Y Acad Sci 915: 54–66
25. Yamamoto N, Sobue K, Miyachi T, Inagaki M, Miura Y, Katsuya H, Asai K (2001) Differential regulation of aquaporin expression in astrocytes by protein kinase C. Brain Res Mol Brain Res 95: 110–116

Correspondence: Anthony Marmarou, Department of Neurosurgery, Virginia Commonwealth University Medical Center, 1101 East Marshall Street, Box 980508, Richmond, VA 23298-0508, USA. e-mail: marmarou@abic.vcu.edu

Astrocytes co-express aquaporin-1, -4, and vascular endothelial growth factor in brain edema tissue associated with brain contusion

R. Suzuki[1], M. Okuda[1], J. Asai[1], G. Nagashima[1], H. Itokawa[1], A. Matsunaga[1], T. Fujimoto[1], and T. Suzuki[2]

[1] Department of Neurosurgery, Showa University Fujigaoka Hospital, Yokohama, Japan
[2] Department of Laboratory of Histochemistry, Showa University Fujigaoka Hospital, Yokohama, Japan

Summary

Introduction. Brain edema may be life threatening. The mechanisms underlying the development of traumatic brain edema are still unclear; however, mixed mechanisms including vasogenic, ischemic, and neurotoxic types of edema may be contributors.

Recent studies indicate that astrocytes, aquaporins (AQPs; a protein family of water channels), and vascular endothelial growth factor (VEGF) may have important roles in the formation and resolution of brain edema. We studied the expression of AQPs and VEGF in the edematous brain.

Methods. We investigated the expression of AQP1, AQP4, and vascular endothelial growth factor (VEGF) in contusional brain tissue surgically obtained from 6 patients. Glial fibrillary acidic protein (GFAP) was also stained to detect astrocytes and to clarify the location of those proteins. The specimens received immunohistological staining and 3-color immunofluorescent staining, and were observed using confocal laser scanning microscopy.

Results. AQP1, AQP4, and VEGF were co-expressed in GFAP-positive astrocytes. AQP1 and AQP4 were expressed strongly in astrocytic end-feet. The astrocytes were located in the edematous tissue, and some cells surrounded cerebral capillaries.

Conclusion. Our results suggest that AQP1, AQP4, and VEGF are induced in astrocytes located in and surrounding edematous tissue. Those astrocytes may regulate the water in- and out-flow in the injured tissue.

Keywords: Brain edema; astrocyte; aquaporin; VEGF; traumatic brain edema.

Introduction

Brain edema may be life-threatening. The mechanisms underlying the development of traumatic brain edema are still unclear; however, mixed mechanisms including vasogenic, ischemic, and neurotoxic types of edema may be contributors [4].

Recent studies indicate that astrocytes, aquaporins (AQPs; a protein family of water channels), and vascular endothelial growth factor (VEGF) may have important roles in the formation and resolution of brain edema [2, 6, 10]. We studied the expression of AQPs and VEGF in the edematous brain.

Materials and methods

Surgical specimens from 6 patients with brain contusion were obtained. The age of the patients ranged from 18 to 67 years, and the period from onset to operation varied from 3 hours to 72 hours.

The specimens were formalin-fixed, embedded in paraffin, then received immunohistological staining and a newly-developed 3-color immunofluorescent staining. The antibodies included anti-VEGF rabbit polyclonal antibody (Oncogene Research Products, Cambridge, MA), anti-glial fibrillary acidic protein (GFAP) mouse monoclonal antibody (clone: 6F-2, DakoCytomation, Copenhagen, Denmark), anti-AQP1 polyclonal antibody (Chemicon, Temecula, CA), and anti-AQP4 polyclonal antibody (Chemicon).

For this study, we developed a novel multicolor immunofluorescent staining method using heat treatment. The details of the staining procedures are described elsewhere [8]. The fluorescent dyes used were fluorescein isothiocyanate (FITC) (DakoCytomation), Cy3 (Chemicon), and Cy5 (Chemicon), which were resistant to heating at $90\,°C$ for 15 minutes, whereas antigenicity of the primary antibodies was lost completely. The specimens were observed using a confocal laser scanning microscope (Olympus FV500, Tokyo, Japan). In the fluorescent images, FITC was seen in yellow-green, Cy3 in orange-red, and Cy5 in blue.

Results

Histological studies, confirmed by hematoxylin and eosin staining, showed brain contusion and many vacuolated spaces compatible with brain edema. Immunohistological studies revealed that VEGF and AQPs were expressed in glial cells located in the edematous tissue and cells located in the white matter where the edema fluid extends.

Color immunofluorescent staining revealed that VEGF and AQP1 were co-expressed in the same

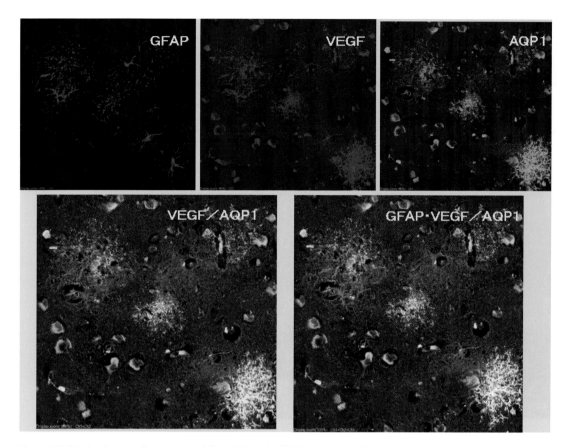

Fig. 1. Multicolor immunofluorescent staining of Case 4. GFAP was stained by Cy5 (blue), VEGF was stained by Cy3 (orange-red), and AQP1 is stained by FITC (yellow-green). Double color of VEGF and AQP1 (VEGF/AQP1) appears yellow, and triple color of GFAP, VEGF, and AQP1 appears white. Staining revealed that VEGF and AQP1 are co-expressed in the same cells that were positive for GFAP. The proteins were widely distributed within the cells, including the fine astrocytic end-feet. GFAP and VEGF were more widely stained than AQP1

GFAP-positive cells. The expression was widely distributed in the cells, including the fine astrocytic end-feet. GFAP and VEGF were more widely stained than AQP1 (Fig. 1). Expression of AQP1 and AQP4 was similarly distributed in the same GFAP-positive cells. GFAP is more widely stained than AQP1 or AQP4 (Fig. 2).

Discussion

Brain edema sometimes contributes to poor prognosis in patients with brain contusion. The mechanisms underlying the formation and resolution of brain edema need to be clarified in order to obtain a better prognosis in patients with cerebral contusion [4].

Recently, the water channel proteins of the AQP family that provide the molecular pathway for water permeability were discovered [11]. Among the AQP family, AQP1, AQP4, AQP5, and AQP9 were detected in brain [2], and AQP1, AQP4, and AQP9 were up-regulated [2, 5, 6] or reduced [3] in various pathological conditions including ischemia, trauma, and brain edema, and are believed to have an important role in the formation and resolution of brain edema in those pathological conditions [2]. Saadoun et al. [5] suggested that the signals that induce AQP expression in pathological conditions might include VEGF, which has been found in edematous tissue [7]. Many AQP water channels are found on astrocytes and their end-feet in contact with brain vessels [2]. Thus, astrocytic AQPs and VEGF may have an important role in the formation and resolution of brain edema [1, 2, 7, 9, 10].

In the present study, newly-developed 3-color immunofluorescent staining methods [8] enabled us to show that VEGF, AQP1, and AQP4 were expressed in the same astrocytes and in astrocytic end-feet located in the edematous tissue. These results confirm that

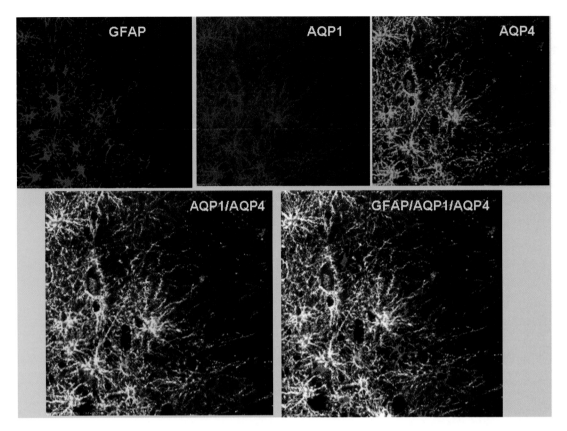

Fig. 2. Multicolor immunofluorescent staining of Case 5. GFAP was stained by Cy5 (blue), AQP1 was stained by Cy3 (orange-red), and AQP4 is stained by FITC (yellow-green). Double color of AQP1 and AQP4 (AQP1/AQP4) appears yellow, and triple color of GFAP, AQP1, and AQP4 appears white. Expression of AQP1 and AQP4 were similarly distributed in the same cells that were GFAP-positive. GFAP was more widely stained than AQP1

astrocytes express AQPs and VEGF to regulate the water in- and out-flow in the injured tissue, and may play an important role in the formation and resolution of brain edema.

Conclusion

VEGF, AQP1, and AQP4 were expressed in the same astrocytes and in astrocytic end-feet located in the edematous tissue. Astrocytes express those proteins to regulate the water in- and out-flow in the injured tissue, and may play an important role in the formation and resolution of brain edema.

References

1. Amiry-Moghaddam M, Frydenlund DS, Ottersen OP (2004) Anchoring of aquaporin-4 in brain: molecular mechanisms and implications for the physiology and pathophysiology of water transport. Neuroscience 129: 999–1010
2. Badaut J, Lasbennes F, Magistretti PJ, Regli L (2002) Aquaporins in brain: distribution, physiology, and pathophysiology. J Cereb Blood Flow Metab 22: 367–378
3. Ke C, Poon WS, Ng HK, Lai FM, Tang NL, Pang JC (2002) Impact of experimental acute hyponatremia on severe traumatic brain injury in rats: influences on injuries, permeability of blood-brain barrier, ultrastructural features, and aquaporin-4 expression. Exp Neurol 178: 194–206
4. Marmarou A (2003) Pathophysiology of traumatic brain edema: current concepts. Acta Neurochir [Suppl] 86: 7–10
5. Saadoun S, Papadopoulos MC, Davies DC, Bell BA, Krishna S (2002) Increased aquaporin 1 water channel expression in human brain tumors. Br J Cancer 87: 621–623
6. Saadoun S, Papadopoulos MC, Davies DC, Krishna S, Bell BA (2002) Aquaporin-4 expression is increased in oedematous human brain tumours. J Neurol Neurosurg Psychiatry 72: 262–265
7. Suzuki R, Fukai N, Nagashijma G, Asai JI, Itokawa H, Nagai M, Suzuki T, Fujimoto T (2003) Very early expression of vascular endothelial growth factor in brain oedema tissue associated with brain contusion. Acta Neurochir [Suppl] 86: 277–279
8. Suzuki T, Tate G, Ikeda K, Mitsuya T (2005) A novel multicolor immunofluorescence method using heat treatment. Acta Med Okayama [in press]
9. Tomas-Camardiel M, Venero JL, Herrera AJ, De Pablos RM, Pintor-Toro JA, Machado A, Cano J (2005) Blood-brain barrier disruption highly induces aquaporin-4 mRNA and protein in perivascular and parenchymal astrocytes: protective effect by es-

tradiol treatment in ovariectomized animals. J Neurosci Res 80: 235–246
10. van Bruggen N, Thibodeaux H, Palmer JT, Lee WP, Fu L, Cairns B, Tumas D, Gerlai R, Williams SP, van Lookeren Campagne ML, Ferrara N (1999) VEGF antagonisms reduces edema formation and tissue damage after ischemia/reperfusion injury in the mouse brain. J Clin Invest 104: 1613–1620
11. Verkman AS, Mitra AK (2000) Structure and function of aquaporin water channels. Am J Physiol Renal Physiol 278: F13–F28

Correspondence: Ryuta Suzuki, Department of Neurosurgery, Showa University Fujigaoka Hospital, 1-30 Fujigaoka, Aoba-ku, Yokohama, 227-8501, Japan. e-mail: ryuta@med.showa-u.ac.jp

Magnesium restores altered aquaporin-4 immunoreactivity following traumatic brain injury to a pre-injury state

M. N. Ghabriel[1], A. Thomas[1], and R. Vink[2]

[1] Department of Anatomical Sciences, University of Adelaide, South Australia, Australia
[2] Department of Pathology, University of Adelaide, South Australia, Australia

Summary

Magnesium reduces edema following traumatic brain injury (TBI), although the associated mechanisms are unknown. Recent studies suggest that edema formation after TBI may be related to alterations in aquaporin-4 (AQP4) channels. In this study, we characterize the effects of magnesium administration on AQP4 immunoreactivity following TBI.

Male Sprague-Dawley rats were injured by impact-acceleration diffuse TBI and a subgroup was administered 30 mg/kg magnesium sulphate 30 minutes after injury. Animals were fixed by perfusion 5 hours later, which corresponded to the time of maximum edema formation according to previous studies. One half of the brain was cut using a Vibratome and the other half blocked in paraffin wax. Wax and Vibratome sections were immunostained for detection of AQP4 by light and electron microscopy, respectively. In untreated animals, AQP4 immunoreactivity was increased in the subependymal inner glia limitans and the subpial outer glia limitans, and decreased in perivascular astrocytic processes in the cerebrum and brain stem. In contrast, animals treated with magnesium sulphate had AQP4 profiles similar to normal and sham control animals. We conclude that magnesium decreases brain edema formation after TBI, possibly by restoring the polarized state of astrocytes and by down-regulation of AQP4 channels in astrocytes.

Keywords: Cerebral edema; astrocytes; aquaporin; magnesium; TBI.

Introduction

The level of intracellular free magnesium declines in the brain following traumatic brain injury (TBI) and the degree of decline correlates with functional outcome [14, 18]. The reduction in brain free magnesium is a ubiquitous feature of brain trauma, and administration of magnesium salts after TBI leads to a significant improvement in neurological outcome [4, 13, 19]. Magnesium administration has also been shown to reduce brain swelling and edema after TBI [2, 3], and to be most effective when administered 30 minutes after injury. A therapeutic window of up to 24 hours may exist after TBI [5]. However, the mechanisms by which magnesium reduces water accumulation in the injured brain are unknown.

During the last decade, a family of integral membrane proteins known as aquaporins have been described, with a specific distribution in various mammalian tissues [15]. Aquaporins act as water channels that facilitate osmotically-induced water movements in and out of cells. In the brain, aquaporin-4 (AQP4) is expressed in astrocytes at the perivascular processes [11]. AQP4 may play an important role in the accumulation of water and the development of brain edema in a variety of neurological conditions [8, 16]. Water homeostasis in the brain is of paramount importance for optimal neuronal function, since alteration in the intracellular and extracellular water content would lead to disruption of ionic homeostasis and perturbation of electrical neuronal conduction.

The aim of this study is to investigate the effects of magnesium administration on AQP4 immunoreactivity in the brain 5 hours following severe brain trauma and to compare the results obtained from injured non-treated animals and control non-injured animals. Previous studies have shown that 5 hours after TBI corresponds to the period of maximum edema formation [20, 21].

Materials and methods

Male Sprague-Dawley rats (400 g) were anesthetized with isoflurane, intubated, and then injured using the 2-meter impact-acceleration model of diffuse TBI [9] as previously described [19]. At 30 minutes after injury, a subgroup of animals (n = 6) was administered magnesium sulphate (30 mg/kg) via the femoral vein, while a second subgroup (n = 7) was untreated. Two more animals served as shams (anesthetized and prepared for injury, but not injured), while 7 additional naïve animals were used as normal controls. Five hours after trauma, animals were killed by cardiac perfusion with 4% paraformaldehyde fixative in 0.1 M phosphate buffer, pH 7.4, and the brains collected. One half of the brain was processed in paraffin and sections were subjected to antigen retrieval in citrate buffer. Vibratome sections (60 mm) were cut from the other half of the brain. Wax and Vibratome sections were immunolabeled for AQP4 using a rabbit polyclonal antibody (Alpha Diagnostic International, Inc., San Antonio, TX) at 1/1000 or 1/2000 dilutions in normal goat serum overnight. Immunolabeling was completed using biotinylated secondary antibody (goat anti-rabbit IgG) and streptavidin-peroxidase complex (1:1000, Rockland Immunochemicals, Gilbertsville, PA). The peroxidase reaction product was developed using liquid diaminobenzidine and substrate-chromogen system (DakoCytomation, Carpinteria, CA) for 10 minutes. Rectangles of tissue were cut from the parietal lobe and pons from immunolabeled Vibratome sections, osmicated, and processed in resin. Ultrathin sections were examined with the electron microscope for the detection of peroxidase reaction product.

Results

Immunolabeling of wax-embedded brain sections from normal and sham control animals showed faint AQP4 immunoreactivity in the neuropil throughout the cerebrum, cerebellum, and brainstem. Immunolabeling showed heterogeneous distribution of AQP4, with the brainstem on the whole showing higher labeling intensity than the cerebrum. In the cerebrum (Fig. 1A), cortical vessels were outlined with immunoreactivity, but the subpial glia limitans was weakly labeled. The epithelium covering the choroid plexus was not labeled. Electron microscopy (Fig. 2A) showed electron-dense reaction product indicative of immunolabeling in astrocytic processes surrounding brain microvessels at the astrocyte-endothelial cell interface.

In injured animals, AQP4 expression 5 hours postinjury showed a range of variation amongst animals and in various parts of the brain. Despite inter-animal variation, there was a general trend for an increase in AQP4 immunoreactivity in the cerebrum (Fig. 1B), cerebellum, and brainstem. Increased immunoreactivity was seen in the neuropil and at the outer and inner glia limitans. However, in 3 of the 7 animals examined, microvessels invariably showed a clear reduction in perivascular immunoreactivity with loss of the linear outline of labeling. Electron microscopy showed a re-

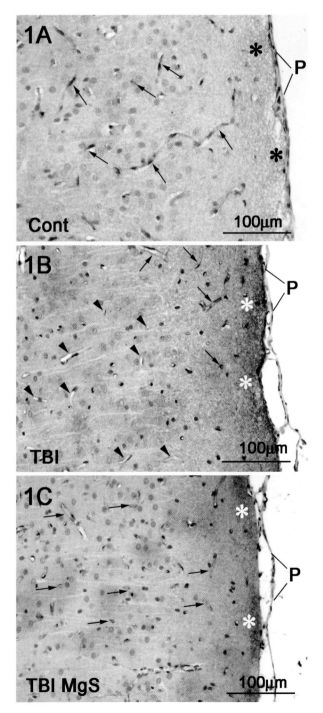

Fig. 1. Light micrographs of brain sections showing the parietal cortex and the pia mater. The sections were immunolabeled for AQP4 using the immunoperoxidase method. Positive labeling appears as brown reaction product. 1A is from a control (*Cont*) animal showing labeling of brain microvessels. Faint labeling is seen in the region of the outer glia limitans (*asterisks*) deep to the pia mater (*P*). 1B is from an injured animal and shows increased labeling in the neuropil and outer glia limitans (*asterisks*) deep to the pia mater (*P*). Microvessels near the surface show faint labeling (*arrows*) but most vessels deep in the cortex are not labeled (*arrowheads*). 1C is from an animal treated with magnesium sulphate (*MgS*) 30 minutes after injury, showing relative reduction of labeling in the neuropil and the outer glia limitans (*asterisks*). Blood vessels show increased labeling (*arrows*). The pia mater in all sections (*P*) is not labeled. All sections were lightly counterstained with hematoxylin

duction or absence of perivascular immunoreactivity (Fig. 2B).

Magnesium sulphate-treated animals showed an apparent reduction in immunolabeling of the outer (Fig. 1C) and inner glia limitans and a reconstitution of the linear perivascular labeling in most vessels (Fig. 1C), thus resembling the pattern of labeling seen in normal and sham animals. Electron microscopy showed perivascular labeling of astrocytic processes (Fig. 2C).

A semi-quantitative approach using a scale of 0 to 4 was used as an objective assessment of the labeling intensity in various regions of the cerebrum and brainstem, which confirmed the qualitative observations of a general increase in AQP4 immunoreactivity after injury and a reduction after magnesium sulphate treatment.

Discussion

In the current study, immunolabeling for AQP4 was used to characterize the distribution and localization of the protein. Control animals showed heterogeneous AQP4 distribution in the brain, with the brainstem showing higher labeling intensity than the cerebrum. Within the brainstem a distinct difference in labeling intensity was noted in the neuropil, where higher intensity was seen in the grey matter nuclei and the transverse pontine fiber, and minimal labeling was seen among the longitudinally directed fibers. A previous immunogold electron microscope study has clearly identified the localization of AQP4 to astrocytic processes [11]. The heterogeneity of AQP4 immunoreactivity detected in the current study may indicate that AQP4 is differentially expressed in certain subsets of astrocytes.

The relationship between AQP4 expression and brain injury has not been conclusively established. An in vitro study reported that hypoxia and reoxygenation leads to an up-regulation in AQP4 in astrocytes [21]. In human edematous brain tumors there is an up-regulation of AQP4 expression [12]. AQP4 knockout mice show less brain edema formation after water intoxication and cerebral ischemia [8]. The conclusion

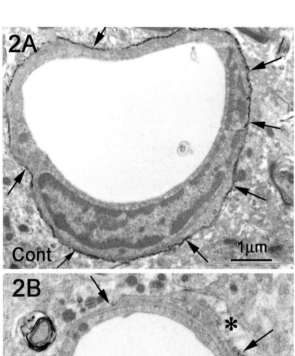
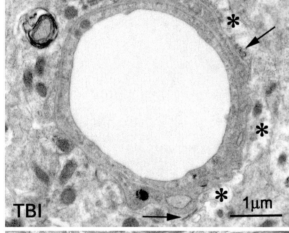
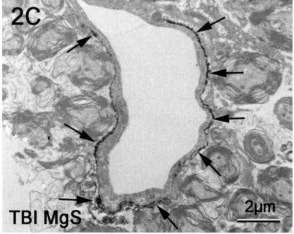

Fig. 2. Electron micrographs of brain vessels from a control animal (*Cont, 2A*), an injured animal (*TBI, 2B*), and an animal treated with magnesium sulphate 30 minutes after injury (*TBI MgS, 2C*). The sections show pre-embedding immunolabeling for AQP1. Positive labeling appears as electron-dense reaction product at the astrocyte-endothelial cell interface. In the control animal, labeling appears around most of the vessel circumference. In the injured animal, labeling is lost from the vessel circumference apart from a few spots (*arrows*). In the treated animals, restoration of labeling is seen around most of the vessel circumference (*arrows*). Asterisks in 2B indicate cellular edema

from the above studies is that brain edema is associated with an up-regulation of AQP4. In contrast, a recent study of cultured astrocytes reported a decrease in AQP4 levels in co-culture with glioma cells [1]. The authors suggested that a decreased expression of AQP4 induced by glioma cells is a mechanism for brain edema in human glioma. Down-regulation of AQP4 mRNA expression was suggested as a mechanism for water homeostasis in a model of TBI with hyponatremia [6]. Similarly, a global reduction in brain AQP4 expression coincides with the development of edema in focal cortical impact injury [7]. We have shown a generalized increase of AQP4 expression 5 hours post-injury. Further studies are needed to establish the relationship between AQP4 expression and the evolution of brain edema.

Our study shows an apparent reduction in immunoreactivity around blood vessels located near central regions of the brainstem and cerebrum associated with increased immunoreactivity in the outer and inner glia limitans. This may indicate that at this time point after TBI, the increase in AQP4 channels occurs by redistribution of the protein between different domains of astrocytes. A recent study reported a reduction in AQP4 immunoreactivity at 1 hour after cerebral hypoxia-ischemia, although a significant difference using Western blotting was detected only at 24 hours but not after 1 hour [10]. This finding may support our hypothesis that in TBI, early changes in the expression of AQP4 may be achieved by shifting AQP4 molecules from one domain of astrocytes to another, while alterations in AQP4 at later time points after brain injury may be accomplished by up-regulation of gene expression.

The administration of magnesium 30 minutes after trauma attenuated the changes in AQP4 immunoreactivity observed in untreated animals. Magnesium is involved in many cellular processes including inhibiting excitotoxicity and apoptosis, and improving the brain bioenergy state [17]. The current study suggests an additional role for magnesium in ameliorating post-traumatic secondary injury, and provides a possible mechanism for the reduction of brain water content by reducing the expression of AQP4 channels in astrocytes.

References

1. Chen YZ, Xu RX, Xu ZJ, Yang ZL, Jiang XD, Cai YQ (2003) Relationship between aquaporin-4 expression in astrocytes and brain edema caused by glioma [in Chinese]. Di Yi Jun Yi Da Xue Xue Bao 23: 566–568, 571
2. Esen F, Erdem T, Aktan D, Kalayci R, Cakar N, Kaya M, Telci L (2003) Effects of magnesium administration on brain edema and blood-brain barrier breakdown after experimental traumatic brain injury in rats. J Neurosurg Anesthesiol 15: 119–125
3. Feng DF, Zhu ZA, Lu YC (2004) Effects of magnesium sulfate on traumatic brain edema in rats. Chin J Traumatol 7: 148–152
4. Heath DL, Vink R (1999) Optimization of magnesium therapy after severe diffuse axonal brain injury in rats. J Pharmacol Exp Ther 288: 1311–1316
5. Heath DL, Vink R (1999) Improved motor outcome in response to magnesium therapy received up to 24 hours after traumatic diffuse axonal brain injury in rats. J Neurosurg 90: 504–509
6. Ke C, Poon WS, Ng HK, Lai FM, Tang NL, Pang JC (2002) Impact of experimental acute hyponatremia on severe traumatic brain injury in rats: influences on injuries, permeability of blood-brain barrier, ultrastructural features, and aquaporin-4 expression. Exp Neurol 178: 194–206
7. Kiening KL, van Landeghem FK, Schreiber S, Thomale UW, von Deimling A, Unterberg AW, Stover JF (2002) Decreased hemispheric Aquaporin-4 is linked to evolving brain edema following controlled cortical impact injury in rats. Neurosci Lett 324: 105–108
8. Manley GT, Fujimura M, Ma T, Noshita N, Filiz F, Bollen AW, Chan P, Verkman AS (2000) Aquaporin-4 deletion in mice reduces brain edema after acute water intoxication and ischemic stroke. Nat Med 6: 159–163
9. Marmarou A, Foda MA, van den Brink W, Campbell J, Kita H, Demetriadou K (1994) A new model of diffuse brain injury in rats. Part I: Pathophysiology and biomechanics. J Neurosurg 80: 291–300
10. Meng S, Qiao M, Lin L, Del Bigio MR, Tomanek B, Tuor UI (2004) Correspondence of AQP4 expression and hypoxic-ischaemic brain oedema monitored by magnetic resonance imaging in the immature and juvenile rat. Eur J Neurosci 19: 2261–2269
11. Nielsen S, Nagelhus EA, Amiry-Moghaddam M, Bourque C, Agre P, Ottersen OP (1997) Specialized membrane domains for water transport in glial cells: high-resolution immunogold cytochemistry of aquaporin-4 in rat brain. J Neurosci 17: 171–180
12. Saadoun S, Papadopoulos MC, Davies DC, Krishna S, Bell BA (2002) Aquaporin-4 expression is increased in oedematous human brain tumours. J Neurol Neurosurg Psychiatry 72: 262–265
13. Smith DH, Okiyama K, Gennarelli TA, McIntosh TK (1993) Magnesium and ketamine attenuate cognitive dysfunction following experimental brain injury. Neurosci Lett 157: 211–214
14. Suzuki M, Nishina M, Endo M, Matsushita K, Tetsuka M, Shima K, Okuyama S (1997) Decrease in cerebral free magnesium concentration following closed head injury and effects of VA-045 in rats. Gen Pharmacol 28: 119–121
15. Takata K, Matsuzaki T, Tajika Y (2004) Aquaporins: water channel proteins of the cell membrane. Prog Histochem Cytochem 39: 1–83
16. Taniguchi M, Yamashita T, Kumura E, Tamatani M, Kobayashi A, Yokawa T, Maruno M, Kato A, Ohnishi T, Kohmura E, Tohyama M, Yoshimine T (2000) Induction of aquaporin-4 water channel mRNA after focal cerebral ischemia in rat. Brain Mol Brain Res 78: 131–137
17. van den Heuvel C, Vink R (2004) The role of magnesium in traumatic brain injury. Clin Calcium 14: 9–14
18. Vink R, McIntosh TK, Demediuk P, Weiner MW, Faden AI (1988) Decline in intracellular free Mg2+ is associated with irreversible tissue injury after brain trauma. J Biol Chem 263: 757–761

19. Vink R, O'Connor CA, Nimmo AJ, Heath DL (2003) Magnesium attenuates persistent functional deficits following diffuse traumatic brain injury in rats. Neurosci Lett 336: 41–44
20. Vink R, Young A, Bennett CJ, Hu X, Connor CO, Cernak I, Nimmo AJ (2003) Neuropeptide release influences brain edema formation after diffuse traumatic brain injury. Acta Neurochir Suppl 86: 257–260
21. Yamamoto N, Yoneda K, Asai K, Sobue K, Tada T, Fujita Y, Katsuya H, Fugita M, Aihara N, Mase M, Yamada K, Miura Y, Kato T (2001) Alterations in the expression of the AQP family in cultured rat astrocytes during hypoxia and reoxygenation. Brain Mol Brain Res 90: 26–38

Correspondence: M. N. Ghabriel, Department of Anatomical Sciences, Medical School, University of Adelaide, Adelaide, South Australia 5005, Australia e-mail: Mounir.Ghabriel@adelaide.edu.au

Neuroprotection and Neurotoxicity

Positive selective brain cooling method: a novel, simple, and selective nasopharyngeal brain cooling method

K. Dohi[1], H. Jimbo[2], T. Abe[2], and T. Aruga[1]

[1] Department of Emergency and Critical Care Medicine, Showa University School of Medicine, Tokyo, Japan
[2] Department of Neurosurgery, Showa University School of Medicine, Tokyo, Japan

Summary

Brain damage is worsened by hyperthermia and prevented by hypothermia. Conventional hypothermia is a non-selective brain cooling method that employs cooling blankets to achieve surface cooling. This complicated method sometimes induces unfavorable systemic complications. We have developed a positive selective brain cooling (PSBC) method to control brain temperature quickly and safely following brain injury.

Brain temperature was measured in patients with a ventriculostomy CAMINO catheter. A Foley balloon catheter was inserted to direct chilled air (8 to 12 L/min) into each side of the nasal cavity. The chilled air was exhaled through the oral cavity. In most patients, PSBC maintained normal brain temperature. This new technique provides quick induction of brain temperature control and does not require special facilities.

Keywords: Brain damage; hypothermia; selective brain cooling.

Introduction

Neurons are more vulnerable to hyperthermia compared to other types of cells [21]. Brain temperature (Tb) elevates during the early phase of severe brain damage caused by cerebral vascular accidents or severe head injury [7] for many reasons, including hypothalamic injury, abnormal release of catecholamines, and production of endogenous pyrogens [7, 8]. Brain cooling mechanisms become unbalanced and dysfunctional from the heat production and loss that occurs during hyperthermia caused by brain damage. Hyperthermia caused by brain damage induces secondary brain damage [6]. Induction of mild brain hypothermia (33 to 35 °C) or normothermia (35 to 37 °C) by body surface cooling blankets has been shown to be neuroprotective in patients [2, 8, 11, 12, 17, 20]. Slight alterations in temperature have profound effects on ischemic cell injury and stroke outcome. Elevated Tb, even if slight, may exacerbate neuronal injury and worsen outcome, whereas hypothermia is potentially neuroprotective. Maintenance of normothermia is the most commonly recommended treatment for stroke and neurotrauma [14, 15]. However, conventional hypothermia is a complicated method that sometimes induces systemic complications [12]. A safe and effective brain cooling method is needed to treat patients with brain damage.

Mammals, including humans [1, 3, 4, 13, 16, 18, 19], have selective brain cooling (SBC) systems [9, 10, 19], one of which is nasopharyngeal cooling. Nasopharyngeal cooling was recently used to reduce cortical and subcortical temperatures rapidly and selectively without affecting the systemic circulation after resuscitation in the rat [10]. In this brief technical note, we describe a simple and selective nasopharyngeal brain cooling method that does not employ cooling blankets.

Methods

Method of positive selective brain cooling (PSBC)

Tb was measured directly in brain-injured patients with a ventriculostomy CAMINO catheter (ICP and Tb sensor; 4HMT, Integra NeuroCare, Hampshire, UK). The PSBC method was performed to cool the brain rapidly. A 16F Foley catheter (Temperature-Sensing Foley Catheter, C.R. Bard, Inc., Murray Hill, NJ) was inserted to blow chilled air (8 to 12 L/min) directly into the nasopharyngeal passage. To direct air from there into the oral cavity, one nasal cavity was occluded by an epistaxis balloon (Eschman Healthcare, Inc., Malaysia) at the inlet. Chilled air (24 to 26 °C) was expelled through

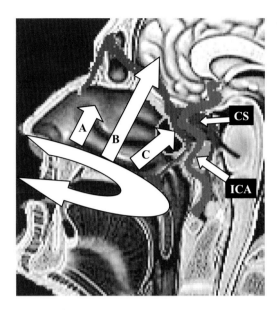

Fig. 1. Positive selective brain cooling (*PSBC*) performed by nasopharyngeal cooling. Artificial nasopharyngeal circulation with chilled air (24 °C, 8 to 12 L/min). Chilled air cooled nasal mucosa and nasal mucosa veins (*A*). Chilled air also cooled the brain and CSF directly (*B*). Cerebral blood is also chilled by heat exchange between the internal carotid artery (*ICA*) and cavernous sinus (*CS*)

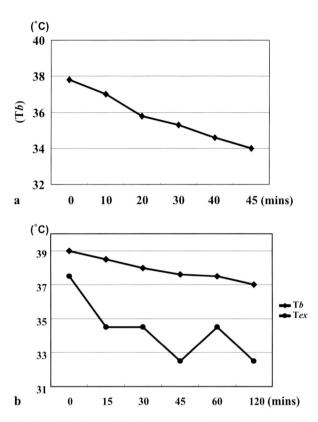

Fig. 2. (a) The change in brain temperature (Tb) after PSBC induction in a tracheal intubated patient with subarachnoid hemorrhage. Tb was decreased to 34 °C without employing surface cooling. (b) The change in brain temperature (Tb) after PSBC induction in a tracheal intubated patient with severe neurotrauma. Tb was elevated to 39 °C and decreased to 37 °C without the use of surface cooling. Exhaust temperature (Tex) was also elevated before the induction of PSBC

the oral cavity (Fig. 1, arrows a and b). Nasal and pharyngeal mucosa were kept clean by irrigation with physiologic saline. We also cooled the patient's head with an electric fan.

Contraindications of PSBC

PSBC is contraindicated in cases of sinusitis or skull base fracture. The procedure should be performed in sedated patients. The procedure may cause an oppressive feeling due to the high volume of circulating air.

Results

Brain temperature

Elevated Tb dropped to within the range of normothermia (37 °C > Tb) in most patients. The initial temperatures of the nasal cavities and exhaust air were very high despite the flow of chilled air. However, these temperatures fell within a short time.

Complications of PSBC

In general, PSBC was performed for a short period. Several patients developed erosion localized to the nasal foramen. Sinusitis, tympanic membrane injury, and dysosmia were not observed.

Case illustrations

Case 1

A 62-year-old woman was admitted to our hospital. She was comatose (Glasgow Coma Scale; E1V1M2). Computed tomography of the brain revealed diffuse, thick, subarachnoid hemorrhage. Nasopharyngeal brain cooling by PSBC was performed immediately. Initial Tb was 37.8 °C, and Tb rapidly decreased to 34.0 °C 45 minutes after induction (Fig. 2a).

Case 2

A 26-year-old man admitted to our hospital was diagnosed with severe neurotrauma. PSBC by nasopharyngeal cooling was performed immediately. Initial Tb was 39.0 °C, which rapidly decreased to 37.0 °C 120 minutes after induction. Exhaust air temperature

(Tex) was also elevated on induction and decreased rapidly (Fig. 2b).

Discussion

SBC system in humans

Mammals have a natural SBC system [9, 10, 19]. Tb is controlled by a balance between heat production/acquisition and heat loss. Heat loss occurs through heat transference from the core of the brain through blood flow, evaporative heat loss through breathing and sweating, and non-evaporative heat loss through breathing. Moreover, heat exchange between the internal carotid artery and venous plexus or cavernous sinus is simple and effectively functions like a radiator. The importance of each cooling mechanism differs among animal species. For example, heat exchange and evaporative heat loss through breathing is the main mechanism in dogs, which have a small number of sweat glands. Mechanisms of SBC in humans are not well-developed in comparison to other mammals. The human brain is believed to have 3 cooling mechanisms. One is cooling of venous blood through the entire skin surface, which in turn cools the arterial blood supply to the brain. A second is cooling by heat loss through the skin via the venous sinuses and diploic and emissary veins. The third mechanism of cooling is heat loss from the upper airways [4, 5, 9, 10, 13, 16, 18, 19].

As stated above, SBC as a treatment for hyperthermia in humans has been proposed but is controversial [1, 3, 5]. Cabanac [5] criticized the use of SBC in humans for the following reasons: 1) SBC masks the error signal which activates defense against hyperthermia; 2) unlike other animals, humans do not pant and thus do not possess a powerful heat sink near the brain; 3) humans do not have a carotid rete, the countercurrent heat exchanger between the arterial and venous bloods flowing in and out of the brain; 4) the high and constant arterial blood flow in the brain is sufficient to cool the brain under all conditions; and 5) the relatively low tympanic temperature recorded in hyperthermic humans is not indicative of SBC but of low head skin temperature. Alternatively, many reports support SBC in humans. The internal carotid artery in humans is surrounded anatomically by a cavernous sinus. The blood temperature of the cavernous sinus decreases in response to chilled venous blood. Arteriolovenular anastomosis is also efficacious in SBC. Emissary and angular veins, the upper respiratory tract, the tympanic cavity, and cerebrospinal fluid are also thought to be components of the SBC system in humans [4, 19]. SBC efficiency is increased by evaporation of sweat from the head and by ventilation through the nose. Moreover, craniofacial features such as thick lips, broad nasal cavity, and large paranasal sinuses, which provide greater evaporating surface, may be anatomical adaptations for effective SBC in hyperthermia [13]. Some of these cooling systems are destroyed by severe brain damage resulting in elevation of Tb. Respiratory ataxia and elevation of venous temperature are the main targets of brain damage, which results in hyperthermia. In addition, tracheal intubation blocks the physiological heat loss systems, thus maintaining the cycle of brain damage and brain hyperthermia. Destruction of the physiological and SBC system worsens thermo-pooling in the brain and secondary brain damage in patients with acute neuronal diseases.

We have demonstrated that SBC mechanisms are important in the control of Tb in humans. We also employed face and head fanning with an electrical fan together with PSBC. Brinnel et al. [4] reported that face fanning maintained tympanic temperature 0.57 °C lower than the esophageal temperature. Face and head fanning might be a safe and effective method of brain cooling. This is the first report which demonstrates SBC in humans by direct measurement of Tb rather than by measurement of tympanic temperature.

Mechanisms of PSBC

We have developed and introduced PSBC, which is an artificial physiological SBC system using circulating chilled air. PSBC promotes heat loss from the nasal cavity and cools through the venous circulation in the nasal mucosa. Chilled venous blood flows into the cavernous sinus through the angular vein and selectively cools the brain by means of a counter-current heat exchange mechanism with the arterial blood. While human nasal mucosa veins are not as developed as in other mammals, the nasal mucosa blood flow increases 3-fold in humans with hyperthermia [22].

Bone in the skull base is only a few millimeters thick. Many important anatomical structures (the pre-optic tract, hypothalamus, hypophysis, and brainstem) control thermoregulation in the basal brain. PSBC prevents hyperthermia in these sites by directly cooling the nasal cavity. Furthermore, CSF chilled at the basal cistern cools the whole brain through the CSF circulation.

In the present study, we cooled the head and face of patients by nasopharyngeal cooling. Thermoradiation through the head and face is an important mechanism of brain cooling in hyperthermic states. Head and face cooling is also an important means of heat exchange between venous and arterial blood. Blood chilled by head and face cooling flows to deep veins in the cavernous sinus and nasal mucosa. Cerebral blood is effectively cooled by heat exchange. Joint use of both methods may result in effective and selective brain cooling.

In conclusion, PSBC is a novel, simple, and selective blood cooling method. PSBC is a safe and effective method expected to be widely applicable in patients who require external control of Tb.

Acknowledgments

This research was partially supported by the Ministry of Education, Science, Sports and Culture, Grant-in-Aid for Scientific Research (C), 16591815, 2004.

References

1. Andrews PJ, Harris B, Murray GD (2004) Randomized controlled trial of effects of the airflow through the upper respiratory tract of intubated brain-injured patients on brain temperature and selective brain cooling. Br J Anesth 94: 330–335
2. Bernard SA, Gray TW, Buist MD, Jones BM, Silvester W, Gutteridge G, Smith K (2002) Treatment of comatose survivors of out-of-hospital cardiac arrest with induced hypothermia. N Engl J Med 346: 557–563
3. Brengelmann GL (1993) Specialized brain cooling in humans? FASEB J 7: 1148–1153
4. Brinnel H, Nagasaka T, Cabanac M (1987) Enhanced brain protection during passive hyperthermia in humans. Eur J Appl Physiol Occup Physiol 56: 540–545
5. Cabanac M (1993) Selective brain cooling in humans: "fancy" or fact? FASEB J 7: 1143–1147
6. Clifton GL, Jiang JY, Lyeth BG, Jenkins LW, Hamm RJ, Hayes RL (1991) Marked protection by moderate hypothermia after experimental traumatic brain injury. J Cereb Blood Flow Metab 11: 114–121
7. Dohi K, Jimbo H, Ikeda Y (2000) Usefulness of the continuous brain temperature monitoring after SAH. Jpn J Crit Care Med 12: 103–104
8. Dohi K, Jimbo H, Ikeda Y, Fujita S, Ohtaki H, Shioda S, Abe T, Aruga T (2005) Pharmacological brain cooling using indomethacin in acute hemorrhagic stroke: anti-inflammatory cytokines and anti-oxidative effects. Acta Neurochir Suppl [in press]
9. Einer-Jensen N, Khorooshi MH (2000) Cooling of the brain through oxygen flushing of the nasal cavities in intubated rats: an alternative model for treatment of brain injury. Exp Brain Res 130: 244–247
10. Hagioka S, Takeda Y, Takata K, Morita K (2003) Nasopharyngeal cooling selectively and rapidly decreases brain temperature and attenuates neuronal damage, even if initiated at the onset of cardiopulmonary resuscitation in rats. Crit Care Med 31: 2502–2508
11. Hypothermia After Cardiac Arrest Study Group (2002) Mild therapeutic hypothermia to improve the neurologic outcome after cardiac arrest. N Engl J Med 346: 549–556
12. Ikeda Y, Matsumoto K (2001) Hypothermia in neurological diseases. No To Shinkei 53: 513–524 [in Japanese]
13. Irmak MK, Korkmaz A, Erogul O (2004) Selective brain cooling seems to be a mechanism leading to human craniofacial diversity observed in different geographical regions. Med Hypotheses 63: 974–979
14. Japanese guidelines for the management of stroke (Provisional Edition) (2003) Rinsho Shinkeigaku 43: 591–708
15. Japanese guideline for severe neurotrauma (2000) Neurotraumatol 23: 15–16
16. Mariak Z, White MD, Lewko J, Lyson T, Piekarski P (1999) Direct cooling of the human brain by heat loss from the upper respiratory tract. J Appl Physiol 87: 1609–1613
17. Marion DW, Obrist WD, Carlier PM, Penrod LE, Darby JM (1993) The use of moderate therapeutic hypothermia for patients with severe head injuries: a preliminary report. J Neurosurg 79: 354–362
18. Mellergard P (1992) Changes in human intracerebral temperature in response to different methods of brain cooling. Neurosurgery 31: 671–677
19. Nagasaka T, Brinnel H, Hales JR, Ogawa T (1998) Selective brain cooling in hyperthermia: the mechanisms and medical implications. Med Hypotheses 50: 203–211
20. Shiozaki T, Sugimoto H, Taneda M, Yoshida H, Iwai A, Yoshioka T, Sugimoto T (1993) Effect of mild hypothermia on uncontrollable intracranial hypertension after severe head injury. J Neurosurg 79: 363–368
21. Sminia P, Haveman J, Ongerboer de Visser BW (1989) What is a safe heat dose which can be applied to normal brain tissue? Int J Hyperthermia 5: 115–117
22. White MD, Cabanac M (1995) Nasal mucosal vasodilatation in response to passive hyperthermia in humans. Eur J Appl Physiol Occup Physiol 70: 207–212

Correspondence: Kenji Dohi, Department of Emergency and Critical Care Medicine, Showa University School of Medicine, 1-5-8 Hatanodai, Shinagawa-ku, Tokyo, Japan. e-mail: kdop@med.showa-u.ac.jp

Mechanism of neuroprotective effect induced by QingKaiLing as an adjuvant drug in rabbits with *E. coli* bacterial meningitis

S. Yue, Q. Li, S. Liu, Z. Luo, F. Tang, D. Feng, and P. Yu

Department of Pediatrics, Xiangya Hospital, Central South University, Changsha, China

Summary

Objective. To explore the neuroprotective effects and underlying mechanism of QingKaiLing (QKL) as an adjuvant treatment for bacterial meningitis.

Method. E. coli bacterial meningitis rabbits were treated with antibiotics (ampicillin) alone or in combination with QKL. The number of leukocytes and the concentration of protein in the cerebrospinal fluid (CSF) of rabbits were determined at 0, 16, and 26 hours after treatment. Brain water, sodium, potassium, and calcium contents were determined at the 26-hour time point. The level of matrix metalloproteinase-9 in the brain was also determined by Western blot.

Result. The average number of leukocytes and the concentration of protein in CSF of the QKL adjuvant treatment group were reduced compared with the ampicillin alone group. Brain water, sodium, and calcium contents were reduced in the QKL adjuvant treatment group. The level of MMP-9 in brain tissue was also reduced in the QKL adjuvant treatment group.

Conclusion. QKL adjuvant treatment alleviates the aggravated inflammatory reaction and partially protects brain tissue from antibiotic-induced injury. The mechanism of this neuroprotective effect of QKL may be due to decreased levels of Ca^{2+} and MMP-9 in the brain.

Keywords: QingKaiLing; bacterial meningitis; brain damage; adjuvant therapy; *Escherichia coli*.

Introduction

Acute bacterial meningitis is a commonly occurring and life-threatening disease of the central nervous system in children. Despite significant advances in antibiotic therapy in the last few decades, bacterial meningitis is still often associated with high mortality and serious neurological sequelae [2, 4]. It is among the 10 most common causes of infection-related death worldwide. Recent studies have demonstrated that antibiotics worsen brain damage and intracranial hypertension in the short-term due to its dissolving effects on bacilli [2].

QingKaiLing (QKL) is a traditional Chinese medicine with mixed components. Recent studies have reported that QKL protects the brain from injury induced by ischemia and infection [14]. We tested the neuroprotective effect and mechanisms of QKL as an adjuvant therapy in *E. coli* bacterial meningitis rabbits.

Materials and methods

Drugs and reagents

QKL injection (composed of cow-bezoare, buffalo horn, scutellaria root, honeysuckle flower, and zhizi) was purchased from Beijing University of Traditional Chinese Medicine, Beijing, China. Ampicillin was purchased from the Huabei Pharmaceutical Company, Huabei, China. *E. coli* was supplied by the Institute for Drug and Biological Product Control of Chinese Public Health Ministry; strain number 44607; serotype: O7: K1: H. Rabbit polyclonal anti-MMP-9 antibody (primary antibody), Caprine anti-rabbit IgG antibody (second antibody), and DBA developer were obtained from Wuhan Boster Biological Technology Co. Ltd., Wuhan, China. Nitrocellulose filter was supplied by Shanghai Biological Technology Co. Ltd., Shanghai, China.

Experimental model preparation and medication

Thirty-seven rabbits, 1.7 to 2.1 kg in weight, were randomly divided into 5 groups: 1) control group (n = 7), 2) meningitis group (n = 8), 3) ampicillin group (n = 8), 4) QKL-treated group (n = 7), and 5) QKL-pretreated group (n = 7). An animal model of meningitis was prepared according to an established method [7]. Rabbits were anesthetized by urethane (1 g/kg), and cisternal puncture was performed to withdraw 0.5 mL cerebrospinal fluid (CSF). The rabbits were then inoculated intracisternally with 0.5 mL 4×10^7 colony-forming units of antigen-k1 positive *E. coli*. The control group was injected with equal volumes of 0.9% sodium chloride. The rabbits were closely observed for any symptoms and signs and 0.5 mL of CSF was withdrawn 16 and 26 hours after inoculation, respectively. Ampicillin (100 mg/kg) was intravenously injected in ampicillin, QKL-treated, and QKL-pretreated groups, 16 and 20 hours after inoculation (a total dose of 200 mg/kg). In the QKL-treated group, 15 mL/kg QKL was intravenously injected with the ampicillin and 5 mL/kg was injected 1 hour later. In the QKL-pretreated group, 15 mL/kg of QKL was intravenously injected 30 minutes before inoculation with *E. coli*. Ampicillin treatment was the same in this group as in the ampicillin and QKL groups. One hour after ampicillin injection, another 5 mL/kg QKL was injected in the QKL-pretreated group. The dosage and medications of QKL and ampicillin were chosen according to previous reports [7, 15]. All rabbits were decapitated 26 hours after inoculation with *E. coli*. The number of leukocytes, concentration of protein, lactate, and nitric oxide in CSF of rabbits were detected at 0, 16, and 26 hours after inoculation, respectively. Brain water, sodium, potassium, and calcium contents were detected at 26 hours after inoculation. The level of matrix metalloproteinase-9 in brain tissue was determined by Western blot analysis.

Parameter determination

Determination of the leukocyte counts in CSF

The numbers of leukocytes in CSF samples (10 μL) were determined using a leukocyte counting plate under a light microscope. The number of leukocytes in 1 liter of CSF was calculated.

Determination of CSF protein concentration

CSF samples were centrifuged at 4000 rpm for 5 minutes, and 0.1 mL of the supernatant fluid was used to determine CSF protein concentration using an automated biochemical analyzer (Hitachi 7170A, Japan).

Determination of brain water content

One hundred mg of fresh brain tissue was dried at 105 °C for 48 hours, with the weight difference before and after the procedure reflecting water content. The water content in brain tissue was calculated by the Elliott equation [(wet weight − dry weight)/wet weight] × 100.

Determination of brain sodium, potassium, and calcium content

Dried brain tissue was dissolved in nitric acid and perchloric acid. Sodium, potassium, and calcium content was determined by atom absorption spectrometry separately.

Determination of concentration of lactate and nitric oxide

Lactate and nitric oxide content in CSF was measured using commercially available kits (Nanjing Jiancheng Bioengineering Institute, Nanjing, China) according to the manufacturer's protocols.

Determination of matrix metalloproteinase-9 (MMP-9) in brain by Western blot

After electrophoresis of total protein from brain tissue in a 6% polyacrylamide gel, proteins were electro-transferred to a nitrocellulose filter and primary antibody (1:400) and second antibody incubations were carried out. After 2 washes, the membranes were exposed to diaminobenzidine and photographed.

Statistical analysis

All data are expressed as mean ± SD. Differences between the groups were analyzed with one-way analysis of variance (ANOVA) and Student-Newman-Keuls test. Values of $p < 0.05$ were considered significant.

Results

Effects of QKL on CSF leukocyte count in E. coli bacterial meningitis

Before antibiotic treatment, the average number of leukocytes in CSF in the infected 4 groups (2, 3, 4, 5) was higher than that of the control group ($p < 0.01$, Table 1). The average number of leukocytes in the QKL-pretreated group was significantly lower than the meningitis, ampicillin, or QKL-treated groups ($p < 0.01$, Table 1). After ampicillin treatment, the average number of leukocytes in the 4 infected groups was higher than that of the control group ($p < 0.01$, Table 1). Furthermore, in the ampicillin group, the average number of leukocytes became higher than before treatment and the other groups at 26 hours after inoculation. The number of leukocytes was, therefore, decreased by treatment with QKL, particularly in the pretreatment group ($p < 0.01$, Table 1).

Effects of QKL on protein content of CSF on E. coli bacterial meningitis

Before antibiotic treatment, the CSF protein content in the 4 infected groups (2, 3, 4, 5) was significantly higher than that of the control group ($p < 0.01$, Table 2). The CSF protein content in the QKL-pretreated group was lower than meningitis, ampicillin, or QKL-treated groups ($p < 0.01$, Table 2). After ampicillin treatment, the CSF protein content in the infected groups (2, 3, 4, 5) was higher than that of the control group ($p < 0.01$). The CSF protein content of the QKL-pretreated group was the lowest among the infected groups after treatment (Table 2). The CSF protein content in the ampicillin group was the highest among the infected groups after treatment (Table 2). These results indicate that pretreatment with QKL reduces CSF protein levels in *E. coli* bacterial meningitis.

Effects of QKL on the brain water, sodium, potassium, and calcium contents in E. coli bacterial meningitis rabbits

Brain water, sodium, and calcium contents from infected groups 26 hours after inoculation with *E. coli*. were higher than those in the control group ($p < 0.05$ and $p < 0.01$, Table 3). The brain potassium content in the infected groups was lower than the control group ($p < 0.01$, Table 3). Brain water, sodium, and calcium

Table 1. Effects of QKL on CSF leukocyte count in E. coli bacterial meningitis (×10⁶/L, mean ± SD)

Group	n	Before inoculation (0 h)	Before treatment (16 h after inoculation)	After treatment (26 h after inoculation)
Control	7	7.4 ± 1.9	8.2 ± 1.5	8.2 ± 2.2
Meningitis	8	7.3 ± 1.2	5040.8 ± 590.3*	4342.5 ± 143.0*□
Ampicillin	8	9.0 ± 1.4	5018.3 ± 585.3*	6517.8 ± 327.1*△△□
Treated	7	9.2 ± 1.9	5002.0 ± 203.2*	4019.0 ± 220.1*△▲□
Pretreated	7	8.6 ± 1.5	3513.2 ± 882.3*△△▲○	2052.0 ± 242.8*△△▲○□

vs. control group: *$p < 0.01$, vs. meningitis group: △$p < 0.05$, △△$p < 0.01$, vs. ampicillin group: ▲$p < 0.01$, vs. treated group: ○$p < 0.01$, vs. pretreated group: □$p < 0.01$.
CSF Cerebrospinal fluid, QKL QingKaiLing.

Table 2. Effects of QKL on CSF protein content in E. coli bacterial meningitis (×g/L, mean ± SD)

Group	n	Before inoculation (0 h)	Before treatment (16 h after inoculation)	After treatment (26 h after inoculation)
Control	7	0.17 ± 0.03	0.17 ± 0.02	0.19 ± 0.04
Meningitis	8	0.19 ± 0.03	2.51 ± 0.19*	1.39 ± 0.15*□
Ampicillin	8	0.18 ± 0.04	2.38 ± 0.23*	1.74 ± 0.24*△□
Treated	7	0.18 ± 0.05	2.53 ± 0.18*	1.42 ± 0.11*▲□
Pretreated	7	0.20 ± 0.03	1.55 ± 0.16*△▲○	0.68 ± 0.08*△▲○□

vs. control group: *$p < 0.01$, vs. meningitis group: △$p < 0.01$, vs. ampicillin group: ▲$p < 0.01$, vs. treated group: ○$p < 0.01$, vs. pretreated group: □$p < 0.01$.
CSF Cerebrospinal fluid, QKL QingKaiLing.

Table 3. Effects of QKL on brain water, sodium, potassium, and calcium content in E. coli bacterial meningitis (mean ± SD)

Group	n	Water (%)	Na (mg/g)	K (mg/g)	Ca (mg/g)
Control	5	78.49 ± 0.30	5.46 ± 0.09	17.23 ± 0.64	0.29 ± 0.03
Meningitis	6	81.42 ± 0.19**	5.73 ± 0.05**	15.73 ± 0.69**	0.49 ± 0.06**
Ampicillin	6	82.10 ± 0.21**△△	5.87 ± 0.04**△△	11.74 ± 0.33**△△	1.62 ± 0.06**△△
Treated	5	81.13 ± 0.25**△▲	5.69 ± 0.03**▲	15.83 ± 0.49**▲	0.44 ± 0.05**▲
Pretreated	5	80.65 ± 0.28**△△▲○○	5.61 ± 0.04**△▲○	16.67 ± 0.45*△▲○	0.37 ± 0.05**△▲○

vs. control group: *$p < 0.05$, **$p < 0.01$, vs. meningitis group: △$p < 0.05$, △△$p < 0.01$, vs. ampicillin group: ▲$p < 0.01$, vs. treated group: ○$p < 0.05$, ○○$p < 0.01$.
QKL QingKaiLing.

contents in the meningitis, QKL-pretreated, and QKL-treated groups were lower than that of the ampicillin group ($p < 0.01$, Table 3), while the contents in QKL-pretreated group were the lowest among all the infected groups ($p < 0.01$, Table 3).

Effects of QKL on CSF lactate content in E. coli bacterial meningitis

Before treatment with ampicillin, the CSF lactate content in the infected groups (2, 3, 4, 5) was higher than that of the control group ($p < 0.01$, Table 4). After treatment with ampicillin, the CSF lactate content in the infected groups was higher than that of the control group ($p < 0.01$). The average CSF lactate content in the QKL-pretreated and QKL-treated groups was decreased after treatment ($p < 0.01$, Table 4). In the ampicillin group, the average CSF lactate content did not change after treatment, while the value of the average CSF lactate content was the highest among all the infected groups 26 hours after inoculation ($p < 0.01$).

Table 4. *Effects of QKL on CSF lactate content in E. coli bacterial meningitis (mmol/L, mean ± SD)*

Group	n	Before inoculation (0 h)	Before treatment (16 h after inoculation)	After treatment (26 h after inoculation)
Control	7	1.42 ± 0.02	2.08 ± 0.27	2.33 ± 0.38
Meningitis	8	1.49 ± 0.07	11.94 ± 1.89*	5.03 ± 0.56*□
Ampicillin	8	1.50 ± 0.13	11.83 ± 0.16*	11.88 ± 1.41*△
Treated	7	1.43 ± 0.03	11.54 ± 0.22*	5.46 ± 0.25*▲□
Pretreated	7	1.46 ± 0.07	3.77 ± 0.61△▲○○	4.46 ± 0.23*▲○□

vs. control group: *$p < 0.01$, vs. meningitis group: △$p < 0.01$, vs. ampicillin group: ▲$p < 0.01$, vs. treated group: ○$p < 0.05$, ○○$p < 0.01$, vs. pretreated group: □$p < 0.01$.
CSF Cerebrospinal fluid, *QKL* QingKaiLing.

Table 5. *Effects of QKL on CSF nitric oxide content in E. coli bacterial meningitis (mmol/L, mean ± SD)*

Group	n	Before inoculation (0 h)	Before treatment (16 h after inoculation)	After treatment (26 h after inoculation)
Control	7	43.0 ± 3.17	49.9 ± 6.12	45.6 ± 5.03
Meningitis	8	43.7 ± 3.43	105.1 ± 7.54*	74.8 ± 3.66*□
Ampicillin	8	43.0 ± 2.79	104.0 ± 8.87*	85.9 ± 2.84*△△□
Treated	7	43.4 ± 2.79	104.0 ± 6.98*	69.0 ± 3.23*△▲□
Pretreated	7	43.9 ± 3.17	88.2 ± 5.32*△△▲○○	64.19 ± 4.56*△△▲○○

vs. control group: *$p < 0.01$, vs. meningitis group: △$p < 0.05$, △△$p < 0.01$, vs. ampicillin group: ▲$p < 0.01$, vs. treated group: ○$p < 0.05$, ○○$p < 0.01$, vs. pretreated group: □$p < 0.01$.
CSF Cerebrospinal fluid, *QKL* QingKaiLing.

Effects of QKL on CSF nitric oxide content in E. coli bacterial meningitis

Before treatment with ampicillin, the CSF nitric oxide content in infected groups was higher than that of the control group ($p < 0.01$, Table 5). The CSF nitric oxide content in the QKL-pretreated group was lower than that in the meningitis, ampicillin, and QKL-treated groups ($p < 0.01$, Table 5). After treatment with ampicillin, CSF nitric oxide content in the infected groups was higher than that in the control group ($p < 0.01$, Table 5). After treatment with ampicillin, the CSF nitric oxide content in QKL-pretreated and QKL-treated groups was significantly lower ($p < 0.01$, Table 5) than that in the other groups. After treatment with ampicillin, the CSF nitric oxide content in the ampicillin group was elevated compared with the other infected groups 26 hours after inoculation ($p < 0.01$, Table 5).

Effects of QKL on the MMP-9 content on brain tissue in E. coli bacterial meningitis

Levels of MMP-9 in brain tissue were measured by Western blot analysis. In the infected groups, MMP-9 levels were higher than those of the control group. Levels of MMP-9 in the QKL-pretreated group were lower than those of the other 3 infected groups. These results indicate that the level of MMP-9 is generally low in normal brain tissue and increased in the inflamed brain. QKL-pretreatment reduces the level of MMP-9 induced by inflammation.

Discussion

Acute bacterial meningitis is a common and serious disease of the central nervous system, and effective antibiotic therapy is an essential component of clinical treatment. However, bacterial lysis induced by early antibiotic therapy facilitates the release of tumor necrosis factor α (TNF-α), interleukins, nitric oxide, and matrix metalloproteinases. These factors aggravate local inflammation, cause blood-brain barrier (BBB) disruption, and disturb the brain microcirculation, leading to transient aggravation of brain edema and intracranial hypertension [2, 4]. In this study, we found an increase in CSF leukocyte count, protein, lactate, nitric oxide, an increase in brain water, sodium, and calcium content, and a decrease in brain potassium content early during antibiotic treatment of *E. coli* meningitis in rabbits. These results are consistent with aggravation of the inflammation-induced reaction and brain edema in the short-term after early antibiotic therapy. Therefore, in order to reduce mortality and neurological sequelae of bacterial meningitis, therapy designed to prevent brain injury is under intensive investigation.

QKL injection is a prepared and mixed pure traditional Chinese medicine, derived from the Angon Niuhuang Pill. The effects of QKL have been reported to clear heat, detoxify, tranquilize and allay excitement, eliminate phlegm, and remove stasis to induce resuscitation. QKL has been widely used in treatment of a number of diseases such as the common cold since 1975, and is especially used to treat central nervous system inflammations [14]. In animal experiments, QKL inhibits the production of inflammatory factors such as IL-1, IL-8, TNF-α [12], and protects the brain from ischemic damage [10]. Studies show that QKL affects vascular endothelial cells to maintain normal microvessel permeability, to improve cerebral blood flow, to reduce brain edema, and to improve survival after brain hypoxia-ischemia [14]. In addition, QKL improves the absorption of necrotic tissue during tissue repair [14]. It is therefore hypothesized that QKL may prevent antibiotic-induced brain damage in meningitis. The results in this study demonstrated that QKL reduces CSF leukocyte count and protein content, and attenuates the antibiotic-induced transient aggravation of meningitis and brain edema in bacterial meningitis. It is particularly effective in the QKL-pretreated group. We provide the first indication that adjuvant QKL treatment has neuroprotective benefit in the management of bacterial meningitis.

In bacterial meningitis, hypoxia-ischemia in brain tissue and excessive release of excitatory amino acids induces opening of calcium channels leading to a subsequent increase in intracellular calcium, which is toxic to the affected cells [3]. This study shows that QKL inhibits the increase in brain calcium content during early antibiotic therapy for *E. coli* meningitis. This suggests the possible involvement of an inhibition of calcium influx via calcium channels. This finding is consistent with our previous reports that QKL effectively inhibits the increase of synaptic calcium induced by glutamate to attenuate cortical and hippocampal neuronal injury [13, 16].

Nitric oxide (NO) is a free-radical gas produced by the conversion of L-arginine to L-citrulline catalyzed by a group of isoenzymes called NO synthases (NOSs). Three different forms have been identified so far: endothelial NOS (eNOS, or NOS III), neuronal NOS (nNOS, or NOS I), and inducible NOS (iNOS, or NOS II) [11]. eNOS and nNOS are expressed constitutively in the adult brain and believed to regulate major physiological functions, including vascular hemostasis and neurotransmission. iNOS is absent in strictly resting cells and strongly induced by cytokines and other immunological stimuli [4]. NO levels in the CSF are elevated in humans with bacterial meningitis and in experimental animal models [4]. Moreover, inducible NOS knockout mice exhibit low levels of inflammatory mediators such as IL-1β, IL-6, TNF-α, and MIP-2, and attenuated BBB disruption in pneumococcal meningitis. Thus, NO derived from inducible NOS plays a major role in bacterial meningitis. This report shows that QKL effectively inhibits the generation of NO during bacterial meningitis. This may be part of the mechanism underlying QKL neuroprotective effects.

Treatment studies of bacterial meningitis suggest that MMP-9 up-regulation mediates BBB breakdown, leukocyte invasion into CSF, brain edema, and brain tissue damage [5]. MMP-9 is not normally present in CSF, but it is markedly increased in bacterial meningitis. High concentrations of MMP-9 are associated with neurological complications, suggesting high CSF concentrations of MMP-9 may significantly increase the risk of bacterial meningitis in humans [6]. Broad-spectrum MMP inhibitors dramatically attenuate neuronal injury and reduce mortality of animals in experimental bacterial meningitis [9]. The result of this study shows that QKL significantly reduces the level of MMP-9 in brain in bacterial meningitis, which suggests that the neuroprotective effects of QKL are likely due to the inhibition of MMP-9 expression. According to the report by Meli *et al.* [8], in bacterial meningitis, brain-resident cells and primarily invading polymorphonuclear cells act as sources of MMP-9. Investigations have shown that the MMP-9 promoter is in a 2-kb 5′ flanking region that contains AP-1, AP-2, and SP-1 factor-binding sites [1]. The transcription factor AP-1 is composed of a mixture of heterodimeric protein complexes derived from the Fos and Jun families, including c-Fos, FosB, Fra-1, Fra-2, c-Jun, JunB, and JunD. QKL not only effectively inhibits the expression of the c-Fos gene induced by glutamate [16], but also reduces the leukocyte count of CSF in bacterial meningitis. Therefore, QKL may reduce the level of

MMP-9 in brain by several different signaling pathways.

In conclusion, QKL adjuvant therapy has the potential to alleviate the aggravated inflammatory reaction and partially protect brain tissue from antibiotic-induced injury. The neuroprotective effects of QKL may be due to decreased levels of Ca^{2+}, NO, and MMP-9 in the brain, and is more efficacious when pretreated.

Acknowledgments

This work was supported by grants from the National Natural Science Foundation of China (39700193) and by the Hunan Traditional Medicine and Drug Research Foundation (204048).

References

1. Atkinson JJ, Senior RM (2003) Matrix metalloproteinase-9 in lung remodeling. Am J Respir Cell Mol Biol 28: 12–24
2. Chaudhuri A (2004) Adjunctive dexamethasone treatment in acute bacterial meningitis. Lancet Neurol 3: 54–62
3. Ko SY, Shim JW, Kim SS, Kim MJ, Chang YS, Park WS, Shin SM, Lee MH (2003) Effects of MK-801 (dizocilpine) on brain cell membrane function and energy metabolism in experimental Escherichia coli meningitis in the newborn piglet. J Korean Med Sci 18: 236–241
4. Koedel U, Scheld WM, Pfister HW (2002) Pathogenesis and pathophysiology of pneumococcal meningitis. Lancet Infect Dis 2: 721–736
5. Lee KY, Kim EH, Yang WS, Ryu H, Cho SN, Lee BI, Heo JH (2004) Persistent increase of matrix metalloproteinases in cerebrospinal fluid of tuberculous meningitis. J Neurol Sci 220: 73–78
6. Leppert D, Leib SL, Grygar C, Miller KM, Schaad UB, Holländer GA (2000) Matrix metalloproteinase (MMP)-8 and MMP-9 in cerebrospinal fluid during bacterial meningitis: association with blood-brain barrier damage and neurological sequelae. Clin Infect Dis 31: 80–84
7. Liu YX (1993) Effects of anisodamine and ampicillin on cerebrospinal fluid in experimental E. coli meningitis. Zhonghua Yi Xue Za Zhi 73: 289–291 [in Chinese]
8. Meli DN, Christen S, Leib SL (2003) Matrix metalloproteimase-9 in pneumococcal meningitis: activation via an oxidative pathway. J Infect Dis 187: 1411–1415
9. Meli DN, Loeffler JM, Baumann P, Neumann U, Buhl T, Leppert D, Leib SL (2004) In pneumococcal meningitis a novel water-soluble inhibitor of matrix metalloproteinases and TNF-alpha converting enzyme attenuates seizures and injury of the cerebral cortex. J Neuroimmunol 151: 6–11
10. Song A, Zhu W (2003) The protective effect of Qingkailin on forebrain neuron in intracranial hemorrhage in rat. J Baot Med Coll 19: 8–10
11. Southan GJ, Szabo C (1996) Selective pharmacological inhibition of distinct nitric oxide synthase isoforms. Biochem Pharmacol 51: 383–394
12. Sun F (2001) Therapeutic effect of Qinkailing injection on serum cytokine in chronic obstructive pulmonary disease. Pharmacol Clin Chin Mat Med 11: 37–41
13. Tao Y, Yue S, Yu Y, Yang Y (2000) Effect of Qingkailin on free calcium of synaptosomes in rats with neurotoxic brain edema induced by L-glutamate. Chin J Contemp Pediatr 2: 326–328
14. Yang J, Cheng H (2001) Advance of QingKaiLing injection in clinic application and its pharmacology. Chin Hosp Pharm J 21: 494–497
15. Yue S, Luo Z, Yu P, Deng S, Xia L (1998) The protective effect of Qingkailin on neurotoxic brain edema induced by glutamate. J Beijing Univ TCM 21: 32–33
16. Yue S, Zeng Q, Zhou J, Tao Y, Zeng S, Wu G, Yu P, Luo Z (2000) Experimental study of protective effect of Qingkailing on glutamate induced neurotoxic damage of brain. Chin J Integrate Tradition West Med 20: 842–845

Correspondence: Shaojie Yue, Department of Pediatrics, Xiangya Hospital, Xiangya Road 141, Changsha, Hunan, P.R. China 410008.
e-mail: shaojieyue@163.com

Acceleration of chemokine production from endothelial cells in response to lipopolysaccharide in hyperglycemic condition

K. Kinoshita, M. Furukawa, T. Ebihara, A. Sakurai, A. Noda, Y. Kitahata, A. Utagawa, and K. Tanjoh

Department of Emergency and Critical Care Medicine, Nihon University School of Medicine, Tokyo, Japan

Summary

Chronic hyperglycemia is an established risk factor for endothelial damage. It remains unclear, however, whether brief hyperglycemic episodes after acute stress alter the function of vascular endothelial cells in response to endotoxin. We hypothesize that brief hyperglycemic episodes enhance the production of interleukin-8 (IL-8) after lipopolysaccharide (LPS) stimulation.

Methods. Human umbilical vein endothelial cells (HUVECs; 1×10^5 cells/mL, cells from subcultures 2–5, n = 6) were cultivated in various concentrations of glucose (200, 300, 400, and 500 mg/dL) with or without LPS stimulation (1 µg/mL) for 24 hours. After culture, IL-8 levels in the supernatant were measured using ELISA.

Results. HUVECs cultured at glucose concentrations of 300 and 400 mg/dL produced more ($p < 0.01$) IL-8 than control cells (200 mg/dL). HUVECs cultured at glucose concentrations of 300 and 400 mg/dL also produced more ($p < 0.01$) IL-8 than those cultured in the absence of LPS.

Conclusions. Hyperglycemic conditions enhance IL-8 production by vascular endothelial cells, and this response is augmented by LPS. Infections may foster neutrophil accumulation at injury sites. These results suggest that it is important to manage even short-term increases in blood glucose after acute stress.

Keywords: Hyperglycemia; endothelium; LPS; IL-8; brain injury.

Introduction

Severe brain damage [5] is associated with an increased prevalence of endothelial cell dysfunction. The mechanisms leading to alterations in endothelial cell function are poorly understood. The strict management of blood glucose has recently been shown to improve outcome in a surgical intensive care unit [6]. Chronic hyperglycemia (diabetes mellitus) is an established risk factor for endothelial damage. It remains unclear, however, whether brief hyperglycemic episodes alter the function of vascular endothelial cells in response to endotoxin. A high incidence of aspiration pneumonia has been reported after severe brain damage and this medical complication is a leading cause of late morbidity and mortality in comatose patients.

We hypothesize that brief hyperglycemic episodes enhance the production of interleukin (IL)-8 after lipopolysaccharide (LPS) stimuli. We evaluated the changes in chemokine IL-8 production from endothelial cells in various hyperglycemic conditions and investigated whether hyperglycemia is associated with an acute inflammatory response. Stress reactions to insults such as severe brain damage could enhance IL-8 production from the endothelial cells.

Materials and methods

Incubation of human umbilical vascular endothelial cells (HUVECs) with glucose

HUVECs were purchased from Sanko Junyaku Co., Ltd. (Tokyo, Japan). HUVECs (1×10^5 cells/mL, cells from subcultures 2–5, n = 6 for each subculture) were cultivated for 24 hours in the 3 systems outlined below. The HUVECs were cultured in RPM11640 medium at 37 °C in a 5% CO_2 atmosphere.

Glucose concentrations in the culture solution were 200, 300, 400, and 500 mg/dL. After 24 hours, IL-8 levels in the supernatant were measured using enzyme-linked immunosorbent assay (ELISA). HUVEC samples from each glucose environment (200, 300, 400, 500 mg/dL) were stimulated with 1 µg/mL LPS and cultured for 24 hours. Changes in IL-8 production in the supernatant were then measured. After incubation for 24 hours at 37 °C, the supernatants were taken from the cultures and stored at −80 °C until cytokine assay.

Measurement of IL-8

The IL-8 levels in the supernatant samples were quantified using a commercially available ELISA kit (IL-8; BioSource International, Inc., Camarillo, CA).

Statistical analysis

Duplicate or triplicate samples were obtained from parallel cultures through the entire experimental protocol. For the bioassays, each sample was plated in triplicate and the results were averaged. Data were expressed as mean values ± SD. Comparisons of cytokine IL-8 data between the groups were conducted using one-way analysis of variance. Fisher's protected least-squares difference analyses were performed to determine differences between each group. Differences were considered significant if the p value was <0.05. Statistical analyses were performed using Stat-View 5.0 (SAS Institute, Inc., Cary, NC).

Results

Among the HUVECs cultured at glucose concentrations of 200, 300, 400, and 500 mg/dL without LPS treatment, IL-8 production was higher (*p < 0.01) in the cells cultured at glucose concentrations of 300 and 400 mg/dL than in the cells cultured at a control glucose concentration of 200 mg/dL (Fig. 1). Among the HUVECs cultured at all glucose concentrations in the presence of LPS, IL-8 production was higher (#p < 0.01) in the cells cultured at glucose concentrations of 300 and 400 mg/dL than in the LPS-treated controls cultured at the glucose concentration of 200 mg/dL (6.6-fold and 6.5-fold, respectively; Fig. 1). The mean IL-8 concentration in the LPS-treated cells cultured at glucose concentrations of 300 and 400 mg/dL was 3.2- and 4.4-fold higher than that in the HUVECs cultured at the same concentrations without LPS treatment, respectively. The mean IL-8 concentration of the LPS-treated cells cultured at a glucose concentration of 200 mg/dL was 2.1-fold higher than that in cells cultured at the same glucose concentration without LPS treatment.

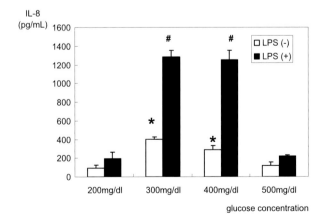

Fig. 1. Changes in IL-8 production from HUVECs treated with or without LPS in various concentrations of glucose (*p < 0.01; #p < 0.01)

Discussion

Hyperglycemia is commonly observed after severe brain damage due to insults such as acute stroke [7]. No reports, however, have discussed whether a brief hyperglycemic episode after a stress reaction can affect secondary tissue damage. One recent report demonstrated that strict management of blood glucose improved outcome in a surgical intensive care unit [6]. Other reports suggest that prolonged hyperglycemia (diabetes mellitus) is a potent risk factor for endothelial damage [2, 3]. It remains unclear, however, whether short episodes of hyperglycemia have any effect on endothelial cells. Endothelial cells play an important role in injury and inflammation through the production and regulation of cytokines, adhesion molecules, free radicals, and vasoactive and chemoattractant mediators [4].

In this study we observed a significant augmentation of IL-8 production by endothelial cells during a 24-hour period of hyperglycemia. Moreover, a similar but significantly stronger augmentation was obtained through LPS treatment. These findings indicate that hyperglycemia associated with infection may create an early window of vulnerability allowing secondary insults to activate deleterious neutrophil functions. We previously reported [1] that a hyperglycemic condition immediately after experimental traumatic brain injury led to significant increases in neutrophil accumulation at the injured site. In patients who have suffered severe brain damage, the high incidence of aspiration pneumonia can provide a nidus for systemic inflammation. The development of such a condition in severely injured patients may leave them at high risk for secondary brain damage. Further studies are needed to elucidate the importance of this data and the consequences of high-glucose conditions over several time courses by investigating the temporal profiles of acute hyperglycemia in patients with severe brain injury.

Conclusions

This study suggests that hyperglycemia may increase the risk of chemokine IL-8 production by endothelial cells. This phenomenon is accelerated by LPS stimulation. Strict control of hyperglycemia may be required after severe brain damage to prevent neutrophils from accumulating in the injured brain and contributing to secondary brain damage processes.

References

1. Kinoshita K, Kraydieh S, Alonso O, Hayashi N, Dietrich WD (2002) Effect of posttraumatic hyperglycemia on contusion volume and neutrophil accumulation after moderate fluid-percussion brain injury in rats. J Neurotrauma 19: 681–692
2. Nishikawa T, Edelstein D, Du XL, Yamagishi S, Matsumura T, Kaneda Y, Yorek MA, Beebe D, Oates PJ, Hammes HP, Giardino I, Brownlee M (2000) Normalizing mitochondrial superoxide production blocks three pathways of hyperglycaemic damage. Nature 404: 787–790
3. Ruderman NB, Haudenschild C (1984) Diabetes as an atherogenic factor. Prog Cardiovasc Dis 26: 373–412
4. Shrotri MS, Peyton JC, Cheadle WG (2000) Leukocyte-endothelial cell interactions: review of adhesion molecules and their role in organ injury. In: Baue A, Faist E, Fry DE (eds) Multiple organ failure: pathophysiology, prevention, and therapy. Springer, New York, pp 224–240
5. Skold MK, von Gertten C, Sandberg-Nordqvist AC, Mathiesen T, Holmin S (2005) VEGF and VEGF receptor expression after experimental brain contusion in rat. J Neurotrauma 22: 353–367
6. van den Berghe G, Wouters P, Weekers F, Verwaest C, Bruyninckx F, Schetz M, Vlasselaers D, Ferdinande P, Lauwers P, Bouillon R (2001) Intensive insulin therapy in critically ill patients. N Engl J Med 345: 1359–1367
7. Weir CJ, Murray GD, Dyker AG, Lees KR (1997) Is hyperglycaemia an independent predictor of poor outcome after acute stroke? Results from a long-term follow up study. BMJ 314: 1303–1306

Correspondence: Kosaku Kinoshita, Department of Emergency and Critical Care Medicine, Nihon University School of Medicine, 30-1 Oyaguchi Kamimachi, Itabashi ku, Tokyo 173-8610, Japan. e-mail: kosaku@med.nihon-u.ac.jp

Photodynamic therapy increases brain edema and intracranial pressure in a rabbit brain tumor model

F. Li[1], G. Zhu[1], J. Lin[1], H. Meng[1], N. Wu[1], Y. Du[2], and H. Feng[1]

[1] Department of Neurosurgery, Southwest Hospital, The Third Military Medical University, Chongqing, China
[2] The Ultrasonic Research Institute, Chongqing University of Medical Sciences, Chongqing, China

Summary

The objective of this study was to evaluate the effect of a single photodynamic therapy (PDT) on brain edema and intracranial pressure (ICP) in a rabbit model of brain tumor.

A total of 57 adult New Zealand rabbits were assigned to 3 groups: the PDT group, the tumor group, and the tumor plus PDT group. Rabbits in the PDT group (n = 9) received PDT but no tumor implantation; rabbits in the tumor group (n = 18) received VX2 carcinoma implantation but no PDT; rabbits in the tumor plus PDT group (n = 30) received tumor implantation with subsequent PDT 16 days later.

Brain edema and ICP levels were then evaluated. We found that ICP in the PDT group was 7.43 ± 0.50 mmHg. After tumor implantation, ICP increased rapidly (18.43 ± 1.10 mmHg, 21 days later). PDT alone did not increase ICP, but compared with that in the tumor group, ICP increased significantly in the tumor plus PDT group (9.55 ± 1.32 vs. 13.31 ± 1.13 mmHg, $p < 0.01$) 24 hours after treatment. Brain water content in the tumor group increased rapidly after tumor implantation. PDT again increased perineoplastic brain edema 24 hours after treatment ($81.09 \pm 0.97\%$ vs. $78.32 \pm 0.49\%$, $p < 0.01$). It should be noted that PDT alone did not induce brain edema.

In conclusion, PDT causes transient brain edema and increases ICP in a rabbit brain tumor model.

Keywords: Photodynamic therapy; VX2 carcinoma; brain edema; intracranial pressure.

Introduction

Photodynamic therapy (PDT) was developed as a modality for malignant tumor treatment, and Perria et al. [7] first used PDT to treat patients with brain tumors. PDT is a potential therapy for brain tumors. Both experimental studies and clinical practice have confirmed that PDT can cause damage to cells and blood vessels. It has been reported that PDT causes brain edema, hypoxic brain injury, cerebral blood vessel injury, and even cerebral herniation [4]. Hence, to provide experimental data for the clinical application of PDT, we examined changes in intracranial pressure (ICP) and brain water content in rabbits following VX2 carcinoma implantation. A single PDT was used and hematoporphyrin derivative (HPD) was the photosensitizer.

Materials and methods

Animals and grouping

A total of 57 adult New Zealand rabbits, weighing 1.5 to 2.0 kg, were assigned to 3 groups: the PDT group, the tumor group, or the tumor plus PDT group. Rabbits in the PDT group (n = 9) received PDT without tumor implantation; rabbits in the tumor group (n = 18) received VX2 carcinoma implantation without PDT; rabbits in the tumor plus PDT group (n = 30) received tumor implantation first and PDT 16 days later.

Hematoporphyrin derivative

This product was developed by the Institute of Pharmaceutical Research of the Chinese Academy of Sciences, and was manufactured by Chongqing Huading Modern Biopharmaceutics Co., Ltd., Chongqing, China.

Instruments

LumaCare LC-051 non-coherent light source (LumaCare, Newport Beach, CA) and a Camino ICP monitor (Integra Neurosciences, Plainsboro, NJ).

Implantation of VX2 carcinoma

Subcutaneous carcinoma mass was obtained from rabbits, and the mass was then sliced into small pieces sized 1.5 mm × 1.5 mm × 1.5 mm. Before tumor implantation, a bone window sized 1.5 cm × 1.5 cm was opened at the site 1 cm lateral of the sagittal suture. A

Table 1. ICP data (Mean ± SD, mmHg)

Day (post-implantation)	Day (post-PDT)	PDT group	Tumor group	Tumor plus PDT group
13	before treatment	7.43 ± 0.50	7.23 ± 0.57	7.23 ± 0.57
17	1	9.55 ± 1.32	7.10 ± 0.55	13.31 ± 1.13*#
18	2	8.96 ± 2.42	10.83 ± 3.91	11.94 ± 1.35*
19	3	7.87 ± 1.59	15.17 ± 1.44	9.84 ± 1.17*#
20	4	7.38 ± 0.36	16.90 ± 0.84	8.63 ± 0.77*#
21	5	7.45 ± 0.46	18.43 ± 1.10	8.35 ± 0.70**#
22	6	–	18.73 ± 0.83	7.89 ± 0.69#
23	7	–	18.77 ± 0.73	7.82 ± 0.66#
24	8	–	18.80 ± 0.81	7.73 ± 0.53#
25	9	–	19.04 ± 0.73	7.47 ± 0.58#

*$p < 0.01$ vs. the PDT group, **$p < 0.05$ vs. of the PDT group, #$p < 0.01$ vs. the tumor group.

piece of carcinoma was then implanted and embedded centrally into brain cortex. Hemostasis with a gelatin sponge and closure of incision was then performed.

Photodynamic therapy

PDT was carried out 16 days after carcinoma implantation. HPD (5 mg/kg, 1 mg/ml) was infused intravenously. The rabbit was anesthetized 24 hours later, and the original incision was reopened. Macroscopic changes in the tumor were then observed. The light spot area was determined according to the size of the tumor, and the time of light exposure was calculated. The output power of the light source was 500 mW, with a total energy of 240 J/cm^2 at a wavelength of 628 nm.

ICP monitoring

Rabbits in the tumor group underwent ICP monitoring at 13 days, 17 days, and 25 days after tumor implantation 3 times a day, while those in the tumor plus PDT group underwent ICP monitoring after tumor implantation 3 times daily.

Pathologic observations

Rabbits in the tumor group were sacrificed without PDT at days 13, 17, 19, 21, 23, and 25 after tumor implantation, while those in the tumor plus PDT group underwent the treatment 16 days after tumor implantation. At days 1, 3, 5, 7, and 9 after treatment, rabbits were sacrificed and pathological observations on brain tissues performed.

Determination of brain water content

After tumors were isolated from the remaining brain tissues, brain water content of right hemisphere was determined by wet/dry method [1].

Experiment with the PDT group

As described above, HPD was intravenously infused into animals in the PDT group; after 24 hours in the dark, the rabbits were anesthetized, the cranial window was opened, and an ICP detector placed. The mean ICP over 30 minutes was regarded as the normal ICP control. Then light exposure time was calculated for a light spot area of 1 cm^2, and light exposure was performed. Pathologic observations, after ICP monitoring, were carried out at days 1, 3, and 5 after light exposure.

Statistical analysis

Data are expressed as mean ± SD. Data were analyzed using ANOVA. Significance levels were measured at $p < 0.05$.

Results

Intracranial pressure

ICP in the PDT group increased 1 and 2 days after HPD infusion and light exposure, and the ICP returned to normal 3 days later. ICP in the tumor group did not change drastically within the first 17 days after tumor implantation, ranging from 7 to 8 mmHg, and was not significantly different from the normal control. With further increase in tumor size, however, ICP increased rapidly. By 21 days, ICP increased to 17 to 19 mmHg.

In the tumor plus PDT group, ICP began to increase after PDT, peaking at 24 hours, and dropping thereafter. ICP was not significantly different from the normal control at day 6 (Table 1).

Brain water content

In the PDT group, brain water content increased 1 day after light exposure and decreased 3 days later (Table 2).

In the tumor group, brain water content was in the normal range at day 13 after tumor implantation, but increased rapidly after day 17 (82.0 ± 0.7% in the day 25 group vs. 78.0 ± 0.5% in the day 13 group, Table 2).

Table 2. Brain water content (%, Mean ± SD)

Day (post-implantation)	Day (post-PDT)	PDT group	Tumor group	Tumor plus PDT group
13	before treatment	–	78.0 ± 0.5	–
17	1	79.7 ± 0.3	78.4 ± 0.5*	81.1 ± 1.0#
19	3	78.3 ± 0.5	79.2 ± 0.5*	79.1 ± 0.7
21	5	77.9 ± 0.5	79.8 ± 0.6*	77.8 ± 0.6#
23	7	–	80.6 ± 0.6*	77.4 ± 0.7#
25	9	–	82.0 ± 0.7*	77.6 ± 0.8#

*p < 0.01 vs. day 0, #p < 0.01 vs. the tumor group.

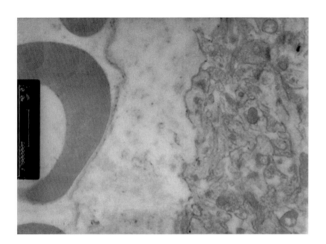

Fig. 1. Perivascular exudation and perineoplastic brain edema 21 days after tumor implantation

In the tumor plus PDT group, brain water content was significantly higher than that in the tumor group 1 day after light exposure (81.1 ± 1.0% vs. 78.4 ± 0.5% in the tumor group, p < 0.01). From day 3, brain water content started to decrease and was lower than that of the tumor group.

Histopathology

In the PDT group, interstitial brain edema developed with mild hyperemia 1 day after light exposure. These changes disappeared within 3 to 5 days.

Seventeen days after tumor implantation, a perineoplastic edema zone developed in the tumor group. A few tumor cells diffusely distributed in the edematous zone, tumor nests were formed, and apparent edema and hyperemia were observed around these nests. Perivascular exudation was observed in tissues adjacent to the tumor using electron microscopy (Fig. 1).

In the tumor plus PDT group, obvious edema adjacent to the tumor was microscopically observed 1 day after light exposure. The edema subsided at day 3, and virtually disappeared 9 days later. The same phenomenon was observed by electron microscopy.

Discussion

It has been reported that PDT causes brain edema, hypoxia of brain tissues, cerebral blood vessel injury, and even cerebral herniation [4]. However, due to relative low doses of photosensitizer used in clinical practice (2–5 mg/kg), few morphological changes of brain can be detected [2, 8]. In glioma patients treated with PDT, Powers et al. [9] found that PDT with HPD led to tumor necrosis in the marginal zone of the tumor. All treated patients had edema in the periphery of the treated zone and clinical symptoms developed 16 hours later. Brain edema disappeared within 1 week. Dehydration agents and glucocorticosteroids could reduce the brain edema.

In our study, rabbits in the PDT group developed brain edema with an increase in ICP at the exposure site 1 day following HPD infusion and light exposure. Brain edema subsided and ICP decreased 3 days later. In the tumor plus PDT group, ICP peaked 1 day after PDT, then declined. Severe edema and vascular injury were found in perineoplastic brain tissues within the first 3 days, and these changes ameliorated 5 days later. Perivascular exudation was also found in perineoplastic brain tissues following PDT using electron microscopy. Brain edema peaked 1 day after treatment, and returned to near normal range 7 days later. These data suggest that PDT results in transient brain edema, which develops immediately after therapy and lasts several days. PDT-induced brain edema is mild with a slight increase in ICP, which may not cause cerebral herniation.

Localization of photosensitizer may also determine the sensitivity of tumor or normal brain tissues to PDT. In normal brain tissues with an intact blood-brain barrier, more photosensitizer distributes on the capillary endothelium, resulting in endothelial necrosis following PDT. In tumor tissue, however, blood-brain barrier permeability is increased. Photosensitizer may pass blood-brain barrier easily, resulting in less vascular injury [3, 5]. Also, PDT produces much necrotic tissue, which results in increased ICP. Although necrotic tissue can be absorbed, a large amount of such tissue cannot be absorbed or discharged swiftly. A marked increase in ICP may cause herniation [10]. Hence, prior to PDT, tumor should be resected as radically as possible.

We treated rabbits with implanted VX2 carcinoma in the brain using the conventional PDT procedure adopted by the Department of Neurosurgery of the Royal Melbourne Hospital of Australia [6]. Our findings indicate that PDT could damage blood vessels in the perineoplastic brain tissues and result in brain edema, but PDT could be a therapy for gliomas. Control of the transient brain edema induced by PDT will be important.

Clinically, PDT for malignant brain tumors should follow surgical treatment and tumor tissues should be resected as radically as possible so as to decompress thoroughly. If necessary, decompression should be attained by removing bone flaps. Within the first few days after PDT, ICP may increase slightly. ICP monitoring and appropriate dehydration treatments immediately after operation are needed.

References

1. Belayev L, Busto R, Zhao W, Ginsberg MD (1996) Quantitative evaluation of blood-brain barrier permeability following middle cerebral artery occlusion in rats. Brain Res 739: 88–96
2. Chen Q, Chopp M, Madigan L, Dereski MO, Hetzel FW (1996) Damage threshold of normal rat brain in photodynamic therapy. Photochem Photobiol 64: 163–167
3. Chopp M, Dereski MO, Madigan L, Jiang F, Logie B (1996) Sensitivity of 9L gliosarcomas to photodynamic therapy. Radiat Res 14: 461–465
4. Goetz C, Hasan A, Stummer W, Heimann A, Kempski O (2002) Experimental research photodynamic effects in perifocal, oedematous brain tissue. Acta Neurochir (Wien) 144: 173–179
5. Karagianis G, Hill JS, Stylli SS, Kaye AH, Varadaxis NJ, Reiss JA, Phillips DR (1996) Evaluation of porphyrin C analogues for photodynamic therapy of cerebral glioma. Br J Cancer 73: 514–521
6. Kaye AH, Popovic EA, Hill JS (1999) Photodynamic therapy. In: Berger MS, Wilson CB (eds) The Gliomas. W.B. Sanders, Philadelphia, pp 619–633
7. Perria C, Capuzzo T, Cavagnaro G, Datti R, Francaviglia N, Rivano C, Tercero VE (1980) Fast attempts at the photodynamic treatment of human gliomas. J Neurosurg Sci 24: 119–129
8. Popovic EA, Kaye AH, Hill JS (1995) Photodynamic therapy of brain tumors. Semin Surg Oncol 11: 335–345
9. Powers SK, Cush SS, Walstad DL, Kwock L (1991) Stereotactic intratumoral photodynamic therapy for recurrent malignant brain tumors. Neurosurgery 29: 688–696
10. Qiu YM, Luo QZ, Xiong WH (1996) Two cases of cerebral hernias due to photodynamic therapy of cerebral glioma. Shanghai Med 19: 183–184

Correspondence: Hua Feng, Department of Neurosurgery, Southwest Hospital, The 3rd Military Medical University, Gaotanyan, Shapingpa, Chongqing, 400038 China. e-mail: fenghua8888@yahoo.com.cn

Whole-body hyperthermia in the rat disrupts the blood-cerebrospinal fluid barrier and induces brain edema

H. S. Sharma[1], J. A. Duncan[2], and C. E. Johanson[2]

[1] Laboratory of Cerebrovascular Research, Institute of Surgical Sciences, Department of Anesthesiology & Intensive Care, University Hospital, Uppsala University, Uppsala, Sweden
[2] Department of Neurosurgery, Brown Medical School, Rhode Island Hospital, Providence, RI, USA

Summary

The present investigation was undertaken to find out whether whole-body hyperthermia (WBH) alters blood-cerebrospinal fluid barrier (BCSFB) permeability to exogenously-administered tracers and whether choroid plexus and ependymal cells exhibit morphological alterations in hyperthermia. Rats subjected to 4 hours of heat stress at 38 °C in a biological oxygen demand (BOD) incubator exhibited a profound increase in the BCSFB to Evans blue and radioiodine. Blue staining of the dorsal surface of the hippocampus and caudate nucleus and a significant increase in Evans blue and [131]Iodine in cisternal cerebrospinal fluid were seen following 4-hour heat stress compared to control. Degeneration of choroidal epithelial cells and underlying ependyma, a dilated ventricular space, and degenerative changes in the underlying neuropil were frequent. Hippocampus, caudate nucleus, thalamus, and hypothalamus exhibited profound increases in water content after 4 hours of heat stress. These observations suggest that hyperthermia induced by WBH is capable of breaking down the BCSFB and contributing to cell and tissue injury in the central nervous system.

Keywords: Hyperthermia; blood-cerebrospinal fluid barrier; edema.

Introduction

Hyperthermia and heat-related illnesses cause large numbers of deaths (about 800 to 2000 cases) during summer months in the United States and in Europe [1, 2, 7–9]. Heat-related death far exceeds that of any other natural calamity such as floods, cyclones, or hurricanes [1, 3, 13, 18, 20, 21, 25, 42, 43]. Recently, the number of heat stroke-induced deaths have increased with global warming and with world-wide increase in the frequency and intensity of heat waves [2, 18, 27, 29, 35, 36].

Heat stress and associated heat stroke are life-threatening illnesses in which body temperature increases above 40 °C causing severe central nervous system (CNS) dysfunction, such as delirium, convulsion, and coma [1–3]. More than 50% of heat stroke victims die within a short time, despite lowering of the body temperature and therapeutic intervention [2, 7–9]. Those who survive heat stroke often show permanent neurological deficit [2, 4, 21, 25].

Interestingly, whole-body hyperthermia (WBH) is commonly used as an adjunct to cytotoxic therapy for cancer [14, 15, 19, 36, 44]. Recently it has been recognized that WBH combined with cytotoxic therapy for cancer causes inhibition of DNA repair, increased drug permeation, and decreased resistance to DNA damaging agents [24]. New clinical and experimental results show that WBH enhances cytotoxic ionizing radiation and chemotherapy [10, 14, 15, 24, 36]. There are reasons to believe that WBH-induced severe side effects, including altered brain function, are probably due to breakdown of BBB function [28, 29, 31].

Previous experiments on WBH in our laboratory suggest that alterations in the brain fluid microenvironment following heat stress are responsible for hyperthermia-induced brain damage [26, 32–34, 38]. However, studies on alterations in the blood-cerebrospinal fluid barrier (BCSFB) in heat stress are still lacking. The BCSFB maintains the composition of the CSF and regulates homeostasis of the CNS within a strict normal limit [6, 16, 22]. Thus, breakdown of the BCSFB adversely influences CNS structure and function.

The present investigation was undertaken to find

out whether WBH alters BCSFB permeability to exogenously-administered tracers and whether choroid plexus and ependymal cells exhibit morphological alterations during hyperthermia.

Materials and methods

Animals

Experiments were carried out on male Sprague-Dawley rats (100 to 150 g; aged 12 to 16 weeks) housed at a controlled room temperature (21 ± 1 °C) on a 12-hour light, 12-hour dark schedule. Food and tap water were supplied ad libitum before the experiments.

Whole-body hyperthermia

Rats were exposed to WBH in a biological oxygen demand incubator (relative humidity 45 to 50%; wind velocity 18 to 25 cm/sec) maintained at 38 °C for 1 to 4 hours [26, 30, 32, 33]. The experiments were conducted according to National Institutes of Health (USA) guidelines for use and care of animals [26, 27, 30]. This experiment was approved by the Ethics Committee of Uppsala University.

BCSFB permeability

The BCSFB was examined in vivo using Evans blue (2%, 0.3 mL/100 g) and [131]Iodine (10 μCi/100 g) tracers [30, 31, 37, 39, 40]. The tracers were administered into the right femoral artery and allowed to circulate for 5 minutes. At the end of the experiment, a CSF sample (about 100 μl) was drawn from the cisterna magna without blood contamination. The animals were then perfused with 0.9% saline through the heart and the brain was dissected out. Extravasation of Evans blue dye was visually examined in the ventricular walls of the lateral, third, and fourth ventricles. Various parts of the brain were then dissected out, weighed, and the radioactivity counted in a gamma counter (energy window 500–800 keV) [30]. Before perfusion, a whole blood sample was withdrawn from the left ventricle by cardiac puncture and the radioactivity determined as above [26, 30, 32]. Extravasation of tracers into the CSF as well as other brain areas around the ventricular system was expressed as percentage increase over the whole blood radioactivity [30]. Evans blue dye that entered some areas of the brain was also measured colorimetrically [26, 30–32].

Brain edema

Brain edema formation was measured using water content calculated from the difference between wet and dry weights of the samples, either in the whole brain or in the several identical brain regions used for radiotracer measurement, as described above [30].

Morphological investigations

At the end of the experiments, rats were perfused transcardially with 4% paraformaldehyde in 0.1 mol phosphate buffer (pH 7.4), preceded by a brief saline rinse [37, 38]. The animals were wrapped in aluminum foil and kept at 4 °C overnight. On the next day, the brain and spinal cord were dissected out and small pieces were embedded in paraffin. About 3 μm thick sections were cut and stained with hematoxylin and eosin or Nissl and examined under a bright field microscope (Leica Microsystems, Bannockburn, IL) for neurodegenerative changes [27]. For semiquantitative analyses of cell injury, rough scores of 0 (no damage), or 1 (least damage) to 4 (maximum damage) were assigned in a blinded fashion [41].

Statistical analyses

Quantitative data were analyzed using ANOVA followed by Dunnet's test for multiple group comparison. The semiquantitative data were analyzed with the chi-square test.

Results

BCSFB permeability

Rats subjected to 4 hours of WBH at 38 °C in a biological oxygen demand incubator exhibited profound alterations in BCSFB to Evans blue and radioiodine tracers. Mild to moderate blue staining of the walls in the lateral, third, and fourth cerebral ventricles was noted. The dorsal surface of the hippocampus and caudate nucleus showed moderate staining. The choroid plexus had deep blue staining. Measurement of Evans blue dye in selected brain regions, such as the hippocampus, caudate nucleus, mid-thalamus (massa intermedia), hypothalamus, dorsal surface of the brain stem, and ventral surface of the cerebellum showed a significant increase compared to the control group (Table 1). A significant increase in Evans blue and [131]Iodine tracers was observed in the CSF samples obtained from the cisterna magna following 4 hours of WBH compared to the control group (Table 1).

On the other hand, blue staining of the ventricular walls and/or surface of the structures within the cerebral ventricles was absent in animals subjected to 1 or 2 hours of heat stress. At these earlier times, no significant increase in Evans blue or radioiodine tracer was noted in various brain areas and/or CSF samples (Table 1).

Brain edema

Measurement of water content in identical brain regions showing leakage of Evans blue or radiotracers, e.g., hippocampus, caudate nucleus, mid-thalamus (massa intermedia), hypothalamus, dorsal surface of the brain stem, and the ventral surface of the cerebellum (sample size 130 to 220 mg wet weight), exhibited a significant increase in water content after 4 hours of WBH (Table 1). However, rats subjected to 1 or 2 hours of heat exposure did not show any increase in the brain water content compared to the control group (Table 1).

Table 1. Changes in BCSFB, brain edema, and structural changes in rats with heat stress for 4 hours

Parameters measured	n	Control	Heat stress 38 °C in a BOD incubator		
			1 h	2 h	4 h
BCSFB permeability#					
[131]Iodine %	5				
Whole brain		0.35 ± 0.06	0.33 ± 0.08	0.42 ± 0.08	1.88 ± 0.24**
Cisternal CSF		0.18 ≈ 0.04	0.12 ± 0.11	0.16 ± 0.12	0.76 ± 0.12**
Hippocampus		0.42 ± 0.12	0.47 ± 0.11	nd	0.84 ± 0.23**
Caudate nucleus		0.28 ± 0.08	0.32 ± 0.14	nd	0.93 ± 0.12**
Cerebellum		0.13 ± 0.08	0.16 ± 0.08	nd	0.65 ± 0.10**
Thalamus		0.48 ± 0.12	0.46 ± 0.08	nd	0.89 ± 0.14**
Hypothalamus		0.54 ± 0.21	nd	nd	0.87 ± 0.23**
Brain Stem		0.18 ± 0.08	nd	nd	0.34 ± 0.14*
Water content#	5				
Whole brain %		76.12 ± 0.18	76.04 ± 0.13	76. ± 4 ± 0.14	80.18 ± 0.24**
Hippocampus		78.43 ± 0.23	78.11 ± 0.21	78.21 ± 0.34	81.56 ± 0.34**
Caudate nucleus		77.43 ± 0.24	77.34 ± 0.32	nd	81.48 ± 0.54**
Cerebellum		74.43 ± 0.21	74.33 ± 0.32	nd	79.34 ± 0.23**
Thalamus		75.21 ± 0.22	75.12 ± 0.33	nd	78.56 ± 0.23**
Hypothalamus		74.54 ± 0.12	nd	nd	76.45 ± 0.23**
Brain Stem		68.54 ± 0.12	nd	nd	69.78 ± 0.12**
Structural changes	5				
Neuronal damage		nil	nil	nil	++++
Glial cell injury		nil	nil	±?	++++
Myelin damage		nil	nil	±?	++++

Values are Mean ± SD of 5 rats in each group.
BCSFB Blood-cerebrospinal fluid barrier; BOD biological oxygen demand; # tissue sample size (135–180 mg); CSF sample 50 to 80 μl; ±? uncertain; ++++ Severe cell damage; nil absent; nd not done; * $p < 0.05$; ** $p < 0.01$ (compared to control); ANOVA followed by Dunnett's test from 1 control.

Morphological alterations

Morphological analysis showed degeneration of choroidal epithelial cells and underlying ependyma in rats subjected to 4 hours of WBH (Fig. 1). The ventricular space appeared to be dilated and the underlying neuropil showed neurodegenerative changes. Neuronal damage, edematous expansion, and edema in hippocampus, cerebral cortex, thalamus, hypothalamus, and brain stem were very common in 4-hour heat-stressed rats (Fig. 1). On the other hand, rats subjected to 1 or 2 hours of WBH did not show structural changes in the brain or spinal cord (results not shown).

Discussion

The salient new finding of the present investigation is a marked increase in BCSFB permeability to Evans blue and radioiodine tracer following 4 hours of WBH in rats, a feature not observed in animals exposed to 1 or 2 hours of heat exposure. These observations suggest that WBH, depending on its duration, is capable of disrupting the BCSFB to large molecule tracers.

Our observations further show that leakiness of the BCSFB is associated with marked cellular changes in several brain regions located within the cerebral ventricles or adjacent regions. Thus, profound cell damage is seen in the hippocampus, caudate nucleus, thalamus, hypothalamus, cerebellum, and brain stem. This indicates that alterations in the BCSFB are somehow contributing to neurodegenerative changes in WBH.

The BCSFB resides in the choroidal epithelial cells that are connected with tight junctions [5, 6, 16, 17, 22]. It is believed that the tightness of the BCSFB is comparable to that of the blood-brain barrier (BBB) located within the cerebral capillary endothelium containing tight junctions [6, 22, 28]. Infusion of hyperosmolar solutions into the internal carotid artery is known to shrink the endothelial cells of the cerebral capillaries and widen the tight junctions leading to breakdown of the BBB [22, 34]. However, it is not known if, under identical conditions, the BCSFB is also compromised.

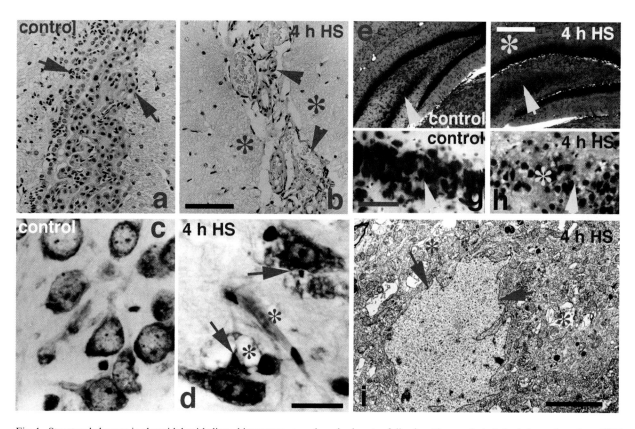

Fig. 1. Structural changes in choroidal epithelium, hippocampus, and cerebral cortex following 4 hours of whole-body hyperthermia at 38 °C. Degeneration of choroidal epithelium (arrowheads) in the lateral ventricle of a 4-hour heat-stressed rat (b) is clearly seen compared to epithelium from a normal animal (a). Damaged nerve cells (arrows) and edema (*) are evident in the cerebral cortex (d) and hippocampus CA4 (f, h) regions in heat-stressed rat compared to control (c, e, g). At the ultrastructural level, the irregular shape of one nerve cell nucleus (arrow, i) in a heat-stressed rat is apparent. Dark and condensed cytoplasm and karyoplasm is clearly visible in the cerebral cortex. Bars: a, b, e, f = 100 μm, c, d = 30 μm, g, h = 20 μm, i = 500 nm

Alterations in the composition of CSF and/or its osmolality are known to occur following lipopolysaccharide-induced fever or heat stress in rabbits [11, 12]. WBH is known to increase plasma viscosity and probably alters the plasma tonicity [2, 4, 44]. Thus, it is possible that hyperosmolality of plasma and/or CSF following WBH could be an important factor in disruption of the BCSFB.

The microvessels supplying choroid plexus are leaky [6, 22]. Thus, choroidal epithelial cells are in direct contact with the hyperosmolar blood plasma [22]. Furthermore, CSF hyperosmolality can also affect the tight junction permeability of the choroidal epithelium from the ependymal side [17]. In addition, the choroid epithelial membrane is subjected to osmotic stress in WBH that could result in increased membrane damage. The structural changes seen in the choroid epithelium and underlying ependymal area are in line with this hypothesis. To confirm these points further, ultrastructural investigations of choroidal epithelium and the tight junctions in WBH are needed.

An increase in Evans blue and radioiodine in the CSF samples obtained from rats subjected to 4 hours of WBH supports the idea of a breakdown of the BCSFB. Evans blue or radioiodine, when injected into the circulation, binds to the endogenous serum proteins [22]. In the present study, administering about 1 molecule of Evans blue binds to 12 molecules of serum albumin in vivo [22, 30]. Therefore, extravasation of Evans blue in the CSF indicates leakage of serum protein complex across the choroid plexus epithelium [30]. Extravasation of serum proteins into the CSF compartment alters the osmotic gradient across the choroid plexus epithelium and the tight junctions resulting in transport of water and other solutes from the vascular compartment leading to edema formation [22, 23, 30].

An increase in the water content of various intracerebral structures following WBH is in line with this hypothesis. Since WBH is known to disrupt BBB in these areas as well [35, 40], a possibility exists that breakdown of the BCSFB will further aggravate regional brain edema formation due to percolation of CSF rich in serum proteins and alterations in CSF to tissue osmotic gradients. Accumulation of serum proteins in the extracellular fluid is likely to initiates a series of cellular and molecular events leading to cell injury and death [28]. The damaged nerve cells, occurrence of sponginess, vacuolation, and edema in many brain areas [28] following 4 hours of WBH are consistent with this idea.

The concept that breakdown of the BCSFB contributes to edema formation and cell injury in WBH is supported by the fact that the short duration of heat exposure (1 or 2 hours) is not associated with leakage of tracers into the CSF or an increase in water content and cell damage. These observations suggest that the magnitude and severity of WBH is primarily responsible for BCSFB damage and brain pathology.

It is unlikely that simple heating of animals following 4 hours of WBH is directly associated with BCSFB leakage [36, 37, 40]. This is evident from the findings that when anesthetized animals are subjected to 4 hours of WBH, no disruption of the BCSFB is observed [Sharma and Johanson, unpublished observation]. Thus, there is reason to believe that WBH-induced alterations in the plasma and CSF composition play an important role in BCSFB disruption. Recent findings in our laboratory suggest that CSF is a conduit of several neurohormones and is capable of transporting several hormones and growth factors in various disease processes [16]. Thus, it is possible that CSF is playing an active role in neurodegeneration and/or neuroprotection. To further explore potential therapeutic strategies involving the CSF microenvironment, the administration of neurohormones, growth factors, or growth hormone into the cerebroventricular spaces should be done in WBH, an approach currently being examined in our laboratory.

Conclusion

In conclusion, our novel observations suggest that hyperthermia induced by WBH is capable of breaking down the BCSFB and contributing to cell and tissue injury in the CNS. It would be important to see whether neuroprotective drugs in heat stress are able to attenuate BCSFB damage in WBH, a subject requiring additional investigation.

Acknowledgments

This investigation was supported by grants from Swedish Medical Research Council (2710, HSS); Astra-Zeneca, Mölndal (HSS), Sweden; National Institutes of Health (NS 27601, CEJ), USA; Alexander von Humboldt Foundation (HSS), Germany; The University Grants Commission (HSS), New Delhi, India; and The Indian Council of Medical Research (HSS), New Delhi, India. The expert technical assistance of Kärstin Flink, Kerstin Rystedt, Franziska Drum, and Katherin Kern, and secretarial assistance of Aruna Sharma are greatly appreciated.

References

1. Belmin J, Golmard JL (2005) Mortality related to the heatwave in 2003 in France: forecasted or over the top? Presse Med 34: 627–628 [in French]
2. Bouchama A (2004) The 2003 European heat wave. Intensive Care Med 30: 1–3
3. Bouchama A, Knochel JP (2002) Heat stroke. N Engl J Med 346: 1978–1988
4. Bouchama A, Roberts G, Al Mohanna F, El-Sayed R, Lach B, Chollet-Martin S, Ollivier V, Al Baradei R, Loualich A, Nakeeb S, Eldali A, de Prost D (2005) Inflammatory, hemostatic, and clinical changes in a baboon experimental model for heatstroke. J Appl Physiol 98: 697–705
5. Bouchaud C, Bouvier D (1978) Fine structure of tight junctions between rat choroidal cells after osmotic opening induced by urea and sucrose. Tissue Cell 10: 331–342
6. Bradbury MWB (1979) The concept of a blood-brain barrier. Wiley, Chichester, England
7. Centers for Disease Control and Prevention (CDC) (2005) Heat-related mortality – Arizona, 1993–2002, and United States, 1979–2002. MMWR Morb Mortal Wkly Rep 54: 628–630
8. Conti S, Meli P, Minelli G, Solimini R, Toccaceli V, Vichi M, Beltrano C, Perini L (2005) Epidemiologic study of mortality during the Summer 2003 heat wave in Italy. Environ Res 98: 390–399
9. Davis RE, Knappenberger PC, Michaels PJ, Novicoff WM (2003) Changing heat-related mortality in the United States. Environ Health Perspect 111: 1712–1718
10. Dewhirst MW, Viglianti BL, Lora-Michiels M, Hanson M, Hoopes PJ (2003) Basic principles of thermal dosimetry and thermal thresholds for tissue damage from hyperthermia. Int J Hyperthermia 19: 267–294
11. Frosini M, Sesti C, Palmi M, Valoti M, Fusi F, Mantovani P, Bianchi L, Della Corte L, Sgaragli G (2000) The possible role of taurine and GABA as endogenous cryogens in the rabbit: changes in CSF levels in heat-stress. Adv Exp Med Biol 483: 335–344
12. Frosini M, Sesti C, Palmi M, Valoti M, Fusi F, Mantovani P, Bianchi L, Della Corte L, Sgaragli G (2000) Heat-stress-induced hyperthermia alters CSF osmolality and composition in conscious rabbits. Am J Physiol Regul Integr Comp Physiol 279: R2095–R2103
13. Gauss H, Meyer KA (1917) Heat stroke: report of one hundred and fifty-eight cases from Cook County Hospital, Chicago. Am J M Sc 154: 554–564
14. Haveman J, Smina P, Wondergem J, van der Zee J, Hulshof MC (2005) Effects of hyperthermia on the central nervous system: What was learnt from animal studies? Int J Hyperthermia 21: 473–487
15. Hildebrandt B, Hegewisch-Becker S, Kerner T, Nierhaus A,

15. Bakhshandeh-Bath A, Janni W, Zumschlinge R, Sommer H, Riess H, Wust P; The German Interdisciplinary Working Group on Hyperthermia (2005) Current status of radiant whole-body hyperthermia at temperatures > 41.5 degrees C and practical guidelines for the treatment of adults. The German 'Interdisciplinary Working Group on Hyperthermia'. Int J Hyperthermia 21: 169–183
16. Jhanson C, Duncan J, Baird A, Stopa E, McMillan P (2005) Choroid plexus: A key player in neuroprotection and neurodegeneration. Int J Neuroprotec Neuroregen 1: 77–85
17. Johanson CE, Foltz FM, Thompson AM (1974) The clearance of urea and sucrose from isotonic and hypertonic fluids perfused through the ventriculo-cisternal system. Exp Brain Res 20: 18–31
18. Kaiser R, Rubin CH, Henderson AK, Wolfe MI, Kieszak S, Parrott CL, Adcock M (2001) Heat-related death and mental illness during the 1999 Cincinnati heat wave. Am J Forensic Med Pathol 22: 303–307
19. Katschinski DM, Wiedemann GJ, Longo W, d'Oleire FR, Spriggs D, Robins HI (1999) Whole body hyperthermia cytokine induction: a review, and unifying hypothesis for myeloprotection in the setting of cytotoxic therapy. Cytokine Growth Factor Rev 10: 93–97
20. Malamud N, Haymaker W, Custer RP (1946) Heat stroke. A clinicopathological study of 125 fatal cases. Milit Surg 99: 397–449
21. Moore R, Mallonee S, Sabogal RI, Zanardi L, Redd J, Malone J (2002) Heat-related deaths – four states, July–August 2001, and United States, 1979–1999. JAMA 288: 950–951
22. Rapoport SI (1976) Blood-brain barrier in physiology and medicine. Raven Press, New York, pp 1–380
23. Reulen HJ, Tsuyumu M, Tack A, Fenske AR, Prioleau GR (1978) Clearance of edema fluid into cerebrospinal fluid. A mechanism for resolution of vasogenic brain edema. J Neurosurg 48: 754–764
24. Robins HI, Peterson CG, Mehta MP (2003) Combined modality treatment for central nervous system malignancies. Semin Oncol 30: 11–22
25. Scoville SL, Gardner JW, Magill AJ, Potter RN, Kark JA (2004) Nontraumatic deaths during US Armed Forces basic training, 1977–2001. Am J Prev Med 26: 205–212
26. Sharma HS (1982) Blood-brain barrier in stress [PhD Thesis] Banaras Hindu University, Varanasi, India, pp 1–85
27. Sharma HS (1999) Pathophysiology of blood-brain barrier, brain edema and cell injury following hyperthermia: new role of heat shock protein, nitric oxide and carbon monoxide. An experimental study in the rat using light and electron microscopy, Acta Universitatis Upsaliensis 830: 1–94
28. Sharma HS (2004) Blood-brain and spinal cord barriers in stress. In: Sharma HS, Westman J (eds) Blood-spinal cord and brain barriers in health and disease. Elsevier Academic Press, San Diego, pp 231–298
29. Sharma HS (2005) Heat-related deaths are largely due to brain damage. Indian J Med Res 121: 621–623
30. Sharma HS (2005) Methods to induce brain hyperthermia. In: Costa E (ed) Current protocols in toxicology, Suppl 23. John Wiley Inc, New York, pp 1–26
31. Sharma HS (2005) Alterations of amino acid neurotransmitters in hyperthermic brain injury. J Neural Transm [in press]
32. Sharma HS, Dey PK (1986) Probable involvement of 5-hydroxytryptamine in increased permeability of blood-brain barrier under heat stress in young rats. Neuropharmacology 25: 161–167
33. Sharma HS, Dey PK (1987) Influence of long-term acute heat exposure on regional blood-brain barrier permeability, cerebral blood flow and 5-HT level in conscious normotensive young rats. Brain Res 424: 153–162
34. Sharma HS, Cervós-Navarro J (1990) Brain oedema and cellular changes induced by acute heat stress in young rats. Acta Neurochir [Suppl] 51: 383–386
35. Sharma HS, Westman J (1998) Brain functions in hot environment. Elsevier, Amsterdam, pp 1–516
36. Sharma HS, Hoopes PJ (2003) Hyperthermia induced pathophysiology of the central nervous system. Int J Hyperthermia 19: 325–354
37. Sharma HS, Cervós-Navarro J, Dey PK (1991) Rearing at high ambient temperature during later phase of the brain development enhances functional plasticity of the CNS and induces tolerance to heat stress. An experimental study in the conscious normotensive young rats. Brain Dysfunction 4: 104–124
38. Sharma HS, Cervós-Navarro J, Dey PK (1991) Acute heat exposure causes cellular alteration in cerebral cortex of young rats. Neuro Report 2: 155–158
39. Sharma HS, Westman J, Cervós-Navarro J, Nyberg F (1997) Role of neurochemicals in brain edema and cell changes following hyperthermic brain injury in the rat. Acta Neurochir [Suppl] 70: 269–274
40. Sharma HS, Westman J, Nyberg F (1998) Pathophysiology of brain edema and cell changes following hyperthermic brain injury. Prog Brain Res 115: 351–412
41. Sharma HS, Lundstedt T, Flardh M, Westman J, Post C, Skottner A (2003) Low molecular weight compounds with affinity to melanocortin receptors exert neuroprotection in spinal cord injury – an experimental study in the rat. Acta Neurochir [Suppl] 86: 399–405
42. Stott PA, Stone DA, Allen MR (2004) Human contribution to the European heatwave of 2003. Nature 432: 559–560
43. Sweeney KG (2002) Heat-related deaths. J Insur Med 34: 114–119
44. Thrall DE, Page RL, Dewhirst MW, Meyer RE, Hoopes PJ, Kornegay JN (1986) Temperature measurements in normal and tumor tissue of dogs undergoing whole body hyperthermia. Cancer Res 46: 6229–6235
45. Wrba E, Nehring V, Chang RC, Baethmann A, Reulen HJ, Uhl E (1997) Quantitative analysis of brain edema resolution into the cerebral ventricles and subarachnoid space. Acta Neurochir [Suppl] 70: 288–290

Correspondence: Hari Shanker Sharma, Department of Surgical Sciences, Anesthesiology & Intensive Care Medicine, University Hospital, Uppsala, Sweden. e-mail: Sharma@surgsci.uu.se

ICP, CSF, and the Cerebrovasculature

Dynamics of cerebral venous and intracranial pressures

E. M. Nemoto

Department of Radiology, University of Pittsburgh, Pittsburgh, PA, USA

Summary

Traumatic brain injury and stroke are both characterized by an ischemic core surrounded by a penumbra of low to hyperemic flows. The underperfused ischemic core is the focus of edema development, but the source of the edema fluid is not known. We hypothesized that flow of edema fluid into the tissue is derived from cerebral venous circulation pressure, which always exceeds intracranial pressure (ICP). As a first step toward testing this hypothesis, the aim of the current study was to determine whether cerebral venous pressure in the normal brain is always equal to or higher than ICP. In studies on 2 pigs, cerebral cortical venous, intracranial (subarachnoid), sagittal sinus, and central venous pressures were monitored with manipulation of ICP by raising and lowering a reservoir above and below the external auditory meatus zero point. The results show that cerebral venous pressure is always higher than or equal to ICP at pressures of up to 60 mmHg. On the basis of these observations, we hypothesize that increased cerebral venous pressure initiated after traumatic brain injury and stroke drives edema fluid into the tissue, which thereby increases ICP and a further increase in cerebral venous pressure in a vicious cycle of brain edema.

Keywords: Traumatic brain injury; intracranial pressure; venous pressure.

Introduction

Both traumatic brain injury and stroke are characterized by an ischemic core with cerebral blood flow values less than 10 ml/100 g/min and a surrounding penumbra with low (10 to 18 ml/100 g/min) to hyperemic (>60 ml/100 g/min) values. The ischemic core is the site of rapid brain edema development partly driven by the increase in tissue osmolality of 80 to 100 mOsm. The source of the edema fluid in the underperfused ischemic core is unknown.

The cortical venous system is shared by both the ischemic core and the ischemic penumbra and could be the source of edema. Because the cortical veins reside in the subarachnoid space and under intracranial pressure (ICP), we hypothesize that cortical venous pressure should always be equal to or exceed ICP. Thus, as ICP rises with the progressive development of brain edema, cortical venous pressure rises pushing more edema fluid into the tissue, thereby creating a vicious cycle of edema and increasing ICP. As the first step in testing this hypothesis, our aim in the present study was to determine whether cortical venous pressure is always equal to or exceeds ICP in the normal brain.

Methods

Animals and anesthesia

Studies were done on 2 male pigs weighing 25 kg. The pigs were anesthetized with ketamine 20 mg/kg, and xylazine 2 mg/kg, i.m. Anesthesia was maintained with fentanyl by continuous infusion.

Surgical procedures

All animals were intubated with cuffed endotracheal tubes and mechanically ventilated on 30% oxygen, balance nitrogen. In both pigs, peripheral venous catheters were inserted for fluid and drug administration, and femoral artery catheters inserted for continuous monitoring of arterial blood pressure and arterial blood samples for blood gases and pH measurements.

The animals were fixed in the prone position and the dorsal calvarium exposed by a U-shaped incision with the bottom of the U anterior. Craniotomies were made over the right parietal hemisphere approximately 2 cm in diameter and over the superior sagittal sinus. Polyethylene catheters were inserted into the sinus and a bridging cortical vein with the tip directed away from the sinus. Two subdural catheters, one to measure ICP and the other for monitoring of ICP, were also placed. The dura around the catheter was sealed with cyanoacrylate glue and the calvarium with cotton sponges soaked in cyanoacrylate glue. The skin was sutured around the catheters. All pressure transducers were zeroed at the level of the external auditory meatus.

Following an equilibration period with arterial blood gas sampling and arterial pressure monitoring to verify physiological variables within normal limits, a reservoir connected to the subarachnoid space was zeroed at the level of the external auditory meatus.

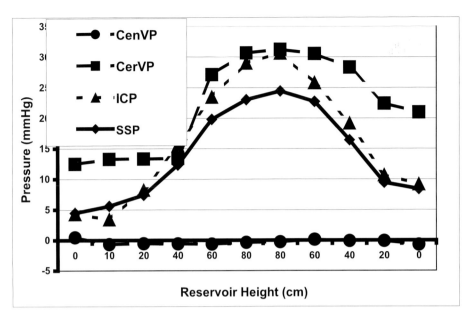

Fig. 1. Cerebral venous (*CerVP*), intracranial (*ICP*), central venous (*CenVP*), and sagittal sinus (*SSP*) pressures measured in a pig. ICP was manipulated by raising and lowering a reservoir filled with mock cerebrospinal fluid zeroed at the level of the external auditory meatus

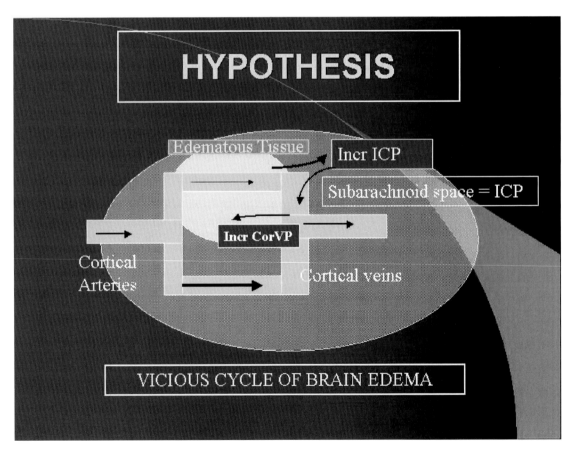

Fig. 2. Hypothetical illustration of the role of CerVP and its relationship to ICP in the generation of cerebral edema in a vicious cycle. Tissue edema initiated by stroke or traumatic brain edema, leads to an increase in CerVP, which drives more fluid into the tissue. The increase in tissue pressure leads to increased ICP, which causes a further rise in CerVP and increased tissue edema in a vicious cycle of edema

The ICP reservoir was then raised and lowered in increments of 10 cm H_2O. ICP was allowed to equilibrate for 5 minutes before pressure readings of cortical vein and superior sagittal sinus were recorded.

Results

In the first pig, a stepwise 10 cm H_2O increase in reservoir height from the zero point at the level of the external auditory meatus caused increases in cerebral venous pressure (CerVP) and always exceeded ICP as the reservoir was moved up and down (Fig. 1). A slight hysteresis was observed when the reservoir was lower and where CerVP remained considerably higher than ICP and sagittal sinus pressure (SSP). SSP also increased but to a less extent than both CerVP and ICP. Similar results were obtained in a second study in a pig where CerVP was always higher than ICP and, again, SSP increased as well. SSP followed ICP closely.

Discussion

Our results show that CerVP is always equal to or higher than ICP. Earlier studies also had shown that CerVP exceeds intracranial or subarachnoid pressure [1, 2]. However, these authors were content with the finding that the waterfall effect between cortical vein pressure and SSP occurred in the lateral lacunae prior to entrance into the sagittal sinus and the source of the resistance might be the arachnoid granulations. Nakagawa, *et al.* [1] and Yada, *et al.* [2] failed to recognize the potential importance of CerVP in the generation of brain edema.

The objective of this study was to determine whether maintenance of CerVP higher than ICP might be the source of edema fluid following stroke and traumatic brain injury. As a first step, we determined whether CerVP is always equal to or higher than ICP, which we have found is the case. On the basis of our findings, we hypothesize that the increase in ICP as a result of brain edema leads to increased CerVP, which drives the increase in tissue edema and a further increase in ICP, thereby further increasing CerVP – a vicious cycle of brain edema (Fig. 2).

References

1. Nakagawa Y, Tsuru M, Yada K (1974) Site and mechanism for compression of the venous system during experimental intracranial hypertension. J Neurosurg 41: 427–434
2. Yada K, Nakagawa Y, Tsuru M (1973) Circulatory disturbance of the venous system during experimental intracranial hypertension. J Neurosurg 39: 723–729

Correspondence: Edwin M. Nemoto, Department of Radiology, University of Pittsburgh, 3950 Children's Main Tower, 200 Lothrop St., Pittsburgh, PA 15213, USA. e-mail: nemotoem@upmc.edu

Effects of angiopoietin-1 on vascular endothelial growth factor-induced angiogenesis in the mouse brain

Y. Zhu[1,4], Y. Shwe[1], R. Du[1,2], Y. Chen[1], F. X. Shen[1], W. L. Young[1,2,3], and G. Y. Yang[1,2,4]

[1] Department of Anesthesia, Center for Cerebrovascular Research, University of California, San Francisco, CA, USA
[2] Department of Neurological Surgery, Center for Cerebrovascular Research, University of California, San Francisco, CA, USA
[3] Department of Neurology, Center for Cerebrovascular Research, University of California, San Francisco, CA, USA
[4] Department of Neurosurgery, Hua-Shan Hospital, Fu-Dan University, Shanghai, China

Summary

A better understanding of angiogenic factors and their effects on angiogenesis in brain is necessary to treat cerebral vascular disorders such as ischemic brain injury. Vascular endothelial growth factor (VEGF) induces angiogenesis and increases blood-brain barrier (BBB) permeability in adult mouse brain. The effect of angiopoietin-1 on BBB leakage during the angiogenesis process is unclear. We sought to identify the effects of combining VEGF with angiopoietin-1 on cerebral angiogenesis and BBB.

Adult male CD-1 mice underwent AdFc (adenoviral vector control), AdAng-1, VEGF protein, VEGF protein plus AdAng-1, or saline (negative control) injection. Brain microvessels were counted using lectin staining on tissue sections after 2 weeks of adenoviral gene transfer. The presence of zonula occludens-1 (ZO-1) was determined by Western blot analysis and immunohistochemistry.

Microvessel count and augmented capillary diameter increased in mice treated with either VEGF protein or AdAng-1 plus VEGF protein compared to saline, AdFc, or AdAng-1 alone ($p < 0.05$). Double-labeled immunostaining demonstrated that ZO-1-positive staining was more complete on the microvessel wall in the AdAng-1 and AdAng-1 plus VEGF protein treated group compared to VEGF protein group. The results of ZO-1 expression from Western blot analysis paralleled that from immunohistochemistry ($p < 0.05$).

We conclude that focal VEGF and angiopoietin-1 hyperstimulation in mouse brain increases microvessel density while maintaining ZO-1 protein expression, suggesting that angiopoietin-1 plays a role in synergistically inducing angiogenesis and BBB integrity.

Keywords: Adenoviral vectors; angiogenesis; angiopoietin-1; blood-brain barrier; VEGF; ZO-1.

Introduction

Vascular endothelial growth factor (VEGF) is an important factor in regulation of the development and differentiation of the vascular system. VEGF affects the integrity of the blood-brain barrier (BBB) by functioning as a capillary permeability-enhancing agent [4]. As primary partners of VEGF, angiopoietins (Ang) play multiple critical roles in vascular development, especially in the maturation of BBB function in the brain [9]. The angiopoietins Ang-1 and Ang-2 are ligands of the endothelial receptor tyrosine kinase Tie-2. Ang-1 stimulates phosphorylation of Tie-2 and stabilizes mature vessels by promoting an interaction between endothelial cells and surrounding support cells, while Ang-2 is an antagonist of Tie-2 [11, 16]. Recent study suggested that angiopoietins and VEGF have been found to have another role in regulating capillary permeability by modulating the expression of tight junction (TJ) proteins in the brain [8]. TJs are essential structural components that maintain the function of the BBB. Zonula occludens-1 (ZO-1), one of the submembranous TJ-associated proteins that are specifically bound to the cytoskeleton, is crucial for the establishment of endothelial polarity and regulation of vascular permeability. Because VEGF stimulates new microvessel growth but promotes vessel leakage, therapy using VEGF alone is not an effective way to accelerate neovascularization to reduce ischemic brain injury and restore impaired function. The aim of our study was to determine whether Ang-1 collaborates with VEGF to affect angiogenesis and BBB structure in the brain and assess their effects on ZO-1.

Materials and methods

Experimental groups

Procedures for the use of laboratory animals were approved by the University of California at San Francisco Committee on Animal Research. Sixty adult male CD-1 mice (Charles River Laboratories, Wilmington, MA) weighing 30–35 g were divided into 5 experimental groups with 12 mice in each group. These mice were injected with AdFc (adenoviral vector control), AdAng-1 (adenoviral vector with Ang-1), VEGF protein, VEGF protein plus AdAng-1, or saline (negative control). Mice were anesthetized with intraperitoneal injections of ketamine/xylazine (Sigma-Aldrich, St. Louis, MO) at 1.5 µL/g body weight. Following anesthesia, mice were placed on a stereotactic frame with a mouse holder (Kopf Instruments, Tujunga, CA). For the adenoviral vector groups (AdFc, AdAng-1, VEGF/AdAng-1, saline), a burr hole was drilled in the frontal skull 2 mm lateral to the sagittal suture and 0.6 mm anterior to the coronal suture. A Hamilton syringe was inserted to a depth of 3.0 mm below the cortical surface into the lateral caudate. Two microliters of adenoviral suspension (AdFc, AdAng-1) containing 2.5×10^9 particles/µL was injected stereotactically into the right caudate/putamen at a rate of 0.2 µL/min. The control animals received the same amount of saline injection. For the protein infusion groups (VEGF, VEGF/AdAng-1), an osmotic mini-pump (Alza Corp., Palo Alto, CA) was implanted between the neck and shoulder, and continuously infused protein into the lateral ventricle for 2 weeks. The tip of the infusion cannula was implanted in the lateral ventricle, 1 mm lateral to bregma, to a depth of 2.5 mm.

Lectin staining and vessel counting

To visualize the vasculature, mice were first anesthetized and injected intravenously with 0.05 mg FITC-labeled tomato lectin (*lycopersicon esculentum*; Vector Laboratories, Burlingame, CA) and then heart-perfused with 4% paraformaldehyde as previously described [15]. Brains were sectioned at 20 µm. Two coronal sections 1 mm anterior and 1 mm posterior to the needle track were chosen. Microvessel counting was carried out at 100× magnification in 3 areas: left, right, and inferior to the needle track.

Double-labeled fluorescent staining

Sections were fixed with acetone at −20 °C for 10 minutes and incubated with 5% normal blocking serum for 1 hour. Vessel staining with lectin was done as described above followed by ZO-1 staining. Slides were incubated with rabbit anti-mouse ZO-1 antibody (1:200 dilution, Zymed Laboratories, South San Francisco, CA) overnight followed by staining with Texas Red anti-rabbit IgG (1:200, Vector Laboratories). Fluorescent immunostaining sections were evaluated using a fluorescence microscope (Nikon Microphoto-SA) with a filter cube (excitation filter, 450 to 490 nm) for fluorescent isothiocyanate labeling, and a filter cube (excitation filter, 515 to 560 nm) for Texas Red. Photomicrographs for double-labeling illustrations were obtained by changing the filter cube without altering the position of the section or focus.

Western blot analysis

Brain tissue was homogenized in lysis buffer and equal amounts of total protein extracts were electrophoresed on 7% Tris-acetate gels (Invitrogen, Carlsbad, CA), and then transferred to a polyvinylidene difluoride membrane (BioRad Laboratories, Hercules, CA) using standard procedures. After blocking with 5% non-fat dry milk, the membrane was incubated at 4 °C overnight with polyclonal rabbit anti ZO-1 antibody (1:500 dilution, Zymed) followed by incubation with peroxidase-conjugated goat anti-rabbit antibody (1:4000, Pierce Biotechnology, Rockford, IL). Finally, the membrane was plastic wrapped and incubated in Femo detection reagent (Pierce Biotechnology) for 5 minutes. Relative densities of the bands were analyzed with NIH Image 1.63 software (National Institutes of Health, Bethesda, MD).

Statistical analysis

Parametric data in the VEGF protein infusion, AdAng-1, VEGF protein/AdAng-1, AdFc, and saline treated groups were compared using analysis of variance (ANOVA) followed by the Scheffe f-test. Means and standard deviation of each group are reported, with p value of less than 5% to indicate statistical significance.

Results

Increase of microvessel counts in VEGF and VEGF/Ang-1 treated mice

Lectin fluorescent staining is a sensitive method for visualizing microvessels in the mouse brain. We analyzed changes in microvessel counts in the 5 groups of mice following 2 weeks of adenoviral vector transduction and protein infusion. We found that AdAng-1 alone did not increase the microvessel counts, but did increase the vessel size. This phenomenon was most prominent around the needle track. The number of microvessels in the VEGF protein infusion and AdAng-1 plus VEGF protein transduction groups was greatly increased compared to the saline treated or AdFc transduced mice ($p < 0.05$, Fig. 1). The microvessel counts were not significantly different between the VEGF and AdAng-1 plus VEGF transduction groups ($p > 0.05$). In the AdAng-1 plus VEGF group, we also observed a similar distribution of microvessel diameters as in the AdAng-1 group. No hemorrhages were detected. These results demonstrate that the combination of VEGF and Ang-1 increased microvessel counting in the adult mouse brain similar to VEGF alone, although for vessel diameter, the VEGF plus Ang1 was similar to Ang-1 alone, not to VEGF alone.

Increase of ZO-1 protein in VEGF/Ang-1 mice

To test whether Ang-1 acts synergistically with VEGF in maintaining BBB structure integrity, we further measured ZO-1 expression using Western blot analysis. We demonstrated that ZO-1 expression decreased in the VEGF treated animals compared to the other 4 groups of mice. Interestingly, the combination of VEGF and Ang-1 increased ZO-1 expression compared to VEGF alone ($p < 0.05$) (Fig. 2a). To identify whether ZO-1 expression was decreased in the vessel

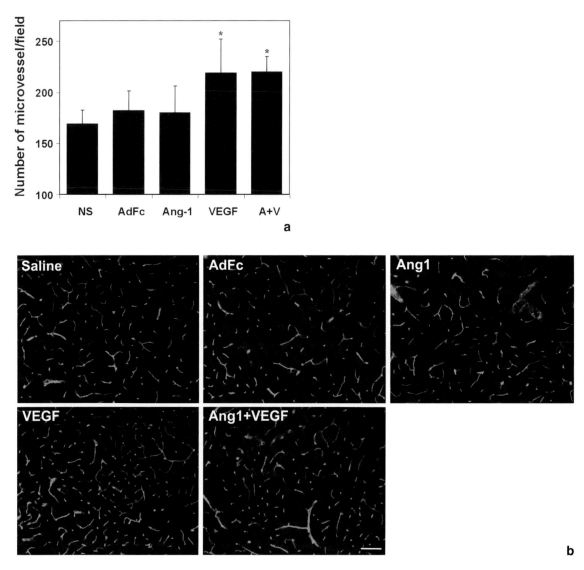

Fig. 1. (a) Bar graph showing microvessel counts in 5 groups of mice following VEGF and Ang-2 treatment. *NS* Saline, *AdFc* AdFc, *Ang-1* AdAng-1, *VEGF* VEGF infusion, *A+V* AdAng-1 plus VEGF infusion. Data are mean ± SD, n = 5 per group. *p < 0.05, VEGF infusion or AdAng-1 plus VEGF infusion vs. Saline. (b) Photomicrographs show lectin staining of microvessls in the 5 treated groups. The numbers of microvessels in the VEGF infusion and AdAng-1 plus VEGF infusion mice are more than that in the other 3 groups. Bar = 100 μm

wall, we performed double-labeled immunostaining. We found that in the normal condition (control group), ZO-1 was expressed around microvessels, suggesting that ZO-1 existed at sites of endothelial cell-to-cell contact and preserved the integrity of the microvessel endothelial cells. However, among the VEGF treated mice, ZO-1 positive staining was much less than the control group. In contrast, ZO-1 positive staining was increased in the Ang-1 and VEGF plus Ang-1 treated mice (Fig. 2b). These data suggest that Ang-1 protects ZO-1 protein in VEGF-induced brain angiogenesis.

Discussion

The use of AdAng-1 gene transfer and the combination of AdAng-1 gene transfer and VEGF protein infusion provides a useful in vivo approach for overexpressing multiple angiogenic proteins in the brain. Employing this method, we demonstrated that: 1) the combination of VEGF and Ang-1 increased microvessel counts, although Ang-1 alone did not stimulate increased microvessel density; 2) the combination of VEGF and Ang-1 prevents ZO-1 protein reduction when compared to VEGF treatment alone. These data

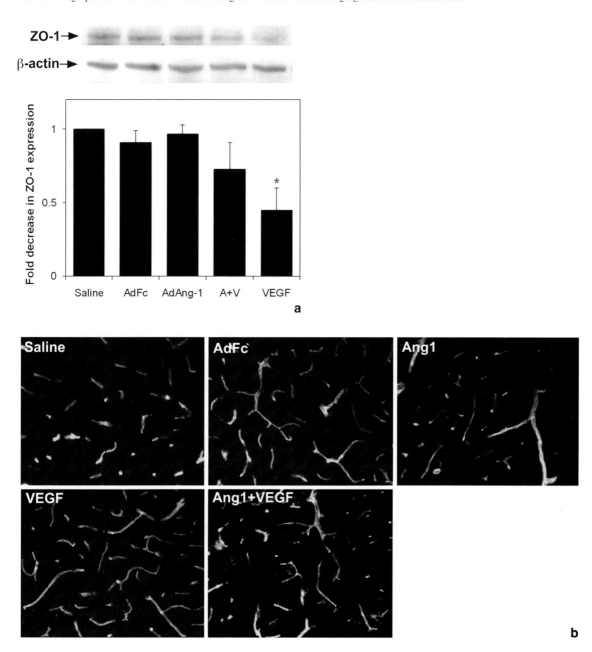

Fig. 2. (a) Upper panel represents one experiment of ZO-1 expression in Western blot analysis. ZO-1 expression is detected in the *NS* saline, *AdFc* AdFc, *Ang-1* AdAng-1, *VEGF* VEGF infusion, *A+V* AdAng-1 plus VEGF infusion. Bar graph shows ZO-1 expression in the mouse brain after the 5 different treatments. Data are mean ± SD, n = 5 per group. *p < 0.05, VEGF infusion vs. other groups. ZO-1 expression decreased in VEGF treated mice. (b) Photomicrographs show double-labeled immunostaining in these 5 groups. Green is lectin staining in brain microvessels, red is ZO-1 positive immunostaining, and yellow is merged staining of lectin and ZO-1. Yellow color is much less in the VEGF treated mice. There is not much difference between the AdAng-1 plus VEGF infusion and the other 3 groups

suggest that Ang-1 may act synergistically with VEGF to promote angiogenesis. Because the combination of VEGF and Ang-1 preserves ZO-1 expression in the microvasculature, Ang-1 may also play a defensive role in BBB integrity during brain angiogenesis.

Angiogenesis is a step-wise process. Necessary steps include an increase in vascular permeability, degradation of surrounding matrix, proliferation and migration of endothelial cells, and stabilization of neo-microvessels [3]. Many angiogenic molecules are involved in this process, including VEGF, Ang-1, and Ang-2. VEGF is essential for angiogenesis and BBB functionality. Our previous work showed that local VEGF transduction promoted capillary angiogenesis [10]. Intraventricular infusion of VEGF protein induces microvascular formation in the whole rodent brain [7]. We used the combination of VEGF and Ang-1 because both angiogenic factors have greatly increased expression during brain angiogenesis. Ang-1 alone did not increase the number of microvessels, however, suggesting that Ang-1 may not directly signal vascular remodeling. The combination of VEGF and Ang-1 still increased the microvessel density, which suggests that Ang-1 may synergize with VEGF during sprouting angiogenesis in vivo [2].

We are particularly interested in the structural features of BBB in neovascularized tissue because it is essential for induction of functional angiogenesis and maintenance of the angiogenic microenvironment. ZO-1 is a peripheral TJ protein that is found on the epithelial and endothelial cell membranes [5]. In the adult brain, most microvessels express ZO-1 [13]. VEGF has been shown to mediate a rapid increase in ZO-1 tyrosine phosphorylation and induce the loss of ZO-1 from endothelial TJs in association with enhanced capillary leakage and BBB permeability, both in vivo and in vitro [1]. However, Ang-1 and VEGF have different effects on vascular function and integrity. A recent study showed that an increase in Ang-1 is likely to account for the ability of SSeCKS to seal the BBB by inducing a linear distribution of ZO-1 and claudin-1 [12]. On the other hand, Ang-1 also up-regulates AHNAK expression, which is widely distributed in endothelial cells with BBB functions and co-localizes with ZO-1 [6]. Since Ang-1 plays a key role in promoting vascular integrity and blocking endothelial cell-to-cell contact leakage [14], it has an enormous effect on upholding ZO-1 expression in the brain microvessel wall. VEGF initially stimulates early angiogenesis, whereas Ang-1 is subsequently required to stabilize the immature vessels and maintain the structural integrity of the endothelial cell-to-cell TJs. There is no doubt that the combination of Ang-1 and VEGF resists the disruptive effects of VEGF on the TJ structure and protects VEGF-induced brain angiogenesis from vascular leakiness.

In conclusion, the present study has demonstrated that focal VEGF and angiopoietin-1 hyperstimulation in the mouse brain increases microvessel density while maintaining ZO-1 protein expression, thereby inducing angiogenesis as well as BBB integrity. Understanding the effects of various angiogenic factors on different aspects of angiogenesis is important for utilizing these factors in the treatment of ischemic brain injury and other pathologic processes.

Acknowledgments

These studies were supported by grants from the National Institutes of Health: R01 NS27713 (WLY), R21 NS45123 (GYY), and PPG NS44144 (WLY, GYY). The authors thank the collaborative support of the staff at the Center for Cerebrovascular Research (http://avm.ucsf.edu/).

References

1. Antonetti DA, Barber AJ, Hollinger LA, Wolpert EB, Gardner TW (1999) Vascular endothelial growth factor induces rapid phosphorylation of tight junction proteins occludin and zonula occluden 1. A potential mechanism for vascular permeability in diabetic retinopathy and tumors. J Biol Chem 274: 23463–23467
2. Asahara T, Chen D, Takahashi T, Fujikawa K, Kearney M, Magner M, Yancopoulos GD, Isner JM (1998) Tie2 receptor ligands, angiopoietin-1 and angiopoietin-2, modulate VEGF-induced postnatal neovascularization. Circ Res 83: 233–240
3. Conway EM, Collen D, Carmeliet P (2001) Molecular mechanisms of blood vessel growth. Cardiovasc Res 49: 507–521
4. Ferrara N, Alitalo K (1999) Clinical applications of angiogenic growth factors and their inhibitors. Nat Med 5: 1359–1364
5. Fischer S, Wobben M, Marti HH, Renz D, Schaper W (2002) Hypoxia-induced hyperpermeability in brain microvessel endothelial cells involves VEGF-mediated changes in the expression of zonula occludens-1. Microvasc Res 63: 70–80
6. Gentil BJ, Benaud C, Delphin C, Remy C, Berezowski V, Cecchelli R, Feraud O, Vittet D, Baudier J (2005) Specific AHNAK expression in brain endothelial cells with barrier properties. J Cell Physiol 203: 362–371
7. Harrigan MR, Ennis SR, Masada T, Keep RF (2002) Intraventricular infusion of vascular endothelial growth factor promotes cerebral angiogenesis with minimal brain edema. Neurosurgery 50: 589–598
8. Iizasa H, Bae SH, Asashima T, Kitano T, Matsunaga N, Terasaki T, Kang YS, Nakashima E (2002) Augmented expression of the tight junction protein occludin in brain endothelial cell line TR-BBB by rat angiopoietin-1 expressed in baculovirus-infected Sf plus insect cells. Pharm Res 19: 1757–1760
9. Jones N, Iljin K, Dumont DJ, Alitalo K (2001) Tie receptors: new modulators of angiogenic and lymphangiogenic responses. Nat Rev Mol Cell Biol 2: 257–267
10. Lee CZ, Xu B, Hashimoto T, McCulloch CE, Yang GY, Young WL (2004) Doxycycline suppresses cerebral matrix metalloproteinase-9 and angiogenesis induced by focal hyperstimulation of vascular endothelial growth factor in a mouse model. Stroke 35: 1715–1719
11. Maisonpierre PC, Suri C, Jones PF, Bartunkova S, Wiegand SJ, Radziejewski C, Compton D, McClain J, Aldrich TH, Papadopoulos N, Daly TJ, Davis S, Sato TN, Yancopoulos GD (1997) Angiopoietin-2, a natural antagonist for Tie2 that disrupts in vivo angiogenesis. Science 277: 55–60
12. Rieckmann P, Engelhardt B (2003) Building up the blood-brain barrier. Nat Med 9: 828–829
13. Song L, Pachter JS (2004) Monocyte chemoattractant protein-1

alters expression of tight junction-associated proteins in brain microvascular endothelial cells. Microvasc Res 67: 78–89

14. Thurston G, Rudge JS, Ioffe E, Zhou H, Ross L, Croll SD, Glazer N, Holash J, McDonald DM, Yancopoulos GD (2000) Angiopoietin-1 protects the adult vasculature against plasma leakage. Nat Med 6: 460–463

15. Xu B, Wu YQ, Huey M, Arthur HM, Marchuk DA, Hashimoto T, Young WL, Yang GY (2004) Vascular endothelial growth factor induces abnormal microvasculature in the endoglin heterozygous mouse brain. J Cereb Blood Flow Metab 24: 237–244

16. Yancopoulos GD, Davis S, Gale NW, Rudge JS, Wiegand SJ, Holash J (2000) Vascular-specific growth factors and blood vessel formation. Nature 407: 242–248

Correspondence: Guo-Yuan Yang, Department of Anesthesia, Center for Cerebrovascular Research, University of California San Francisco, Box 1371, 1001 Potrero Ave, San Francisco, CA 94110, USA. e-mail: gyyang@anesthesia.ucsf.edu

Inflammation and brain edema: new insights into the role of chemokines and their receptors

S. M. Stamatovic[1], O. B. Dimitrijevic[1], R. F. Keep[1,2], and A. V. Andjelkovic[1,3]

[1] Department of Neurosurgery, University of Michigan Medical School, Ann Arbor, Michigan, USA
[2] Department of Physiology, University of Michigan Medical School, Ann Arbor, Michigan, USA
[3] Department of Pathology, University of Michigan Medical School, Ann Arbor, Michigan, USA

Summary

Brain edema is associated with a variety of neuropathological conditions such as brain trauma, ischemic and hypoxic brain injury, central nervous system infection, acute attacks of multiple sclerosis, and brain tumors. A common finding is an inflammatory response, which may have a significant impact on brain edema formation. One critical event in the development of brain edema is blood-brain barrier (BBB) breakdown, which may be initiated and regulated by several proinflammatory mediators (oxidative mediators, adhesion molecules, cytokines, chemokines). These mediators not only regulate the magnitude of leukocyte extravasation into brain parenchyma, but also act directly on brain endothelial cells causing the loosening of junction complexes between endothelial cells, increasing brain endothelial barrier permeability, and causing vasogenic edema. Here we review junction structure at the BBB, the effects of pro-inflammatory mediators on that structure, and focus on the effects of chemokines at the BBB. New evidence indicates that chemokines (chemoattractant cytokines) do not merely direct leukocytes to areas of injury. They also have direct and indirect effects on the BBB leading to BBB disruption, facilitating entry of leukocytes into brain, and inducing vasogenic brain edema formation. Chemokine inhibition may be a new therapeutic target to reduce vasogenic brain edema.

Keywords: Blood-brain barrier; monocyte chemoattractant protein; CCL2; tight junctions.

Introduction

Inflammation and proinflammatory mediators play an essential role in edema development in a variety of neuropathological conditions such as brain trauma, ischemic or hypoxic brain injury, central nervous system (CNS) infection (HIV infection, tuberculosis, or bacterial meningitis), acute attacks of multiple sclerosis, and brain tumors [12, 19, 21, 29, 45, 49, 64]. Brain edema in all of these conditions is mostly classified as vasogenic with extracellular water accumulation resulting from blood-brain barrier (BBB) disruption and massive infiltration of leukocytes. The critical pathophysiological mechanism in vasogenic brain edema formation is BBB disruption, characterized by activation of brain endothelial cells followed by loosening of junctional complexes between those cells and increased barrier permeability. This is accompanied by leukocyte recruitment into the brain parenchyma and extravasation of plasma proteins [64]. We reviewed current knowledge on the ability of proinflammatory mediators to regulate BBB permeability and how they contribute to brain edema formation.

Unique properties of the BBB and cerebral endothelium

Under basal conditions, the BBB acts as a highly specialized structural and biochemical barrier that regulates the entry of blood-borne molecules into brain and preserves ionic homeostasis within the brain microenvironment [53, 67]. The BBB is composed of a specialized tight adherent microvascular endothelium and glial cell elements (astrocytes and microglia) along the entire endothelial abluminal surface [48]. Specific properties of the brain endothelial barrier are the presence of continuous strands of intercellular junction complexes that almost completely seal the paracellular cleft between adjacent endothelial membranes [4, 53]. Two morphologically distinct structural units occur in these intercellular junctional complexes: tight junctions (TJs) and adherens junctions [25].

TJs at the BBB are composed of an intricate combination of transmembrane integral proteins and several cytoplasmic-accessory proteins classified into 2 major groups: postsynaptic density (PDZ) domain containing proteins, and non-PDZ proteins [9, 25]. The major structural proteins of TJs are: (i) claudins (claudin-1, -5, -11), tissue-specific proteins that form the primary seal of TJ, (ii) occludin, an integral membrane protein involved in regulation of electrical resistance across the BBB and paracellular permeability, and (iii) junctional adhesion molecules (JAMs; JAM-1, -2, -3), single membrane-spanning proteins that belong to the immunoglobulin superfamily, which are mostly involved in leukocyte-endothelial cell interaction and leukocyte transmigration [9, 13, 25, 32, 40]. The TJ accessory proteins are multi-domain cytoplasmic molecules that form structural support for the TJ and are involved in signal transduction. The TJ PDZ-containing proteins are zonula occludens proteins (ZO-1, ZO-2, ZO-3) and AF6. The group of non-PDZ TJ proteins contains cingulin, 7H6, and atypical protein kinase C [9, 13, 25, 40].

Although the proteins of the TJ complex ultimately determine the barrier properties of endothelial cells, the adherens junction (AdJ) proteins mediate initial adhesion between endothelial cells and modulate TJ permeability [59]. The adherens junction complex is composed of a cadherin-catenin complex (Ve cadherin bound to β-catenin, plakoglobin, and α-catenin) and associated proteins (e.g., p 120 protein) [42, 59]. Both TJ and AdJ proteins are linked to cytoskeletal proteins. Actin, the primary cytoskeletal protein, has both structural and dynamic roles within cells [57, 62].

BBB opening: morphological aspects

At the functional level, the junctional complexes result in a high transendothelial electrical resistance, typically 1500–2000 $\Omega.cm^2$, making the BBB a unique selective permeability barrier [25]. Alterations in brain endothelial junctional complexes can result in increased in BBB permeability. At the cellular level, BBB "opening" is manifested by intercellular gap formation, changes in cell shape, reorganization of actin microfilament bundles, and redistribution of endothelial junction proteins. The typical morphological pattern of actin reorganization involves increased stress fiber density and a reduction or loss of the cortical actin band. This pattern occurs in parallel with and/or could cause spatial redistribution of both AdJ and TJ structures. There is decreased association of the cadherin/catenin complex with a shift to intracellular pools, disorganization of occludin on the endothelial surface with a loss of ZO-1, and increased association of occludin, ZO-1, and ZO-2 with actin filaments, which shifts these proteins to an insoluble cytosolic pool [15, 20, 63, 66]. The loss of adherence between brain endothelial cells, accompanied by conformational changes in TJ proteins and increased intercellular forces, widens the gap between brain endothelial cells and facilitates extravasation of leukocytes and plasma proteins into the brain parenchyma. Alterations in brain endothelial junctions represent a basic substrate for developing vasogenic brain edema.

Inflammation, proinflammatory mediators, and BBB opening

In conditions such as multiple sclerosis, AIDS-associated encephalopathy, and meningitis/meningoencephalitis, inflammation is considered a primary cause of brain damage [26, 39, 46]. On the other hand, in conditions such as Alzheimer's disease and stroke, inflammation can induce secondary brain injury by aggravating the initial insult [16, 22]. In all of these cases, there are the classic hallmarks of brain inflammation: BBB breakdown, edema formation, tissue infiltration by peripheral blood cells, activation of immunocompent cells, and intrathecal release of numerous immune mediators such as interleukins and chemotactic factors [33]. Common pathological findings are the presence of many leukocytes (monocytes or neutrophils) and activated microglial cells in the brain parenchyma, and a loss of immunostaining for most TJ proteins (occludin, ZO-1, claudin-5, or ZO-2) at the BBB [6, 26, 31, 43]. This may be an indicator of TJ protein redistribution/reorganization at the junction complex.

Proinflammatory mediators play a crucial role in regulation of CNS inflammation as well as in modulating BBB permeability. For example, proinflammatory cytokines (IL-1α, IL-1β, TNF-α, IL-6, GM-CSF) produced by invading leukocytes, activated endothelial cells, or other components associated with the BBB (astrocytes or perivascular macrophages), regulate the magnitude and persistence of inflammatory brain reactions [5, 46, 68]. But those factors also directly through their own receptors, or indirectly via other mediators, participate in BBB disruption. IL-1β and TNF-α increase BBB permeability under in vitro and in vivo

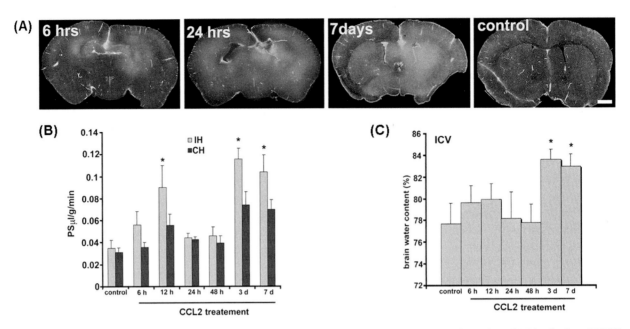

Fig. 1. Intracerebroventricular administration of CCL2 increases BBB permeability and brain edema formation. (A) Distribution of FITC-albumin in brain coronal sections; (B) permeability surface (PS) area products for FITC albumin; and (C) brain water content in CCL2- and saline-treated control mice. Measurements were made 6, 12, 24, and 48 hours after a single CCL2 dose (25 µg) or after a 3-day (5 µg/hr) or 7-day (2.5 µg/hr) chronic infusion. PS products were measured in ipsilateral (IH) and contralateral (CH) hemispheres. Brain water content measurements are of the whole brain. Values are mean ± SD. * indicates significant difference between CCL2-treated mice and control animals at the $p < 0.001$ level. Scale bar = 1000 µm

conditions, altering endothelial cell junction complexes leading to the development of local inflammatory responses and edema formation [5, 18, 68]. Neutralizing antibodies to these cytokines can diminish brain edema formation [24, 44].

Oxidative stress and mediators of oxidative injury (superoxide, hydrogen peroxide, peroxynitrite, nitric oxide, eicosanoids) are a second group of factors implicated in reducing TJ integrity and causing edema formation [23, 27]. Oxidative mediators can directly trigger a signal cascade with one end-point being a redistribution of TJ and AdJ proteins and reorganization of the endothelial actin cytoskeleton. Oxidative mediators can also regulate the expression/activity of other proinflammatory mediators and, in this way, "open" the BBB [18, 27].

Adhesion molecules (ICAM-1, VCAM-1) and selectins can also alter TJ complexes by regulating leukocyte/endothelial cell interactions. Those interactions may trigger expression of other proinflammatory mediators or intracellular signal cascades that regulate BBB integrity [14]. Leukocytes can also contribute to the persistence of increased vascular permeability via interactions with endothelial cells and the release of proinflammatory mediators, or by production of proinflammatory mediators in the brain parenchyma [11, 38, 47]. One particularly interesting group of proinflammatory mediators that contribute to alterations in BBB permeability are chemokines. In recent years, the role of these chemoattractant cytokines in CNS inflammation has emerged.

Chemokine-regulated BBB opening: role in brain edema formation

Chemokines are proinflammatory mediators involved in the selective driving of leukocytes into brain parenchyma. Enhanced perivascular chemokine expression is found in many pathological settings accompanied by inflammation, providing a chemoattractant gradient for leukocyte influx [41, 52].

Chemoattractant cytokines (chemokines) are a novel superfamily of structurally-related proinflammatory peptides (~70 to 90 amino acids) of low molecular weight (8 to 10 kDa). They have been divided into 4 classes on the basis of the positions of the first 2 conserved cysteine residues (C) and the number of intervening amino acids (X) between them (e.g., C,

CC, CXC, and CXXXC subfamilies) [41, 52, 69]. These subfamilies further exhibit functional differences. For example, CXC chemokines target mainly neutrophils, while CC chemokines primarily target monocytes/macrophages, eosinophils, and basophils, while CXC and CC chemokines invoke responses in lymphocytes [51, 69]. All chemokines mediate their effects by binding to 7-transmembrane G protein-coupled receptors related to rhodopsin. Some of the functional consequences of chemokine-receptor interactions are alterations in integrin adhesiveness, cell migration, polarization, and proliferation, as well as in gene expression [36, 51]. The discovery that parenchymal cells (astrocytes, oligodendroglial cells, and microglia in the CNS) and endothelial cells (including brain endothelial cells) express chemokine receptors significantly extended the possible functions of chemokines in inflammation and other (patho)physiological conditions [1, 2]. One of these novel functions is that chemokines can regulate BBB permeability.

The chemokine CXCL8 (IL-8) has already been shown to contribute to brain edema formation during ischemia/reperfusion injury. Thus, Matusmoto and colleagues [35] showed that adding an IL-8 neutralizing antibody significantly reduced edema formation in rabbit. In addition to IL-8, several studies examining the effect of IL-1β on BBB permeability have found a strong correlation between IL-1β-induced expression of chemokines, such as CINC1 and MIP-2 (CXCL2), and increased BBB permeability. Neutralizing antibodies to MIP2 and CINC1 can prevent BBB breakdown [3, 7].

A well-studied example of the effect of chemokines on BBB permeability is the action of monocyte chemoattractant protein-1 (MCP-1, CCL2). CCL2 belongs to the CC subfamily of chemokines and is involved in recruitment of monocytes/macrophages and activated lymphocytes into brain during neuropathological states [37]. CCL2 is highly expressed in the perivascular space and brain parenchyma during CNS inflammation, but it is also present in cerebrospinal fluid in several CNS inflammatory states (stroke, meningitis, multiple sclerosis) [8, 30, 34, 56, 58]. Expression of CCL2 receptor CCR2 was found on components of BBB (brain endothelial cells and astrocytes), suggesting that CCL2 might not only act on leukocyte recruitment but could also participate in regulation of the inflammatory response at the level of the BBB.

Our laboratory has shown that CCL2 (in μmol concentrations) in brain parenchyma or in cerebrospinal fluid can increase BBB permeability several fold (as measured by the permeability surface area product for fluorescein isothiocyanate-labeled albumin) and also induce brain edema formation (Fig. 1). In areas of BBB leakage, immunostaining for TJ proteins (occludin, ZO-1, ZO-2, claudin-5) was extensively reduced and there was intense infiltration of leukocytes [61]. There are 2 possible explanations of how CCL2 changes BBB permeability and causes vasogenic brain edema formation.

The effects of CCL2 on BBB permeability could be direct by acting on brain endothelial cells, or indirect by inducing production of other proinflammatory mediators by endothelial cells, astrocytes, or leukocytes. Evidence indicates that CCL2 acts via both mechanisms. In vitro treatment of brain endothelial cell monolayers or an in vitro model of BBB (co-culture of brain endothelial cells and astrocytes) with recombinant CCL2 exerts the same morphological and biochemical changes. Thus, TJ proteins are redistributed from a TritonX-100 soluble to a TritonX-100 insoluble fraction (potentially reflecting internalization of TJ proteins into a cytoplasmic compartment) and there is a loss or fragmentation of TJ protein staining [60, 61]. These biochemical and morphological alterations are absent if the CCL2 receptor CCR2 is not present on the brain endothelial cells [60, 61]. The effects of CCL2 on BBB permeability in vivo are abolished if the CCR2 receptor is deleted, pinpointing that CCL2 alters BBB permeability directly via its own CCR2 receptor.

Indirect effects of CCL2 on BBB permeability may involve regulation of leukocyte recruitment and/or regulation of the expression of proinflammatory mediators. We have found that depletion of circulating monocytes and activated macrophages by liposomally-encapsulated clodronate attenuates the effect of CCL2 on BBB permeability in vivo [61]. This implies that leukocytes are one of the factors contributing to BBB breakdown. Other factors may include proinflammatory factors such as VEGF and IL-1, which are significantly reduced in CCL2 knockout mice undergoing middle cerebral artery occlusion [28].

These results on CCL2, which may also apply to other chemokines, indicate a potential key role in mediating BBB breakdown during inflammation. CCL2 may not only direct leukocytes to sites of injury, but it may facilitate leukocyte migration into brain parenchyma by enhancing BBB permeability. Such BBB breakdown may also result in vasogenic edema.

Molecular mechanisms underlying regulation of permeability by CCL2

In general, several signal pathways regulate BBB permeability. For alterations in actin cytoskeleton architecture, prominent roles are proposed for: (i) Ca^{2+}/calmodulin- and Rho/Rho kinase-dependent pathways, primarily acting on the activity of myosin light chain kinase in order to facilitate actin and myosin light chain interaction and stress fiber formation. (ii) Phosphorylation of TJ and AdJ proteins by protein kinase C isoforms, tyrosine kinase lyn, serine kinase, and protein tyrosine phosphatase. (iii) Under some circumstances, proteolytic degradation of junctional constituents by matrix metalloproteinases (MMP-2, MMP-9) leading to the loss of TJs and AdJs and increased BBB permeability [10, 17, 18, 20, 50, 54, 65].

Chemokines, like other proinflammatory mediators, have a prolonged effect on the brain endothelial barrier denoted as *thrombin type* and characterized by intercellular widening for more than 30 minutes and specific activation of Rho/Rho kinase pathways. Two independent studies on 2 different types of model systems have shown that chemokines CXCL8 and CCL2 induce activation of Rho/Rho kinase with stress fiber formation [55, 60]. Further, Stamatovic et al. [60] found that Rho/Rho kinase not only altered the actin cytoskeletal organization but also had a significant impact on TJ protein complexes between brain endothelial cells. Diminishing Rho activity by transfection with dominant negative mutant of Rho stabilized brain endothelial barrier integrity. This led to the conclusion that Rho is a "nodal point" in brain endothelial intercellular signaling directed to alter junction complex of brain endothelial cells. Future investigation is needed to elucidate how Rho induces the phosphorylation and redistribution of TJ proteins.

Conclusion

Proinflammatory mediators, including chemokines, have a significant impact on brain edema formation in a variety of neuropathological conditions. Experiments examining CCL2, studied as a prototype for chemokine activity, show effects on leukocyte extravasation not only through directing leukocytes into the brain parenchyma but also by enhancing BBB disruption. CCL2 affects brain endothelial cells directly (via CCR2 receptors) and indirectly (via production of other proinflammatory agents). CCL2-induced BBB disruption may contribute to plasma protein extravasation and vasogenic brain edema as well as aggravating leukocyte influx. Chemokines are emerging as powerful factors controlling inflammation-induced brain edema. Elimination of chemokine activity via newly developed chemokine receptor antagonists offers a new potential avenue for the treatment of vasogenic brain edema.

Acknowledgments

This study was supported by grant NS-044907 (A.V.A.) from the National Institutes of Health, Bethesda, MD.

References

1. Andjelkovic AV, Kerkovich D, Shanley J, Pulliam L, Pachter JS (1999) Expression of binding sites for β chemokines on human astrocytes. Glia 28: 225–235
2. Andjelkovic AV, Spencer DD, Pachter JS (1999) Visualization of chemokine binding sites on human brain microvessels. J Cell Biol 145: 401–413
3. Anthony D, Dempster R, Fearn S, Clements J, Wells G, Perry VH, Walker K (1998) CXC chemokines generate age-related increases in neutrophil-mediated brain inflammation and blood-brain barrier breakdown. Curr Biol 8: 923–926
4. Bauer HC, Bauer H (1999) Neural induction of the blood brain barrier: still an enigma. Cell Mol Neurobiol 20: 13–28
5. Blamire AM, Anthony DC, Rajagopalan B, Sibson NR, Perry VH, Styles P (2000) Interleukin-1beta-induced changes in blood-brain barrier permeability, apparent diffusion coefficient, and cerebral blood volume in the rat brain: a magnetic resonance study. J Neurosci 20: 8153–8159
6. Brown H, Hien TT, Day N, Mai NT, Chuong LV, Chau TT, Loc PP, Phu NH, Bethell D, Farrar J, Gatter K, White N, Turner G (1999) Evidence of blood-brain barrier dysfunction in human cerebral malaria. Neuropathol Appl Neurobiol 25: 331–340
7. Campbell SJ, Wilcockson DC, Butchart AG, Perry VH, Anthony DC (2002) Altered chemokine expression in the spinal cord and brain contributes to differential interleukin-1beta-induced neutrophil recruitment. J Neurochem 83: 432–441
8. Chen Y, Hallenbeck JM, Ruetzler C, Bol D, Thomas K, Berman NE, Vogel SN (2003) Overexpression of monocyte chemoattractant protein 1 in the brain exacerbates ischemic brain injury and is associated with recruitment of inflammatory cells. J Cereb Blood Flow Metab 23: 748–755
9. Citi S, Cordenonsi M (1998) Tight junction proteins. Biochim Biophys Acta 1448: 1–11
10. Coghlan MP, Chou MM, Carpenter CL (2000) Atypical protein kinases C-λ and -ζ associate with the GTP-binding protein Cdc42 and mediate stress fiber loss. Mol Cell Biol 20: 2880–2889
11. Couraud PO (1998) Infiltration of inflammatory cells through brain endothelium. Pathol Biol (Paris) 46: 176–180
12. Davies DC (2002) Blood-brain barrier breakdown in septic encephalopathy and brain tumours. J Anat 200: 639–646
13. Denker BM, Nigam SK (1998) Molecular structure and assembly of the tight junction. Am J Physiol 274: F1–F9
14. Dietrich JB (2002) The adhesion molecule ICAM-1 and its regu-

lation in relation with the blood-brain barrier. J Neuroimmunol 128: 58–68
15. Farshori P, Kachar B (1999) Redistribution and phosphorylation of occludin during opening and resealing of tight junctions in cultured epithelial cells. J Membr Biol 170: 147–156
16. Feuerstein, GZ, Wang X, Barone FC (2000) Inflammatory gene expression in cerebral ischemia and trauma. Ann New York Acad Sci 24: 179
17. Fujimura M, Gasche Y, Morita-Fujimura Y, Massengale J, Kawase M, Chan PH (1999) Early appearance of activated matrix metalloproteinase-9 and blood brain barrier disruption in mice after focal cerebral ischemia and reperfusion. Brain Res 842: 92–100
18. Garcia JG, Schaphorst KL (1995) Regulation of endothelial cell gap formation and paracellular permeability. J Invest Med 43: 117–126
19. Gerriets T, Stolz E, Walberer M, Muller C, Kluge A, Bachmann A, Fisher M, Kaps M, Bachmann G (2004) Noninvasive quantification of brain edema and the space-occupying effect in rat stroke models using magnetic resonance imaging. Stroke 35: 566–571
20. Gloor SM, Wachtel M, Bolliger MF, Ishihara H, Landmann R, Frei K (2001) Molecular and cellular permeability control at the blood brain barrier. Brain Res Brain Res Rev 36: 258–264
21. Gray F, Belec L, Chretien F, Dubreuil-Lemaire ML, Ricolfi F, Wingertsmann L, Poron F, Gherardi R (1998) Acute, relapsing brain oedema with diffuse blood-brain barrier alteration and axonal damage in the acquired immunodeficiency syndrome. Neuropathol Appl Neurobiol 24: 209–216
22. Halliday G, Robinson SR, Shepherd C, Kril J (2000) Alzheimer's disease and inflammation: a review of cellular and therapeutic mechanisms. Clin Exp Pharmacol Physiol 27: 1–8
23. Heo JH, Han SW, Lee SK (2005) Free radicals as triggers of brain edema formation after stroke. Free Radic Biol Med 39: 51–70
24. Hosomi N, Ban CR, Naya T, Takahashi T, Guo P, Song XY, Kohno M (2005) Tumor necrosis factor-alpha neutralization reduced cerebral edema through inhibition of matrix metalloproteinase production after transient focal cerebral ischemia. J Cereb Blood Flow Metab 25: 959–967
25. Huber JD, Egleton RD, Davis TP (2001) Molecular physiology and pathophysiology of tight junctions in the blood-brain barrier. Trends Neurosci 24: 719–725
26. Kirk J, Plumb J, Mirakhur M, McQuaid S (2003) Tight junctional abnormality in multiple sclerosis white matter affects all calibres of vessel and is associated with blood-brain barrier leakage and active demyelination. J Pathol 201: 319–327
27. Koedel U, Pfister HW (1999) Oxidative stress in bacterial meningitis. Brain Pathol 9: 57–67
28. Kumai Y, Ooboshi H, Takada J, Kamouchi M, Kitazono T, Egashira K, Ibayashi S, Iida M (2004) Anti-monocyte chemoattractant protein-1 gene therapy protects against focal brain ischemia in hypertensive rats. J Cereb Blood Flow Metab 24: 1359–1368
29. Lorenzl S, Koedel U, Pfister HW (1996) Mannitol, but not allopurinol, modulates changes in cerebral blood flow, intracranial pressure, and brain water content during pneumococcal meningitis in the rat. Crit Care Med 24: 1874–1880
30. Losy J, Zaremba J (2001) Monocyte chemoattractant protein-1 is increased in the cerebrospinal fluid of patients with ischemic stroke. Stroke 32: 2695–2696
31. Mark KS, Davis TP (2002) Cerebral microvascular changes in permeability and tight junctions induced by hypoxia-reoxygenation. Am J Physiol 282: H1485–H1494
32. Martin-Padura I, Lostaglio S, Schneemann M, Williams L, Romano M, Fruscella P, Panzeri C, Stoppacciaro A, Ruco L, Villa A, Simmons D, Dejana E (1998) Junctional adhesion molecule, a novel member of the immunoglobulin superfamily that distributes at intercellular junctions and modulates monocyte transmigration. Cell Biol 142: 117–127
33. Martiney JA, Cuff C, Litwak M, Berman J, Brosnan CF (1998) Cytokine-induced inflammation in the central nervous system revisited. Neurochem Res 23: 349–356
34. Mastroianni CM, Lancella L, Mengoni F, Lichtner M, Santopadre P, D'Agostino C, Ticca F, Vullo V (1998) Chemokine profiles in the cerebrospinal fluid (CSF) during the course of pyrogenic and tuberculous meningitis. Clin Exp Immunol 114: 210–214
35. Matsumoto T, Ikeda K, Mukaida N, Harada A, Matsumoto Y, Yamashita J, Matsushima K (1997) Prevention of cerebral edema and infarct in cerebral reperfusion injury by an antibody to interleukin-8. Lab Invest 77: 119–125
36. Mellado M, Rodriguez-Frade JM, Manes S, Martinez-A C (2001) Chemokine signaling and functional responses: the role of receptors dimerization and TK pathway activation. Ann Rev Immunol 19: 397–421
37. Menicken F, Maki R, de Souza EB, Quirion R (1999) Chemokines and chemokine receptors in the CNS: a possible role in neuroinflammation and patterning. Trends Pharmacol Sci 20: 73–77
38. Merrill JE, Murphy SP (1997) Inflammatory events at the blood brain barrier: regulation of adhesion molecules, cytokines, and chemokines by reactive nitrogen and oxygen species. Brain Behav Immun 11: 245–263
39. Miller RJ, Meucci O (1999) AIDS and the brain: is there a chemokine connection? Trends Neurosci 22: 471–476
40. Mitic LL, Aderon JM (1998) Molecular architecture of tight junctions. Ann Rev Physiol 60: 121–142
41. Murphy PM (1994) The molecular biology of leukocyte chemoattractant receptors. Ann Rev Immunol 12: 593–633
42. Nagafuchi A (2001) Molecular architecture of adherens junctions. Curr Opin Cell Biol 13: 600–603
43. Ng I, Yap E, Tan WL, Kong NY (2003) Blood-brain barrier disruption following traumatic brain injury: roles of tight junction proteins. Ann Acad Med Singapore 32: S63–S66
44. Oprica M, Van Dam AM, Lundkvist J, Iverfeldt K, Winblad B, Bartfai T, Schultzberg M (2004) Effects of chronic overexpression of interleukin-1 receptor antagonist in a model of permanent focal cerebral ischemia in mouse. Acta Neuropathol (Berl) 108: 69–80
45. Ozates M, Kemaloglu S, Gurkan F, Ozkan U, Hosoglu S, Simsek MM (2000) CT of the brain in tuberculous meningitis. A review of 289 patients. Acta Radiol 41: 13–17
46. Paul R, Koedel U, Winkler F, Kieseier BC, Fontana A, Kopf M, Hartung HP, Pfister HW (2003) Lack of IL-6 augments inflammatory response but decreases vascular permeability in bacterial meningitis. Brain 126: 1873–1882
47. Petty MA, Lo EH (2002) Junctional complexes of the blood-brain barrier: permeability changes in neuroinflammation. Prog Neurobiol 68: 311–323
48. Prat A, Biernacki K, Wosik K, Antel JP (2001) Glia cell influence on the human blood brain barrier. Glia 36: 145–155
49. Reidel MA, Stippich C, Heiland S, Storch-Hagenlocher B, Jansen O, Hahnel S (2003) Differentiation of multiple sclerosis plaques, subacute cerebral ischaemic infarcts, focal vasogenic oedema and lesions of subcortical arteriosclerotic encephalopathy using magnetisation transfer measurements. Neuroradiology 45: 289–294
50. Ridley AJ (1997) Signaling by rho family proteins. Biochem Soc Trans 25: 1005–1010

51. Rodriguez-Frade JM, Mellado M, Martinez-A C (2001) CCR2. In: Oppenheim J, Feldmann M, Durum SK (eds) Cytokine reference: a compendium of cytokines and other mediators of host defense, vol 2: receptors. Academic Press, London, pp 2041–2052
52. Rollins BJ (1997) Chemokines. Blood 90: 909–928
53. Rubin LL, Staddon JM (1999) The cell biology of the blood-brain barrier. Ann Rev Neurosci 22: 11–28
54. Sakakibara A, Furuse M, Saitou M, Ando-Akatsuka Y, Tsukita S (1997) Possible involvement of phosphorylation of occludin in tight junction formation. J Cell Biol 137: 1393–1401
55. Schraufstatter IU, Chung J, Burger M (2001) IL-8 activates endothelial cell CXCR1 and CXCR2 through Rho and Rac signaling pathways. Am J Physiol 280: L1094–L1103
56. Sindern E, Niederkinkhaus Y, Henschel M, Ossege LM, Patzold T, Malin JP (2001) Differential release of beta-chemokines in serum and CSF of patients with relapsing-remitting multiple sclerosis. Acta Neurol Scand 104: 88–91
57. Small VJ, Rottner K, Kaverina I (1999) Functional design in the actin cytoskeleton. Curr Opin Cell Biol 11: 54–60
58. Sorensen TL, Ransohoff RM, Strieter RM, Sellebjerg F (2004) Chemokine CCL2 and chemokine receptor CCR2 in early active multiple sclerosis. Eur J Neurol 11: 445–449
59. Staddon JM, Rubin LL (1996) Cell adhesion, cell junctions and the blood-brain barrier. Curr Opin Neurobiol 6: 622–627
60. Stamatovic SM, Keep RF, Kunkel SL, Andjelkovic AV (2003) Potential role of MCP-1 in endothelial cell tight junction 'opening': signaling via Rho and Rho kinase. J Cell Sci 116: 4615–4628
61. Stamatovic SM, Shakui P, Keep RF, Moore BB, Kunkel SL, Van Rooijen N, Andjelkovic AV (2005) Monocyte chemoattractant protein-1 regulation of blood-brain barrier permeability. J Cereb Blood Flow Metab 25: 593–606
62. Sutherland JD, Witke W (1999) Molecular genetic approaches to understanding the actin cytoskeleton. Curr Opin Cell Biology 11: 142–151
63. Tsukamato T, Nigam SK (1999) Role of tyrosine phosphorylation in the reassembly of occludin and other tight junction proteins. Am J Physiol 276: F737–F750
64. Unterberg AW, Stover J, Kress B, Kiening KL (2004) Edema and brain trauma. Neuroscience 129: 1021–1029
65. Wachtel M, Frei K, Ehler E, Fontana A, Winterhalter K, Gloor SM (1999) Occludin proteolysis and increased permeability in endothelial cells through tyrosine phosphatase inhibition. J Cell Sci 112: 4347–4356
66. Wang AJ, Pollard TD, Herman IM (1983) Actin filaments stress fibers in vascular endothelial cells in vivo. Science 219: 867–869
67. Wolburg H, Risau W (1995) Formation of the blood-brain barrier. In: Kettenmann H, Ransom BR (eds) Neuroglia. Oxford University Press, Oxford, pp 763–776
68. Yang GY, Gong C, Qin Z, Liu XH, Betz AL (1999) Tumor necrosis factor alpha expression produces increased blood-brain barrier permeability following temporary focal cerebral ischemia in mice. Brain Res Mol Brain Res 69: 135–143
69. Yoshie O, Imai T, Nomiyama H (1997) Novel lymphocyte-specific CC chemokines and their receptors. J Leukoc Biol 62: 634–644

Correspondence: Anuska V. Andjelkovic, Departments of Neurosurgery and Pathology, University of Michigan Medical School, 5550 Kresge I Bldg., Ann Arbor, MI 48109-0532, USA. e-mail: anuskaa@umich.edu

Atrial natriuretic peptide: its putative role in modulating the choroid plexus-CSF system for intracranial pressure regulation

C. E. Johanson[1], J. E. Donahue[2], A. Spangenberger[1], E. G. Stopa[1,2], J. A. Duncan[1], and H. S. Sharma[3]

[1] Department of Clinical Neuroscience, Brown Medical School, Rhode Island Hospital, Providence, RI, USA
[2] Department of Pathology, Brown Medical School, Rhode Island Hospital, Providence, RI, USA
[3] Department of Surgical Sciences, Anesthesiology & Intensive Care Medicine, University Hospital, Uppsala, Sweden

Summary

Evidence continues to build for the role of atrial natriuretic peptide (ANP) in reducing cerebrospinal fluid (CSF) formation rate, and thus, intracranial pressure. ANP binds to choroid plexus (CP) epithelial cells. This generates cGMP, which leads to altered ion transport and the slowing of CSF production. Binding sites for ANP in CP are plentiful and demonstrate plasticity in fluid imbalance disorders; however, specific ANP receptors in epithelial cells need confirmation. Using antibodies directed against NPR-A and NPR-B, we now demonstrate immunostaining not only in the choroidal epithelium (including cytoplasm), but also in the ependyma and some endothelial cells of cerebral microvessels in adult rats (Sprague-Dawley). The choroidal and ependymal cells stained almost universally, thus substantiating the initial autoradiographic binding studies with ^{125}I-ANP. Because ANP titers in human CSF have previously been shown to increase proportionally to increments in ICP, we propose a compensatory ANP modulation of CP function to down-regulate ICP in hydrocephalus. Further evidence for this notion comes from the current finding of increased frequency of "dark" epithelial cells in CP of hydrocephalic (HTx) rats, which fits our earlier observation that the "dark" choroidal cells, associated with states of reduced CSF formation, are increased by elevated ANP in CSF. Altogether, ANP neuroendocrine-like regulation at CSF transport interfaces and blood-brain barrier impacts brain fluid homeostasis.

Keywords: CSF homeostasis; natriuretic peptide receptors; hydrocephalus; cGMP; brain natriuretic peptide.

Introduction

Overview of peptidergic effects on fluid balance: choroid plexus (CP) vs. kidney

Atrial natriuretic peptide (ANP) is well known for down-regulating or "unloading" excessive plasma volume in systemic hypertensive states [6]. In response to fluid overload in the plasma compartment, the ANP in atrial myocardial cells is up-regulated and secreted. The resultant elevated titer of ANP in the plasma exerts inhibitory effects on the kidney tubule transport of ions, causing natriuresis and enhanced urinary outflow. This compensatory renal excretion of fluid reduces plasma hypervolemia and blood pressure.

We suspect that ANP synthesized centrally, in response to cerebrospinal fluid (CSF) retention (ventriculomegaly) or elevated intracranial pressure (ICP), exerts inhibitory effects on CP epithelial cells to down-regulate CSF production and reduce ICP. Cumulative evidence supports this kind of a "feedback" compensatory system.

Raichle [20] originally proposed a neuroendocrine hypothesis for the regulation of ions and volume in the central nervous system (CNS). The literature is replete with data that the CP, blood-brain barrier, and various CSF-bordering tissues (ependyma, circumventricular organs, and arachnoid membrane) are involved in neurohormonal signaling [18] to adjust volume/pressure parameters associated with fluid expansion in brain extracellular compartments [11, 28]. Thus, the augmented ICP or ventriculomegaly "stretch stimulus" reflex should increase the synthesis and release of neuropeptides into CSF. Accordingly, ANP and other neuropeptides manufactured and secreted by cells within the CNS should stimulate their respective receptors at the blood-CSF, blood-brain, and CSF-brain interfaces. We propose that these peptidergic actions lead to attenuation of fluid generation and build up in the ventriculo-subarachnoid spaces and brain interstices, thereby decreasing pathologically-elevated ICP.

The CP manufactures the preponderance of extracellular fluid in the CNS. Depending upon species, as much as 90% of CSF originating from the plasma is secreted at choroidal sites of the blood-CSF interfaces in the lateral and fourth ventricles. Collectively, CPs act as "kidneys" for the brain [23]. The plexuses mediate the transfer of a relatively large volume of CSF to maintain brain extracellular fluid integrity through a mechanism similar to proximal renal tubule regulation of chemical balance in plasma. Just as diuretic agents (e.g., acetazolamide) inhibit ion and water transport in the kidney, they also curtail the turnover of fluid from blood to CSF. Consequently, drugs such as acetazolamide suppress CSF production and can relieve elevated ICP. Similarly, ANP and arginine vasopressin, which alter the fluxes of ions and water along various segments of the nephron, also modify the hydrodynamics of the CP-CSF system.

We have developed a model of neuroendocrine-like "dark" cells in the plexus, which regulate CSF volume and hence, ICP [15]. Although several hormones and peptides regulate fluid turnover at multiple sites in the CNS, this paper focuses on ANP modulation of the blood-CSF barrier. The discussion below treats several aspects of ANP in relation to CP receptors, epithelial ultrastructure, ion transporters, CSF formation, hydrocephalus, and ICP adjustments.

Materials and methods

The rodent model has been widely used to explore ANP-cyclic guanosine monophosphate (cGMP) relationships in the CP in the context of CSF dynamics and pressure phenomena [2, 4, 8, 25, 27, 31, 32]. These extensive rodent studies constitute a useful database with which to expand CSF modeling. Therefore we used rats to probe receptors for ANP in the CP, and to analyze the neuroendocrine-like dark cells, inducible by ANP [19], in the plexus of hydrocephalic (HTx) animals. HTx rats are of particular interest because their ICP is reducible by intracerebroventricular injections of ANP [9].

Immunohistochemistry

Fixed-frozen brain specimens from adult Sprague-Dawley rats were sectioned at 10 microns and incubated with antibodies against natriuretic peptide receptor A (NPR-A) or natriuretic peptide receptor B (NPR-B) (Santa Cruz Biotechnology, Santa Cruz, CA). Immunohistochemical procedures, utilizing avidin-biotin-peroxidase, were carried out with Ventana automated equipment with antibody dilutions of 1:50 (corresponding to ~1:100 by manual methodology). Control tissues were run with primary antibody omitted.

Electron microscopy

The number of dark epithelial cells in the CP was assessed in perinatal HTx rats to test the hypothesis that the predicted reduction in CSF formation is associated with increasing numbers of dark cells. Brains were removed from fetal (E20) and postnatal (P3) rats and prepared for electron microscopy using procedures previously described [15]. Light epithelial cells (normal or typical) and dark epithelial cells were counted in choroidal tissues from 10 rats: 5 with ventriculomegaly and 5 littermate controls without hydrocephalus. The occurrence of dark cells was expressed as a percent = 100 × [number of dark cells/specimen] ÷ [total number of epithelial cells (dark + light) in that specimen].

Results and discussion

Binding sites vs. receptors for ANP in CP

Choroid epithelial cells are replete with high affinity binding sites for ANP. Numerous earlier autoradiographic studies using ^{125}I-ANP as ligand described the plasticity of these choroidal natriuretic peptide binding sites. In hypertension and in spaceflight, for example, there is down-regulation [32] and up-regulation [10], respectively, of these binding sites. Antibodies against specific NPRs are now available. Using an antibody directed against NPR-A, we demonstrate the presence of NPR-A in rat CP epithelium (Fig. 1A). NPR-A immunostaining, as well as that of NPR-B, occurred in most of the epithelial cells in the plexus. NPR-A and NPR-B are long forms of NPR which generate cGMP, a cyclic nucleotide associated with lowering the rate of CSF formation by CP [7, 12].

Both ANP and brain natriuretic peptide (BNP) bind to and activate NPR-A [6]. Porcine BNP generates cGMP in the amphibian CP [27], the epithelial cells of which contain guanylate cyclase particulate (GCp) at all surfaces: basal, lateral, and apical [25]. Such widespread GCp distribution on the external limiting membranes of the choroidal epithelium suggests a similarly extensive presence of NPRs on the cells.

ANP induction of dark epithelial cells in CP

Neuroendocrine-like dark cells are widely extant in CP at various stages of life. Their number varies with the physiological state, e.g., dehydration or water overload. In baseline (non-activated) conditions, dark cells represent about 3% to 5% of the total epithelial population. Dark cells are induced by nanomolar (or less) concentrations of ANP and arginine vasopressin (AVP) [15, 17, 19]. A 2- to 3-fold increase in the number of dark cells occurs after choroidal exposure to increasing CSF levels of ANP [19] and AVP [15, 17], both in vitro and in vivo. Dark cells in CP are shrunken and have condensed cytoplasm, but they contain normally-appearing organelles [15]. They are usually associated with states of decreased CSF formation [28].

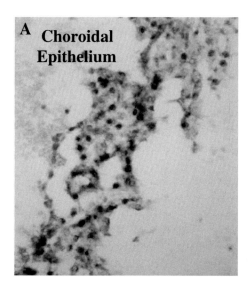 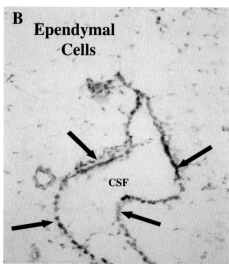

Fig. 1. Immunostaining of NPR at choroidal and extrachoroidal sites of fluid regulation in the adult rat CNS. (A) Widespread staining of NPR-A occurred in CP epithelial cells of lateral ventricle tissue, as indicated by the dark reaction product (peroxidase). NPR-B (not shown) also extensively stained in the choroidal epithelium. Omission of the primary antibody greatly reduced the staining intensity of the NPRs. Magnification is ×400. (B) Immunostaining of NPR-A in ependymal cells (1 cell-layer thick) lining the third ventricle. There was nearly universal staining of the ependyma. Arrows point to individual ependymal cells. There was negligible staining of NPR-A in the surrounding brain parenchyma. CSF = cerebrospinal fluid of the third ventricle. Magnification is ×200

ANP-induced reduction of Na uptake by CP

CSF formation is intimately coupled to net Na transport from plasma to CP to ventricular fluid. Na uptake into the CP-CSF system is reduced by acetazolamide, which slows basolateral Na-H exchange via increased epithelial cell pH. ANP also reduces Na uptake in the in vitro CP [14]. This inhibition may result from reduced Na-H exchange secondary to elevated choroid cells (Na) following ANP stimulation of the reabsorptive arm of the Na-K-Cl cotransporter [16]. Thus, coordination of basolateral and apical ion transporters involved in CSF formation may be disrupted by ANP. Elevation of cGMP in the choroidal cytoplasm slows down the apically-located Na pump [7]. However, more information is needed on how ANP-generated cGMP and nitric oxide are linked to other apical membrane transport phenomena, including aquaporin-1 channels [1], during the inhibition of CSF secretion.

ANP and CSF formation rate

Direct measurements of CSF production, as assessed by dilution of non-actively transported markers administered into the ventricular system, indicate that ANP can reduce fluid output by the plexus. An inhibitory effect of 35% on CSF formation was found in one investigation of rabbits [24], but a smaller reduction was observed in another [21]. The reduction in rabbit CSF formation by ANP was presumably due to a peptidergic effect at the blood-CSF barrier. On the other hand, no alteration in CSF formation was observed during ventriculo-cisternal perfusion of sheep with exogenous ANP [5]. ANP dosage, anesthesia, species differences, physiologic state, and hormonal levels all need to be factored into an analysis of the responses of various animal preparations to ANP.

Intracerebroventricular infusion of ANP to lower ICP

Attenuated CSF formation is often associated with reduced ICP. Because ANP seems to slow down CSF production by the CP, one would expect that an intraventricular injection of ANP might lower ICP. In fact, intracerebroventricular administration of 2 μg of α-hANP decreases ICP by 25 to 30 mmH$_2$O in HTx rats with ventriculomegaly and elevated CSF pressure at 3 to 4 weeks postpartum [9]. This raises the possibility of a compensatory mechanism involving ANP to adjust ICP increments by "unloading" CSF volume via inhibition of CSF formation.

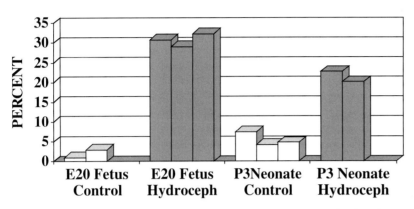

Fig. 2. "Dark" neuroendocrine-like epithelial cells in lateral ventricle CPs of perinatal HTx rats (grey bars) and in non-HTx littermate controls (white bars). Each bar represents counts of at least 300 epithelial cells in the plexus tissue from an HTx animal. The percent calculation (y-axis plot) is defined in the text. In non-HTx rats, the baseline control value of the frequency of occurrence of dark cells is about 2% to 5% of the total epithelial population. In HTx animals, the number of dark cells in the plexus increased by 4- to 6-fold over control values. Dark cells typically increase in number during states of decreased CSF formation, in some cases being induced by ANP [19] or AVP [15]

Relationship between ICP and CSF concentration of ANP in human disorders

A key question is whether increments in ICP are followed by proportional elevations in titers of ANP in CSF. This is evidently the case, at least in humans. Yamasaki and colleagues [30] examined ANP levels in CSF of patients with various neurological diseases, with and without intracranial hypertension. They observed that an increase in CSF (ANP) correlated positively with ICP augmentation, while CSF (ANP) rose with ICP, serum (ANP) did not. Similarly in pediatric patients, Tulassay et al. [26] found that CSF ANP concentration (which was elevated in hydrocephalus) increased with ICP but was not affected by serum levels. Collectively, these findings constitute evidence for ANP-mediated regulation of CP-CSF hydrodynamics and consistently point to a central modulation of CSF (ANP) that is independent of plasma.

Congruence of morphological and functional evidence from hydrocephalus models: ANP, epithelial shrinkage, and curtailment of CSF formation

Given that ANP exerts suppressant effects on the CP-CSF system, and that CSF ANP levels rise with ICP, it seems worthwhile to cultivate the model for a compensatory reduction of CSF formation in hydrocephalus. In this regard, morphological evidence is consistent with functional and biochemical findings. Microscopic analyses reveal an expansion of intercellular spaces between choroid epithelial cells in a variety of hydrocephalus models [28]. This is significant because paracellular dilations occur in CP epithelium when CSF formation decreases [3].

We found that the number of dark cells in CP increases in the HTx rat model (Fig. 2). Because dark cells are usually associated with curtailed CSF formation, this finding raises the possibly of a compensatory decrease in CSF production in this genetic model of hydrocephalus. Under various conditions of hydrocephalus in cats, dogs, and even humans, there have been other observations of lower CSF formation rates [22, 29]. Clearly, there are reasons to link hydrocephalus-induced alterations of CP structure and function with CSF-mediated neuroendocrine responses.

The ANP model of CSF regulation: progress and challenges

Many "pieces of the puzzle" are coming together to form a larger picture of the system that mediates natriuretic peptide effects on brain fluid dynamics. Due to the complexity of this natriuretic regulatory system, however, and its interaction with other neurohumoral control systems, more information needs to be procured for the paradigm. Titers of ANP, BNP, and C-type natriuretic peptide (CNP) in the CSF of hydrocephalic animals and humans have yet to be established. Also, the site of synthesis of these neuropeptide ligands in the CNS awaits clarification. Periventricular regions, circumventricular organs and hypothalamic nuclei are possible regions that synthesize and secrete ANP, BNP, and CNP to stimulate distant NPR receptors in CP, ependyma, and endothelial microvessels. It would also be helpful to gain additional insight on the effects of natriuretic peptides on CSF formation in rodent models.

The blood-CSF vs. other CNS transport interfaces: toward an integrated model of neuroendocrine-mediated fluid balance

Although CP has the major role in CNS fluid homeostasis, there are regulated fluxes of ions and water at other interfaces which also affect ICP. We found heavy staining of NPR-A in the ependyma (Fig. 1B) and on the walls of some cerebral microvessels. A comprehensive model of fluid balance among brain compartments should integrate the Na and water movements across the ependyma and cerebral microvessels with the corresponding fluxes across the plexuses. Because there are receptors for ANP and AVP in ependymal and endothelial cells, it is therefore essential to appreciate neuroendocrine regulation at these other interfaces that are "gateways" to the CSF [13] and brain interstitial fluid. This will provide greater understanding of how natriuretic and vasopressinergic inhibition of CSF and ISF flow alters ICP.

Acknowledgments

We thank R. Tavares for carrying out the Ventana immunostaining analyses of natriuretic peptide receptors and N. Johanson for graphics. Also, we gratefully acknowledge research support from Lifespan at Rhode Island Hospital, the Department of Neurosurgery at Brown Medical School, and National Institutes of Health R01 NS 27601 (CEJ).

References

1. Boassa D, Yool AJ (2005) Physiological roles of aquaporins in the choroid plexus. Curr Top Dev Bio 67: 181–206
2. Brown J, Zuo Z (1993) C-type natriuretic peptide and atrial natriuretic peptide receptors of rat brain. Am J Physiol 264: R513–R523
3. Burgess A, Segal MB (1970) Morphological changes associated with inhibition of fluid transport in the rabbit choroid plexus. J Physiol 208: 88P–91P
4. Carcenac C, Herbute S, Masseguin C, Mani-Ponset L, Maurel D, Briggs R, Guell A, Gabrion JB (1999) Hindlimb-suspension and spaceflight both alter cGMP levels in rat choroid plexus. J Gravit Physiol 6: 17–24
5. Chodobski A, Szmydynger-Chodobska J, Cooper E, McKinley MJ (1992) Atrial natriuretic peptide does not alter cerebrospinal fluid formation in sheep. Am J Physiol 262: R860–R864
6. D'Souza SP, Davis M, Baxter GF (2004) Autocrine and paracrine actions of natriuretic peptides in the heart. Pharmacol Ther 101: 113–129
7. Ellis DZ, Nathanson JA, Sweadner KJ (2000) Carbachol inhibits Na^{+}-K^{+}-ATPase activity in choroid plexus via stimulation of the NO/cGMP pathway. Am J Physiol Cell Physiol 279: C1685–C1693
8. Grove KL, Goncalves J, Picard S, Thibault G, Deschepper CF (1997) Comparison of ANP binding and sensitivity in brains from hypertensive and normotensive rats. Am J Physiol 272: R1344–R1353
9. Hashimoto K, Kikuchi H, Ishikawa M. Yokoi K, Kimura M, Itokawa Y (1990) The effect of atrial natriuretic peptide on intracranial pressure in a congenital hydrocephalic model. No To Shinkei 42: 683–687 [in Japanese]
10. Herbute S, Oliver J, Davet J, Viso M. Ballard RW, Gharib C, Gabrion J (1994) ANP binding sites are increased in choroid plexus of SLS-1 rats after 9 days of spaceflight. Aviat Space Environ Med 65: 134–138
11. Hertz L, Chen Y, Spatz M (2000) Involvement of non-neuronal brain cells in AVP-mediated regulation of water space at the cellular, organ, and whole-body level. J Neurosci Res 62: 480–490
12. Hise MA, Johanson CE (1978) Inhibition of cerebrospinal fluid flow by dibutyryl guanosine-3'-5'-cyclic monophosphoric acid. Fed Proceed 37: 514
13. Johanson C (2003) The choroid plexus-CSF nexus: gateway to the brain. In: Conn PM (ed) Neuroscience in medicine. Humana Press, Totowa, NJ, pp 165–195
14. Johanson CE, Sweeney SM, Parmelee JT, Epstein MH (1990) Cotransport of sodium and chloride by the adult mammalian choroid plexus. Am J Physiol 258: C211–C216
15. Johanson CE, Preston JE, Chodobski A, Stopa EG, Szmydynger-Chodobska J, McMillan PN (1999) AVP V1 receptor-mediated decrease in Cl-efflux and increase in dark cell number in choroid plexus epithelium. Am J Physiol 276: C82–C90
16. Johanson C, McMillan P, Tavares R, Spangenberger A, Duncan J, Silverberg G, Stopa E (2004) Homeostatic capabilities of the choroid plexus epithelium in Alzheimer's disease. Cerebrospinal Fluid Res 1: 3
17. Liszczak TM, Black PM, Foley L (1986) Arginine vasopressin causes morphological changes suggestive of fluid transport in rat choroid plexus epithelium. Cell Tissue Res 246: 379–385
18. Nilsson C, Lindvall-Axelsson M, Owman C (1992) Neuroendocrine regulatory mechanisms in the choroid plexus-cerebrospinal fluid system. Brain Res Brain Res Rev 17: 109–138
19. Preston JE, McMillan PN, Stopa EG, Nashold JR, Duncan JA, Johanson CE (2003) Atrial natriuretic peptide induction of dark epithelial cells in choroid plexus: consistency with the model of CSF downregulation in hydrocephalus. Eur J Pediatr Surg 13: S40–S42
20. Raichle ME (1981) Hypothesis: a central neuroendocrine system regulates brain ion homeostasis and volume. Adv Biochem Psychopharmacol 28: 329–336
21. Schalk KA, Faraci FM, Williams JL, VanOrden D, Heistad DD (1992) Effect of atriopeptin on production of cerebrospinal fluid. J Cereb Blood Flow Metab 12: 691–696
22. Silverberg GD, Huhn S, Jaffe RA, Chang SD, Saul T, Heit G, Von Essen A, Rubenstein E (2002) Downregulation of cerebro-

spinal fluid production in patients with chronic hydrocephalus. J Neurosurg 97: 1271–1275
23. Spector R, Johanson CE (1989) The mammalian choroid plexus. Sci Am 261: 68–74
24. Steardo L, Nathanson JA (1987) Brain barrier tissues: end organs for atriopeptins. Science 235: 470–473
25. Tei S, Vagnetti D, Secca T, Santarella B, Roscani C, Farnesi RM (1995) Response of guanylate cyclase to atrial natriuretic factor in epithelial cells of the frog choroid plexus. Tissue Cell 27: 233–240
26. Tulassay T, Khoor A, Bald M, Ritvay J, Szabo A, Rascher W (1990) Cerebrospinal fluid concentrations of atrial natriuretic peptide in children. Acta Paediatr Hung 30: 201–207
27. Vagnetti D, Tei S, Secca T, Santarella B, Roscani C, Farnesi RM (1995) Biochemical and cytochemical analyses of BNP-stimulated guanylate cyclase in frog choroid plexus. Brain Res 705: 295–301
28. Weaver CE, McMillan PN, Duncan JA, Stopa EG, Johanson CE (2004) Hydrocephalus disorders: their biophysical and neuroendocrine impact on the choroid plexus epithelium. In: Hertz L (ed) Advances in molecular and cell biology. Elsevier Press, Greenwich, CT, pp 269–293
29. Welch K (1975) The principles of physiology of the cerebrospinal fluid in relation to hydrocephalus including normal pressure hydrocephalus. Adv Neurol 13: 247–332
30. Yamasaki H, Sugino M, Ohsawa N (1997) Possible regulation of intracranial pressure by human atrial natriuretic peptide in cerebrospinal fluid. Eur Neurol 38: 88–93
31. Zorad S, Alsasua A, Saavedra JM (1991) A modified quantitative autoradiographic assay for atrial natriuretic peptide receptors in rat brain. J Neurosci Methods 40: 63–69
32. Zorad S, Alsasua A, Saavedra JM (1998) Decreased expression of natriuretic peptide A receptors and decreased cGMP production in the choroid plexus of spontaneously hypertensive rats. Mol Chem Neuropathol 33: 209–222

Correspondence: Conrad E. Johanson, Department of Neurosurgery, Rhode Island Hospital, 593 Eddy Street, Providence, RI 02903, USA. e-mail: Conrad_Johanson@Brown.edu

Author index

Abe, T. 57, 409
Akimoto, H. 168
Alm, P. 288, 309
Amorini, A. M. 393
Andjelkovic, A. V. 444
Aruga, T. 57, 249, 409
Asai, J. 398
Aygok, G. 24, 348
Azzini, C. 81

Badgaiyan, R. D. 288
Balestreri, M. 114
Barnes, J. 7
Beaumont, A. 30, 171
Beiler, C. 177
Beiler, S. 177
Bhattathiri, P. S. 61, 65
Binder, D. K. 389
Blumbergs, P. C. 263
Borrelli, M. 81

Cannon, J. R. 222
Casey, K. 177
Cavallo, M. 81
Cernak, I. 121
Ceruti, S. 81
Chambers, I. R. 7
Chan, M. T. V. 21
Chang, C. W. 74
Chen, H. 244
Chen, Y. 438
Chiba, N. 69
Chieregato, A. 53, 81, 85
Citerio, G. 7
Colohan, A. R. T. 188
Compagnone, C. 53, 85
Corteen, E. 11, 17
Corwin, F. 171
de Courten-Myers, G. M. 177
Czosnyka, M. 11, 17, 103, 108, 114

Daley, M. L. 103
Dimitrijevic, O. B. 444
Dohi, K. 57, 249, 409
Donahue, J. E. 451
Du, R. 438
Du, Y. 422
Dunbar, J. G. 258, 303, 393
Duncan, J. A. 426, 451

Ebihara, T. 37, 44, 48, 69, 97, 419
Enblad, P. 7
Endo, S. 144, 254, 272, 279, 299
Ennis, S. R. 276, 295
Eymann, R. 343, 364

Fainardi, E. 53, 81, 85
Fatouros, P. 24, 171, 348
Fazzina, G. 303, 393
Felt, B. T. 183
Feng, D. 413
Feng, H. 422
Fujimoto, M. 74, 283
Fujimoto, T. 74, 283, 398
Fujita, S. 57
Furukawa, M. 37, 44, 48, 69, 97, 419

Gennarelli, T. 30, 171
Ghabriel, M. N. 402
Glisson, R. 258, 303, 393
Gong, Y. 232
Gordh, T. 309, 335
Gräwe, A. 373
Gregson, B. 65
Gregson, B. A. 61
Guendling, K. 108

Helps, S. C. 263
Hiler, M. 11, 114
Hodoyama, K. 249
Hoff, J. T. 194, 199, 218, 227, 232
Howells, T. 7
Hua, Y. 183, 194, 199, 203, 218, 222, 227, 232
Huang, F. 78
Hutchinson, P. J. 11, 17, 108, 114

Ikeda, Y. 57, 157, 163
Inada, K. 97
Ishibashi, S. 144, 272, 279, 299
Ishii, N. 157, 163
Ishikawa, Y. 163
Itano, T. 134
Ito, U. 239
Itokawa, H. 398

James, H. E. 125
Januszewski, S. 267
Jimbo, H. 57, 409
Johanson, C. E. 426, 451

Katayama, Y. 3, 40, 130
Katsumata, N. 144, 254, 272, 279, 299
Kawai, N. 212
Kawakami, E. 239
Kawamata, T. 3, 40, 130
Keep, R. F. 134, 194, 199, 203, 207, 218, 222, 227, 232, 276, 295, 444
Kelley, R. 91
Kiefer, M. 343, 358, 364
Kiening, K. 7
Kiening, K. L. 139
Kinoshita, K. 33, 37, 44, 48, 69, 97, 419
Kintner, D. B. 244
Kirkman, J. 177
Kirkpatrick, P. J. 11, 17
Kishimoto, K. 283
Kitahata, Y. 48, 69, 97, 419
Kleindienst, A. 258, 303, 393
Kudo, Y. 157
Kuroiwa, T. 144, 168, 239, 254, 272, 279, 299

Larnard, D. 177
Leffler, C. W. 103
Lemcke, J. 352, 358, 368, 373, 377, 381
Lewis, P. 108
Li, C. 163
Li, F. 422
Li, Q. 413
Li, S. 144, 299
Lin, J. 422
Linke, M. J. 177
Liu, S. 413
LiYuan, S. 299
Lodhia, K. R. 207
Luo, J. 244
Luo, Z. 413

Maeda, T. 130
Manley, G. T. 389
Marmarou, A. 24, 171, 258, 303, 348, 393
Matsunaga, A. 398
Matsunaga, M. 249
Mattern, J. 7
Meier, U. 343, 352, 358, 368, 373, 377, 381
Mendelow, A. D. 17, 61, 65
Meng, H. 422
Menon, D. K. 11, 17
Mitchell, P. 17, 61

Miyamoto, O. 134
Miyasakai, N. 168
Mizusawa, H. 272, 299
Mohanty, S. 151, 288, 316, 322
Mori, T. 40, 130
Moriya, M. 74, 283
Moriya, T. 33, 37, 44, 48, 69, 97
Morota, S. 157, 163
Murray, G. 17

Nagao, K. 69
Nagao, S. 134, 194, 212, 218
Nagaoka, T. 168
Nagasao, J. 239
Nagashima, G. 398
Nakamachi, T. 249
Nakamura, T. 134, 194, 203, 212, 218, 222, 227
Nakano, I. 239
Nanda, A. 91
Nawashiro, H. 148
Neher, M. 139
Nemoto, E. M. 435
Neumann, U. 352, 358
Ng, S. C. P. 21
Nilsson, P. 7
Noda, A. 44, 48, 419
Noda, E. 37
Nomura, N. 148
Nortje, J. 108
Nyberg, F. 309

O'Connor, C. A. 121
Ohno, K. 144, 168, 254, 279
Ohsumi, A. 148
Ohtaki, H. 57, 249, 283
Okuda, M. 74, 398
Okuno, K. 33, 37, 48, 97, 258
Ooigawa, H. 148
Ostrowski, R. P. 188
Otani, N. 148
Oyanagi, K. 239

Pascarella, R. 53, 85
Patnaik, R. 151, 316
Patnaik, S. 151, 316
Pickard, J. D. 11, 17, 103, 108, 114
Piper, I. 7
Pluta, R. 267
Poon, W. S. 21
Portella, G. 24
Prasad, K. S. M. 61, 65

Ragauskas, A. 7
Ravaldini, M. 53
Ray, A. K. 151
Reyer, T. 373
Richardson, R. J. 222
Rickels, E. 17
Robinson, T. 177
Rodgers, K. M. 263
Russo, M. 81

Sahuquillo, J. 7, 17
Sakowitz, O. W. 139
Sakurai, A. 33, 37, 44, 48, 69, 97, 419
Saletti, A. 81
Sarpieri, F. 85
Sato, T. 283
Schallert, T. 183
Schardt, C. 139
Schivalocchi, R. 81
Servadei, F. 17
Shakui, P. 207
Shao, J. 183
Sharma, H. S. 151, 288, 309, 316, 322, 329, 335, 426, 451
Shen, F. X. 438
Shibasaki, F. 157, 163
Shigemori, Y. 130
Shima, K. 148
Shioda, S. 57, 249, 283
Shull, G. E. 244
Shwe, Y. 438
Siesjö, B. K. 157
Signoretti, S. 24
Sims, N. R. 263
Sjöquist, P. O. 322
Smielewski, P. 108, 114
Spangenberger, A. 451
Stamatovic, S. M. 444
Steudel, W. I. 364
Stopa, E. G. 451
Stover, J. F. 139
Sun, D. 244
Sun, L. 144, 272
Suzuki, R. 74, 398
Suzuki, T. 398

Tagliaferri, F. 53, 85
Takahashi, T. 157, 163
Tamarozzi, R. 81
Tamura, A. 168
Tanabe, F. 168
Tanaka, Y. 279
Tanfani, A. 53, 85

Tang, F. 413
Tanjoh, K. 33, 37, 44, 48, 69, 97, 419
Targa, L. 53, 85
Teasdale, G. M. 17
Thomas, A. 402
Tian, H. 232
Timofeev, I. 11, 17, 108
Tominaga, Y. 69
Toyooka, T. 148
Turner, R. J. 263

Uchino, H. 157, 163
Ulamek, M. 267
Unterberg, A. 17, 33, 37, 44, 48, 69, 97, 419
Unterberg, A. W. 139

Vannemreddy, P. 91, 151, 316
Verkman, A. S. 389
Vink, R. 121, 263, 402

Wagner, K. R. 177
Wan, S. 199
Wang, M. 218
Wiklund, L. 288, 322
Willis, B. 91
Wu, G. 78
Wu, N. 422

Xi, G. 78, 183, 194, 199, 203, 218, 222, 227, 232

Yamada, I. 254, 279
Yamashita, S. 134
Yang, G. Y. 438
Yang, S. 203, 227
Yano, A. 148
Yao, X. 389
Yau, Y. H. 7
Yofu, S. 249
Yoshitake, A. 37
Young, H. 348
Young, W. L. 438
Younger, J. G. 227
Yu, P. 413
Yue, S. 413

Zhang, J. H. 188
Zhao, F. Y. 168
Zhu, G. 422
Zhu, Y. 438
Zuccarello, M. 177

Index of keywords

ADC 279
Adenoviral vectors 438
Adjuvant therapy 413
Alzheimer's disease 267
β-amyloid peptide 267
γ-aminobutyric acid 222
Analysis 7
Aneurysm 91
Angiogenesis 283, 438
Angiopoietin-1 438
Antioxidants 322
AP sites 194
APE/Ref-1 194
Apparent diffusion coefficient 81
AQP water channel 303
AQP4 393
Aquaporin 398, 402
Aquaporin-4 389
Astrocyte 398
Astrocytes 402
Autoregulation 108

Bacterial meningitis 413
Behavior 134, 183, 199, 279, 299
Blood flow velocity 21
Blood-brain barrier 130, 151, 171, 258, 267, 288, 295, 438, 444
Blood-cerebrospinal fluid barrier 426
Blood-spinal cord barrier 309, 322, 329, 335
Brain damage 409, 413
Brain edema 3, 11, 30, 74, 78, 171, 194, 203, 212, 218, 227, 258, 303, 389, 393, 398, 422
Brain injury 419
Brain ischemia 267
Brain natriuretic peptide 451
Brain swelling 121
Brain-derived neurotrophic factor 329

Calcineurin 157
Calcium 244
Cardiac arrest 44
Cardiopulmonary arrest 69
Carotid artery occlusion 276
3CB2 134
CCL2 444
Cellular edema 24
Cerebral blood flow 85
Cerebral contusion 3

Cerebral edema 402
Cerebral hemorrhage 199, 218, 227
Cerebral infarction 254
Cerebral ischemia 239, 279, 295
Cerebral perfusion pressure 108
Cerebral vasoreactivity 21
Cerebrovascular pressure transmission 103
Ceruloplasmin 203
c-fos 322
cGMP 451
Choroid plexus 276
Clinical characteristics 48
Closed head injury 151
$CMRO_2$ 97
Coagulation 69
Cocaine 91
Cold injury 163
Cold-induced brain injury 134
Comorbidity 364
Complement C3 227
Computed tomography 30, 78
Computer monitoring 108
Connexin 148
Contusion 30
Cortical atrophy 272
Cortical cold lesioning 168
Cortical layer 3 and 6 272
CPP 37
Craniectomy 373
Cranio-cervical trauma 139
Crossover 61
CSF absorption 207
CSF drainage 37
CSF homeostasis 451
Cyclooxygenase inhibitor 57
Cyclophilin D 157
Cytokines 177

Data collection 7
D-dimer 69
Decompressive craniectomy 11, 17
Deferoxamine 199
Diffuse axonal injury 144
Diffusion tensor MR imaging 168
Diffusion-weighted imaging 81
Dimethyl sulfoxide 258
DNP 194, 295
Dopamine 222
Dynorphin A (1-17) 309

Edema 121, 125, 151, 183, 244, 263, 288, 316, 322, 373, 426
Elasticity 254
Endothelium 419
Eosinophilic neuron 272
Epilepsy 389
Escherichia coli 413
Estrogen 121
Estrogen receptor 218
Evans blue extravasation 258

Factor VII 212
Fibrinogen degradation products 69
Fibrinolysis 69
Fluid management 37
Fluid percussion injury 144
FR901495 157
Free radical scavenger 57

Gap junction 148
Gender 121
Gene expression 163, 322
Gerbil 144, 299
Glasgow Outcome Scale 11
Glial cells 389
Glial-derived neurotrophic factor 329
Global ischemia 249, 272, 283, 299
Glutamate 139
Gravitational valve 377
Growth factor 139

H-290/51 322
Head injury 7, 11, 17, 24, 108, 114, 373
Hematoma 212
Hematoma growth 78
Hematoma lysis 78
Hematoma removal 74
Heme oxygenase 151
Heme oxygenase inhibitor 232
Hemeoxygenase-1 227
Hemispheric swelling 40
Highest modal frequency 103
Histamine 316
HOE 642 244
Horseradish peroxidase 267
Hydrocephalus 65, 207, 343, 348, 352, 358, 451
Hyperbaric oxygen 188
Hyperglycemia 419
Hyperthemia 426

Hypoperfusion 283
Hypotension 48, 276
Hypothermia 33, 44, 57, 177, 249, 409

ICP 11
Idiopathic 377, 381
IL-8 419
Immunoreactivity 335
Immunosuppression 163
Indentation method 254
Inflammation 263
Interleukin-1β 57
Intracerebral haemorrhage 61, 65, 69, 74, 78, 81, 177, 183, 194, 212, 222, 232
Intracranial hematoma 48
Intracranial hypertension 11
Intracranial pressure 17, 108, 114, 125, 373, 422, 435
Intraventricular hemorrhage 65, 207
Inulin 295
Iron 194, 199, 203
Iron deficiency anemia 183
Ischemia 53, 85, 263
Ischemic brain damage 157

Jugular vein temperature 97

Ku-proteins 194

Leukoaraiosis 267
Lipid peroxidation 188
LPS 419

Magnesium 402
Magnetic resonance imaging 171
Mannitol 125
Matrix metalloproteinase 130
Maturation phenomenon 239
Metabolic uncoupler 295
Mice 134, 227
Middle cerebral artery occlusion 258, 303, 393
Miethke Dualswitch-Valve 352
Mitochondrial permeability transition 157
Mongolian gerbil 272
Mongolian gerbils 254
Monocyte chemoattractant protein 444
Motor dysfunction 329
Mouse 249
Multi-parametric MRI 254

Na^+/Ca^{2+} exchange 244
NADPH oxidase 188

Natriuretic peptide receptors 451
Necrosis 3
Nestin 134
Network 7
Neuronal cell death 249
Neuronal nitric oxide synthase 288
Neuronal remodeling 239
Neuropeptides 263
Neuroprotection 139
Neurotrauma 121
Neurotrophins 329
Nitric oxide 288
Nitric oxide synthase 309
Normal pressure hydrocephalus 343, 364, 377, 381
NPH 348

8-OHdG 194
Organ dysfunction syndrome 33
Osmolality 3, 125
Outcome 65, 114, 358, 364, 377, 381
Outflow resistance 352
Overdrainage 343
Oxidative DNA injury 194
Oxidative stress 322
Oxygen and glucose deprivation 244

Pain 335
PANT 194
Periventricular lucency 348
Permeability 295
Phorbol ester 393
Photodynamic therapy 422
Plasmin inhibitor complex 69
Plateau waves 103
PMA 393
Preconditioning 203
Pressure autoregulatory response 21
Pressure reactivity 114
Pressure/volume index 364
Progesterone 121
Programmable valves 381
PTF1+2 69
Pulmonary artery temperature 97

QingKaiLing 413

Remodeling 134
Repetitive head injury 40
Resistance to outflow 364
Return of spontaneous circulation 69

Selective brain cooling 409

Sensory motor functions 288
Serotonin 151
Shunt 343, 352, 364
Shunt operation 358
SjO_2 97
Sodium 244
Spatial neglect 144
Spinal cord edema 329
Spinal cord injury 309, 316, 322, 329
Spinal cord pathology 329
Spinal nerve lesion 335
Sports 40
Stroke 57, 91, 276
Subarachnoid hemorrhage 53, 69, 85, 91, 188
Subdural hematoma 33
Surgery 61

T2 279
TBI 402
Thin subdural hematoma 40
Thrombin 203
Tight junctions 444
Transferrin 183
Traumatic brain edema 24, 398
Traumatic brain injury 11, 17, 21, 30, 33, 37, 44, 48, 130, 139, 148, 163, 171, 288, 435
TUC-4 134

Ultrastructure 168

Validation 7
Valves 358
Vascular endothelial growth factor 283
Vasogenic brain edema 168
Vasogenic edema 24, 348
Vasopressin 303
VEGF 398, 438
Venous pressure 435
Viscosity 254
VX2 carcinoma 422

Water transport 389
White matter 177
White matter lesions 267

Xenon CT 53
Xenon-CT 85

Zinc protoporphyrin 151, 232
ZO-1 438

SpringerNeurosurgery

Klaus . R. H. von Wild (ed.)

Re-Engineering of the Damaged Brain and Spinal Cord

Evidence – Based Neurorehabilitation

In cooperation with Giorgio A. Brunelli.
2005. XVI, 232 pages. 55 figures.
Hardcover **EUR 120,–**
Reduced price for subscribers to "Acta Neurochirurgica": **EUR 108,–**
(Recommended retail prices)
Net-prices subject to local VAT.
ISBN 3-211-24150-7
Acta Neurochirurgica, Supplement 93

Traumatic Brain Injury (TBI) can lead to loss of skills and to mental cognitive behavioural deficits. Paraplegia after Spinal Cord Injury (SCI) means a life-long sentence of paralysis, sensory loss, dependence and in both, TBI and SCI, waiting for a miracle therapy. Recent advances in functional neurosurgery, neuroprosthesis, robotic devices and cell transplantation have opened up a new era. New drugs and reconstructive surgical concepts are on the horizon. Social reintegration is based on holistic rehabilitation. Psychological treatment can alleviate and strengthen affected life. This book reflects important aspects of physiology and new trans-disciplinary approaches for acute treatment and rehabilitation in neurotraumatology by reviewing evidence based concepts as they were discussed among bio and gene-technologists, physicians, neuropsychologists and other therapists at the joint international congress in Brescia 2004.

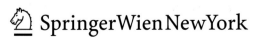

P.O. Box 89, Sachsenplatz 4–6, 1201 Vienna, Austria, Fax +43.1.330 24 26, books@springer.at, **springer.at**
Haberstraße 7, 69126 Heidelberg, Germany, Fax +49.6221.345-4229, SDC-bookorder@springer.com, springer.com
P.O. Box 2485, Secaucus, NJ 07096-2485, USA, Fax +1.201.348-4505, service@springer-ny.com, springer.com
Prices are subject to change without notice. All errors and omissions excepted.

SpringerNeurosurgery

Yasuhiro Yonekawa, Yoshiharu Sakurai,
Emanuela Keller, Tetsuya Tsukahara (eds.)

New Trends of Surgery for Stroke and its Perioperative Management

2005. IX, 187 pages. With partly coloured figures.
Hardcover **EUR 110,-**
Reduced price for subscribers to "Acta Neurochirurgica": **EUR 98,-**
(Recommended retail prices)
Net-prices subject to local VAT.
ISBN 3-211-24338-0
Acta Neurochirurgica, Supplement 94

In July 2004 specialists in neurosurgery, neuroradiology, neurology and neurointensive care discussed recent trends at the 2nd Swiss Japanese Joint Conference on Cerebral Stroke Surgery, held in Zurich, Switzerland. New concepts were worked out during the conference and are published in this volume. The book starts with the topic intracranial aneurysms, discussing microsurgical and endovascular treatment modalities, as well as new surgical approaches. Further chapters deal with the management of unruptured aneurysms and with subarachnoid hemorrhage. Practical guidelines for vasospasm treatment are given. Together with contributions about arteriovenous malformations and fistulas, cerebral revascularization techniques and surgery related to the intracranial venous system a comprehensive overview about stroke surgery is given with an interdisciplinary approach. The book will be of interest for all specialists involved in therapy of cerebrovascular disease.

SpringerWienNewYork

P.O. Box 89, Sachsenplatz 4–6, 1201 Vienna, Austria, Fax +43.1.330 24 26, books@springer.at, **springer.at**
Haberstraße 7, 69126 Heidelberg, Germany, Fax +49.6221.345-4229, SDC-bookorder@springer.com, springer.com
P.O. Box 2485, Secaucus, NJ 07096-2485, USA, Fax +1.201.348-4505, service@springer-ny.com, springer.com
Prices are subject to change without notice. All errors and omissions excepted.

SpringerNeurosurgery

W. S. Poon, C. J. J. Avezaat, M. T. V. Chan,
M. Czosnyka, K. Y. C. Goh, P. J. A. Hutchinson,
Y. Katayama, J. M. K. Lam, A. Marmarou,
S. C. P. Ng, J. D. Pickard (eds.)

Intracranial Pressure and Brain Monitoring XII

2005. XIV, 484 pages. With partly coloured figures.
Hardcover **EUR 172,–**
Reduced price for subscribers to "Acta Neurochirurgica": **EUR 145,–**
(Recommended retail prices)
Net-prices subject to local VAT.
ISBN 3-211-24336-4
Acta Neurochirurgica, Supplement 95

88 short papers originating from the 12th International Symposium on Intracranial Pressure and Brain Monitoring held in August 2004 in Hong Kong present experimental as well as clinical research data on invasive and non-invasive intracranial pressure and brain biochemistry monitoring. The papers have undergone a peer-reviewing and are organized in nine sections: ICP management in head injury, neurochemical monitoring, intracranial hypertension, neuroimaging, hydrocephalus, clinical trails, experimental studies, brain compliance and biophysics.

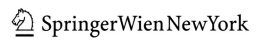

P.O. Box 89, Sachsenplatz 4–6, 1201 Vienna, Austria, Fax +43.1.330 24 26, books@springer.at, **springer.at**
Haberstraße 7, 69126 Heidelberg, Germany, Fax +49.6221.345-4229, SDC-bookorder@springer.com, springer.com
P.O. Box 2485, Secaucus, NJ 07096-2485, USA, Fax +1.201.348-4505, service@springer-ny.com, springer.com
Prices are subject to change without notice. All errors and omissions excepted.

Springer and the Environment

WE AT SPRINGER FIRMLY BELIEVE THAT AN INTERnational science publisher has a special obligation to the environment, and our corporate policies consistently reflect this conviction.

WE ALSO EXPECT OUR BUSINESS PARTNERS – PRINTERS, paper mills, packaging manufacturers, etc. – to commit themselves to using environmentally friendly materials and production processes.

THE PAPER IN THIS BOOK IS MADE FROM NO-CHLORINE pulp and is acid free, in conformance with international standards for paper permanency.